DIGITAL ELECTRONICS

Logic and Systems

Second Edition

John D. Kershaw
ITT Telecom Network Systems Division
formerly of West Virginia Northern Community College

Breton Publishers
A Division of Wadsworth, Inc.
North Scituate, Massachusetts

Breton Publishers
A Division of Wadsworth, Inc.

Library of Congress Cataloging in Publication Data

Kershaw, John D.
Digital electronics; logic and systems.

Includes index.
1. Digital electronics. 2. Logic circuits.
I. Title.
TK7868.D5K47 1983 621.381 82-17891
ISBN 0-534-01471-2

Printed in the United States of America
4 5 6 7 8 9—87 86 85 84

Digital Electronics, Second Edition was prepared for publication by the following people:

Sponsoring editor: *George Horesta*
Production supervisor: *Sylvia Dovner*
Production editor: *Jean T. Peck*
Art coordinator: *Mary S. Mowrey*
Copy editor: *Nancy Blodget*
Interior designer: *Breton Staff*
Cover designer: *Steven Wm. Snider*

Illustrations were done by Jay's Publishers Services, Inc. The book was typeset by Jay's Publishers Services, Inc.; printing and binding was by The Maple-Vail Book Manufacturing Group.

Contents

Preface

For the first edition of this text, it was thought that TTL and CMOS circuits would be the dominant logic family through the end of the 1970s. Back then, both RTL and DTL logic were yielding to TTL as the ideal logic family. In retrospect, that judgment has proven correct, for although microprocessor circuitry, gate array, and other custom LSI circuits have taken over large segments of the hardware, these advances have merely brought about the economies necessary for a massive expansion in the amount of digital circuitry that could be put to practical use.

At the time of writing the second edition, most of the basic data and information pertaining to TTL and CMOS logic families are still relevant. Although for mass-scale production, gate array and other forms of custom circuits may supply ever increasing portions of the circuitry in use, the TTL and CMOS standard family of circuits will continue to be used as the essential "glue" that binds many custom devices together. They will also be used in small-scale production and in prototyping many of the custom devices. The term *TTL compatible* continues to be a keyword in integrated circuit design and will continue to be through the next decade.

Significantly, improved versions of both TTL and CMOS circuits are now being released to the marketplace. They provide speed and power improvements without any radical change in other operating parameters. In most cases, the old pin-outs are still available. Therefore, for the second edition of this text, Chapter 9 has been expanded to provide a detailed discussion of the AC and DC parameters of both current and newer versions of TTL and CMOS families.

In the area of memory circuits, change has been even more dramatic. Accordingly, this new edition includes the latest in memory technology. The parallel development of three classes of integrated circuits—RAMs, ROMs, and microprocessors—has led to a virtual explosion in the application of digital electronics. Each year, new generations of these devices reach the marketplace, and each generation is an improvement. The new devices are likely to be faster, cheaper, of higher capacity, and more reliable and less power consuming than the devices they replace.

To bring users of this book up to date with the recent improvements in memory technology, the latter chapters have been extensively revised. Furthermore, a chapter has been added on the important subject of multiplexers and demultiplexers, which are commonly used in the handling of microprocessor buses. A relatively new and promising family of devices, the programmable array logic, has also been added with this edition.

The application of devices such as these places a renewed emphasis on the use of Boolean algebra. The dispersed and somewhat incomplete coverage of this topic in the first edition has been expanded and is now included in a single section. Moreover, in recognition of the fact that microprocessor op codes and address numbering are commonly expressed in hexadecimal, coverage of this number system has been added to Chapter 2.

Acknowledgments

During the preparation of the manuscript for the second edition, a number of people provided helpful comments and suggestions, for which I am grateful. In particular, thanks are expressed to John R. Pelong, Henry Ford Community College, Dearborn, Michigan; Harold R. Morgan, Northeast Mississippi Junior College, Booneville, Mississippi; Peter Holsberg, Mercer County Community College, Trenton, New Jersey; and William W. Kleitz, Tompkins-Cortland Community College, Dryden, New York. The editorial and production staff at Breton Publishers—especially George Horesta, Sylvia Dovner, Jean Peck, Mary Mowrey, and Betty Slinger—are also to be thanked for their efforts and their many contributions to the second edition.

Finally, thanks are due to numerous individuals and firms within the electronics industry for providing up-to-date information that was helpful in preparing the second edition and, in many cases, for giving their permission to reprint copyrighted material in the book. My appreciation is extended to the following firms:

Advanced Micro Devices
B & K, Product of Dynascan
Burroughs Corporation
Data I/O
Fairchild Semiconductor, Inc.
Harris Semiconductor Products Group
Hewlett-Packard Company
IBM Corp.
Intel Corp.
International Crystal Manufacturing Company
Monolithic Memories, Inc.
Mostek Corporation
Motorola, Inc.
National Semiconductor Corporation
OKI Semiconductor
Precision Monolithics Incorporated
PRO-LOG Corporation
RCA Solid State Division
Signetics Corporation
Solid State Scientific, Inc.
Sprague Electric Company
Synertek
Tektronix, Inc.
Texas Instruments Incorporated
Toledo Scale, Division of Reliance Electric

Preface to the First Edition

This text is aimed at providing students with an easy understanding of digital logic and digital logic systems. Although I discuss some circuitry in introducing the logic elements, there is no attempt to present detailed mathematical circuit analysis. To make effective use of this text the reader needs only knowledge of basic algebra and basic electricity. A knowledge of semiconductor theory would be highly desirable but is not an absolute prerequisite.

Digital electronics, like other areas of electronic technology, presents both student and instructor with a problem of too much: too much to learn, too much to teach, in too little time. Unless the course hours allotted this subject are increased, we face hard choices about what to include and what to leave out. A count of the new application sheets issued by circuit manufacturers each year shows how much new and perhaps necessary information is being added. Ideally, culling obsolete information from the program will create the time to discuss new developments. In practice, though, the pressure to include more information means that some basic principles of logic systems may receive inadequate coverage. Mindful of this problem, I have left out of this text what is obviously obsolete, and where doubt exists I leave the choice to the user. The four basic logic gates and the flip-flop storage elements are given full coverage, with a wealth of applications. Although individual gates and flip-flops are currently used only to tie together large- or medium-scale integrated circuits into larger logic units, one must still master their application.

For the sake of simplicity, I explain logic subassemblies such as adders, counters, and shift registers in their simplest logic form before presenting them as medium-scale integrated circuits. The circuits presented are primarily TTL and CMOS, except in the area of memory circuits, where large-scale ECL and MOS have a substantial share of the market.

The timing diagram has grown in importance as an instrument for analyzing digital logic circuits, and therefore it is introduced early in the text and is given continual emphasis thereafter.

Each chapter is preceded by a set of objectives to be accomplished by students as a result of reading the chapter and performing the many problems and exercises. As an additional study aid, summaries and glossaries are included at the end of the chapters.

The glossaries are written to support the material in the chapter and may not provide all possible meanings of the terms. They are arranged in logical rather than alphabetical order. In many chapters, the glossary can provide an effective review of the material covered. Students should be encouraged to refer to the glossaries whenever they have difficulty understanding something in the text.

I have relied heavily on drawings of logic symbols, waveforms, and wiring diagrams to make the exercises more meaningful. Two kinds of exercises have been provided, allowing the instructor greater flexibility in making assignments. The "Questions" are easier and less time consuming. They provide a test of the student's understanding of the material. The "Problems" are more time consuming and test performance as well as understanding. They require the student to perform calculations, make simple logic drawings, select the correct logic unit for an application, indicate the proper wiring of logic circuits, and complete timing diagrams. The wiring diagram problems are ideal for assembly and checkout in the lab. Only a few of the problems require the student to do extensive drawings; these should be assigned when practice in logic drawing is desirable.

Students will find it useful to have a drawing template MIL-STD-806C (available at most drafting supply stores), which contains the most widely accepted set of logic symbols used in the field. It would also be useful for students to have access to catalogues of standard TTL and CMOS circuits. These are available from numerous manufacturers. The majority of the circuits called for in exercises, however, have been presented as figures in the text.

In the formation of this text I have received frequent advice and guidance from colleagues and reviewers, for which I am grateful. John Bakum (Middlesex County College), Frederick Driscoll (Wentworth Institute), John Nagi (Hudson Valley Community College), and Robert Shapiro (College of San Mateo) read the manuscript at various stages; Robert Coughlin (Wentworth Institute) provided especially valuable technical consultation. Thanks are due also to those members of the electronics industry whose generous distribution of current information kept this writing up to date; to my wife Katherine, son Terry, and daughter Jackie for many hours of typing, drawing, and reading; and, above all, to the editors and staff at Breton, whose skill and enthusiasm converted a rough manuscript into this finished product.

John D. Kershaw

CHAPTER

1

Introduction to Digital Machines

Objectives

Upon completion of this chapter, you will be able to:

- Name some benefits derived from digital electronics.
- Identify some job market opportunities open to technicians trained in digital electronics.
- Draw a block diagram of the general-purpose stored-program digital computer.
- Identify other noncomputer types of digital equipment.
- Use system drawings to describe digital equipment.
- Draw the three basic symbols of flow diagrams and state their use for describing the construction and functioning of digital equipment.

1.1 Introduction

Machines that can duplicate human motion have been with us now for many generations. The degree to which they benefit us may be a subject for debate, but there is no doubt they are responsible for a major reduction in the number of hours in our work week. Nevertheless, individual workers, like the fabulous John Henry, often feel the anxiety of being in competition with the machine.

Until recently, workers may have taken comfort from the fact that the machine could not think, make decisions, choose a course of action based on its decisions, or learn. But these elements of human superiority over the machine may no longer be valid. In 1962, a computer was programmed to play the game of checkers. In the beginning, it lost most of the time, but after playing thousands of games it became practically unbeatable. Here, obviously, was a machine that could make decisions, choose its alternatives, and learn.

In the past decade, computers of much greater capacity have been developed. Where now do we stand in light of our newest machines? What can humans do that machines cannot do better? Do we, like John Henry, persist in valiant competition with the machine, or do we conclude that the computer is superhuman and look to it to solve all our problems?

The computer is not superhuman. As complex as it may be, it is only a tool, and, as with any other tool, human beings play the key role in using it. Our key role requires knowledge of what the computer is, how it functions, and what it can and cannot do for us. That is the object of this book—to explain the surprisingly simple devices that can be assembled to produce these automatic machines and to show that sensible employment of computers can substantially enrich the quality of human life.

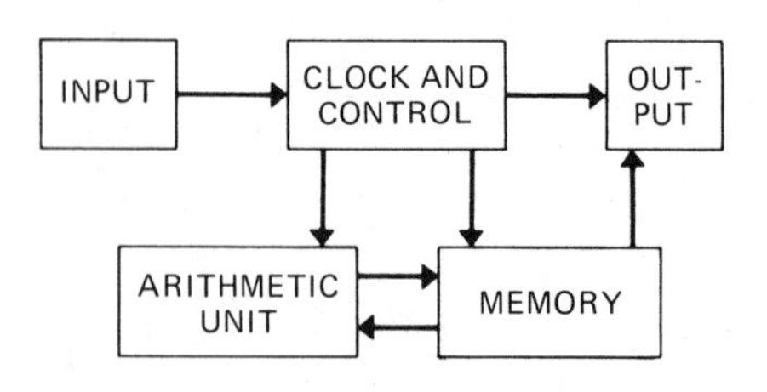

FIGURE 1-1

Block diagram of a general-purpose stored-program digital computer.

1.2 Digital Electronic Equipment

1.2.1 The General-Purpose Stored-Program Digital Computer

In size, complexity, and usefulness, the general-purpose stored-program digital computer is at the top of the list. The first generation of digital computers was produced in the 1940s. They employed vacuum tube and relay switches, consumed large amounts of power, and occupied considerable space. Yet, even with those components, they displayed a surprising degree of reliability and showed promise of a high level of usefulness in science, statistics, and accounting. With the introduction of semiconductor switches, other component and hardware improvements, and improvements in software (programming techniques), those promises have been fulfilled beyond the wildest expectations. The term *computer* is often used rather loosely, particularly in advertising, where even a simple cooking timer may be called a computer. To be termed a computer, a device should have at least the units shown in Figure 1-1.

Input-Output (I/O). The most versatile input-output (I/O) device is the teletypewriter. Depressing a key on the teletypewriter not only prints the letter or character on paper, but simultaneously sends the same character to the computer in digital code. Conversely, when the computer sends that digital code to the teletypewriter, it causes that same letter to print on the paper. Other input devices are the punched tape reader and the card reader, with their corresponding output devices, the tape punch, and the card punch.

Memory. Elaborate mathematical computations usually require mathematicians either to remember or to write down subtotals, partial answers, and so forth. To do those same functions, the computer must also have a memory. Memories have been constructed from acoustic tubes, magnetic drums, ferrite cores, and, more recently, semiconductor circuits. The terms used to describe the many different types of memories are explained in detail in Chapter 22. The time required to obtain a number from memory and deliver it to the arithmetic unit or other location in the computer is called the memory cycle time. Measured in microseconds or less, it is one key to the speed of computer operations.

Arithmetic Unit. This unit is the main functional section for most computer operations. It is essentially a high-speed digital adder with sufficient register circuits to perform all the basic arithmetic functions. The add cycle time—the time required to enter two numbers into the adder and produce a sum—is another measure of the speed of a computer.

Clock and Control Unit. A digital clock and its supporting circuits provide timing, cycling, and sequencing signals for the other units of the computer. The control unit contains digital circuitry that, when selected by number, will cause certain simple, logical, or mathematical routines to occur. Complex functions can be made to occur as a result of a stored program that produces many of those routines in a logical and useful sequence at extremely high speed. The software, or programming, provides the wealth of simple, routine combinations behind the powerful capabilities of today's computers.

1.2.2 Special-Purpose Computers

Another class of digital machines may be called special-purpose computers. The program they perform is hard wired and cannot readily be changed. They receive variable input data and subject it to a fixed mathematical routine for such purposes as navigation and industrial testing. They are seldom as elaborate as general-purpose computers, but they contain similar components.

1.2.3 Accounting Machines

Accounting machines, used to keep cash and inventory records, are becoming increasingly more digital. The price of the digital calculator has decreased from several hundred dollars to as low as ten dollars. The functions of the cash register have been expanded from merely counting cash taken in to computing change and transmitting data to a central terminal that keeps track of inventory and automatically prints out a purchasing list.

Even the cashier's operations have been speeded up by the use of a wand that senses a digitally coded machine-readable tag and transmits the description to the register. The universal product code (Figure 1-2) now appears on the labels of many grocery items. The bar code is scanned electronically, and the product description is sent in digital form to a small computer. The price is drawn from the computer memory rather than from the memory of a cashier or from a price stamped on by a store clerk.

FIGURE 1-2
Universal product code: a machine-readable code used on grocery and other products.

1.2.4 Digital Test Equipment

It is natural that the electronics industry should be the forerunner in using digital equipment. Many items of electronic test equipment that have long been analog in nature are now available in digital form. Figure 1-3 compares an analog volt ohmmeter with a digital VOM. The analog meter requires the user to select and interpret the correct scale. The digital meter can be read more rapidly and with less confusion.

Instruments that measure time and frequency are primarily digital circuits. Figure 1-4 shows the HP 5345A counter, an instrument that can count frequencies accurately as high as 500 MHz and time intervals as short as 50 nanoseconds. Its range can be extended with adapters and converters to as high as 18 GHz.

The cathode ray tube, which for some time has been used to display analog waveforms, can today be used to display printed data from a computer or digital information system. Figure 1-5 shows the Tektronix 4023 computer display terminal. This device uses digital and analog techniques to store and read out the data it receives for display.

Digital techniques have been successfully used to automate electronic testing. Many minor electronic components that were tested by hand a decade ago are today tested automatically and at much higher speed. Digital test terminals,

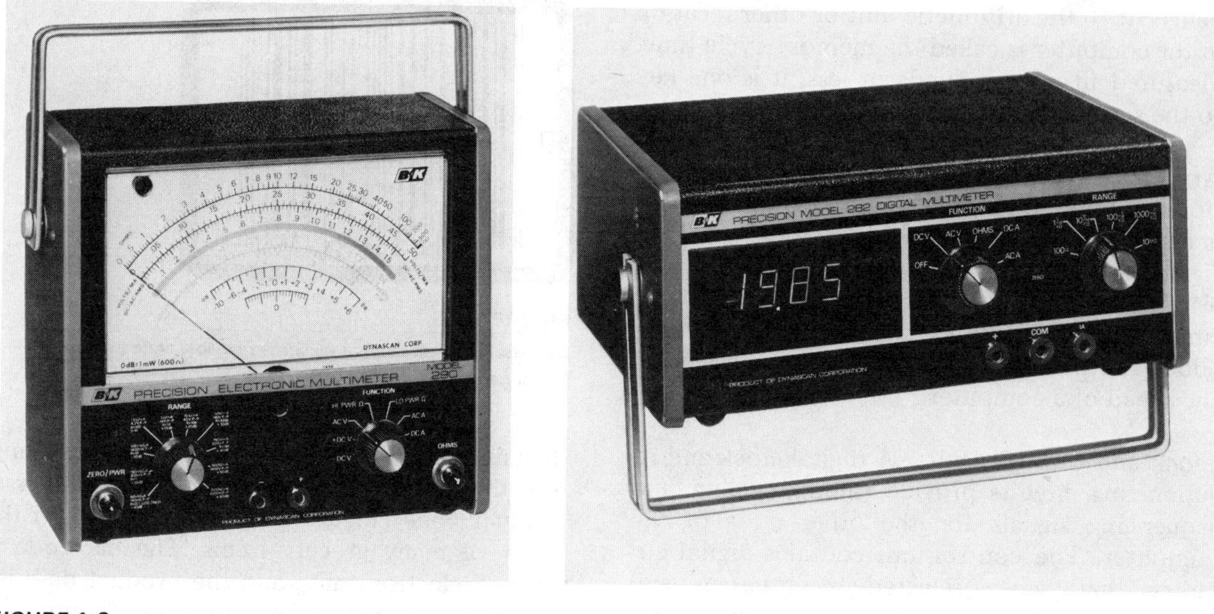

FIGURE 1-3

Dynascan analog meter (left) and digital meter (right).

which can be programmed for a wide variety of tests, do most of today's mass production testing.

Modern aircraft contain so much electronic equipment that automated, programmable test systems that are primarily digital in nature have been devised to check the air-frame periodically.

1.2.5 Digital Scales

Today even the scale used at the meat or produce counter at the grocery store may be digital. Figure 1-6 shows the Toledo 8201 digital scale, which displays not only the weight, but also the price per pound and the total price. It also provides both clerk and customer with an easy-to-read display.

1.2.6 Digital Communication

It has long been recognized that a trained operator can copy CW or Morse code through severe conditions of noise and interference with greater accuracy than can be accomplished by voice transmission. CW code is digital in nature, having

FIGURE 1-4

The Hewlett-Packard 5345A digital counter.

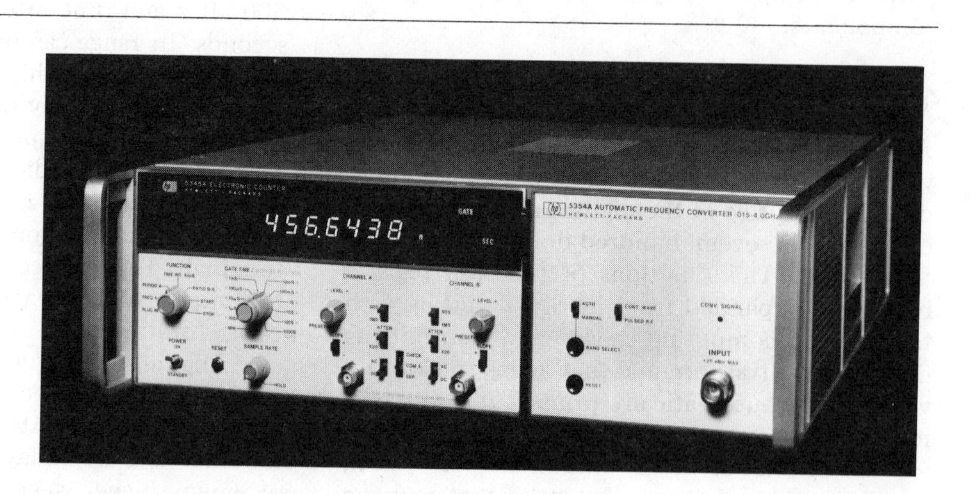

FIGURE 1-5
Tektronix 4023 computer display terminal.

only two levels—on and off. It is not surprising, then, that high-speed mass communications systems are using digital techniques to considerable advantage. Two of the systems used are pulse width modulation and pulse code modulation. These modulation systems make possible the sending of numerous messages simultaneously on the same communication channel by a technique known as multiplexing. Like CW code, these modern digital systems require the receiver to recognize only two distinct levels (ON and OFF), which thus makes them less susceptible to errors caused by noise.

There are numerous digital information systems that use computer-type input-output devices and memory storage. Messages may be transmitted by punching them on IBM cards, punched paper tape, or magnetic tape. These systems have the added advantage of being able to work directly with computer terminals.

FIGURE 1-6
The Toledo 8201 digital scale.

1.3 Employment Opportunities

Over the past 30 years, employment opportunities in the electronics industry have grown rapidly. This industry has provided a job market that offers easy placement and rapid advancement for those who are willing and able to keep up with continual change. For a long time, application of electronics was confined to the single-channel AM radio. After World War II, application was rapidly expanded into the new fields of radar, television, and the computer, while the industry was still limited by a hard-wired vacuum tube technology. Technicians had to increase their knowledge of circuits and systems to benefit from the expanding job opportunities. Along with knowledge of new circuits, they needed skill in using new and more elaborate forms of electronic test equipment.

The printed circuit board, which replaced the bulk of the hand wiring required in electronic systems, certainly reduced the amount of labor needed for assembly and testing. But, for every job lost at lower skill levels, a new and more challenging opportunity became available at a higher skill level.

The progress of the electronics industry and the employment opportunities it presents can be described as a series of overlapping and repetitive cycles; that is, new developments, shrinking costs, increasing reliability, and expanded applications.

New developments and expanded applications do not necessarily create opportunities for the same skills or for application of old knowledge. But, for those who are willing to add knowledge of new devices and applications to their knowledge of basic principles, there appears to be sustained growth in challenging job openings.

At this writing, we are seeing the latter stage of one of the above cycles. The integrated circuit, and in particular the digital integrated circuit, was developed in the 1960s. The cost per circuit has shrunk. Its reliability has been developed to a superior level. Today applications are expanding. Instruments that were strictly analog a decade ago are available today in all or partially digital form—digital television, digital volt ohmmeters, even digital organs, to mention a few. Even the automobile industry, which once confined its electronics to the car radio, uses electronics in both analog and digital form.

A consumer advocate accused the automotive industry of having lagged in efforts to develop safety equipment by asking, "How do you explain the fact that the air bag is available in the automobile today and not ten years ago?" An automotive executive replied, "Ten years ago we did not have the remarkable little electronic black boxes that are available to us today."

There is no way to predict how many things "little black boxes" will be doing for us in the next decade.

1.4 Drawings and Instructions

1.4.1 Drawing Systems

Most manufacturing plants use a clearly defined system to identify the many drawings that are needed to assemble and test a complex digital system. The main drawing is the assembly drawing, which shows the mechanical layout of the main unit. It includes a parts list that itemizes individual parts and subassemblies, including printed circuit cards.

Other drawings, such as the system block diagram, schematic diagram, wiring diagram, and logic diagrams, may also be listed by drawing numbers on the parts list. These drawings, like the drawings for the circuit cards and other subassemblies, must be requested as a separate package from the print room. The subassembly drawings will again include a parts list, which may list the logic and schematic diagrams for the subassembly.

In the field, most of these drawings are found in the back of the instruction book.

1.4.2 The Flow Diagram

The flow diagram is relatively new to the electronic scene. The engineer uses it to design a digital system, but it is equally valuable to the technician in understanding the systems operation. It is very difficult to troubleshoot a defective digital system without first gaining a thorough understanding of how it functions. This understanding may come from a written description in the instruction book, or, in manufacturing, from a specification. As part of such descriptions, the flow diagram gives a symbolic illustration of the unit's functions. Flow diagrams are also used extensively by computer

FIGURE 1-7
Flow diagram symbols used to illustrate the functioning of digital electronic systems.

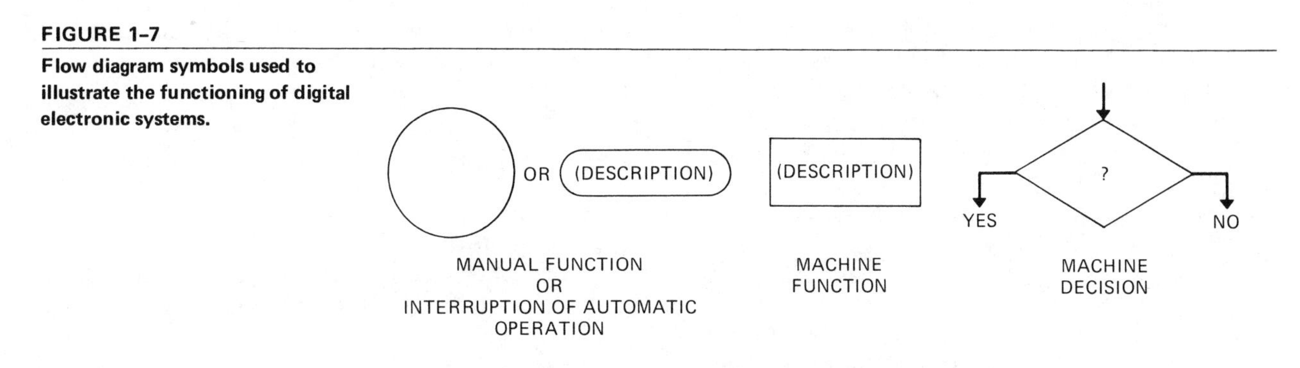

programmers. Many highly descriptive symbols are used for programming, but the electronic technician need be concerned with only three symbols (see Figure 1-7).

A circle is used to describe a manual function that the operator must perform or to indicate an interruption in the automatic functioning of the machine. The circle is sometimes flattened to make it easier to write in the description. If the machine is fully automatic, these symbols may appear only at the beginning and end of the diagram.

A rectangle is used to describe the machine functions. Several functions may appear in one block, but they must appear in the order in which they occur.

The diamond is used as a decision block. Most automatic machines have cycles of operation, or loops, that are repeated over and over again. The flow diagram is particularly useful in describing these loops. It is at the decision diamond that a machine decides either to continue in a loop or to break out of it. It is common for the decision diamond to have several inputs and as many as three outputs.

To understand the use of a flow diagram, let us first look at a specification or description of a digital machine that can grade a multiple-choice test paper:

1. The answer sheet must be in machine-readable form.

FIGURE 1-8
Flow diagram of a machine that automatically marks test papers.

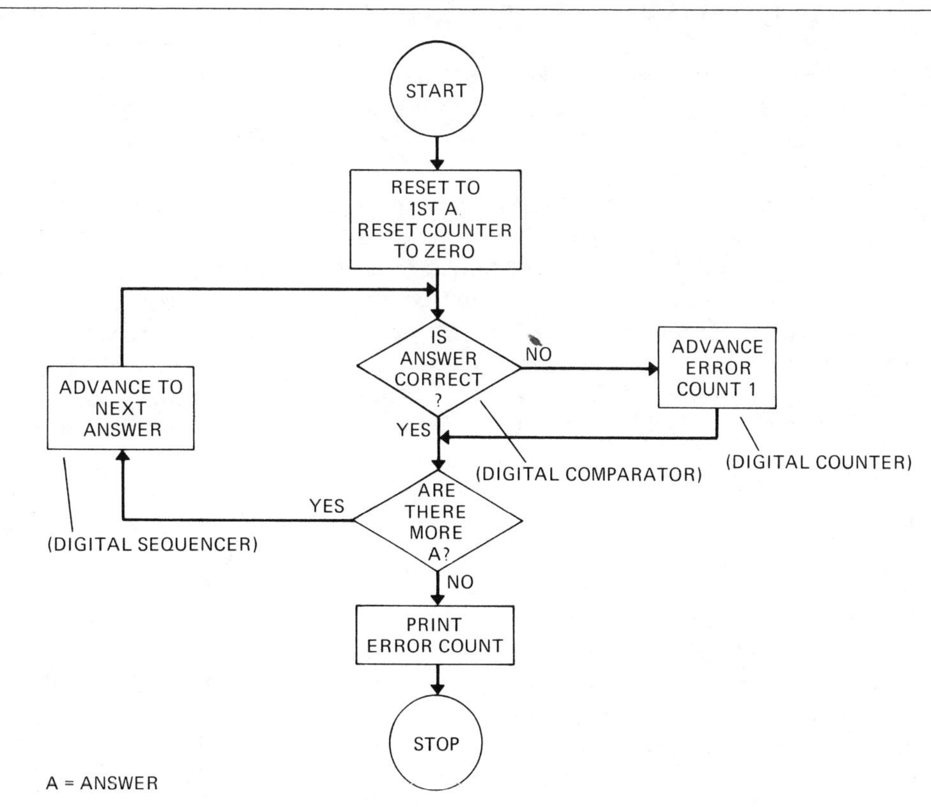

2. It must be programmable for the correct answers to any set of 20 test questions.
3. It must examine each question in sequence and count an error each time the answer on the sheet differs from the correct answer.
4. When all 20 questions have been checked, it will print the number of errors on the answer sheet and stop.

The flow diagram of Figure 1-8 shows how the device will operate. These functions are identical to the mental and physical functions teachers perform in comparing student answers with their own answer key; that is, counting the errors and writing the number of errors after checking the last question.

After having completed this text, you will be able to determine from specifications and, in particular, from flow diagrams what digital subassemblies are required to perform the set of functions indicated by a flow diagram. The components and subassemblies implied by the diagram in Figure 1-8 are listed in parentheses for future reference. At this point, it is necessary only to analyze the machine by its basic functions.

1.5 Summary

As complex as computers may be, they are only tools, and, as with any other tools, human beings play the key role in using them. This book describes the devices that can be assembled to produce computers. The first generation of computers, produced in the 1940s, used vacuum tubes and relay switches, and it consumed large amounts of power and space. Today's microcomputers are hundreds of times smaller, yet much more powerful than their early predecessors. All computers have some common characteristics, such as input-output devices, memory, arithmetic unit, and clock and control unit. Computers are often classified as either general purpose or special purpose.

Computers are not the only pieces of electronic equipment that use digital devices. Some other examples are digital VOMs, counters, scales, and communication equipment.

Flow diagrams (Figure 1-7) are often used to show how a system or program operates. Each flow diagram symbol illustrates a particular function. Notes inside and outside give further explanation of how the system operates.

Glossary

Digital Circuit: A circuit in which voltage levels or other conditions are given a numerical value. Usually they are binary, having only two levels, 1 and 0. Distinct voltage levels are established for 1 and 0, and the system is designed to avoid levels in between.

Analog Circuit: A circuit designed to accept and respond correctly to all voltage levels between the circuit minimum and maximum. If used, numerical values are determined by measurement.

Software: Written programs and instructions that are provided to enhance the usefulness of a general-purpose digital computer.

Ferrite Core: A small metal washer made by compressing a powered magnetic material into doughnut-like shape. The clockwise and counterclockwise directions of magnetization are given digital values of 1 and 0. Thousands of ferrite cores are arranged to form memory circuits to store numbers within a digital computer.

Magnetic Drum Memory: A cylinder coated with ferrite material that can be magnetized in 1 and 0 directions and used to store numbers in a digital computer. Memory locations are located by the motion of the magnetic read-head along the axis of the drum and by the rotation of the drum.

Memory Cycle: The operation of writing a number into, or reading a number out of, a memory circuit—normally a continuously occurring cycle that is used when needed and that continues in an inactive or ineffective state when not needed.

Memory Cycle Time: The time required to write a number into memory or to read a number out of memory.

Arithmetic Unit: The section of a computer designed to perform high-speed arithmetic operations—normally including a digital adder, storage registers, and occasionally scratch pad or buffer memories for short-term storage of the numbers being operated on.

Add-Cycle Time: The time required to enter two numbers into the adder and produce a sum.

Questions

1. What are the main elements of a general-purpose stored-program digital computer? Draw the block diagram.
2. In what way does the special-purpose computer differ from the general-purpose stored-program digital computer?
3. Describe four types of drawings that may be used to describe a digital electronic system.
4. Draw the three general flow diagram symbols and give their meanings.

Problem

1-1 The specifications for a digital teaching machine, a mechanical description, and a set of flow diagram functions are shown in Figure 1-9. You are to redraw the symbols into a correct flow diagram for the specified machine.

DIGITAL TEACHING MACHINE
SPECIFICATIONS

The teaching machine will be constructed of relays and other digital components, and its capabilities will be as follows:

1. It will operate a standard slide projector.
2. It will have four push buttons to give the operator a choice of four answers presented on a slide with the question.
3. It will reprimand the operator for wrong answers.
4. It will accept only the operator's first answer to a question.
5. It will tell the operator the correct answer after an error has been made.
6. It will allow the operator to advance to the next question after having had time to consider the correction.
7. It will praise the operator for giving the correct answer.
8. It will advance automatically to the next question when the operator gives a correct answer.
9. It will count the number of wrong answers given to any set of questions programmed.

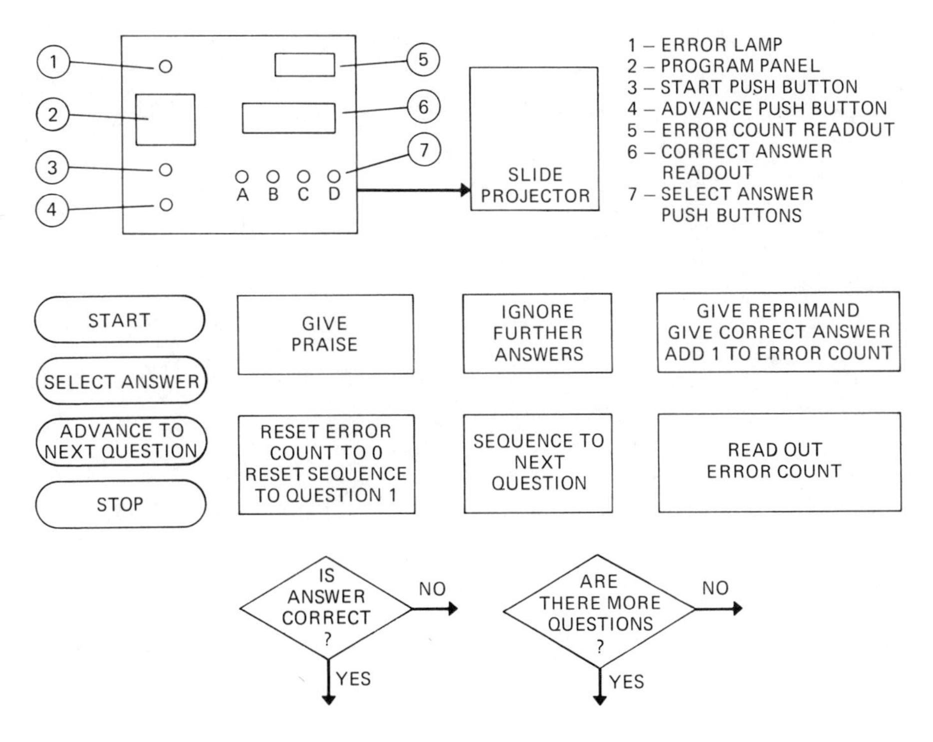

FIGURE 1-9. Problem 1-1

CHAPTER

Number Systems

Objectives

Upon completion of this chapter, you will be able to:

- Identify number systems other than the decimal system (by tens).
- Convert between decimal and binary numbers.
- Cite the advantage of the binary number system for machine operation.
- Cite the special advantages of using the octal and hexadecimal number systems.
- Convert between octal and binary number systems.
- Convert between decimal and binary-coded decimal numbers.
- Convert between binary and hexadecimal number systems.
- Convert between decimal and hexadecimal number systems.
- Convert between octal and decimal number systems.

2.1 Introduction

The decimal system is so ingrained in our culture that it is hard to accept the fact that there are other number systems that would enable us to perform the same operations of arithmetic we usually do in decimal numbers. As a matter of experience, we develop some concept of the quantity represented by a number such as 1523. We do not stop to remind ourselves that this number means $1 \times 10^3 + 5 \times 10^2 + 2 \times 10^1 + 3 \times 10^0$. If we did, perhaps it would be easier to accept the fact that quantities might be represented by powers other than 10. This chapter develops two other very useful number systems that are essential to an understanding of digital electronics.

2.2 General Number System

The general number system can be written as shown in this equation:

$$A_nX^n + \cdots + A_3X^3 + A_2X^2 + A_1X^1 + A_0X^0. + A_aX^{-1} + A_bX^{-2} + A_cX^{-3} + \cdots + A_nX^{-n}$$

Figure 2-1 shows a number system table and the effect of making X = 10. The A numbers would be digits limited to (X - 1) in value. There is never any need for a digit of value equal to or greater than X, for when A = X it can then be represented by 1 times the next power of X, because $XX^n = X^{n+1}$. This example shows why digits 0 through 9 only are used in the decimal system. Ten is not a single digit; it indicates 1 times the next power of ten.

In the general number system, X is referred to as the radix or base of the system. If we accept that the positive powers of X increase by 1 for each term to the left of X^0 and the negative powers of X increase by 1 for each term to the right of X^0, then the number can be expressed by a representation of the coefficients (A). Therefore, a four-digit decimal number with the radix or base understood as 10 is written (as we are accustomed) with only the coefficients shown: A_3 A_2 A_1 A_0. If there is some possibility the radix will be mistaken, a subscript to the number is used, such as 1523_{10}. We must realize, of course, that 1523_{10} is not equal to 1523_8, nor is it equal to 1523_6. The same set of figures will represent different quantities if the radix is different. A numerical quantity can be represented in many different number systems, but, except for some single-digit numbers, the same quantities will have different sets of digits depending on the radix. We will soon be able to determine that $1523_{10} = 2763_8 = 5F3_{16}$.

2.3 Binary Number System

2.3.1 Importance to Electronic Computers

The binary number system is the language of the digital computer. Its primary advantage is that it requires only two digits, 1 and 0. In the binary system, the base or radix is (X) = 2. Therefore, as we specified, the coefficients $A \leqslant (X - 1)$ cannot be greater than 1. Thus, only 1 and 0 are digits.

Many electrical devices have two stable states that can be easily distinguished. A switch

FIGURE 2-1
Table for the general number equation, with values progressively substituted for the decimal number 1523.

	A_3X^3	A_2X^2	A_1X^1	A_0X^0	•	A_aX^{-1}	A_bX^{-2}	A_cX^{-3}
IF X = 10	$A_3 10^3$	$A_2 10^2$	$A_1 10^1$	$A_0 10^0$	•	$A_a 10^{-1}$	$A_b 10^{-2}$	$A_c 10^{-3}$
	$A_3 1000$	$A_2 100$	$A_1 10$	$A_0 1$	•	$A_a .1$	$A_b .01$	$A_c .001$
IF Q = 1523	1 × 1000	5 × 100	2 × 10	3 × 1	•	0 × .1	0 × .01	0 × .001
	1000	500	20	3	•	0	0	0

$A \leqq (X - 1)$

is either OFF or ON. A relay is either energized or de-energized. A diode may be reverse-biased or forward-biased. A transistor can be in cutoff or in saturation. These electrical states can be assigned numerical values of 1 or 0 and the devices connected to perform arithmetic operations.

Although the binary system is highly advantageous for computer operations, it poses difficulty for the people using the computer. The decimal number 1623 is one that a person of reasonable aptitude can read and both remember and conceive of as a size of number. Without much thought, we recognize it as being larger than the number 1298.

The binary number of value equal to 1623_{10} is 11001010111_2, while 1298_{10} is 10100010010_2. It took 11 binary places to represent both four-place decimal numbers, and if the two binary numbers fell on separate pages in a book, we would have to count places and check carefully to be sure which was the larger. For this reason, the results of a binary computation must be converted to decimal before being read out to the operator.

The operator putting numbers into the computer will probably do so through an input device that converts from decimal to binary.

2.3.2 Decimal-to-Binary Conversion

Beginning at X^5, the general number equation $A_5X^5 + A_4X^4 + A_3X^3 + A_2X^2 + A_1X^1 + A_0X^0$ in binary becomes $A_5 2^5 + A_4 2^4 + A_3 2^3 + A_2 2^2 + A_1 2^1 + A_0 2^0. = A_5 32 + A_4 16 + A_3 8 + A_2 4 + A_1 2 + A_0$. The values of the coefficients are either 0 or 1.

There are several methods of converting a number from decimal to binary. The most straightforward is to determine first the highest power of 2 in the decimal number. Using the decimal number 83 as an example, we proceed as Figure 2-2 shows. The highest power of 2 in 83 is 64, and 64 is 2^6. Therefore, our binary number would have a 1 in the 2^6 column. Subtract 64 from 83. Since 2^5 will not divide into the remainder, that place has a 0 digit. Since 2^4 will divide into the remainder, we place a 1 digit in that place and subtract 16. Now 2^3 will not divide into the remainder; nor will 2^2. We use 0 digits in these places. Since 2^1 will divide into the remainder with 1 left over, the two (least significant) digits are 1s.

The binary number for 83 is 1010011. It is not correct to refer to this as one million ten thousand eleven. That implies a decimal number. The binary number is correctly expressed as one "oh" one "oh" "oh" one one.

In the decimal number system, the lowest-value digit, or digit farthest to the right, is referred to as the least significant digit (abbreviated LSD). The highest-value digit in the number is the most significant digit (MSD). The same references may be used in other number systems, including the binary, but in the binary number system the binary places are referred to as bits, and the abbreviations LSB, for least significant bit, and MSB, for most significant bit, may be used.

Another method of conversion from decimal to binary is to divide successively by 2 and

FIGURE 2-2

Table for the general number equation, with values substituted for a base 2 number the equivalent of 83_{10}.

	A_6X^6	A_5X^5	A_4X^4	A_3X^3	A_2X^2	A_1X^1	A_0X^0
IF $X = 2$	$A_6 2^6$	$A_5 2^5$	$A_4 2^4$	$A_3 2^3$	$A_2 2^2$	$A_1 2^1$	$A_0 2^0$
	$A_6 64$	$A_5 32$	$A_4 16$	$A_3 8$	$A_2 4$	$A_1 2$	$A_0 1$
IF $Q = 83_{DEC}$	1×64	0×32	1×16	0×8	0×4	1×2	1×1
	64	0	16	0	0	2	1

83	1	0	1	0	0	1	1
−64							
19							
−16							
3							
−2							
1							

$\therefore 1010011_2 = 83_{10}$

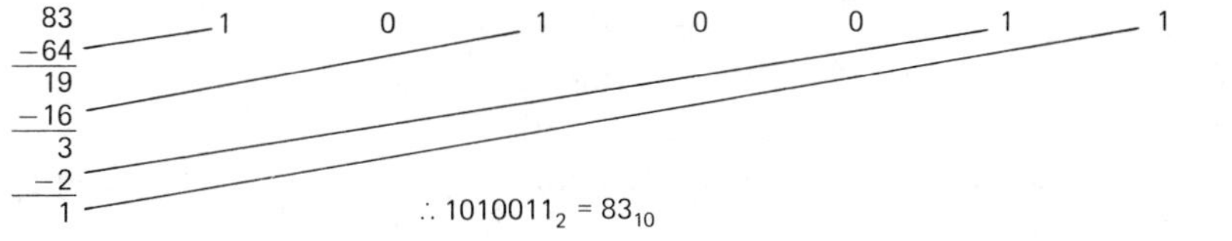

use the remainders as the digits for the binary number. Although the logic of this is less obvious, it is correct and, for most people, easier. Here is an example:

$83 \div 2 = 41$ with 1 over (LSB)
$41 \div 2 = 20$ with 1 over
$20 \div 2 = 10$ with 0 over
$10 \div 2 = 5$ with 0 over
$5 \div 2 = 2$ with 1 over
$2 \div 2 = 1$ with 0 over
$1 \div 2 = 0$ with 1 over (MSB)

This procedure, which produces the LSB first and the MSB last, gives the same 1010011 for the binary number, equal to 83 decimal.

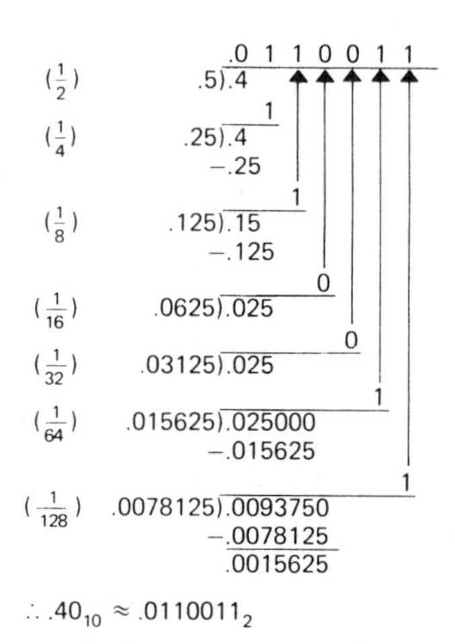

FIGURE 2–3

Conversion of a decimal fraction to a binary number.

■ EXERCISES

1. Determine the highest power of 2 in the following decimal numbers: 1, 625, 10, 144, 49, 96, 27, 275, 3.
2. Convert the decimal numbers of Exercise 1 to binary.
3. Convert the following decimal numbers to binary: 511, 7, 63, 255, 15, 127, 31.
4. Convert the following decimal numbers to binary: 33, 9, 129, 513, 65, 17, 257.

The binary number system is not limited to whole numbers. To the right of the binary point are the places $.A_a 2^{-1} + A_b 2^{-2} + A_c 2^{-3} + A_d 2^{-4} + \cdots + A_m X^{-m}$. The coefficients are still limited to the values 1 and 0. From algebra, $2^{-x} = 1/2^x$. Therefore, $2^{-1} = 1/2$, $2^{-2} = 1/2^2 = 1/4$, $2^{-3} = 1/8$. The decimal values of the places to the right of the binary point are therefore:

$$.A_a \frac{1}{2} + A_b \frac{1}{4} + A_c \frac{1}{8} + A_d \frac{1}{16} + A_e \frac{1}{32}$$

$$(.5) \quad (.25) \quad (.125) \quad (.0625) \quad (.03125)$$

The decimal .6875 can be converted by first comparing the number with 1/2 or .5. It is larger than .5. Therefore, place a 1 in the first column to the right of the binary point and subtract .5. Next try .25 or 1/4. It doesn't go into the remainder, so place a 0 in the second column. Try .125 or 1/8. It goes. Place a 1 in the third column and subtract .125. Next divide 1/16 or .0625 into the remainder. It goes, with no remainder. Place a 1 in the fourth column and the conversion is complete. This conversion went to the fourth binary place, or 1/16. Decoding the whole number 1011 gives 11. It, therefore, represents 11/16. Diagrammatically, this conversion can be shown as follows:

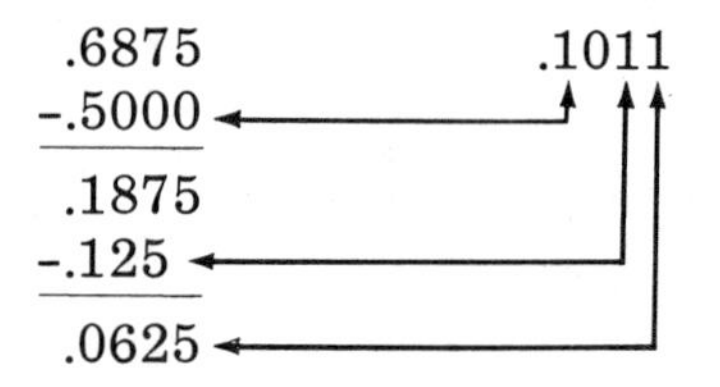

Some precise decimal fractions with only a few decimal places may require many binary places or may prove to be irrational when expressed in binary. The usual practice is to convert to a given number of binary places and truncate or drop the remainder. For example, .4 decimal has only one decimal place, but, as Figure 2–3 shows, we have gone to six binary places and still have a remainder of $.0015625_{10}$.

If the denominator of a decimal fraction is a power of 2, we know the LSB of the binary fraction. Thus, it is necessary only to convert the numerator to binary and be sure to place its LSB far enough to the right of the binary point. For example, to convert the fraction 7/16, first convert the numerator:

$$7_{10} = 111_2$$

The LSB must appear in the 1/16 (2^{-4}) column.

Thus,

$$\frac{7}{16} = .0111_2$$

To convert 5/8,

$$5_{10} = 101_2$$

or, $\frac{5}{8} = .101$

To convert 11/64,

$$11_{10} = 1011_2$$

and 1/64 is 2^{-6}. Therefore,

$$\frac{11}{64} = .001011$$

Even mixed numbers may be converted. For example, to convert 10–3/8, the whole number is converted first:

$$10_2 = 1010_2$$

The numerator of the fraction, 3/8, is converted next:

$$3_{10} = 11_2$$

In this case, the LSB is in the 1/8 column. Therefore,

$$10\frac{3}{8} = \underbrace{1010}.\underbrace{011}$$

10 + 3 times the LSB

■ EXERCISES

5. Convert the following decimals to binary to six significant bits (truncate after 2^{-6}): .654, .075, .175, .325, .675, .825.
6. Convert the following fractions to binary to six significant bits (truncate after 2^{-6}): 1/2, 1/3, 1/4, 1/5, 1/6, 1/7, 1/8, 1/9, 1/10.
7. Convert the following fractions to binary: 15/32, 21/64, 3/8, 13/16, 3/4.
8. Convert the following mixed decimals to binary: 5-11/16, 15-5/32, 12.625, 9.75, 3.9375.

2.3.3 Binary-to-Decimal Conversion

Converting from a binary number to decimal is a simple matter of adding up the decimal values of each binary place in which a 1 appears. To determine the value of a number such as 101101.0101, merely add up the place values of the one bits in the number, as in the table of Figure 2-4.

■ EXERCISES

9. Convert the following binary numbers to decimal: 101101, 1110111, 10110, 1000101, 10101, 1111, 10001.
10. Convert the following binary numbers to fractions: .001, .11, .101, .0101, .111, .1101, .1.
11. Convert the following binary numbers to decimal: .1001, .011, .1101, .011, .1111.
12. Convert the following binary numbers to decimal: 1.011, 1000.1, 11.11, 10.101, 111.0111, 101.101, 100.001.

2.4 Octal Number System

2.4.1 General

Of the many number systems that could be used in digital machines, four are used most often—the decimal, binary, octal, and hexadecimal. The

FIGURE 2-4

Binary number table for the number 101101.0101_2, showing it to equal 45.3125_{10}.

1×2^5	0×2^4	1×2^3	1×2^2	0×2^1	1×2^0	•	0×2^{-1}	1×2^{-2}	0×2^{-3}	1×2^{-4}
1×32	0×16	1×8	1×4	0×2	1×1	•	$0 \times \frac{1}{2}$	$1 \times \frac{1}{4}$	$0 \times \frac{1}{8}$	$1 \times \frac{1}{16}$
32	0	8	4	0	1	•	0	$\frac{1}{4}$	0	$\frac{1}{16}$

32 + 8 + 4 + 1 + 1/4 + 1/16 = 45 5/16 or 45.3125

decimal system is used often because it is the operator's number system. The binary system is used because of the two stable states common to electrical components. The octal and hexadecimal systems are used because it is more convenient and economical to group binary numbers using one of these formats.

In the octal number system, the radix is 8. Therefore, the general number

$$A_3X^3 + A_2X^2 + A_1X + A_0X^0. + A_aX^{-1} + A_bX^{-2} + A_cX^{-3}$$

becomes

$$A_38^3 + A_28^2 + A_18 + A_08^0. + A_a8^{-1} + A_b8^{-2} + A_c8^{-3}$$

or, $A_3(512) + A_2(64) + A_1(8) + A_0(1).$

$$+ A_a\left(\frac{1}{8}\right) + A_b\left(\frac{1}{64}\right) + A_c\left(\frac{1}{512}\right)$$

The coefficient values are limited ($A \leqslant X - 1$) to the digits 0 through 7. The digits 8 and 9 never occur, because $8_{10} = 10_8$ and $9_{10} = 11_8$.

2.4.2 Decimal-to-Octal Conversion

A decimal number such as 1623_{10} can be converted to octal by first determining the highest power of 8 that will divide into the decimal number. Since $8^4(4096)$ is too large, we start with 8^3, but 8^3 will divide into 1623_{10} three times. Our MSD of the octal number is, therefore, 3, and we proceed as Figure 2-5 shows.

There is an easier method of conversion from decimal to octal. Its logic is more difficult to perceive. We divide by 8 successively, and the remainders that occur form the octal digits, beginning with the LSD. Figure 2-6 shows how to convert 1623_{10} to octal by this method.

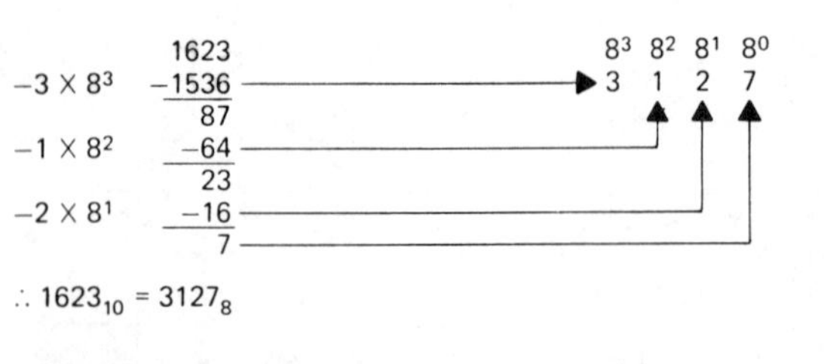

FIGURE 2-5

Conversion of 1623 decimal number to its equivalent octal number, 3127.

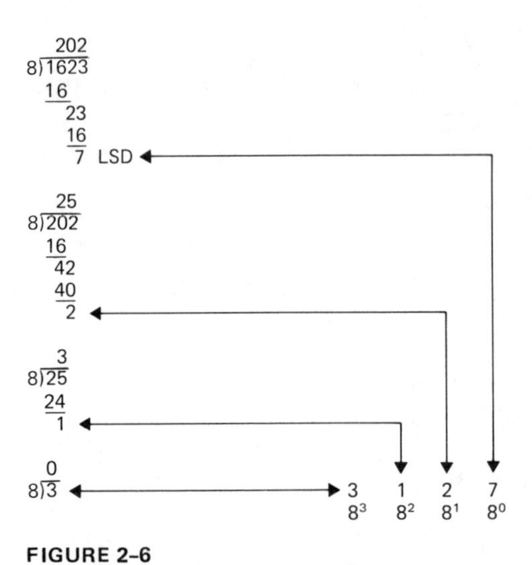

FIGURE 2-6

Simplified method of converting a decimal number to its equivalent octal number.

■ EXERCISE

13. Convert the following decimal numbers to octal: 76, 15, 724, 1322, 162, 17, 285.

2.4.3 Octal-to-Decimal Conversion

An octal number can be converted back to decimal merely by substituting the appropriate values in the general number equations. An octal number, such as 372, can be converted as shown:

$$A_2X^2 + A_1X + A_0X_0$$

where: $X = 8$

$$3 \times 8^2 + 7 \times 8^1 + 2 \times 8^0 =$$
$$3 \times 64 + 7 \times 8 + 2 \times 1 =$$
$$192 + 56 + 2 = 250$$

Therefore,

$$372_8 = 250_{10}$$

In abbreviated form, let us convert the whole number 3527_8 to decimal:

$$\begin{aligned} 3 \times 8^3 &= 3 \times 512 = 1536 \\ 5 \times 8^2 &= 5 \times 64 = 320 \\ 2 \times 8^7 &= 2 \times 8 = 16 \\ 7 \times 8^0 &= 7 \times 1 = 7 \\ & \overline{1879_{10}} \end{aligned}$$

Therefore,

$$3527_8 = 1879_{10}$$

■ EXERCISE

14. Convert the following octal numbers to decimal: 2571, 721, 32, 111, 13.

2.4.4 Binary-to-Octal Conversion

The octal number is similar to the decimal number in that, for the usual magnitude of number, it seldom is more than one digit larger than its equivalent decimal number. Let us consider the three numbers below:

17642 21643 16235

As they have neither 8 nor 9 as digits, they may be either decimal or octal. Although they do not represent the same magnitude in both systems, the larger number in the decimal system is also the larger in the octal system. The smaller in the decimal system is also the smaller in the octal system. An operator could easily list them in ascending or descending order.

Binary numbers of equivalent magnitude, such as these

1111110100010
10001110100011
1101010011101

are not nearly so easily placed in order of magnitude. When an operator must deal with these numbers regarding only relative magnitudes and does not have to apply them arithmetically, conversion from binary to either decimal or octal is equally useful.

As we shall learn, it costs less to convert electrically to octal than to convert to decimal. For this reason, the azimuth and elevation readouts of some radar computers are in octal rather than decimal form, and some computers have their memory locations numbered in octal rather than decimal.

Converting from binary to octal is a simple matter of dividing the number into three place groups and converting each of the three place groups to the digits 0 through 7. Here are some

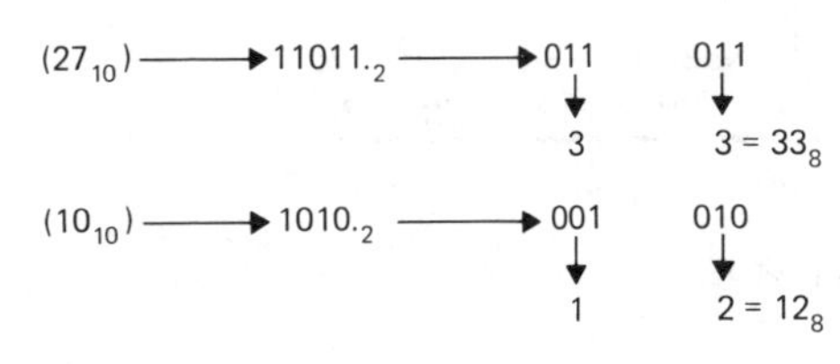

FIGURE 2–7
Conversion of a binary number to an octal number by separating the binary number into three-bit groups.

examples:

1	111	110	100	010	Binary
1	7	6	4	2	Octal

The digits required for octal are 0 through 7, which is the exact capacity of a three-bit or three-place binary number. Numbers 0 through 9 are needed for decimal; a three-place or three-bit binary number is too small. Four-bit division overlaps to the next decimal place.

Division into three-bit groups begins at the binary point and moves to both left and right. Zeros may be added to the end to fill out the groups, as shown in Figure 2–7.

■ EXERCISE

15. Convert the following binary numbers to octal: 1110101, 1101101, 1010001, 11011001101, 111101011.

2.5 Binary-Coded Decimal (BCD)

2.5.1 Decimal-to-BCD Conversion

In the binary-coded decimal number code, each digit of a decimal number is converted individually to its binary equivalent. To do so, we must allow four binary places for each decimal digit. A four-bit binary number can represent quantities as high as 15, but the BCD uses only 0 through 9. The BCD code is, therefore, less efficient in the use of circuits than straight binary. Yet, it is widely used because of the ease with which it can be coded and decoded. The number

For BCD

$\underbrace{2^3\ 2^2\ 2^1\ 2^0}_{\times 10^3}\ \underbrace{2^3\ 2^2\ 2^1\ 2^0}_{\times 10^2}\ \underbrace{2^3\ 2^2\ 2^1\ 2^0}_{\times 10^1}\ \underbrace{2^3\ 2^2\ 2^1\ 2^0}_{\times 10^0}$

For Binary

$2^{10}\ 2^9\ 2^8\ 2^7\ 2^6\ 2^5\ 2^4\ 2^3\ 2^2\ 2^1\ 2^0$
1 1 0 0 1 0 1 0 1 1 1

FIGURE 2-8

Conversion from decimal numbers to binary-coded decimal numbers by converting each decimal digit to a four-bit binary equivalent.

1623_{10}, for example, is changed as shown below:

1	6	2	3	
0001	0110	0010	0011	16 bits (BCD)

$1623_{10} = 11001010111_2$ 11 bits (binary)

Sixteen bits are needed to represent this number in BCD, but only 11 to represent it in straight binary. The place values of the bits have, however, different meaning, as shown in Figure 2-8.

■ EXERCISE

16. Convert the following decimal numbers to BCD: 237, 3224, 15.25, 6572, 16.625.

2.5.2 BCD-to-Decimal Conversion

If a binary-coded decimal number is to be decoded to decimal, it should be divided into sets of four bits each on each side of the binary point. Then each set of bits should be decoded to form the digits of the decimal number, as shown below:

1001010000110101 (binary number)

1001	0100	0011	0101	(BCD)
9	4	3	5	(decimal)

■ EXERCISE

17. Convert the following BCD numbers to decimal: 1111001, 1011010000, 100011, 1100101, 101010101.

2.6 Hexadecimal Number System

As was demonstrated in Section 2.4 (octal), a number system that allows conversion from binary by first separating the binary number into groups of bits can be very useful. For the octal system, the binary number is separated into three-bit groups. The modern-day microprocessor, however, has address and data buses that are divisible by four. This fact has favored the hexadecimal number system, which permits conversion in groups of four binary bits. The hexadecimal number system is, in this respect, similar to BCD, except that the bits being converted are true binary rather than BCD.

The four-bit groups of the BCD number can convert to values between 0 and 9. The four-bit groups of the hexadecimal, on the other hand, convert to values between 0 and 15. The place values of the hexadecimal, instead of being powers of 10, are powers of 16. In terms of a four-digit general number,

$$A_3X^3 + A_2X^2 + A_1X^1 + A_0X^0$$

where: $X = 16$

$$A_3(16^3) + A_2(16^2) + A_1(16^1) + A_0(16^0) =$$
$$A_3(4096) + A_2(256) + A_1(16) + A_0(1)$$

Since the values of the coefficients are 0 through 15, and our human experience has accustomed us only to digits 0 through 9, a set of six new digits had to be established. In early years, special characters such as punctuation marks were used. Today, the letters A through F are more commonly used, which is fortunate for the beginner because the human mind has already established an order for these symbols. Figure 2-9 shows the value assigned to these digits.

Although a hexadecimal number may some-

BINARY	DECIMAL	HEXADECIMAL
1010	10	A
1011	11	B
1100	12	C
1101	13	D
1110	14	E
1111	15	F

FIGURE 2-9

Hexadecimal digits used in addition to the digits 0 through 9.

times be distinguished from decimal or octal by the presence of letters, there are hexadecimal numbers whose integers do not exceed 9. Such numbers may be mistaken for decimal or, if no digits exceed 7, octal. To avoid this confusion, a subscript 16 or an H may be used to signify a hexadecimal number.

2.6.1 Binary-to-Hexadecimal Conversion

Converting binary numbers to hexadecimal is a simple matter of separating the binary number into four-bit groups and converting each group individually to form the hexadecimal digits. The binary number 11101101111110_2 converts to hexadecimal as follows:

0011	1011	0111	1110	Binary
3	B	7	E	Hexadecimal
	(11)		(14)	

■ EXERCISE

18. Convert the following binary numbers to hexadecimal:

    1011001100101100
    1100100011101111
    100010110010
    111010010110
    101111101001011

2.6.2 Hexadecimal-to-Binary Conversion

Conversion from hexadecimal to binary is no more difficult than binary to hexadecimal. Merely change each digit of the hexadecimal number to its four-bit equivalent binary number. The hexadecimal number 5DA7 converts to binary as follows:

5	D	A	7	Hexadecimal
101	1101	1010	0111	Binary

■ EXERCISE

19. Convert the following hexadecimal numbers to binary: 31B4, C70E, 2F3A, AC2D, 1EB6, F900.

2.6.3 Hexadecimal-to-Decimal Conversion

Occasionally, it is useful to convert hexadecimal to decimal numbers. To do so, we multiply each hexadecimal place value by its coefficient and add the results. The hexadecimal number $4C2B_H$ converts to decimal as follows:

$$A_3X^3 + A_2X^2 + A_1X^1 + A_0X^0$$

where: $X = 16$

$$A_3 16^3 + A_2 16^2 + A_1 16 + A_0 =$$

$$4(4096) + C(256) + 2(16) + B =$$

$$16384 + 3072 + 32 + 11 = 19499$$

Therefore,

$$4C2B_H = 19499_{10}$$

■ EXERCISE

20. Convert the following hexadecimal numbers to decimal: 1FE5, A72C, 36B4, 8D94, 6743, 41BC2D.

2.6.4 Decimal-to-Hexadecimal Conversion

The conversion of decimal to hexadecimal numbers can be accomplished in several ways. The most straightforward way is to determine the number of successively lower powers of 16 needed to produce a hexadecimal number of equivalent value. We convert the decimal number $42,627_{10}$ to a hexadecimal as shown in Figure 2-10. Since 16^4 is 65,536, it obviously will not divide into 42,627. However, 16^3 or 4096 will divide into 42,727. It will go A times.

	42627
$A \times 16^3$	−40960
	1667
6×16^2	−1536
	131
8×16^1	−128
3×16^0	3

∴ 42627 = A683

FIGURE 2-10

Conversion of a decimal number to its equivalent hexadecimal number.

FIGURE 2-11
Simplified method of converting a decimal number to a hexadecimal number.

The other method is to divide the decimal number successively by 16 and to continue the division until the dividend is zero. As shown in Figure 2-11, the remainders form the hexadecimal digits. This method produces the LSD first.

■ EXERCISE

21. Convert the following decimal numbers to hexadecimal: 26247, 63999, 9827, 14386, 39154, 10673.

2.7 Comparison of Number Systems

Figure 2-12 shows the five number systems most widely used with digital circuits. The need for fewer places for a high value number obviously favors hexadecimal, decimal, and octal numbers. The need for fewer digits favors binary and BCD numbers. Hexadecimal, decimal, and octal numbers are the same up to the value 7. Binary and BCD do not differ until the value 10. Only the hexadecimal system contains digits A through F, which are not familiar to us as numbers.

2.8 Summary

The decimal number system, which we are most familiar with, is unworkable for digital components. The reason is that noise would prevent electronic components from maintaining ten distinct steps in the decimal system. Therefore, digital circuits operate using two logic levels—0 and 1. This number system is called the binary system. All information in computers is represented by these two levels. Section 2.3.2 shows how decimal numbers, both integers and fractions, are converted to binary. The conversion from binary to decimal is shown in section 2.3.3.

Long strings of binary patterns are very cumbersome for people to use. More convenient systems are needed to group the binary digits. These systems are called codes. Three of the most common codes are octal, hexadecimal, and

FIGURE 2-12
Comparison of five different number systems used in digital circuits.

DECIMAL	BINARY	OCTAL	BCD	HEXADECIMAL
0	0	0	0000	0
1	1	1	0001	1
2	10	2	0010	2
3	11	3	0011	3
4	100	4	0100	4
5	101	5	0101	5
6	110	6	0110	6
7	111	7	0111	7
8	1000	10	1000	8
9	1001	11	1001	9
10	1010	12	0001 0000	A
11	1011	13	0001 0001	B
12	1100	14	0001 0010	C
13	1101	15	0001 0011	D
14	1110	16	0001 0100	E
15	1111	17	0001 0101	F

binary-coded decimal (BCD). The octal system has eight digits—0 through 7. The hexadecimal system has 16—0 through 9 and A through F. The BCD system has ten—0 through 9. Sections 2.4–2.6 describe these number systems and show how to convert from one system to another. A comparison summary is given in section 2.7.

Glossary

Number: A figure or group of figures used to represent a quantity or collection of units, whether of persons, things, or abstract units.

Radix: A number used as the base for a system of numbers, such as 10 in the decimal system, 2 in the binary system, 8 in the octal system.

Number System: A system for representing quantities by a summation of successive powers of a given base or radix. The general equation of a number system is:

$$A_nX^n + \cdots + A_3X^3 + A_2X^2 + A_1X^1$$
$$+ A_0X^0 + A_aX^{-1} + \cdots + A_mX^{-m}$$

X: The radix or base of the number system (in the decimal system X = 10).

A: The coefficient that multiplies the given power of the radix ($A \leqslant X - 1$).

The coefficient may be any integer needed between 0 and (X – 1). (In the decimal system, coefficient values are 0 through 9.) In the usual case, the radix is understood and only the coefficients are expressed, such as for a four-digit decimal number. For instance,

$$A_3X^3 + A_2X^2 + A_1X^1 + A_0X^0 \text{ becomes}$$
$$A_3\ A_2\ A_1\ A_0$$

A quantity such as 2 9 4 7 need not be expressed as 2000 + 900 + 40 + 7, as would be derived from the equation.

X^n. The highest power of the radix that does not exceed the quantity being represented.

Radix point or base point is referred to as decimal point for base 10 or binary point for base 2.

$A_0X^0 = A_0$. Any number to the 0 power is equal to 1. Therefore, the first term to the left of the radix point is equal to its coefficient, regardless of the base of the number system.

$A_aX^{-1} = A_a/X$. To the right of the radix point are terms with increasing negative powers of X. These terms can also be expressed as powers of the radix dividing the coefficients. In the decimal system,

$$.625 = \frac{6}{10} + \frac{2}{100} + \frac{5}{1000}$$

Digit: The places needed to represent a quantity in the decimal number system. The decimal number 3762 is a four-digit number. The leftmost digit (3) is the most significant digit (MSD). The rightmost digit is the least significant digit (LSD).

Binary System: A number system that represents quantities by powers of 2. It has the advantage of having coefficients of 0 and 1 only, instead of the 0 through 9 used in decimal.

Bit: The places needed to represent a quantity in the binary number system. The binary number 1010 is a four-bit number. The leftmost bit is the most significant bit (MSB). The rightmost bit is the least significant bit (LSB).

Binary-Coded Decimal (BCD): A number system in which each decimal digit has had its coefficient individually converted to binary, requiring four binary bits for each decimal digit.

Octal System: A number system that represents quantities by powers of 8. This system has only eight integers: 0, 1, 2, 3, 4, 5, 6, and 7.

Hexadecimal System: A number system that represents quantities by powers of 16. This system has the integers 0 through 9 along with A, B, C, D, E, and F.

Questions

1. List the highest value integer for each of the following number systems: binary, octal, decimal, hexadecimal, BCD.
2. What number system does a digital computer use?
3. How many dozen are there in each of the

quantities listed: 1000_{10}, 1000_2, 1000_H, 1000_8, 1000_{BCD}.?

4. Define the terms LSD and MSB.
5. What is the lowest power of 2 that exceeds 1000_{10}.?
6. Can the digits 8 and 9 ever occur in the octal system?
7. What power of 2 has the same value as an equal power of 10?
8. Of the following numbers: 1010, 1001, 1B39, 1827, 1764.
   Which could be binary?
   Which could not be octal?
   Which could not be decimal?
   Which could be hexidecimal?
   Which could be BCD?
9. What binary numbers do not exist in the BCD system?
10. Will it take more bits to express a decimal number in straight binary or BCD?
11. Why has the hexadecimal system become popular with microprocessor systems?
12. Write the first 32 hexadecimal digits.

Problems

Convert the following decimal numbers to binary:

2-1 527, 27, 327, 1278, 73

2-2 0.6, .625, .84375, .390625

2-3 17.5, 51.63, 39.175, 61.375

2-4 1/2, 11/16, 5/8, 13/32, 7/64, 3/4

2-5 7-15/16, 13-3/8, 63-13/16, 3-23/64, 14-7/32

Convert the following binary numbers to decimal:

2-6 1111, 101101, 110011, 10101, 11101

2-7 0.101, 0.1101, 0.001011, 0.0111, 0.10101

2-8 11.1011, 101.101, 111.111, 10.10101, 100.001

Convert the following binary numbers to octal:

2-9 110110100, 11001011, 101011

2-10 110011, 10010111, 1001110101, 110011100110

Convert the following binary-coded decimal numbers to decimal:

2-11 10011, 1100101, 10010110, 10011, 10010100, 10010101, 10010110, 11101000110, 100111001

2-12 Of the numbers listed below, only eight are octal, and the remainder are decimal. Convert the octal to binary and the decimal to BCD:
2376, 5872, 3129, 4735, 6297, 3846, 3265, 5829, 2134, 6833, 4276, 1492, 1066, 2732, 1776, 1945

2-13 Of the numbers listed below, four are binary and the rest are BCD. Convert the binary to octal, and the BCD to decimal:
1101101, 10100101, 101101011, 1101101011, 10111011, 1100110011, 10000011, 10000101

2-14 Convert the following hexadecimal numbers to binary and decimal: 56A, C14, BA2, 4E8, 3A7F, 19DO, 42FE, FFFF.

2-15 Convert the binary numbers in problems 2-26, 2-29, 2-10 to hexadecimal.

2-16 Convert the following hexadecimal numbers to binary:
A3C6, 5B72, 31F9, E753, 2679, 6083, 56A3, C2B5, 43E2, 3386, 672F, 2CB1, 6E09, B372, 67A1, 549C.

2-17 Convert the numbers given in problem 2-15 to decimal.

2-18 Convert the following decimal numbers to hexadecimal:
3549, 1671, 3722, 5609, 2321, 6672, 5432, 3252, 1563, 4683, 7535, 6319, 2572, 3361.

CHAPTER

Binary Arithmetic

Objectives

Upon completion of this chapter, you will be able to:

- Add with binary numbers.
- Convert binary numbers to ones and twos complements.
- Perform three methods of subtraction with binary numbers.
- Multiply and divide with binary numbers.
- Subtract decimal numbers by nines complement.

3.1 Introduction

We have emphasized the importance of the binary number system to the digital machine. We will now consider how a machine uses its dual-state components to accomplish the tasks of arithmetic. To understand the machine operations in this new number system, we must first master for ourselves the low-speed, pencil-and-paper operations that the machine must duplicate, in its own style, at extremely high speed.

A_0	0	1	0	1
$+B_0$	0	0	1	1
$S_0 + C_0$	0 + 0	1 + 0	1 + 0	0 + carry 1 = (2^1)

A_0	B_0	S_0	C_0
0	0	0	0
1	0	1	0
0	1	1	0
1	1	0	1

FIGURE 3-1

Addition of the LSB. In adding the LSB of binary numbers, one of four possible conditions will occur to produce the sum (S_0) and the carry (C_0), as shown in the right-hand column of the truth table.

3.2 Binary Addition

Let us add two numbers, A and B, to get the sum S. If A and B are four-bit numbers, S will be either four or five bits, depending on whether there is a carry from the addition of the MSB. Let us represent the individual bits of the numbers with subscript letters:

$$\begin{array}{rl} \text{A} & A_3 A_2 A_1 A_0 \\ +\text{B} & B_3 B_2 B_1 B_0 \\ \hline \text{S} & S_3 S_2 S_1 S_0 \end{array}$$

Each subscript letter represents a binary 1 or 0 times the power of 2 of the subscript number. As in the decimal system, we begin adding with the LSB. As the LSB addition does not involve a carry-in, there are only four possible combinations, as shown in Figure 3-1. The addition of A_0 and B_0 produces two separate outputs, the sum, S_0, and the carry, C_0. Figure 3-1 also shows a truth table. The truth table is a device used to analyze digital circuits by showing every possible input combination and the outputs resulting from each one. A digital circuit used to perform LSB addition would necessarily have such a truth table.

Three of the four possible combinations are identical to decimal: 0 + 0 = 0, 0 + 1 = 1, 1 + 0 = 1. The condition 1 + 1 = 2 in decimal, but in binary 2_{10} is 10_2. Therefore, the statement for this condition is 1 plus 1 is 0 carry 1.

The second significant digit has a third input. The carry-out of the first significant digit (C_0) is the carry-in to the second significant digit (C_{in}). In this digit and higher ones, there are eight possible combinations, as shown in Figure 3-2.

C_{in}	0	1	0	0	0	1	1	1
A_1	0	0	1	0	1	0	1	1
B_1	0	0	0	1	1	1	0	1
$+S_1 + C_0$	0 + 0	1 + 0	1 + 0	1 + 0	0 + 1	0 + 1	0 + 1	1 + 1

C_{in}	A_1	B_1	S_1	C_0
0	0	0	0	0
1	0	0	1	0
0	1	0	1	0
0	0	1	1	0
0	1	1	0	1
1	0	1	0	1
1	1	0	0	1
1	1	1	1	1

FIGURE 3-2

Addition of the higher order bits of a binary number. In adding the higher order bits, one of eight possible conditions will occur to produce the sum (S_1) and the carry (C_1), as shown in the right-hand column of the truth table.

The first four of these combinations are the same as decimal in that adding three zeros produces zero, and a single one plus any number of zeros equals one. Regardless of which input happens to be zero, two ones produce a zero carry one. The final condition is 1, plus 1, plus 1 from carry equals 1 carry 1.

Let us begin by comparing decimal and binary additions of numbers between 1 and 3. These numbers involve both LSB and MSB addition.

In decimal, 2 + 1 = 3. In binary, 0 plus 1 is 1. One plus 0 is 1. The answer is 11, which equals 3 in decimal, as we see below:

$$\begin{array}{r} 2 \\ +1 \\ \hline 3 \end{array} \longrightarrow \begin{array}{r} 10 \\ +01 \\ \hline 11 \end{array}$$

In decimal, 2 + 2 equals 4. In binary, (LSB) 0 plus 0 equals 0. (MSB) 1 plus 1 equals 0 carry 1. The answer 100 is 2^2 or (4_{10}). This addition is shown below:

```
         1
  2     10
 +2 →  +10
 --    ---
  4    100
```

In decimal, 3 plus 1 equals 4. In binary, (LSB) 1 plus 1 is 0 carry 1. (MSB) 0 plus 1 plus 1 from carry equals 0 carry 1. The answer is 100_2 or (4_{10}), as shown here:

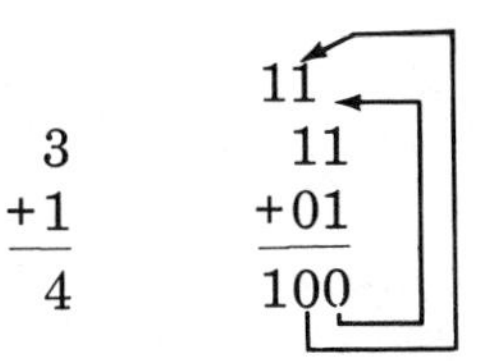

In decimal, 3 plus 2 equals 5. In binary, (LSB) 1 plus 0 is 1. (MSB) 1 plus 1 is 0 carry 1. The answer is 101_2 or (5_{10}), as we see below:

```
        1
  3    11
 +2   +10
 --   ---
  5   101
```

In decimal, 3 plus 3 equals 6. In binary, (LSB) 1 plus 1 is 0 carry 1. (MSB) 1 plus 1 plus 1 from carry is 1 carry 1. The answer is 110_2 or (6_{10}), as shown here:

```
       11
  3    11
 +3   +11
 --   ---
  6   110
```

These same procedures can be extended to larger numbers, as shown in Figure 3-3.

Fractional binary numbers, as well as whole numbers, can be added:

```
  1
 .0101     5/16
+.0110   + 3/8
------   -----
 .1011    11/16
```

```
   1110111  ← Cin
   1010101     85
 + 1110011    115
 ---------    ---
  11001000    200
                LSB  1 + 1                is 0 carry 1
 128            2SB  0 + 1 + 1 from carry is 0 carry 1
  64            3SB  1 + 0 + 1 from carry is 0 carry 1
   8            4SB  0 + 0 + 1 from carry is 1
 ---            5SB  1 + 1                is 0 carry 1
 200            6SB  1 + 0 + 1 from carry is 0 carry 1
                MSB  1 + 1 + 1 from carry is 1 carry 1
```

FIGURE 3-3

Addition of two seven-bit binary numbers. The binary numbers equivalent to 85 and 115 are added to produce the binary equivalent of 200.

Mixed numbers present no special problem, as we see here:

```
   111111          Cin
     11.1011      3 11/16
 +  110.0110     +6  3/8
 -----------     --------
   1010.0001     10  1/16
```

■ EXERCISES

1. Add the following pairs of binary numbers using the method A + B = S:

```
  1101      10101
 +1010     + 1011

  11011     10111
 +10010    +11011
```

Convert A, B, and S to decimal and confirm your answers.

2. Add the following pairs of binary numbers:

```
  .1011      .111      .10101
 +.00101    +.1001    +.01011
```

3. Add the following pairs of binary numbers:

```
  11011.101
 +  101.1011

  11100.011
 + 1011.101

     11.1101
 +1101.101
```

3.3 Binary Subtraction

3.3.1 Direct Subtraction

There is only one basic procedure for binary addition. However, for subtraction, there are at least three methods. First we will discuss the straightforward method, which is similar to the decimal procedure: subtract X from A to obtain a remainder, R. Again, we use subscripts to represent a subtraction of one four-bit number from another, as shown by the following:

$$\begin{array}{r} A \\ -X \\ \hline R \end{array} \longrightarrow \begin{array}{r} A_3A_2A_1A_0 \\ -X_3X_2X_1X_0 \\ \hline R_3R_2R_1R_0 \end{array}$$

$$\begin{array}{lcccc} A_0 & 0 & 1 & 0 & 1 \\ -X_0 & -0 & -0 & -1 & -1 \\ \hline R_0 - B_0 & 0 & 1 & 1 \text{ borrow } 1 & 0 \end{array}$$

A_0	X_0	R_0	B_0
0	0	0	0
1	0	1	0
0	1	1	1
1	1	0	0

FIGURE 3-4

Subtraction of the LSB. In subtracting the LSB of binary numbers, one of four possible conditions will occur to produce the remainder (R_0) and the borrow (B_0), as shown in the right-hand column of the truth table.

As in addition, we first examine the operations needed for the LSB. As shown in Figure 3-4, there are four possible combinations: two outputs, a remainder, and a borrow. The truth table shown in Figure 3-4 would be used to represent these functions.

Three of the operations in Figure 3-4 are identical to decimal subtraction. Let us look closely at the one that is different:

$$\begin{array}{r} 0 \\ -1 \\ \hline \end{array}$$

For this operation, we must borrow 1. After we borrow, it becomes:

$$\begin{array}{r} 10 \\ -\ 1 \\ \hline \end{array}$$

In decimal, this problem would be 10 minus 1 equals 9. In binary, it is one zero minus one. In binary, 10 equals a decimal 2. Therefore, 10 minus 1 in binary equals 1.

For the second significant digit and higher, there are three inputs and, therefore, eight possible combinations, as shown in Figure 3-5. The borrow has a value of only 1 in the column *from* which it is borrowed, but it has a value of 10 or 2_{10} in the column *for* which it is borrowed:

$$\begin{array}{r} 10 \\ -01 \\ \hline 01 \end{array} \quad \begin{array}{r} 2 \\ -1 \\ \hline 1 \end{array}$$

It is sometimes necessary to borrow from several columns over:

FIGURE 3-5

Subtraction of the higher order bits of a binary number. In subtracting the higher order bits, one of eight possible conditions will occur to produce the remainder (R_1) and the borrow (B_1), as shown in the right-hand column of the truth table.

$$\begin{array}{lcccccccc} -B_{in} & 0 & 0 & -1 & 0 & -0 & -1 & -1 & -1 \\ A_1 & 0 & 1 & 0 & 0 & 1 & 0 & 1 & 1 \\ -X_1 & 0 & 0 & 0 & -1 & -1 & -1 & 0 & -1 \\ \hline R_1 - B_0 & 0 & 1 & 1 \text{ borrow } 1 & 1 \text{ borrow } 1 & 0 & 0 \text{ borrow } 1 & 0 & 1 \text{ borrow } 1 \end{array}$$

A_1	X_1	B_{IN}	R_1	B_0
0	0	0	0	0
1	0	0	1	0
0	1	0	1	1
0	0	1	1	1
1	1	0	0	0
1	0	1	0	0
0	1	1	0	1
1	1	1	1	1

4	100
-1	-001
3	11

In this case, we must borrow from two columns over. In doing so, we take the one from the third column for the second column. The third column one (2^2) has a value of 2 in the second column (2^1). We leave one in the second column (2^1), and carry one to the first column (2^0), where it has a value of 2.

When we borrow from several columns over, the column from which the one is borrowed becomes zero. The column for which the one is borrowed receives a value equal to 2. All columns in between become one. An example, 32 minus 1, is shown in Figure 3-6.

The subtraction of 1 from 32, in binary, is similar to subtracting 1 from 100,000 in decimal:

```
 0 9 9 9 9(10)→ten
 1 0 0 0 0 0
-          1
  9 9,9 9 9
```

Some other examples of subtraction follow:

	-12		- 12
5	101	6	110
-2	- 10	-5	-101
3	11	1	1

■ EXERCISES

4. Subtract the following binary numbers A − X = R:

1010	1001	101011
- 111	- 11	-110101

3.3.2 Subtraction by Complement

Clearly, subtraction of binary numbers is more difficult than addition. There are, fortunately, methods of converting binary numbers to complements, which changes the operation to addition. That is, the subtraction can be done by

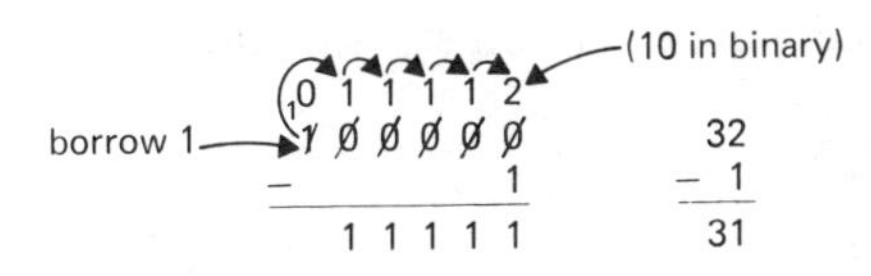

FIGURE 3-6
Borrowing 1 from five columns over.

complementing the larger number, adding, and then complementing the sum. The complement of a binary number is obtained by converting the ones to zeros and the zeros to ones. For example, to find the complement of A = 1011001, change the 1s to 0s and 0s to 1s with the following result:

$\overline{A} = 0100110$

Note that the notation for the complement of A is $\overline{A}$.

Another method for finding the complement of a number is to subtract it from another number having the same number of bits that are all ones. With this method, we thus subtract to find the complement of A = 1011001:

```
 1111111
-1011001
 0100110
```

Therefore,

$\overline{A} = 0100110$

To subtract A − X (when A > X) by complementing, use the method:

$$\begin{array}{l} \overline{A} \\ +X \\ \hline S = \overline{R}, \therefore R = \overline{S} \end{array}$$

With this method, we must first convert the larger number, A, to its complement, $\overline{A}$, before adding. After adding, the sum, S, is equal to the complement of the remainder, $\overline{R}$. R is, therefore, equal to $\overline{S}$.

Let us try an example in which A = 13 and X = 10. First convert the decimal numbers to binary:

A	13	1101
-X	-10	-1010

Find the complement of A and substitute:

$$\begin{array}{cr} \bar{A} & 0010 \\ +X & +1010 \\ \hline S & 1100 = \bar{R} \end{array}$$

Therefore,

$$R = \bar{S} = 0011_2 = 3_{10}.$$

■ EXERCISES

5. Determine the complements of the following binary numbers:

   A = 10110101

   B = 1110011

   C = 101.101

6. Subtract by complementing the larger number:

$$\begin{array}{r} 1101011 \\ -1001101 \\ \hline \end{array} \qquad \begin{array}{r} 110101 \\ -\ 11011 \\ \hline \end{array} \qquad \begin{array}{r} 111000 \\ -\ 11101 \\ \hline \end{array}$$

3.3.3 Ones Complement

In the preceding method for subtraction, the larger of the two numbers in the subtraction was complemented. In every case, the sum was complemented to obtain the remainder. The sign of the remainder would, of course, be predetermined by the mathematician. In operating a digital machine, however, it is not always convenient to have to determine the larger of two numbers. This operation may take as long as the subtraction itself.

If we use the ones complement method, the subtrahend is complemented regardless of the relative size of the numbers:

$$\begin{array}{c} A \\ -X \\ \hline R \end{array} \qquad \begin{array}{c} A \\ +\bar{X} \\ \hline S \end{array}$$

The problem occurs in handling the sum when there is a carry beyond the highest significant digit of the numbers in the subtraction. The carry, instead of forming the MSB, is added to the LSB of the sum. This type of carry is referred to as an *end-around carry* (*EAC*). An EAC occurs only when A > X. The occurrence of an EAC in a digital machine indicates a positive remainder. Here is an example of subtraction with an EAC:

$$\begin{array}{crr} A & 13 & 1101 \\ -X & -10 & -1010 \\ \hline \end{array}$$

Use the ones complement method:

$$\begin{array}{cr} A & 1101 \\ +\bar{X} & +0101 \\ \hline S & (1)0010 \\ & \rightarrow 1 \quad \text{EAC} \\ \hline & 0011 = +3 \end{array}$$

If the subtrahend (X) has fewer bits than the minuend (A), complement as many zeros to the left of the MSB as are needed to give X the same number of bits as A, as shown here:

$$\begin{array}{crr} A & 13 & 1101 \\ -X & -\ 3 & -0011 \\ \hline \end{array}$$

Then use the ones complement method:

$$\begin{array}{cr} A & 1101 \\ +\bar{X} & +1100 \\ \hline S & (1)1001 \\ & \rightarrow 1 \quad \text{EAC} \\ \hline & 1010 = +10 \end{array}$$

If no EAC occurs, A is less than X, and the operation is identical to that in Section 3.3.2. The following is an example:

$$\begin{array}{crr} A & 10 & 1010 \\ -X & -13 & -1101 \\ \hline \end{array}$$

Here the sum is equal to the complement of the remainder:

$$\begin{array}{cr} A & 1010 \\ +\bar{X} & +0010 \\ \hline S = \bar{R} & 1100 \quad \text{(no EAC)} \end{array}$$

Therefore,

$$R = \bar{S} = -0011_2 = -3_{10}$$

The machine, seeing no EAC, recognizes it as a negative answer in complement form and automatically complements it.

■ EXERCISES

7. Subtract by ones complement:

(pos. R)

A	11011	11001
-X	-10101	-10100
R		

(add zeros)

10011	1100101
- 1001	- 11101

(neg. R)

101101	100111
-1011010	-111001

Fractional numbers can also be handled by the ones complement method, as we see in this example in which

A = 7/8 = (28/32)

X = 17/32

In this problem,

A	28/32	.11100
-X	-17/32	-.10001

Use the ones complement method:

A	.11100	
$+\overline{X}$	+.01110	
S	(1).01010	
	→ 1	EAC
	.01011 = +11/32	

In the following example, A = 3/4 and X = 15/16, which must be converted to binary numbers:

A	12/16	.1100
-X	-15/16	-.1111

Then, the ones complement method is used:

A	.1100	
$+\overline{X}$	+.0000	
$S = \overline{R}$	.1100	(no EAC)

Therefore,

$R = \overline{S} = -.0011 = -3/16$

Mixed numbers present no special difficulty, as can be seen in the following example, in which

A = 9 1/2 (2/4)

X = 6 3/4

In this problem,

A	9 2/4	1001.10
-X	-6 3/4	-0110.11

Then:

A	1001.10	
$-\overline{X}$	+1001.00	
S	(1)0010.10	
	→ 1	EAC
	$10.11_2 = +2\ 3/4$	

If A = 7 5/16 and X = 11 1/2, the problem is solved as shown below:

A	7 5/16	0111.0101	
-X	-11 8/16	-1011.1000	← Equalize bits by adding zeros

Then use the ones complement method:

A	0111.0101	
$-\overline{X}$	-0100.0111	
$S = \overline{R}$	1011.1100	(no EAC)

Therefore,

$R = \overline{S} = -100.0011 = -4\ 3/16$

Proof: R + A = X	111.0101
	100.0011
	1011.1000 = X

■ EXERCISES

8. Subtract by ones complement:

```
  .10110        .10101
 -.10010       -.11011

 1010.1        101.0111
- 111.01      -1001.11
```

3.3.4 Twos Complement Subtraction

For one type of digital machine, the ones complement is the preferred method of subtraction; in other machines, the principle of twos complement is better. The twos complement of a number is a ones complement plus one. The twos complement of $A = \bar{A} + 1$. It can be obtained, as shown below, by subtracting A from the next power of 2 higher than A, which is a 1 followed by all zeros.

$A = 21_{10} = 10101_2$

```
A̅ =  01010
   +     1
     01011 (twos complement)
```

```
Or, 100000
   - 10101
     01011 (twos complement)
```

It can also be accomplished by complementing only those bits to the left of the least significant one bit. As shown in Figure 3-7, the identical results are obtained using the ones complement plus 1.

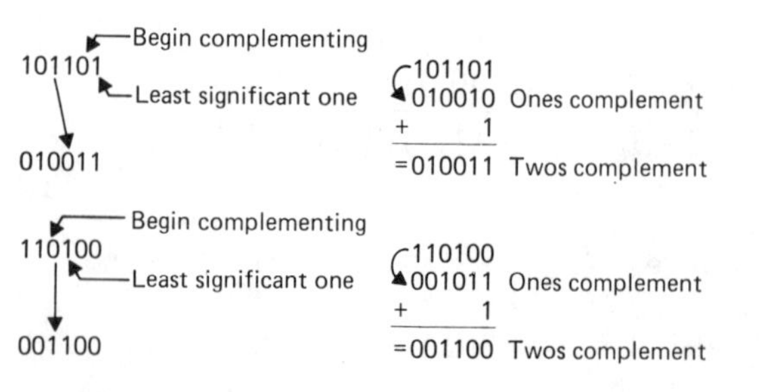

FIGURE 3-7

Alternate methods of producing the twos complement of binary numbers. The examples on the left show complementing after the least significant one bit; the examples on the right show the ones complementing plus 1 method.

■ EXERCISES

9. Using three of the above methods on each number, convert the following numbers to 2s complement:

```
101101      .10101     11.1011
1110110      .10101
101.01001
```

Here is an example of subtraction by twos complement, where A is greater than X:

```
 A     13     1101
-X    -10    -1010
```

Use the twos complement method:

```
   A              1101
+(X̅ + 1)        +0110
   S           (1)0011 = +11₂ = 3₁₀
               EAC
```

In this operation, an end carry or a one in a column higher than the MSB column of the numbers in the subtraction indicates a positive remainder. This end carry, however, is not a part of the remainder. It only indicates a positive sign.

The following is an example of subtraction by twos complement, where A is less than X:

```
 A     10     1010
-X    -13    -1101
```

Then:

```
   A              1010
+(X̅ + 1)        +0011
S = R̅ + 1        1101 (no EAC)
```

In this case, there is no end carry. The absence of an end carry indicates a negative remainder in the twos complement form. The correct answer can be obtained by taking the twos complement of S ($R = \bar{S} + 1$):

$$R = \bar{S} + 1 = 0010 + 1 = 0011_2 = -3_{10}$$

Fractional numbers can be subtracted using the twos complement method. In the following example:

A = 5/8 (10/16)

X = 7/16

In this problem,

$$\begin{array}{lrr} A & 10/16 & .1010 \\ -\overline{X} & -\ 7/16 & -.0111 \\ \hline \end{array}$$

Use the method:

$$\begin{array}{lr} A & .1010 \\ +(\overline{X}+1) & +.1001 \\ \hline S & (1).0011 = 3/16 \\ & \text{EAC} \end{array}$$

In the following example, A = 5/16 and X = 7/8:

$$\begin{array}{lrr} A & 5/16 & .0101 \\ -X & -14/16 & -.1110 \\ \hline \end{array}$$

Use:

$$\begin{array}{lrl} A & .0101 & \\ +(\overline{X}+1) & +.0010 & \\ \hline S = \overline{R}+1 & .0111 & \text{(no EAC)} \end{array}$$

Therefore,

$$R = (\overline{S}+1) = -.1001_2 = -9/16$$

Mixed numbers may also be subtracted with this method. If A = 3 5/8 and X = 2 3/4, the problem is solved as follows:

$$\begin{array}{lrr} A & 3\ 5/8 & 11.101 \\ -X & -2\ 6/8 & -10.110 \\ \hline \end{array}$$

Use the twos complement method:

$$\begin{array}{lr} A & 11.101 \\ +(\overline{X}+1) & +01.010 \\ \hline S & (1)00.111 \\ & \text{EAC} \end{array}$$

Therefore,

$$R = +.111_2 = +7/8$$

In a problem where A = 3 3/4 and X = 2 7/16, we proceed as follows:

$$\begin{array}{lrr} A & 3\ 9/16 & 011.110 \\ -X & -2\ 7/16 & -101.001 \\ \hline \end{array}$$

Then:

$$\begin{array}{lrl} A & 011.110 & \\ +(\overline{X}+1) & +010.111 & \\ \hline S = \overline{R}+1 & 110.101 & \text{(no EAC)} \end{array}$$

Therefore,

$$R = (\overline{S}+1) = -001.011 = -1\ 3/8$$

■ EXERCISES

10. Using the twos complement method, subtract the following numbers:

$$\begin{array}{rr} 1011 & 10110 \\ -\ 111 & -11001 \\ \hline \\ .01001 & .0101 \\ -.01101 & -.001 \\ \hline \\ 101.01 & 10.1 \\ -1000.101 & -110.001 \\ \hline \end{array}$$

3.3.5 Nines Complement Subtraction

Subtraction by adding complements can be applied to number systems other than the binary. The nines complement is sometimes used in operations with binary-coded decimal numbers. The nines complement is obtained by subtracting a decimal number from all nines. If A = 764,

$$\begin{array}{r} 999 \\ -764 \\ \hline 235 = \overline{A} \end{array}$$

If A = 23.675,

$$\begin{array}{r} 99.999 \\ -23.675 \\ \hline 76.324 = \overline{A} \end{array}$$

■ **EXERCISES**

11. Convert the following numbers to nines complement:

    7254 .875 16.725 138.62

Subtraction by nines complement proceeds as follows when A is greater than X:

```
 A    225
-X   -162
```

Use the nines complement method:

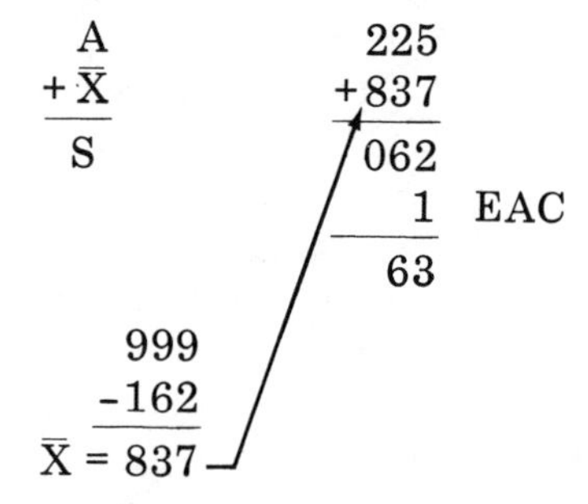

Subtraction by nines complement proceeds as follows when A is less than X:

```
 A    162
-X   -225
```

Then:

```
  A         162
 +X̄        +774
  S = R̄     936  (no EAC)

  999
 -225
X̄ = 774
```

Therefore, since $R = \bar{S}$,

```
 999
-936
 -63
```

■ **EXERCISES**

12. Subtract the following by the nines complement method:

```
 1762      154
-  521    -273

 23.65     .75
-54.7    -.875
```

3.4 Binary Multiplication

Binary multiplication is simpler than decimal multiplication in that it is only a matter of shifting and adding the multiplicand. Let us take 3 × 5 = 15, for example:

```
  101
×  11
  101   Multiply by 1
 101    Shift and multiply by 1
 1111₂ = 15₁₀
```

$1111_2 = 15_{10}$

For another example, let us consider the problem 10 × 26 = 260:

```
     11010
   ×  1010
     00000  Multiply by 0
    11010   Shift and multiply by 1
   00000    Shift and multiply by 0
  11010     Shift and multiply by 1
 100000100₂ = (256 + 4) = 260₁₀
```

$100000100_2 = (256 + 4) = 260_{10}$

Binary multiplication is possible with fractions. For example, to multiply 5/16 by 3/4, we proceed as follows:

```
  .0101 }  6 binary places
 × .11  }
   0101
  0101
.001111
   15  ↖(1/64)
```

Therefore,

.001111 = 15/64

Multiplication of mixed numbers is also possible, as illustrated in the example 4 3/4 × 3 5/8:

```
     100.11 }  5 binary places
  ×  11.101 }
      10011
     00000
    10011
   10011
  10011
 10001.00111
  17      7 ↖(1/32)
```

Therefore,

10001.00111 = 17 7/32

■ **EXERCISES**

13. Perform the following binary multiplications, convert to decimal, and verify your answers:

```
   1101       10001
× 10101      ×  1011

  11011       11110
×  1001      ×  1100

   1101        1010
×  .011      × 10.01

 1011.01      10.011
×  1.101     ×  11.01
```

3.5 Binary Division

Like multiplication, binary division is easier than decimal division. Let us divide 5 into 45, or 101 into 101101:

```
       Quotient
Divisor)Dividend
```

```
  9.            1    .
5)45.     101)101101.
             -101
                0
```

If the first three figures of the dividend is a binary number greater than or equal to the three figures of the divisor, place a one in the quotient, as shown above, and subtract the divisor from the first three digits of the dividend. Then bring down the next (lower) bit of the dividend to form the LSB of the remainder; 101 is larger than 01. Therefore, put a zero for the next lower bit of the quotient:

```
       10   .
  101)101101.
     -101
       01
```

Bring down the next lower bit of the dividend to the remainder; 101 is larger than 010. Therefore, put another zero bit in the quotient:

```
       100 .
  101)101101.
     -101
       010
```

Bring down the next lower (1) bit from the dividend to the remainder and 101 will divide into the remainder exactly once. Place the LSB (1) in the quotient and the division is complete:

```
       1001.              1001. = 9₁₀
  101)101101.       101)101101.
     -101               -101
       0101               0101
                         - 101
                             0
```

(The quotient above equals 9_{10}.)

As in decimal, binary division does not always come out even. Divide 12 into 57, or 1100 into 111001:

```
    4.75
 12)57.00
   -48
     90
    -84
     60
```

As with the decimal point, we must be careful to locate the binary point properly. The divisor goes into the first four bits of the dividend, leaving a remainder of 10:

```
            1.
  1100)111001.
      -1100
         10
```

Bring down the next lower bit of the dividend to the LSB of the remainder; 1100 is larger than 100. Place a zero in the next lower bit of the quotient:

```
           10 .
  1100)111001.
      -1100
        100
```

Bring down the next lower bit of the dividend to the LSB of the remainder; 1100 is larger than

1001. Place a zero in the next lower bit of the quotient:

```
        100.
1100)111001.
     -1100
       1001
```

Bring down the next lower bit of the dividend, now a zero, to the right of the binary point. This alerts us that any additional bits generated for the quotient will now be to the right of the binary point. Since 1100 will divide into 10010, place a 1 in the quotient to the right of the binary point:

```
        100.1
1100)111001.0
     -1100
      10010
```

Subtract the divisor from the remainder. Bring down the next zero bit from the dividend to the LSB of the remainder:

```
        100.1
1100)111001.00
     -1100
      10010
     - 1100
        1100
```

Since 1100 will divide evenly into this remainder, place the LSB 1 bit in the quotient:

```
        100.11 = 4 3/4
1100)111001.00
     -1100
      10010
     - 1100
        1100
       -1100
           0
```

Binary division, like decimal division, does not always come out even. In such cases, the machine will carry out to a set number of places and then truncate, or cut off the remainder.

■ **EXERCISES**

14. Perform the binary divisions shown below:

11)1111   1001)101101

1100)10010000   1010)100010

10.1)101.0   110.11)111.01

10110)110111

3.6 Summary

This chapter dealt with how the arithmetic operations—addition, subtraction, multiplication, and division—are performed using the binary number system.

When two binary numbers are added together, three of the four possible combinations are identical to decimal. The exception is 1 + 1 = 2 in decimal. In binary, 1 + 1 = 0 with a carry of 1.

Unlike binary addition, there are three commonly used procedures for binary subtraction—straightforward, ones complement, and twos complement. A complement of a binary number is obtained by changing all the 0s to 1s and all the 1s to 0s. Most computers now use the twos complement form of subtraction. In straightforward binary subtraction, three of the four operations are identical to decimal subtraction. The exception in binary is 0 - 1 = 1 with a borrow of 1.

In ones complement subtraction, the subtrahend is complemented and added to the minuend. If there is a carry beyond the highest significant digit, the carry is added to the least significant bit (LSB) of the sum. The carried number is referred to as an end-around carry (EAC). An EAC occurs only if the minuend is greater than the subtrahend.

The twos complement of a number is a ones complement plus one. In twos complement subtraction, the subtrahend is twos complemented and added to the minuend. In this operation, a carry is never added to the LSB. A carry is generated, though, if the minuend is greater than the subtrahend.

Binary multiplication requires only shifting and adding the multiplicand. It is usually easier than decimal multiplication.

Although binary division may be easier than decimal division, it is possible to get a remainder that does not come out even in binary, even though it may if the division is done in decimal. In these cases, the remainder is cut off after a set number of places.

Glossary

Ones Complement: The ones complement of a binary number is accomplished by converting all the 1 bits to 0 and all the 0 bits to 1s: The ones complement of number A is expressed as $\bar{A}$. If A = 1011001, then $\bar{A}$ = 0100110. The ones complement can also be obtained by subtracting the number from another number with equal binary places that are all ones:

$$\text{If } A = 1011001, \text{ then } \bar{A} = \begin{array}{r} 1111111 \\ -1011001 \\ \hline 0100110 \end{array}$$

Twos Complement: The twos complement of a number is equal to the ones complement plus 1. If A = 1011001, then the twos complement of

$$A = \bar{A} + 1 = \begin{array}{r} 0100110 \\ + \quad\quad 1 \\ \hline 0100111 \end{array}$$

Nines Complement: The nines complement of a decimal number can be accomplished by subtracting that number from another number having equal decimal places that are all nines. If A = 236, the nines complement of

$$\begin{array}{r} 999 \\ A = -236 \\ \hline 763 \end{array}$$

End Carry: A carry having a significance (power of 2) greater than the MSB of either term in the addition. In adding numbers of like sign, the end carry forms the MSB of the sum. In subtraction by complements, an end carry indicates a positive remainder.

End-Around Carry: An end carry generated during ones complement subtraction is taken around and added to the LSB of the sum.

Questions

1. For the three-input addition, what combination yields a binary 1 for the sum and carry?
2. What is the truth table for a two-input binary subtraction?
3. Is an end-around carry generated in every ones complement subtraction operation? If your answer is no, which combination produces an EAC?
4. Is an end-around carry generated in twos complement subtraction? If your answer is yes, give an example.
5. What type of complement subtraction is used with BCD numbers?
6. What two major functions are characteristic of binary multiplication?
7. In some binary division problems, does the remainder have to be truncated?

Problems

3-1 Add the following binary numbers:

$$\begin{array}{r} 10011 \\ +\ 1101 \\ \hline \end{array} \quad \begin{array}{r} 101110 \\ +\ 11011 \\ \hline \end{array} \quad \begin{array}{r} 11011011 \\ +\ 1001001 \\ \hline \end{array}$$

$$\begin{array}{r} .01011 \\ +.1101 \\ \hline \end{array} \quad \begin{array}{r} .11001 \\ +.00101 \\ \hline \end{array} \quad \begin{array}{r} .001011 \\ +.100111 \\ \hline \end{array}$$

$$\begin{array}{r} 11.011 \\ +101.1101 \\ \hline \end{array} \quad \begin{array}{r} 1011.011 \\ +\ 110.1011 \\ \hline \end{array} \quad \begin{array}{r} 111.11 \\ +1010.101 \\ \hline \end{array}$$

3-2 Convert to both ones and twos complements:

10.011 1.101 1011.1 11.011 1011 1101

11001 10111 11.111 1011.101 .11011 .1001

3-3 Subtract by the ones complement method:

101101 − 11001 101101 − 110110 101111 − 1111001

10.1101 − 1.101 111.11 − 101.101 11011.111 − 100.01

3-4 Subtract by the twos complement method:

10011 − 1110 10011 − 11001 110001 − 111010

101.1011 − 11.11 11.1001 − 1.1101 1.0011 − 11.01

3-5 Determine the nines complement of the following numbers:

1536 2379 516 372 .175 .325

.675 .876 11.25 15.625 34.5 55.75

3-6 Subtract by nines complement:

625 − 317 2237 − 378 6254 − 89

587 − 8312 15.62 − 27.35 3.875 − 5.625

3-7 Multiply the following binary numbers:

111101 × 111 11001 × 10101 101101 × 101010

11.101 × 1.11 1011.1011 × 110.11 1011.011 × 10.1

3-8 Divide the following binary numbers:

101)111100 1001)1001000 1100)10000100

1011)10001111 1100)1111000

CHAPTER

Digital Signals and Switches

Objectives

Upon completion of this chapter, you will be able to:

- Identify the electrical nature of a digital signal.
- State the importance of timing and how it is accomplished in the digital system.
- Show the importance of clock cycles to a digital system operation.
- Identify binary numbers in serial and parallel form.
- Identify the characteristics of manual and automatic switches used in digital machines.
- Write and simplify Boolean equations.
- Identify logic gates and symbols.
- Draw digital circuits from Boolean equations.
- Apply DeMorgan's theorem to simplify Boolean equations.
- Apply Boolean functions for parallel operation.

4.1 Introduction

In Chapter 3, we discussed the binary system of arithmetic and discovered that the arithmetic operations—addition, subtraction, multiplication, and division—can be accomplished in the binary number system as accurately as in the decimal system. The results are the same with respect to quantities in either system; that is, A + B = X, whether it be 6 + 8 = 14 or $110_2 + 1000_2 = 1110_2$.

In the decimal system, we have fewer places for the same quantities, but must deal with 0 through 9 digits. Electrically, we could represent the digits of the decimal system by voltage levels such as 0 through 9 volts or 1 volt per digit. The electronic device needed to differentiate between these voltage levels is an elaborate one. The devices needed to transmit, receive, and store these decimal voltage levels without shifting them up or down in value are expensive and difficult to keep in alignment.

In the binary system, on the other hand, the electrical quantities 1 and 0 can be represented by two distinct electrical states or two voltage levels. The two voltages representing binary 1 and 0 can be degraded as much as 50 percent during transmission, reception, and storage and still be recognized. The devices handling the binary signals may even restore them to ideal levels as they are processed.

Boolean algebra can be used to predict what the output will be after these digital signals pass through a circuit. Section 4-7 introduces Boolean algebra and the digital logic gates that conform to the Boolean rules.

4.2 Binary States

Mathematically speaking, there are two binary states, the digits 1 and 0. In the majority of digital systems, binary 1 and 0 are represented by two distinct voltage levels—zero volts for 0, and a positive voltage for 1. If the 1-level voltage is the more negative voltage, as is true for the -4.5V and 0V system in column 1 of Figure 4-1, then it is referred to as negative logic. When the 1-level voltage is more positive than the 0 level, as in the other three columns of Figure 4-1, it is called positive logic. Figure 4-1 shows some typical logic levels. For the sake of simplicity, we will use positive logic.

To avoid possible ambiguity in using the terms *one* and *zero*, because zero volts may not always be binary 0, and the term *one input* may refer to a single input and not that which has a binary 1 level on it, the terms *high level* and *low level* are often used.

The terms *true* and *false* are sometimes used in place of 1 and 0. These are logical terms taken from Boolean algebra, and they are occasionally used by engineers and technicians. Figure 4-2 compares these terms with positive logic. Logically, or mathematically, positive and negative logic systems are the same, or are analyzed in the same fashion.

POS. LOGIC	LOGIC TERMS IN USE		
+5V	1	HIGH	TRUE
0V	0	LOW	FALSE

FIGURE 4-2
Alternative logic terms for one and zero.

4.3 Digital Signals

In any operating digital system, there is constant transmission, reception, and processing of signals between various components and subsystems—most of which we might expect to be representations of binary numbers. Often, however, they are control signals, and in digital communications they may represent letters or punctuation. The term *signal* here refers to intelligence transmitted in the form of periodic changes between the 1 and 0 levels. These are generally pictured as they would appear on an oscilloscope, a voltage-versus-time waveform. There are two general terms used for signals with respect to duration—*level* and *pulse*, as shown in Figure 4-3. The

FIGURE 4-1
Binary levels.

	TYPE LOGIC			
DIGIT	NEGATIVE	POSITIVE	POSITIVE	POSITIVE
1	−4.5V	0V	+5V	+6V
0	0V	−12V	0V	−6V

FIGURE 4-3

Typical digital signal waveforms.

FIGURE 4-4

Typical negative-going waveforms.

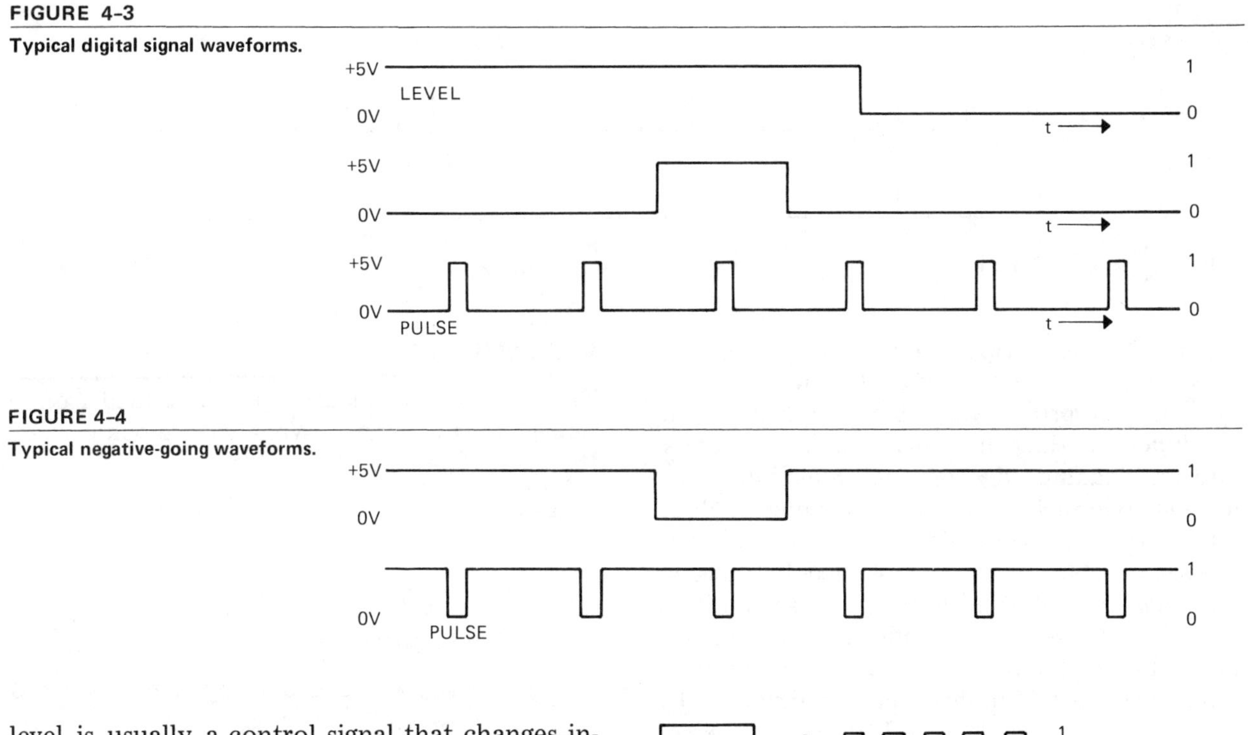

level is usually a control signal that changes infrequently and remains at 1 and 0 levels for a long time.

The pulse is a change in level of very short duration. In defining these terms, we face the question, How long is long and how short is short? The terms are relative, but in a system controlled by a clock, *pulse* usually means a level change of less duration than a clock period. The digital clock is a pulse generator that generates periodic voltage pulses.

The same general terms would apply if the resting state of the signal were 1 and the changes occurred to 0, as Figure 4-4 shows.

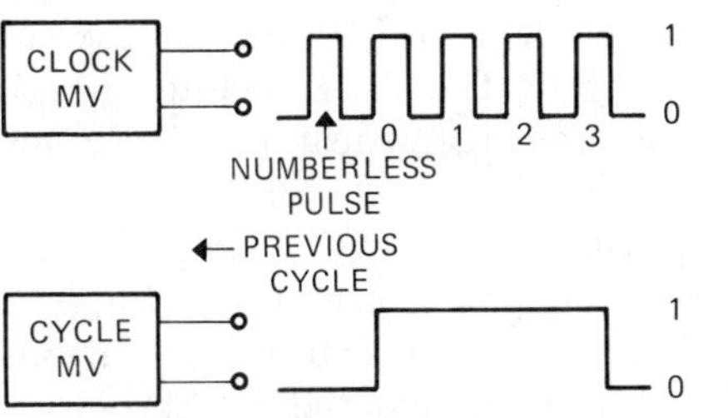

FIGURE 4-5

Multivibrator outputs used to provide timing within a digital machine.

4.4 Timing

The majority of digital machines require some exact degree of timing. Timing is generally supplied from a digital clock, which provides either a squarewave or train of pulses with exact frequency. Multivibrators, uni-junction pulse generators, or crystal-controlled oscillators can be used.

A device that puts out continuous clock pulses alone is as worthless as a mechanical clock without a face. Most digital operations occur in cycles, and, as Figure 4-5 shows, another timing device may be needed to set up those cycles. The read-write cycle of a computer memory occurs automatically in cycles of so many clock pulses. The add cycle of the arithmetic unit is another. Somehow we must designate when these cycles begin. As Figure 4-5 shows, the first clock pulse occurring after the cycle MV goes high may be designated as the zero pulse. Every pulse thereafter is numbered like the numbers on the face of a clock. A new zero pulse is generated each time the cycle MV goes high and the numbering begins anew.

Another method, described in Chapter 16, uses a digital counter that counts the necessary number of pulses that form a cycle, starting over with a zero pulse at the end of each count. This system provides two all-important timing signals, as Figure 4-6 shows.

The term *leading edge* is the change between 1 and 0 that occurs at the beginning of

FIGURE 4-6

Timing Signals.

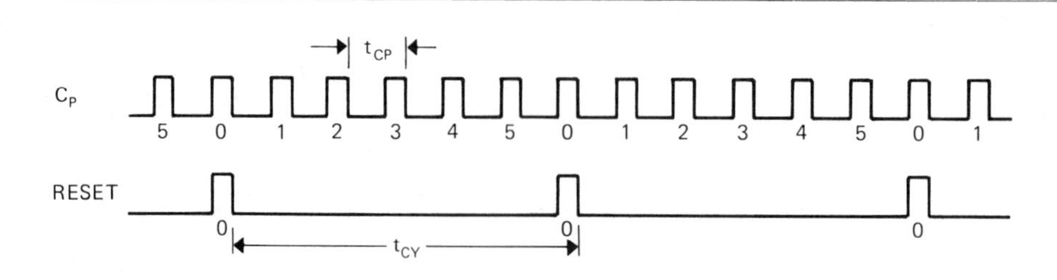

a pulse. The term *trailing edge* is the change between 1 and 0 that occurs at the end of a pulse.

An important time measurement is the *clock period*—the time, usually in microseconds, between leading edge of one clock pulse and leading edge of the next. If we assume all clock pulses are of the same width, as is the usual case, the same value should be obtained by measuring between trailing edge of one pulse and trailing edge of the next. A time measurement between like edges of two zero pulses is the *cycle time.* The cycle time can also be obtained by multiplying the clock period by the number of clock pulses per cycle. The clock period can also be obtained by the reciprocal of the clock frequency.

Later we will learn how to isolate single clock pulses on a separate line and send them to other parts of the machine, causing operations to occur at the time designated by the clock pulse.

The clock pulse is directly or indirectly used in generating the signals within a digital system. It is not surprising that many of the signals we examine show edges that coincide with the leading or trailing edge of a clock pulse. The clock and the reset, or zero, pulse lines are the two most commonly used timing signals. The clock period (t_{CP}) and the cycle time (t_{CY}) are important time measurements that indicate the operating speed of the machine.

■ **EXAMPLE 4-1**

The clock line in Figure 4-6 is obtained from a 2-megahertz generator. What is the clock period? the cycle time?

Solution:

$$t_{CP} = \frac{1}{\text{clock frequency}} = \frac{1}{2 \times 10^6\text{Hz}}$$
$$= (0.5\mu\text{sec})$$

$$t_{CY} = t_{CP} \times 6 \text{ pulses per cycle} = .5\mu\text{sec} \times 6$$
$$= (3\mu\text{sec})$$

4.5 Signals Representing Binary Numbers

4.5.1 Serial Numbers

Signals representing binary numbers occur in both parallel and serial form. In serial numbers, the binary places or powers of 2 are represented by clock or time periods. They may occur in either pulse trains or level trains. Figure 4-7

FIGURE 4-7

Binary number 10011 represented by a serial pulse train. Time periods between clock pulses are designated as 2^0 through 2^4 in value. The presence of a pulse is a 1; the absence of the pulse, a 0.

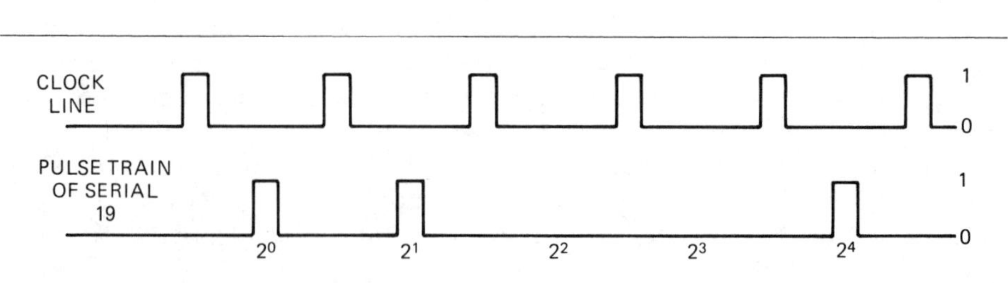

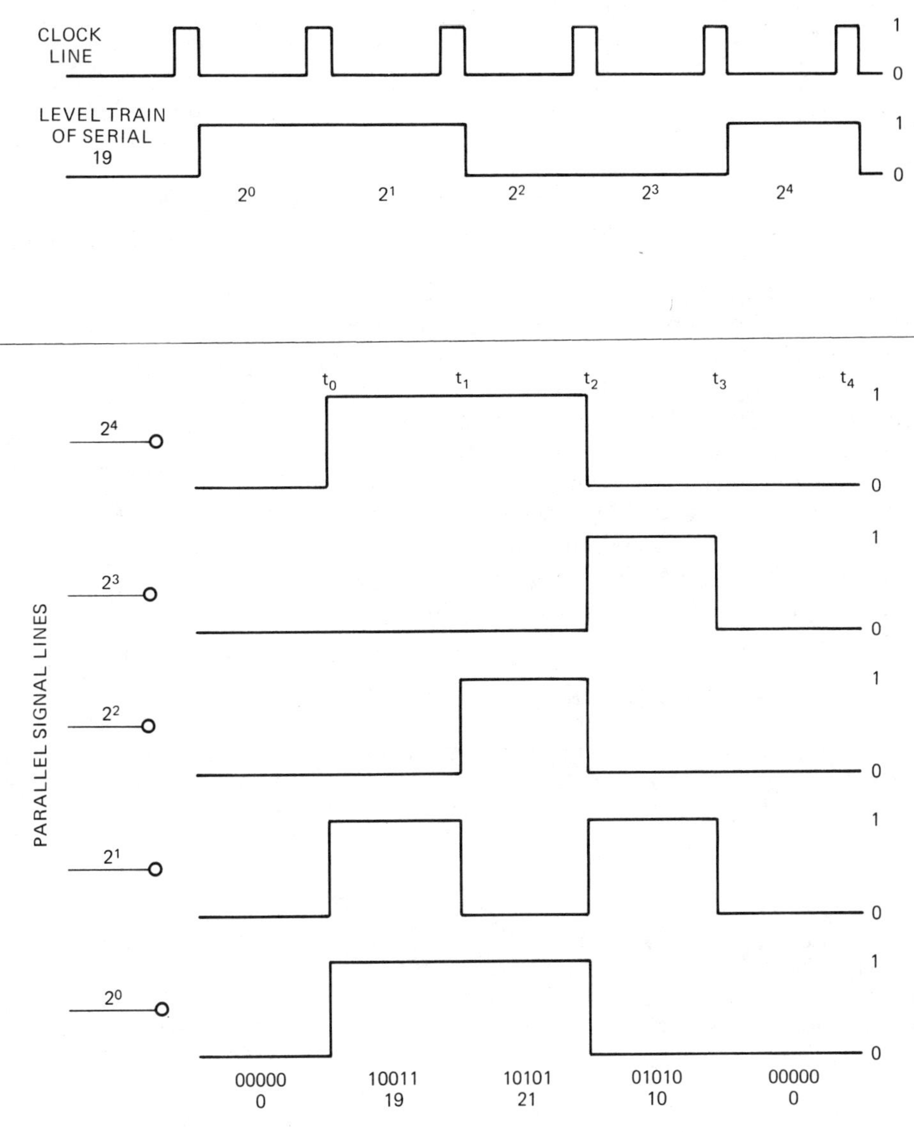

FIGURE 4-8

Binary number 10011 represented by a serial level train. A 1 is represented by a high level that lasts the entire period; a 0 is represented by a low level that lasts the entire period.

FIGURE 4-9

Waveforms resulting from a five-bit parallel number. Each waveform is one line showing only how that individual bit is changing, but the entire number can be transferred in one clock period.

shows a pulse train representing the binary number for 19. The LSB occurs on the left because it is first in time and the oscilloscope will present it on the left. The same number might also appear as a level train, as Figure 4-8 shows. In this signal, the 1 level remains for the full period. If two or more adjacent ones occur, there is no change in level between them.

The serial numbers have the advantage of requiring only one line or channel for processing them, but they require a separate clock period for each binary place, and for some applications this makes them too slow.

4.5.2 Parallel Numbers

A parallel number, which requires a separate wire or channel for each binary place, is expensive with regard to circuits. The entire number can be transferred in one clock period, making it many times faster than the serial number. Figure 4-9 shows the waveforms that would occur simultaneously on the five lines or parallel channels of a five-bit number that is changing from 0 to 19 at t_0, to 21 at t_1, to 10 at t_2, and back to 0 at t_3. A parallel number may occur in pulse form as well as levels.

4.6 Switching of Digital Signals

4.6.1 Binary State of Electrical Devices

The 1 and 0 levels of voltage are not the only electrical states that may be assigned binary values. A lamp may represent a 1 when turned on and a 0 when turned off. A switch may be in the 1 state when on, the 0 state when off. A relay may be a 1 when energized, a 0 when de-energized. Figure 4-10 lists electrical devices and the states that may be assigned a binary number value. Of the devices in Figure 4-10, only one is not a form of switch. These devices are of particular interest to us because it is complex, high-speed switching that accomplishes the many remarkable functions occurring in digital machines.

4.6.2 The Ideal Switch

The ideal switch for digital machines must have the following characteristics:

1. Be automatic (turned off and on with binary 1 or 0 levels).
2. Be high-speed (instantaneous transition from off to on state).
3. Have infinite off resistance.
4. Have 0 on resistance.
5. Be in isolation from the signal turning it on and off.
6. Consume minimum power in the off and on states.

Figure 4-11 compares electrical switches with respect to these characteristics.

The manual switch and relay have two other disadvantages—contact bounce and large size in comparison to diodes and transistors, particularly if the diodes and transistors are in integrated circuit form. Contact bounce is the creation of sporadic and irregular voltage spikes during the instant that contact is made or broken. These spikes may appear like a multitude of pulses, which will cause erroneous operation of some types of digital circuits.

Because of its simplicity, we will use the manual switch to discuss some general switching

FIGURE 4-10

Electrical devices and their binary states. The seven electrical devices have distinct operating states, which can be assigned the binary values 1 and 0.

VALUE *	LAMP	SWITCH	RELAY	DIODE	VACUUM TUBE	TRANSISTOR	FET
1	ON	CLOSED	ENERGIZED	FORWARD BIASED	SATURATION	SATURATION	SATURATION
0	OFF	OPEN	DE-ENERGIZED	REVERSE BIASED	CUT OFF	CUT OFF	CUT OFF

*Depending on construction, transistor switches may produce an inversion from 1 to 0.

FIGURE 4-11

Comparison of ideal switching characteristics in the five most important electrical switches.

CHARACTERISTICS	TOGGLE SWITCH	RELAY	DIODE	TRANSISTOR	FET
AUTOMATIC	NO	YES	YES	YES	YES
SPEED	SEC.	mSEC.	nSEC.	nSEC.	μ SEC.
OFF RESISTANCE	∞	∞	MΩ	MΩ	MΩ
ON RESISTANCE	0Ω	0Ω	OFFSET V_D	VOLTAGE $V_{CE\,SAT}$	50 TO 200Ω
ISOLATION	PERFECT	EXCELLENT	NONE	POOR	GOOD
POWER CONSUMPTION	NONE	HIGH	SIGNAL PWR. ONLY	LOW	VERY LOW

functions before proceeding to more complex devices.

4.6.3 The Manual Switch

The manual toggle switch in Figure 4-12 has input and output and two states, OFF and ON. These two states can be designated 0 and 1, as marked, which gives the switch a numerical value, to be represented by the algebraic term A. If we assume the switch to be a part of a digital system, the input will have either a 1 or 0 level and can be represented by an algebraic term, B. The output, which we will designate with the algebraic term X, depends on both A and B and is, in fact, equal to A times B.

Figure 4-13 shows the algebraic representation of the switch. This equation may seem trivial in terms of ordinary algebra, but with both A and B limited to the values 1 and 0, it becomes quite useful. This particular form of algebra is known as Boolean algebra, and in this application may be referred to as switching algebra. It follows the rules of ordinary algebra, but there are some special conditions because values are limited to 1 and 0.

The term $X = A \cdot B$ is referred to as an AND function or AND multiplication. It has only four possible conditions.

A chart known as a truth table is often used to plot the possible conditions for a given switching function. As can be seen from the truth table of Figure 4-14, of the four conditions, only where a 1 exists on both A and B does the value X equal 1. This condition is called an AND function. Let it be understood, however,

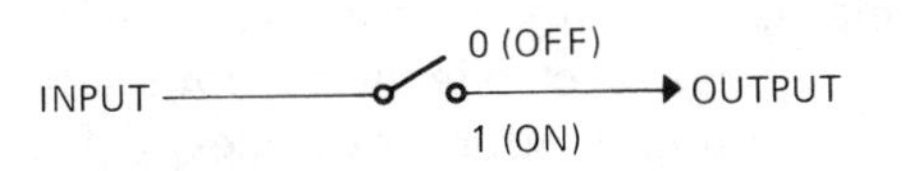

FIGURE 4-12
Toggle switch showing binary values assigned to off-on positions.

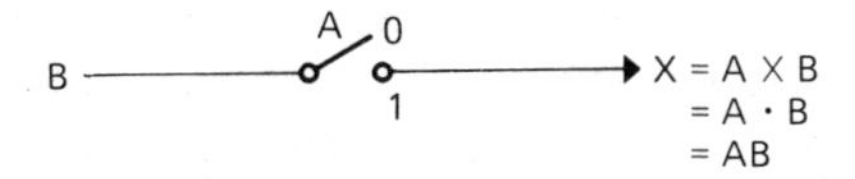

FIGURE 4-13
Algebraic representation of a toggle switch. Such an algebraic term may be part of a Boolean algebra equation.

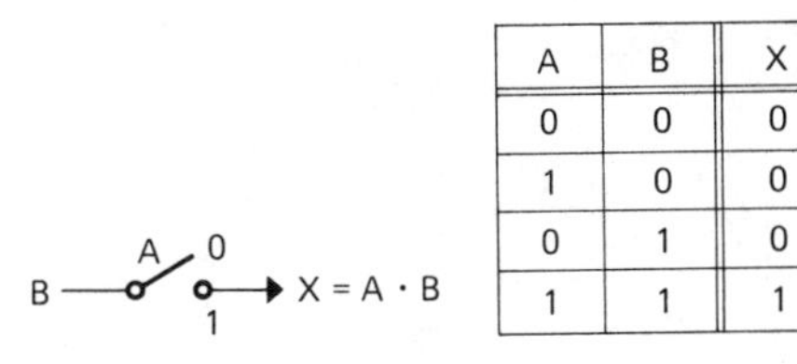

A	B	X
0	0	0
1	0	0
0	1	0
1	1	1

FIGURE 4-14
Truth table listing all the possible conditions for the given Boolean expression.

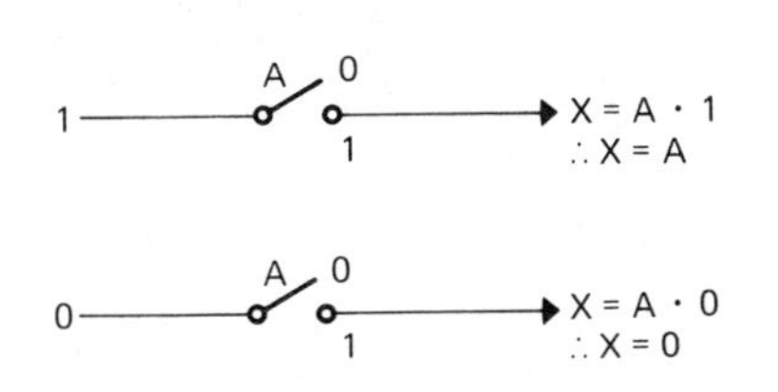

FIGURE 4-15
Simplification of Boolean expressions when the variables are made constant (a fixed 1 or 0).

that the switch (A) alone is not an AND function. It is an AND function only in conjunction with the variable (B).

If the input of the switch were connected to the power supply (a constant 1), the function would be $X = A \cdot 1$. As in ordinary algebra, anything multiplied by 1 equals itself, and so $X = A$. If the input of the switch is terminated at ground (a constant 0), $X = A \cdot 0$. As in ordinary algebra, anything multiplied by 0 equals 0. Figure 4-15 shows these conditions.

If several switches are combined in the same circuit, the Boolean functions and a truth table can still apply, as Example 4-2 shows.

■ **EXAMPLE 4-2**

Using Figure 4-16 (a), determine the Boolean equation for X′ and X. Construct the truth table for X.

Solution:

$$X' = A \cdot 1 = A$$

Therefore,

$$X = X' \cdot B = A \cdot B$$

The truth table is shown in Figure 4-16(b).

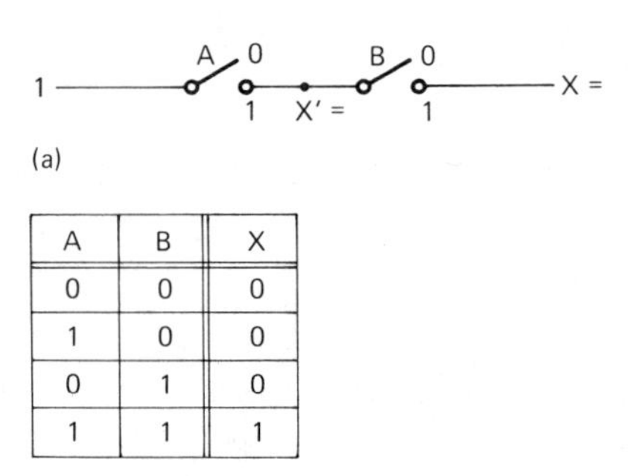

A	B	X
0	0	0
1	0	0
0	1	0
1	1	1

FIGURE 4-16
Example 4-2.

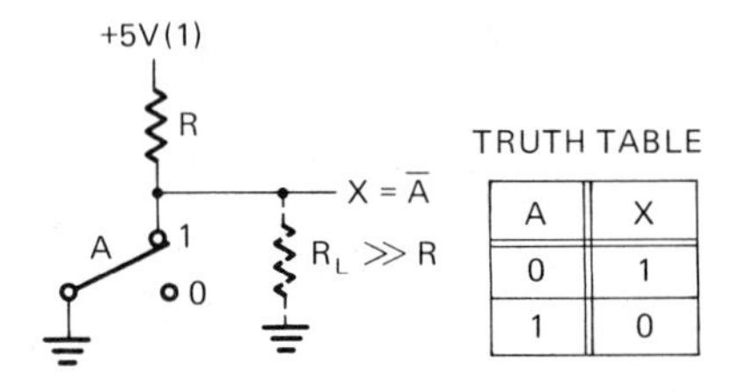

A	X
0	1
1	0

FIGURE 4-17
Inverting switch. The 1 level need not always pass through the switch; it may be applied or removed from the output by the action of the switch.

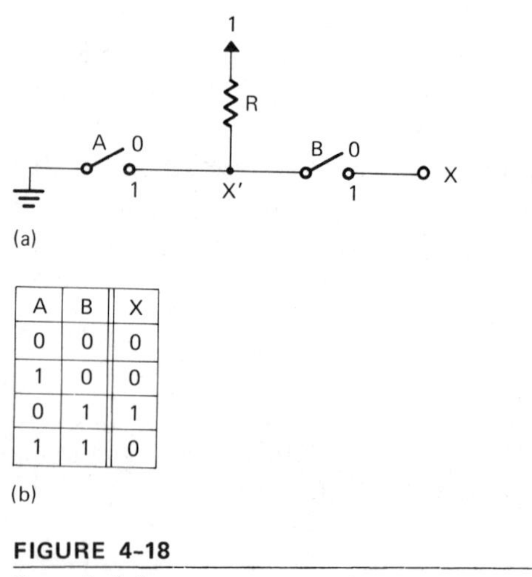

A	B	X
0	0	0
1	0	0
0	1	1
1	1	0

FIGURE 4-18
Example 4-3.

It may seem strange to apply such elaborate analysis to a simple single-pole switch. However, as we develop complex arrangements of series and parallel switching circuits, these procedures become quite useful. It is advisable to gain an understanding of Boolean notations and truth tables while the circuits are still simple.

4.6.4 Shorting Switch

Another switching function is accomplished by shorting an output line to ground with a switch. As Figure 4-17 shows, a resistor is necessary to prevent shorting the 1-level voltage bus. When this switch is turned on, the X is connected to ground, causing all the 1-level voltage to drop across the resistor, R. When the switch is open, the 1 level will appear at X, provided $R_L \gg R$. The switch is an inverting switch in that A and X always have opposite values. When A is 0, X is 1; when A is 1, X is 0. X is the complement of A ($X = \bar{A}$).

If inverting and noninverting switches are combined in the same circuit, Boolean equations and truth tables still apply, as Example 4-3 shows.

■ **EXAMPLE 4-3**

In Figure 4-18(a), determine the Boolean equation for X. Construct the truth table.

Solution:

$$X' = \bar{A}$$

Therefore,

$$X = \bar{A} \cdot B$$

The truth table is shown in Figure 4-18(b).

4.6.5 The Relay as a Switch

The relay was one of the earliest automatic switching devices. The telephone system reached its present state of high speed and reliability by the use of relays, and only recently, and with some reluctance, has the telephone company begun to change to semiconductor devices.

Figure 4-19(a) is a schematic diagram of a two-pole relay. The sets of contacts are similar to double-pole, double-throw switches. Their terminals are usually designated as common, normally open, normally closed. In a complex switching assembly, it is too awkward to use the schematic symbols that locate contacts and coil together. Such a drawing uses the logic symbols shown in Figure 4-19(b). The individual contacts or the coil may appear at separate points on the drawing, as long as they are marked. When the energizing voltage is suddenly removed from a relay coil, it creates a reverse high-voltage spike, which is shorted out by the diode placed across the coil.

When relays and their contacts are wired into a logic circuit along with switches, the outputs can be expressed in terms of a Boolean equation.

■ EXAMPLE 4-4

Using the circuit shown in Figure 4-20(a), prove that the output X can be expressed in terms of A and B. Construct its truth table.

Solution: X = K2, but each relay is a function of the switches and contacts needed to turn it ON. Therefore,

$$K1 = A \qquad K2 = \overline{K1} \cdot B$$

Therefore,

$$K2 = \overline{A} \cdot B \text{ and } X = \overline{A} \cdot B$$

A 1 level appears at X only if A is OFF and B is ON. The truth table is shown in Figure 4-20(b).

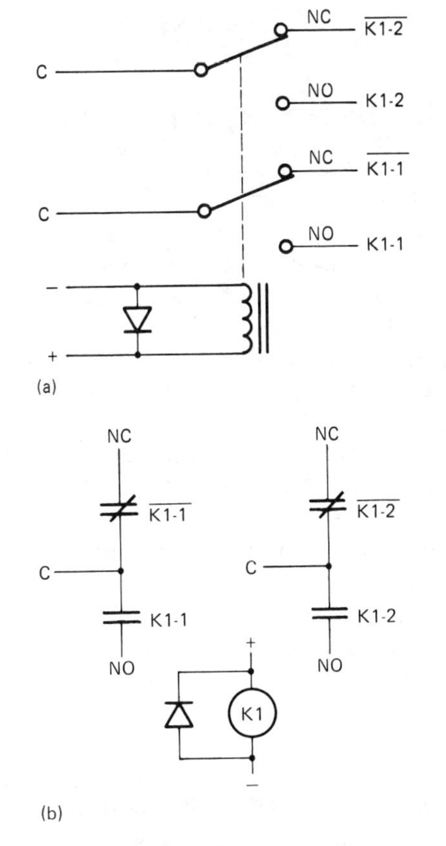

FIGURE 4-19

(a) Schematic diagram of relay designated K1. K1-1 and K1-2 have values of 1 when K1 is energized; $\overline{K1\text{-}1}$ and $\overline{K1\text{-}2}$ have values of 1 when K1 is de-energized. (b) Logic symbol for the same relay.

The relay is, of course, automatic in that the logic 1 level is usually the voltage level needed to energize the relay. It is high-speed in comparison to human reaction time and, therefore, adequate for semiautomatic equipment. But the relay pull-in and drop-out time is normally measured in milliseconds, as compared to microsecond or even nanosecond delay times for semiconductor switches. Although miniature relays with low power consumption have been developed, relays require more power than semiconductors and occupy considerably more space. Entire computing systems, such as missile-launching pads, have been built with relays, but they were developed before the availability of low-priced reliable semiconductors.

Relay assemblies may still be advantageous in less complicated devices using high current or high voltage, or for use in conjunction with semiconductors where contact closures are needed to drive high-current solenoids and motors, or for switching signals that are not at the logic level.

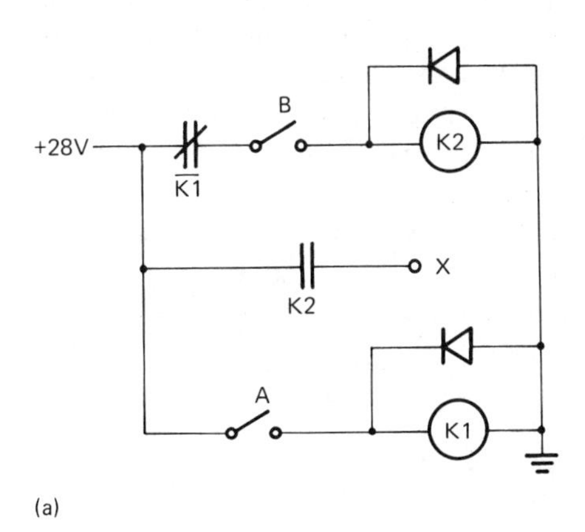

(a)

A	B	X
0	0	0
0	1	1
1	0	0
1	1	0

(b)

FIGURE 4-20

Example 4-4.

4.7 Boolean Algebra

4.7.1. The Boolean Equation

As the foregoing demonstrates, the outcome of passing digital levels through a network of switches can be predicted with an algebraic equation. This form of algebra was developed by an English mathematician named George Boole around 1854, which was some years before electrical switching was even considered.

Boolean algebra was aimed at the dual state of true-false logic, an assumption that any proposition can be proven with correct answers to a specific number of true-false questions. (With each question expressed as an independent variable and the conclusion as a dependent variable in an algebraic equation, it was shown that logic could be handled as a mathematical operation.)

The Boolean operation, in addition to the truth of a theorem, may point out the redundancy of variables. Since Boolean algebra deals with the dual states (true and false), it lends itself readily for use on binary machines with the dual states 1 and 0. In this application, it can be used to determine the correctness of circuit functions and possible redundancies in circuitry.

Boolean algebra uses only the operators multiply and add. Multiply is referred to as an AND function $A \cdot B = AB$. Add is referred to as an OR function $A + B$. An equation such as $AX + BY = C$ could be meaningful as either a Boolean algebra equation or a mathematical algebraic equation. However, $5X + 9Y = 15$ could not be a Boolean algebra equation because, in Boolean algebra, both constants and variables are limited to the values 1 and 0. In spite of being restricted to integers 1 and 0, the laws of normal algebra apply.

Commutative Law

$$A + B = B + A$$
$$AB = BA$$

Associative Law

$$A + (B + C) = (A + B) - C$$
$$A(BC) = (AB)C$$

Distributive Law

$$A(B + C) = AB + AC$$

4.7.2. Logic Gates and Symbols

A standard set of symbols are used to describe switching circuits in terms of Boolean functions. Figure 4-21(a) shows the logic symbol for an AND function of three variables. Figure 4-21(b) is the AND function produced with manual switches. In order for a 1 voltage level to appear at the output on either, a 1 must exist on inputs A and B and C. The truth table in Figure 4-21(c) shows all eight possible combinations that may exist on the inputs, and the only set that results in a 1 at the output is 1 on all three inputs. The proof of this is simple when one considers that if, in a multiplication of three variables, any one variable is zero, the product is zero. On the other hand,

$$1 \cdot 1 \cdot 1 = 1$$

The three-variable OR function is shown in Figure 4-22. The OR function provides a 1 at the output if any one of the inputs is a 1. Only

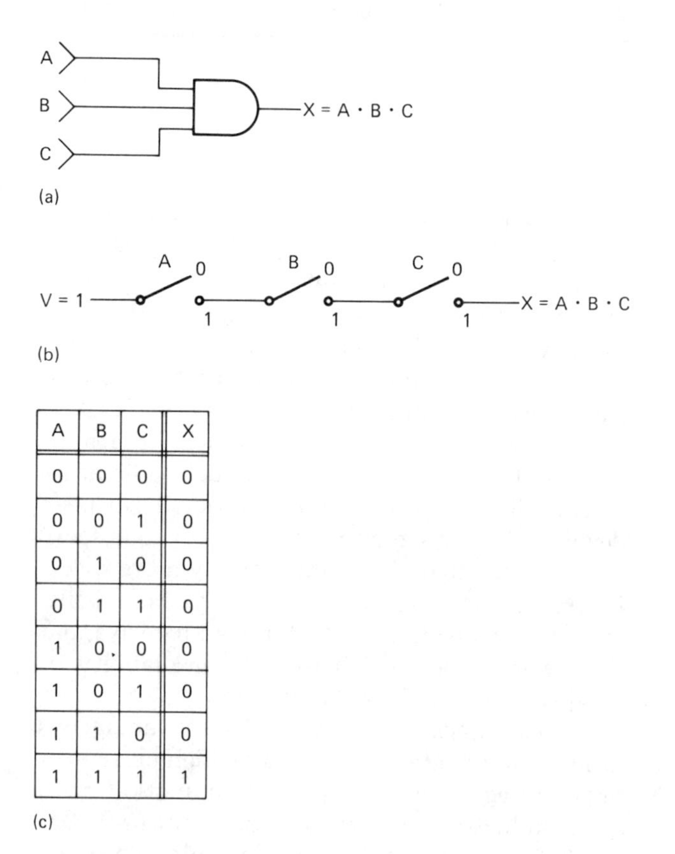

A	B	C	X
0	0	0	0
0	0	1	0
0	1	0	0
0	1	1	0
1	0	0	0
1	0	1	0
1	1	0	0
1	1	1	1

FIGURE 4-21

(a) Logic symbol, (b) mechanical switch equivalent, and (c) truth table for a three-input AND gate.

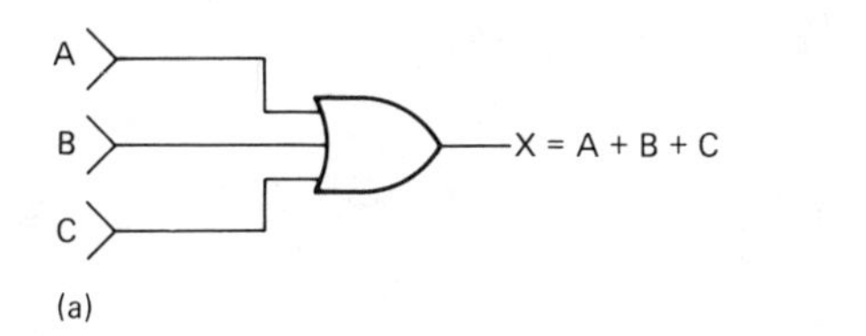

(a)

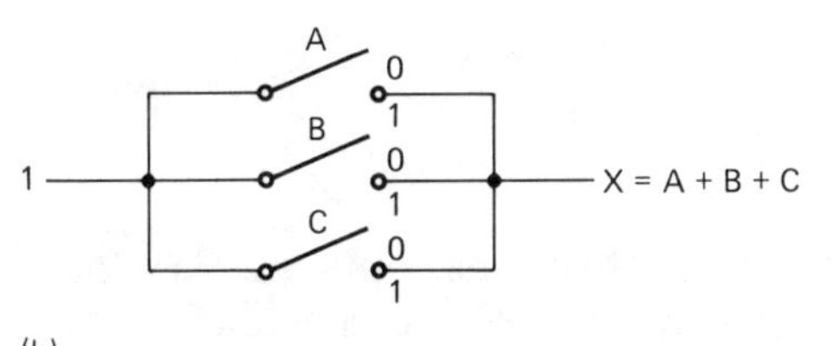

(b)

A	B	C	X
0	0	0	0
0	0	1	1
0	1	0	1
0	1	1	1
1	0	0	1
1	0	1	1
1	1	0	1
1	1	1	1

(c)

FIGURE 4-22

(a) Logic symbol, (b) mechanical switch equivalent, and (c) truth table for a three-input OR gate.

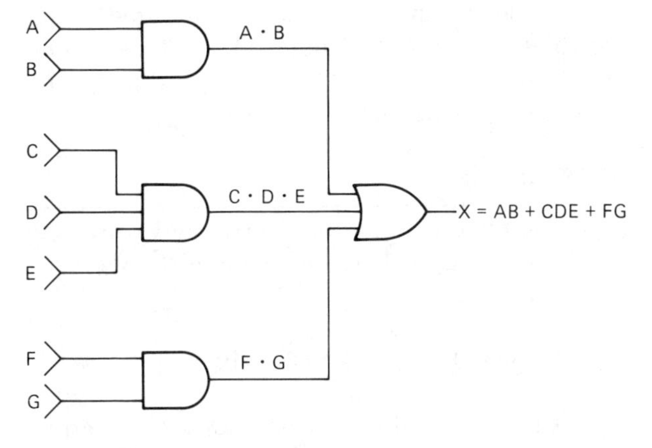

FIGURE 4-23

Three logic gates interconnected to form the desired Boolean function.

when all inputs are zero will the output be 0. Figure 4-22(a) shows the logic symbol. As the manual switch version of the OR function in Figure 4-22(b) shows, if any one of the switches is in a 1 state, the output will be 1. As in other forms of algebra, adding any number of zeros to a sum does not change its value. However, unlike other forms of algebra, ORing any number of functions with the value 1 does not produce anything higher than 1. As Figure 4-22(c) shows, the 1 will appear at the output whether one, two, or three switches are closed.

The logic symbols for AND and OR functions appear on logic drawings which are used to design and document complex switching networks. The symbols apply whether the functions are produced with manual switches or with high-speed electronic circuits. (When used in electronic circuits, these functions are called gates because, like gates, they can be opened and closed for the passage of digital signals.) This function is explained in Section 6.5. Figure 4-23 shows a set of three interconnected logic gates with the Boolean functions expressed at the output of each gate.

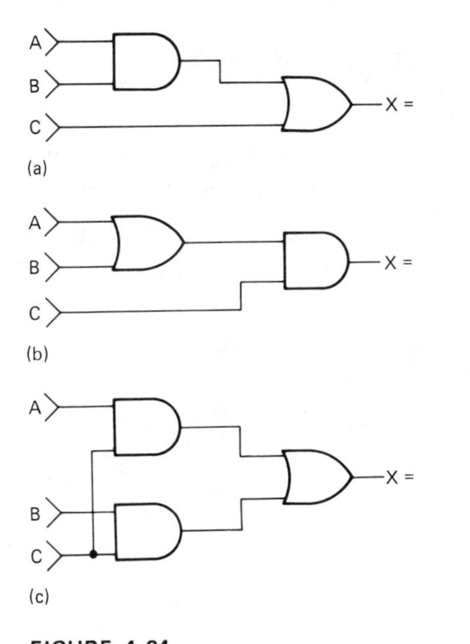

FIGURE 4-24

Interconnection of logic gates for Exercise 1.

■ EXERCISES

1. Determine the Boolean functions for the output of each gate in Figure 4-24.
2. From the Boolean equations below draw the implied logic diagrams:

   X = D + ABC

   X = CD (A + B)

Boolean equations can be factored using the rules of standard algebra. The factored equations imply other methods of producing the same function, as Figure 4-25 shows.

■ EXERCISES

3. Factor the equation below and draw the circuit implied by each step of the factored equation:

   $X = ACD + AEF + BCD + BEF$

4. Show by factoring that the Boolean equation at the output in Figure 4-24(b) produces the same function as that in Figure 4-24(c).

4.7.3 Operations with 1 and 0

The only constants in Boolean algebra are 1 and 0. Boolean functions between these values result in the following:

$$0 \cdot 0 = 0$$

$$1 \cdot 0 = 0$$

$$1 \cdot 1 = 1$$

$$0 + 0 = 0$$

$$1 + 0 = 1$$

$$1 + 1 = 1$$

Of the preceding theorems, only the last one, $1 + 1 = 1$, is peculiar to Boolean algebra.

The theorems covering functions between constants and variables are as follows:

$$0 \cdot A = 0$$

$$1 \cdot A = A$$

$$0 + A = A$$

$$1 + A = 1$$

Again, only the last one is unique to Boolean algebra.

FIGURE 4-25

Equivalent logic circuits. Logic diagrams produce an equivalent Boolean equation at the output. Therefore, factoring a Boolean equation can result in fewer gates being needed.

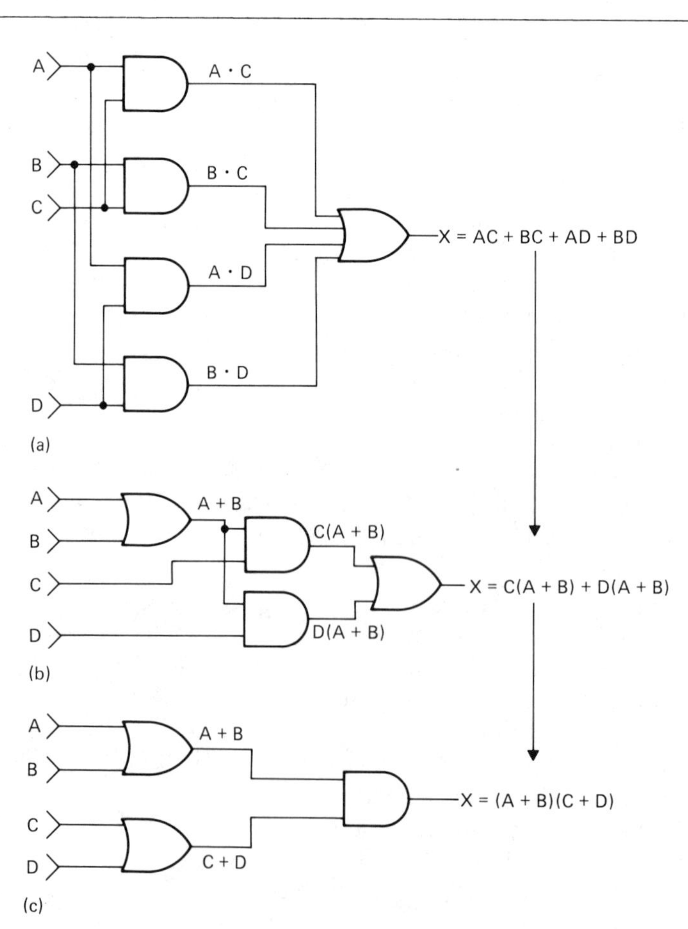

As the following examples show, these theorems can prove useful in reducing or simplifying Boolean equations.

■ **EXAMPLE 4-5**

Figure 4-26 is a logic circuit consisting of a two-input AND gate and a two-input OR gate. The resulting Boolean equation at the output is X = C + BC. Simplify the Boolean equation.

Solution: On simplifying the equation, we discover that the output is a function of only one input. That is, the circuit is redundant:

$$X = C + BC = C\,(1 + B)$$
$$= C\,(1) = C$$

Output X will be a function of C only.

■ **EXERCISES**

5. Simplify the following Boolean equations:

   1 - X = AC + B + BD

   2 - X = BA + (C + B)

6. Draw the circuits implied by both the original and the reduced version of the equations in Exercise 5.

4.7.4 Powers and Multiples of a Boolean Variable

Another advantage unique to Boolean algebra occurs with the results of raising variables to a power or adding to produce multiples of the same variable, as follows:

$$A \cdot A = A$$

$$A + A = A$$

Variables are restricted to 1 or 0, and these integers raised to any power do not change in value. A + A can be factored, as shown here:

$$A + A = A\,(1 + 1) = A\,(1) = A$$

■ **EXAMPLE 4-6**

Simplify the logic circuit diagram of Figure 4-27.

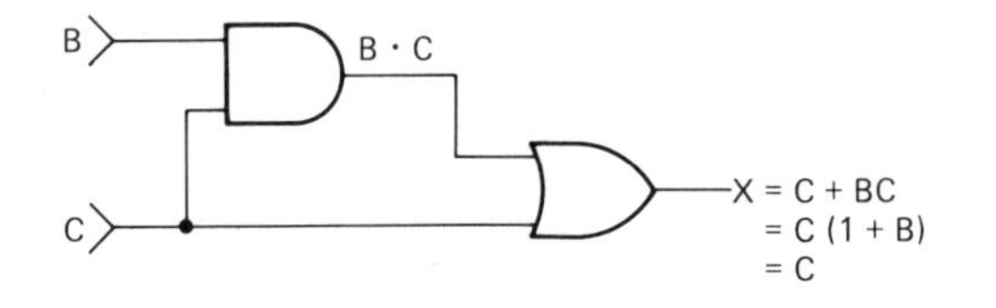

FIGURE 4-26
Example 4-5.

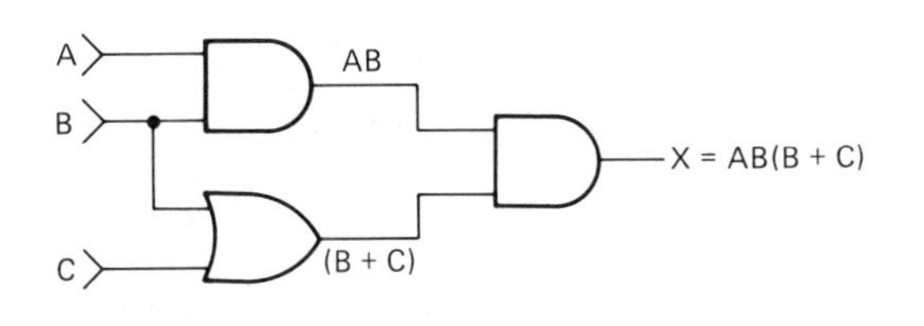

FIGURE 4-27
Example 4-6.

Solution:

$$X = AB\,(B + C) = ABB + ABC = AB + ABC$$
$$= AB\,(1 + C) = AB\,(1) = AB$$

The reduction of the equation indicates that the function gives results identical to a two-input AND gate. The input C has no effect on the output X.

■ **EXERCISES**

7. Prove the Boolean identities given below and draw the logic diagrams implied by each side of the identity:

$$(A + B)(B + C + D) \equiv A(C + D) + B$$

$$(AB + CDE)(BC + BCD) \equiv BC(A + AD + DE)$$

$$(AB + CD)(BC + BD)(CD + AC) \equiv BC(AD + A + D)$$

4.7.5 The Inversion (Complement)

In logic, some propositions are found to be true only if a condition is false. These can be expressed as Boolean algebra variables:

A and $\overline{A}$ (A not)

As was shown in Figure 4-17, a switch that will

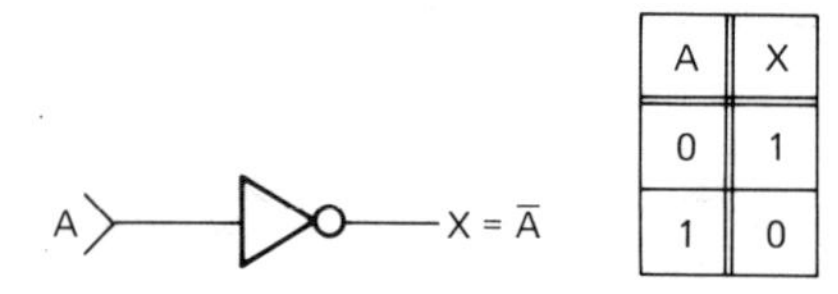

A	X
0	1
1	0

FIGURE 4-28

Logic symbol and truth table for an inverter gate.

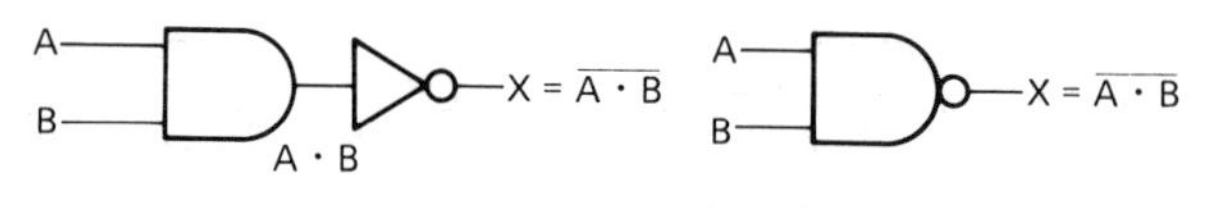

FIGURE 4-29

Equivalent symbols. When an inversion exists as an integral part of the input or output of the gate circuit, the symbol is simplified by showing only the bubble.

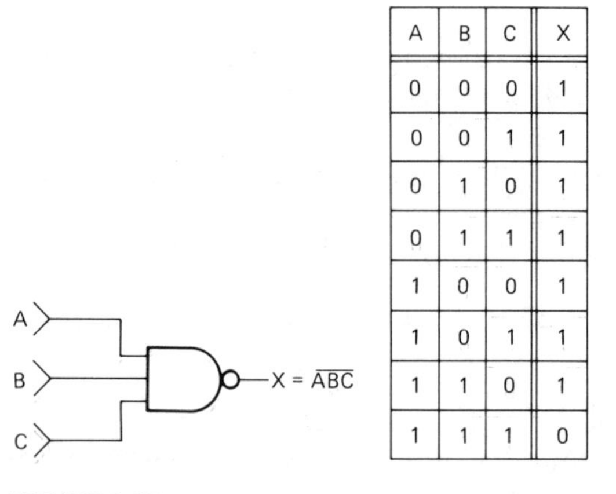

A	B	C	X
0	0	0	1
0	0	1	1
0	1	0	1
0	1	1	1
1	0	0	1
1	0	1	1
1	1	0	1
1	1	1	0

FIGURE 4-30

Logic symbol and truth table for a three-input NAND gate.

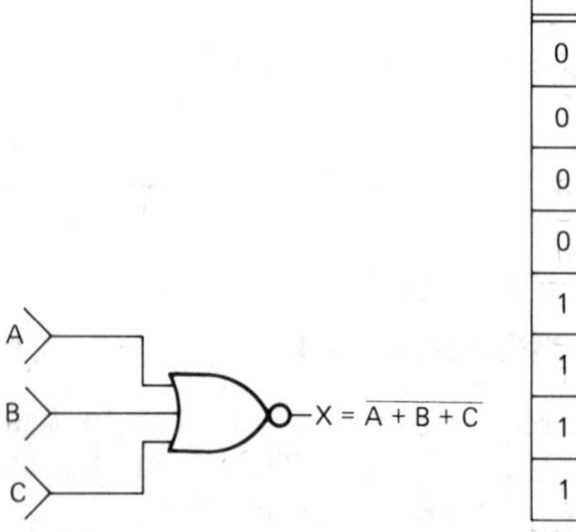

A	B	C	X
0	0	0	1
0	0	1	0
0	1	0	0
0	1	1	0
1	0	0	0
1	0	1	0
1	1	0	0
1	1	1	0

FIGURE 4-31

Logic symbol and truth table for a three-input NOR gate.

cause an inversion can be constructed. The logic symbol and truth table for such a circuit is given in Figure 4-28.

Important relationships between Boolean variables and their inversions are:

$$A \cdot \overline{A} = 0$$

$$A + \overline{A} = 1$$

When we consider the two possibilities existing for each of these equations

$A \cdot \overline{A} = 0$ must be either $1 \cdot 0 = 0$ or $0 \cdot 1 = 0$

$A + \overline{A} = 1$ must be either $1 + 0 = 1$ or $0 + 1 = 1$

the proof is obvious. In addition, if a function is given an even number of inversions, the inversion is canceled:

$$\overline{\overline{A}} = A$$

When inversions occur in a circuit as a part of an AND or OR gate, the symbol is abbreviated and only the bubble is shown, as Figure 4-29 depicts.

4.7.6 The NAND and NOR Functions

An AND gate with output inverted results in a NAND gate. The symbol and truth table for a three-input NAND gate is shown in Figure 4-30. Any zero input to a NAND gate results in a 1 output. Only when all inputs are 1 will the output be zero.

An OR gate with output inverted results in a NOR gate. The symbol and truth table for a three-input NOR gate is shown in Figure 4-31. If any input to a NOR gate is 1, the output will be 0. Only when all inputs are 0 will the output be 1.

Boolean equations may include inverted functions for which all of the foregoing theorems apply.

■ EXERCISES

8. Draw the circuits implied by the following Boolean equations:

$$X = \overline{AB} + \overline{CD}$$

$$X = \overline{A + B} \cdot \overline{C + D}$$

$$X = \overline{A + B} + \overline{C + D}$$

When factoring with inverted terms, care must be taken in handling terms under inversion bars. Terms can be factored under inversion bars, as we see here:

$$X = \overline{AC + AB} = \overline{A\ (C + B)}$$

But special conditions exist when factoring terms under inversion bars or separating the inversion bars, as shown below:

$$\overline{A + B} \neq \overline{A} + \overline{B}$$

$$\overline{A \cdot B} \neq \overline{A} \cdot \overline{B}$$

In order to factor these inverted terms, we need to apply a theorem known as DeMorgan's theorem.

4.7.7 DeMorgan's Theorem

Another important theorem of Boolean algebra is DeMorgan's theorem, which says that the complement of an OR function is equal to the AND function of the complements. Therefore,

$$\overline{A + B + C} = \overline{A} \cdot \overline{B} \cdot \overline{C}$$

The theorem also states that the complement of an AND function is equal to the OR function of the complements. Therefore,

$$\overline{A \cdot B \cdot C} = \overline{A} + \overline{B} + \overline{C}$$

From this, we can conclude that there are two valid symbols for the NOR gate as shown in Figure 4-32(a). The two valid symbols for the NAND gate are shown in Figure 4-32(b).

DeMorgan's theorem often proves useful in simplifying circuit functions or finding alternate methods for producing a function.

■ **EXAMPLE 4-7**

Prove that the function

$$X = CD \cdot \overline{A + B + C} = 0$$

Solution:

$$X = CD \cdot \overline{A + B + C} = C \cdot D \cdot \overline{A} \cdot \overline{B} \cdot \overline{C}$$

$$C \cdot \overline{C} = 0$$

Therefore,

$$X = 0$$

■ **EXERCISES**

9. Prove that Figure 4-33(a) is the Boolean identity to Figure 4-33(b).
10. Use DeMorgan's theorem to simplify the following equations:

$$X = \overline{(A + B)} + \overline{B}$$

$$X = \overline{(A\,B\,C)}\,C$$

$$X = \overline{(A + B) + (B + C)}$$

11. Prove the Boolean identity that follows:

$$\overline{A \cdot B} \cdot \overline{A} \cdot \overline{B} \equiv (A + B)\overline{(A \cdot B)}$$

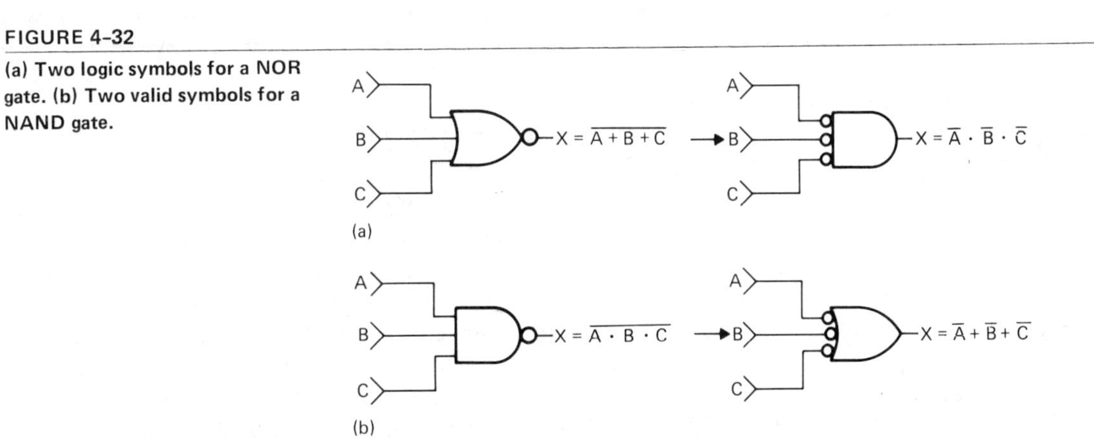

FIGURE 4-32
(a) Two logic symbols for a NOR gate. (b) Two valid symbols for a NAND gate.

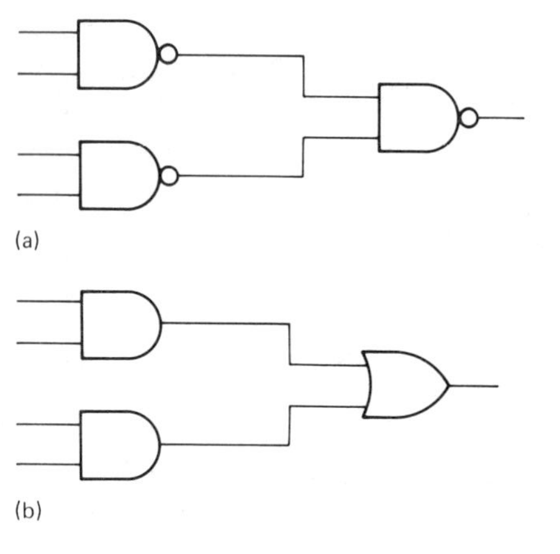

FIGURE 4-33
Logic diagram for Exercise 9.

4.7.8 Boolean Functions in Parallel

Similar to addition and subtraction, binary numbers can be subjected to Boolean operations in parallel, as shown in the following examples.

■ **EXAMPLE 4-8**

AND 10111001 with 11010101.

Solution:

```
10111001
11010101
--------
10010001
```

■ **EXAMPLE 4-9**

OR 10111001 with 11010101.

Solution:

```
10111001
11010101
--------
11111101
```

In both the AND and OR operations, each bit from LSB to MSB in the answers is determined separately. There are no carry or borrow bits.

■ **EXERCISES**

11. Determine the AND and OR functions of the following sets of numbers:

```
10101110     01101011     10110010
10011011     11000101     01010110
```

Masking Operations. Boolean functions in parallel are often used in microprocessor programs. Frequently a program will require that certain bits of a number be examined separately or individually by masking out unwanted bits. If the unwanted bits are to be reduced to zero, an AND function with a mask can be used.

■ **EXAMPLE 4-10**

Operate on the number 10100101 using the four least significant bits first, followed by the four most significant bits.

Solution: For the four LSBs, use mask 00001111, which results in:

```
00001111
10100101
--------
00000101
```

For the four MSBs, use mask 11110000, resulting in:

```
11110000
10100101
--------
10100000
```

■ **EXERCISES**

12. Show the masking operation needed to examine the three least significant bits of the following eight-bit numbers.

    11011010   10010011   11010001

13. Show the masking operation needed to examine the four most significant bits of the above numbers leaving the unwanted bits high instead of low. (Use OR with a mask.)

4.8 Summary

The binary number system is ideal for digital machines because it has two integers, 1 and 0. A number of electronic devices have two stable states between which they can be switched with a high degree of reliability. These two states, designated 1 and 0, can usually be identified by a voltage level on the terminals of the device. The voltage levels are also designated 1 and 0. A typical system has 0 represented by 0 volts and

1 represented by +5V. Figure 4-1 shows other logic levels. These levels are referred to as 1 and 0, but sometimes the terms *high* and *low* are used to avoid ambiguities.

In the operation of a digital system, there is a constant transmission of signals between the various components and subsystems. These signals are in the form of voltage changes between the 1 and 0 levels of the system. In some cases, they represent binary numbers, but in many cases, they are control signals. Two general terms that are used to describe these signals are *level* and *pulse.* The signals are often examined with the aid of an *oscilloscope*, which presents them as voltage-amplitude-versus-time graphs—referred to as waveforms. Figure 4-3 shows a typical set of digital signal waveforms.

An important part of the digital system is the clock circuit. The *digital clock* generates pulses that give timing and sequencing control to the system. The clock pulses are generally numbered in sets, which periodically begin with a 0 pulse. When the required number have passed, a new 0 pulse occurs and the numbering starts over. Figure 4-6 shows a typical *cycle* of pulses. The *clock period* and *cycle time* are two important indicators of the speed of a digital system.

Binary numbers appear in various forms. They may be in serial form. Serial form needs only one line or channel to transmit the number, but it requires a separate time period for each bit in the number. Figure 4-7 shows a serial pulse train representing the binary number 10011 (19). Presence of a pulse during the bit time is a 1; absence of a pulse is 0. Note that the LSB occurs on the left. The LSB occurs first in time, and the oscilloscope will present the first events on the left. Figure 4-8 shows the same number 10011 (19) as a serial-level train. The level train differs from the pulse train in that a 1 level occupies the full width of the time period. Adjacent 1 levels blend together, causing a markedly different appearance from that of the pulse train.

Numbers may also occur in parallel. In parallel, each bit of the number needs a separate line or channel. The number 10011 (19) would require five channels. The advantage of parallel transmission is speed. The entire number can be transmitted in one time period.

Most of the logic circuits through which digital signals must pass are a form of switch. The ideal digital switch has the following characteristics:

1. Must be automatic (turned off and on with binary 1 or 0 level).
2. Must be high-speed (instantaneous transition between off and on states).
3. Must have infinite off resistance.
4. Must have zero on resistance.
5. Must have isolation from the signal turning it on and off.
6. Must consume minimum power in the off and on states.

Figure 4-10 shows electronic devices used as digital switches, with the state of these devices given 1 and 0 designations. These designations may, however, be inverted, depending on the supporting circuit. Figure 4-11 compares the characteristics of these devices as switches. A digital system today will contain some number of each of the devices compared in Figure 4-11, but the bulk of the switches in the system will be either bipolar or field-effect transistors.

Boolean algebra deals with true and false states and, hence, lends itself readily for use on binary machines. Boolean algebra uses only the operations of multiply and add. Multiply is the AND function, while addition is the OR function. The commutative, associative, and distributive laws of algebra also apply to Boolean algebra.

Figures 4-21 and 4-22 show the standard logic symbols for the AND and OR functions, respectively. Figure 4-23 illustrates the concept that, given the logic diagram, a Boolean equation can be written. The reverse can also be done; that is, given the Boolean equation, a logic diagram can be drawn.

The powers and multiples described in Section 4.4.4 are unique to Boolean algebra because the variables are restricted to 1 or 0. The complement or inversion is another unique property of Boolean algebra.

An AND gate followed by an inverter yields a NAND gate. Similarly, an OR gate followed by an inverter produces a NOR gate.

Two important theorems of Boolean algebra are DeMorgan's theorems. These theorems often prove useful in simplifying circuit functions.

In computers, Boolean operations are performed in parallel.

Glossary

Logic: A correct process of reasoning or calculating.

Logic Level: A voltage level used to represent

binary 1 or 0 in a mathematical sense, true or false in a logical sense. See Figure 4-1.

Digital Signal: A form of intelligence or control transmitted through an electrical channel in the form of voltage changes between logic 1 and 0 levels. See Figure 4-3.

Boolean algebra: A type of algebra for manipulating logic variables. It was developed by George Boole.

Gate: A digital circuit with one or more inputs and a single output. The output is a direct result of the input conditions.

AND Gate: A digital circuit whose output is a logic 1 only when all inputs are a logic 1. See Figure 4-21.

OR Gate: A digital circuit whose output is a logic 1 when any input is a logic 1. See Figure 4-22.

Inverter: A digital circuit whose output is the complement of the input. See Figure 4-28.

NOR Gate: A digital circuit whose output is a logic 1 when all inputs are a logic 0. See Figure 4-31.

NAND Gate: A digital circuit whose output is a logic 1 when any input is a logic 0. See Figure 4-29.

DeMorgan's Theorem: Two rules which are useful in simplifying Boolean expressions. The rules are: (1) An OR gate is equal to a NAND gate with its inputs inverted; (2) An AND gate is equal to a NOR gate with its inputs inverted.

Squarewave: A rectangular voltage waveform of abrupt periodic changes between 1 and 0 levels for which the 0 and 1 levels are of equal duration. See Figure 4-5.

Pulse: A voltage waveform composed of a change in digital level of only short duration. See Figure 4-3.

Digital Clock: A circuit that produces a squarewave or periodic set of pulses that are used to provide timing in a digital system.

Clock Period: The time, usually measured in μsec or nsec, between leading edge of one clock pulse and leading edge of the next. See Figure 4-6.

Serial Number: A binary number transmitted on a single-wire circuit line or channel. A separate period is required for each bit in the binary number. See Figures 4-7 and 4-8.

Parallel Number: A binary number transmitted with each bit occupying a separate wire, circuit line, or channel. All bits of the number are transmitted simultaneously in one period. See Figure 4-9.

Voltage Waveform: A variation of voltage amplitude with respect to time. Translated to an amplitude-versus-time graph by use of an oscilloscope.

Voltage Spike: Very narrow pulse. Pulse of very short duration.

Cycle: A series of events that are repeated periodically.

Questions

1. A digital machine uses positive logic with levels of 0 volts and 5 volts. What six logic terms might be used to describe these voltage levels?
2. A pulse is generally considered to be a change in level narrower than ___?
3. A digital system uses 0 volts as logic 1 and -4.5V as logic 0. Is this positive or negative logic?
4. An eight-bit digital number is transmitted in serial. How many transmission channels are needed? How many periods are needed to complete the transmission?
5. A 12-bit digital number is transmitted in parallel. How many transmission channels are needed? How many periods are needed to complete the transmission?
6. List six important characteristics of an ideal digital switch.
7. If X is 0 when A is 1, and A is 0 when X is 1, express X as a function of A. (X = ?)
8. The diode connected across the coil of a relay should be ___-biased.
9. What is the purpose of the diode across the relay coil?
10. How does the speed of a relay compare with the speed of a transistor switch?
11. For what application are relays advantageous?
12. What are the dual states in Boolean algebra?
13. What are the two mathematical operators in Boolean algebra?
14. Can the commutative law be expanded to three variables?
15. What is the Boolean expression for (a) a four-input AND gate; (b) a four-input OR gate? Inputs are A, B, C, and D.
16. What are two theorems unique to Boolean algebra?
17. What does the term NAND represent?
18. Write DeMorgan's theorems.

Problems

4-1 The clock line in Figure 4-6 is every other pulse from 1 MHz multivibrator. (Every other pulse has been removed.) What is the clock period? the cycle time?

4-2 Above the clock line in Figure 4-34, draw a waveform representing the binary number for 21 in serial pulse train form. Pulses representing the one-bits will occur in the middle of the clock period having the same width as the clock pulse.

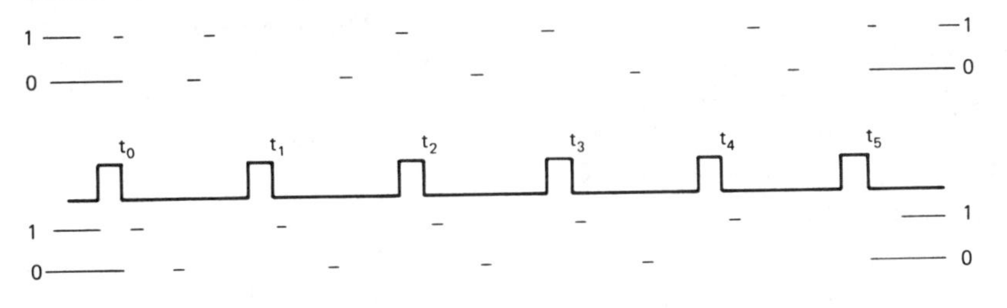

FIGURE 4-34. Problems 4-2 through 4-4

4-3 At the bottom of Figure 4-34, draw a waveform representing the binary number for 25 in serial level train form.

4-4 If the clock frequency in Figure 4-34 is 500 kHz, how long does it take to complete the numbers in Problems 4-12 and 4-13?

4-5 On the five lines designated in Figure 4-35, draw the waveforms that would occur as the numbers changed from 0 to 25 at t_0, to 10 at t_1, to 21 at t_2, to 9 at t_3.

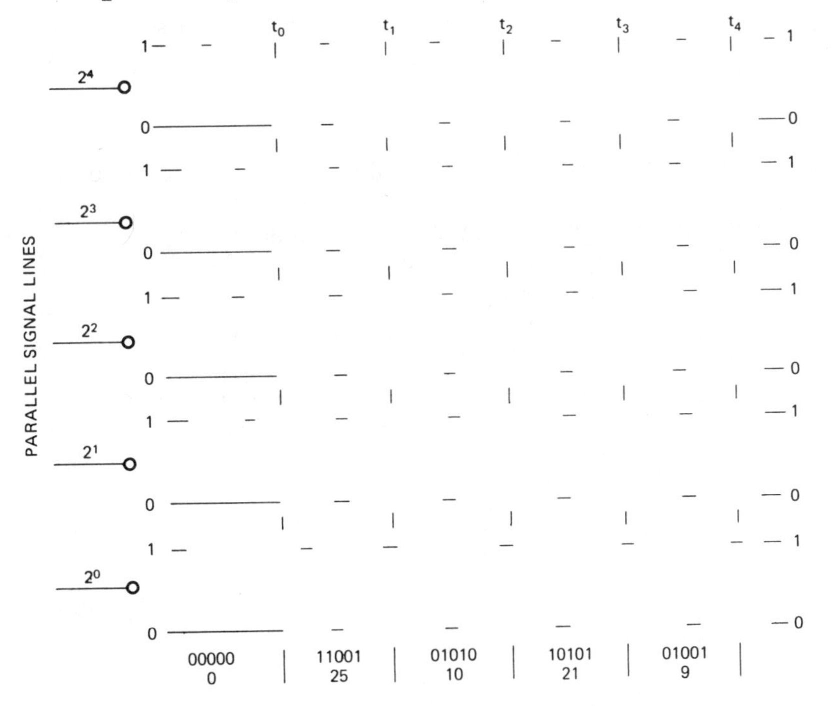

FIGURE 4-35. Problem 4-5

4-6 If the clock frequency is 500 kHz, how long does it take to complete the number 25? How long does it take to complete all four numbers? At a given clock frequency, which system provides for faster transmission of numbers, the serial system of Problem 4-14 or the parallel system of Problem 4-5?

4-7 In Figure 4-36, determine the Boolean function at X and complete the truth table.

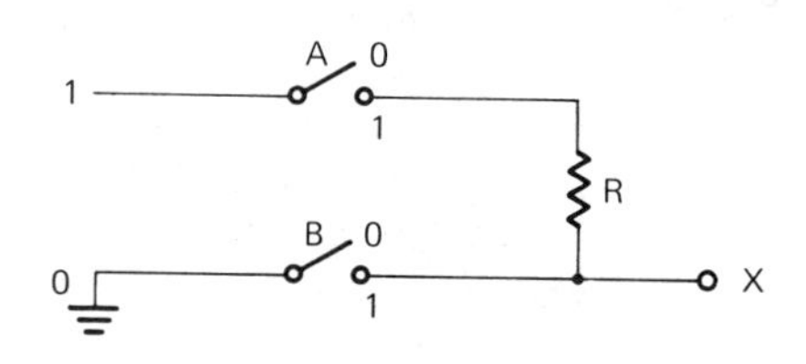

FIGURE 4-36. Problem 4-7

4-8 In Figure 4-37, the +28V relay bus is the binary 1 level. Write the Boolean AND function of X and fill in the truth table.

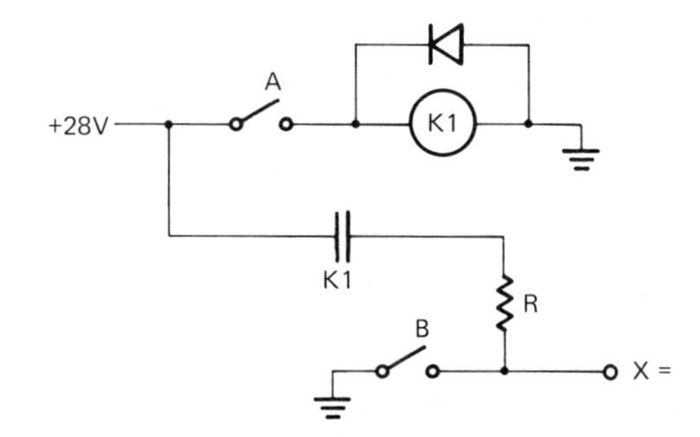

TRUTH TABLE

(HINT) K1 = A

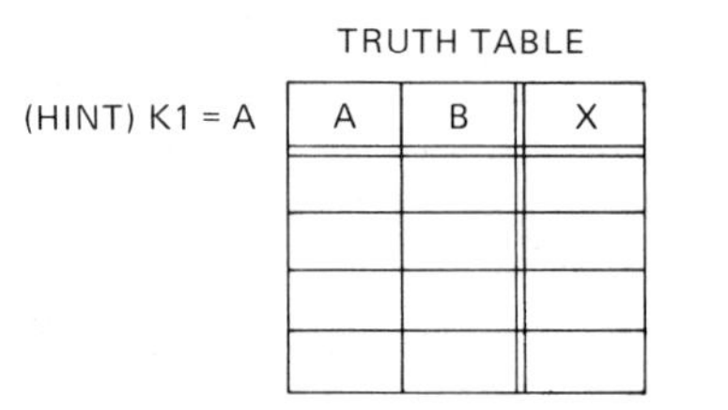

A	B	X

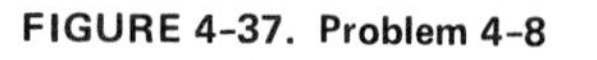

FIGURE 4-37. Problem 4-8

4-9 The clock waveform shown in Figure 4-38 is the output of a 250 kHz clock. What is the clock period in μsec? The reset pulses drawn below mark the beginning of cycles. What is the cycle time in μsec?

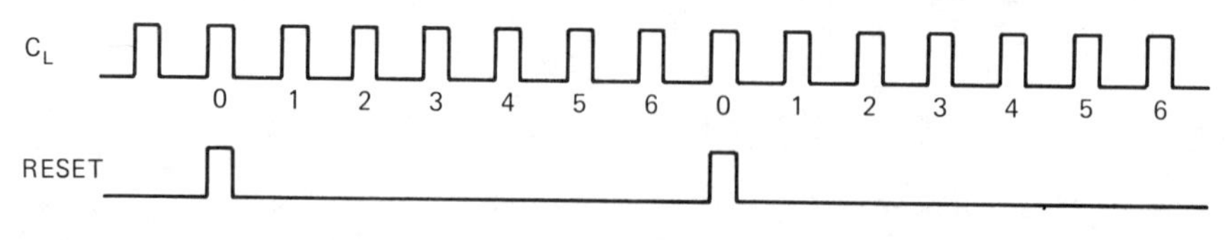

FIGURE 4-38. Problem 4-9

4–10 If the bit time in the serial system of Figure 4–39 was one clock period and the clock rate was 100 kHz, what time in μsec would be needed to transmit the entire four-bit number?

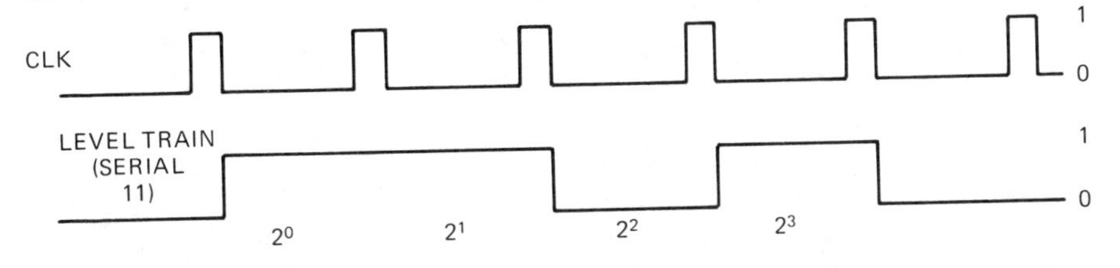

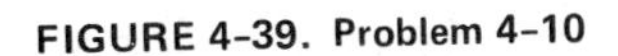

FIGURE 4–39. Problem 4–10

4–11 If the bit time in the parallel system of Figure 4–40 was one clock period and the clock rate was 100 kHz, what time in μsec would be needed to transmit each four-bit number?

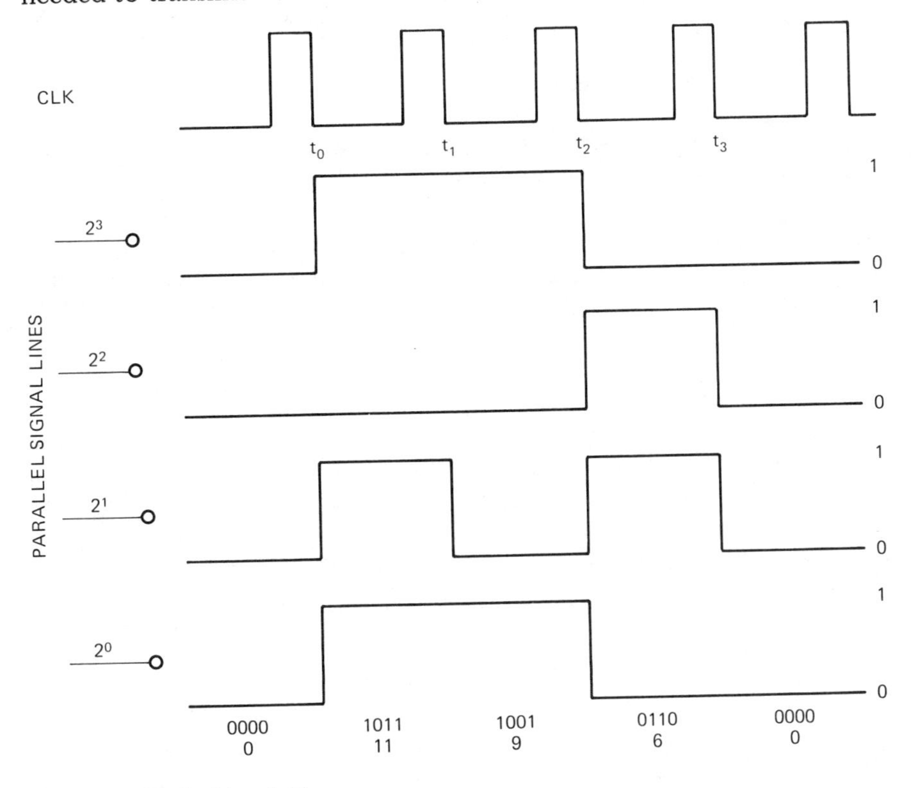

FIGURE 4–40. Problem 4–11

4–12 What clock rate would be needed in the serial system of Figure 4–39 to transmit four-bit numbers at the same rate as the parallel system of Figure 4–40?

4–13 If the size of the numbers were increased to eight bits, what should be added to the parallel system of Figure 4–40? What clock rate would then be needed in the serial system to match the speed of the parallel system?

4-14 In Line a of Figure 4-41, draw the voltage waveform of a pulse train representing the binary number for 28. The LSB starts at t_0; the pulse width is $T_c/2$.

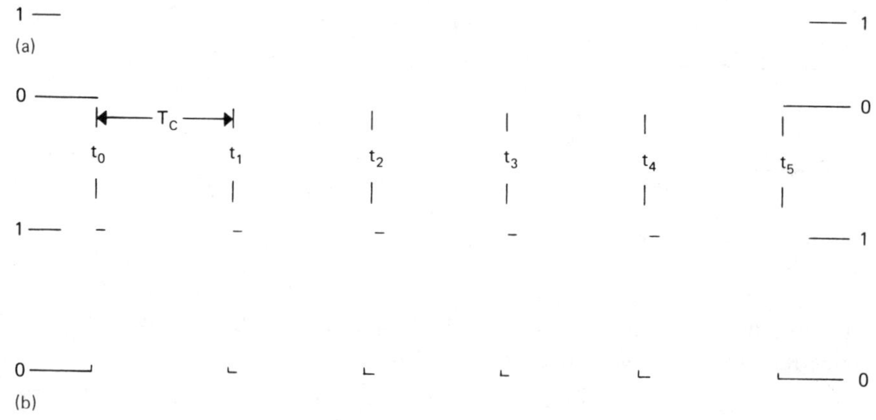

FIGURE 4-41. (a) Timing diagram for Problem 4-14. (b) Timing diagram for Problem 4-15

4-15 In Line b of Figure 4-41, draw the voltage waveform of a level train representing the binary number for 23. The LSB starts at T_0.

4-16 In Figure 4-42, draw the voltage waveform of each parallel output line below as the binary number changes from 0 to 12 at t_0, back to 0 at t_1, to 15 at t_2, and back to 0 at t_3.

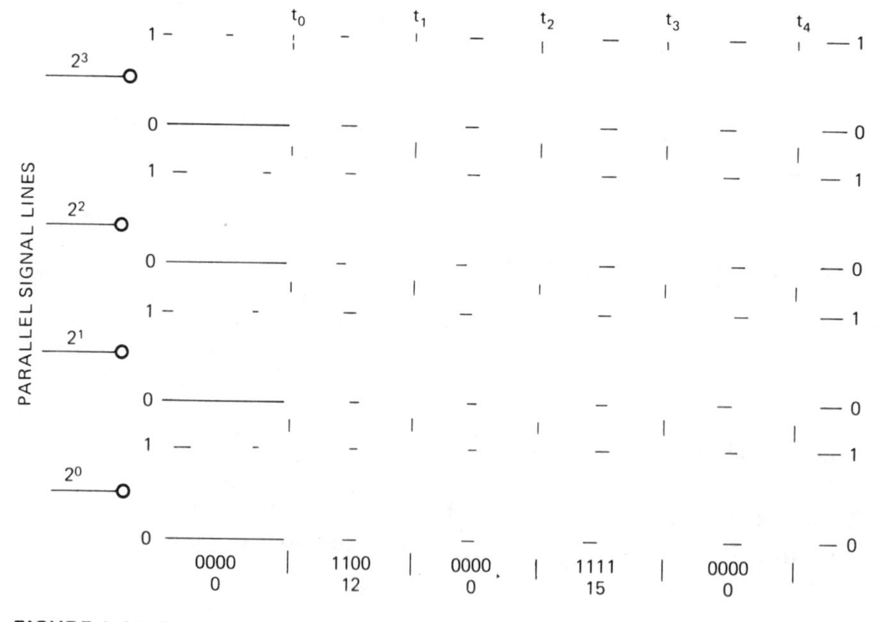

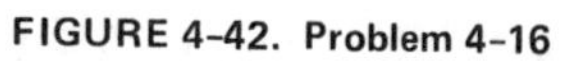
FIGURE 4-42. Problem 4-16

4-17 In Figure 4-43, K1 and K2 are single-pole, double-throw relays. Express the Boolean function at X_1, X_2, and X_3. Fill in the truth tables for X_1 and X_3.

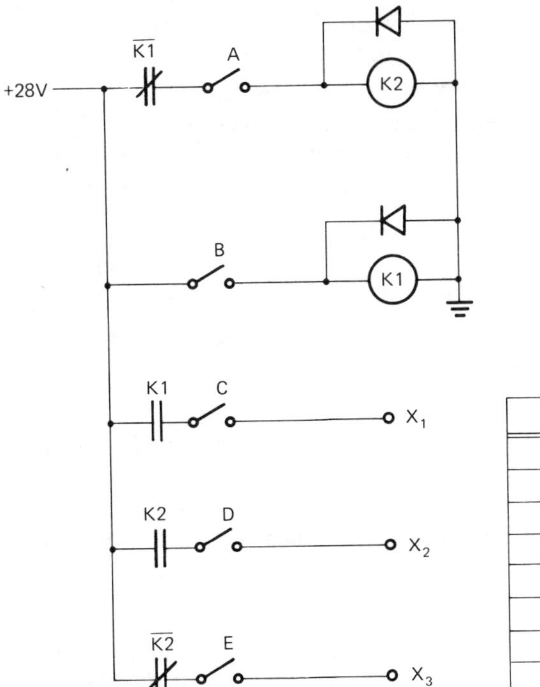

TRUTH TABLE

		X_1

TRUTH TABLE

			X_3

FIGURE 4-43. Problem 4-17

4-18 Determine the output Boolean equation for each logic diagram given in Figure 4-44.

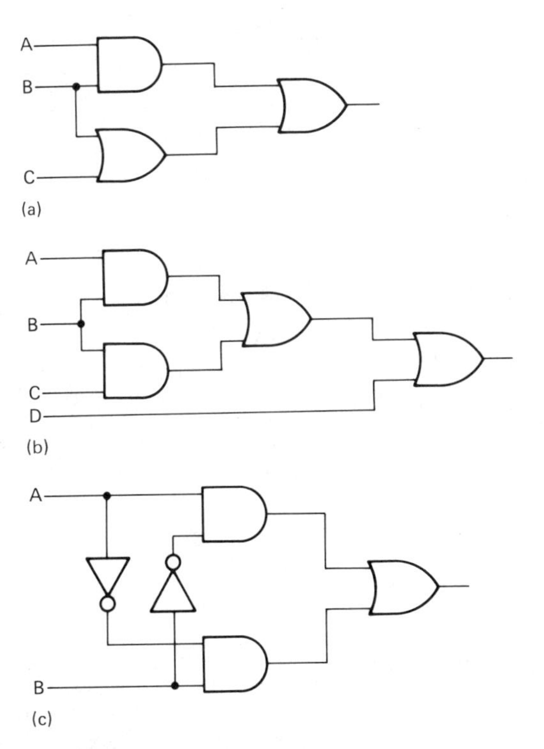

FIGURE 4-44. Problem 4-18

4–19 Draw the logic diagram for each of the following Boolean equations:

$$X = A + BCD$$
$$X = A\overline{B} + BC$$
$$X = C + AB(C + D)$$

4–20 Can any of the expressions of Problem 4–19 be simplified? If your answer is yes, draw the equivalent simplified logic diagram.

4–21 Prove the identities listed below and show each step:

$$\overline{\overline{A} + \overline{B} + C} \equiv AB\overline{C}$$

$$\overline{A\,\overline{B} + C} \equiv \overline{C}(\overline{A} + B)$$

$$\overline{(A + \overline{B})\,(\overline{C} + \overline{D})} \equiv \overline{A}B + CD$$

4–22 Draw the DeMorgan equivalent of the following gates: (a) NAND (b) NOR

4–23 Determine the AND function for each of the following 8 bit numbers:

11001101	00011110
01111100	10110111

4–24 Determine the OR function for each of the eight-bit numbers in Problem 4-23.

CHAPTER

Electronic Switches: Discrete and Integrated

Objectives

Upon completion of this chapter, you will be able to:

- Describe the functioning of a diode switch.
- Describe the functioning of a bipolar transistor switch.
- Describe the functioning of a field effect transistor switch.
- Describe the functioning of integrated circuit RTL, TTL, MOS, and CMOS inverters.

5.1 Introduction

Except for the fact that a toggle switch must be operated manually, it has ideal electrical parameters. Its ON resistance is a relatively short circuit, or near zero ohms resistance. In the OFF state, its resistance is the air between its contacts, which thus makes it an open circuit of near infinite resistance.

There are many applications that do not need such perfect conditions. In some applications, an ON resistance of several ohms, or even several hundred ohms, will do as well as a short circuit. For most digital applications, an OFF resistance of 100 Kohms may work as well as a hundred megohms. In such a situation, a rheostat and a switch would be interchangeable.

As Figure 5-1 shows, rheostats and switches have two comparable states—a very high resistance (OFF) and a very low resistance (ON). There are electronic devices whose resistance can be varied automatically and almost instantaneously by an incoming signal. For example, the electronic amplifier varies its output resistance in proportion to an incoming signal. It is, therefore, an automatic rheostat. But if we used it only at its maximum and minimum levels, it would, for most applications, act like an automatic switch, with resistance very low at saturation and very high at cutoff.

There are various methods of producing high-speed automatic switches with electronic amplifier components. One method involves the assembly of resistors, diodes, and transistors on a printed circuit board. These are referred to as *discrete component circuits.* The computers and digital machines having this type of construction are gradually being replaced by machines containing integrated circuits.

The *integrated circuit switch* contains all its resistors, diodes, and transistors as an integral part of a minute silicon chip. This concentration has numerous advantages, such as substantially increased circuit density, lower power consumption, and a superior production method for batch fabrication. The integrated circuit switch can be explained partially in terms of an equivalent discrete component switch, but it is often more complicated.

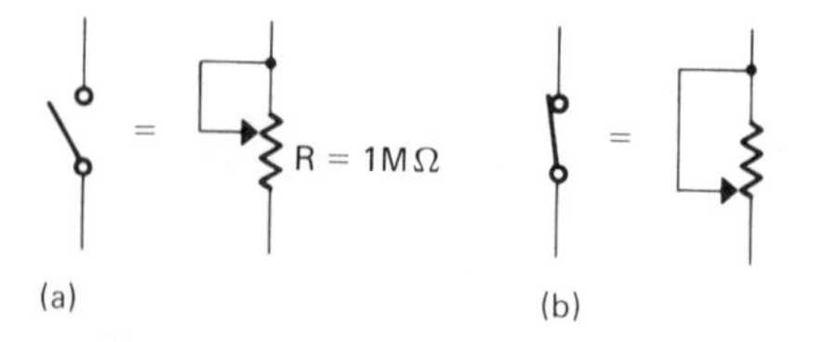

FIGURE 5-1

Comparison of rheostats and switches. (a) A high-resistance rheostat in maximum resistance is like an open switch. (b) The rheostat in 0 resistance position is like a closed switch.

The discrete component switch normally contains one transistor. The equivalent integrated circuit switch usually contains several transistors, because of manufacturing advantages as well as improved circuit performance. In discrete form, a transistor may cost ten times more than a resistor. In integrated form, a transistor may be cheaper than a given size of resistor, and, therefore, will often replace resistors within the integrated circuit.

5.2 The Diode as a Switch

It may seem far-fetched to call a semiconductor diode a switch. It is a unidirectional device in that it allows current to flow in one direction. Figure 5-2 shows the effect of a diode in series with a 12-volt, 50-milliampere lamp. The lamp has a hot resistance of 240 ohms.

When the battery is connected with the negative terminal on the anode as in Figure 5-2(a), the diode is reverse-biased and places a high resistance of more than 1 megohm in series with the lamp. This condition is the equivalent of an open switch, and the lamp will not light.

When the battery is in the circuit with the plus terminal toward the anode as in Figure 5-2(b), the diode is then forward-biased. It provides a low resistance of 1 to 10 ohms, which, in this circuit, is the equivalent of a closed switch, and the lamp will light.

The diode is typically used in digital circuits to allow a point in a circuit to be energized from two or more lines without those lines being shorted together. If the voltages in the circuit are such that the more positive potentials are toward the anode, the diode is forward-biased and looks like a closed switch. If the voltages are such that the more positive voltage is on the cathode side, the diode is reverse-biased and looks like an open switch.

The diode is a passive device that degrades the signal passing through it. Thus, many logic systems rule that a diode switch may not receive the output of another diode switch and can

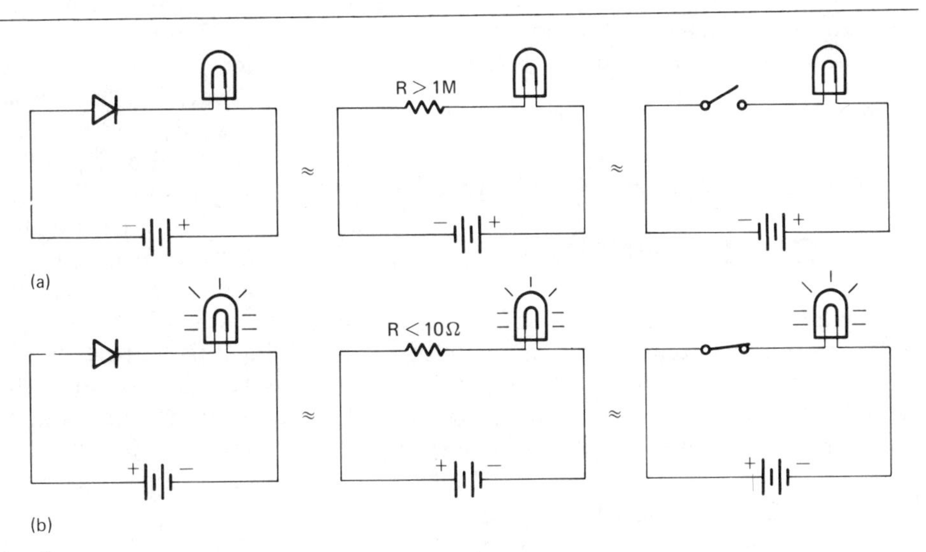

FIGURE 5-2

The diode functioning as a switch. (a) When the diode is reverse-biased, it has a high resistance like an open switch. (b) When the diode is forward-biased, it has a low resistance like a closed switch.

drive only an active device, like the bipolar or field effect transistor switch, which is able to restore the degraded signal level.

The diode as a switch is automatic, but the signal must be such that the diode can be turned off and on. It cannot be turned off or on by a signal other than the one passing through it. While applications are thus limited, there are many uses for which it is ideal. It is high-speed and can be changed from forward to reverse bias in a matter of nanoseconds.

Figure 5-3 shows the typical volt ampere curve of a diode and its equivalent circuit. In the reverse bias quadrant of Figure 5-3(a) the diode as a switch is turned off. The reverse resistance is in the megohms, which is more than adequate. In the forward direction, however, there is very significant resistance until a voltage of about 0.3V for germanium diodes and 0.7V for silicon is dropped across the diode. The signal is, therefore, degraded by that amount as it passes through a diode switch. Figure 5-4 shows the typical result.

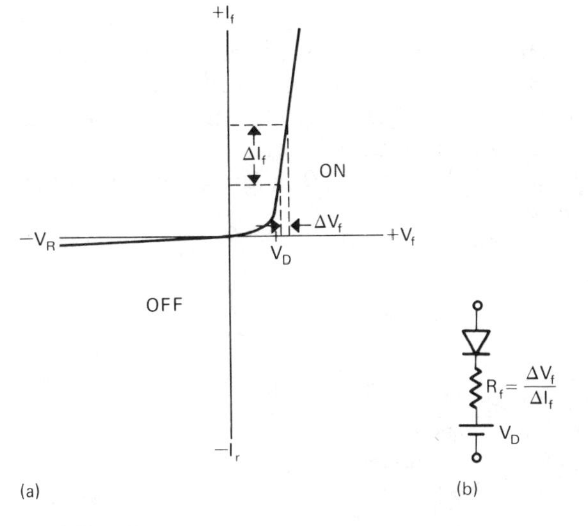

FIGURE 5-3

(a) Volt ampere curve of a diode as a switch. The upper right quadrant describes the ON conditions; the lower left quadrant, the OFF conditions. (b) Equivalent circuit of a diode as a switch, with a very small resistor and small offset voltage.

5.3 The Vacuum Tube as a Switch

The vacuum tube as a switch is of historical interest only. It once had the advantage of high speed over the relay, but it was rapidly replaced by transistors in the mid-1950s. The transistor provides the same high-speed functions, but uses low voltages—12 volts or less—as compared to 50

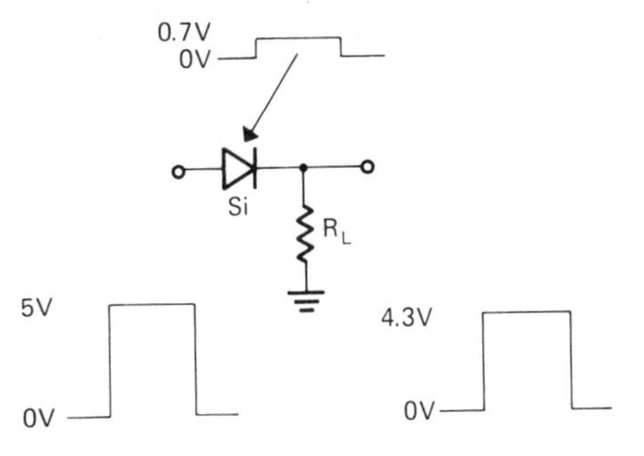

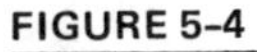

FIGURE 5-4

Signal degradation passing through a silicon diode switch. The output waveform of the signal shows the signal degraded by the amount of the offset voltage (0.7V).

to several hundred volts for the vacuum tube. The transistor consumes less power and is many times smaller, cheaper, and more reliable.

5.4 The Bipolar Transistor as a Switch

The majority of digital switching devices in use today are bipolar transistor circuits in either discrete or integrated form. The transistor as a switch may be either a common emitter switch or a common collector (emitter follower) switch.

5.4.1 Discrete Component Common Emitter Switch

The common emitter switch operates as Figure 5-5 shows. The binary value of the output coming from the collector is controlled by the application of 1 and 0 levels to the input or base lead. The common emitter operates like the shorting switch of Figure 4-19 and is, therefore, an inverter.

When the input, A, is at 0, as in Figure 5-5(a), there is no base current, I_B. With 0 base current, I_B, there will also be 0 collector current, I_C. If no collector current flows through R_C, then the (V_{CC}) binary 1 level will appear at the output, X. The voltage divider relationship between R_C and R_L dictates some reduction in the V_{CC} level at X, but if R_L remains very large in comparison to R_C, that reduction will be small.

If a 1 level is applied to the input, A, as in Figure 5-5(b), the base current, I_B, will result in a collector current sufficient to drop most of the 1-level voltage across R_C. This provides a binary 0. It is imperfect in that a small voltage, the saturation voltage, that is necessary to operate the transistor is still present. But it is close enough to 0 volts to function as a binary 0. To work properly, I_C must reach a level equal to V_{CC}/R_C.

Because of the current gain, β, of the transistor, the input base current, I_B, needed to produce a saturating collector current, $I_C = V_{CC}/R_C$, is relatively small. Thus, the input impedance of the switch is high and numerous inputs can be connected to the output of one common emitter switch without exceeding the condition of $R_L \gg R_C$. This affects the fan-out of the switch—that is, the number of inputs that the output of one switch can drive.

Single-input inverting switches are usually

FIGURE 5-5

(a) Common emitter switch with logic 0 input and equivalent manual switch circuit. (b) Common emitter switch with logic 1 input and equivalent manual switch circuit.

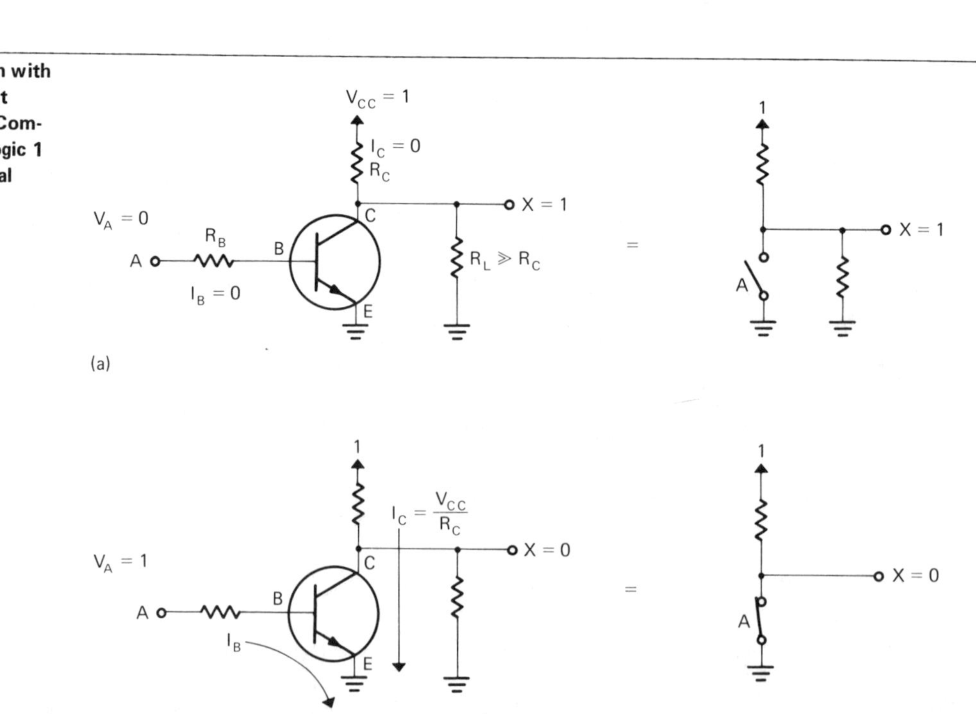

called inverters. Figure 5-6 shows the logic symbol and truth table of the inverter. The logic symbol of Figure 5-6 is much simpler to draw than the schematic drawing of Figure 5-5(a), and since all inverters in a given logic system are likely to have the same schematic, it is necessary to draw it only once. The use of logic symbols provides a drawing that is cheaper to do and easier to read. The only two permitted input levels to an inverter are 1 and 0. As the truth table indicates, a 1 input produces a 0 out, and a 0 in produces a 1 out. The Boolean equation for this is $X = \overline{A}$.

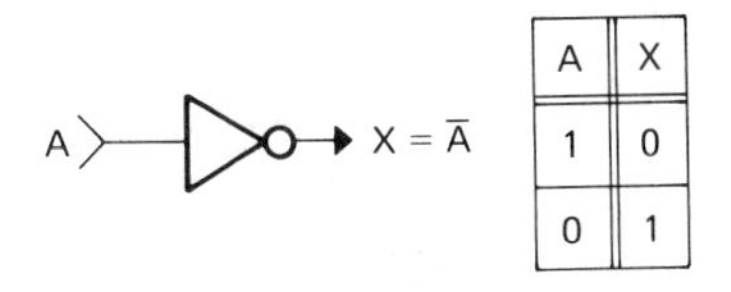

A	X
1	0
0	1

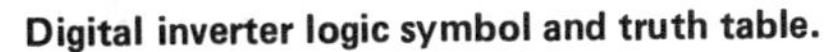

FIGURE 5-6

Digital inverter logic symbol and truth table.

5.4.2 Discrete Component Emitter Follower Switch

The transistor used in a common collector, or emitter follower, configuration functions like the switch shown in Figure 5-7. A 1 level on the base of the transistor in effect closes the switch. The emitter follower switch of Figure 5-7 degrades the 1 level by the amount V_{BE}, resulting in output 1 levels 0.6V lower than the input 1 level. The restriction $R_L \gg R_C$ specified for the common emitter switch is not necessary for R_E of the emitter follower. The emitter follower switch is an excellent driver and can supply much more current at the 1 level than it draws from the input signal.

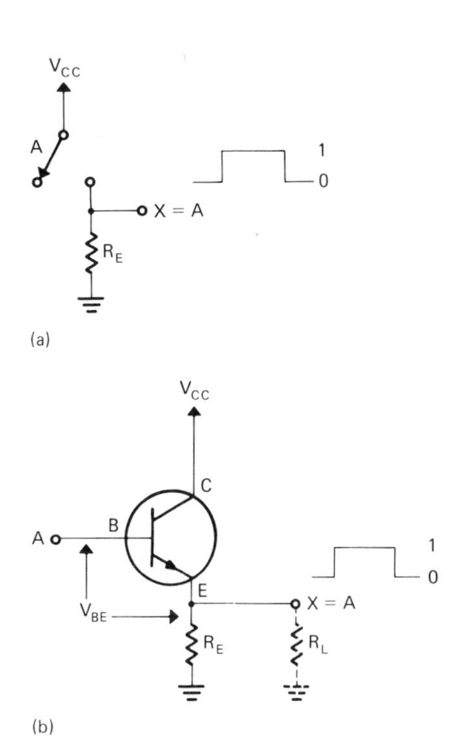

FIGURE 5-7

Emitter follower switch. (a) A common collector, or emitter follower, switch can be simulated by a manual switch that applies a 1 level to the top of the load resistor when it is in the 1 position. (b) The output of the emitter follower switch is not inverted; the output signal is degraded by the amount V_{BE}.

5.4.3 Integrated Circuit RTL Switch

One of the simplest integrated circuit switches is the RTL (resistor-transistor logic) switch. Figure 5-8 shows the Motorola MCL quad inverter, a type of RTL switch. The 14-pin DIP (dual-in-line package) provides four inverting switches, each very much like the discrete switch in Figure 5-5. The V_{CC} voltage for this circuit is 3.6V. The 1 output level at full load may be as low as 1.1V. The 0-level output never exceeds 0.4V. This narrow spread between the 1 and 0 levels is the main disadvantage of the RTL circuit.

5.4.4 Integrated Circuit TTL Switch

The integrated circuit TTL (transistor-transistor logic) switch is now the most widely used integrated circuit switch. It has the advantages of high noise immunity over the RTL switch and high speed over the MOSFET switches, to be discussed in Section 5.5.4. They are normally supplied in a 14-pin DIP package. Figure 5-9 shows one switch from a Sprague 5404/7404 Hex Inverter. There are six such inverters in a 14-pin package.

Improved switching characteristics result from using four transistors to provide the switching capabilities that used to be provided by only one transistor. The reason can be understood if we consider the problem of selecting an ideal

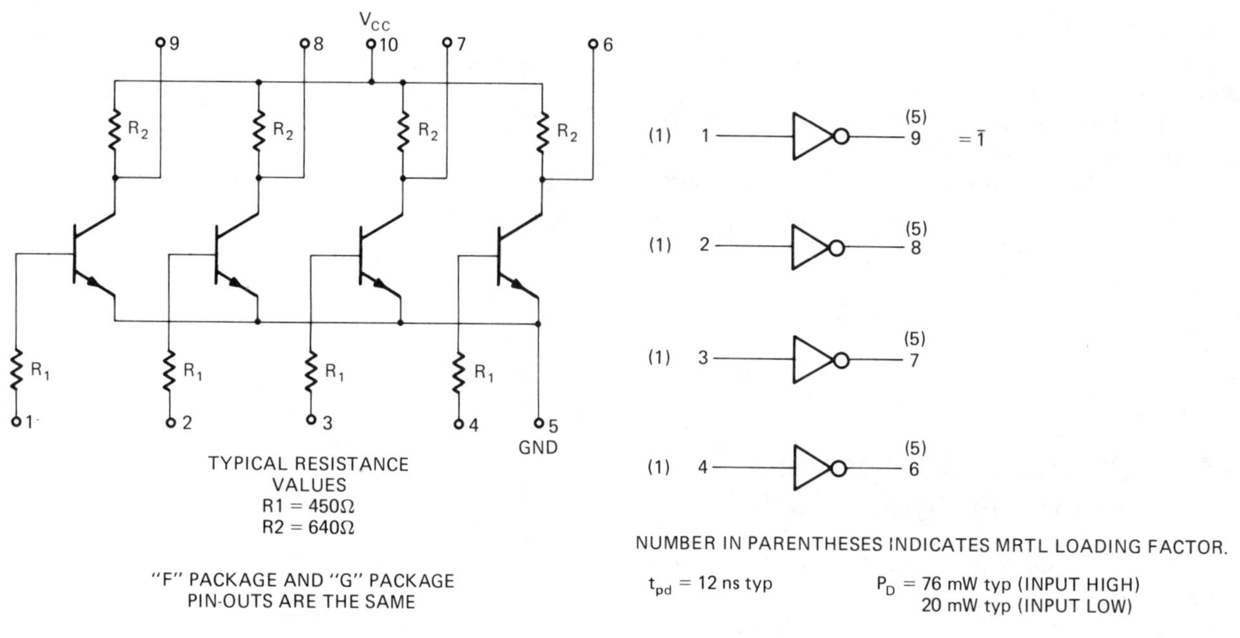

FIGURE 5-8

RTL integrated circuit switch: Motorola MCL Quad Inverter MC927-MC827.

FIGURE 5-9

TTL integrated circuit switch: Sprague 7404 Hex Inverter.

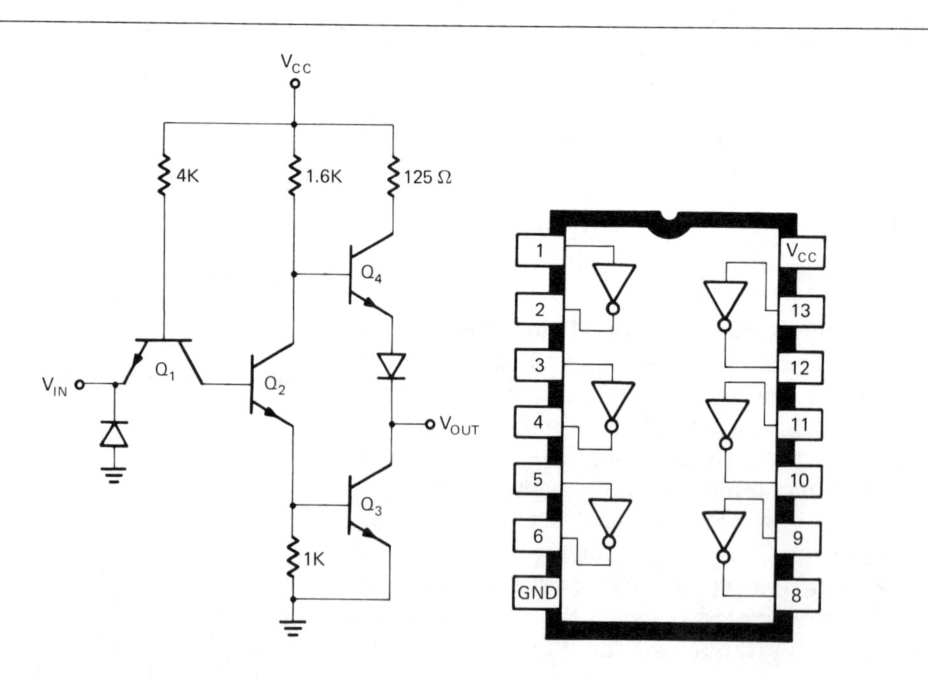

value for the collector resistor, R_C, in Figure 5-10(a). When the transistor is turned off, it would be ideal to have the 1-level V_X as nearly equal to V_{CC} as possible. This is accomplished if R_C is very small and R_L is very large, but such an arrangement would limit the number of circuits the switch could drive.

When the transistor is turned on, as Figure 5-10(b) shows, the amount of current, I_C, needed to drop the V_{CC} voltage across R_C is $I_C = V_{CC}/R_C$. If R_C is small, the power consumption of the switch becomes excessive and calls for an R_C of very high resistance. If the resistor could function as shown in Figure 5-11(a) and change resistance as the switch is turned off and on, it would be ideal for both conditions. The

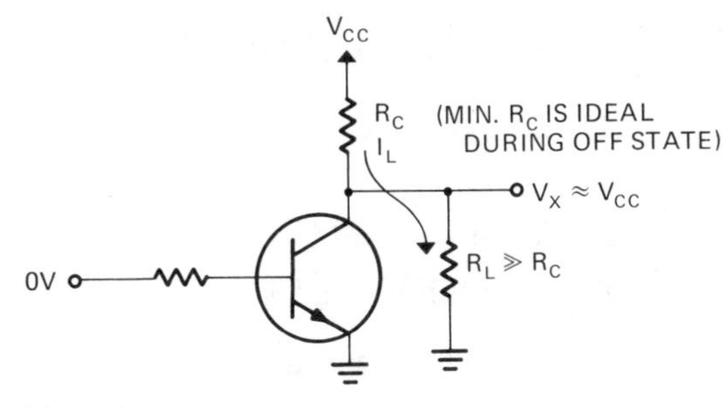

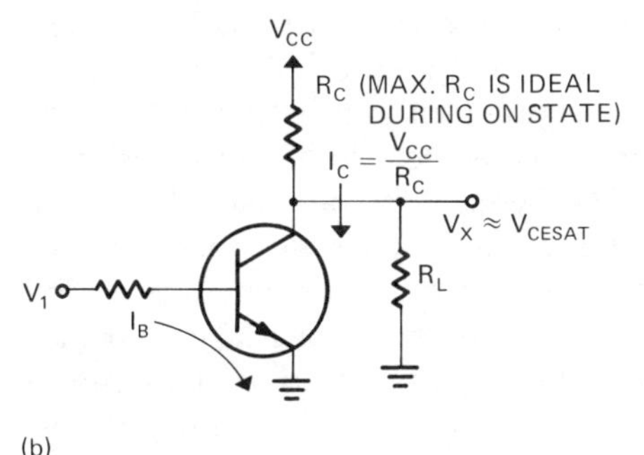

FIGURE 5-10

Selection of an ideal value for R_C. (a) With a binary 0 input, no collector current flows through R_C, and the output 1 level is reduced from V_{CC} only by the voltage I_L R_C. (b) With a binary 1 input, collector current must reach V_{CC}/R_C in order that all of V_{CC} will drop across R_C to produce a 0 output.

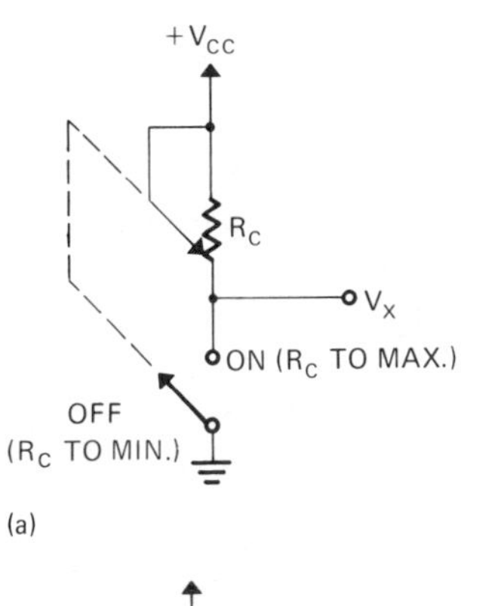

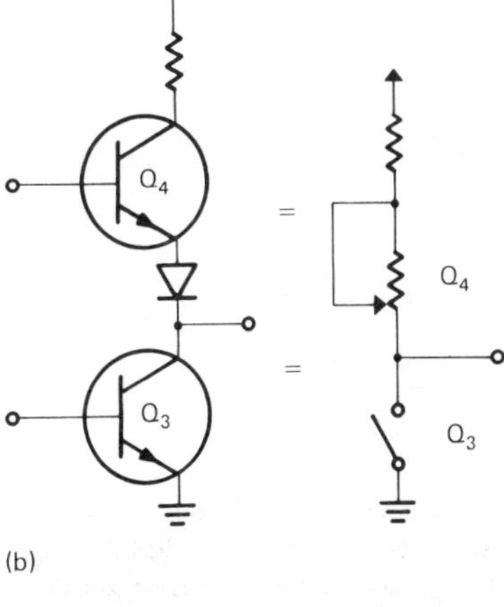

FIGURE 5-11

Operation of "totem pole" transistors. (a) If R_C is raised to maximum at the same time the switch is turned on and lowered to minimum as the switch is turned off, it would be ideal for both OFF and ON states. (b) Using one transistor as a switch and the other as a variable resistor provides an automatic condition as in circuit 5-11(a).

switch could be turned on and the resistance R_C increased simultaneously by the 1-level input. The two output transistors shown in Figure 5-11(b) function in this fashion and provide an ideal output condition. This arrangement of the transistors is commonly referred to as a "totem pole."

The input is on the emitter of Q_1. With a 1-level input, the circuit appears as in Figure 5-12(a). The transistor Q_1, seen as two diodes, has its base-to-emitter diode reverse-biased. The base-to-collector diode, however, is forward-biased, sending a base current through Q_2. A 1 on the input, therefore, appears as a 1 on Q_2 and Q_3, and a 0 on Q_4. With a 0-level input, the circuit appears as in Figure 5-12(b). The base-to-emitter diode is forward-biased, dropping most of V_{CC} across the 4K resistor. A 0 level appears on the base of Q_2 and Q_3, and a 1 on Q_4. The reverse-biased diode on the input protects the circuit from ringing, which often occurs with high-speed switching.

The four-transistor switch can be divided into three sections, as Figure 5-13 shows, and they function as follows: When the control transistor, Q_2, is turned OFF by a 0 level on its base, a high-voltage level on the base of Q_4 reduces its collector to emitter resistance. With Q_2 turned OFF, Q_3 is also turned OFF. This provides the ideal 1-level output condition, with a low value of R_C and an open circuit (turned-off transistor) to ground, as in the equivalent circuit of Figure 5-11.

When Q_2 is turned ON by a 1 level on its base, the current through Q_2 drops voltage across the 1.6K, reducing the base voltage on Q_4 and increasing its collector to emitter resistance. At the same time, the current through Q_2 produces a voltage drop across the 1K resistor sufficient to turn on Q_3. This provides the ideal

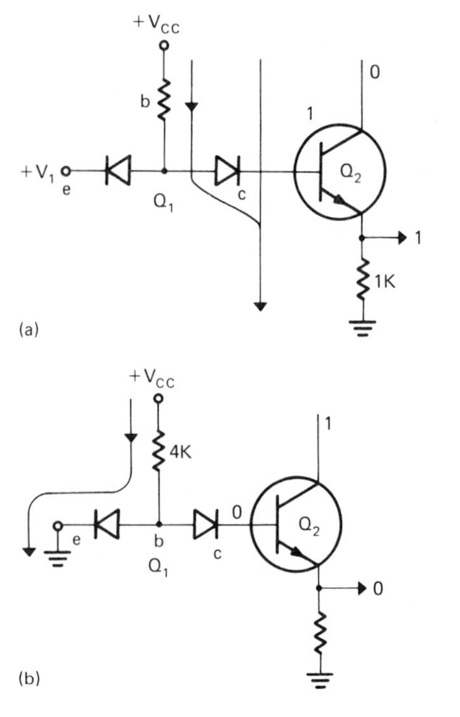

FIGURE 5-12

Input transistor of TTL inverter drawn as two diodes. (a) A 1 level at the input causes base current through to Q_2. (b) A 0 level at the input drops the $+V_{CC}$ across the 4K resistor, turning OFF Q_2.

0-level output condition, with a high value of R_C and the output shorted to ground through a turned-on transistor.

5.5 The Field Effect Transistor as a Switch

5.5.1 Types of Discrete Component FETs

Explanation of the FET switch is complicated by the fact that there are six types available. If we accept that N channel differs from P channel only by polarity of power supply, then we are left with three available types of FET switches. One of the three, the junction FET, is seldom employed as a digital switch in either discrete or integrated circuit form because of the superiority of the other two types. On the basis of positive voltage-1 levels, the two types we must consider are N-channel depletion-mode insulated-gate FET and N-channel enhancement-mode insulated-gate FET. It should be kept in mind that there are also available P-channel FETs of the same types. The abbreviation MOSFET, or in most cases just MOS, is commonly used to refer to either of these devices.

5.5.2 Depletion-Mode MOS Field Effect Transistors

Figures 5-14(a) and (b) show a cross section of the depletion-mode MOSFET and its schematic

FIGURE 5-13

Four-transistor switch of the TTL inverter divided into three sections.

FIGURE 5-14

(a) Cross section of an N-channel depletion-mode MOSFET. (b) Schematic symbol of an N-channel depletion-mode MOSFET.

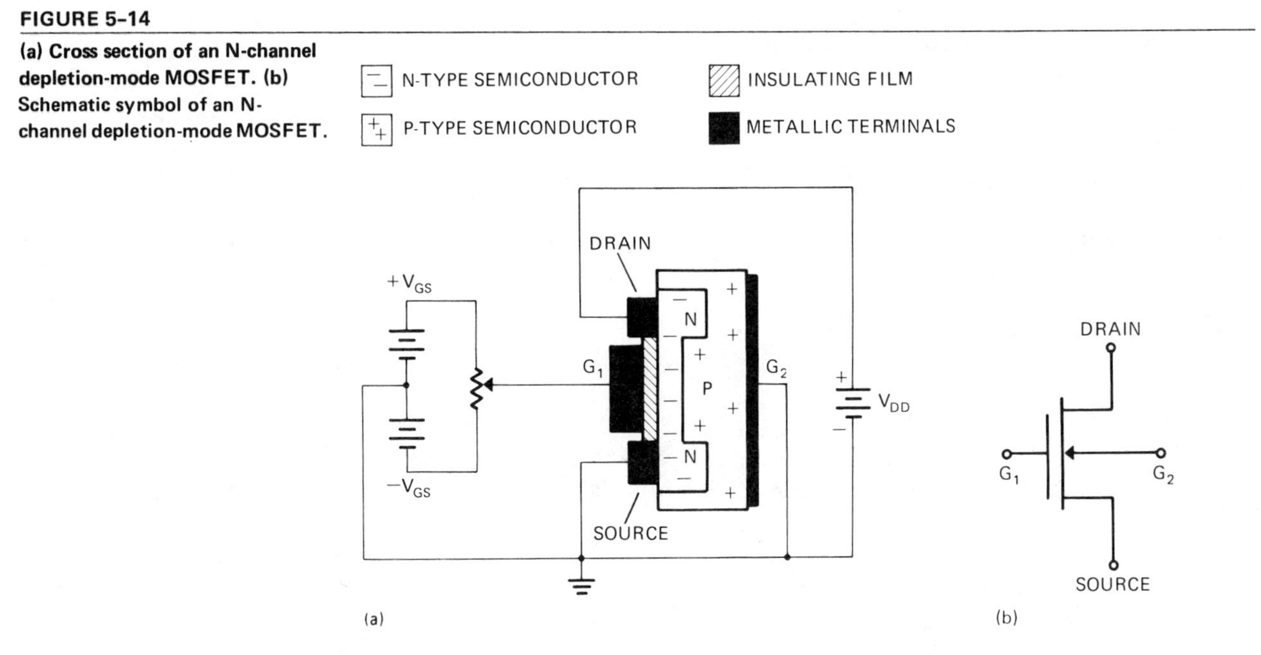

symbol. Embedded in the P material of Figure 5-14 are two rather large wells of N material connected by a narrow channel of N material. The gate (G_1) parallels this channel but is separated from it electrically by a thin insulating film.

The drain and source terminals are connected directly to the two wells of N material. If these terminals are biased, as in Figure 5-14, with the potentiometer set at $V_{GS} = 0$, a current (I_{DS}) will flow between drain and source. If, however, a negative gate voltage, $-V_{GS}$, is applied, the negative electric field will repel the carriers from the narrow N channel into the source well, reducing its conductivity and reducing the current, I_{DS}. If the negative gate voltage is increased to $-V_P$, then the channel will be completely depleted of carriers and the drain current, I_{DS}, will be shut off. This level, $-V_P$, is called the pinchoff voltage.

If the gate voltage is raised to a positive level, however, the positive gate voltage will attract more electrons into the narrow channel, increasing the drain current beyond what it was at $V_{GS} = 0$. The depletion-mode MOSFET can, therefore, be switched on with a positive voltage and off with a negative voltage, as Figure 5-15 shows.

As shown by the waveforms of Figure 5-15, there are three voltage levels to contend with. To obtain cutoff and saturation of the switch, the input levels must be positive (1) and negative ($-V_P$). The outputs, however, are positive and 0. The 0V level will not cut off a succeeding switch of the same depletion-mode type. Although techniques have been found to overcome this in

FIGURE 5-15

Inverting switch using a depletion-mode MOSFET.

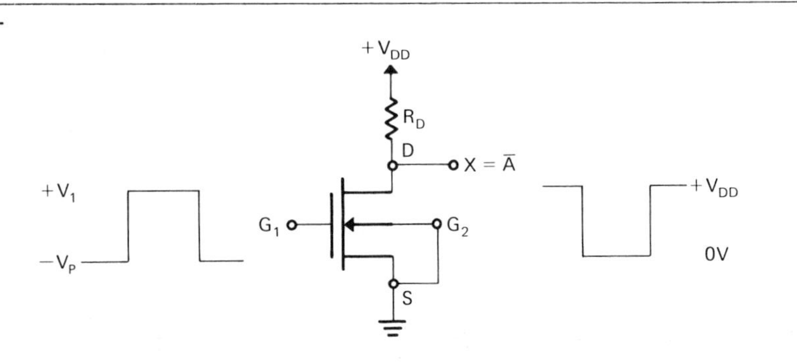

integrated circuit application, the depletion-mode MOSFET is less widely used than is the enhancement-mode MOSFET.

5.5.3 Enhancement-Mode MOS Field Effect Transistor

The enhancement-mode MOSFET is constructed as shown in Figure 5-16. Two relatively large wells of N material are embedded at opposite ends of a block of P-type material. Unlike the depletion-mode MOSFET, they are not joined by a narrow channel of N-type material. If a voltage is applied between drain and source, no current will flow. Of the three types of FETs available, this is the only one that has no drain current with 0 gate voltage. For this reason, it is called a normally OFF FET.

If an increasing positive gate voltage is applied, the positive electric field repels the holes of the P-type material. The channel under the gate is now depleted of both positive and negative charge carriers. But if the positive gate voltage increases beyond a point, $+V_P$, it draws negative charge carriers from the source well into the channel, creating a complete N channel between drain and source. Thus, a drain current, I_D, will flow. Increasing the positive gate voltage will increase the drain current. Reducing the gate voltage to 0 will return the drain current, I_D, to 0.

A digital switch can be constructed as shown in Figure 5-17. Of the three FET devices, enhancement-mode MOS is the only one that does not require a negative voltage bias network to translate the 0-level voltage of an output downward to the pinchoff voltage. The binary 0-level output of this switch is low enough to

FIGURE 5-16

(a) Cross section of N-channel enhancement-mode MOSFET. (b) Schematic symbol of an N-channel enhancement-mode MOSFET.

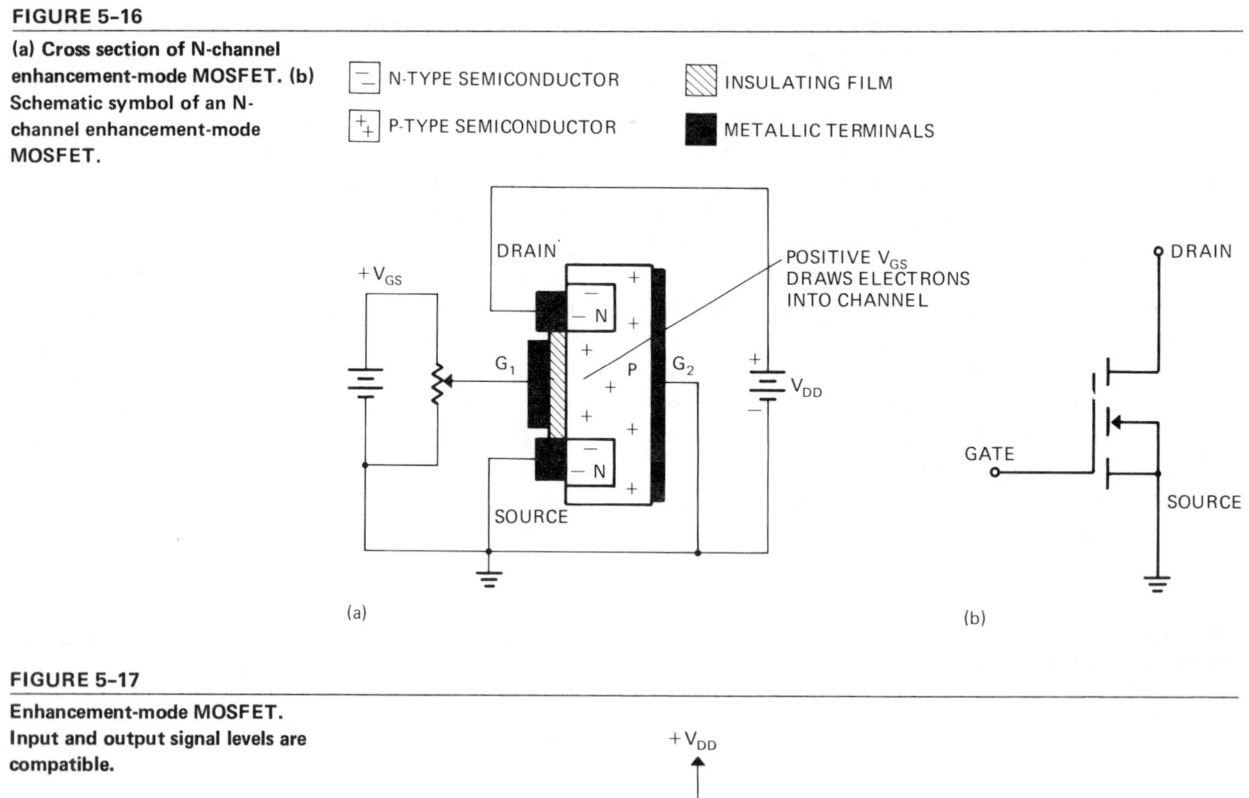

FIGURE 5-17

Enhancement-mode MOSFET. Input and output signal levels are compatible.

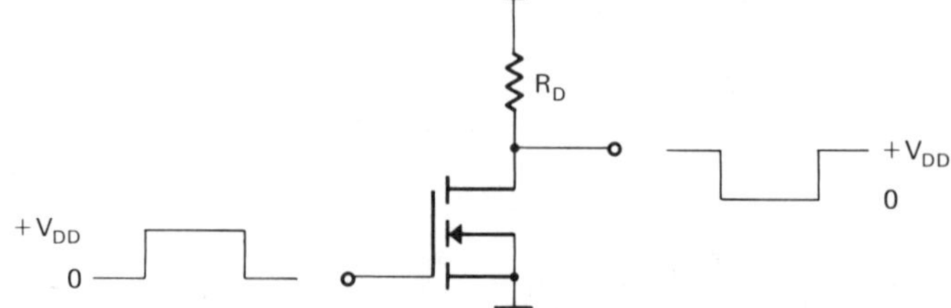

provide a cutoff gate voltage at the input to another enhancement-mode MOSFET. The 1-level output can be made high enough to saturate the next switch without any design difficulty.

5.5.4 MOS Integrated Circuit Switches

The discrete component field effect transistor switches discussed in Sections 5.5.1 and 5.5.2 were never put to widespread use because the FET and, in particular, the MOSFETs were under development during the same period that integrated circuit technology was being developed. Today the MOSFET integrated circuit is generally used with a popularity rivaling that of TTL integrated circuits. The MOS integrated circuit switch generally uses no other components. The drain resistor, R_D, shown in Figures 5-15 and 5-17, rarely appears in an integrated circuit MOS inverter. As shown in Figure 5-18(a), the drain resistor is normally replaced by a second transistor. This load transistor may be saturated by having the gate and drain connected to V_{DD}, as in 5-18(a), or it may be operated in the triode region by having the gate connected to a voltage lower than drain voltage, as in 5-18(b).

A four-letter designation is used to classify MOS inverters. As Figure 5-18(c) shows, the first letter is N or P, describing the channel. The second is D or E, for depletion or enhancement mode. The last two describe the load transistor as S or T, for saturated or operating in triode region. Eight configurations are possible: The two shown in Figure 15-18(a) and 15-18(b)—NELS and NELT—and six others—PELS, PELT, NDLS, NDLT, PDLS, and PDLT.

A ninth configuration for MOSFET switches uses both P- and N-type enhancement-mode transistors. This configuration, designated CMOS, has been increasing in popularity. The CMOS switch shown in Figure 5-19(a) consists of a complementary pair of MOSFETs, the top one a P-channel, the bottom one an N-channel. As can be seen in the equivalent circuit of Figure 5-19(b), when the input is at binary 1 ($+V_{DD}$), the N-channel MOSFET turns ON while the P-channel MOSFET turns OFF (or goes to maximum resistance). This applies $-V_{SS}$ or binary 0 to the output. When the input is at binary 0 ($-V_{SS}$), the N-channel CMOS turns OFF while the P-channel CMOS turns ON (goes to minimum resistance). This applies a binary 1 to the output. This operation is like that of the two transistor switches of Figure 5-11 except that using complementary devices eliminates need for a control transistor.

5.6 Summary

A manual switch has nearly zero ON resistance and an open circuit OFF resistance. Many electronic devices have the two states of low resistance (ON) and very high resistance (OFF). Although these are slightly imperfect in terms of OFF and ON resistance, their use provides us with switches that operate automatically and at high speed.

FIGURE 5-18

Integrated circuit MOS inverters. (a) Inverter using saturated load transistors in place of R_D. (b) Inverter using load transistor in triode region. (c) Four-letter classification of MOS inverters.

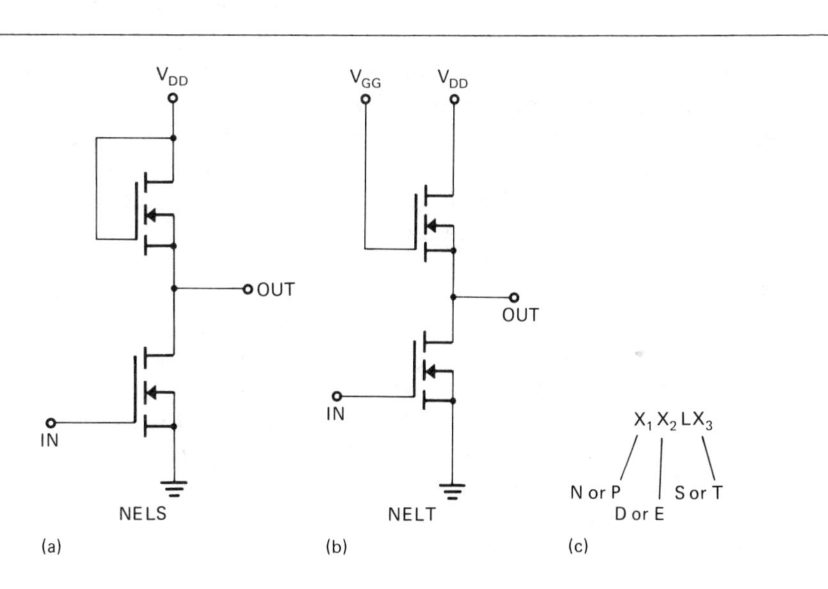

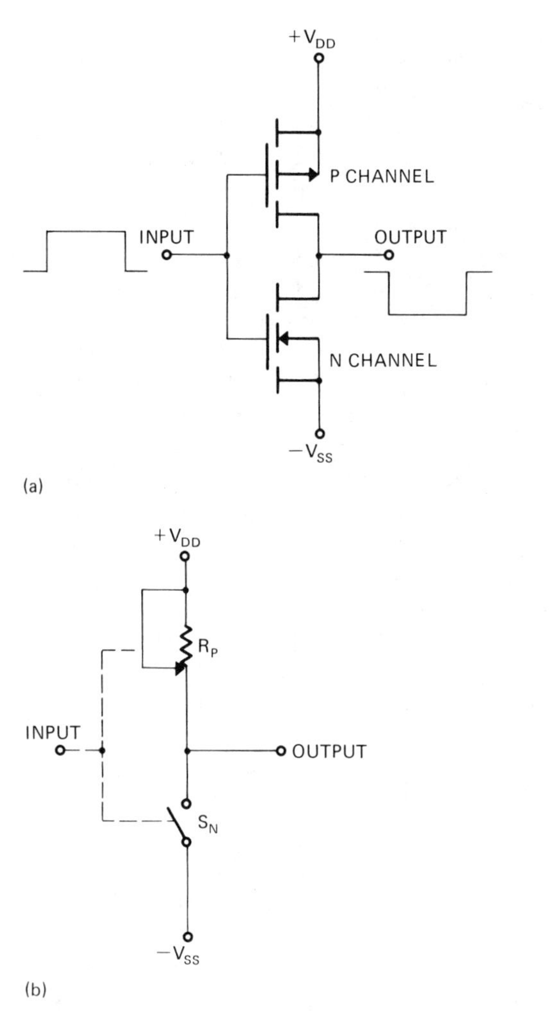

FIGURE 5-19

CMOS integrated circuit switch. (a) Complementary enhancement-mode MOSFETs at the P channel and at the N channel. (b) Equivalent circuit shorting-type switch. Here we consider the N-channel MOSFET a switch and the P-channel MOSFET a rheostat.

The *semiconductor diode* can function as a switch. When *forward-biased* by more than 0.7V, it has a low resistance (ON). When *reverse-biased*, it has a very high resistance (OFF).

The *bipolar transistor* can be used as a switch. A 1-level voltage applied to the base lead of the transistor turns the transistor ON. In the ON state, a heavy collector current flows. This current is high enough to cause *saturation* of the circuit. When a 0 is applied to the base lead, it turns the transistor off. In the OFF state, or *cut-off*, no base or collector current flows. A saturated or turned-ON transistor will produce a 1 or 0 at the output of the transistor switch circuit depending on whether it is operating in common emitter or in emitter follower configuration.

In emitter follower configuration, the high saturation current caused by the 1-level voltage on the base or input produces a 1 level at the output, as Figure 5-7 shows. In the common emitter configuration, shown in Figure 5-5, the high saturation current caused by the 1 level on the base or input drops the supply voltage across the collector resistor, producing a 0 at the output. The common emitter switch is, therefore, called an *inverter* because a 1 level at the input produces a 0 on the output, and a 0 on the input produces a 1 on the output. This is shown in the truth table of Figure 5-6.

The field effect transistor can be used as a switch. The enhancement-mode *MOSFET* is most widely used for this purpose. A 1-level voltage on the gate input of a MOSFET turns it ON. In the ON state, a heavy drain current flows. This current is high enough to cause saturation of the circuit. When a 0 is applied to the gate lead, it turns the FET off and no drain current flows. The 0 level used must be lower than the pinchoff voltage of the MOSFET.

The usual configuration of a MOSFET switch is an *inverter.* Figure 5-17 shows an enhancement-mode MOSFET inverter. A 1-level voltage applied to the gate input causes a saturation-level drain current. This current drops all the drain supply voltage across the drain resistor, producing a 0 level at the output. A 0 level on the gate input results in no drain current. With no drain current flowing through the drain resistor, the supply voltage is not dropped and a 1 level will appear at the output.

Most transistor inverters in use today are in *integrated circuit* form. Integrated circuit inverters are available with four to eight inverters on a single 14-pin circuit that occupies less than a square centimeter of space. Figure 5-8 shows a quad *RTL* inverter. These individual circuits are like the inverter explained in Figure 5-5, in that each inverter uses a single transistor. The TTL integrated circuit is more complicated, as Figure 5-9 shows. It uses four transistors for each inverter. The two output transistors, Q_3 and Q_4, are known as a *totem pole.* When Q_3 turns ON, Q_4 becomes a high resistance. This reduces the power consumed during the 0 output state. When Q_3 turns OFF, Q_4 becomes a low resistance. This provides a low impedance during the 1 output.

The MOS integrated inverter usually contains no resistors. The drain resistor is usually a FET, operating either in *saturation*, as in Figure

5-18(a), or in the *triode region*, as in Figure 5-18(b). The *CMOS* integrated inverter uses an enhancement-mode N-channel switch transistor with a P-channel load transistor. Figure 5-19 shows that the input is connected to both gates. A 1 level applied to the input turns the N-channel transistor ON, shorting the output to ground, while the P-channel transistor turns OFF, preventing an excessive drain on the power supply. A 0 level applied to the input turns the N-channel transistor OFF, isolating the output from ground, while the P-channel transistor turns ON, providing a low-resistance connection between the output and the drain voltage. The CMOS inverter has the advantage of very high input impedance and very low power consumption.

Glossary

Semiconductor: Chemical elements with a valence of four having an ability to conduct electricity midway between the conductor elements and insulator elements.

N-Type Semiconductor: A semiconductor material doped with an impurity that provides excess electrons not needed in the crystal valence structure. The presence of these loosely bound electrons makes this material a good conductor.

P-Type Semiconductor: A semiconductor material doped with an impurity that creates holes in the crystal valence structure. The holes act like positive-charge carriers, making the material a good conductor.

Semiconductor Diode: A diffused junction of P- and N-type semiconductor material that will conduct electron flow effectively in only one direction. The P material forms the anode, and the N material the cathode. When forward-biased—positive on the anode, negative on the cathode—it will conduct a high level of current. When reverse-biased—positive on the cathode, negative on the anode—it will conduct very little current.

Forward Bias: The application of DC voltage to a semiconductor PN junction in a direction that produces a major current flow. The high current flow occurs when the more positive voltage is applied to the anode lead (P material). The more negative voltage is applied to the cathode (N material). When forward-biased, the diode has very low resistance. See Figure 5-2.

Reverse Bias: The application of DC voltage to a diode or semiconductor PN junction in a direction that produces a minor current flow. Only a minute amount of current flows when the more negative voltage is applied to the anode (P material). The more positive voltage is applied to the cathode (N material). When reverse-biased, the diode has a very high resistance. See Figure 5-2.

Bipolar Transistor: A three-terminal semiconductor device made by the diffusion of three segments of N- and P-type semiconductor material to form emitter, base, and collector leads. They are produced in NPN or PNP form. This device exhibits the useful characteristic of current gain in that a small forward-bias current between base and emitter leads produces a major current flow between collector and emitter. In switching applications, a forward bias applied to base emitter leads turns the transistor ON. Removing the forward bias turns the transistor OFF. See Figure 5-5.

Cutoff: Three zones exist in the operation of a transistor amplifier: cutoff, saturation, and the zone between cutoff and saturation, called transition or triode zone. Cutoff occurs when the input is taken so low that no output current flows. The output voltage will be minimum or maximum (1 or 0) depending on the circuit configuration.

Saturation: In the operation of a transistor amplifier, an increase at the input causes a proportional increase in output current. As the output current increases, the amount of the supply voltage dropped across the resistors in the output circuit increases. Saturation occurs when all the available supply voltage is dropped across the resistors, and a further increase at the input has no effect.

Triode Region: The transition zone between cutoff and saturation of an amplifier. When a transistor amplifier is used as a switch, the triode region is normally avoided. When a transistor is used in place of a load resistor of a switching circuit, the load transistor may be biased in the triode zone to provide (see Figure 5-18) a resistance somewhere between the low resistance of saturation and the high resistance of cutoff.

Inverter: A single-input logic circuit that produces a 1 output when the input is 0 and a 0 output when the input is 1. See Figure 5-6.

MOSFET: A three-terminal device useful as an amplifier or a switch. The output terminals are the drain and source leads. The output current is conducted through a semiconductor channel existing between drain and source. Devices are manufactured with either N- or P-channel material. Figure 5-16 shows an N-channel enhancement-mode MOSFET. For the N-channel enhancement-mode MOSFET, the channel between drain and source must be enhanced by free electrons attracted into the channel by a positive charge on the gate lead. No drain current will flow without this positive charge on the gate lead. When used as a switch, a 0 voltage on the gate turns the channel OFF, and a positive 1 voltage turns it ON.

Integrated Circuit: A method of producing electronic circuits in which all components are part of a semiconductor chip. Resistors, transistors, diodes, and even small capacitors are produced as minute semiconductor devices.

Discrete Component Circuit: A method of producing electronic circuits in which resistors, transistors, diodes, and capacitors as individual components are assembled on a circuit board or other circuit assembly.

RTL: Abbreviation for resistor-transistor logic, a method of producing integrated logic circuits. The input is a resistor connected to the base of an NPN transistor. The output is the collector of the transistor. See Figure 5-8.

TTL: Abbreviation for transistor-transistor logic, a method of producing integrated logic circuits. The input or inputs are the emitter of a transistor. The output is in most cases a totem pole circuit. One or more control transistors are used between the input transistor and the totem pole output. See Figure 5-9.

Totem Pole: A pair of transistors connected with their collector-to-emitter circuits in series. When used in TTL logic circuits, the lower transistor or switch is turned ON while the upper or load transistor is switched to high resistance. When the switch transistor is turned OFF, the load transistor is switched to a low resistance. See Figure 5-11.

N-Channel MOSFET: A MOSFET in which the source and drain wells are N-type semiconductor. A positive gate voltage is needed to enhance conduction through an N-channel MOSFET.

P-Channel MOSFET: A MOSFET in which the source and drain wells are P-type semiconductor. A negative gate voltage is needed to enhance conduction through a P-channel MOSFET.

CMOS: Abbreviation for *complementary MOSFET*, a method of producing MOS integrated logic circuits using an N-channel switch transistor and a P-channel load transistor. With a positive 1 level at the input, the N-channel switch transistor is ON and the P-channel load transistor is at high resistance. With 0-level input, the N-channel transistor is OFF and the P-channel load transistor is at low resistance. See Figure 5-19.

Questions

1. Under what condition is a diode like an open switch?
2. Under what condition is a diode like a closed switch?
3. What loss occurs when a signal passes through a diode switch?
4. What limitation must be considered in the loading of a common emitter switch?
5. What advantage has the totem pole output over a single transistor with a collector resistor?
6. How does a depletion-mode MOSFET differ from an enhancement-mode MOSFET?
7. Compare the polarities of pinchoff voltage (V_P) for the following MOSFETS: N-channel depletion-mode, P-channel depletion-mode, N-channel enhancement-mode, P-channel enhancement-mode.
8. Which inverter has the highest resistance load transistor: NELS, PELT, or PELS?
9. In what important characteristic is the CMOS inverter like the totem pole circuit of Figure 5-11(b)?
10. The top RTL inverter of Figure 5-8 has its input connected to a 0 (gnd). Consider all resistor values as typical. Compute the output voltage with 1, 2, and 3 of the remaining inverter inputs connected to Pin 9.

Problems

5-1 Consider the diode ideal. In Figure 5-20(a), what voltage will appear at V_X? In Figure 5-20(b), what voltage is at V_X? Redraw both circuits using a manual switch in place of the diode.

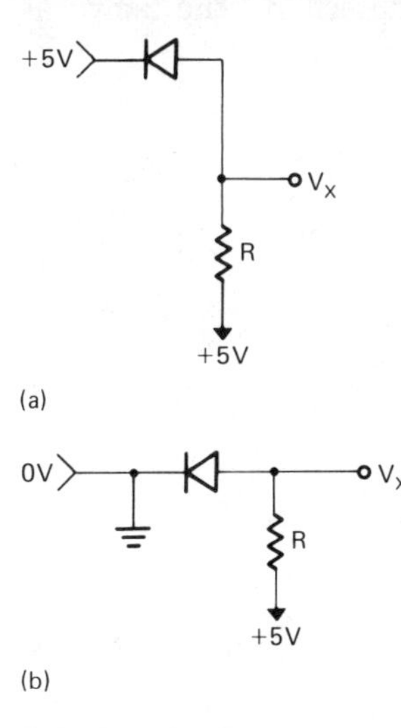

FIGURE 5-20. Problem 5-1

5-2 In the circuit of Figure 5-21, which diodes are forward-biased? Redraw the circuit using open or closed switches to represent the diodes.

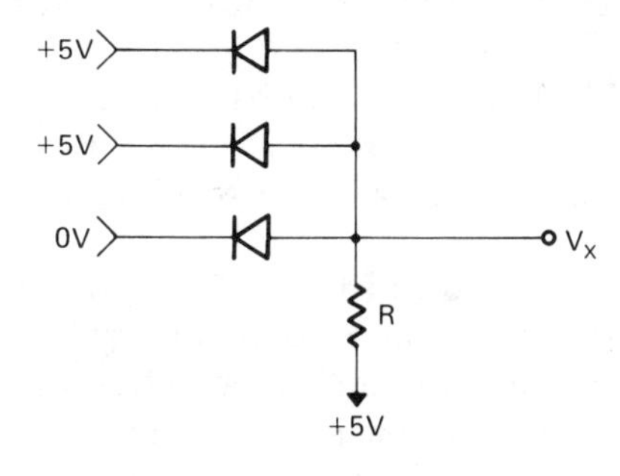

FIGURE 5-21. Problem 5-2

5-3 In the circuit of Figure 5-22, redraw the circuit using open or closed switches to replace the diodes.

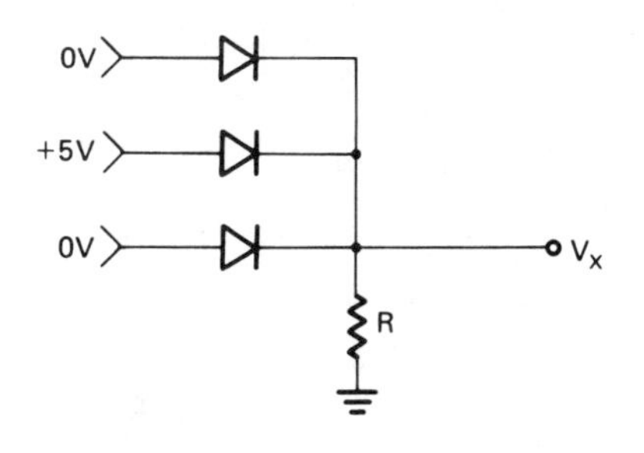

FIGURE 5-22. Problem 5-3

5-4 In the circuit of Figure 5-23, voltages are applied as in each line of the table. Complete the table by describing the lamps as ON or OFF.

TRUTH TABLE

A	B	C	L_1	L_2
1	0	0		
0	1	0		
0	0	1		

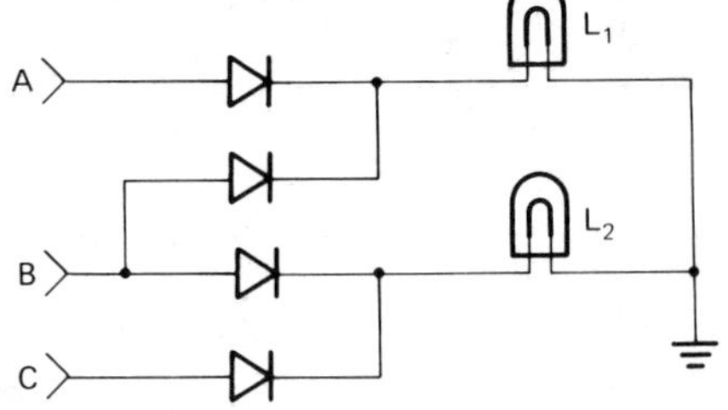

FIGURE 5-23. Problem 5-4

5-5 The diodes in Figures 5-21 and 5-22 are silicon. What voltages will appear at V_X?

5-6 If the value of R_C in Figure 5-5 were 4.7K with a V_{CC} of 5V, approximately how much collector current, I_C, would be needed to saturate the transistor and produce the output 0 level?

5-7 If a 1 level on the base of the transistor of Problem 5-6 produces an (I_B) input current of 50μA, how many inputs can be connected to a single output without reducing the 1 output level below 4V?

5-8 In the circuit of Figure 5-24, the three lamps will light with 4 to 6 volts. Six volts is applied to one input while the remaining lines are at 0V (ground). List in the table the forward-biased diodes and the lamps that will light for each line energized with 6 volts.

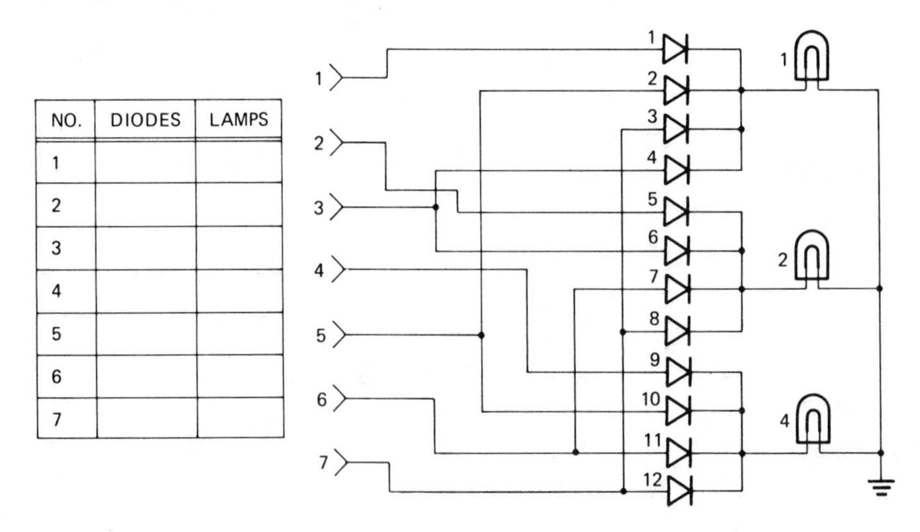

NO.	DIODES	LAMPS
1		
2		
3		
4		
5		
6		
7		

FIGURE 5-24. Problem 5-8

5-9 In the circuit of Figure 5-25, what value of collector current should flow to produce a 0 output (R_C = 2.2K)?

5-10 If the input-1 level to the switch in Figure 5-25 is 3.7V, what will be the value of I_B (R_B = 50K) (V_{BE} = 0.7V)?

5-11 If four such inputs are connected to the output of the switch of Figure 5-25, how will it affect the output-1 level?

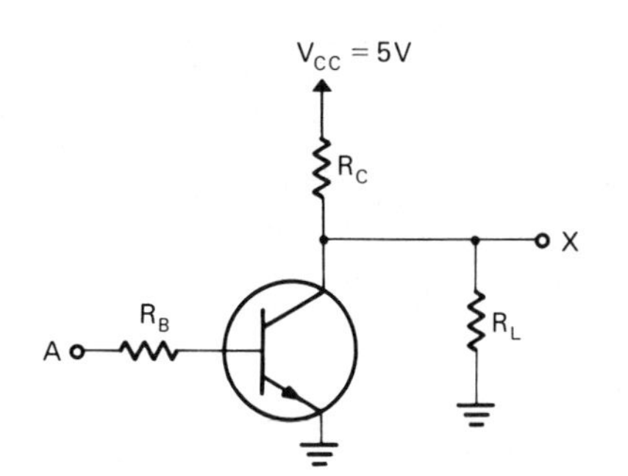

FIGURE 5-25. Problems 5-9 through 5-11

CHAPTER

Logic Gate Circuits

Objectives

On completion of this chapter, you will be able to:

- Identify and use the logic symbols of AND and OR gates.
- Draw the truth table of AND and OR gates.
- Write and simplify the Boolean equations of circuits composed of AND and OR gates.
- Assemble diodes and resistors to form diode AND and OR gates.
- Assemble transistors and resistors to form transistor AND and OR gates.
- Identify correct pin connections for TTL/SSI integrated circuit AND and OR gates.
- Use AND and OR gates to enable or inhibit the passage of a digital signal.
- Draw and analyze timing diagrams for the operation of AND and OR gates.
- Use AND and OR gates to encode and decode digital data.
- Form special waveforms by proper connection of shift counter output to AND or OR gates.
- Isolate individual pulses or sets of pulses from the clock pulse line using AND and OR gates in conjunction with shift counter waveforms.

6.1 Introduction

The digital signals described in Chapter 4 are generated and used by logic circuits. These logic circuits give a particular response to certain combinations of binary 1 and 0 levels appearing on their inputs. There are four simple logic functions—AND, OR, NOR, and NAND—that must be thoroughly understood before we proceed with our analysis of digital machines. Figure 6-1 shows the four basic logic gate symbols. These terms, particularly AND and OR, have their origins in Boolean algebra. The reader who has already mastered Boolean algebra will perhaps be at a slight advantage. However, in this day of integrated circuits, Boolean algebra need not be mastered to understand logic circuits. Boolean expressions and equations will, however, be valuable to us in identifying and describing the logic functions. For this reason, the Boolean notations will be described and used in this chapter where helpful in explaining material.

In this chapter, we will discuss the two noninverting gates, the AND and OR gates. The inverting gates, NAND and NOR, will be discussed in Chapter 7.

It is not the object of this text to examine in detail the numerous discrete-component circuits that can be used to form the four basic logic functions. A few typical circuits will be described, but we will proceed on the assumption that the logic gates are provided in integrated circuit form. At this writing, the state of the art in integrated circuit construction is such that even the complex systems described herein may be available in large-scale integrated circuit (LSI) form; but it is still essential to understand the logic within the LSI for proper interfacing of LSI chips into complete logic systems.

6.2 The AND Gate

Figure 6-2 shows the symbol for a three-input AND gate. Let us specify this gate, and unless otherwise stated all the gates described in this

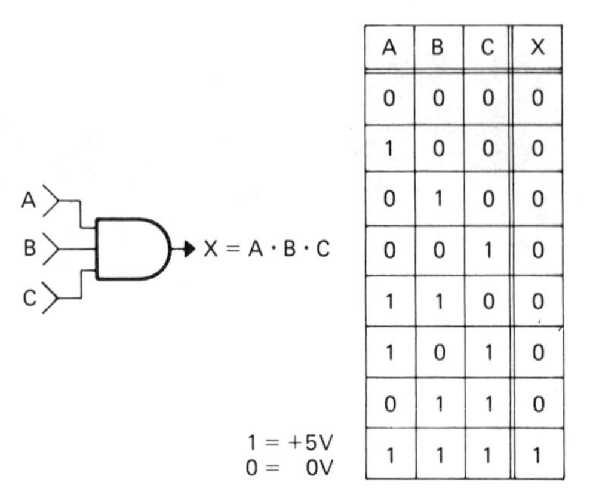

A	B	C	X
0	0	0	0
1	0	0	0
0	1	0	0
0	0	1	0
1	1	0	0
1	0	1	0
0	1	1	0
1	1	1	1

FIGURE 6-2

Three-input AND gate and its truth table.

chapter, as operating at two distinct electrical levels, binary 1 and binary 0. We will assume binary 1 to be +5V and binary 0 to be 0V. As we discuss some of the circuits in this chapter, it will become clear that losses from loading, junction potentials, and so forth will result in variations of the 1 level below 5V and increase in the 0 level above 0V. For the sake of simplicity, these variations will be dealt with in Chapter 9. Until then we will assume ideal circuits and levels.

For the three-input AND gate in Figure 6-2, a binary 1 level will appear on the output, X, of the AND gate only if there is a binary 1 level on inputs A, B, and C. If any one of the inputs is at the binary 0 level, the output will be 0. The Boolean algebra notation for this is $A \cdot B \cdot C = X$. The dot, which is a multiplication operator sign in ordinary algebra, indicates AND function in Boolean algebra. The AND function defines X as being 1 only when A, B, and C are all 1. In the electrical sense, it means that the output, X, is 1 (at 5V) only when 5V is applied to all three inputs. If any input is at 0V, the output will be at 0V.

The truth table of Figure 6-2 shows every combination of 1 and 0 that could possibly occur on the three inputs, and all result in a 0 output except the condition of all ones. The truth table of a two-input AND gate would have

FIGURE 6-1

The four basic logic gate symbols.

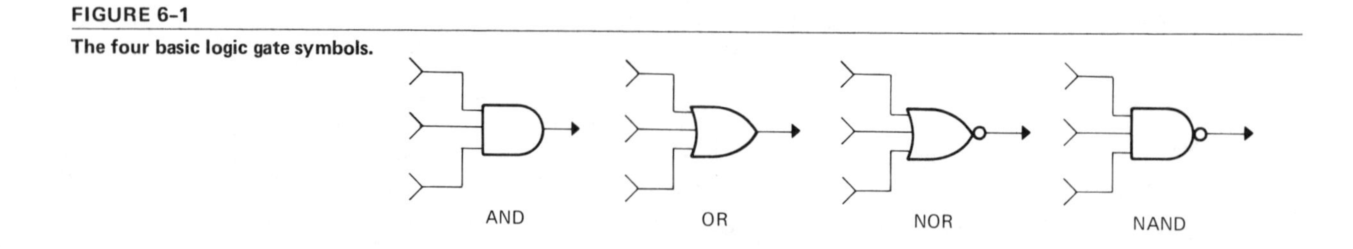

FIGURE 6-3

Three-input AND gate with manual switches in a series.

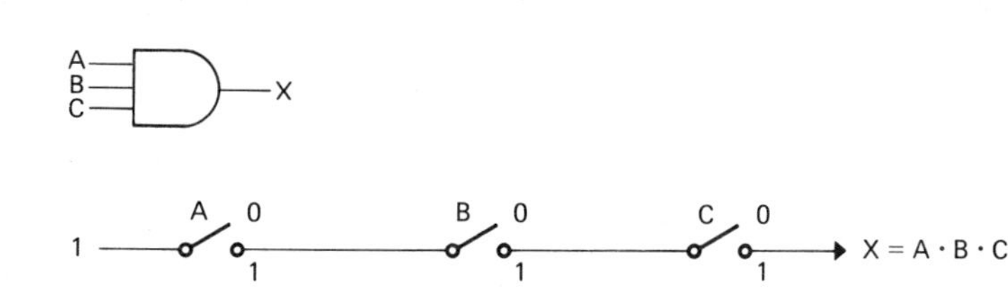

fewer input combinations. The truth table of a four-input AND gate would have more combinations. Both would have an output of 1 only when all inputs were 1.

A simple electrical AND function could be made with the three switches of Figure 6-3. The switches are in series, and only when switches A, B, and C are closed (in the 1 state) will the 1 level reach X.

Figure 6-4 shows another method of producing an AND function with manual switches. In this circuit, the switches are in the 1 state when open. If either switch is closed or in the 0 state, the top of resistor R is at ground 0 and all of the +5V is dropped across R_L. Both switches must be in the 1 state for the output, X, to be at 1 level. The load circuit resistance, R_L, must be high in proportion to R so that load current will drop only a minimal amount of the 1-level voltage.

High-speed, solid-state logic devices are cheaper and smaller than switches and require much less power to operate. Accordingly, the simple circuit of Figure 6-4 can be converted to a high-speed automatic AND gate by replacing the manual switches with semiconductor diode switches. A diode appears like an open switch when it is reverse-biased and like a closed switch when forward-biased.

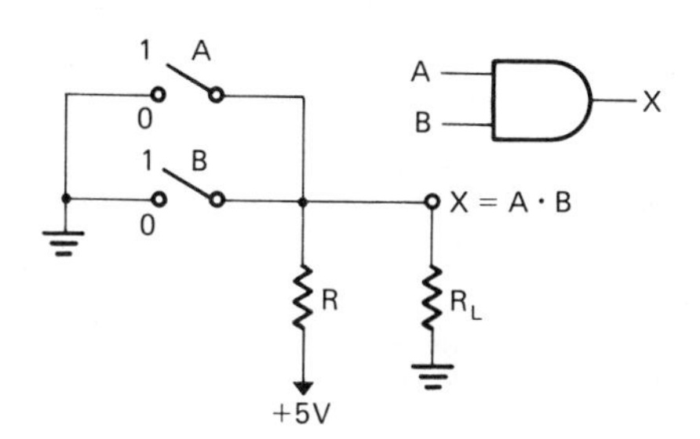

FIGURE 6-4

Two-input AND gate with manual switches and a resistor.

If inputs A and B in Figure 6-5 are connected to the outputs of other logic gates, then a 0 level on either input will result in its diode being forward-biased. A current flow through resistor R, the forward-biased diode, and the driving circuit results in a 0 level at X. Any 0 in will result in a 0 out. Only when both A and B are in the 1 state will the output be 1. The truth table for the two-input AND gate in Figure 6-5 gives every possible input combination and the resulting output.

The two-input diode gate of Figure 6-5 can be expanded to a gate of a larger number of inputs merely by putting more diodes in parallel with D_1 and D_2. The anodes of all such diodes must connect to the resistor. Figure 6-6 shows the two-input gate expanded to three inputs. The dotted lines indicate how this expansion is accomplished. The resulting changes to logic symbol and truth table are also indicated.

AND gates may also be constructed from transistors. One method for doing so is shown in Figure 6-7. A 0V level on the base of one or more of the transistors will forward-bias the respective base junction, causing an emitter current to flow through R, dropping all but a few tenths of a volt of V_E across R. The transistor

FIGURE 6-5

Two-input AND gate with semiconductor diodes for switches.

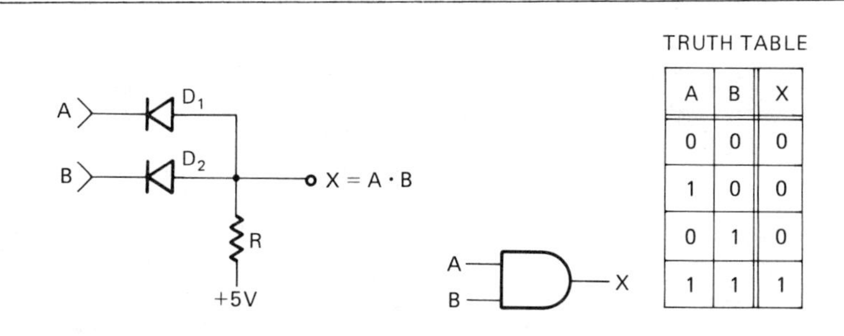

TRUTH TABLE

A	B	X
0	0	0
1	0	0
0	1	0
1	1	1

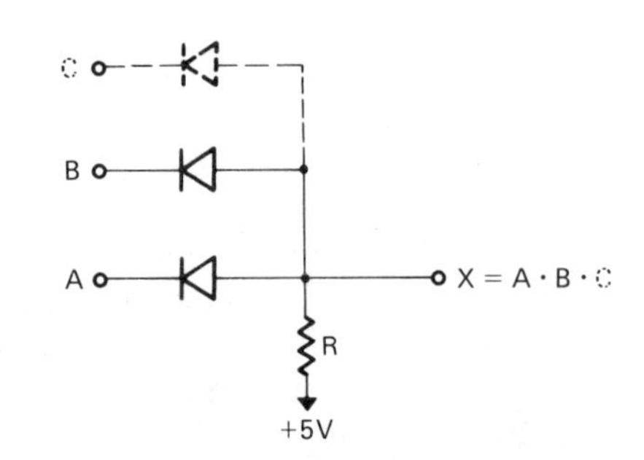

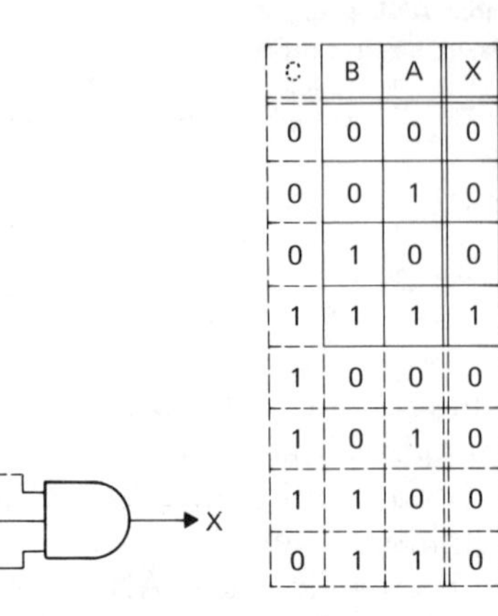

C	B	A	X
0	0	0	0
0	0	1	0
0	1	0	0
1	1	1	1
1	0	0	0
1	0	1	0
1	1	0	0
0	1	1	0

FIGURE 6-6

Expansion of a two-input AND gate and the necessary change of the logic symbol and truth table.

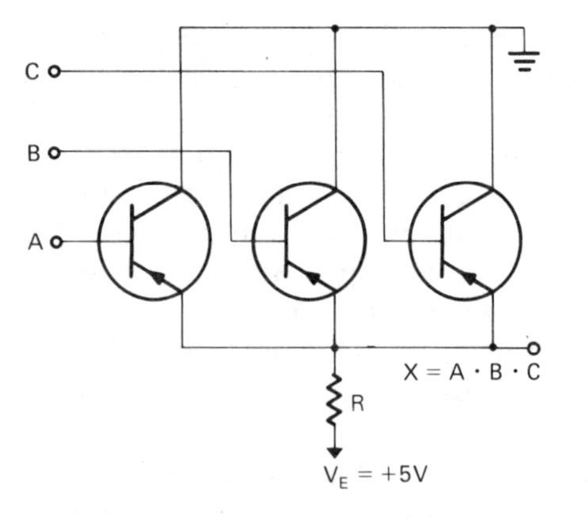

FIGURE 6-7

Three-input AND gate with transistors for switches.

AND gate is essentially an emitter follower circuit with a very high input impedance and a low output impedance.

The most widely used form of AND gate is the integrated circuit TTL (transistor-transistor logic), like the Fairchild TTL/SSI 7411 of Figure 6-8. One 14-pin package contains three identical three-input AND gates, as shown in Figure 6-8(a). The schematic diagram of one of the gates appears in Figure 6-8(b). These are modifications of the TTL switch explained in Section 5.4.4, except that the input transistor, Q_1, has three emitters. The transistor, Q_1, alone forms the actual AND function, while the output transistors, Q_6 and Q_7, form the "totem pole"-type

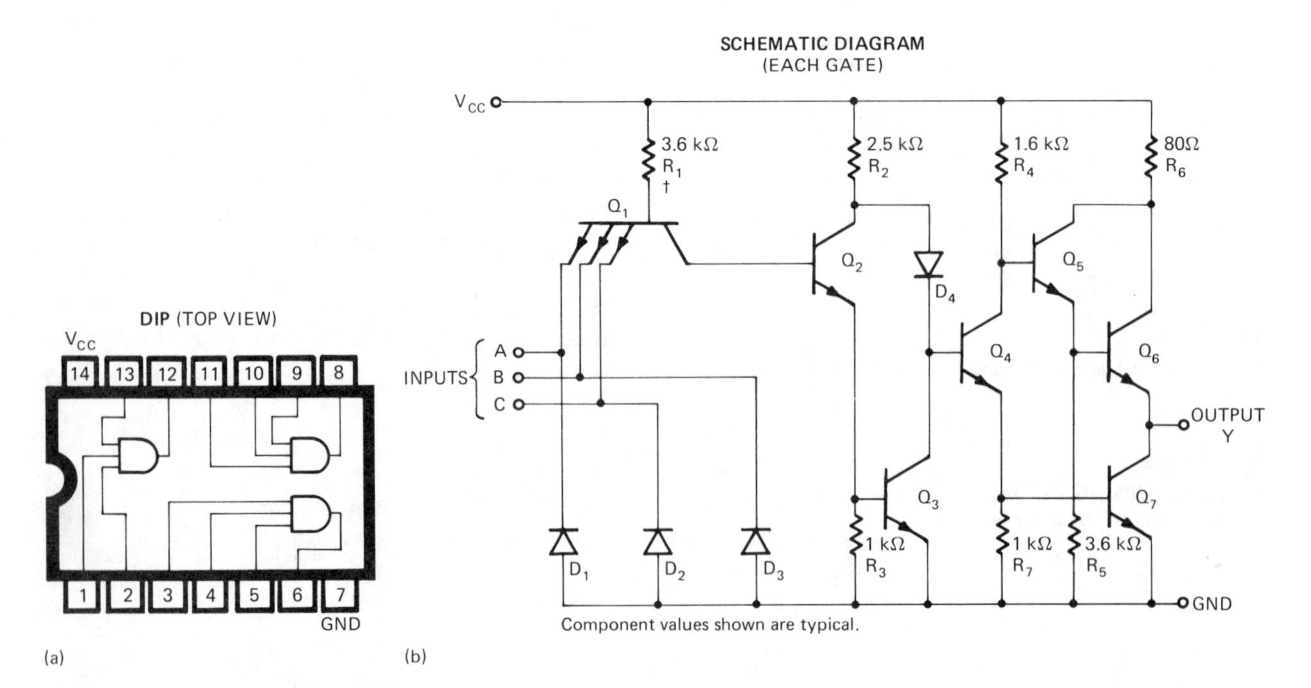

FIGURE 6-8

Fairchild TTL/SSI 7411 three-input AND gates.

output circuit explained in Section 5.4. The remaining transistors act as a driver circuit between Q_1 and the output "totem pole."

Figure 6-9 shows the diode analysis of Q_1, which functions like the diode AND gate of Figure 6-6. A ground or 0 level on any one of the three inputs will draw current from V_{CC} through the 3.6K resistor, removing the base current from Q_2. Therefore, all three inputs must be in the 1 state to cause Q_2 to turn on. Turn-on of Q_2 produces a 1 level at the output, Y.

Figure 6-10 demonstrates the functioning of the output circuit. This two-transistor output configuration is often referred to as a "totem pole." As shown in Figure 6-10(a), a binary 1 output occurs when the base of Q_6 is high and the base of Q_7 is low. As in the equivalent manual circuit, Q_6 (rheostat) goes to its minimum resistance, and Q_7 appears like an open switch. This results in a low-resistance path between the V_{CC} (1 level) and the output, Y. In Figure 6-10(b), a binary 0 output occurs when the base of Q_6 is low and the base of Q_7 is high. As in the equivalent manual circuit, Q_6 (rheostat) goes to its maximum resistance, and Q_7 appears like a closed switch, shorting the output to ground (0).

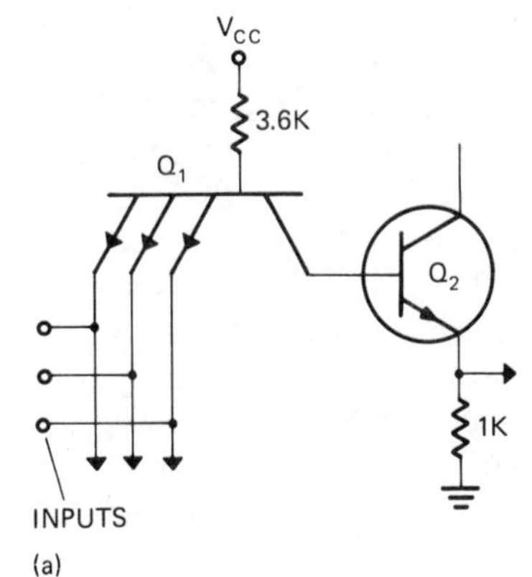

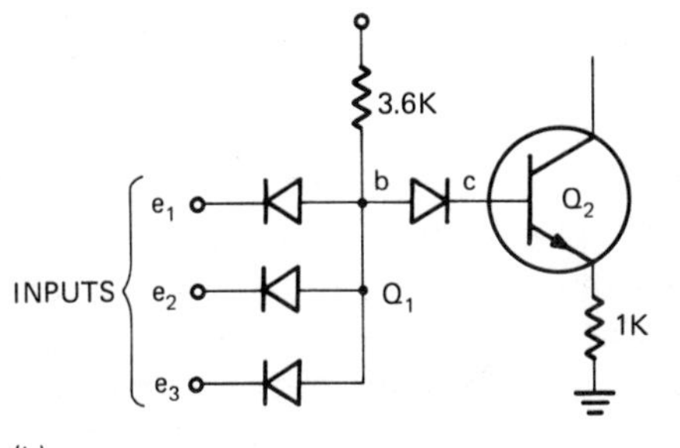

FIGURE 6-9

(a) Input transistor of the TTL AND gate. (b) Diode AND gate.

6.3 The OR Gate

Another logic function used in digital machines is the OR gate. Figure 6-11 shows the symbol for a three-input OR gate. If a digital 1 level occurs on input A *or* B *or* C, a 1 level will appear on the output, X. As the truth table in Figure

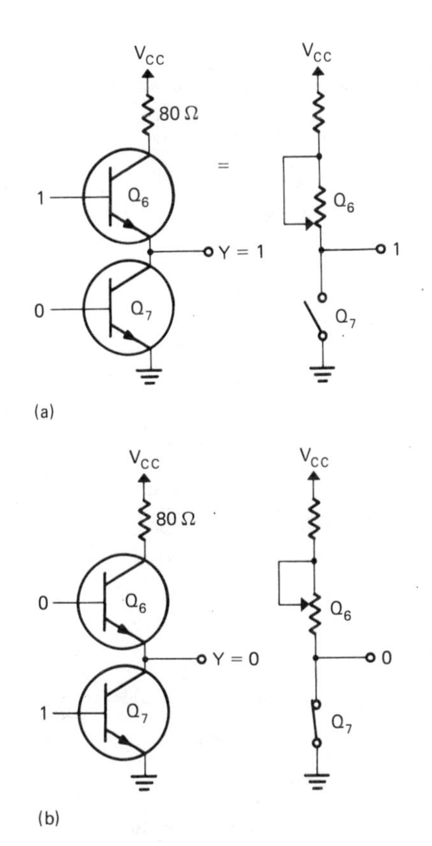

FIGURE 6-10

TTL integrated circuit "totem pole" output transistors and equivalent manual circuits for 1 and 0 outputs.

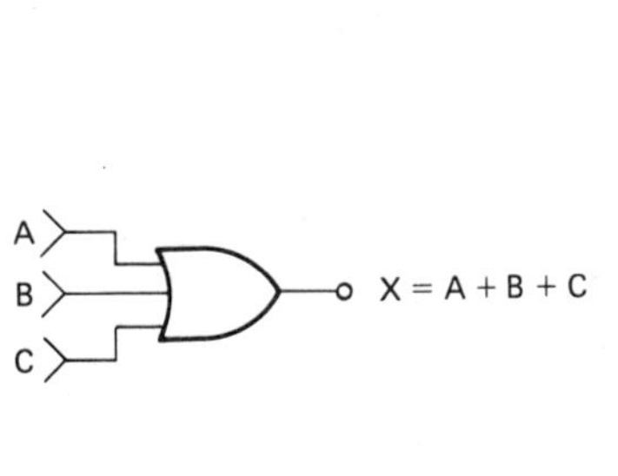

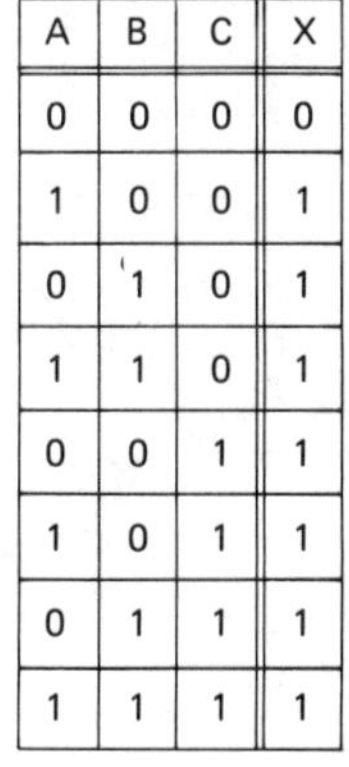

A	B	C	X
0	0	0	0
1	0	0	1
0	1	0	1
1	1	0	1
0	0	1	1
1	0	1	1
0	1	1	1
1	1	1	1

FIGURE 6-11

Three-input OR gate and its truth table.

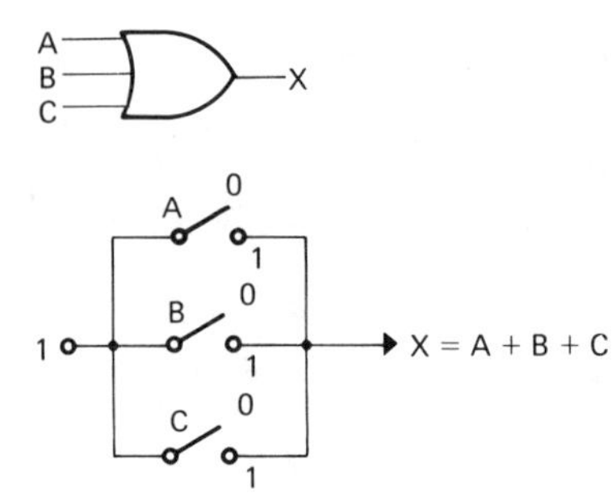

FIGURE 6-12

Three-input OR gate with manual switches wired in parallel.

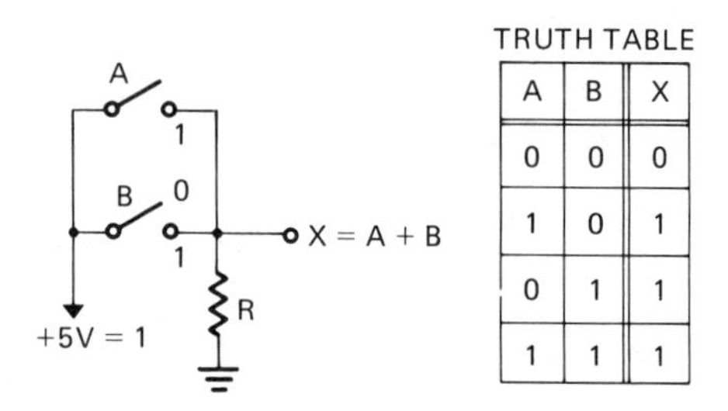

A	B	X
0	0	0
1	0	1
0	1	1
1	1	1

FIGURE 6-13

Two-input OR gate with manual switches and a resistor.

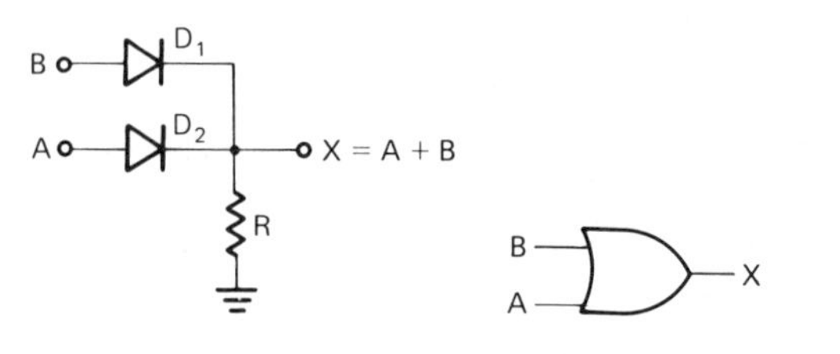

FIGURE 6-14

Two-input OR gate with semiconductor diodes for switches.

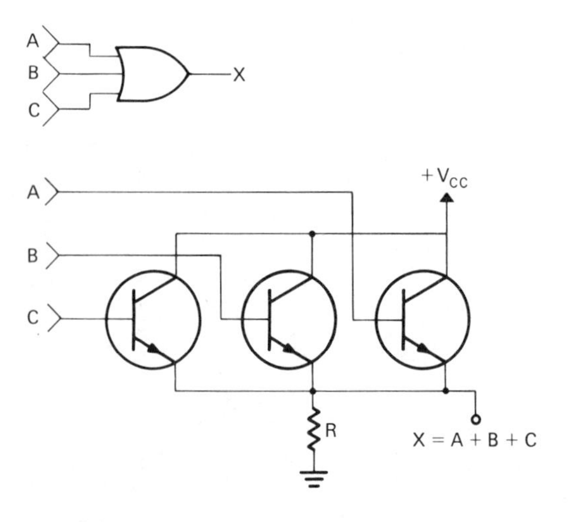

FIGURE 6-15

Three-input OR gate with transistors for switches.

6-11 shows, only when all inputs are at 0 will 0 occur at the output, X.

The three-input OR function can be accomplished electrically by wiring switches in parallel, as Figure 6-12 shows. The + sign, an operator sign for addition in ordinary algebra, stands for the OR function in Boolean algebra. Relay contacts can be wired in parallel to provide an automatic, although relatively low-speed OR function.

Figure 6-13 shows an alternative method of producing the manual OR function. In this circuit, the switches are in the 1 state when closed. If either switch is closed, the +5V 1 level will be applied to the top of resistor R, producing a 1 level at the output. Only when both switches are in the 0 state will there be a 0 at X.

As we did with the manual AND gate, we can replace the switches used in Figure 6-13 with a pair of diodes. These diodes appear like closed switches when forward-biased and like open switches when reverse-biased. The resulting OR gate is automatic and capable of operating at high speed.

Figure 6-14 shows a two-input diode OR gate. In normal use, the inputs, A and B, will be connected to other logic gates that will apply combinations of 1 and 0 levels to them. The positive 1 level on either input will forward-bias the diode, making it appear as a short circuit, causing the positive 1 level to appear at the top of R and at the output, X. If one input is 0 while the other is 1, the diode at the input having a 0 applied to it will be reverse-biased, isolating it from the 1-level output. Only when both inputs are at 0 will the output be 0.

The diodes in Figure 6-14 can be replaced by transistors, which provide higher-speed operation along with improved fan-out advantages. In Chapter 9, we will discuss in detail fan-in-fan-out problems. Figure 6-15 shows a typical direct-coupled transistor OR gate. A positive 1 level on a transistor base lead will forward-bias the base-to-emitter junction, causing a flow of base current, I_B.

Because of the gain of the transistor, this small amount of base current will cause a large emitter current. The circuit operates like an emitter follower, and the emitter voltage is only slightly lower than the input 1 level. If more than one transistor is turned on, it will merely bring the 1 output a little closer to highest input 1 level. Only when all three inputs are 0 will there be no current through R, resulting in a 0 at

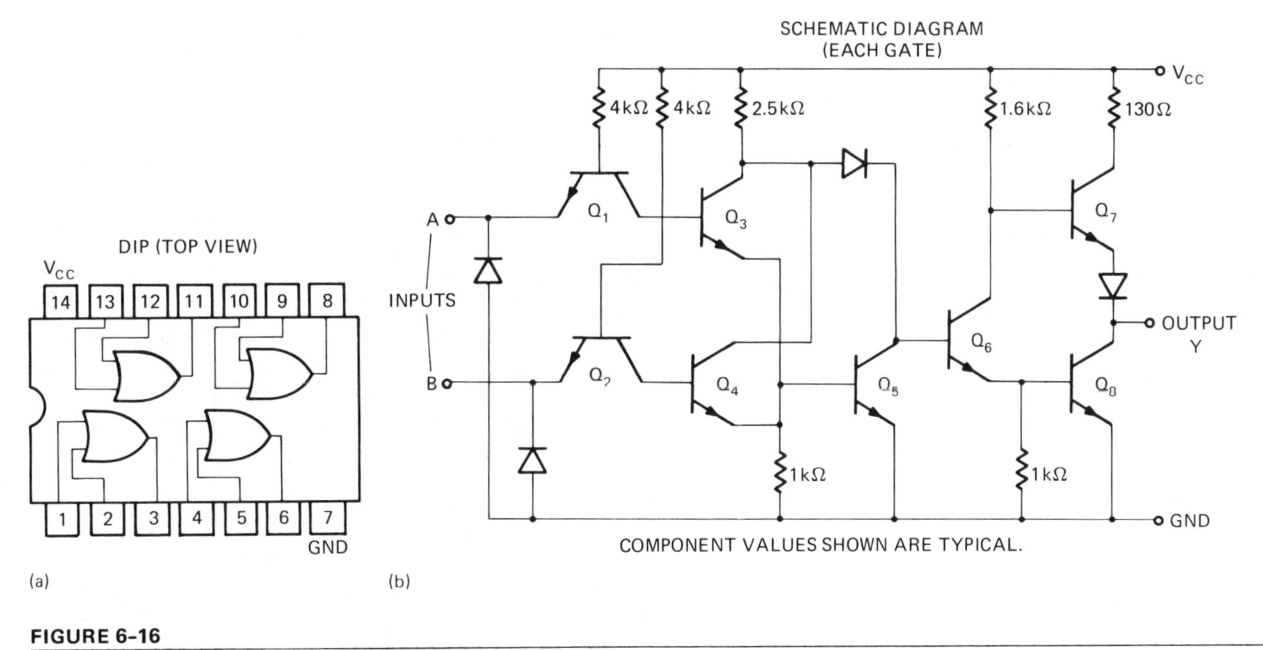

FIGURE 6-16
Fairchild TTL/SSI 7432 quad two-input OR gates.

X. Figure 6-15, therefore, has the same truth table as in Figure 6-11.

A widely used OR gate is the TTL integrated circuit. Figure 6-16 shows a Fairchild TTL/SSI 7432 two-input OR gate. One 14-pin package contains four identical two-input OR gates, as shown in Figure 6-16(a). The schematic diagram for one such gate appears in Figure 6-16(b). The output transistors function exactly like those explained in Section 5.4.4 and Figure 6-10. As is standard for TTL gates, the 1-level output occurs with the top transistor turned ON and the bottom transistor turned OFF. The 0-level output occurs with the top transistors OFF and the bottom one saturated.

The input transistors, Q_1 and Q_2, function like back-to-back diodes, passing the input 1 or 0 level through to the base leads of transistors Q_3 and Q_4. This provides an input compatible with that of other TTL gates. The parallel transistors, Q_3 and Q_4, form the actual OR function. These function like the direct-coupled transistor OR gate of Figure 6-15. Their output emitter leads are connected to the base of Q_5. Transistors Q_5 and Q_6 act as drivers between the actual OR function and the output totem pole.

6.4 Boolean Manipulation

A combination of logic gates can be assembled for a given Boolean equation. The circuitry can sometimes be simplified by algebraic manipulation of the equation.

■ EXAMPLE 6-1

Assemble a circuit producing the function ABC + ABD. Simplify if possible.

Solution: As grouped, the output of the two three-input AND functions apply to a two-input OR gate, as Figure 6-17 shows. The function $X = ABC + ABD$ can be accomplished with two three-input AND gates connected to one two-input OR gate, as shown in Figure 6-17(a). But, by factoring $ABC + ABD = AB(C + D)$, it becomes obvious that only two gates are needed for the equivalent function, shown in Figure 6-17(b).

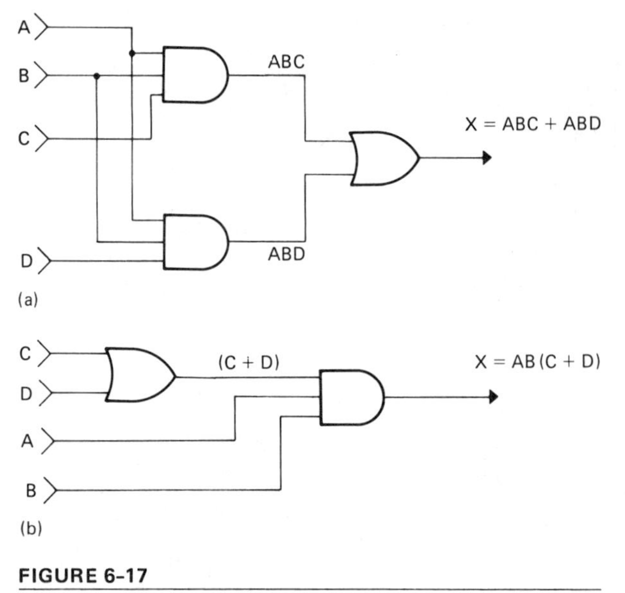

FIGURE 6-17

Solution for Example 6-1.

The Boolean equation often simplifies more easily than standard algebra because the values are limited to 1 and 0. Therefore, the following special conditions exist:

Anything ORed with 1 = 1
  $(A + B + 1 = 1)$
Anything AND with 1 = itself
  $(A \cdot B \cdot 1 = A \cdot B)$
Anything ORed with 0 = itself
  $(A + B + 0 = A + B)$
Anything AND with 0 = 0
  $(A \cdot B \cdot 0 = 0)$

Along with these conditions, a term AND with itself = itself:

$$A \cdot A = A$$

Since the values of A are limited to 1 or 0, we can raise them to any power without changing these values.

Similarly, a term ORed with itself = itself:

$$A + A = A$$

That is,

$$A + A = A(1 + 1) = A \cdot 1 = A$$

■ EXAMPLE 6-2

Draw the logic circuit of the equation X = ABC + BC + AD. Simplify the equation and draw the more economical circuit.

Solution: The obvious circuit is three AND gates connected to a three-input OR gate, as shown in Figure 6-18(a). But:

$$ABC + BC + AD = BC\,(A + 1) + AD$$
$$= B \cdot C \cdot (1) + AD = BC + AD$$

After the equation has been simplified, we need an equivalent circuit with only two AND gates and one two-input OR gate. This proves that the top three-input AND gate adds nothing to the function. Therefore, three two-input gates are enough, as Figure 6-18(b) shows.

6.5 Enable or Inhibit Functions by AND Circuits

If we look at the two-input AND circuit in Figure 6-19, we see that it has the Boolean notation $X = A \cdot B$, which is valid for any random combination of ones and zeros at the two inputs. The truth table shows these combinations and their resulting outputs, X.

Let us consider now two special conditions for the same AND circuit: (1) A input permitted to vary between 1 and 0 with B fixed at 0, $X = A \cdot 0$; and (2) A input permitted to vary between 1 and 0 with B fixed at 1, $X = A \cdot 1$.

In Figure 6-20, a switch has been provided for the first conditions. With the switch in the 0 position, we have restricted the AND gate to the

FIGURE 6-18

Solution for Example 6-2.

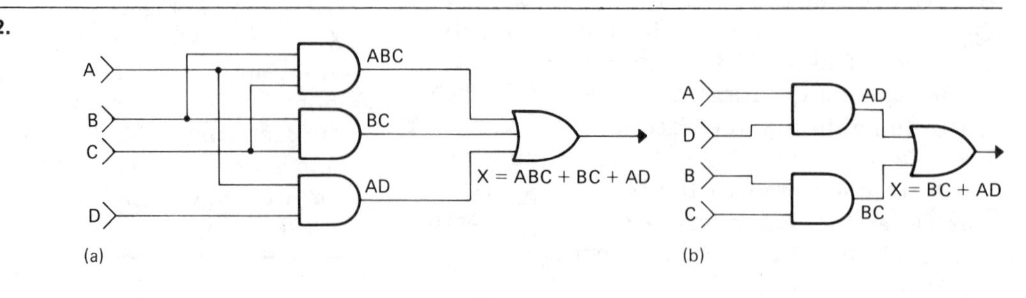

top two conditions of the truth table of Figure 6-20, and 0 is the only possible output. If we refer to A as the signal input and B as the control input, then the 0 level on B is termed an *inhibit*, for it inhibits the signal, A, from passing through the gate.

Figure 6-21 shows the second condition—that is, with switch B in the 1 position. This creates the condition of the bottom half of the truth table in Figure 6-19. As long as B remains at 1, the signal at A will get through to the output. For this reason, the 1 level of thc control input, B, is termed an *enable*, for it enables the signal, A, to pass through the gate.

The same principle can be applied to a three-input gate having one signal input and two control inputs. Figure 6-22 shows a digital clock and counter. The clock puts out pulses used to time the operations of digital machines. After a designated time 0, each pulse can be given a number, much as minutes on a mechanical clock face are numbered. If, during a particular digital program, we wanted to wait until after time 3 and then count up to five, we could connect the clock pulses to the counter after first running it through the AND gate, which would enable them after time 3 and inhibit them after time 8. The timing diagram in Figure 6-22 shows the relative timing of each signal change.

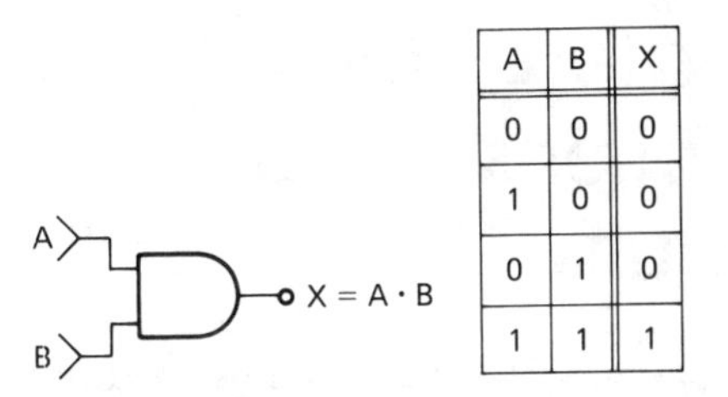

A	B	X
0	0	0
1	0	0
0	1	0
1	1	1

FIGURE 6-19

Two-input AND gate and its truth table.

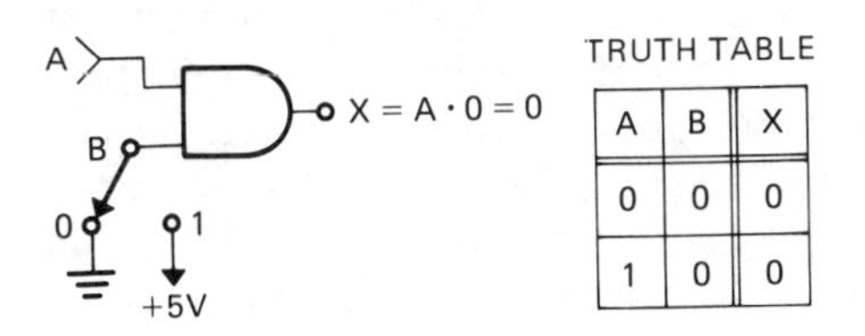

A	B	X
0	0	0
1	0	0

FIGURE 6-20

Two-input AND gate in the inhibit state.

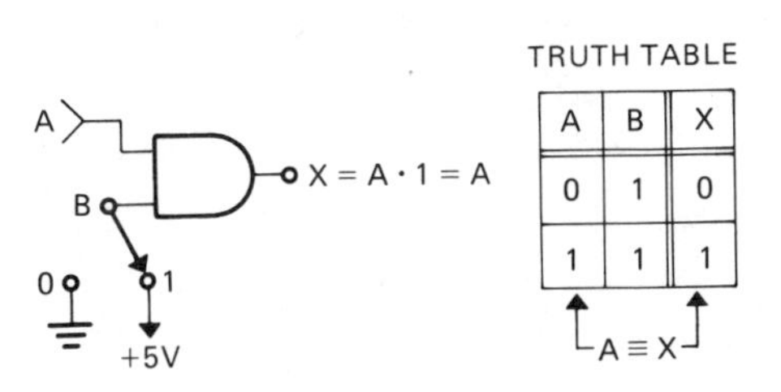

A	B	X
0	1	0
1	1	1

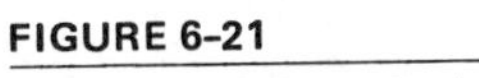

FIGURE 6-21

Two-input AND gate in the enable state.

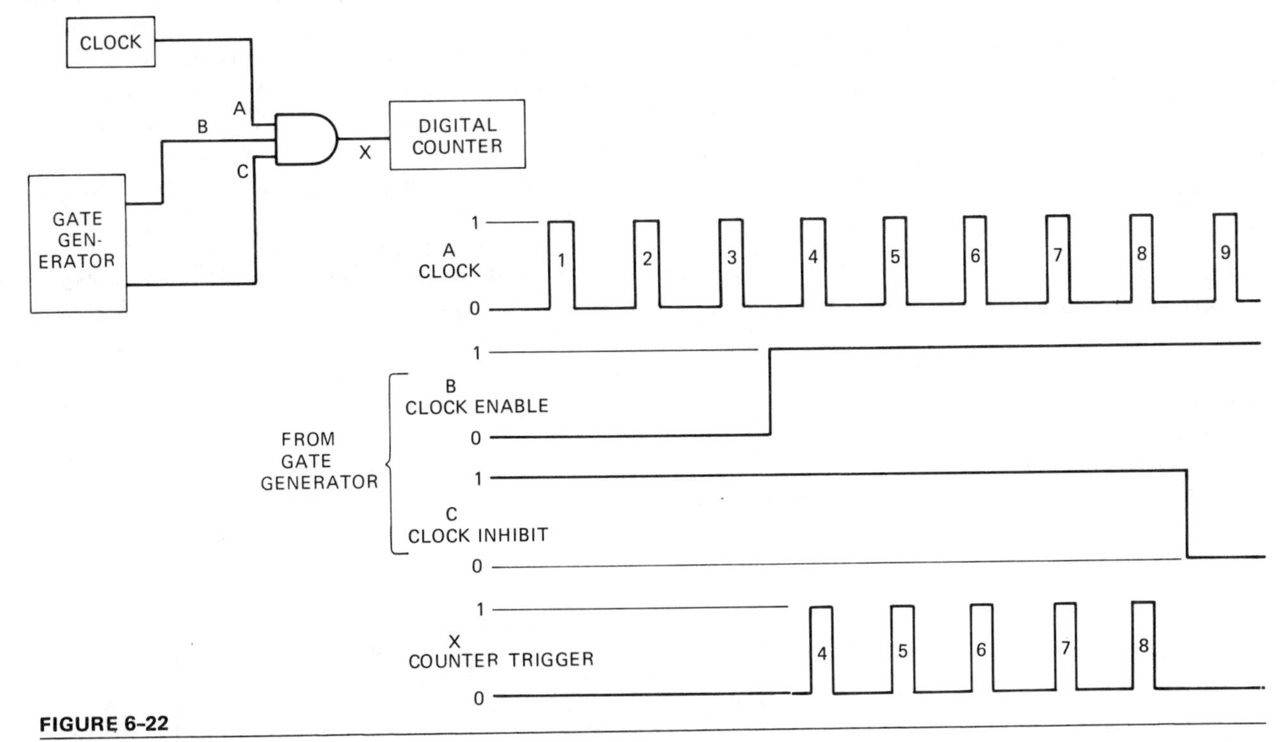

FIGURE 6-22

AND gate used to control the clock pulses applied to a digital counter.

6.6 Enable or Inhibit Functions by OR Circuits

The OR gate, like the AND gate, can be used for enabling or inhibiting passage of a digital signal. Let us consider as the signal the A input of the two-input OR gate in Figure 6-23 and allow it to vary, while the B input enables or inhibits its passage through the gate. The A + 0 input produces an output identical to the A input in the truth table in Figure 6-23(a). The 0 level on the control input may thus be referred to as an enable. When a 1 level occurs on the control input, giving the Boolean function A + 1 = 1, as in Figure 6-23(b), the output is 1 regardless of any variation of the A input. The 1 level on this control input may be referred to as an inhibit.

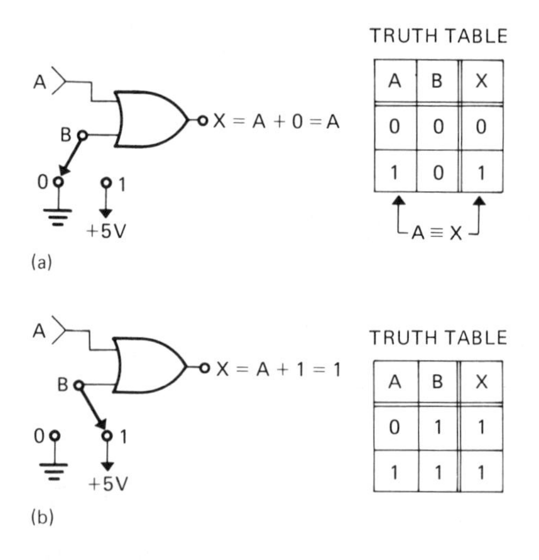

A	B	X
0	0	0
1	0	1

A	B	X
0	1	1
1	1	1

FIGURE 6-23

(a) OR gate in the enable state. (b) OR gate in the inhibit state.

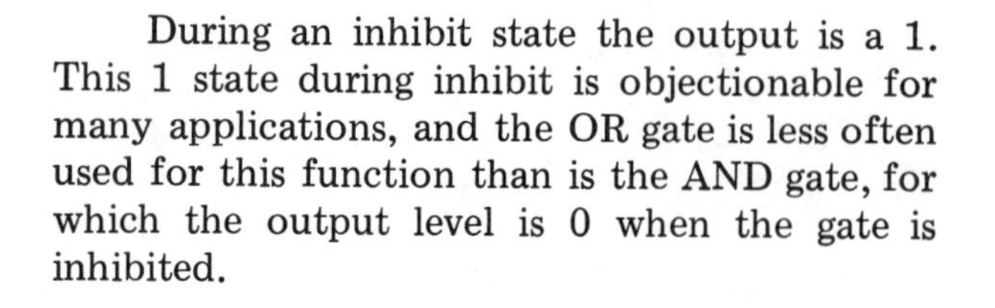

During an inhibit state the output is a 1. This 1 state during inhibit is objectionable for many applications, and the OR gate is less often used for this function than is the AND gate, for which the output level is 0 when the gate is inhibited.

6.7 Analysis by Timing Diagram

Plotting the actual operation of a logic gate is not always as easy as simply considering one input a signal and the other a control input. Often all the inputs are changing with considerable frequency, and it becomes difficult to determine the output that should occur. In this instance, the timing diagram is the best instrument for plotting gate operation.

Let us take a case in which the first input is changing from 1 to 0 and from 0 back to 1 every microsecond. The second input is changing every two microseconds, and the third input, every four microseconds. These signals can be plotted as voltage-versus-time waveforms on a time scale with one-microsecond increments for analysis to determine the gate output. The timing may be based on a clock pulse line, as in Figure 6-22. Figure 6-24 shows the three input waveforms. Their effect on the AND gate output is shown above the input waveforms. The output waveform can be determined by finding the periods during which all ones coincide on the inputs. During these periods, the AND gate output will be 1. During all other periods, there is a 0 on at least one of the inputs, and any 0 input will result in a 0 out.

If the same inputs are applied to an OR gate, as in Figure 6-25, the output waveform can be determined by finding the periods during

FIGURE 6-24

Timing diagram showing the input and output waveforms of an AND gate.

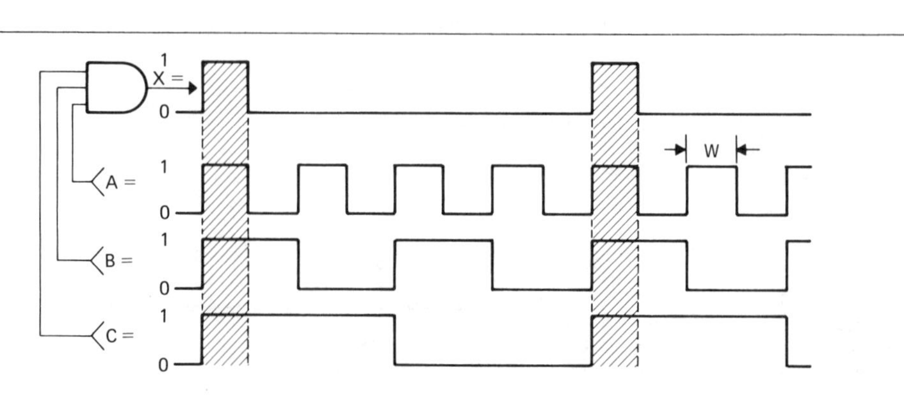

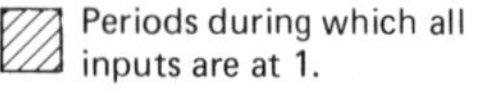

Periods during which one or more inputs are at 0.

which all zeros coincide on the inputs. During these periods, the output will be 0. During all other periods, there is a 1 level on at least one of the inputs, and any 1 input will result in a 1 out of an OR gate.

To illustrate the use of timing diagrams in developing special waveforms, we will take an advanced look at a logic assembly that is commonly used for waveform generation. This assembly, known as a *shift counter*, receives a clock pulse input, and its outputs are a set of waveforms that begin and end on the trailing edge of clock pulses.

The clock generator of Figure 6-26 provides the clock pulses shown at the top line C_P. It also provides a delayed clock C'_P. The shift counter provides the outputs A through E and their complements, $\bar{A}$ through $\bar{E}$. These outputs can be used in conjunction with AND and OR gates to provide various logic signals.

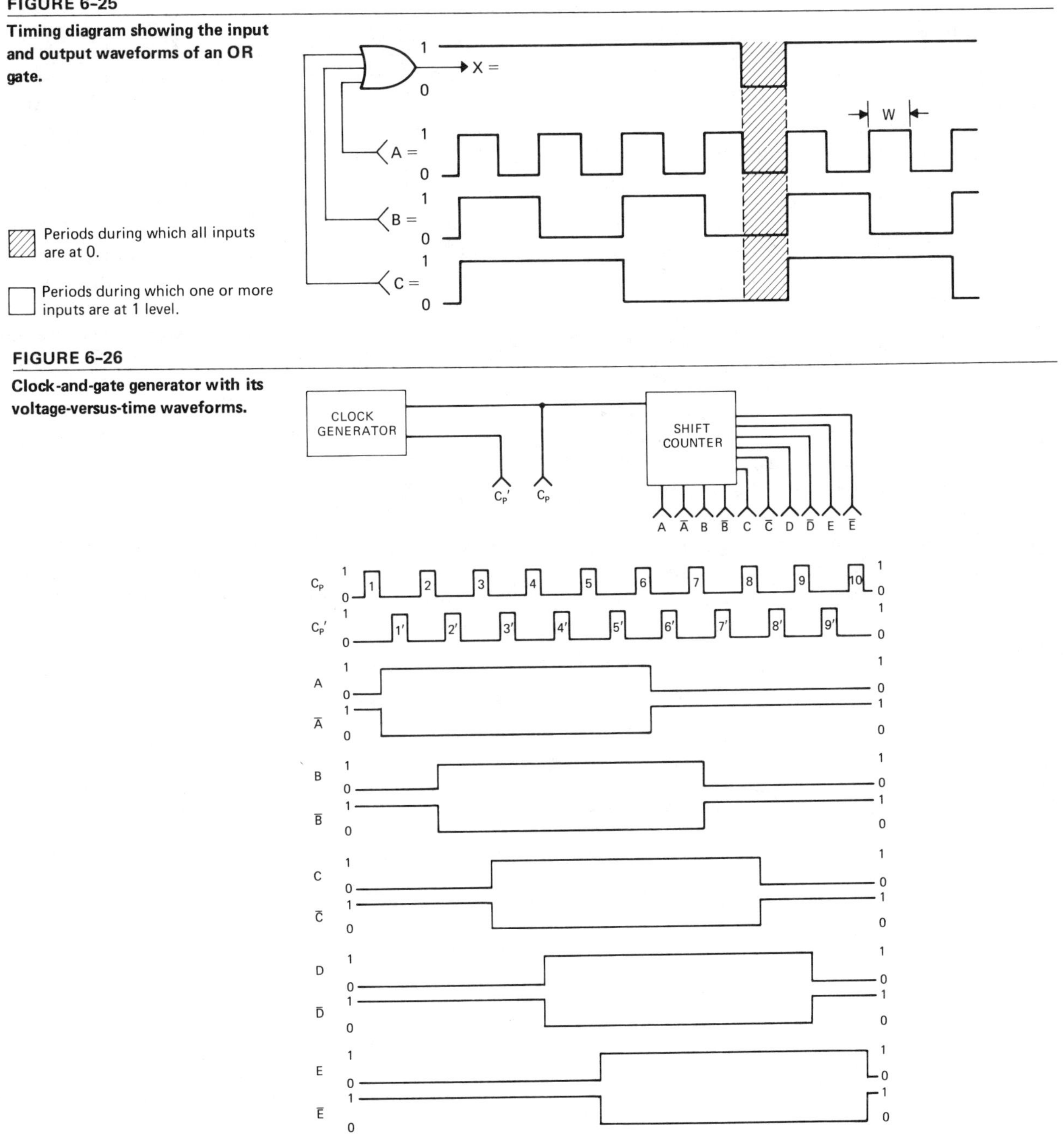

FIGURE 6-25
Timing diagram showing the input and output waveforms of an OR gate.

FIGURE 6-26
Clock-and-gate generator with its voltage-versus-time waveforms.

FIGURE 6-27
Example 6-3.

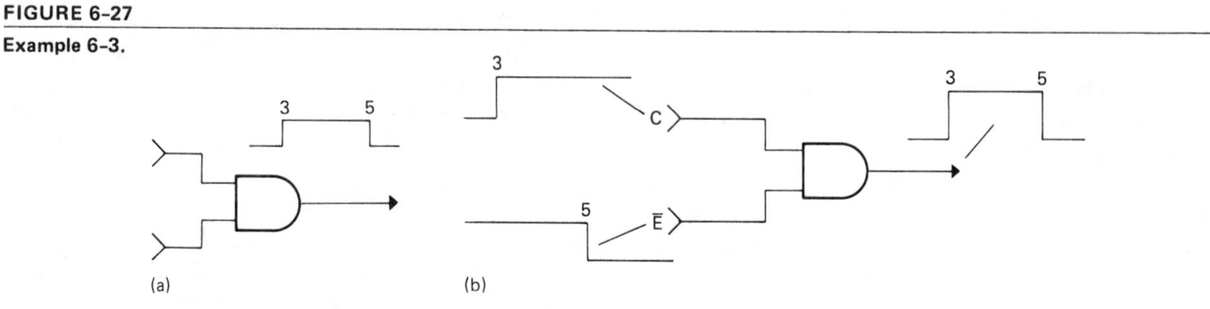

■ EXAMPLE 6-3

Connect the two-input AND gate of Figure 6-27(a) to the shift counter outputs of Figure 6-26 to provide the indicated output. (Label the inputs with the necessary shift counter connection.)

Solution: The AND gate output goes high at time 3, so one of the inputs must do this. A straight line down from C_P 3 on Figure 6-26 indicates that waveform C is the only one going high at time 3. Therefore, one input must be C. The output goes low at time 5. Waveform C is still in the 1 state at time 5, so the other input must be going low at time 5. A straight line down from clock pulse 5 on Figure 6-26 indicates that waveform $\overline{E}$ would have to be used for this. As Figure 6-27(b) shows, the shift counter waveform C and $\overline{E}$ are the only pair for which both are in the 1 state between times 3 and 5. C input turns the gate ON at time 3. $\overline{E}$ input turns gate OFF at time 5.

■ EXAMPLE 6-4

The AND gate in Figure 6-28(a) has shift counter waveforms B and $\overline{C}$ at its inputs. Draw the output waveform and label the waveform edges with clock times.

Solution: If we look first at waveform B, it goes from 0 to 1 at clock time 2. At time 2, $\overline{C}$ is already high, so the output will go high at 2. $\overline{C}$ goes low at time 3, so the output goes low at 3. Between no other times are both inputs high. The solution is the waveform in Figure 6-28(b).

■ EXAMPLE 6-5

Label the three-input AND gate of Figure 6-29(a) with the gate and clock generator inputs needed to provide the single pulse shown at the output.

FIGURE 6-28
Example 6-4.

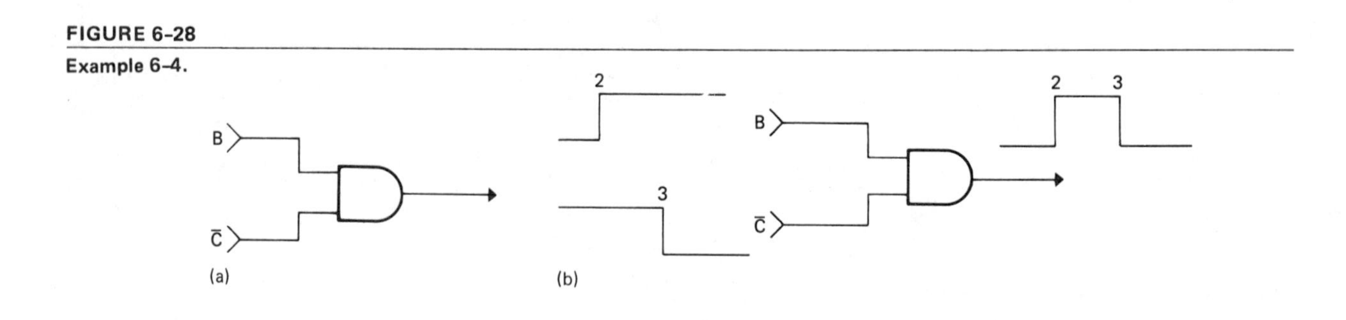

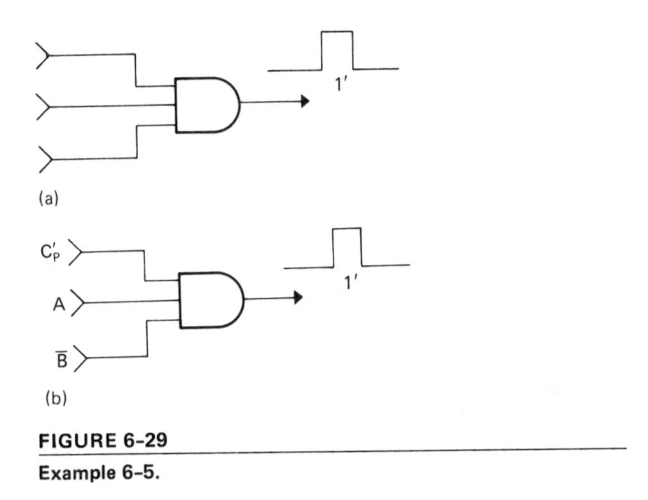

FIGURE 6-29
Example 6-5.

Solution: To get the 1′ pulse, one connection must be the C'_P line. Another must enable the gate at time 1. The third must inhibit the gate at time 2, so that the gate is enabled only long enough for the single 1′ pulse to pass. A straight line down from clock pulse 1 indicates waveform A will enable the gate at time 1, while a straight line down from clock pulse 2 indicates $\overline{B}$ can be used to inhibit the gate at time 2. As shown in Figure 6-29(b), input A enables gate at time 1, and input B inhibits gate at time 2. The 1′ pulse alone is passed through the gate.

6.8 Decimal-to-BCD Conversion Using OR Gates

The OR gate is widely used in encoding and decoding circuits. A typical example is conversion from decimal to BCD. In their decimal form, the numbers may exist as separate lines for each of the digits 1 through 9. A 1 level will exist on whichever line the decimal value is equal to at any particular time. The remaining eight lines will be at 0 (a decimal 0 will be represented by all lines being at 0). Figure 6-30(a) shows the "black box" representation of the converter. Figure 6-30(b) shows the table of this conversion. Note that the 2^0 line goes high (to 1 state) on all the odd numbers. It can, therefore, be expressed as an OR function: $2^0 = 1 + 3 + 5 + 7 + 9$. Examination of the 2^1 column shows a function of $2^1 = 2 + 3 + (6 + 7)$. The 2^2 column indicates a function $2^2 = 4 + 5 + (6 + 7)$, while the 2^3 column is a simple OR function, $2^3 = 8 + 9$. These connections are made as shown in Figure 6-31.

The function for a $1_{BCD} = 1 + 3 + 5 + 7 + 9$ could be accomplished by using a single five-input OR gate, but two- and three-input gates are the most common. OR functions can be expanded by connecting the output of one gate to the input of another. As is done for the 1_{BCD} function in Figure 6-31, two three-input gates are connected to produce a five-input OR function. In doing this, economies are often possible, such as occur with the partial function (6 + 7), which is used on both the 2_{BCD} and 4_{BCD} gates. The output of a gate can drive a number of inputs, so once a partial function is produced, it can be applied to a number of inputs if needed.

Inputs: 0, 1, 2, 3, 4, 5, 6, 7, 8, 9 → DECIMAL-TO-BCD ENCODER → outputs 2^0 (1), 2^1 (2), 2^2 (4), 2^3 (8)

(a)

DEC	BCD			
INPUT	2^0	2^1	2^2	2^3
0	0	0	0	0
1	1	0	0	0
2	0	1	0	0
3	1	1	0	0
4	0	0	1	0
5	1	0	1	0
6	0	1	1	0
7	1	1	1	0
8	0	0	0	1
9	1	0	0	1

(b)

FIGURE 6-30
(a) Block diagram: decimal-to-BCD encoder. (b) Decimal-to-BCD table.

FIGURE 6–31

Decimal-to-BCD encoder using two- and three-input OR gates.

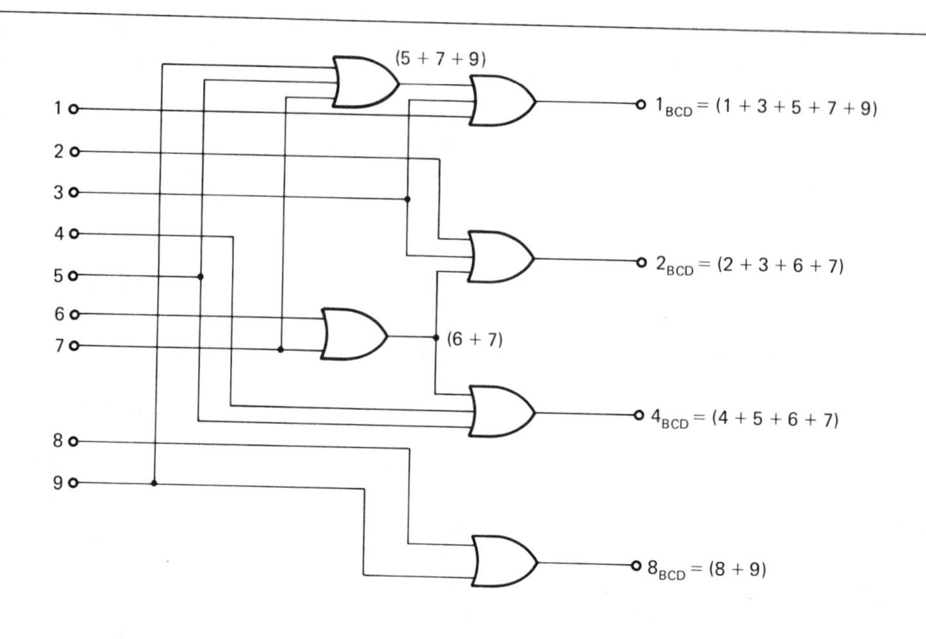

6.9 BCD-to-Decimal Conversion

In Section 6.8, we saw how OR gates are used to encode decimal number lines into binary-coded decimal. It is often necessary to reverse that process or decode from BCD to decimal. Figure 6–32 shows the "black box" representation of the decoder. BCD 1 will be high (in the 1 state) for every odd decimal value, but the decimal 1 should be high (in the 1 state) only when BCD 1 is high, while BCD 2, 4, and 8 are 0. This can be expressed as the AND function $1_{DEC} = (1 \cdot \overline{2} \cdot \overline{4} \cdot \overline{8})_{BCD}$. The NOT values can be obtained by connecting an inverter to each of the BCD lines, as Figure 6–33 shows.

Integrated circuit inverters were previously explained in Section 5.4, which shows the RTL quad inverter (Figure 5–8), the TTL hex inverter

FIGURE 6–32

(a) Block diagram: BCD-to-decimal decoder. (b) BCD-to-decimal table.

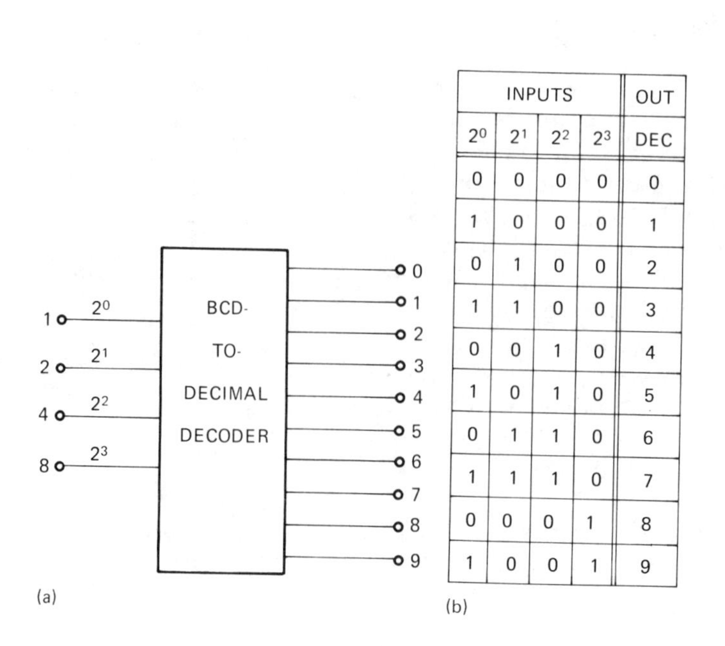

INPUTS				OUT
2^0	2^1	2^2	2^3	DEC
0	0	0	0	0
1	0	0	0	1
0	1	0	0	2
1	1	0	0	3
0	0	1	0	4
1	0	1	0	5
0	1	1	0	6
1	1	1	0	7
0	0	0	1	8
1	0	0	1	9

FIGURE 6-33

BCD-to-decimal decoder.

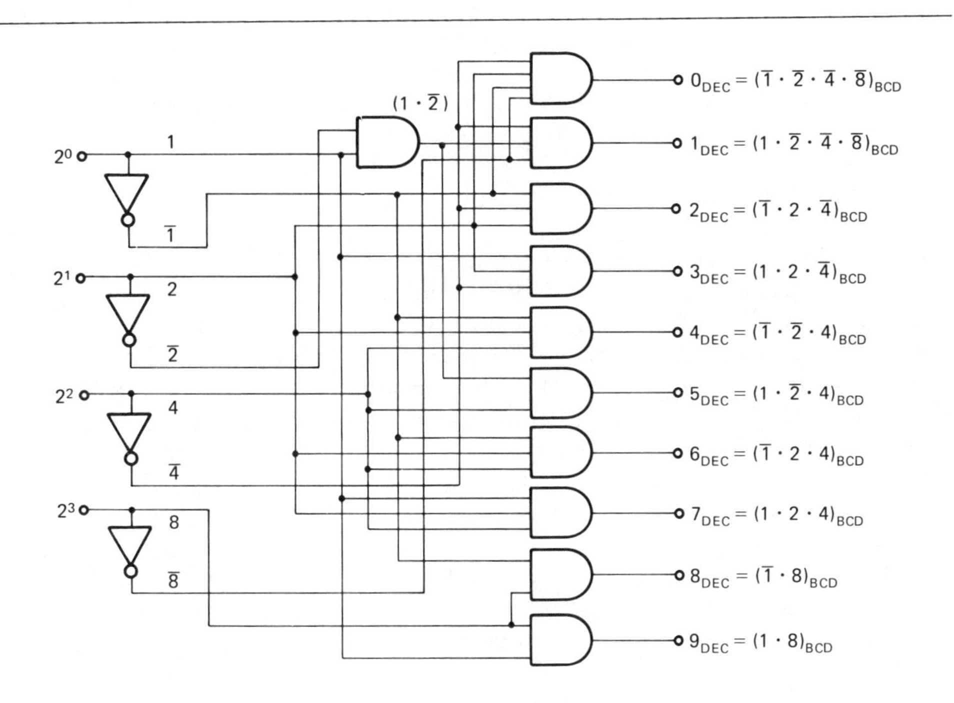

(Figure 5-9), and the basic CMOS inverter (Figure 5-19). An inverter is often referred to as a NOT circuit, and the outputs of the four inverters ($\overline{1}$, $\overline{2}$, $\overline{4}$, $\overline{8}$) may be referred to as "not one, not two, not four, not eight." This means that the output of the AND gate producing $2_{DEC} = \overline{1} \cdot 2 \cdot \overline{4}$ may be called either "not one and two and not four" or "one not and two and four not."

In BCD code, 8 never occurs with 2 or 4. Therefore, the $\overline{8}$ is required only on the 1_{DEC} function and the $\overline{2}$ and $\overline{4}$ are not required for 8 and 9. This would not be true for binary-to-decimal conversion.

6.10 Summary

The two basic noninverting *logic gates* are the *AND gate* and the *OR gate*. Unlike the inverter, described in Chapter 5, the logic gates have two or more inputs. The most widely used are two- and three-input. The output of the logic gate is a logic 1 or 0 level, depending on the combination of 1 and 0 levels applied to the inputs.

The AND gate produces a logic 1 on its output only when all its inputs are at logic 1. If any one or more inputs are at logic 0 level, the output will be 0. Figure 6-2 shows the logic diagram and truth table of a three-input AND gate.

The *truth table* shows every possible combination of ones and zeros that can occur at the inputs, and the level that will result at the output. Note that for eight possible combinations, a single condition produces a 1-level output. The Boolean equation for an AND gate with three inputs, A, B, and C, is $X = A \cdot B \cdot C$ or just $X = ABC$.

AND gates can be constructed by the connection of diodes and a resistor, as in Figure 6-6, or by the connection of transistors, as in Figure 6-7.

The most widely used AND gate in present-day equipment is the integrated circuit AND gate, such as the TTL gate shown in Figure 6-8. This gate is very much like the TTL inverter shown in Figure 5-9 except that the input transistor, Q_1, has three emitters that supply the input leads. If the transistor, Q_1, is drawn as three emitter-base diodes and one base collector diode, as in Figure 6-9, it indicates a circuit very much like the diode AND gate of Figure 6-6. The two output transistors are the totem pole circuit explained in Figure 6-10.

Regardless of the construction of the AND gate circuit, the logic symbol is the same, and its function is still that of producing a 1 output only when all inputs are 1. If any input is 0, the output will be 0.

The OR gate, on the other hand, produces a 1 on the output lead if any one or more of its inputs is 1. All inputs must be 0 before a 0 will occur on the output. Figure 6-11 shows the logic diagram and truth table of a three-input OR gate. Note that for the eight possible input conditions listed in the truth table, a single condition produces a 0 output. The Boolean equation for an OR gate with three inputs, A, B, and C, is $X = A + B + C$.

OR gates can be constructed by interconnection of diodes and a resistor, as in Figure 6-14, or by interconnection of transistors and a resistor, as Figure 6-15 shows. In present-day equipment, OR gates are more likely found in integrated circuit form, such as the TTL gate in Figure 6-16. One 14-pin circuit provides four two-input OR gates. Each input of the TTL OR gate is the emitter of a separate transistor. The output is a totem pole circuit.

A Boolean equation can be formed to describe the output of any combination of AND and OR gates. The correct algebraic manipulation of a Boolean function can often show the existence of simpler circuits that produce the same output function, as in Figure 6-17.

The AND gate can be used to *enable* or *inhibit* the passage of a digital signal. If a digital signal is applied to input A of a two-input AND gate, the signal will be passed through the gate if input B is at logic 1 level. If input B is at 0, the signal will be inhibited and there will be a fixed 0 level at the output.

Figure 6-22 shows a three-input AND gate used to control the passage of a signal. Input B enables or turns the signal on. Input C inhibits or turns the signal off. The OR gate can be used to enable or inhibit signals. A 0 level will enable a signal through an OR gate; a 1 level will inhibit the signal, producing a fixed 1 output. Having a 1 level during inhibit is not always acceptable. Therefore, OR gates are less often used for this function than are AND gates.

In many applications of logic gates, inputs change too frequently to be analyzed statistically, and a *timing diagram* must be used to track the resulting output. The timing diagram shows the changes on both inputs and outputs with respect to time.

Figure 6-24 shows a set of input waveforms and the resulting outputs for a three-input AND gate. The waveforms are drawn as they would appear on an oscilloscope with zero time on the left. In Figure 6-24, the output is 1 only when all three inputs are 1. Figure 6-25 shows a similar set of waveforms applied to an OR gate. In this case, the output is 0 only when all three inputs are 0.

AND gates and OR gates have many functions. Figure 6-26 shows a clock and shift counter with the variety of output waveforms produced by this circuit. These outputs can be applied as inputs to AND gates to produce a variety of output waveforms that begin and end on the edges of clock pulses, as demonstrated in Examples 6-3 and 6-4. With three input gates, individual clock or delayed clock pulses can be isolated from the clock pulse line by using these same shift counter waveforms.

A typical use for AND and OR gates is found in encoding and decoding circuits. Figures 6-30 and 6-31 demonstrate the use of OR gates to convert from decimal number lines to binary-coded decimal (BCD) lines. Figure 6-32 and 6-33 demonstrate use of AND gates to convert BCD lines to decimal.

Glossary

Logic Circuit: Circuit that produces the electrical equivalent of a logical function, such as a Boolean algebra operation. The correct combination of logic circuits can be assembled to give the electrical analogy of any Boolean algebra equation. This includes the capacity for mathematics in binary form.

Logic Gate: A logic circuit with two or more inputs and a single output. There are two voltage levels occurring at inputs and output. These are defined with limits and used to represent binary 1 and 0. There are various logic gate circuits in use. They differ by the Boolean operation or function they produce. In general, the output of a logic gate will be 1 or 0, depending on the combination of 1 and 0 levels applied to its inputs. The term *gate* is derived from the fact that a digital signal can be applied to one input of a logic gate, and the remaining input or inputs can be used to enable or stop the passage of the signal, like opening and closing a gate.

AND Gate: A logic gate that produces a 1 at its output only when all its inputs are 1. If any one or more inputs is 0, the output will be 0. Figure 6-2 shows the logic symbol, truth table, and Boolean function for a three-input AND gate.

OR Gate: A logic gate that produces a 1 at its output if any one or more of its inputs is 1. A 0 occurs at its output only when all its inputs are 0. Figure 6-11 shows the logic symbol, truth table, and Boolean function for a three-input OR gate.

Truth Table: A table showing every possible combination of 1 and 0 levels that can occur on the inputs to a logic circuit and the output or outputs that result from each. A logic circuit with N inputs can have no more than 2^N possible combinations of 1- and 0-level input combinations. In special instances, some of the 2^N possible combinations may be known not to occur and are, therefore, not included in the table.

Enable Function: A control waveform applied to one input of a logic gate to allow the passage of a signal or part of a signal occurring on a different input of the same gate. See Figure 6-23.

Inhibit Function: A control waveform applied to one input of a logic gate to prevent the passage of a signal or part of a signal occurring on a different input of the same gate. See Figure 6-23.

Timing Diagram: A voltage-versus-time plot of waveforms appearing at the inputs and outputs of logic circuits. Timing is usually based on the clock pulse line. The time reference usually starts with the leading edge of the 0 pulse, and changes in level on the inputs being compared are referenced to clock pulse time. These waveforms can be registered on an oscilloscope by externally synchronizing the "scope" with the 0 pulse and connecting the logic circuit point to the oscilloscope vertical input. Timing diagrams, however, are often drawn on paper and used to predict, from a theoretical knowledge of the circuits, the output voltage waveform that will result from a given set of input voltage waveforms.

Shift Counter: A logic assembly that produces a variety of waveforms that change level on the trailing edge of the clock pulses operating it. Each output of the shift counter has its level changes occurring on different clock pulses. This makes them ideal for use in conjunction with logic gates to produce control and timing waveforms. See Figure 6-26.

Questions

1. Draw the logic symbol and truth table of a three-input AND gate.
2. In Figure 6-5, the inputs are the second line of the truth table. Which diode is forward-biased?
3. Draw the logic symbol and truth table for Figure 6-7.
4. Draw the logic symbol and truth table of a three-input OR gate.
5. The schematic diagrams of Figures 6-5 and 6-14 show the same components. How do they differ?
6. What single condition produces a 1 level out of an AND gate? What single condition exists for the OR gate?
7. A 0 applied to any input of an AND gate results in what output?
8. A 1 applied to any input of an OR gate results in what output?
9. Select the correct word or words: If a zero is applied to the control input of a two-input AND gate, the signal on the other input (will will not) get through. The gate is (enabled inhibited).
10. Select the correct word or words: If a 1 is applied to the control input of a two-input OR gate, the signal on the other input (will will not) get through the gate. The gate is (enabled inhibited).
11. When two or more inputs to a gate are changing with considerable frequency, what method may be used to predict its output?
12. In Figure 6-24, if the signal on input A were shifted (in time) one-half W, how would this affect the output? Draw the timing diagram for this. (*Note:* Input A keeps the same waveform, but the leading and trailing edges are shifted to the right W/2 with respect to inputs B and C.)
13. In Figure 6-25, if input B were shifted (in

time) one-half W, how would this affect the output? Draw the timing diagram.

14. If the waveforms in Figure 6-26 repeat themselves every 100 microseconds, how often will the output of the AND gate in Figure 6-27(a) repeat itself?
15. In Figure 6-30, can more than one input line be high at any given time? Why? Can more than one output line be high at one time?
16. If only two input gates were available, what would be the minimum number required for the decimal-to-BCD encoder of Figure 6-31?
17. If only two input gates were available, what would be the minimum number required for the BCD-to-decimal decoder of Figure 6-33? Draw the circuit.

Problems

6-1 A six-bit parallel binary number is to drive a pair of octal readouts. Figure 6-34 is the block diagram. Draw the logic diagram of the converter using AND gates and inverters.

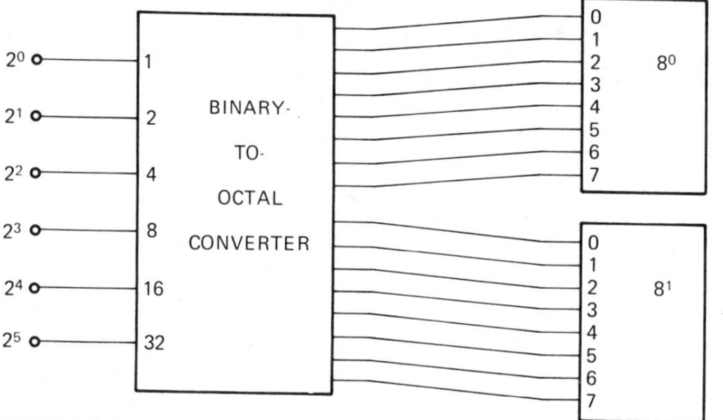

FIGURE 6-34. Problem 6-1

Note: For Problems 6-2 through 6-9, the clock generator of Figure 6-26 provides the clock pulses shown at the top line, C_P. It also provides a delayed clock, C'_P. The shift counter provides the outputs A through E and their complements, $\overline{A}$ through $\overline{E}$.

6-2 Connect the two-input AND gates in Figure 6-35 to the clock-and-gate generator outputs of Figure 6-26 to provide the indicated output signals. (Label the input with the necessary shift counter connection.)

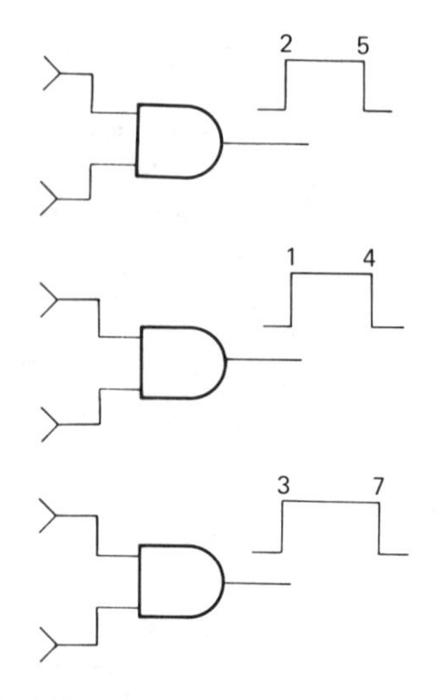

FIGURE 6-35. Problem 6-2

6-3 For the gates in Figure 6-36, draw the voltage waveforms that will occur at the output between clock times 1 and 10.

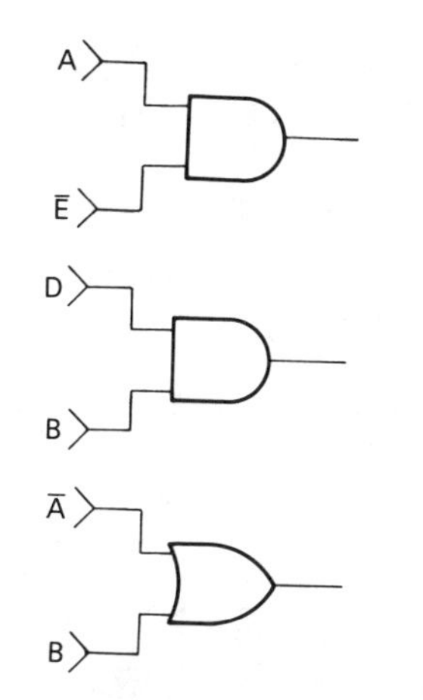

FIGURE 6-36. Problem 6-3

6-4 Connect the three-input AND gates in Figure 6-37 to the clock-and-gate generator of Figure 6-26 to provide the single pulses shown at the outputs. (Label the inputs with the necessary shift counter or clock connection.)

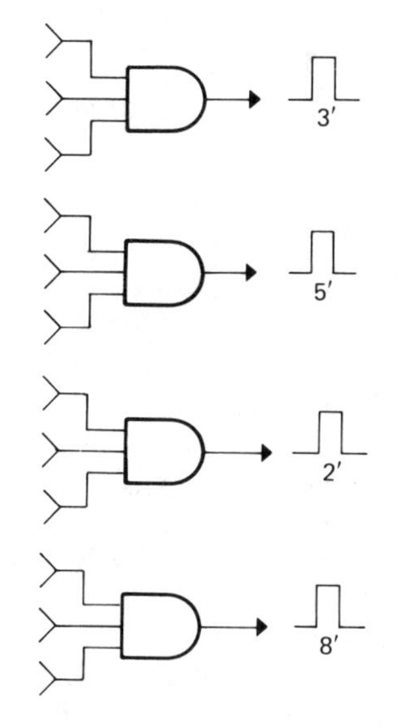

FIGURE 6-37. Problem 6-4

6-5 For the gates in Figure 6-38, draw the voltage waveforms that will occur at the output between clock time 1 and 10.

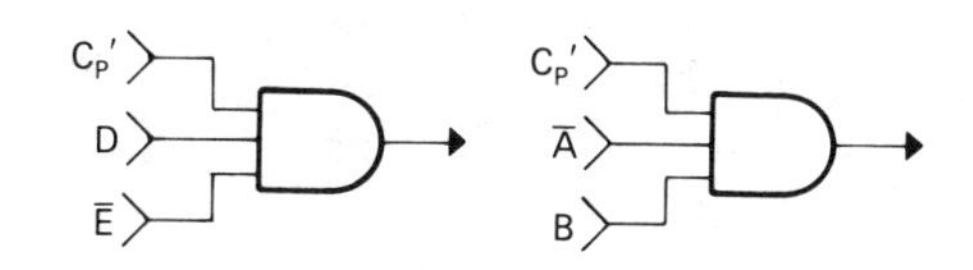

FIGURE 6-38. Problem 6-5

6-6 Connect the gate inputs of Figure 6-39 to the clock-and-gate generator outputs of Figure 6-26 to provide the indicated outputs from the OR gates.

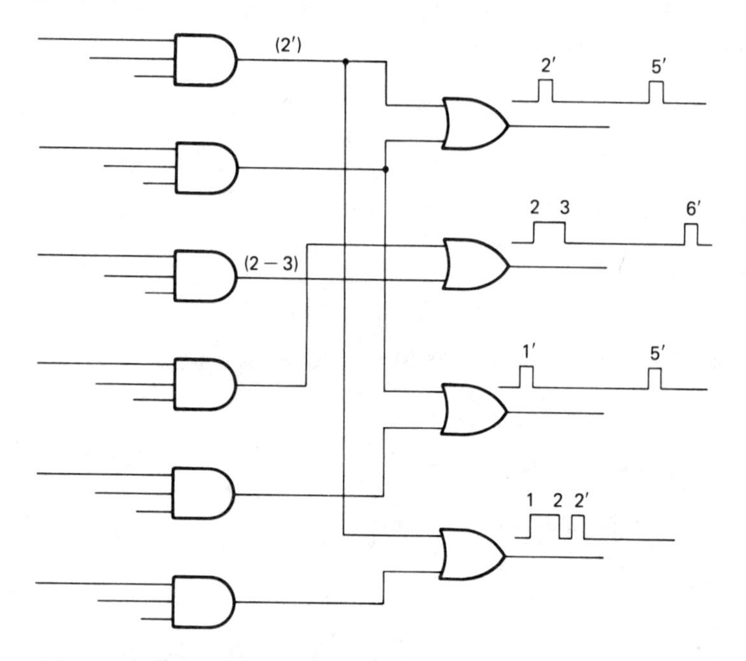

FIGURE 6-39. Problems 6-6 and 6-20

6-7 Label the two AND gates in Figure 6-40 with the shift counter waveforms needed to produce the given output.

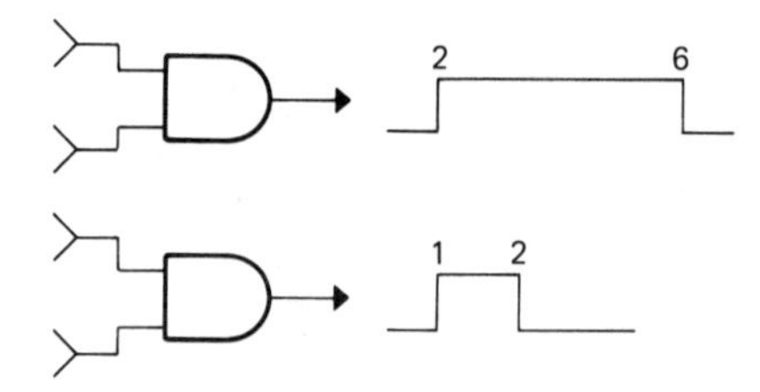

FIGURE 6-40. Problem 6-7

6-8 The gates in Figure 6-41 have shift counter inputs as labeled. Draw the output waveforms. (Label waveform edges with clock time.)

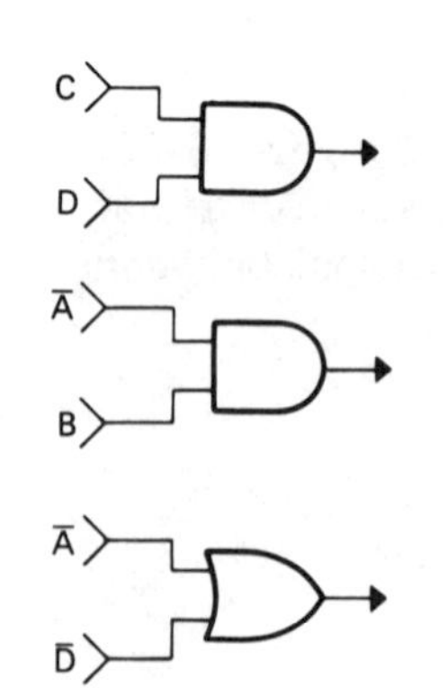

FIGURE 6-41. Problem 6-8

6-9 Label the three inputs of the AND gates of Figure 6-42 with the waveforms needed to produce the given outputs.

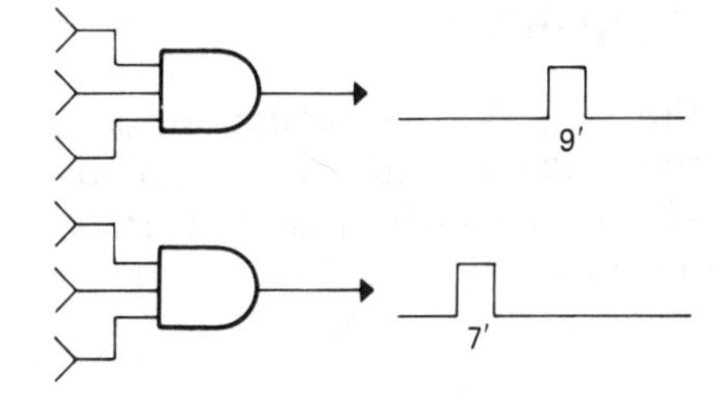

FIGURE 6-42. Problem 6-9

Boolean Algebra (Problems 6–10 through 6–15)

6-10 Give the Boolean equation for the circuit of Figure 6-43. Draw the truth table.

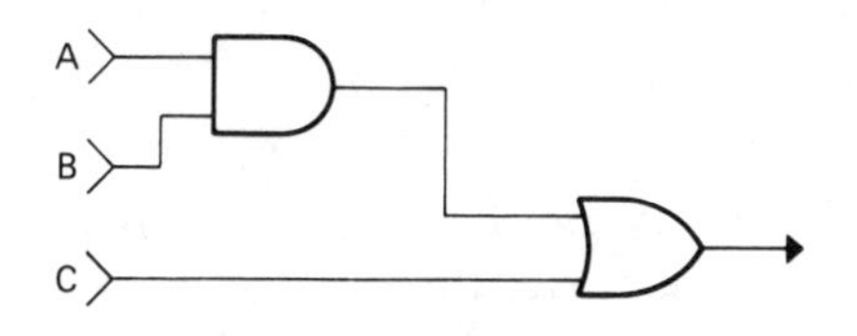

FIGURE 6-43. Problem 6-10

Note: Boolean equations can be multiplied and factored similar to algebraic equations, with several important exceptions, such as:

$$A + A = A \text{ (not } 2A)$$
$$A \cdot A = A \text{ (not } A^2)$$
$$A + 1 = 1$$

But, similar to algebra:

$$A \cdot B = AB$$
$$A(B + C) = AB + AC$$
$$(A + B) \cdot (C + D) = AC + AD + BC + BD$$

Useless Circuits

6-11 Prove by Boolean equations that the logic circuits of Figure 6-44(a) and (b) both equal B.

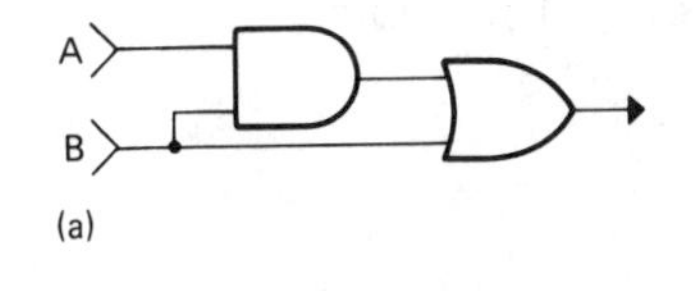

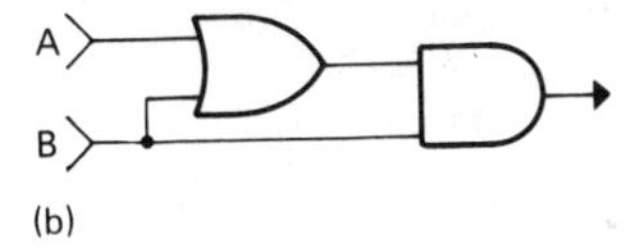

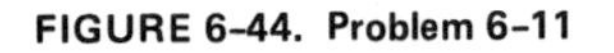

FIGURE 6-44. Problem 6-11

Simplify

6-12 Prove by Boolean equations that the logic circuits of Figure 6-45(a) and (b) have identical outputs.

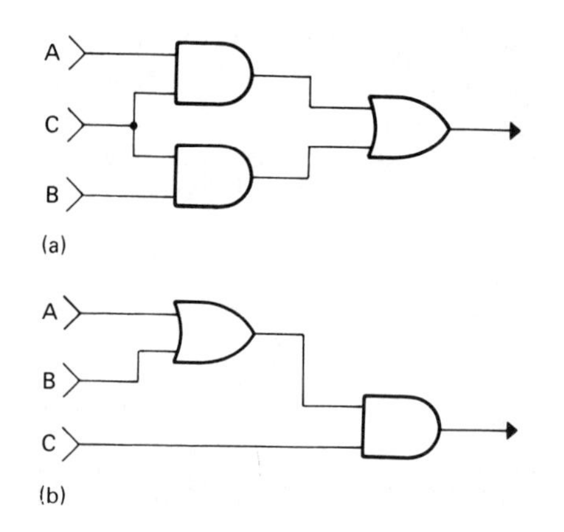

FIGURE 6-45. Problem 6-12

6-13 Prove by Boolean equation that the logic circuits of Figure 6-46(a) and (b) have identical outputs.

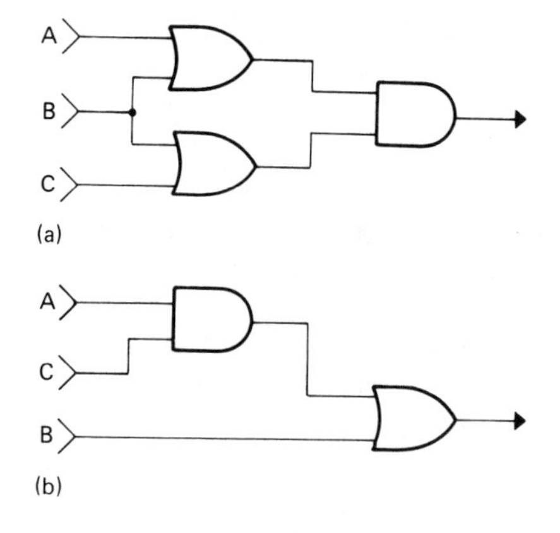

FIGURE 6-46. Problem 6-13

6-14 Using three AND gates and one OR gate, assemble the logic circuit for the function X = AB + BC + CD. Factor the equation and assemble the logic, using all two-input gates.

6-15 Draw the logic circuit needed for the function X = ABC + CDE + CD + D. Simplify if possible.

6–16 The circuits in Figure 6–47 are the Fairchild TTL/SSI 7411 and 7432 integrated circuits in Figures 6–8 and 6–16. Connect the internal leads of the logic circuits to the correct pins.

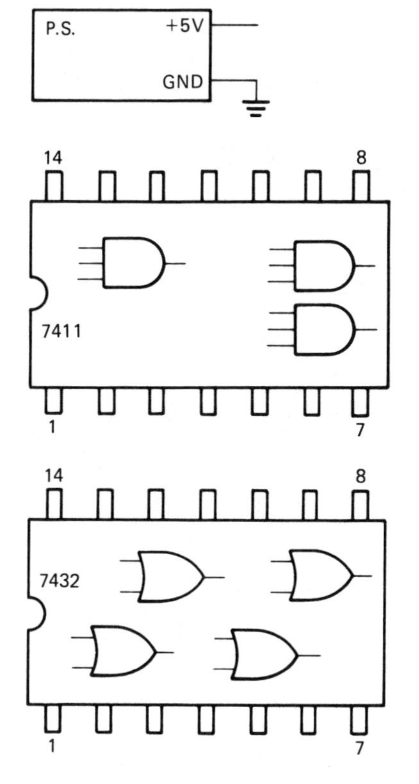

FIGURE 6–47. Problems 6–16 through 6–18

6–17 Draw a wiring diagram that connects the power supply to the circuits of Figure 6–47 and also provides the logic of Figure 6–17(b). Label the pins for A, B, C, D, and X. (Unused inputs should be connected to a used input of the same circuit.)

6–18 From the remaining gates of Figure 6–47, connect a circuit of Figure 6–18(b).

6-19 The circuits in Figure 6-48 are the Fairchild TTL/SSI 7411 and 7432 integrated circuits in Figures 6-8 and 6-16. Connect the internal leads of the logic circuits to the correct pins.

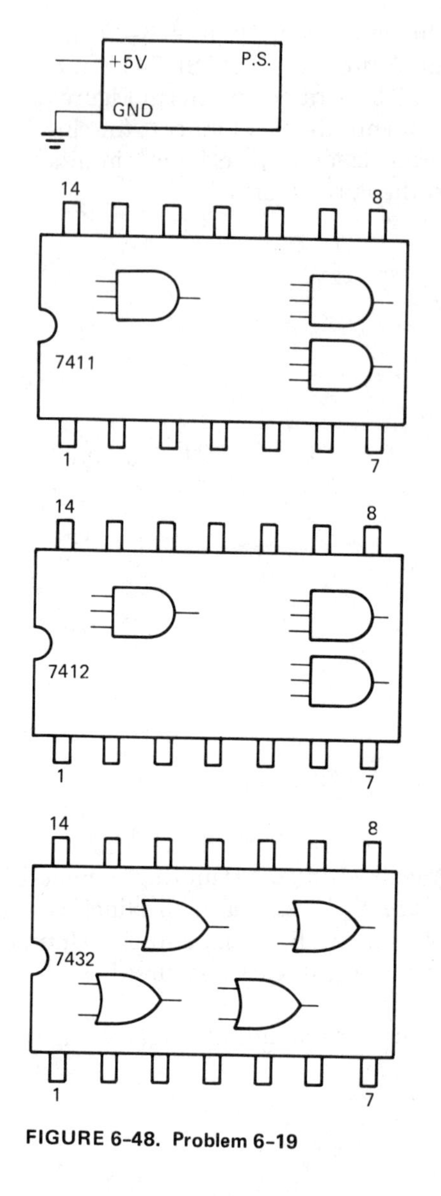

FIGURE 6-48. Problem 6-19

6-20 Draw a wiring diagram to connect the power supply to the circuits and also provide the logic of Figure 6-39. (Unused inputs should be connected to used inputs of the same circuit.)

CHAPTER

7

Inverting Logic Gates (NOR, NAND)

Objectives

Upon completion of this chapter, you will be able to:

- Identify and use the logic symbols of NAND gates and NOR gates.
- Draw the truth tables of NAND and NOR gates.
- Write and simplify Boolean equations of circuits including NAND and NOR gates.
- Assemble transistors and resistors to form NAND and NOR gates.
- Identify correct pin connections for TTL integrated circuit NAND and NOR gates.
- Identify correct pin connections for CMOS NAND and NOR gates.
- Use NAND and NOR gates to enable or inhibit passage of a digital signal.
- Assemble inverting logic gates of one type to form AND and OR functions.
- Analyze the functioning of NAND and NOR gates by timing diagram.
- Use NAND and NOR gates in conjunction with clock and shift counter to generate special waveforms.
- Use NAND and NOR gates in conjunction with clock and shift counter to isolate individual clock pulses from the clock pulse line.
- Use De Morgan's theorem to find alternative methods of producing a digital logic function.

7.1 Introduction

In Chapter 6, we discussed the AND and OR gates. When those gates are enabled to pass a digital signal, the output is essentially the same as the input. When the inverting gates, NOR and NAND, are enabled to pass a signal, the output is the inversion or complement of the input signal. The inverting gates are more complicated to use, but they are essential in producing many functions. In addition, inverting-type gates are superior electrically to noninverting gates. For this reason, their use is preferred in most logic systems.

7.2 The NOR Gate

7.2.1 NOR Gate Symbol and Truth Table

Figure 7-1 shows one of several symbols for the NOR gate. As the symbol indicates, it functions like an OR gate, for which the output is inverted. The Boolean notation is that of an OR function with an inversion over the entire function. The truth table of the NOR gate has outputs that are the complement of those shown in the OR gate truth table of Figure 6-11(b).

7.2.2 NOR Gate Discrete Circuits

A NOR gate circuit can be provided by connecting the output of either a diode or transistor OR gate to the input of an inverter circuit. Figure 7-2 shows the schematic of a diode transistor (DTL) NOR gate.

The same function can be produced by direct connection of transistors in a common emitter circuit, as in Figure 7-3. If a 1 level is applied to an input (base lead), the base-to-emitter junction will be forward-biased, causing a base current, I_B, to flow. The resulting collector current, I_C, will draw enough current through R_C to drop all the V_{CC} voltage, leaving a 0-level voltage at X. The only condition that will result in a 1 level at X is a 0 level on all three inputs. This exclusive condition is shown in the truth table of Figure 7-3.

7.2.3 TTL Integrated Circuit NOR Gate

At present, the most widely used NOR gate is the TTL, like the Sprague US7402 quad dual-input NOR gate shown in Figure 7-4. It is very much like the AND and OR TTL gates previously discussed. It has the same "totem pole" dual-transistor output. The inputs are the emitters of

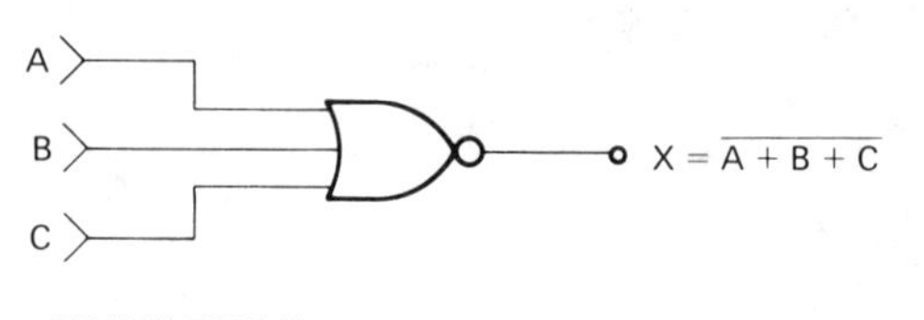

TRUTH TABLE

A	B	C	X
0	0	0	1
1	0	0	0
0	1	0	0
0	0	1	0
1	1	0	0
1	0	1	0
0	1	1	0
1	1	1	0

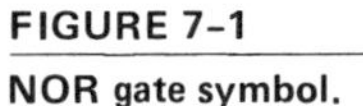

FIGURE 7-1
NOR gate symbol.

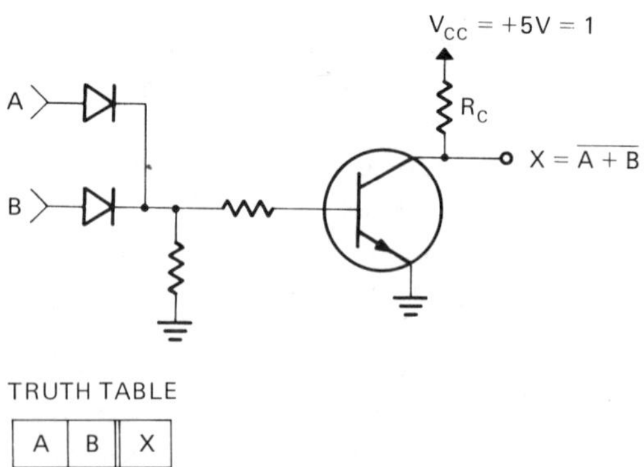

TRUTH TABLE

A	B	X
0	0	1
1	0	0
0	1	0
1	1	0

FIGURE 7-2
Two-input NOR gate composed of a diode OR gate connected to a transistor inverter.

separate transistors, which serve to standardize the inputs with those of the other TTL gates. The actual NOR function is formed by the two middle transistors, which are connected in parallel like the transistors of the NOR gate in Figure 7-3.

7.2.4 CMOS Integrated Circuit NOR Gate

NOR gates are produced in CMOS integrated circuits by using complementary enhancement-mode MOSFETs, which operate as explained in Section 5.5.3. The enhancement-mode FET is normally OFF, which means that it conducts no current with 0V V_{GS}. It is therefore turned off by an OV or negative voltage if N-channel, or an OV or positive voltage if P-channel. The logic system uses $-V_{SS}$ as a logic 0. A logic 0 turns off an N-type MOSFET. The 1 level is $+V_{DD}$. A logic 1 turns off a P-type MOSFET.

To produce a NOR function, a 1 input must result in a 0 at the output. This can be accomplished with parallel N-channel MOSFETs, as in

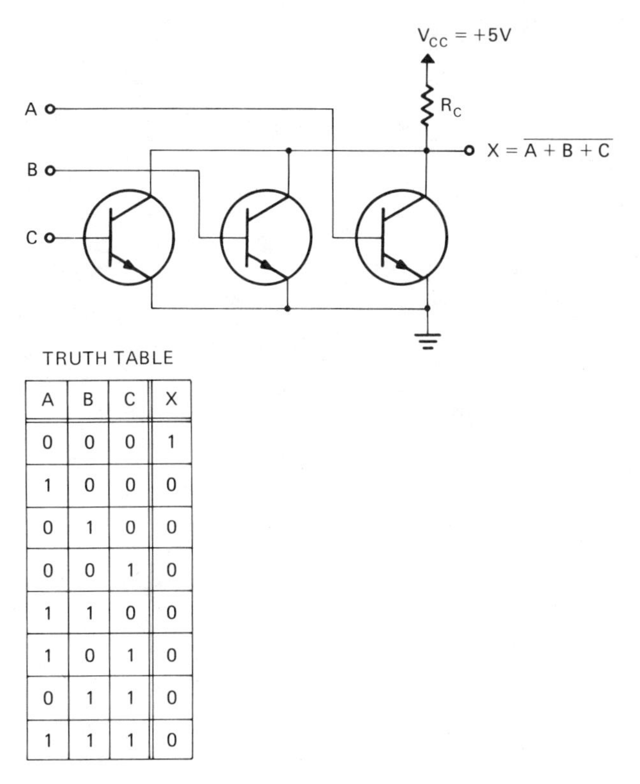

A	B	C	X
0	0	0	1
1	0	0	0
0	1	0	0
0	0	1	0
1	1	0	0
1	0	1	0
0	1	1	0
1	1	1	0

FIGURE 7-3
Three-input transistor NOR gate.

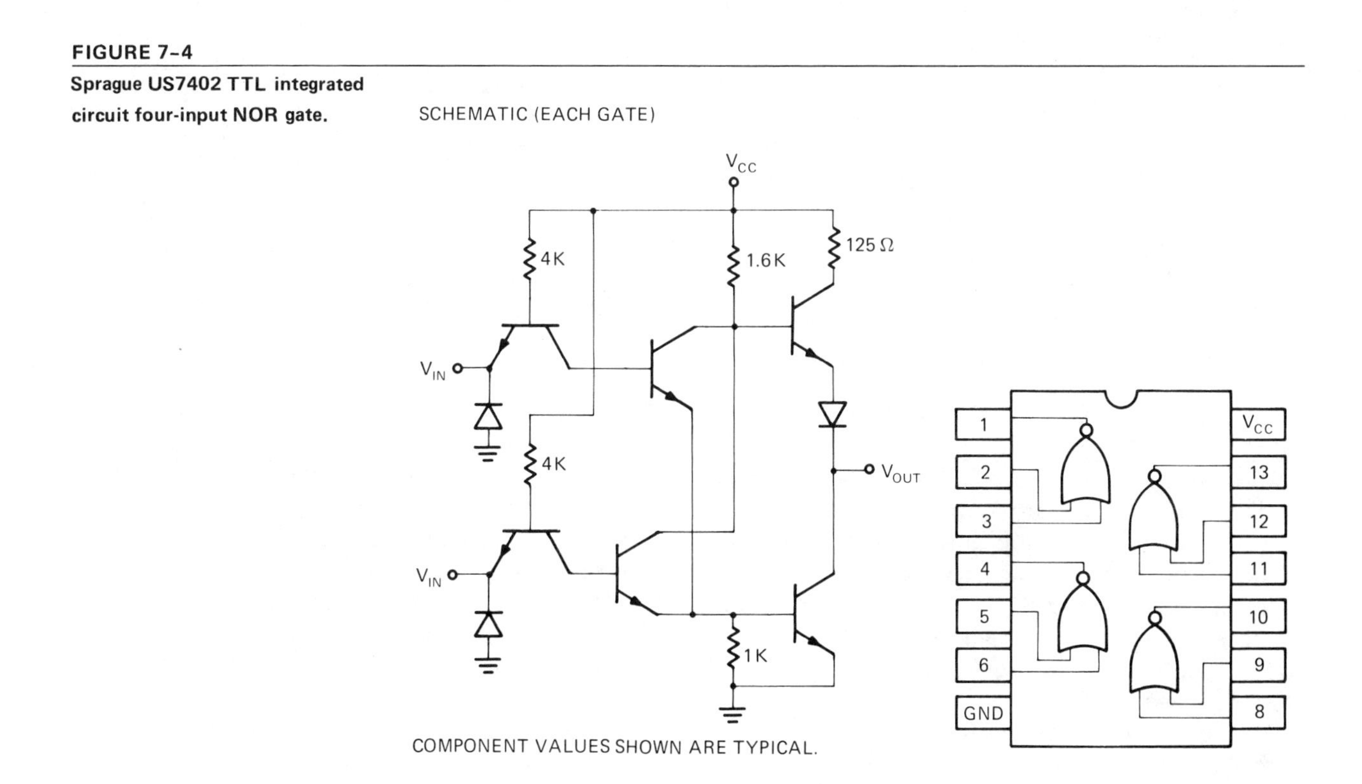

FIGURE 7-4
Sprague US7402 TTL integrated circuit four-input NOR gate.

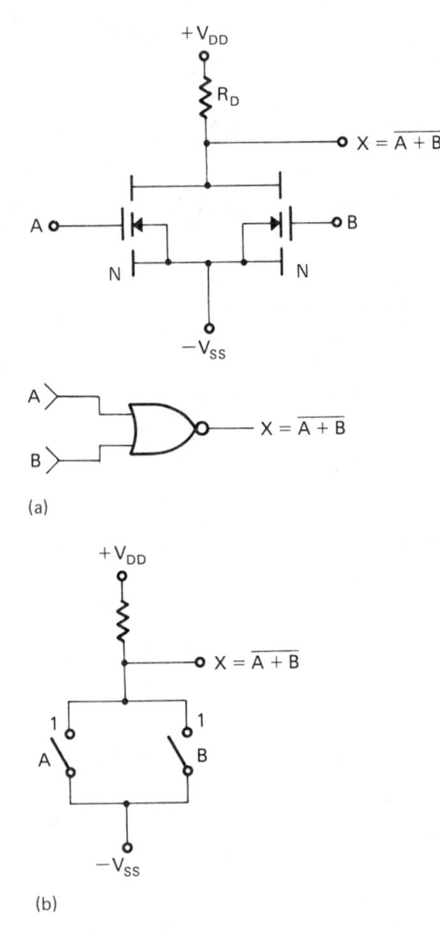

FIGURE 7-5

(a) N-channel enhancement-mode MOSFET NOR gate. (b) Equivalent manual switch circuit. A 1 level (+V_{DD}) on one or more inputs results in a 0-level output.

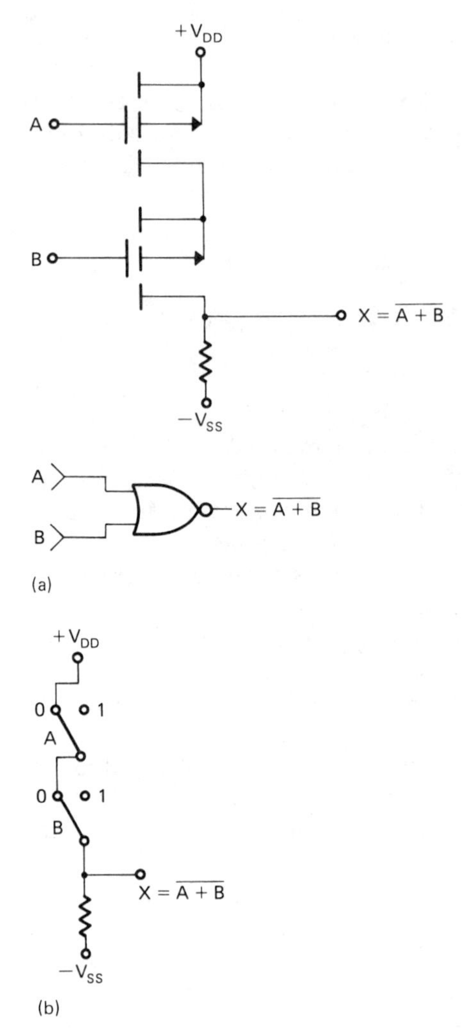

FIGURE 7-6

(a) P-channel enhancement-mode MOSFET NOR gate. (b) Equivalent manual switch circuit. A high level (+V_{DD}) on any input turns the P-channel FET off, thereby removing the 1 level from the output.

Figure 7-5. The two MOSFETs in parallel work like a pair of switches in parallel. Turning ON one or both results in application of the -V_{SS} to the output. Only with both siwtches in the 0 state will a 1 level appear at the output. Although this circuit functions correctly as a NOR gate, it does not give the ideal low resistance during a 1 out, or very high resistance during a 0 out, such as we obtained from the "totem pole" circuits used in TTL.

To use another approach: A 1 level, +V_{DD}, should appear at the output only when both inputs are 0. Figure 7-6 shows two P-channel MOSFETs in series. These function like two switches in series. Both must be turned ON before +V_{DD} will occur at the output. The P-channel MOSFETs turn ON with -V_{SS} on the gates. Therefore, a 1 will occur at the output only when both inputs are binary 0.

The NOR gates of Figures 7-5 and 7-6 are valid NOR gates, but both lack the ideal conditions provided by the totem pole circuit. An ideal circuit results from removing the resistors and combining the two to form a CMOS NOR gate, as Figure 7-7 shows. CMOS gates consume very little power and have a much higher fan-out

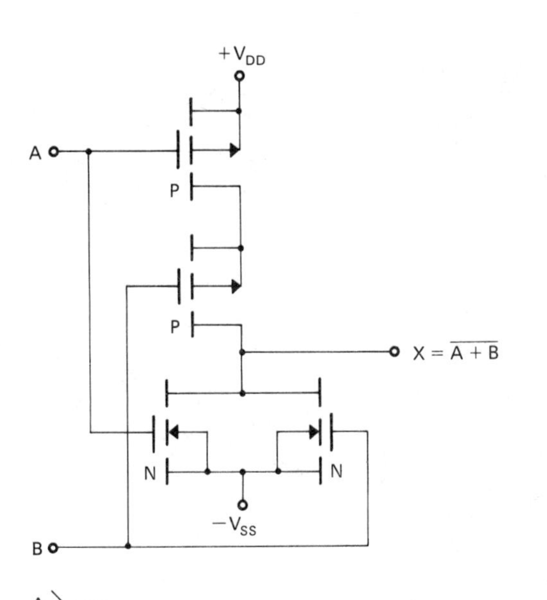

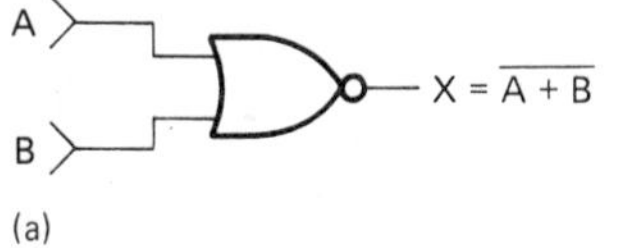

(a)

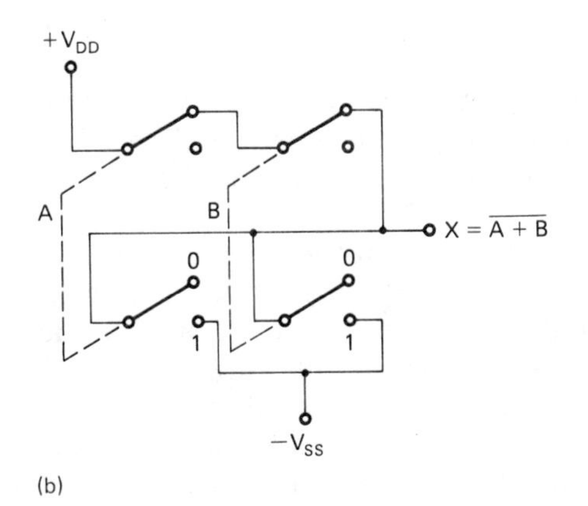

(b)

FIGURE 7-7

(a) CMOS NOR gate. (b) Equivalent manual switch circuit. Combining the N-channel and P-channel circuits of Figures 7-5 and 7-6 results in a NOR circuit that needs no resistors and has low power consumption with high fan-out capabilities.

capability, which means that one output can drive a large number of inputs to subsequent gates.

Figure 7-8 shows three integrated circuits by Solid State Scientific, Inc. The circuits provide two-, three-, and four-input CMOS NOR gates. The CMOS gates can be operated with the same voltage levels as TTL gates by using ground or 0V for $-V_{SS}$ and $+5V$ for $+V_{DD}$.

7.3 Boolean Functions with NOR Gates

Circuits containing AND, OR, and NOR gates can be expressed in Boolean functions.

■ **EXAMPLE 7-1**

Draw the circuit described by the Boolean function $X = \overline{A + B} \cdot \overline{C + D}$.

Solution: See Figure 7-9.

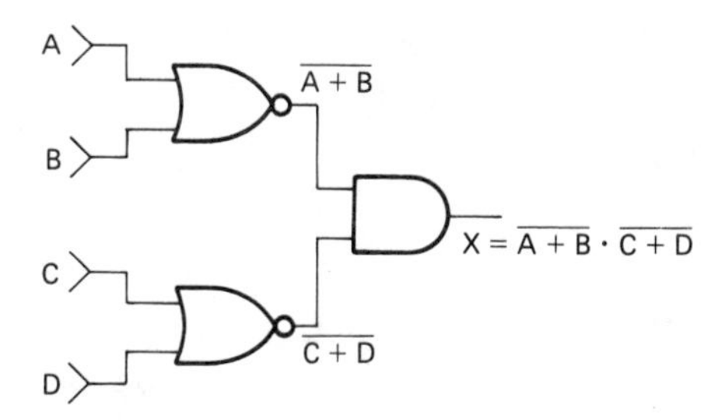

FIGURE 7-9

Solution for Example 7-1.

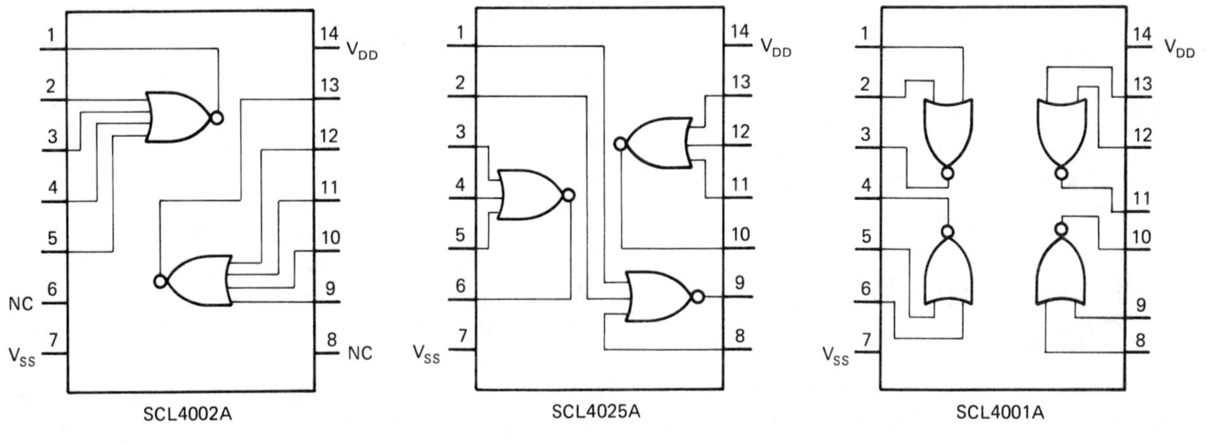

FIGURE 7-8

Solid State Scientific, Inc., integrated circuits with two-, three-, and four-input CMOS NOR gates.

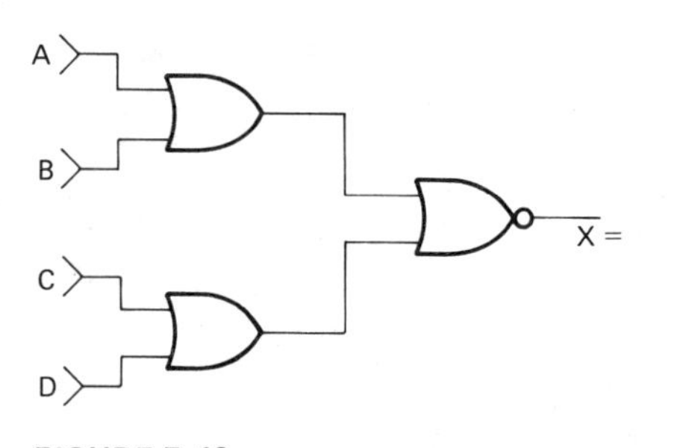

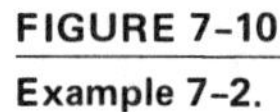

FIGURE 7-10

Example 7-2.

■ **EXAMPLE 7-2**

Write the Boolean function of the circuit shown in Figure 7-10. What single gate is it equal to?

Solution: The functions are A + B and C + D. Therefore,

$$X = \overline{A + B + C + D}$$

This function is equivalent to a single four-input NOR gate.

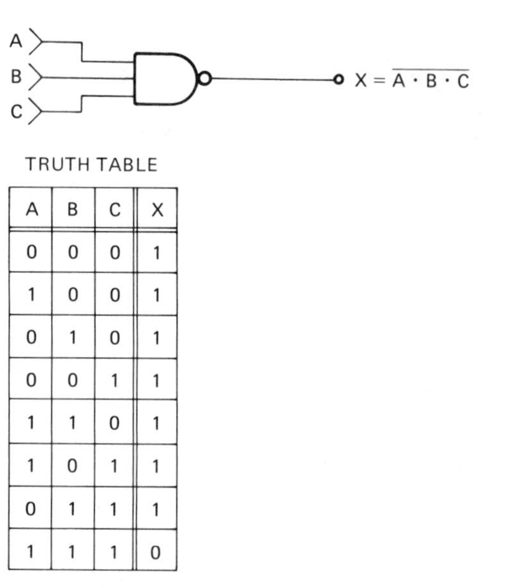

TRUTH TABLE

A	B	C	X
0	0	0	1
1	0	0	1
0	1	0	1
0	0	1	1
1	1	0	1
1	0	1	1
0	1	1	1
1	1	1	0

FIGURE 7-11

Three-input NAND gate logic symbol.

7.4 The NAND Gate

7.4.1 NAND Gate Symbol and Truth Table

Figure 7-11 represents a three-input NAND gate. The symbol is that of an AND gate with the output inverted. The Boolean notation for a NAND function is that of an AND function with an inversion line drawn over the entire function, as follows:

$$X = \overline{A \cdot B \cdot C}$$

Comparison of its truth table (in Figure 7-11) with the truth table of the AND gate in Figure 6-2 shows the output of the NAND to be the inversion or complement of the AND. The NAND gate produces a 0 on its output only when there are all ones on the inputs. Any 0 input will result in a 1 on the output.

7.4.2 NAND Gate Discrete Circuits

As the symbol indicates, the NAND circuit can be produced by connecting the output of an AND circuit to the input of an inverter (common emitter circuit), as shown in Figure 7-12. Although it appears that two separate symbols are involved, an AND gate connected to an inverter, the logic symbol is contracted by deleting the triangle.

Figure 7-13 shows another transistor NAND circuit. In this circuit, the transistors are in series, and all three inputs must have 1s on them in order to produce a saturation current through R_C. All 1 inputs, therefore, result in a 0 output at X. Any transistor that has a 0-level input acts like an open switch to produce the 1 output at X.

7.4.3 TTL Integrated Circuit NAND Gate

The TTL NAND gate is very much like the TTL gates already described except that it is the most economical of the TTL gates with respect to the number of transistor elements needed to form the gate. Figure 7-14 is a Sprague US7410 three-input NAND gate. One 14-pin integrated circuit contains three such NAND gates. The input appears like a common base with three emitters, but a drawing in diode form indicates a diode AND function like that shown in Figure 6-9. The output is the same highly effective totem pole circuit explained in Figure 6-10. Note that one control transistor is used between the input and the totem pole output.

FIGURE 7-12

Diode transistor NAND gate composed of a diode AND gate followed by a transistor inverter.

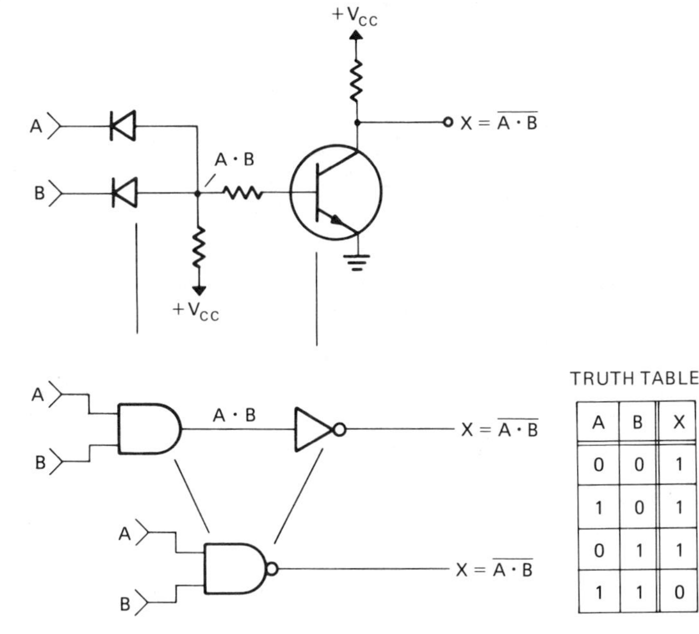

TRUTH TABLE

A	B	X
0	0	1
1	0	1
0	1	1
1	1	0

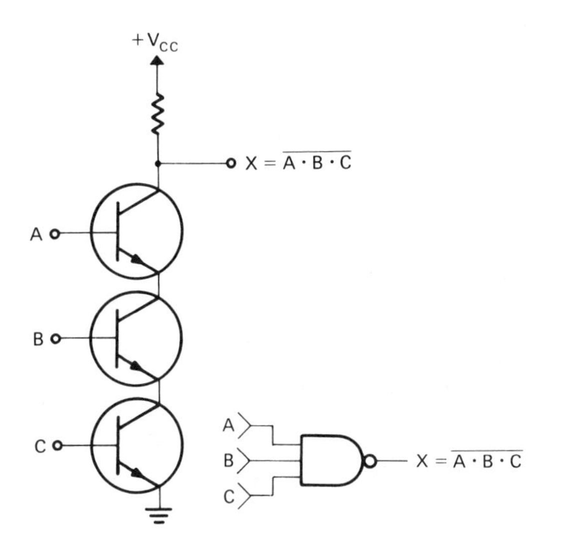

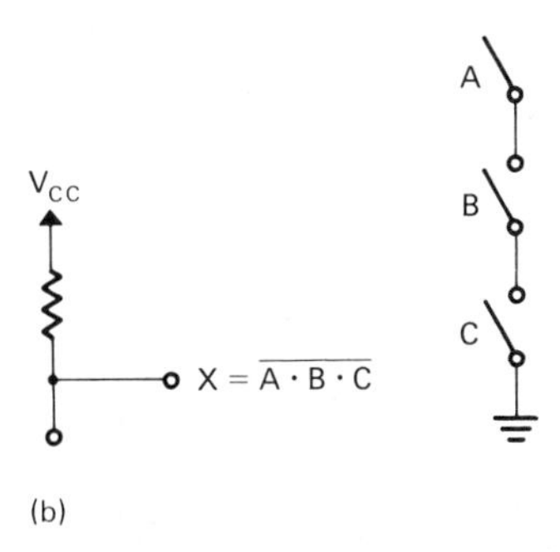

FIGURE 7-13

(a) Direct-coupled transistor NAND gate composed of three transistors in series. (b) Equivalent manual switch circuit. Only when all three transistors are turned on will the output drop to 0.

One control transistor in the NAND gate compares with as many as four for an equivalent AND gate, and five for an equivalent OR gate. Thus, the NAND is cheaper to produce and, therefore, is given preference in TTL design.

7.4.4 CMOS Integrated Circuit NAND Gate

The CMOS NAND gate is very much like the CMOS NOR gate except that the N- and P-channel devices are reversed. In the NAND function, any binary 0 input results in a binary 1 output. This can be accomplished with two or more P-channel MOSFETs in parallel, as Figure 7-15 shows. Like a pair of switches in parallel, if either one or both are turned on, the $+V_{DD}$ will appear at the output. The P-channel MOSFET is ON with a $-V_{SS}$, or 0-level input. This satisfies the NAND condition that any 0 input produces a 1 output.

If we use the viewpoint that all 1 inputs must produce a 0 output, a NAND circuit can be produced with two N-channel enhancement-mode MOSFETs in series. As Figure 7-16 shows, the 1 level, $+V_{DD}$, will turn on the N-channel FET, but both must be turned on to produce a 0 at X. Therefore, the circuit forms a valid NAND gate in that all 1 inputs are needed in order to produce a 0 output. Although both Figures 7-15 and 7-16 form valid NAND gates, they use resistors and do not have the ideal output conditions and low power consumption of a CMOS device.

FIGURE 7–14

Sprague US7410 TTL integrated circuit three-input NAND gate.

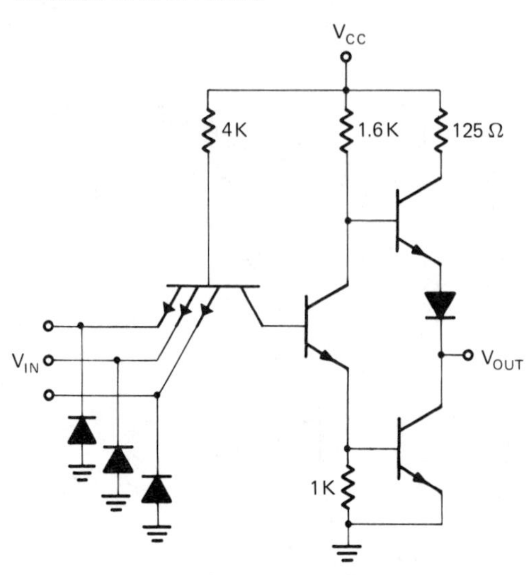

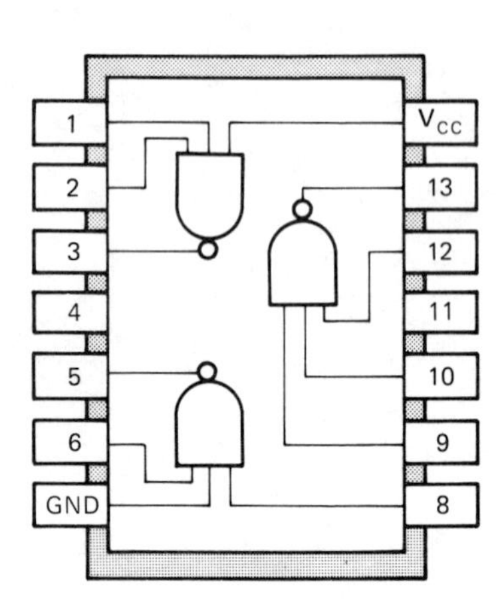

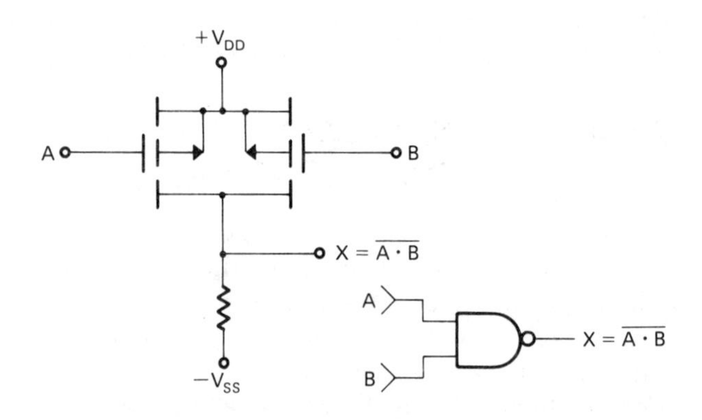

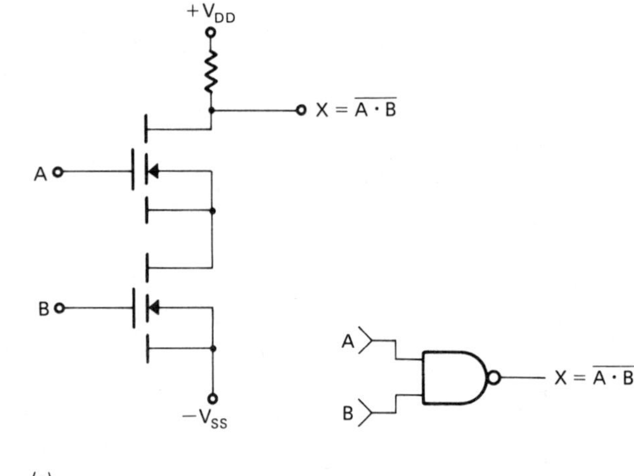

FIGURE 7–15

(a) P-channel enhancement-mode MOSFET NAND gate. (b) Equivalent manual switch circuit. A 0 on either input will produce a 1-level output.

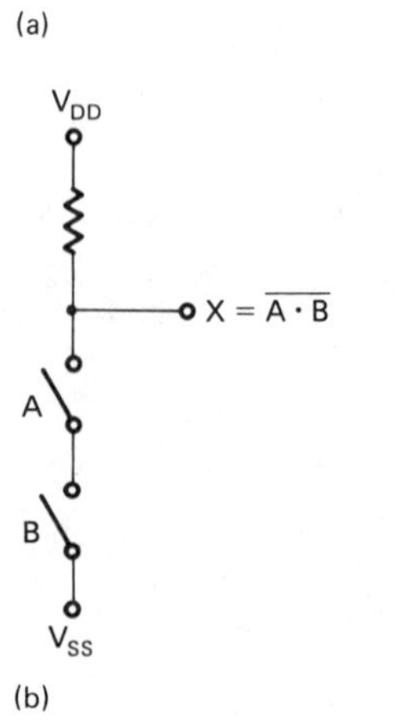

FIGURE 7–16

(a) N-channel enhancement-mode MOSFET NAND gate. (b) Equivalent manual switch circuit. A 1 on both inputs is required to produce a 0 output.

As we did with the NOR gate, we again remove the resistors and combine the two circuits to form the complete CMOS circuit shown in Figure 7-17. The fourth column of the truth table lists the devices that are turned on for the given input conditions. This type of CMOS NAND circuit is provided by the Solid State Scientific, Inc., circuits in Figure 7-18.

7.5 Boolean Functions with NAND Gates

The outputs of circuits containing AND, OR, and NAND gates can be expressed in Boolean functions.

FIGURE 7-17

CMOS NAND gate and equivalent manual switch circuit. Combining the N-channel and P-channel circuits of Figures 7-18 and 7-19 results in a NAND gate that needs no resistors and has low power consumption and high fan-out capability.

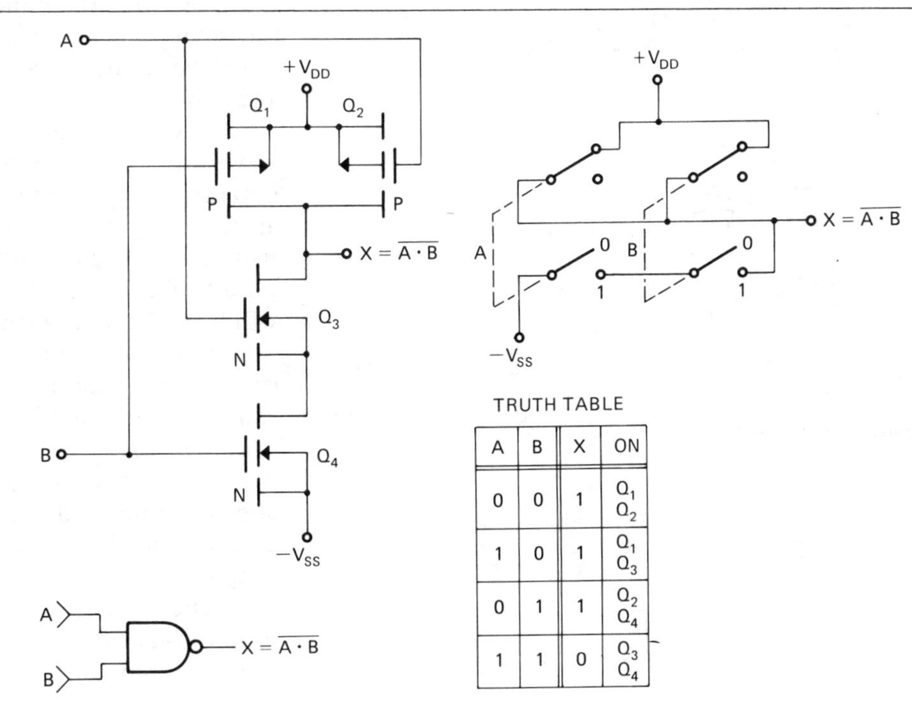

A	B	X	ON
0	0	1	Q_1 Q_2
1	0	1	Q_1 Q_3
0	1	1	Q_2 Q_4
1	1	0	Q_3 Q_4

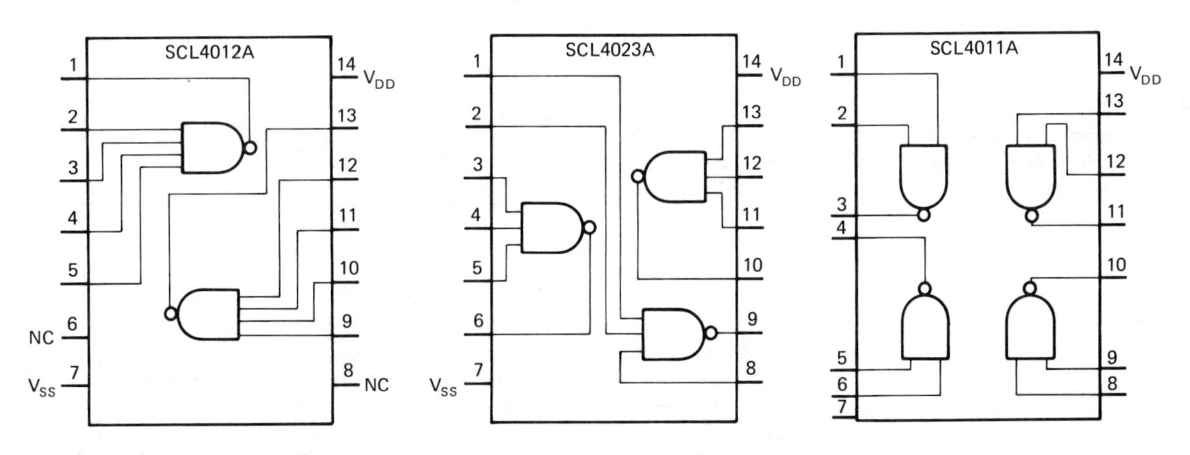

FIGURE 7-18

Solid State Scientific, Inc., integrated circuits, with two-, three-, or four-input CMOS NAND gates.

■ **EXAMPLE 7-3**

Draw the circuit described by the Boolean function $X = \overline{A \cdot B} + \overline{C \cdot D}$.

Solution: See Figure 7-19.

■ **EXAMPLE 7-4**

Write the Boolean function of the circuit shown in Figure 7-20.

Solution: The NAND functions are $\overline{A \cdot B}$ and $\overline{C \cdot D}$. Therefore,

$$X = \overline{A \cdot B} \cdot \overline{C \cdot D}$$

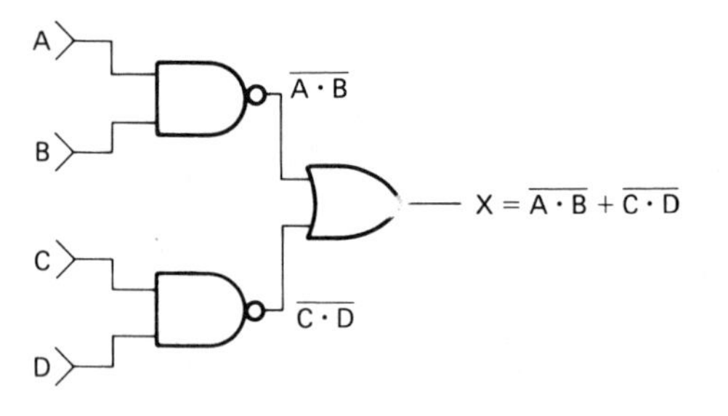

FIGURE 7-19

Solution for Example 7-3.

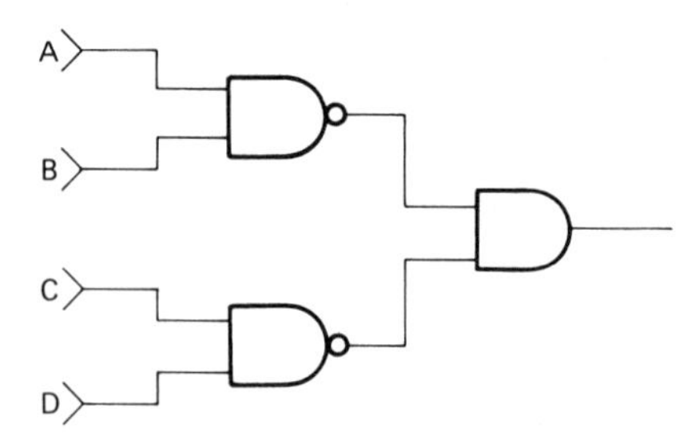

FIGURE 7-20

Example 7-4.

7.6 Analysis by Timing Diagram

Analyzing the outputs of inverting gates is not always as easy as considering one input a signal and the others a control input. Often all the input signals are changing with a frequency and irregularity for which the timing diagram is the only practical means to analyze the output function. If we use the example of Section 6.7, let the first input change from 0 to 1 and back to 0 once each microsecond, the second input make the same changes every two microseconds, and the third input every four microseconds. Figure 7-21 shows the three input waveforms.

The resulting NOR gate output voltage waveform is shown at the top. This output is determined by finding the periods during which all three inputs are 0. Only during those periods will the NOR gate output be at 1 level. At every other point along the graph, one or more inputs will be at 1 level, resulting in a 0 output.

If the same inputs are applied to the NAND gate, as in Figure 7-22, the output waveform can be determined by finding the periods during which all three inputs are 1 coincidentally. During those periods only, the output will be 0. At any other point along the graph, one or more inputs will be at 0 level, resulting in a 1 level at the output.

The clock-and-gate generator at the top of Figure 7-23 generates the waveforms shown in the lower portion of the figure.

FIGURE 7-21

Timing diagram used to plot the output waveform of a NOR gate when inputs A, B, and C are changing.

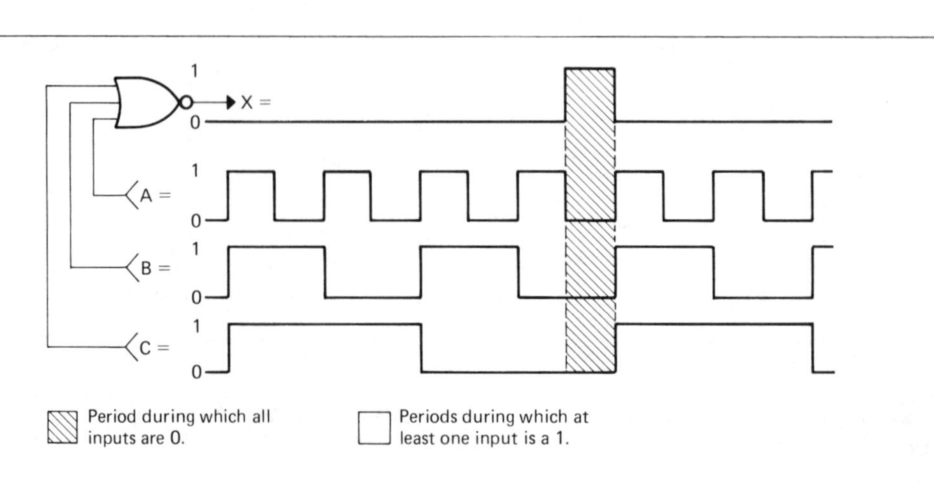

FIGURE 7-22

Timing diagram used to plot the output waveform of a NAND gate when inputs A, B, and C are changing.

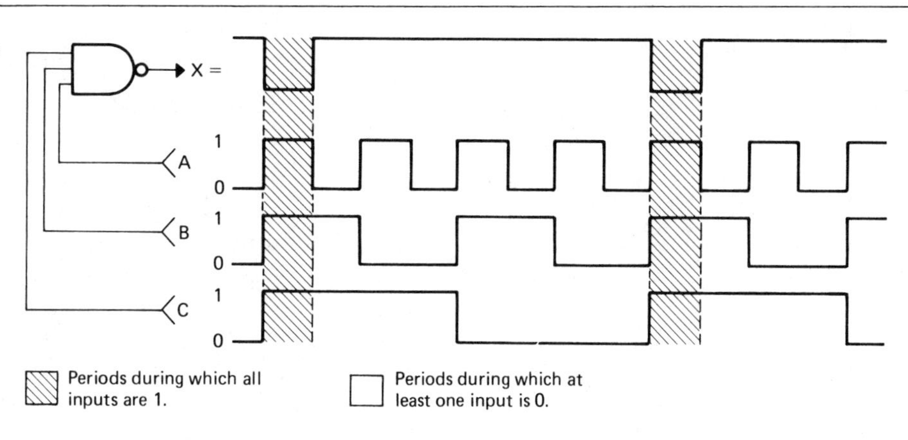

FIGURE 7-23

Clock-and-gate generator with its voltage-versus-time waveforms.

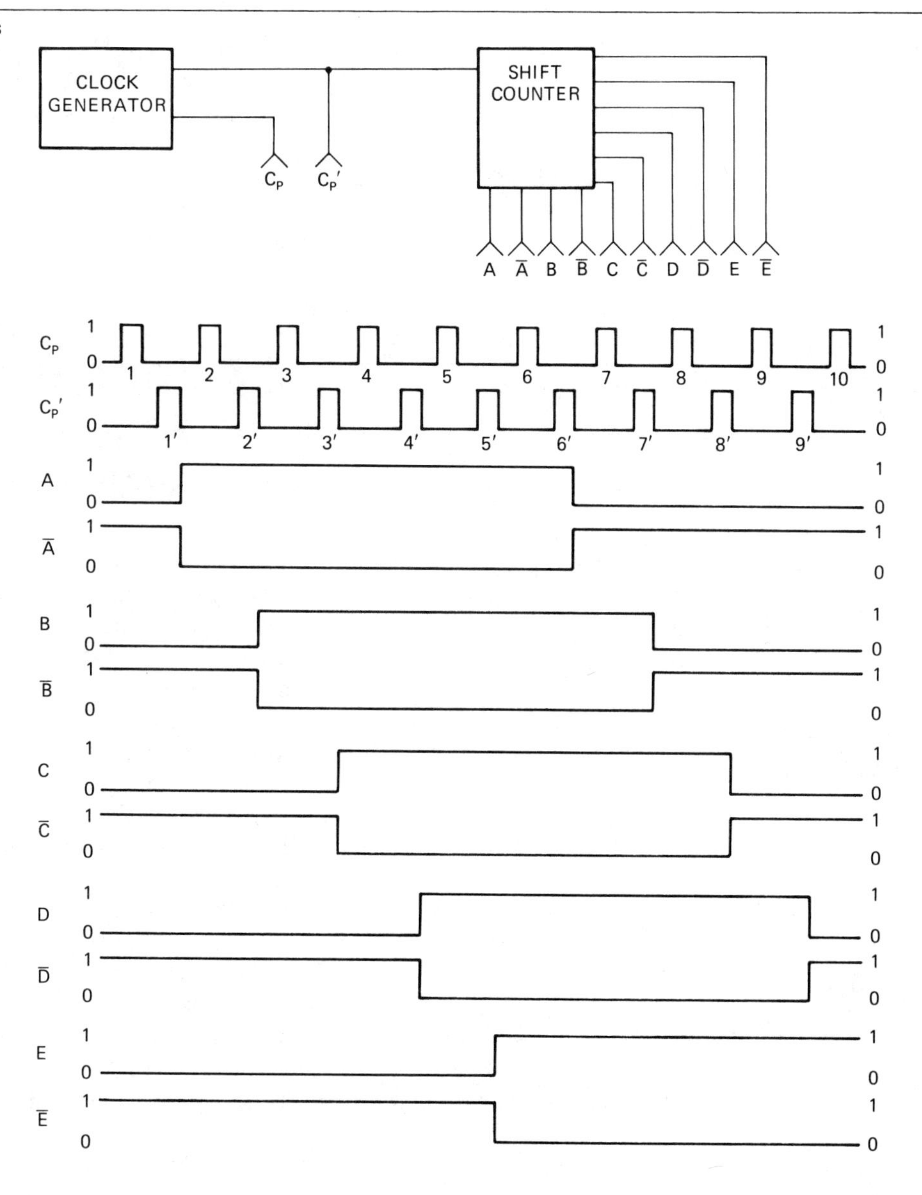

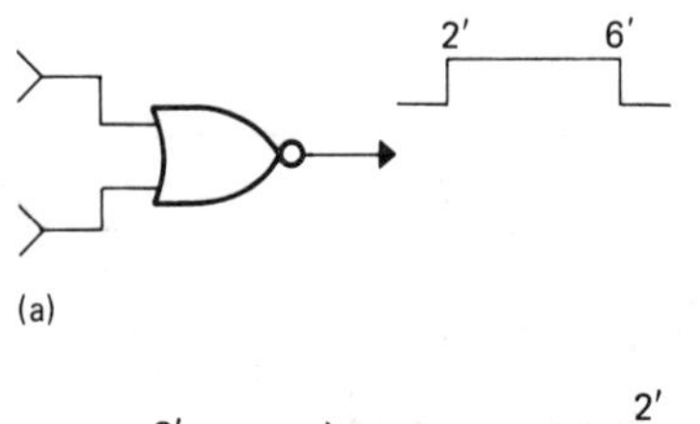

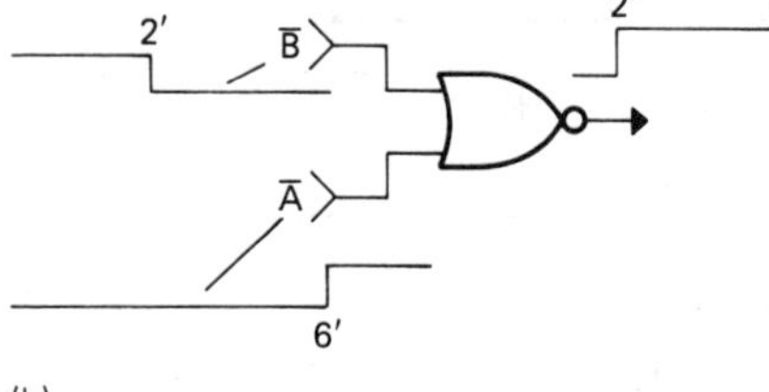

FIGURE 7-24
Example 7-5.

■ EXAMPLE 7-5

Connect the two-input NOR gate of Figure 7-24(a) to the shift counter outputs of Figure 7-23 to provide the indicated output signal.

Solution: A straight edge down from the trailing edge of the 2′ pulse on Figure 7-23 indicates that only waveform B is going to 0 at 2′. This input would be needed to turn the gate on at that time. The straight edge down from pulse 6′ indicates that A is going high at time 6′ and must be used to turn the gate off, as shown in Figure 7-24(b).

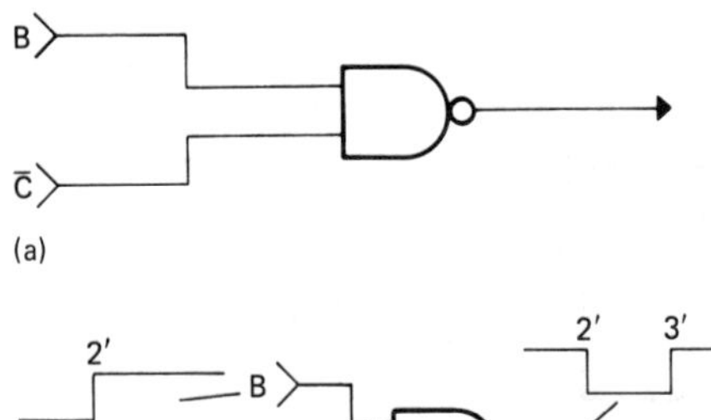

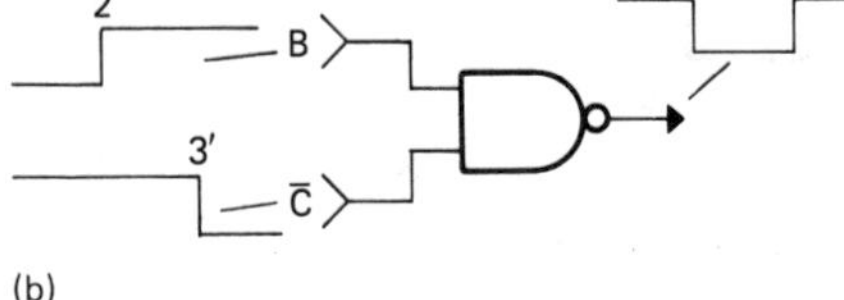

FIGURE 7-25
Example 7-6.

■ EXAMPLE 7-6

The NAND gate in Figure 7-25(a) is connected to the indicated shift counter outputs. Draw the output waveform.

Solution: The NAND gate goes to 0 when all inputs are 1. Using the straight edge again on Figure 7-23, we find that input B goes to 1 at time 2′, at which time C is already 1. Two 1s take the output to 0. C goes to 0 at time 3′. Any 0 into a NAND gate produces a 1 out. The output is shown in Figure 7-25(b).

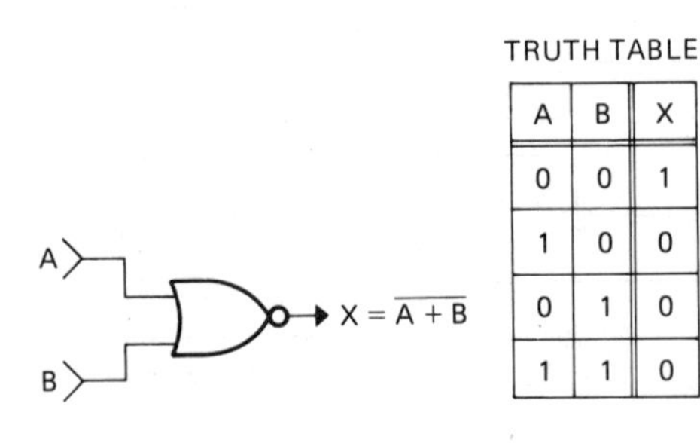

A	B	X
0	0	1
1	0	0
0	1	0
1	1	0

FIGURE 7-26
Two-input NOR gate.

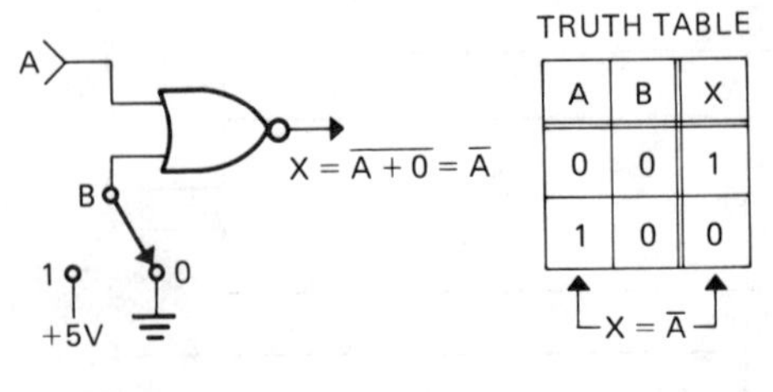

A	B	X
0	0	1
1	0	0

FIGURE 7-27
Two-input NOR gate enabled by input B to pass signal A.

7.7 Enable or Inhibit Functions of the NOR Gate

In Section 5.4, we discussed the use of a two-input AND gate in which a signal is applied to one input while the other input controls the passage of the signal through the gates. The NOR gate can be employed in this fashion except for the problem that the signal is inverted. Figure 7-26 shows a two-input NOR gate and its truth table. Let us call input A the signal input, which will vary between 1 and 0 levels, and input B the control input. When the switch in Figure 7-27 is in the 0 position, the gate functions according to the top two lines of the truth table. Under this condition, the output is the inversion of the signal input: $X = \overline{A}$. Even though the signal output is inverted, the NOR gate is considered *enabled* with a 0 on its control input.

In Section 6-5, we manipulated the Boolean function of the two-input OR as shown:

$$X = A + B$$

$$B = 0$$

Therefore,

$$X = A + 0 = A$$

The NOR function, therefore, is the same function under the inversion sign, as shown here:

$$X = \overline{A + B}$$

$$B = 0$$

Therefore,

$$X = \overline{A + 0} = \overline{A}$$

When the switch B is in the 1 state, as Figure 7-28 shows, then the NOR gate is operating according to the bottom half of the truth table of Figure 7-26. This tells us that no matter how the input varies, the output will be 0. Therefore, with a 1 on the control input, the gate is *inhibited.*

Let us again review a manipulation of the Boolean OR function used in Section 6.5:

$$X = A + B$$

$$B = 1$$

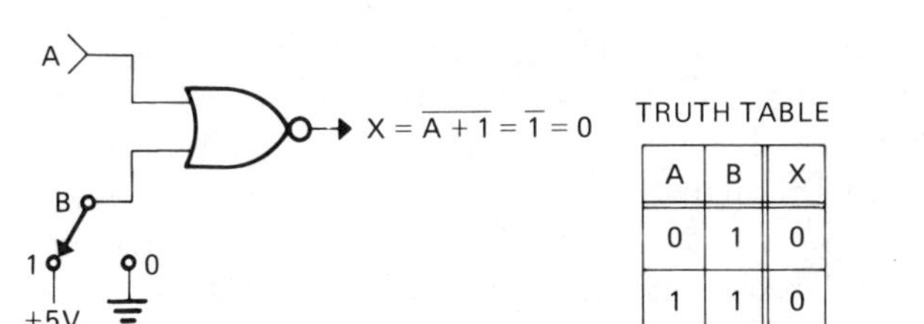

A	B	X
0	1	0
1	1	0

FIGURE 7-28

Two-input NOR gate inhibited by input B from passing signal A.

Therefore,

$$X = A + 1 = 1$$

The NOR, therefore, is the same function under an inversion sign:

$$X = \overline{A + B}$$

$$B = 1$$

Therefore,

$$X = \overline{A + 1} = \overline{1} = 0$$

Because of the inversion of the signal, using the NOR gate to control a signal takes a little more care. In the example in Figure 7-29(a), the control voltage is being used to isolate clock pulse 3,

FIGURE 7-29

NOR gate used to isolate a number 3 pulse.

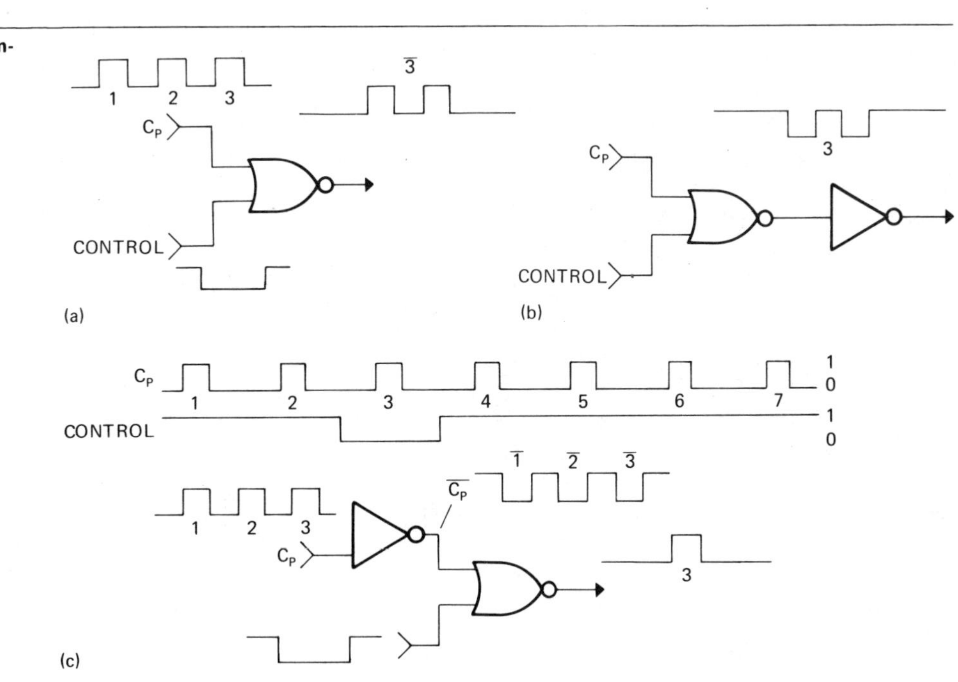

but the pulse is inverted and the baseline on each side of the pulse is distorted. It might seem that it is necessary only to use an inverter either on the signal input or on the output. However, as we see in Figure 7-29(b), inverting the output does result in a positive pulse, but the baseline is distorted. In Figure 7-29(c), inverting the input results in a positive pulse with no baseline distortion.

For most applications, the input inverter would be correct. If an inverted signal is already available within the system, the inverter will, of course, be unnecessary.

■ **EXAMPLE 7-7**

Using the clock-and-gate generator waveforms of Figure 7-23, indicate connections to the NOR gate of Figure 7-30(a) that will produce the clock pulse 5 only on the output. Connect inverters to the inputs or outputs as needed.

Solution: The gate must be enabled between 4′ and 5′. The C_P must be inverted to $\overline{C}_P$ before connecting to the input. A straight edge down from 4′ indicates that the gate must be enabled by $\overline{D}$. The straight edge down from 5′ indicates that the remaining input must be inhibited by E. Figure 7-30(b) shows the result.

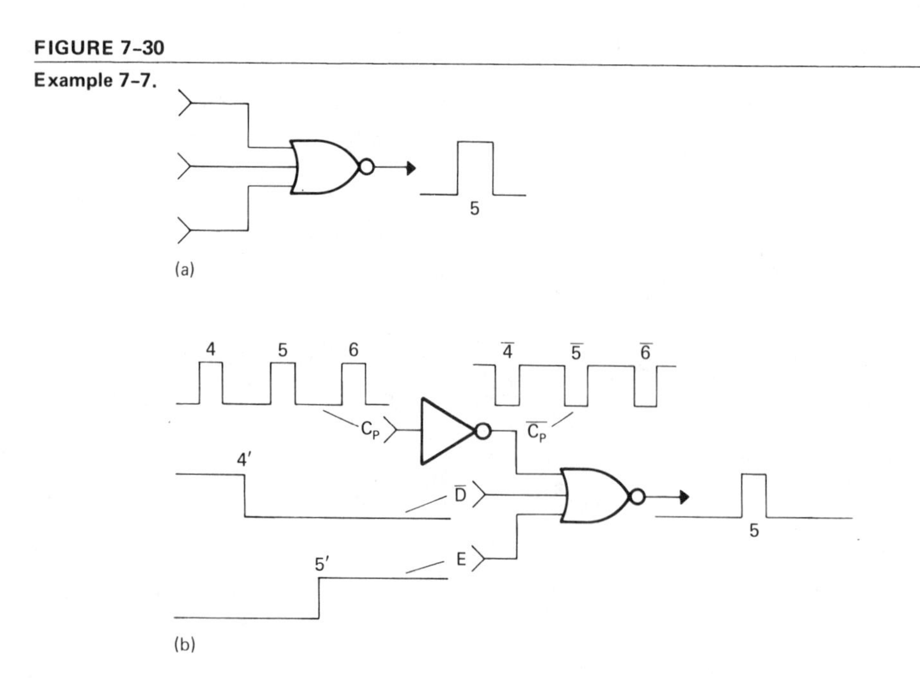

FIGURE 7-30
Example 7-7.

7.8 Enable or Inhibit Functions of the NAND Gate

Figure 7-31 shows again the two-input NAND gate and its truth table. Let us consider the A input as the signal and allow it to vary, while the B input enables or inhibits its passage through the gate. In Figure 7-32, there is a 1 on the B input, and the output is the inversion of A. Despite

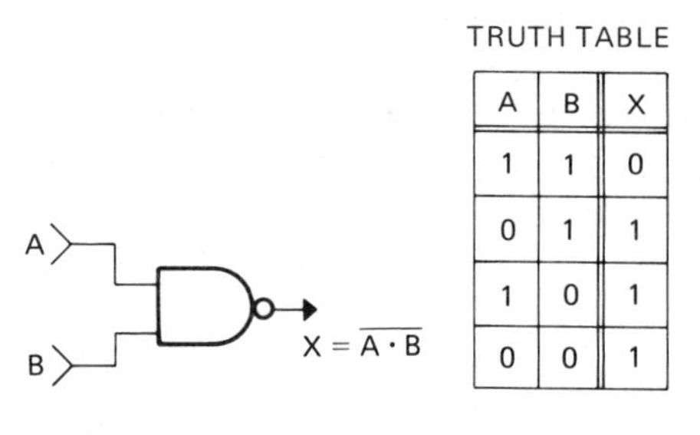

TRUTH TABLE

A	B	X
1	1	0
0	1	1
1	0	1
0	0	1

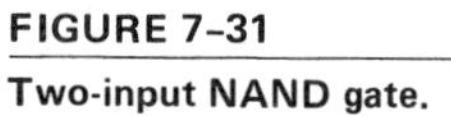

FIGURE 7-31
Two-input NAND gate.

the inversion, the gate is considered enabled. After switching the B input to 0, as in Figure 7-33, the bottom half of the truth table of Figure 7-31 applies and the output will remain at 1 regardless of the signal variation. With a 0 on the control input, the NAND gate is inhibited and the output is fixed at a 1 level, regardless of A.

In using the NAND gate to control a signal, it might seem necessary only to invert either input or output. However, as the example in Figure 7-34 shows, there is a difference. In Figure 7-34(a), the clock line and control voltage are used to obtain only a number 3 pulse, which appears inverted. In Figure 7-34(b), inverting the input results in a positive pulse but distorted baseline. Inverting the output, as in Figure 7-34(c), results in a correct pulse with no distortion. For most applications, then, the output inverter would be correct.

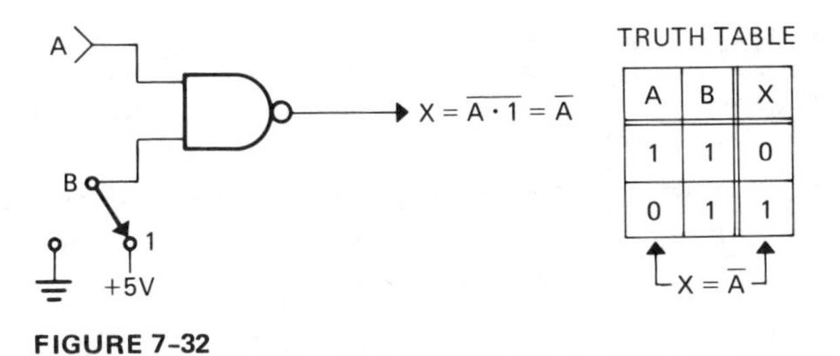

FIGURE 7-32

Two-input NAND gate enabled by input B to pass signal A.

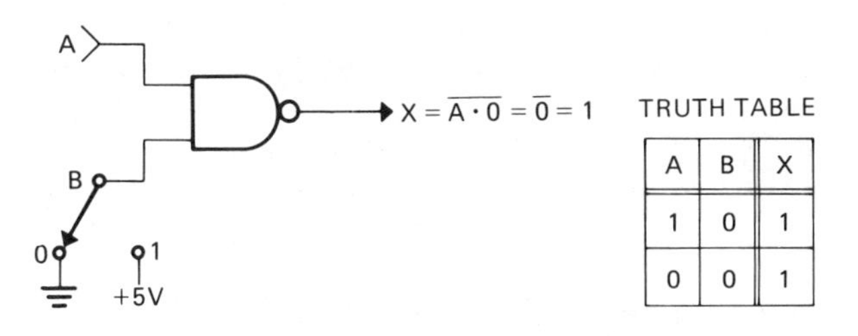

FIGURE 7-33

Two-input NAND gate inhibited by input B from passing signal A.

FIGURE 7-34

NAND gate used to isolate a number 3 pulse.

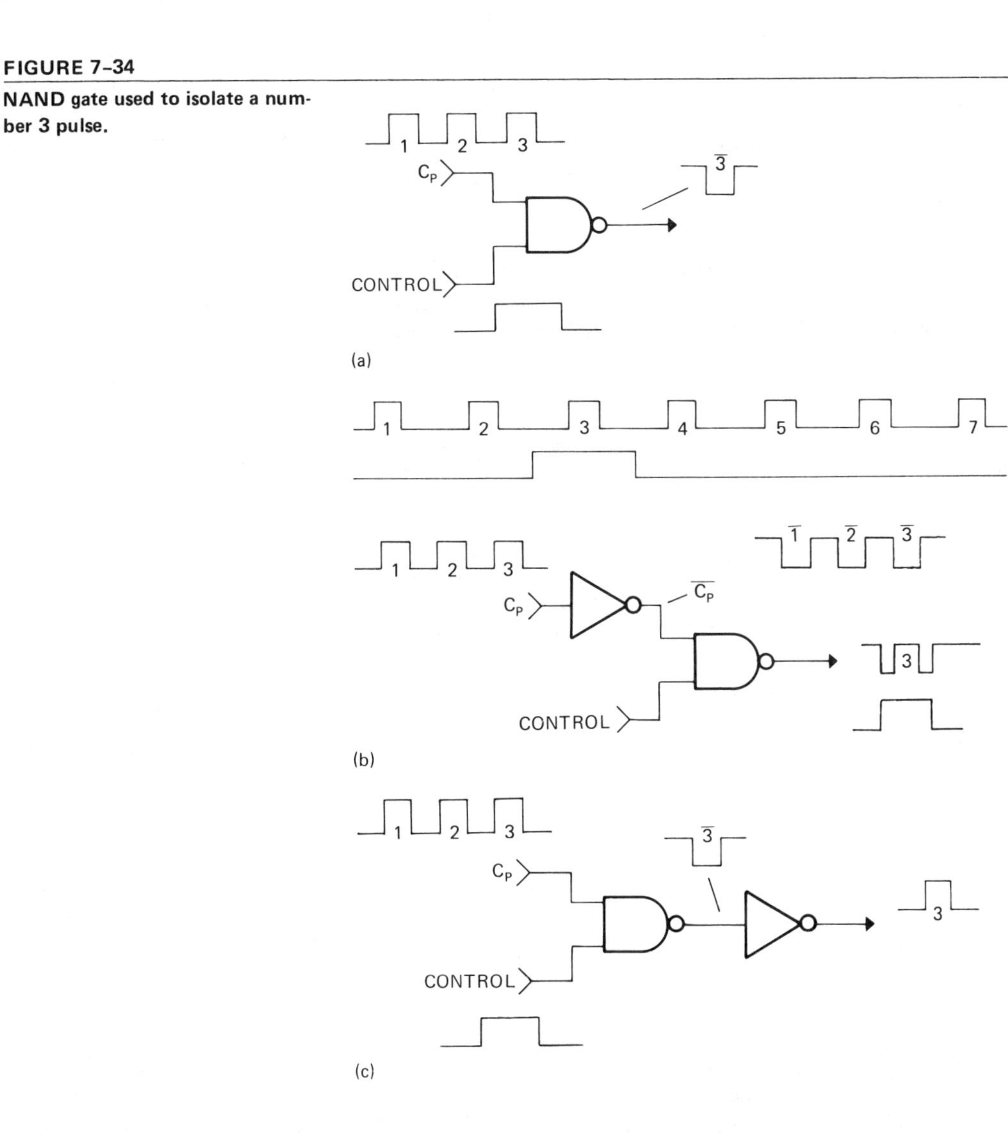

■ **EXAMPLE 7-8**

Using the clock-and-gate generator waveforms of Figure 7-23, indicate the necessary connections to a three-input NAND gate that will produce the clock pulse 8 on the output. Connect inverters to the inputs or outputs as needed.

Solution: The gate must be enabled between 7′ and 8′. A straight edge down from 7′ indicates that the gate must be enabled by B. The straight edge down from 8′ indicates that the remaining input must be inhibited by $\overline{C}$. To obtain an upright pulse, an inverter must be connected to the output. Figure 7-35 shows the result.

7.9 De Morgan's Theorem

One of the more important theorems of Boolean algebra is De Morgan's theorem, which states that the complement of an OR function is equal to the AND function of the complements:

$$\overline{A + B + C} = \overline{A} \cdot \overline{B} \cdot \overline{C}$$

This theorem gives the identity shown in Figure 7-36. It also implies that the NOR symbol we have been using, an OR gate with the output inverted, can be replaced by an AND symbol with the inputs inverted.

We can prove De Morgan's theorem if we take an OR gate truth table, keep the inputs the same, but invert the outputs. As shown in Figure 7-37(a), the OR gate table becomes a NOR gate truth table. As we see in Figure 7-37(b), if we take the AND gate truth table, keep the output the same, and invert the inputs, the result is also a NOR gate truth table.

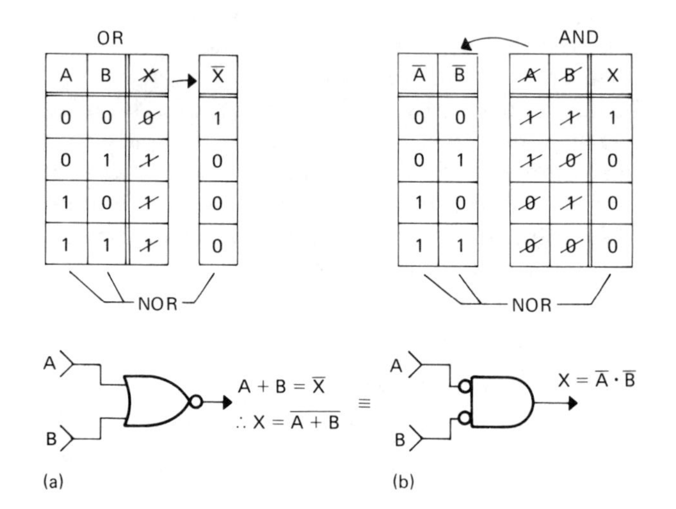

FIGURE 7-37

(a) OR gate truth table with output inverted. (b) AND gate truth table with inputs inverted. In both cases, the inversions change the tables to NOR gate truth tables.

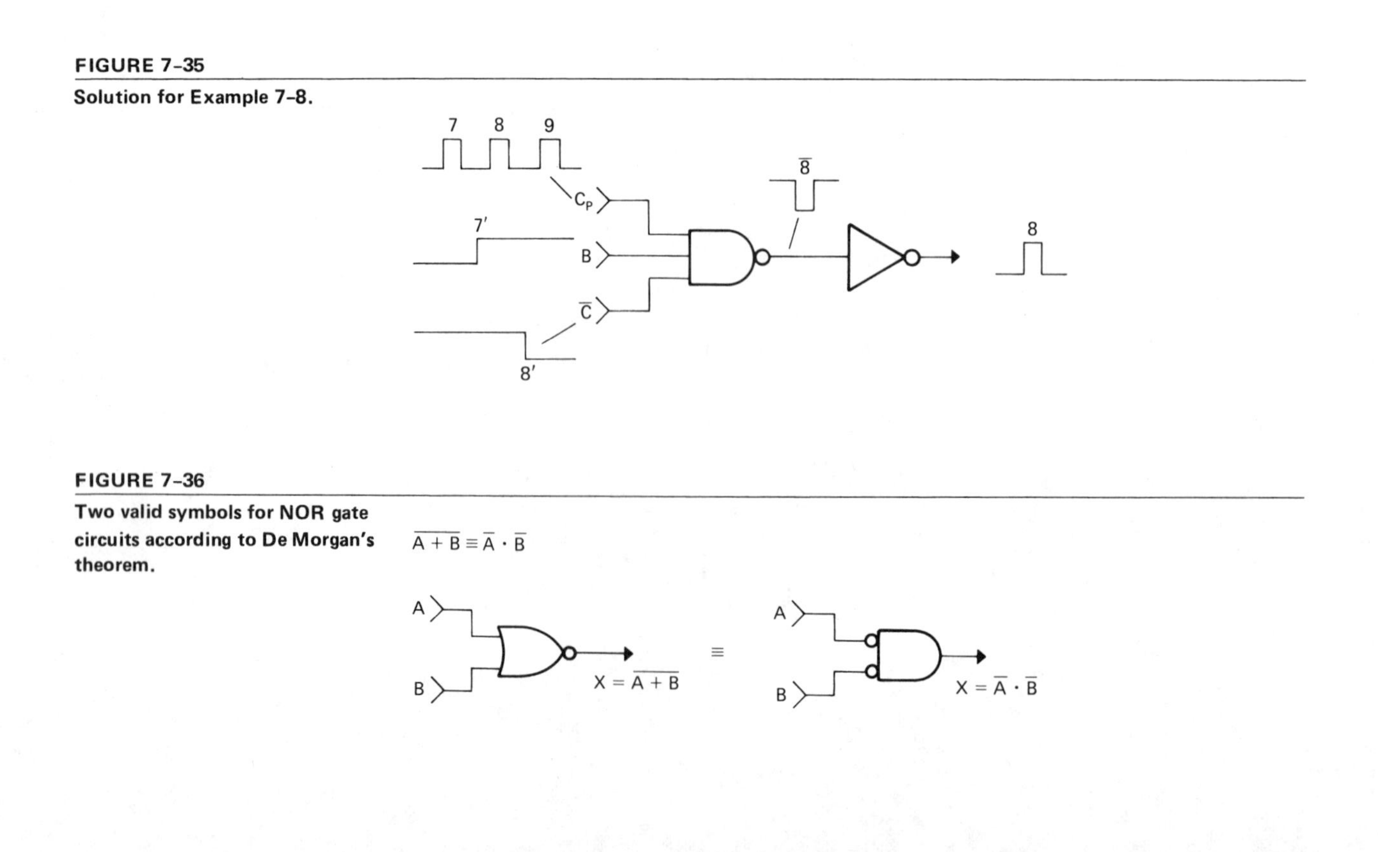

FIGURE 7-35

Solution for Example 7-8.

FIGURE 7-36

Two valid symbols for NOR gate circuits according to De Morgan's theorem.

FIGURE 7–38

Two valid symbols for NAND gate circuits according to De Morgan's theorem.

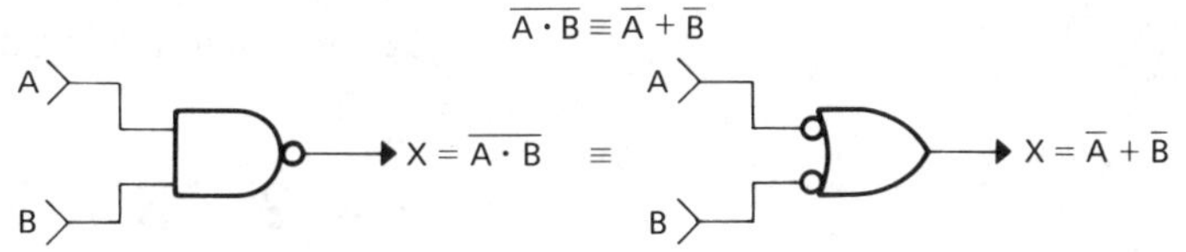

$\overline{A \cdot B} \equiv \overline{A} + \overline{B}$

A, B → NAND → $X = \overline{A \cdot B}$ ≡ A, B → (inverted inputs) OR → $X = \overline{A} + \overline{B}$

FIGURE 7–39

(a) AND gate truth table with output inverted. (b) OR gate truth table with inputs inverted. In both cases, the tables become NAND gate truth tables.

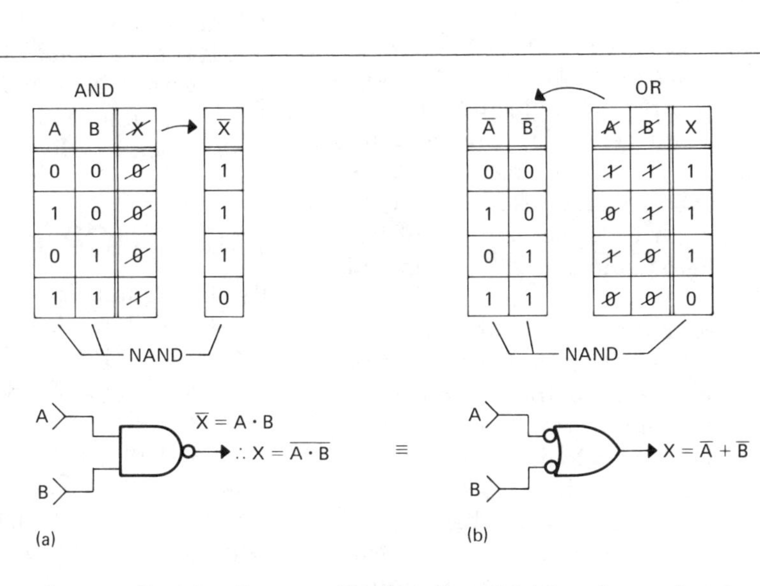

De Morgan's theorem also applies to the NAND function. It states that the complement of an AND function is equal to the OR function of the complements:

$$\overline{A \cdot B \cdot C} = \overline{A} + \overline{B} + \overline{C}$$

It implies that the NAND gate has two valid symbols, as shown in Figure 7–38. Again, manipulation of the truth tables helps to verify the theorem. The first symbol implies that an AND truth table with the inputs held constant and the output inverted is a NAND truth table, as shown in Figure 7–39(a). As the second symbol implies, an OR gate truth table with the output held constant and the inputs inverted is also a NAND gate truth table, as shown in Figure 7–39(b). In short, an AND function with the output inverted is identical to an OR function with the inputs inverted; and an OR function with the output inverted is identical to an AND function with the inputs inverted.

■ EXAMPLE 7–9

Prove that the circuits in Figure 7–40 are Boolean identities.

Solution: If equal,

$$\overline{\overline{A} \cdot \overline{B}} \cdot \overline{A \cdot B} = \overline{(A \cdot B) + \overline{A + B}}$$

Change the right-hand member by De Morgan's theorem until it becomes identical to the left-hand member:

$$\overline{(A \cdot B) + \overline{A + B}} = \overline{(A \cdot B)} \cdot \overline{\overline{A + B}}$$
$$= \overline{A \cdot B} \cdot \overline{\overline{A} \cdot \overline{B}}$$

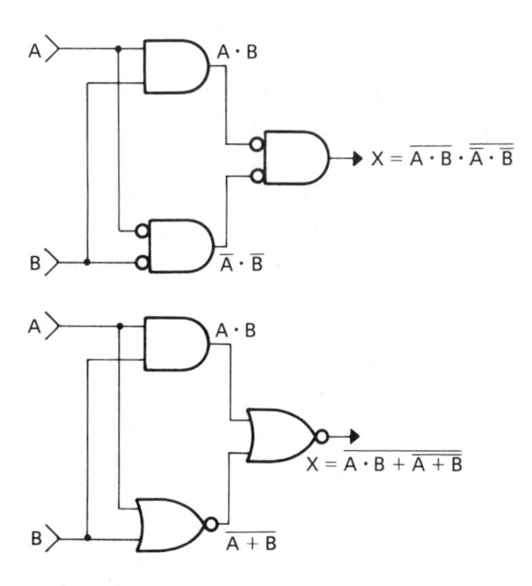

FIGURE 7–40

Example 7–9.

7.10 Single-Gate Logic

7.10.1 Complete Logic Capability

A system of logic gates is complete if it has the AND, OR, and either of the two inverting gates. That is, with these three types of gates, any logic function can be provided using the minimum number of gates. There is nothing to be gained by mixing NOR gates and NAND gates in the same logic system. In fact, the presence of both NAND and NOR gates in a logic system only adds confusion.

If all inverting-type gates were dropped from our logic system, many logic functions could not be produced. On the other hand, if one or both of the noninverting gates (AND and OR) were excluded from a logic system, all logic functions could still be produced. This becomes obvious from the fact that a number of NOR gates can be connected to produce an AND gate, and only two NOR gates are required to produce an OR gate. Similar conversions exist for the NAND gate.

There is, of course, an advantage to using a single inverting-type logic gate throughout an entire logic system. If there weren't such an advantage, it would be pointless to discuss using several inverting gates in place of one noninverting gate. There is some economic advantage in that it is cheaper to produce or purchase items in larger quantities; but over and above the economic advantage is the superior electrical quality of the inverting-type gate. The noninverting gates previously explained were either diode or emitter follower-type circuits, for which output-1 levels are always lower than input-1 levels. The output-0 levels are always higher than the input-0 levels. The signals are degraded as they pass through each gate. The inverting gates are common emitter circuits, and with this type of circuit, the levels are restored at the output of each gate.

7.10.2 NOR to OR

When we look at the NOR symbol in Figure 7-41, it appears like an OR symbol with the output inverted. A second inversion can be produced by using a second NOR gate at the output to invert back to OR.

7.10.3 NOR to AND

By De Morgan's theorem, the NOR function $\overline{A + B + C} = \bar{A} \cdot \bar{B} \cdot \bar{C}$ implies that both symbols of Figure 7-42(a) are valid for the NOR gate. If three additional NOR gates are connected as inverters on each input, the result is a double inversion and the output is equivalent to the AND function, as in Figure 8-42(b). If inverters are not available, NOR gates with all inputs shorted will substitute for inverters, as in Figure 7-42(c).

7.10.4 NAND to AND

When we look at the NAND symbol in Figure 7-43, it appears like an AND symbol with the output inverted. A second inversion can be produced by using a second NAND gate at the output to invert back to AND.

7.10.5 NAND to OR

By De Morgan's theorem, the NAND function $\overline{A \cdot B \cdot C} = \bar{A} + \bar{B} + \bar{C}$ indicates that the symbol of Figure 7-44(a) is identical to the symbol of 7-44(b). Connecting three additional NAND gates, one to each of the three inputs, results in a double inversion—an OR gate—as in Figure 7-44(c).

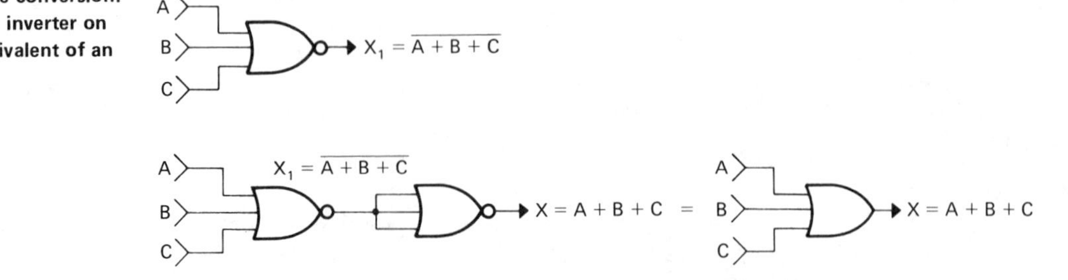

FIGURE 7-41
NOR gate to OR gate conversion. A NOR gate with an inverter on its output is the equivalent of an OR gate.

FIGURE 7–42

NOR gate to AND gate conversion. (a) According to De Morgan's theorem, both symbols are valid for the NOR gate. (b) Inverting the inputs to a NOR gate results in an AND gate. (c) A NOR gate with all inputs shorted is equivalent to an inverter.

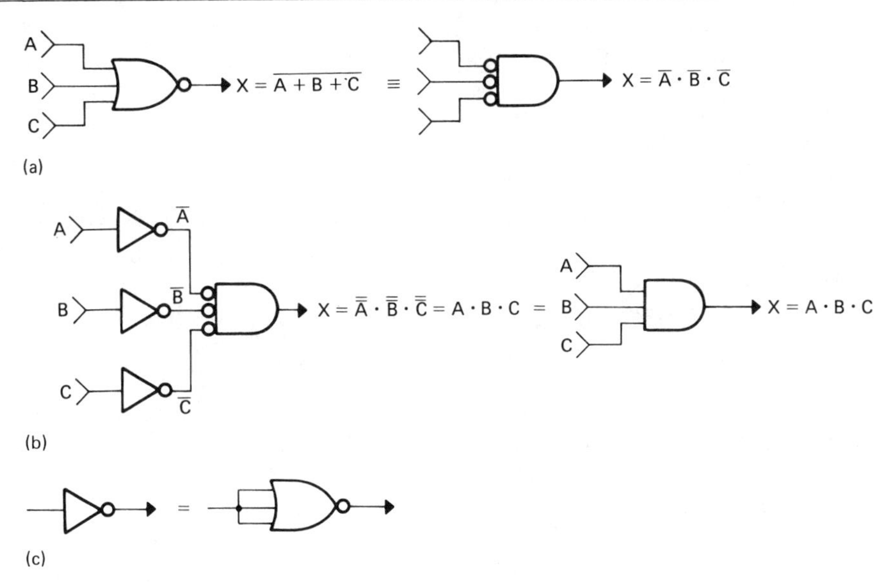

FIGURE 7–43

NAND gate to AND gate conversion. A NAND gate with an inverter on its output is the equivalent of the AND gate.

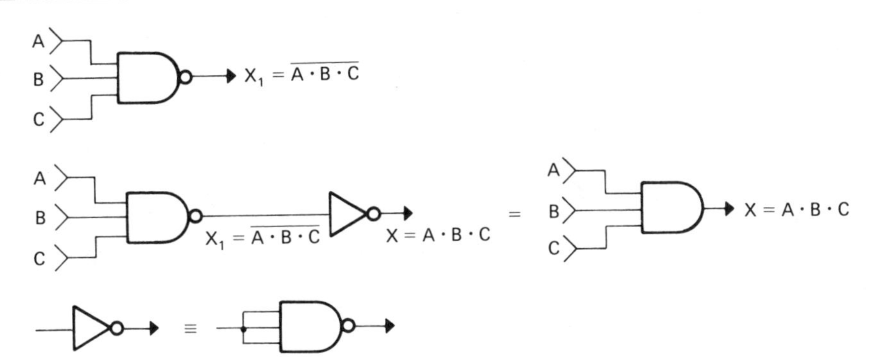

FIGURE 7–44

NAND gate to OR gate conversion. (a) Identities by De Morgan's theorem. (b) Identities by De Morgan's theorem. (c) Inverted inputs on a NAND gate resulting in an OR gate.

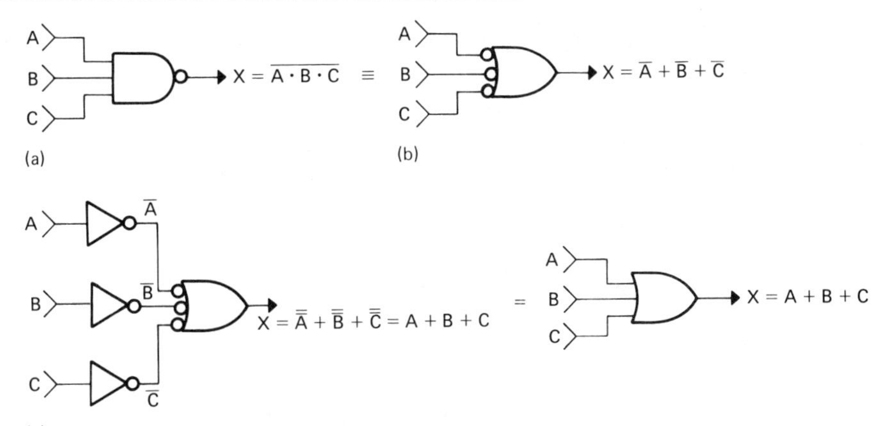

■ EXAMPLE 7-10

Convert the circuit of Figure 7-45(a) to an identical function using all NOR gates.

Solution: See Figure 7-45(b), which shows a NOR circuit with all inputs connected together forming an inverter.

7.10.6 Conversion by Complementary (Inverted) Inputs

It is, without question, a serious disadvantage to use two or more inverting-type gates to provide the same function as one noninverting gate. Fortunately, direct conversions are not necessary too often. A careful look at the logic system often shows that inverted or complementary values for a given input are already available; and, as Figure 7-46 shows, it removes the need for input inverters.

7.11 Summary

The two primary inverting logic gates are the NOR and NAND gates. The most widely used logic symbol for the NOR gate is an OR gate symbol with a small *inversion circle* on the output. This implies that the NOR gate functions like an OR gate with the output inverted or *complemented*. Only when all inputs are 0 will the output of a NOR gate be 1. If any input is 1, the output will be 0. Figure 7-1 shows the logic symbol and truth table of a three-input NOR gate. The Boolean equation for a NOR gate with three inputs, A, B, and C, is $X = \overline{A + B + C}$. Note that this function differs from the OR gate function only by the *inversion bar* over the entire function.

A NOR gate can be constructed by tying the output of a diode OR gate directly to the input of a transistor inverter. As Figure 7-2 shows, this technique is known as diode transistor logic (DTL).

FIGURE 7-45
Example 7-10.

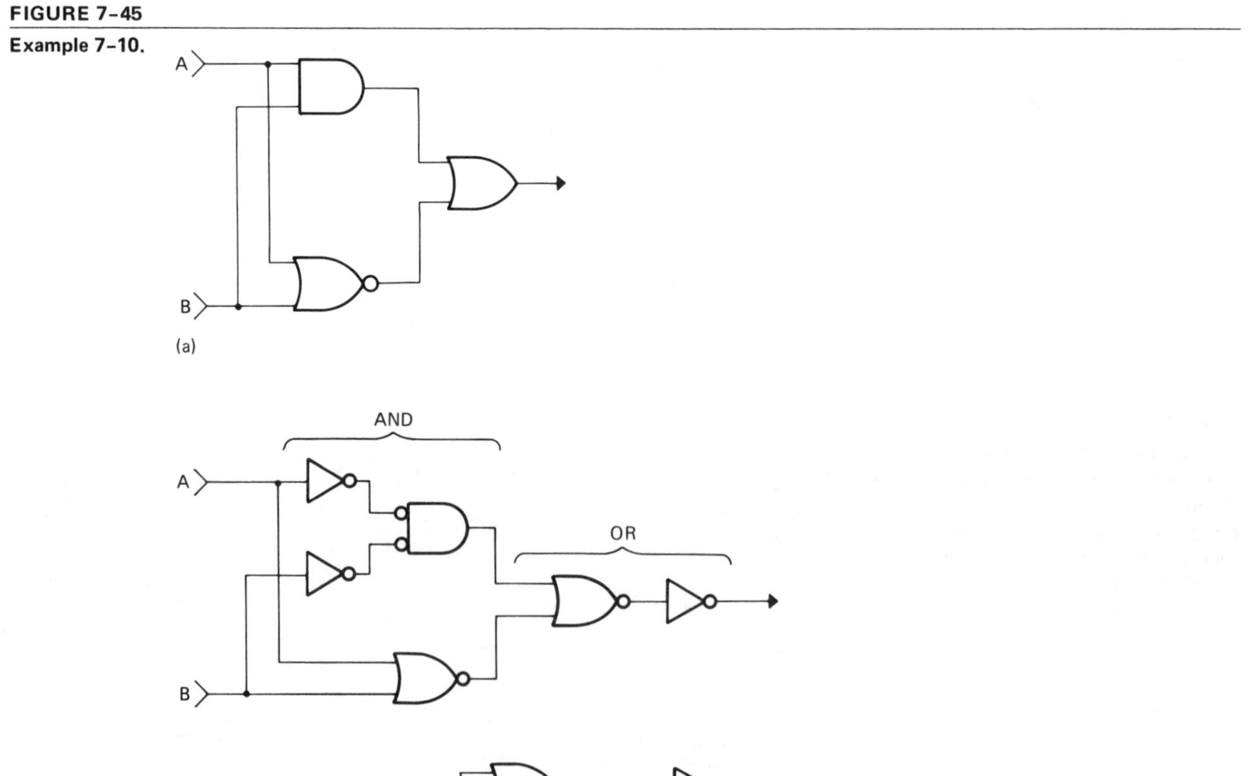

FIGURE 7-46

NOR and NAND gate conversion by inverted inputs. (a) Inverted inputs to a NOR gate produce the same results as their complements applied to an AND gate. (b) Inverted inputs to a NAND gate produce the same results as their complements applied to an OR gate.

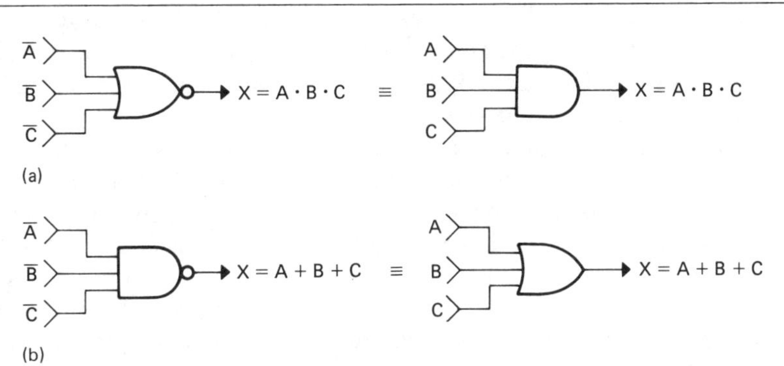

A NOR gate can be constructed by connecting resistors and transistors as shown in Figure 7-3. This method is called resistor transistor logic (RTL).

At present, NOR gates are more widely used in TTL or CMOS integrated circuit form. Figure 7-4 shows a quad dual-input NOR gate, which differs from the 7432 OR gate in that it does not have two transistors in each gate. This elimination of the two transistors in each gate allows for the output inversion. As in the TTL OR gate, each input is a separate transistor and the output a "totem pole" circuit.

The CMOS NOR gate consists of two or more N-channel enhancement-mode MOSFETs connected with drain and source leads in parallel. In place of a load resistor, two or more P-channel enhancement-mode MOSFETs are connected in series. As Figure 7-7 shows, the gates of the N- and P-channel FETs are connected so that the inputs operate complementary pairs of N- and P-channel devices. The result is that a 1 applied to an input turns the N-channel device ON and the P-channel device OFF. This shorts the output to $-V_{SS}$ (usually ground). At the same time, the series P-channel FET opens, reducing the power drain on $+V_{DD}$. One or more inputs having a 1 applied to them will cause a 0 output. To obtain a 1 level on the output, all P-channel devices must be turned ON, which requires all inputs to be 0. This applies $+V_{DD}$ or 1 level to the output. When all P-channel FETs are ON, N-channel FETs are OFF, isolating the output from ground (or $-V_{SS}$).

The most widely used logic symbol for the NAND gate is an AND gate symbol with a small *inversion circle* on the output. This implies that the NAND gate functions like an AND gate with the output inverted or complemented. Only when all inputs are 1 will the output of a NAND gate be 0. If any input is 0, the output will be 1. Figure 7-11 shows the logic symbol and truth table of a three-input NAND gate. The Boolean equation for a NAND gate with inputs A, B, and C is $X = \overline{A \cdot B \cdot C}$ or $\overline{ABC}$.

A NAND gate can be constructed by tying the output of a diode AND gate directly to the input of a transistor inverter, as Figure 7-12 shows.

A NAND gate can be constructed by interconnecting resistors and transistors, as Figure 7-13 shows.

At present, TTL or CMOS integrated circuit NAND gates are more widely used than those previously mentioned. Figure 7-14 shows a triple three-input NAND gate TTL integrated circuit. The input transistor with its three emitters can be drawn in diode form, as was done in Figure 6-10. This is identical to a diode AND gate. The remainder of the circuit is an inverter that converts the input AND function to a NAND function. The output transistors are again the "totem pole" circuit. The TTL NAND gate has the fewest transistors per logic gate of all the TTL gates. For that reason, the TTL NAND gates are the most economical and are preferred for use in TTL design.

The CMOS NAND gate is very similar in construction to the CMOS NOR gate except that, as Figure 7-17 shows, the N-channel devices are in series and the P-channel devices are in parallel. To obtain a 0 output, both N-channel MOSFETs must be turned ON. Since a positive 1 level is required to turn the N-channel MOSFETs ON, the result is a NAND function, requiring all 1s on the input to produce a 0 out. If any input is 0, one or more of the N-channel MOSFETs will be OFF, isolating the output from ground, and at least one P-channel MOSFET will be ON, applying $+V_{DD}$ 1 level to the output.

When the inputs to an inverting gate are changing rapidly, the most effective means of predicting the nature of the signal coming from the output is the timing diagram. Figure 7-21 shows a timing diagram of typical waveforms applied to a three-input NOR gate. Note that the output goes to 1 level only during the period when all three inputs are 0. Figure 7-22 shows the same set of inputs applied to a NAND gate, and the output goes to 0 only when all three inputs are 1.

NAND and NOR gates can be used in conjunction with clock and shift counter waveforms to produce control signals of various clock pulse widths, as demonstrated in Examples 7-5 and 7-6. Three-input gates can be used to isolate one or more clock pulses from the clock pulse line, as demonstrated in Example 7-7.

The NOR gate can be used to control passage of a digital signal. If the signal is applied to one input of the gate, a 1 level on a second or control input will inhibit the signal from passing. A 0 level on the control input will allow the signal to pass, but it will appear at the output in inverted form. The signal will also be subjected to a *baseline distortion*, as in Figure 7-29(a). If the signal is inverted at the input, both signal inversion and baseline distortion are corrected. If the inverter is connected to the output, signal inversion will be corrected but baseline distortion will still be evident.

The NAND gate can also be used to enable or inhibit passage of a signal. A 0 level is used to inhibit the signal, during which time the output is 1. A 1 level is used to enable the signal, but, as Figure 7-34 shows, the signal is inverted at the output. Unlike the NOR gate, however, the baseline is not distorted. An inverter on the output will correct the inversion without baseline distortion. As Figure 7-34(b) shows, inverting the input produces baseline distortion.

One of the most important theorems of Boolean algebra is *De Morgan's theorem.* Application of this theorem points out greater versatility for inverting gates than exists for non-inverting gates. De Morgan's theorem tells us that an inverted OR function is equivalent to an AND function of the inverted terms, ($\overline{A + B} = \overline{A} \cdot \overline{B}$). It also tells us that an inverted AND function is equal to an OR function of the terms inverted, ($\overline{A \cdot B} = \overline{A} + \overline{B}$). In terms of logic symbols, the NOR gate symbol, as shown in Figure 7-36, can be drawn as an OR symbol with the inversion circle on the output, or as an AND symbol with inversion circles on the input. Also, the NAND gate symbol, as shown in Figure 7-38, can be drawn as an AND symbol with the inversion circle on the output, or as an OR symbol with inversion circles on the inputs.

It is usual to have logic systems composed of AND gates, OR gates, and one of the inverting gates, NOR or NAND. NAND gates and NOR gates are seldom used together. It is less confusing when only one inverting gate is used. It is even possible to construct the entire logic system using a single inverting-type gate. If a system consists of all NOR gates, an inverter can be obtained by shorting together all the inputs of a NOR gate. An OR gate can be accomplished by connecting an inverter to the output of a NOR, as shown in Figure 7-41.

The AND function can be accomplished by connecting an inverter to each input of a NOR gate, as we see in Figure 7-42.

In a system composed of all NAND gates, an inverter can be constructed by shorting together all the inputs to a NAND gate. The AND function can be accomplished by connecting an inverter to the outputs of the NAND, as in Figure 7-43. The OR function can be accomplished by connecting inverters to the inputs of a NAND gate, as in Figure 7-44.

Glossary

NOR Gate: A logic gate that produces a 1 on its output only if all inputs are 0. If a 1 level exists on any of its inputs, the output will be 0. The Boolean equation for the output of a NOR gate with inputs A, B, and C is $X = \overline{A + B + C}$. Figure 7-1 shows the logic symbol and truth table of a three-input NOR gate.

NAND Gate: A logic gate that produces a 0 at its output only when all inputs are 1. If a 0 level exists on any of its inputs, the output will be 1. The Boolean equation for the output of a NAND gate with inputs A, B, and C is $X = \overline{A \cdot B \cdot C}$. Figure 7-47 shows the logic symbol and truth table of a three-input NAND gate.

Complement Bar or Inversion Bar: The bar over the top of a term in Boolean algebra, which means that the value is 1 when the term is 0 and 0 when the term is 1. That is, $\overline{A}$ is 1 if A is 0, and $\overline{A}$ is 0 if A is 1. The same is true for a function, so that $\overline{A + B + C}$ is 1 if A

+ B + C is 0, and $\overline{A + B + C}$ is 0 if A + B + C is 1.

Inversion Circle: A small circle drawn on an output lead of a logic symbol to indicate that this connection supplies an output that is the complement of what the logic symbol would normally indicate. For example, the circle on the output of an OR gate logic symbol indicates the complement of an OR function and makes a NOR symbol. Inversion circles may also occur on inputs, indicating that the circuit provides the functions indicated by this symbol if the inputs are complemented. For example, an OR gate logic symbol with inversion circles on each input is used as a symbol for a NAND gate.

De Morgan's Theorem: The complement of an AND function equals the OR function of the complements.

$$\overline{A \cdot B \cdot C} = \overline{A} + \overline{B} + \overline{C}$$

The complement of an OR function equals the AND function of the complements.

$$\overline{A + B + C} = \overline{A} \cdot \overline{B} \cdot \overline{C}$$

In terms of logic gates, this theorem indicates that an OR symbol with the output inverted is the identity of an AND symbol with the inputs inverted, as Figure 7-36 shows, and also that an AND symbol with the output inverted is the identity of an OR symbol with the inputs inverted, as Figure 7-38 shows.

Fan-Out. The number of inputs that an output can drive and still maintain its correct 1 and 0 logic levels is called "fan-out." (The output of a logic gate is often connected to the inputs of several other logic gates.)

Questions

1. Draw the logic symbol and truth table of a three-input NOR gate.
2. In Figure 7-7, the output of the NOR gate is at logic 1. Which of the four transistors are turned ON?
3. Draw that portion of the TTL schematic of Figure 7-4 that forms the actual NOR function.
4. Draw a truth table for the logic circuit of Figure 7-9.
5. Draw the logic symbol and truth table of a three-input NAND gate.
6. Describe briefly the differences between the schematics of Figures 7-7 and 7-17.
7. Of the four TTL logic gates described thus far, why is the NAND gate likely to be preferred?
8. Draw the truth table of the logic circuit of Figure 7-19.
9. Apply the waveforms A and B in Figure 7-21 to a two-input NOR gate and draw a timing diagram of all three waveforms (including the output).
10. Apply the waveforms B and C in Figure 7-22 to a two-input NAND gate and draw a timing diagram of all three waveforms (including the output).
11. What single condition produces a 1 level out of a NOR gate?
12. What single condition produces a 0 level out of a NAND gate?
13. Which level must be applied to the control input of a NOR gate to inhibit passage of a signal?
14. Which level must be applied to the control input of a NAND gate to inhibit passage of a signal?
15. A 0 applied to any input of a NAND gate produces what output level?
16. A 1 applied to any input of a NOR gate produces what output level?
17. Select the correct word or words: To enable a signal through a two-input NOR gate, a (0 level 1 level) must be applied to the control input. To avoid inversion and distortion of the signal, an inverter should be used at the signal (input output).
18. Select the correct word or words: To enable a signal through a two-input NAND gate, a (0 level 1 level) must be applied to the control input. To avoid inversion and distortion of the signal, an inverter should be used at the signal (input output).
19. Draw the two logic symbols for a two-input NOR gate. Explain their relationship by De Morgan's theorem.
20. Draw the two logic symbols for a two-input NAND gate. Explain their relationship by De Morgan's theorem.

Problems

7-1 Draw the circuit described by the Boolean function

$$X = \overline{(A \cdot B) + (C \cdot D)}$$

7-2 Write the Boolean function for the circuit in Figure 7-47.

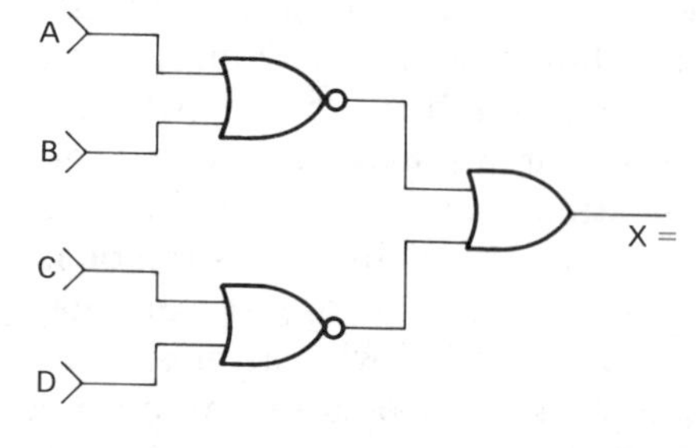

FIGURE 7-47. Problem 7-2

7-3 Write the Boolean function for the circuit in Figure 7-48.

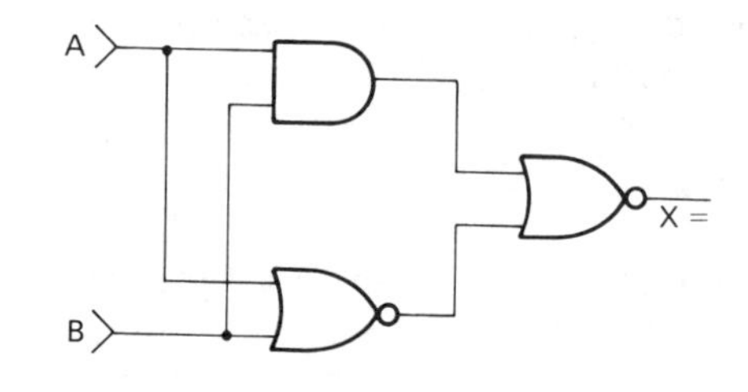

FIGURE 7-48. Problems 7-3 and 7-6

7-4 Draw the logic circuit equivalent to the Boolean function $A \cdot B + \overline{C \cdot D}$.

7-5 The 14-pin DIP integrated circuits in Figure 7-49 are the TTL AND circuit 7408 and the TTL NOR circuits of Figure 7-4. Draw the internal pin connections.

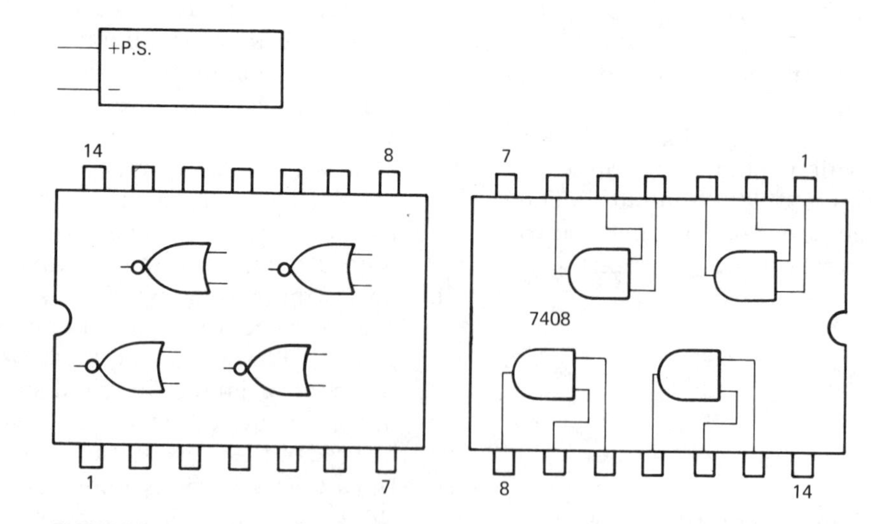

FIGURE 7-49. Problems 7-5 and 7-6

7-6 In Figure 7-49, draw in the external connections to power supply and draw the connection necessary to produce Figure 7-48.

7-7 Draw the circuit described by the Boolean function

$$X = \overline{(A + B) \cdot (C + D)}$$

7-8 Write the Boolean function of the circuit in Figure 7-50.

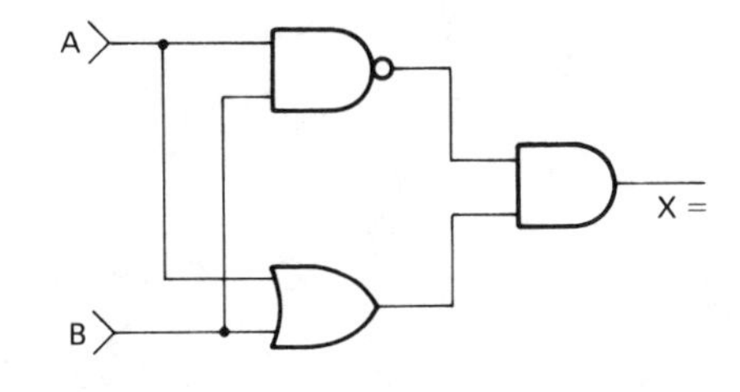

FIGURE 7-50. Problem 7-8

7-9 The 14-pin DIP integrated circuits in Figure 7-51 are the TTL OR circuits of Figure 6-16 and the TTL NAND circuits of Figure 7-14. Draw the internal pin connections.

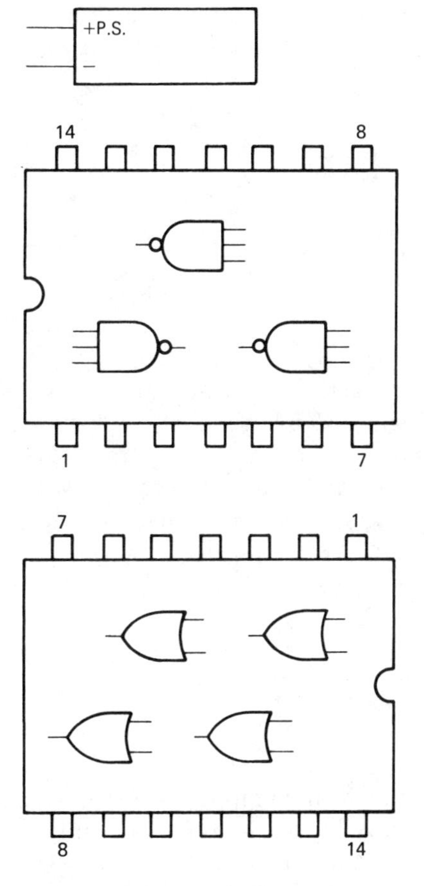

FIGURE 7-51. Problems 7-9 and 7-10

7-10 In Figure 7-51, draw the connections to power supply and draw in the connections needed to produce the circuit of Figure 7-19.

7-11 Connect the inputs of the gates in Figure 7-52 to the shift counter outputs of Figure 7-23 needed to provide the indicated outputs.

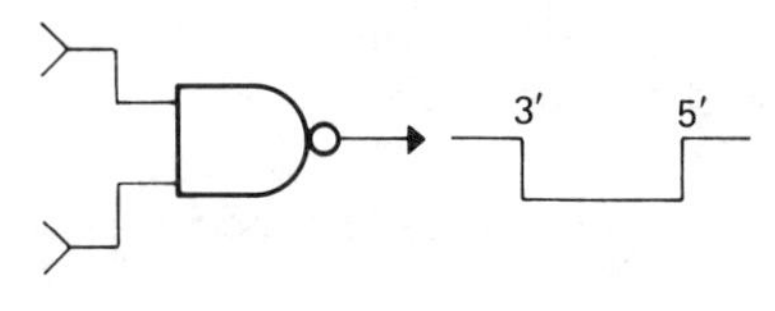

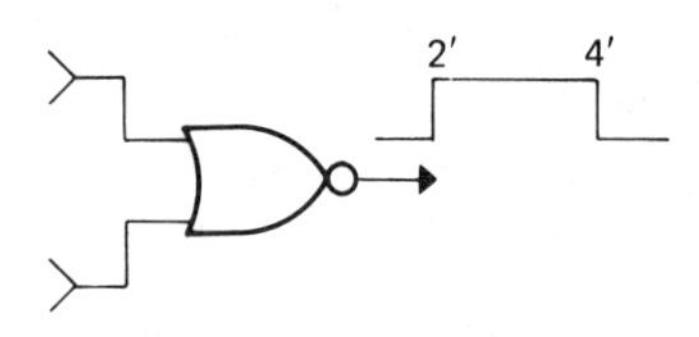

FIGURE 7-52. Problem 7-11

7-12 Determine the output waveform of the gates shown in Figure 7-53. The inputs are connected to the shift counter of Figure 7-23 as indicated by the letters.

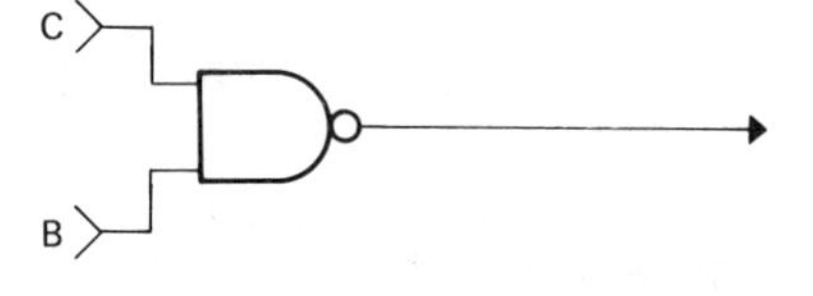

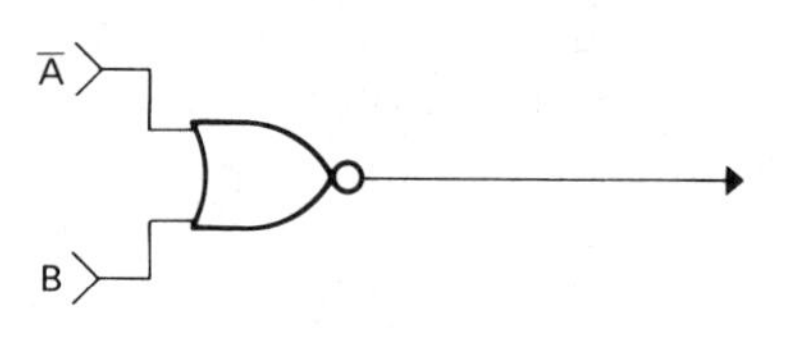

FIGURE 7-53. Problem 7-12

7-13 Connect the three-input gates of Figure 7-54 to the clock-and-gate generator of Figure 7-23 so that the indicated pulse will occur on the outputs. Use inverters where needed.

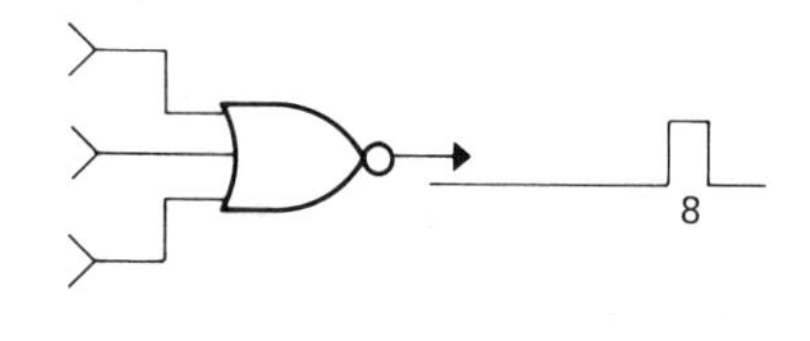

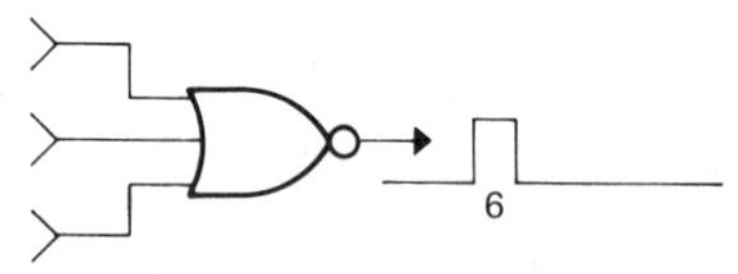

FIGURE 7-54. Problem 7-13

7-14 Draw a three-input NAND gate connected to the clock-and-gate generator of Figure 7-23 so that only clock pulse 5 will occur on the output. Use inverters where needed. Draw a second NAND gate connected so that only clock pulse 9 occurs on the output.

7-15 Prove that the circuits of Figure 7-55 are Boolean identities.

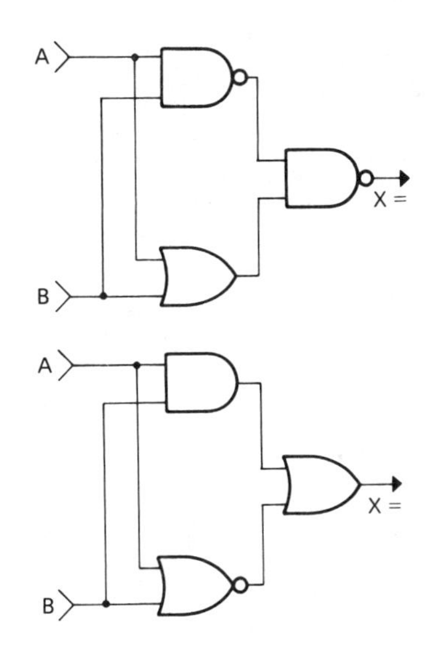

FIGURE 7-55. Problem 7-15

7-16 Convert the circuit of Figure 7-56 to an identical function using all NAND gates.

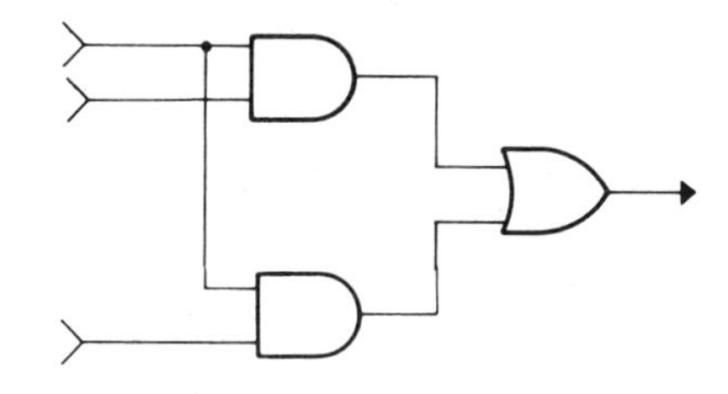

FIGURE 7-56. Problems 7-16 and 7-17

7-17 Convert the circuit of Figure 7-56 to an identical function using all NOR gates (the complements of A and B are available).

7-18 Using all NOR gates, draw the logic diagram of a BCD-to-decimal converter. (See Section 6.9.)

7-19 Using all NAND gates, draw the logic diagram of a decimal-to-BCD converter. (See Section 6.8.)

Note: For Problems 7-20 through 7-23, use the pulses and waveforms generated by the clock-and-gate generator of Figure 7-23.

7-20 Connect the two-input gates shown in Figure 7-57 to the shift counter of Figure 7-23 to provide the indicated output signals.

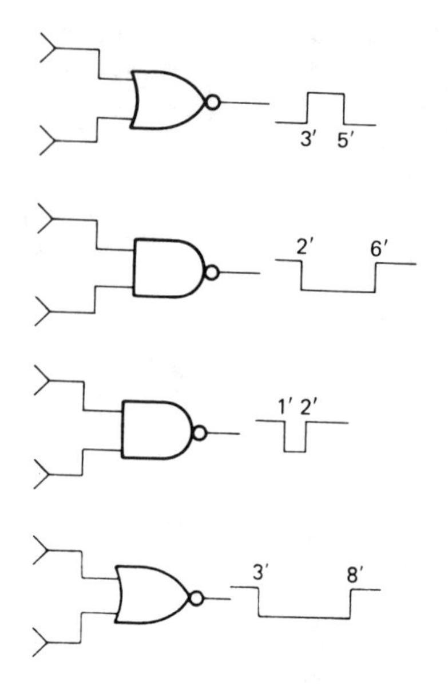

FIGURE 7-57. Problem 7-20

7-21 The gates shown in Figure 7-58 are connected to the shift counter outputs as labeled. Draw the resulting output waveforms.

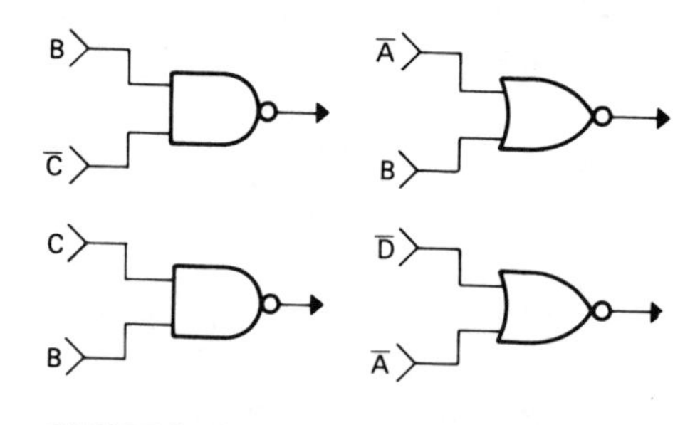

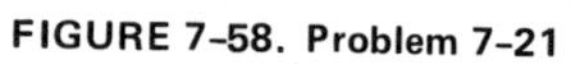
FIGURE 7-58. Problem 7-21

7-22 Connect the three-input gates of Figure 7-59 to shift-counter-and-clock outputs so that the signal pulses shown will occur at the outputs. Connect inverters to the inputs or outputs as needed.

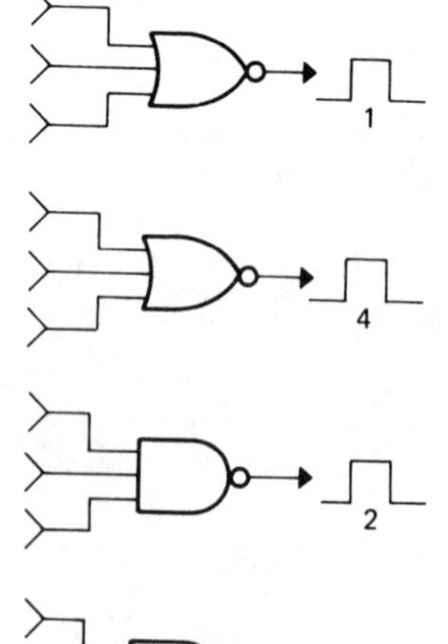

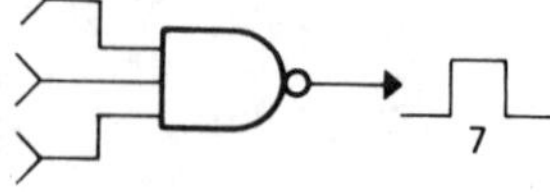

FIGURE 7-59. Problem 7-22

7-23 Connect the gate inputs of Figure 7-60 to clock-and-shift-counter outputs of Figure 7-23 to provide the indicated outputs from the NOR gates.

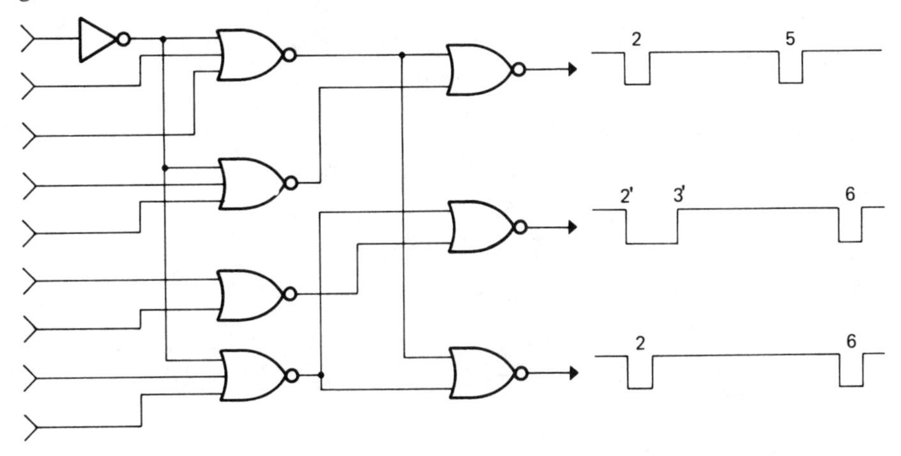

FIGURE 7-60. Problem 7-23

Boolean Algebra (Problems 7-24 through 7-26)

7-24 Give the Boolean equation for the circuits of Figure 7-61. Draw the truth tables.

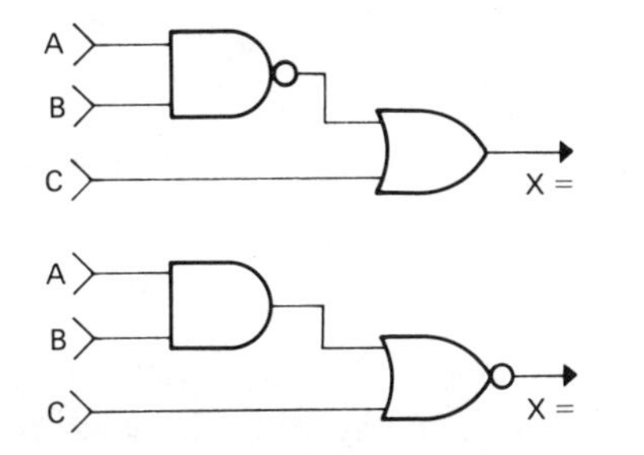

FIGURE 7-61. Problem 7-24

7-25 Give the Boolean equation for the circuit of Figure 7-62. Draw the truth table.

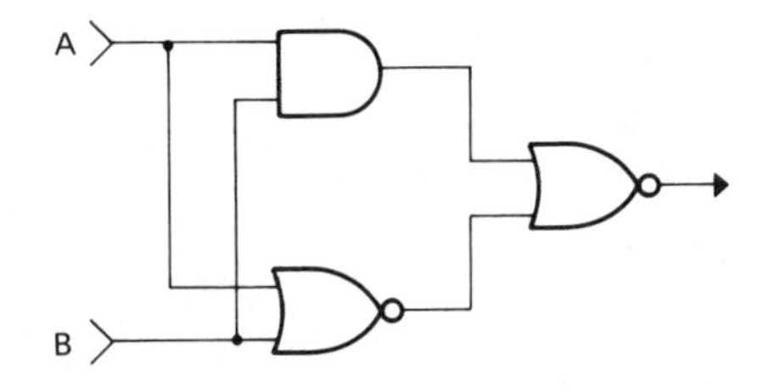

FIGURE 7-62. Problems 7-25 and 7-26

7-26 Prove that the circuit of Figure 7-62 produces identical outputs to those of Figure 7-63.

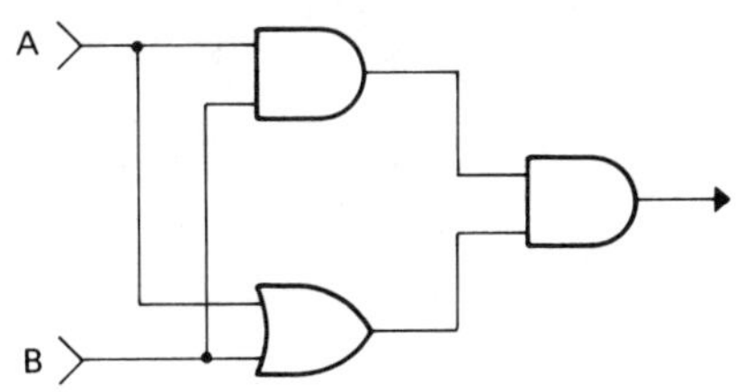

FIGURE 7-63. Problem 7-26

7-27 Convert the circuit of Figure 7-64 to one using all NAND gates.

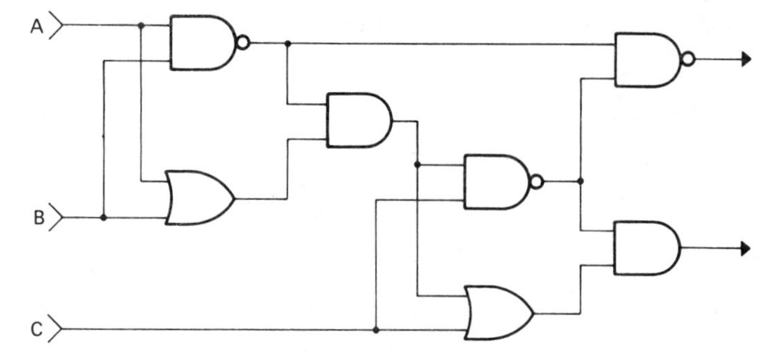

FIGURE 7-64. Problems 7-27, 7-29, and 7-31

7-28 Convert the circuit of Figure 7-65 to one using all NOR gates.

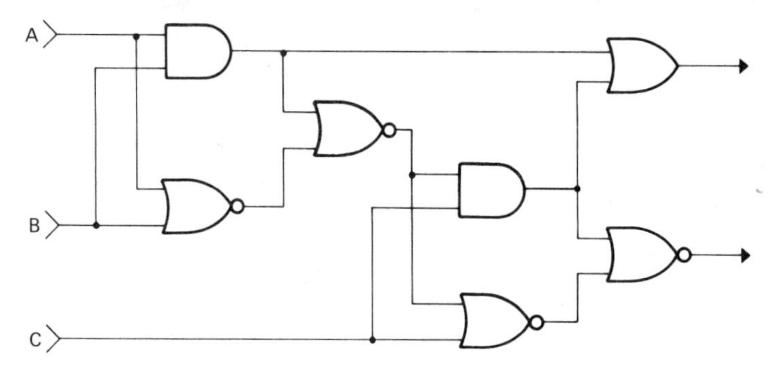

FIGURE 7-65. Problems 7-28 and 7-33

7-29 For the circuit of Figure 7-64, the complements of A, B, and C are available in the system. Convert the circuit to use the minimum number of NAND gates only.

7-30 The circuits in Figure 7-66 are CMOS NAND gates and CMOS NOR gates, described in Figures 7-18 and 7-8. Draw the internal pin connections.

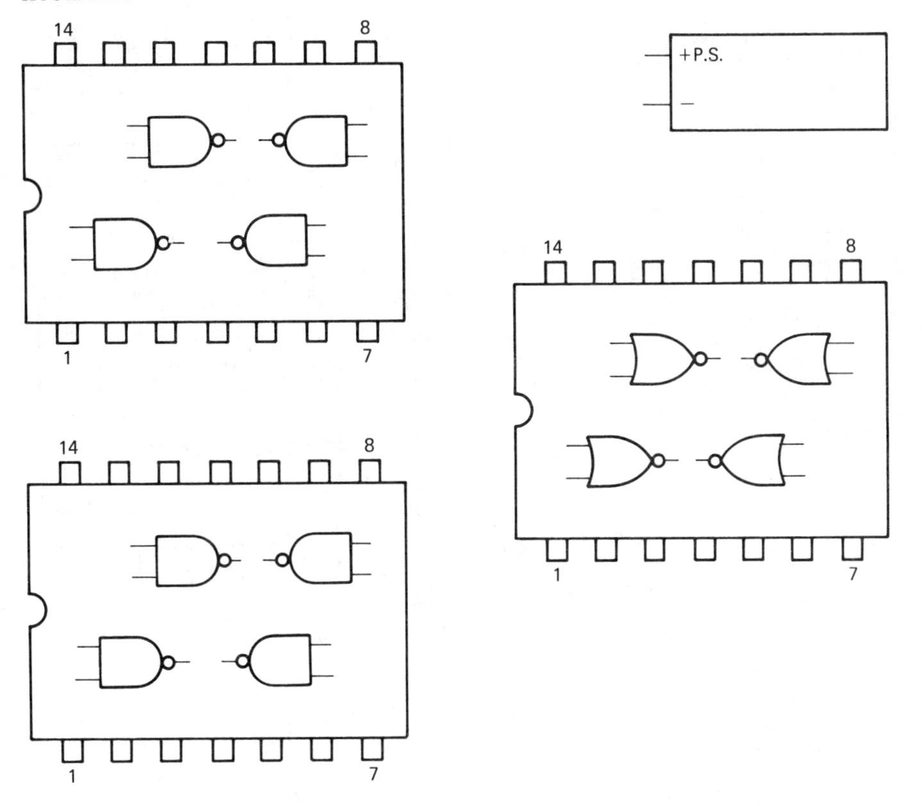

FIGURE 7-66. Problems 7-30 and 7-31

7-31 In Figure 7-66, using some of the gates as inverters, draw the external connections to power supply and the connections needed to form a circuit equivalent to Figure 7-64.

7-32 The circuits shown in Figure 7-67 are CMOS NOR gates and CMOS NAND gates, described in Figures 7-8 and 7-18. Draw the internal pin connections.

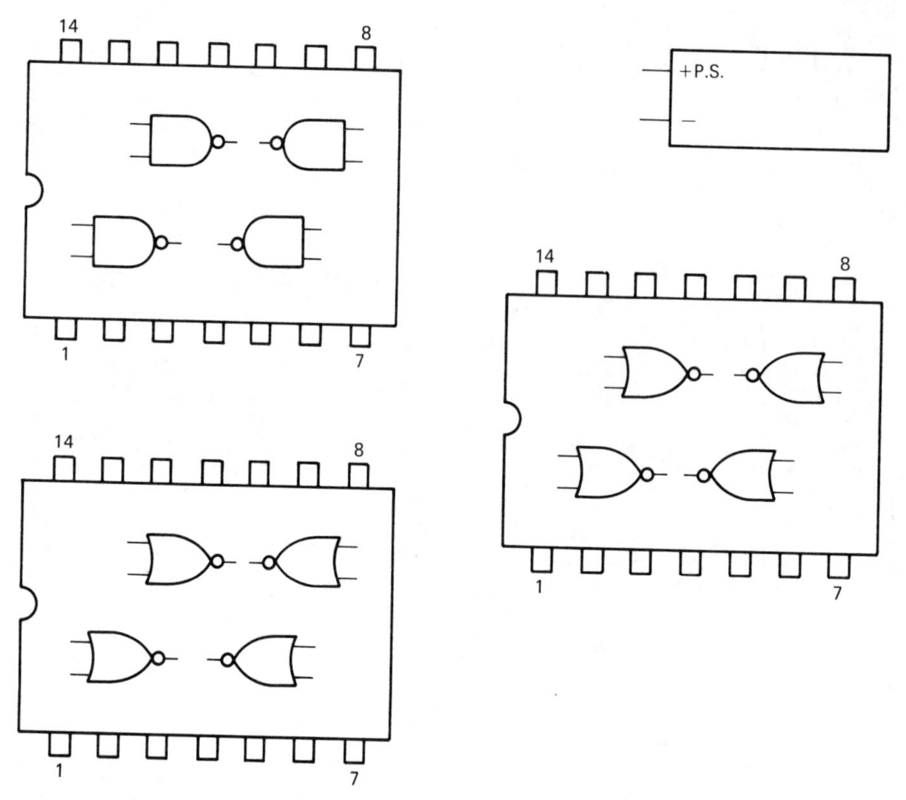

FIGURE 7-67. Problems 7-32 and 7-33

7-33 In Figure 7-67, using some of the gates as inverters, draw the external connections to power supply and the connections needed to form a circuit equivalent to Figure 7-65.

CHAPTER

Exclusive OR and Exclusive NOR Gates

Objectives

Upon completion of this chapter, you will be able to:

- Identify and use logic symbols of the exclusive OR and exclusive NOR gates.
- Draw the truth tables of the exclusive OR and exclusive NOR gates.
- Interconnect other logic gates to produce the exclusive OR gate.
- Use the AND OR invert integrated circuit to form an exclusive OR function.
- Use the exclusive OR gate to form a complementing switch.
- Determine parity of digital numbers.
- Assemble exclusive OR gates to form a parity generator or parity checker.
- Use exclusive OR gates to compare parallel digital numbers and determine if they are the same or different.
- Use exclusive OR gates to complement data.

8.1 Introduction

The logic gates discussed thus far—AND, OR, NOR, and NAND—are single-element gates. They have the common characteristic of producing an output, X (1 or 0), only when there is a coincidence of 1s (for AND and NAND) or 0s (for OR and NOR) on all inputs. Any other condition produces the output $\overline{X}$. Figure 8-1 shows these gates, which are often called coincidence gates, and their truth tables.

There are two other widely used gates—the exclusive OR and exclusive NOR gates—that result from a combination of at least three of the simple gates. They are two-input gates, and their output level depends on whether those inputs are the same or different. Figure 8-2 shows the exclusive OR and exclusive NOR and their truth tables. As we can see from the truth table, an exclusive OR with the output inverted becomes an exclusive NOR, and vice versa. There are many circumstances in logic systems where these gates are needed. The exclusive OR, however, is the more widely used of the two.

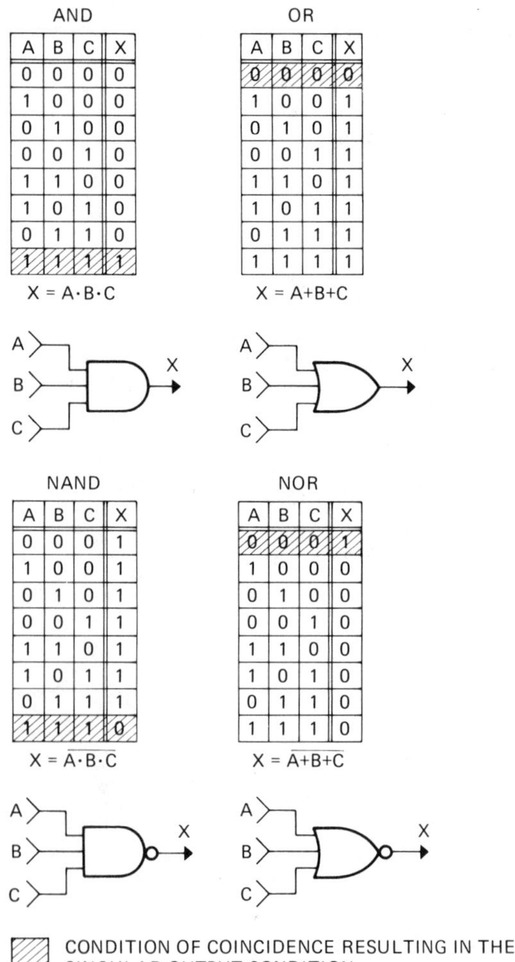

AND

A	B	C	X
0	0	0	0
1	0	0	0
0	1	0	0
0	0	1	0
1	1	0	0
1	0	1	0
0	1	1	0
1	1	1	1

$X = A \cdot B \cdot C$

OR

A	B	C	X
0	0	0	0
1	0	0	1
0	1	0	1
0	0	1	1
1	1	0	1
1	0	1	1
0	1	1	1
1	1	1	1

$X = A+B+C$

NAND

A	B	C	X
0	0	0	1
1	0	0	1
0	1	0	1
0	0	1	1
1	1	0	1
1	0	1	1
0	1	1	1
1	1	1	0

$X = \overline{A \cdot B \cdot C}$

NOR

A	B	C	X
0	0	0	1
1	0	0	0
0	1	0	0
0	0	1	0
1	1	0	0
1	0	1	0
0	1	1	0
1	1	1	0

$X = \overline{A+B+C}$

FIGURE 8-1

Four basic logic gates and their truth tables.

8.2 The Exclusive NOR Gate

The exclusive NOR gate exists in the wiring of many homes—for example, the simple two-way light switch that makes it possible to turn a hallway light on and off from a switch upstairs and downstairs. Figure 8-3 shows a two-way switch. From either switch position, the light can be turned on or off. The light is on when the switches are both up (1 state) or both down (0 state). The light is off when one switch is up and the other is down. These switches have the truth table shown in Figure 8-4. The same function

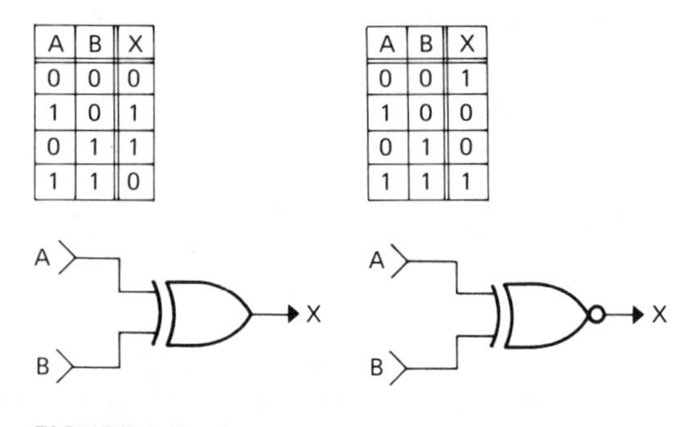

A	B	X
0	0	0
1	0	1
0	1	1
1	1	0

A	B	X
0	0	1
1	0	0
0	1	0
1	1	1

FIGURE 8-2

Exclusive OR and exclusive NOR gates and their truth tables.

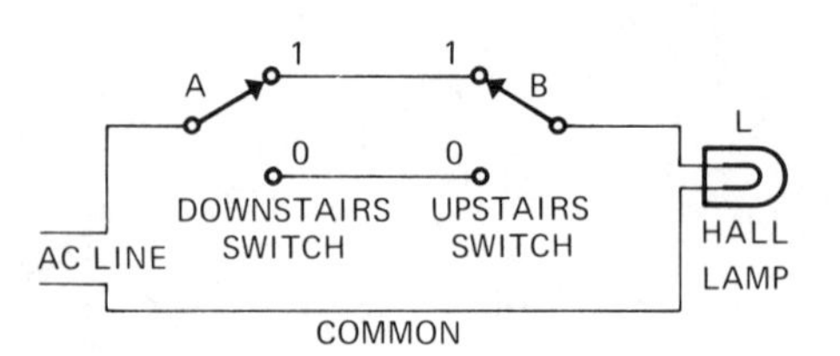

A	B	L
UP	UP	ON
DOWN	UP	OFF
UP	DOWN	OFF
DOWN	DOWN	ON

FIGURE 8-3

Two-way light switch using exclusive NOR function.

can be accomplished with the three logic gates of Figure 8-4. Simply stated: 1 occurs at the output when the inputs are both 1 or both 0. A 0 occurs at the output only when the inputs are different.

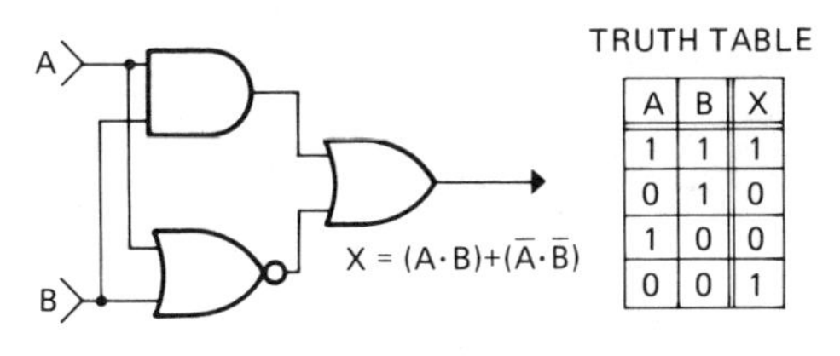

A	B	X
1	1	1
0	1	0
1	0	0
0	0	1

FIGURE 8-4

Exclusive NOR function made from AND, OR, and NOR gates.

8.3 The Exclusive OR Gate

If the two wires connecting the two-way switch are crossed over, as Figure 8-5 shows, the switch still works, but the light goes on only when one switch is up while the other is down. This function, called the exclusive OR gate, has many useful applications in digital logic. Figure 8-6 shows its symbol and truth table. The exclusive OR has only two inputs, and it produces a 1 on the output if these are different, a 0 if they are the same. The Boolean equation shown here is only one of numerous identities that describe this function, and each identity points to a different method of producing the function.

There is no simple combination of diodes and transistors that will result in an exclusive OR gate, but it can be produced by connecting three or more dual-input gates of the four types we have already discussed. The AND-and-two-NOR connection in Figure 8-7 is one way to produce the exclusive OR. It is a favored method because it produces certain economies when used in arithmetic circuits.

An exclusive OR gate may also be produced by an AND, OR, and NAND connection, as in Figure 8-8. If all inverting gates are used, conversions of the noninverting gates will produce the exclusive OR functions; but if the complements of A and B ($\bar{A}$ and $\bar{B}$) are already available, exclusive OR can be made as in Figure 8-9.

These varied methods of producing an exclusive OR seem to produce different Boolean equations. Yet they all agree with the truth table. The outputs can all be proven identities by the theorems and postulates of Boolean algebra.

If the complements of A and B are available within the system, the integrated circuit known as the AND OR INVERT gate can be used.

This circuit is available in several integrated circuit families. Figure 8-10 shows the basic AND OR INVERT circuit connected for an exclusive OR function.

In TTL circuits, the exclusive OR need not be constructed from other SSI gates but may be

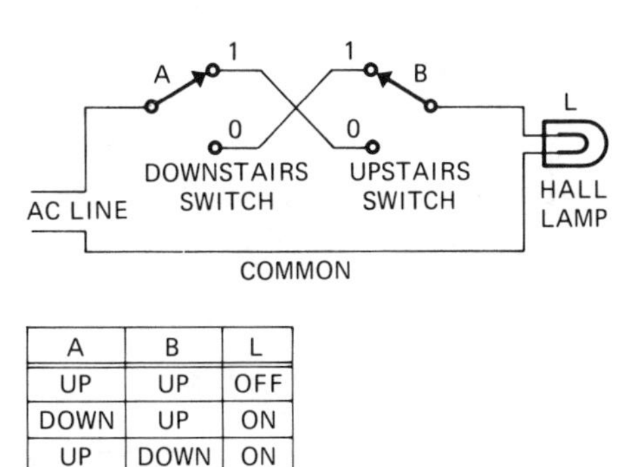

A	B	L
UP	UP	OFF
DOWN	UP	ON
UP	DOWN	ON
DOWN	DOWN	OFF

FIGURE 8-5

Two-way light switch using exclusive OR function.

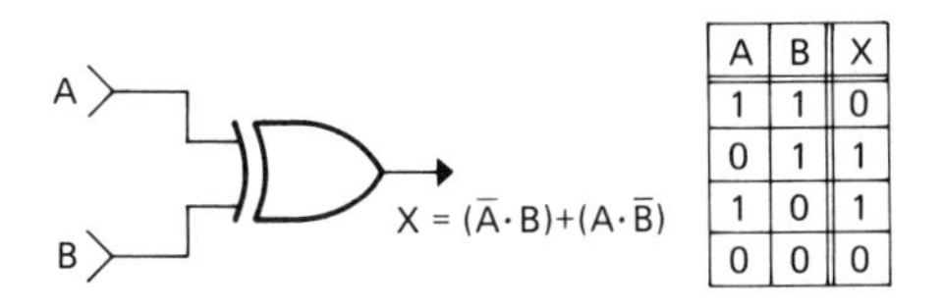

A	B	X
1	1	0
0	1	1
1	0	1
0	0	0

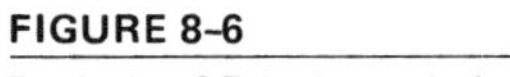

FIGURE 8-6

Exclusive OR logic symbol and its truth table.

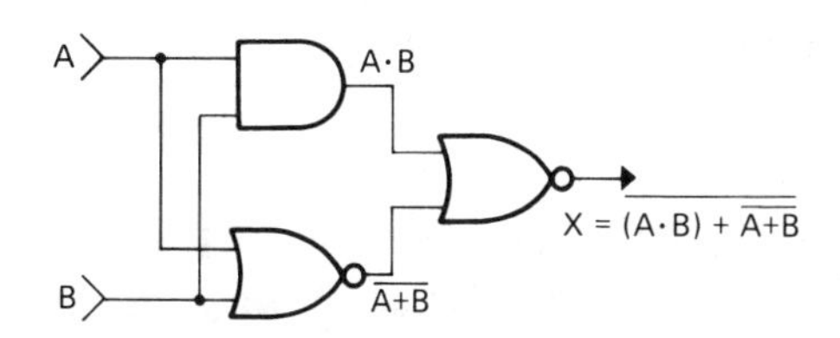

FIGURE 8-7

Exclusive OR function made from one AND and two NOR gates.

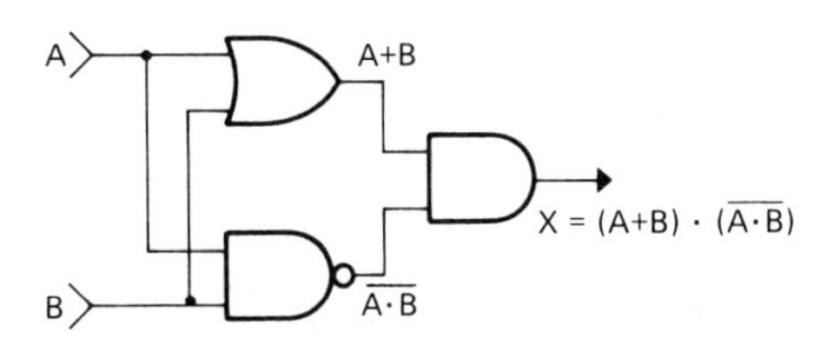

FIGURE 8-8

Exclusive OR function made from NAND, OR, and AND gates.

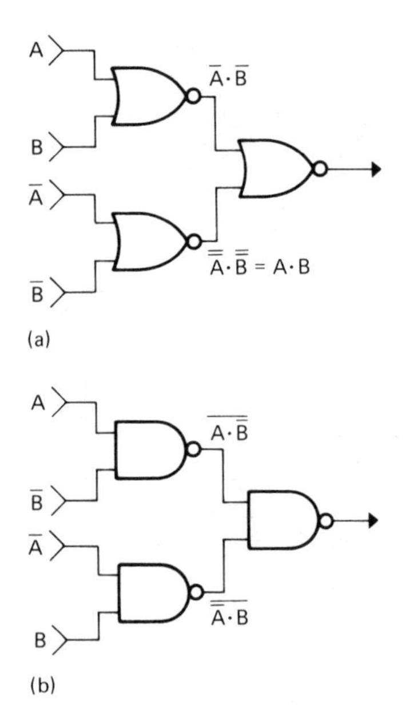

FIGURE 8-9

(a) Exclusive OR function made from NOR gates. (b) Exclusive OR function made from NAND gates. Exclusive OR functions made with inverting gates require complementary inputs.

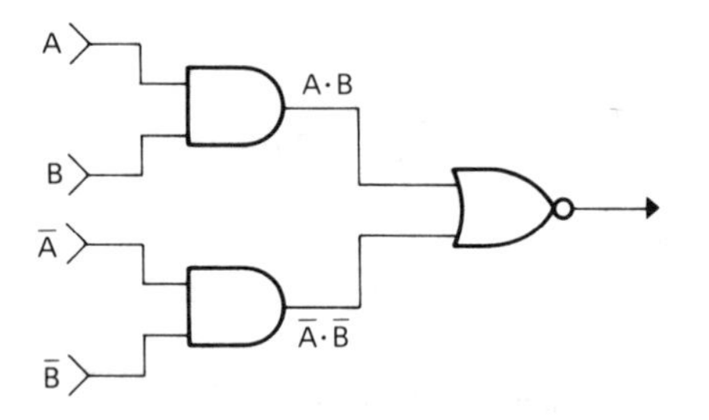

FIGURE 8-10

AND OR INVERT gate, a low-cost integrated circuit that can be used where complementary inputs are available.

obtained directly as a circuit. Figure 8-11 shows the Fairchild 7486 quad exclusive OR gate. It has the totem pole output circuit (see Figure 6-10) typical of TTL gates. A 0 level is needed on the base of Q_9 to have a binary 1 output at Y. This can occur only if one of the transistors Q_7 or Q_8 is saturated and the emitter of one transistor is shorted to ground through a turned-on transistor Q_3 or Q_6, while its base lead simultaneously receives a high-level voltage through D_2 R_5, or D_1 R_2. This condition can occur only if input A or B is high while the other is low. If inputs A and B are both 1, the emitters of both Q_7 and Q_8 will be grounded through transistors, but so will the base leads, resulting in a turnoff of both Q_7 and Q_8. If inputs A and B are both 0, the base leads of both Q_7 and Q_8 will be high, but neither will turn on because their emitters will be high also.

If input A or B has a binary 1 while the other has a binary 0 level, then either Q_7 or Q_8 will receive a turn-on condition—a high level on its base lead simultaneously with an emitter short to ground through Q_3 or Q_6. This results in a binary 1 at the totem pole output.

The exclusive OR is available also in CMOS. Figure 8-12 shows the Solid State Scientific, Inc., SCL 4030A quad exclusive OR gate.

8.4 Complementing Switch

In Chapters 6 and 7, we discussed the operation of the simple gates, in which one input is a signal and the other a control that enables or inhibits passage of the signal through the gate. If we try this same operation with the exclusive OR gate, the results are surprising. Figure 8-13 shows the exclusive OR gate with its truth table. Connecting the signal to input A and allowing it to vary while input B remains at 0 results in the top half of the truth table of Figure 8-13. The truth table of Figure 8-14 and the Boolean identities indicate that the signal will pass through the gate unchanged: X = A. On the other hand, if A is allowed to vary, while input B remains in the 1 state, as in Figure 8-15, the signal passes through the gates but is inverted: $X = \overline{A}$. The exclusive NOR can be used in this same fashion, except that a 0 on the control input produces $X = \overline{A}$. A 1 on the control input produces X = A.

8.5 Analysis by Timing Diagram

The methods of analyzing the exclusive OR operation thus far are of little value if both inputs are subject to frequent variations. If both inputs are likely to vary frequently, the timing diagram is a practical means of plotting the exclusive OR gate output. Let us use the example of a signal on input A varying from 1 to 0 and 0 back to 1 each microsecond, the signal on B having the

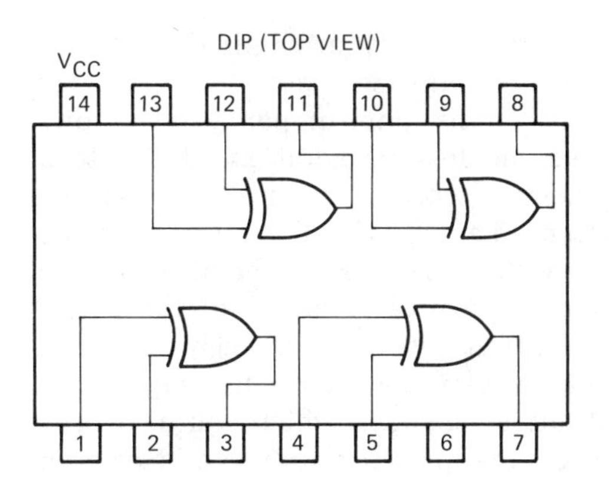

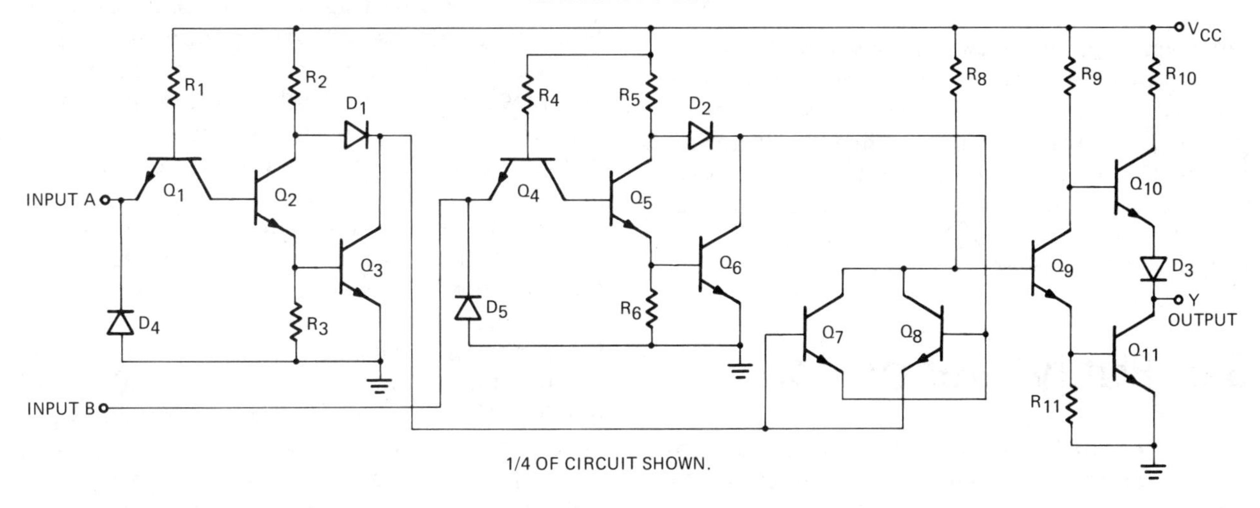

FIGURE 8–11

Fairchild quad exclusive OR gate: integrated circuit TTL/SSI 7486.

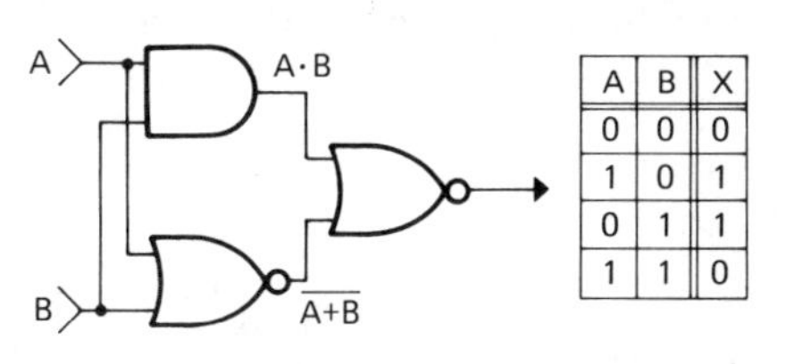

A	B	X
0	0	0
1	0	1
0	1	1
1	1	0

FIGURE 8–13

Exclusive OR gate and its truth table.

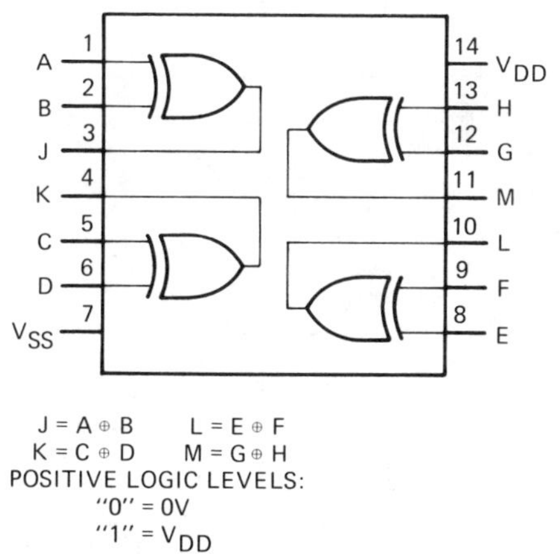

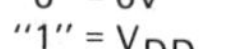

FIGURE 8–12

Solid State Scientific, Inc., quad exclusive OR gate: CMOS integrated circuit SCL4030A. It is also available as a 4070B with the same pinout.

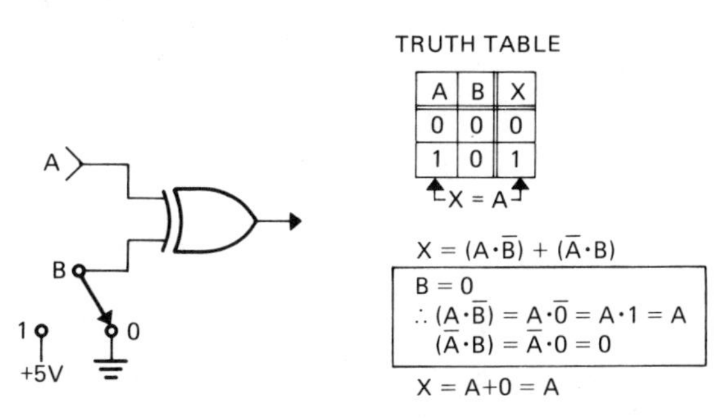

A	B	X
0	0	0
1	0	1

FIGURE 8–14

Exclusive OR gate with B input held at 0. The signal passes without changing.

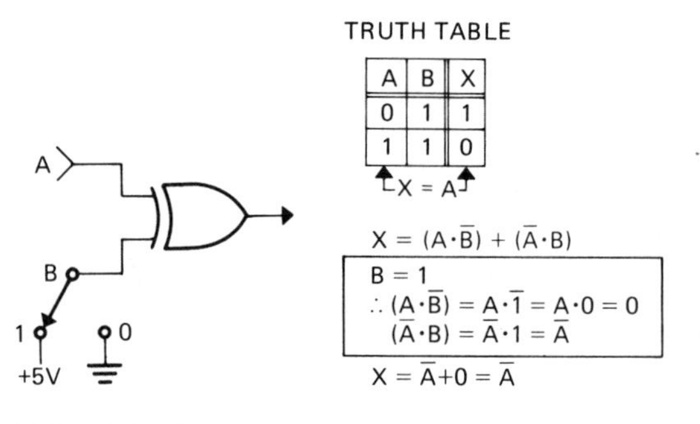

FIGURE 8-15

Exclusive OR gate with B input held at 1. The signal is inverted as it passes.

same variation each two microseconds. Figure 8-16 shows the voltage-versus-time waveform of the two input signals. The output waveform is plotted above them by finding the periods during which the inputs are different. During those periods, the output levels will be 1. During the periods when the input levels are the same, the output level will be 0.

8.6 Parity Generator

A complex, high-speed digital machine is susceptible to errors. With tens of millions of operations being performed through thousands of inches of wires and circuit lines every hour, the probability of an error due to noise or other factors is disturbingly high. Occasional errors of this type become less disturbing if they can be detected the moment they occur.

Use of a parity line, or parity bit, allows detection of the loss or addition of a 1 to a digital number or word as it travels through the digital system. The additional bit is positioned adjacent to either the LSB or the MSB. Figure 8-17(a) shows possible locations of the parity bit in parallel transmission of an eight-bit word. In a parallel system, the parity bit requires an additional line or channel, which will be 1 or 0, depending on the need to produce odd or even parity.

Figure 8-17(b) shows the possible locations of a parity bit in serial transmission of an eight-bit word. In a serial system, the parity bit occupies a time period either before the LSB or after the MSB. A pulse will occur or not occur, depending on the need to produce odd or even parity.

Parity may be designated as odd or even. In an odd parity system, the total number of bits that are (high) 1 levels must be odd. If the word bits themselves are even, then a 1 is placed on the parity line, or parity position, to make the count odd. If the count of the 1s in the word bits is odd, then the parity is kept low (or 0), to keep the total odd. For an even parity system, 1s are placed in parity line or position to make the total count even. The parity bit is generated at the transmitting or input end of a digital system and examined for loss or addition of a bit at the receiving or using end of the system.

FIGURE 8-16

Timing diagram used to plot the output waveform when inputs are frequently changing.

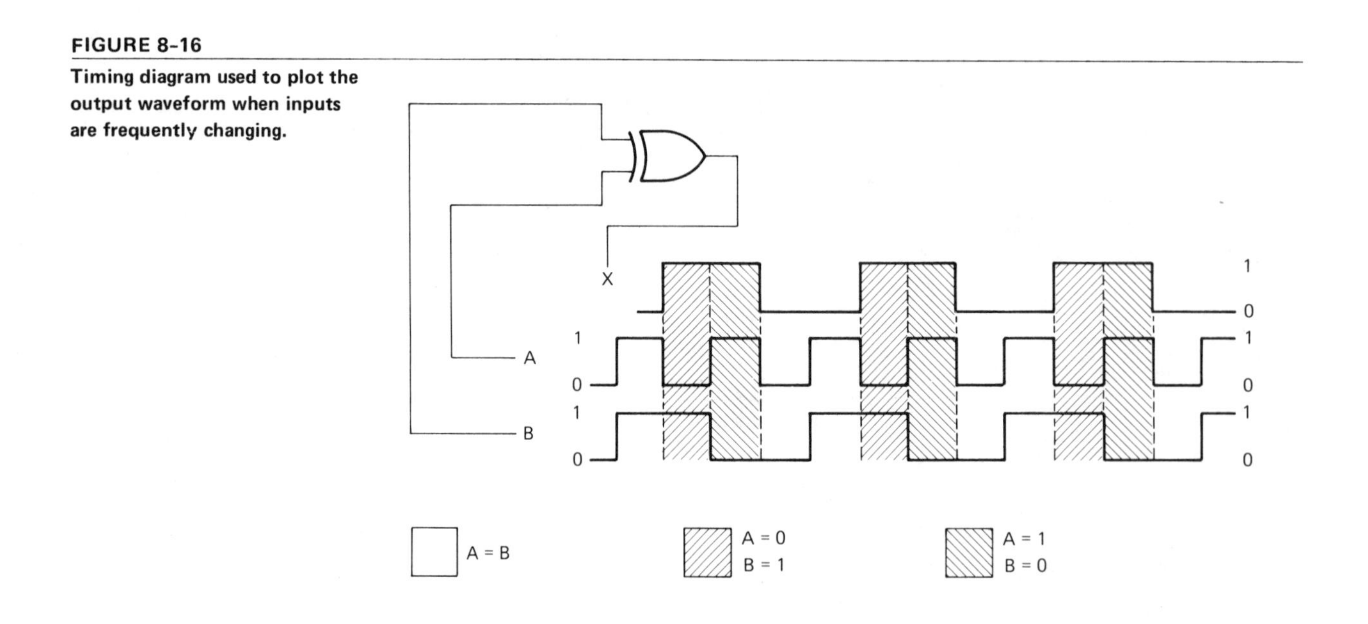

FIGURE 8-17

Parity bits used to detect the loss or addition of a 1 to a digital number or word as it travels through digital systems.

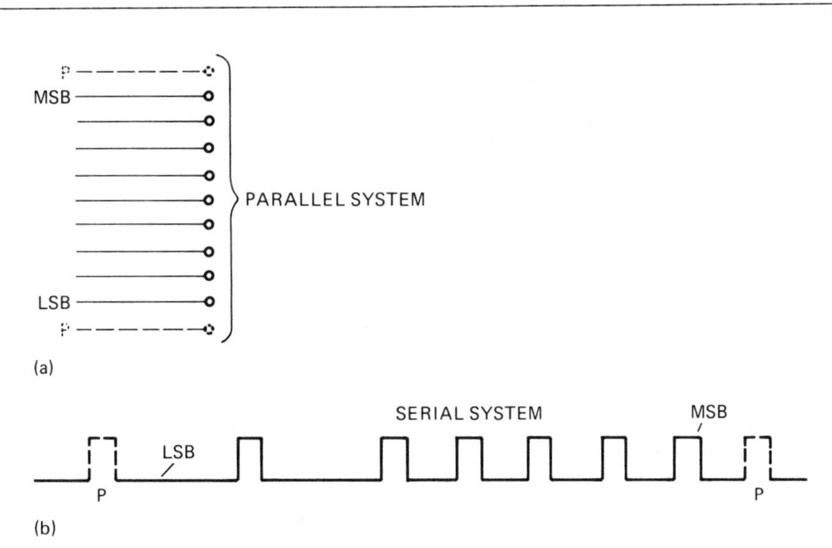

■ **EXAMPLE 8-1**

Place an odd parity bit adjacent to the MSB of the eight-bit numbers listed below:

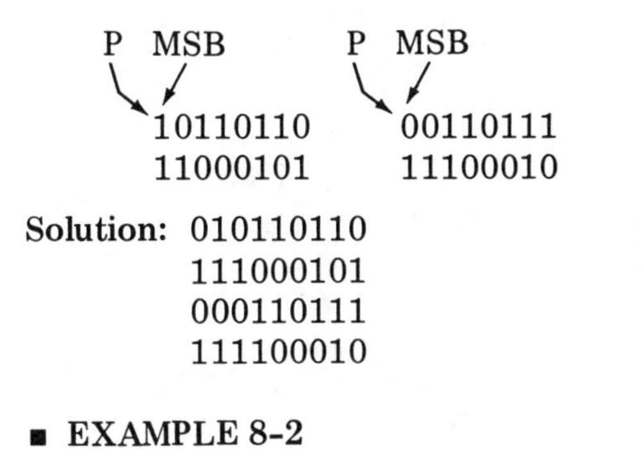

P MSB P MSB

10110110 00110111
11000101 11100010

Solution: 010110110
111000101
000110111
111100010

■ **EXAMPLE 8-2**

A tape reader is often used as an input device to a digital machine. Figure 8-18(a) is a segment of eight-level punched paper tape showing the numbers 15386 in seven-bit ASCII code without parity. The top row, which is not punched, is often used for parity. Draw in the punched holes needed for odd parity. Sprocket holes are not included in the parity count.

Solution: See Figure 8-18(b), which shows the punched tape with parity.

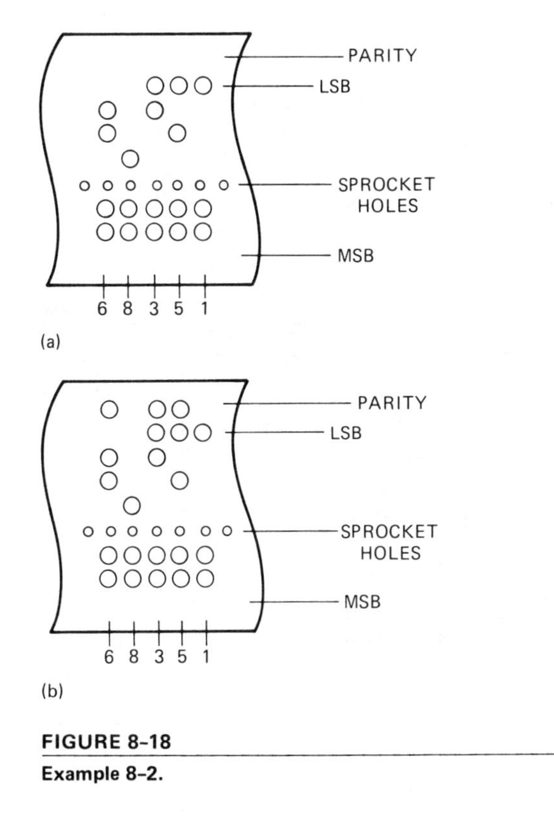

FIGURE 8-18

Example 8-2.

Figure 8-19 shows an odd parity system applied to a seven-bit parallel word. The seven lines are fed to a parity generator, which immediately decides whether the number of lines in the 1 state is odd or even. If they are odd, it keeps the parity line at 0. If they are even, it applies a 1 to the parity line, thereby making the eight lines odd in parity. After the word has been processed, and just before it is to be used, the eight lines are checked for parity. If the check still indicates an odd number of 1s, the operation, is allowed to continue. But if the check

FIGURE 8-19

Block diagram: parity check system.

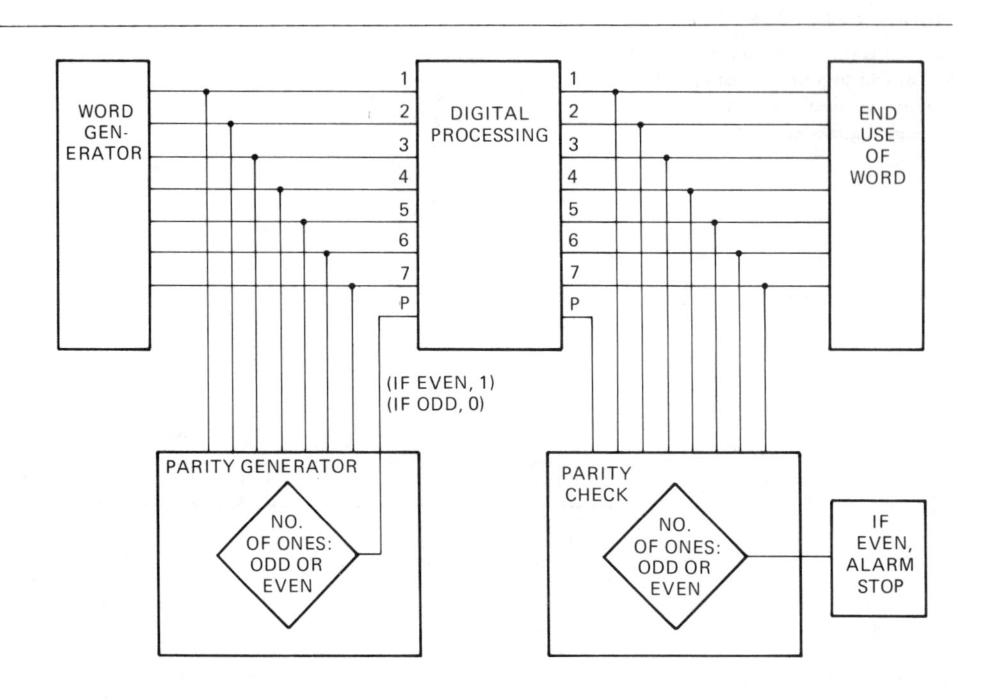

indicates even parity from loss or addition of a 1, the operation will be stopped and an alarm set off to notify the operator.

The exclusive OR gate discussed in Section 8.3 can determine whether a pair of lines is odd or even. The digital lines can be divided into pairs and each pair connected to an exclusive OR. Remembering that the sum of an odd and an even number is odd, and that the sum of two odd or two even numbers is even, one can again compare the outputs of the exclusive OR in pairs. Figure 8-20 shows a seven-bit parity generator and the resulting outputs for the number 1011011. The method shown in Figure 8-20 is known as a parity tree. Identical results can be obtained by the method shown in Figure 8-21. In both methods, the number of exclusive OR circuits is the same and an output of 1 indicates odd parity, a 0 even parity. This can be reversed by connecting an inverter at the output.

If, in the parity tree of Figure 8-20, we were to use the exclusive NOR gate in place of the exclusive OR, it would still result in an odd parity generator (a high output indicating an odd number of 1s at the inputs). It could be converted to even parity by an inverter on the output or on a single input line. Using the exclusive NOR for a serial circuit like the one in

FIGURE 8-20

Logic diagram: seven-bit parity generator (tree).

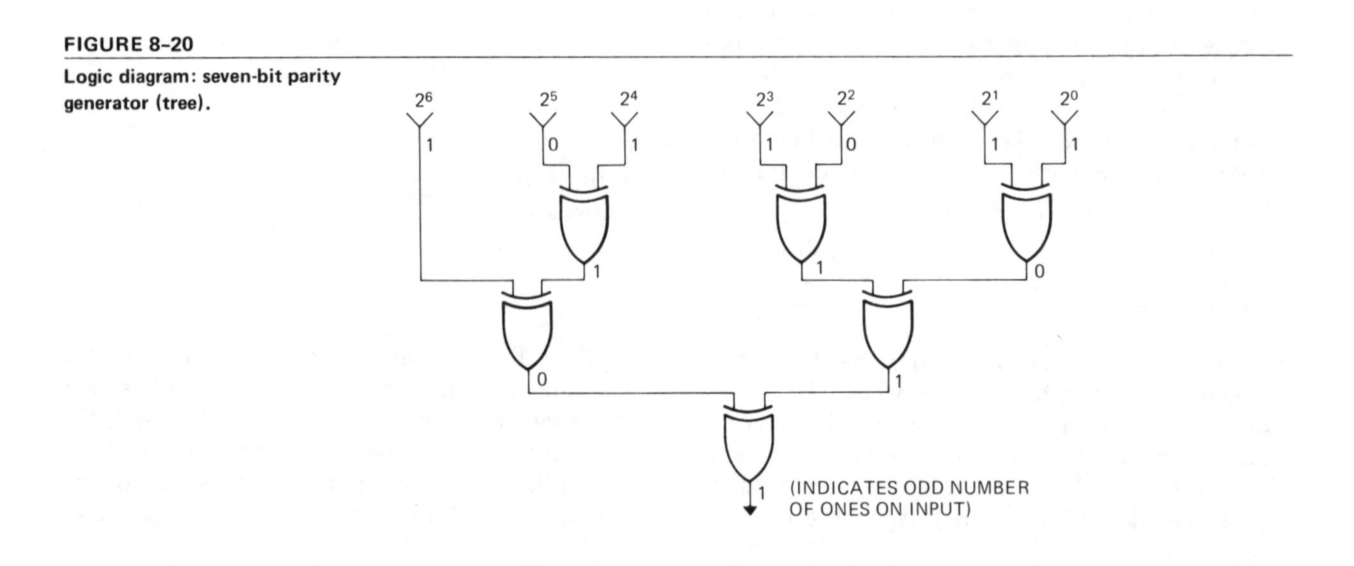

Figure 8-21 would result in an inversion from even to odd parity at the output of each gate. An odd number of gates would result in even parity. The final result could, of course, be changed with an inverter.

Figure 8-22 shows an integrated circuit parity tree. It uses both exclusive OR and exclusive NOR gates and provides complementary outputs that can be switched from odd to even parity. As the truth table indicates, a high

FIGURE 8-21

Logic diagram: seven-bit parity generator (serial).

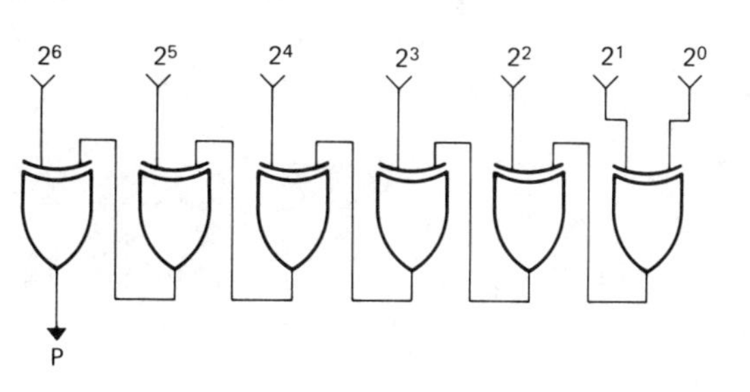

FIGURE 8-22

Fairchild TTL MSI eight-bit parity tree.

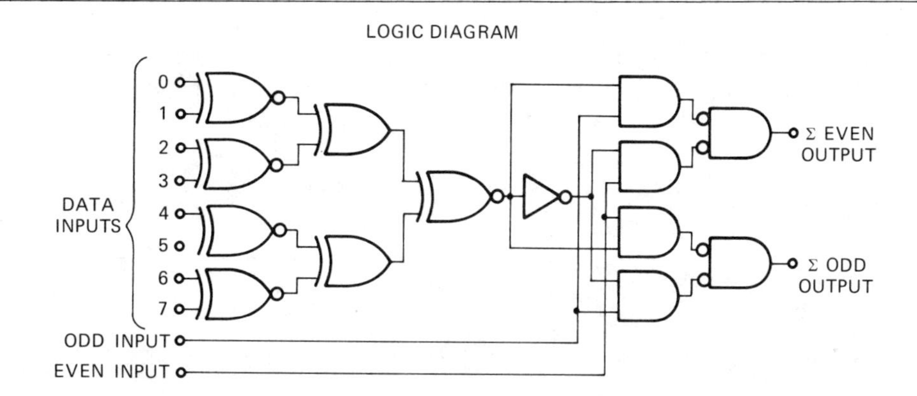

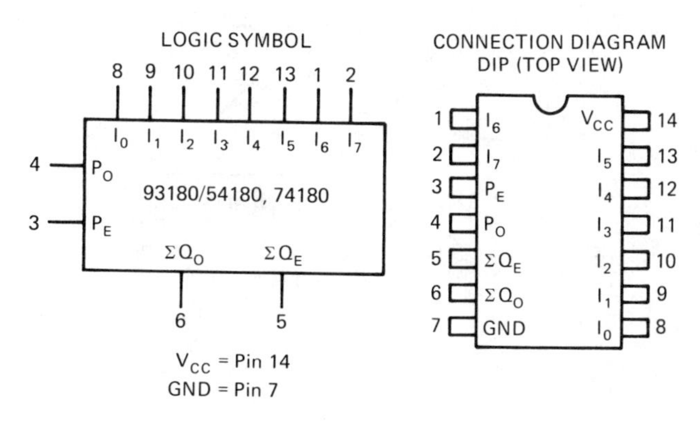

TRUTH TABLE

INPUTS			OUTPUTS	
Σ OF 1'S AT 0 THRU 7	EVEN	ODD	Σ EVEN	Σ ODD
EVEN	H	L	H	L
ODD	H	L	L	H
EVEN	L	H	L	H
ODD	L	H	H	L
X	H	H	L	L
X	L	L	H	H

X = IRRELEVANT

on the even input and a low on the odd input results in an even parity generator. A high on the odd input and a low on the even input results in an odd parity generator. The output can also be inhibited in either low or high state. The switching of the output is accomplished with a pair of AND OR invert gates like the circuit in Figure 8-10.

8.7 The Parallel Comparator

In operating, and particularly in testing digital machines, there is often a need to compare two parallel numbers to determine whether they are the same or different. A parallel comparator, shown in Figure 8-23(a), can be used to compare the numbers.

With the four-bit digital comparator shown in Figure 8-23(b), each binary-place bit of the four-bit number A is compared with the corresponding binary bit of B. If all A and B bits are the same, they will produce only zeros out of the exclusive OR gate. The OR gate will produce a zero at the output, X. If the numbers A and B differ on any bit, the OR gate will put out a 1. This can be reversed by using an inverter on the output or a NOR gate at the output in place of the OR. The number of bits compared can be increased by adding other exclusive OR gates.

8.8 Complementing Switch for Parallel Data

Another useful application of the exclusive OR gate is the complementing of digital numbers or data. As Figure 8-24 shows, parallel inputs can be passed through the gates true, or complemented,

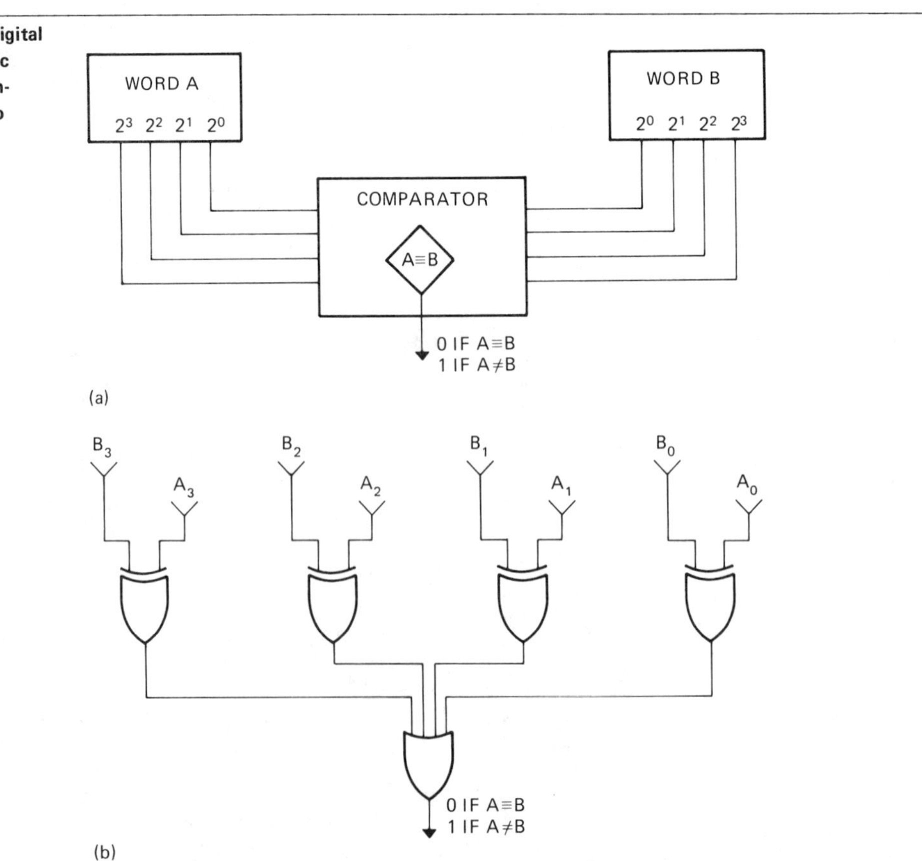

FIGURE 8-23
(a) Block diagram: parallel digital comparator system. (b) Logic diagram: four-bit digital comparator used to compare two four-bit parallel numbers, A and B.

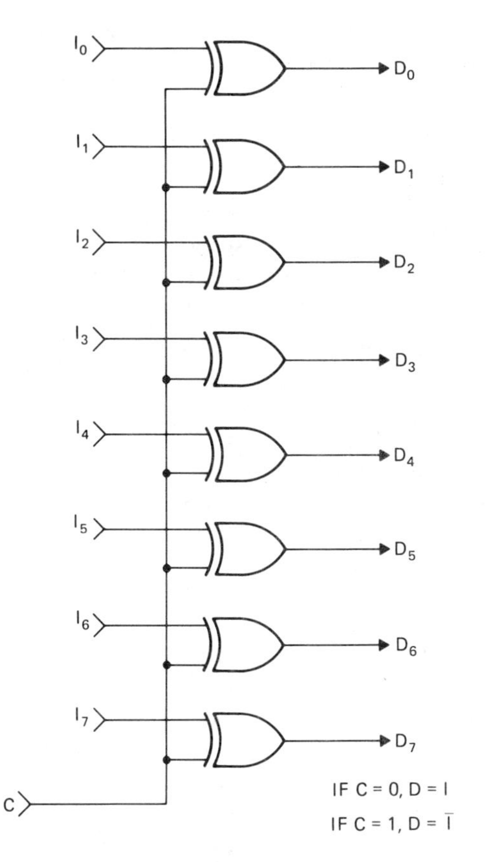

FIGURE 8-24

Parallel complementing circuit. The level on input C determines whether the input will pass through the gates in a complemented or uncomplemented state.

by changing the single control input level. This circuit is useful in ones and twos complement adders and many similar applications that require parallel signals to be complemented under some conditions and not complemented under other conditions.

8.9 Summary

The *exclusive OR gate* is a widely used logic circuit. It has only two inputs and one output. The output logic level is determined by whether the inputs are the same or different. If both inputs are 0, or if both are 1, the output will be 0. If the inputs are different, the output will be 1. Figure 8-6 shows the logic symbol and truth table of the exclusive OR gate.

A similar circuit, the *exclusive NOR gate*, produces a 1 at the output when the inputs are the same, a 0 when the inputs are different. The symbol for this gate is the exclusive OR symbol with an inversion circle at its output. The exclusive OR circuit is more widely used than the exclusive NOR. The two-way switch demonstrates a simple exclusive OR function. If we consider the up position of the switch as 1 and the down position as 0, then the switches of Figure 8-5 form an exclusive OR gate. Figure 8-5 shows a wiring method by which the hall lamp is on only when one switch is on and one off, and is out when both switches are in the same position. The advantage to this system is that we have the capacity to control the hall lamp from either the upstairs or downstairs position.

The Boolean equation of the exclusive OR can take numerous forms, all of which can be proven identities by the theorems of Boolean algebra, including De Morgan's theorem. Some of these are: $X = (\overline{A} \cdot B) + (A \cdot \overline{B}) = \overline{(A \cdot B) + (\overline{A + B})} = (A + B) \cdot (\overline{A \cdot B})$. Each of these forms of the equation points to a method of producing the exclusive OR by an assembly of the simple logic gates such as shown in Figures 8-7 through 8-10. The exclusive OR, however, is available in integrated circuit form—for example, the 7486 quad exclusive OR gate shown in Figure 8-11—or in CMOS, as shown in Figure 8-12. In the operation of a digital arithmetic unit, numbers must be complemented for some operations and not for others. This can be accomplished with the exclusive OR gate by applying the data to one input and a control level to the second input. When the control input is at 1 level, the output of the exclusive OR will be the complement of the input data. When the control input is at 0 level, the output of the exclusive OR will be the same as the data input. This is shown in Figures 8-14 and 8-15.

A complex, high-speed digital machine is susceptible to producing errors. With tens of millions of operations being performed through thousands of inches of wires and circuit lines every hour, the probability of an error being introduced by noise or other factors is disturbingly high. Sometimes errors of this type become less disturbing if they can be detected the moment they occur. Use of a parity line or *parity bit* makes it possible to detect the loss or addition of a 1 to a digital number or word as it travels through the digital system.

An eight-bit digital number has an even number of bits, but it will be odd or even parity, depending on whether the number of bits in the 1 state is odd or even. We can control the parity by adding a parity bit next to the LSB or MSB. Figure 8-17 shows typical locations for the parity bit for serial and parallel transmission of numbers. If a digital system uses odd parity, a 1 level will appear in parity only when a count of the 1 bit of the number is even, thus maintaining the odd parity at all times regardless of the number being generated or processed.

The *parity bit* is generated at the point of transmission or at the point where the number is generated. At the point of reception, the parity is checked. If an extra 1 level has been introduced into the line, or a 1 bit has been lost, the parity will have changed. The parity check circuit detecting a *parity error* would shut down the machine and notify the operator by alarm. Figure 8-19 shows a block diagram of an odd parity system.

The parity generator and parity check circuit can both be made from exclusive OR gates. Several methods of connection are used. Figure 8-20 shows six exclusive OR gates connected to form a seven-bit parity tree. A serial connection of the same number of gates produces the same results, as Figure 8-21 shows. An eight-bit parity tree is available in TTL integrated circuit, as shown in Figure 8-22. Two such circuits can be connected in serial to provide 15 bits. Each additional circuit connected will increase the capacity another seven bits.

Another application for the exclusive OR gate is the parallel *comparator.* The parallel comparator is used to determine whether two digital numbers are the same or different. Figure 8-23(a) shows a block diagram of a comparator. As shown in Figure 8-23(b), like bits of each number are connected to an exclusive OR gate. This requires one exclusive OR gate per bit. If the characters being compared are identical, the output of each gate will be 0. The outputs of all exclusive OR gates are tied to the inputs of a single OR gate. If the numbers (or characters) being compared differ, one or more of the exclusive OR gates will produce a 1, causing the output of the OR gate to be 1. The output of the OR gate will be 0 only if the two numbers are the same. Substituting a NOR gate for the OR gate will invert the output.

Figure 8-24 shows yet another application for exclusive OR gates—the parallel complementing circuit. The control input, C, allows this circuit to pass the input data in a complemented or uncomplemented form.

Glossary

Exclusive OR Gate: A two-input logic gate producing a 1 on its output when the inputs are different, and a 0 on its output when both inputs are the same. Figure 8-6 shows the logic symbol and truth table of the exclusive OR gate.

Exclusive NOR Gate: A two-input logic gate producing a 1 on its output when the inputs are the same, and a 0 on its output when the inputs are different.

AND OR Invert Gate: A connection of the outputs of two AND gates to a two-input NOR gate. Figure 8-10 shows an AND OR invert circuit used with complementary inputs to form an exclusive OR function.

Digital Word: Digital data that may represent numbers, letters, or special characters. It has a dimension expressed in bits and can be in serial or parallel form.

Odd Parity: The situation when the number of 1-level bits in a digital word *cannot* be divided into pairs without having 1 left over.

Even Parity: The situation when the number of 1-level bits in a digital word *can* be divided into pairs without having 1 left over.

Parity Bit: A bit that accompanies a digital word through its processing and which has no meaning. Used solely to give the word a parity (odd or even, depending on which parity is correct for the system).

Parity Error: A parity error exists when a digital word (including parity bit) is checked at the end of its transmission or processing and the parity does not agree with the parity of the system. In such cases, the system stops and an operator alarm is generated.

Comparator: A digital circuit that examines two digital words and determines if they are the same or different.

Complementing Circuit: A digital circuit whose output is the complement of the input. Most complementing circuits have a control switch which allows the complementing function to be enabled or disabled. When the complementing function is disabled, the complementing function is disabled, the input data are uncomplemented.

Questions

1. Draw the logic symbol and truth table of an exclusive OR gate.
2. Prove by Boolean algebra (including De Morgan's theorem) that the circuits of Figures 8-7 and 8-8 are logic identities.

$$\overline{(A \cdot B) + (\overline{A + B}})= A + B \cdot \overline{A \cdot B}$$

3. Use the alternative symbol for the bottom NOR gate in Figure 8-9(a) and prove that the output of that circuit is identical to the circuit of Figure 8-7. (See Section 7.9.)
4. Use the alternative symbol for the output NAND gate in Figure 8-9(b) (see Section 7.9) and prove that that circuit is identical to two AND gates and an OR gate.
5. Draw the logic symbol and truth table of a nonexclusive OR gate.
6. Quad exclusive OR gates in TTL and CMOS have identical 14-pin DIP packages. Are the pin connections the same for the power connections? Are they the same for the gate input and output leads?
7. If an exclusive NOR gate is used as an inverting switch, what change must be made to the control input to obtain the same inverting function as with the exclusive OR?
8. Which of the statements below are incorrect?
   (a) All odd decimal numbers converted to binary will have odd parity.
   (b) All four-, six-, and eight-bit numbers have even parity.
   (c) Eight-bit numbers can be in odd or even parity.
   (d) A digital system processes numbers in eight bits plus odd parity bit.
   (e) The quantity 0 will have a 1-level parity bit.
9. A digital system processes seven-bit numbers. How many exclusive OR gates are needed for its parity generator if odd parity is used? How many if even parity is used? How many are needed for the parity checker?
10. What is the largest number of bits available from a parity generator made by interconnecting a single quad exclusive OR gate? Draw the logic diagram.
11. What is the largest number of bits that can be obtained by interconnecting one TTL eight-bit parity tree and one quad exclusive OR gate?
12. What change or changes can be made to the circuit of Figure 8-23 to obtain a 1, if A = B and 0? If A = B?
13. Describe briefly the differences between the logic circuit of the parity generator and the comparator.
14. Assuming the complements are available for the numbers A and B, draw the logic circuit of the comparator of Figure 8-23 using the AND OR invert gate to form the exclusive OR.
15. Consider the circuit of Figure 8-24 and assume it is built using exclusive NOR gates. Complete the following table.

If C = 0, D = 10110101
If C = 1, D =

Problems

8-1 (a) For the parallel numbers in the left-hand column, install the correct odd parity bit to the left of the MSB. (b) For the parallel numbers in the right-hand column, install the correct even parity bit to the left of the MSB.

P	P
110101	111011
011010	111001
110011	100111
101010	100101

8-2 The section of paper tape in Figure 8-25 is punched with the ASCII letters RECORD without parity. Draw in the punched holes needed for even parity.

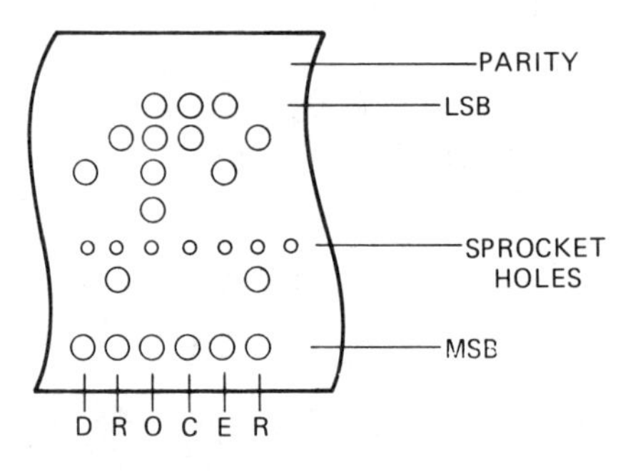

FIGURE 8-25. Problem 8-2

8-3 The circuits in Figure 8-26 are quad exclusive OR gates of Figure 8-11. Draw the internal pin connections. Draw the external connections to the power supply and the connections needed to produce the parity generator of Figure 8-20.

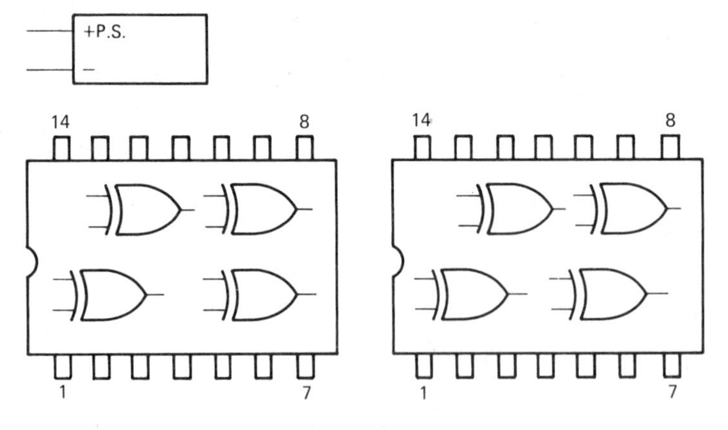

FIGURE 8-26. Problem 8-3

8-4 The circuits shown in Figure 8-27 are the SCL 4030A of Figure 8-12 and the SCL 4002A of Figure 7-8. Draw in the internal pin connections. Draw in the external connections to the power supply and the connections needed to produce the circuit of Figure 8-23.

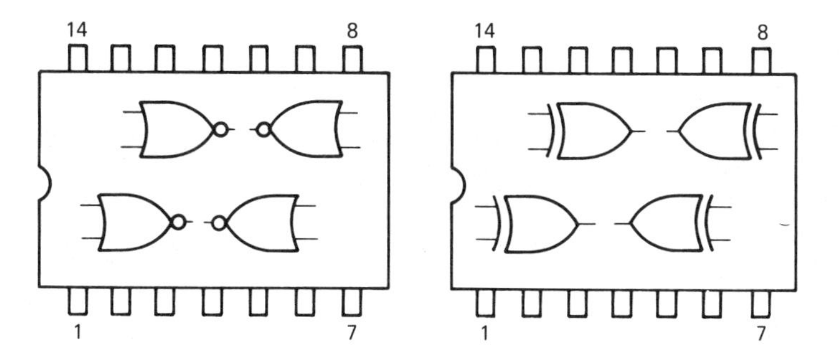

FIGURE 8-27. Problem 8-4

8-5 The numbers listed below are to be processed in binary-coded decimal plus an odd parity bit next to the MSB position. List the numbers plus parity bit in BCD form.

52	29	94
37	85	63

8-6 The binary numbers listed below are binary-coded decimal plus even parity bit next to the MSB position. Convert to decimal and circle those with parity errors.

100110110	101011000	101110011
101001001	110010010	000100101

8-7 The circuits of Figure 8-28 are TTL MSI eight-bit parity trees identical to those shown in Figure 8-22. Show the connection needed to produce a 14-bit odd parity generator.

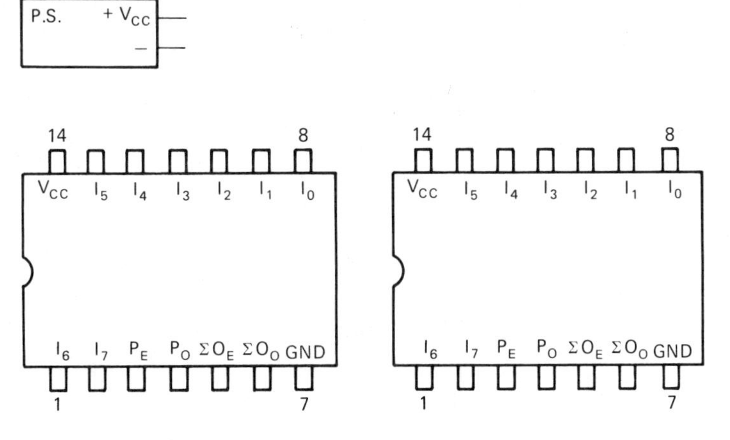

FIGURE 8-28. Problem 8-6

8-8 The binary numbers listed below are the input data to Figure 8-24. What is the output for each number when C = 0 and C = 1?

11001011	00001111
01110101	10100111

CHAPTER

Logic Gate Specifications

Objectives

Upon completion of this chapter, you will be able to:

- Determine if a logic circuit output is correctly loaded.
- Determine the minimum 1 output and maximum 0 outputs acceptable for a given logic circuit.
- Determine the minimum 1 and maximum 0 inputs acceptable for a given logic circuit.
- Compute the noise immunity levels of a logic circuit from input and output specifications.
- Measure delay time parameters and predict their effect on logic circuits.
- Compute the amount of power consumed by a logic circuit.
- Determine power supply requirements for a given number of logic gates.
- Select the ideal logic circuits for the conditions, speed, and power drain requirements of a digital system.
- Expand the fan-in of a logic gate.
- Use an open-collector circuit for wired OR connection.
- Study the advantages of CMOS devices.
- Compare the DC and AC parameters of CMOS and TTL circuits.
- Determine CMOS noise immunity, driver and fan-out impedance, crosstalk, power line and ground noise, and power requirements.
- Know the types of CMOS outputs.
- Determine the trade-offs between speed versus power for CMOS gates.
- Use precautions to protect the input circuits of CMOS gates.
- Observe the absolute maximum ratings of CMOS devices.
- Interface TTL-to-CMOS.
- Interface CMOS-to-TTL.

SCHEMATIC (each gate)

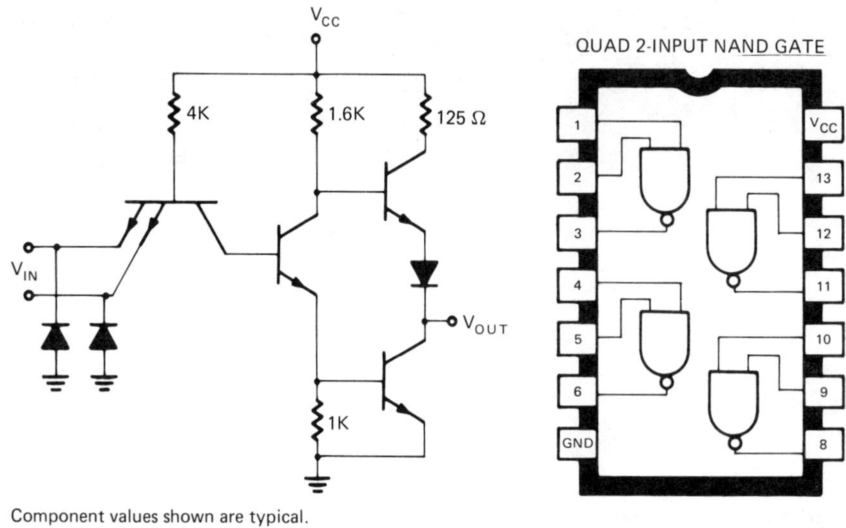

Component values shown are typical.

RECOMMENDED OPERATING CONDITIONS

		MIN.	NOM.	MAX.	UNIT
Supply Voltage (V_{CC}):	US5400	4.5	5.0	5.5	V
	US7400	4.75	5.0	5.25	V
Operating Temperature Range:	US5400	−55	25	+125	°C
	US7400	0	25	+70	°C

Fan-Out from each output (N): .... 1 to 10

ELECTRICAL CHARACTERISTICS: (over operating temperature range unless otherwise noted)

Characteristic	Symbol	Test Conditions						Limits				Notes
		Test Fig.	Temp.	V_{CC}	Driven Input	Other Input	Output	Min.	Typ.	Max.	Units	
"1" Input Voltage	$V_{in(1)}$	1A		MIN				2.0			V	
"0" Input Voltage	$V_{in(0)}$	1B		MIN						0.8	V	
"1" Output Voltage	$V_{out(1)}$	1B		MIN	0.8V	V_{CC}	−400μA	2.4	3.3		V	1
"0" Output Voltage	$V_{out(0)}$	1A		MIN	2.0V	2.0V	16mA		0.22	0.4	V	1
"0" Input Current	$I_{in(0)}$	1C		MAX	0.4V	4.5V				−1.6	mA	2
"1" Input Current	$I_{in(1)}$	1D		MAX	2.4V	0V				40	μA	2
				MAX	5.5V	0V				1	mA	
Output Short Circuit Current	I_{OS}	1E		MAX	0V	0V	0V	−20		−55	mA	3
"0" Level Supply Current	$I_{CC(0)}$	1F		NOM	5.0V	5.0V			3	4.8	mA	1, 4
"1" Level Supply Current	$I_{CC(1)}$	1G		NOM	0V	0V			1	2.6	mA	1, 4
Input Clamp Voltage	V_{clamp}	1J	NOM	MIN	−20mA				−1	−1.5	V	1, 2

SWITCHING CHARACTERISTICS: $V_{CC} = 5.0V$, $T_A = 25°C$, N = 10

Characteristic	Symbol	Test Fig.	Test Conditions	Limits				Notes
				Min.	Typ.	Max.	Units	
Turn-On Delay Time	t_{pd0}	13	C_L = 15 pF, R_L = 400Ω	3	8	12	ns	
Turn-Off Delay Time	t_{pd1}	13	C_L = 15 pF, R_L = 400Ω	5	13	20	ns	
Output Rise Time	t_r	13	C_L = 15 pF, R_L = 400Ω	3	12	18	ns	
Output Fall Time	t_f	13	C_L = 15 pF, R_L = 400Ω	1	5	8	ns	

NOTES:

1. Typical values are at V_{CC} = 5.0V, T_A = 25°C.
2. Each input tested separately.
3. Not more than one output should be shorted at one time.
4. Each gate.

FIGURE 9-1

Typical logic unit specifications for a TTL NAND gate.

9.1 Introduction

In the preceding chapters, we dealt with logic gates as if they were perfectly uniform devices delivering precise logic 1 or 0 levels at the output and responding correctly to those precise levels at their inputs. This chapter deals with logic levels and specifications for both TTL and CMOS gates. Trade-offs between speed and power for these devices are covered. CMOS circuits provide a distinct advantage in some designs, while TTL gates are a better choice in other systems. Some digital applications require both types of circuitry, and Section 9.15 explains how the two are interfaced.

Integrated circuits, being batch-fabricated, are far from uniform. Their input and output levels vary by a significant amount. This variation is, however, held within carefully controlled limits, information on which is supplied to the user in the form of a specification. The specification covers the supply voltage, 1 and 0 levels at the outputs, 1 and 0 levels needed to produce the correct results at the inputs, and loading and switching characteristics. Figure 9-1 shows typical logic unit specifications for a TTL NAND gate.

9.2 Unit Load

The first characteristic to be described for any system of logic gates is the unit load—a load representative of the typical logic unit input. An input is typically a *sink load*, or a *source load*. The sink load tends to degrade the 1-level output of the circuit that is driving it but has an insignificant effect on the 0 output level. The source load tends to degrade the 0-level output of the circuit that is driving it, pulling it upward. It has an insignificant effect on the 1 output level.

9.3 Unit Sink Load

The unit sink load is often expressed as a positive current that an input will draw when driven by a specified 1 output level of the gate that drives it. For discrete circuit logic and some older types of integrated circuit logic, the unit sink load is represented by a resistor (R_U) to ground, as in Figure 9-2(a). More exact approximations may include ideal diode and diode potentials, as Figure 9-2(b) shows. If the units are to be used at high speed, shunt capacitance of the input may be specified, as in Figure 9-2(c).

Loading calculations are greatly simplified by having the output of each type of gate, register, or other logic unit rated in the number of unit loads it can drive. This makes it possible to count, rather than compute, the loading of the output circuits. Figure 9-3 shows a typical rating of a NOR gate. The usual gate input is a single-unit load, but other logic circuits discussed in later chapters may have inputs of higher unit

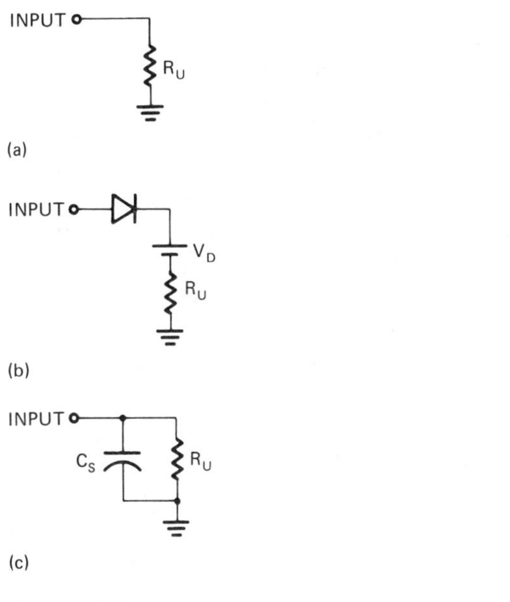

FIGURE 9-2
Schematic representations of unit sink load approximations. (a) and (b) Medium frequencies. (c) High frequency.

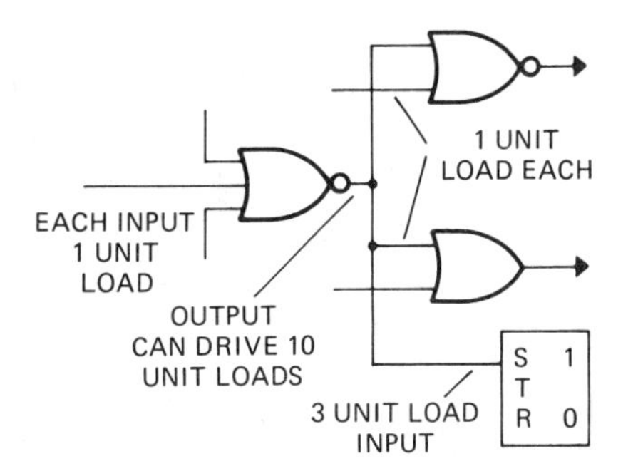

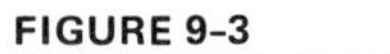

FIGURE 9-3
Typical use of unit load rating of a NOR gate.

load rating. The NOR gate in Figure 9-3 is rated for ten unit loads. We can be sure it is not overloaded by merely adding up the unit load ratings of the inputs connected to the NOR gate and comparing the total with the output rating of the gate. In this case, only five are connected, which is well within its capability.

9.4 Unit Source Load

A source load does not reduce or degrade the 1 level of the output driving it. It resembles a high-resistance generator, and its disadvantage is that it tends to bring up the 0 level of the driver. It may be represented by a resistance in series with a voltage source usually equal to V_{CC} or the 1 level, as in Figure 9-4. Many logic systems have only sink loads; for those having both sink and source loads, inputs will be rated in either sink or source load, but an output will be rated in the number of sink loads and source loads it can drive separately or simultaneously.

The input of TTL circuits is a typical source load, as Figure 9-5 shows. For the driving circuit to maintain a 0 level on the input, it must be able to receive enough current to drop the voltage (V_{CC} - V_{BE}) across the 4K base resistor. During a 1-level out, however, the input tends to aid the development of the 1 level by the driver.

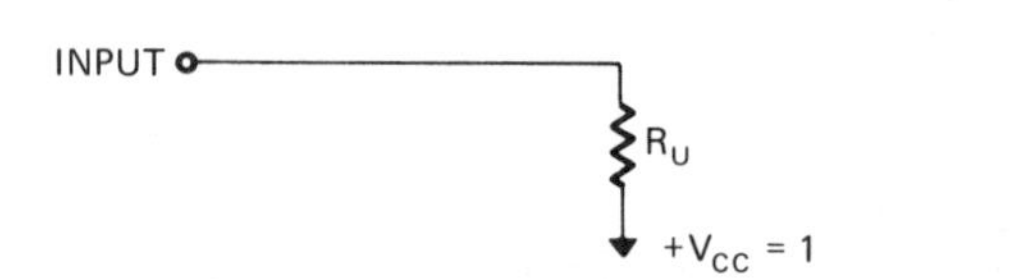

FIGURE 9-4

Schematic representation of a unit source load approximation.

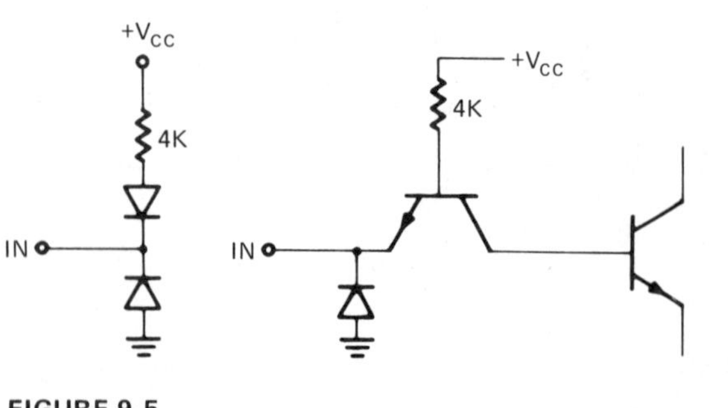

FIGURE 9-5

TTL input and its equivalent resistor-diode circuit.

■ EXAMPLE 9-1

Determine whether the gates in the following figures are sink or source loads:

Figure 6-6	Figure 6-15
Figure 7-14	Figure 7-2

Solution: Figures 6-15 and 7-2 are sink loads, as they return to ground through forward-biased junctions. Figures 6-6 and 7-14 are source loads, as they draw current from V_{CC} through forward-biased junctions.

9.5 Output-Level Parameters

The unit load concept just discussed is a general working specification that may be computed for convenience from the manufacturer's specifications or from the schematic drawing of the circuit. It is neither precise enough nor complete enough for the overall specification.

Another important aspect of the specification is the output level. If the switch in Figure 9-6(a) was opened at time t_o and closed at time t_c, the voltage waveform appearing at X would have the ideal rectangular shape and go exactly from 0V to the full V_{CC}, as shown in Figure 9-6(b). But, if we apply a number of unit loads to X as shown in Figure 9-7(a), the 0 level will still be 0V, but the 1 level will now be reduced, as in Figure 9-7(b).

The more unit loads that are attached, the lower the output 1 level will be. If the mechanical switch is replaced by a transistor, as in Figure 9-8(a), both 0 and 1 levels are degraded slightly from the ideal, as shown in Figure

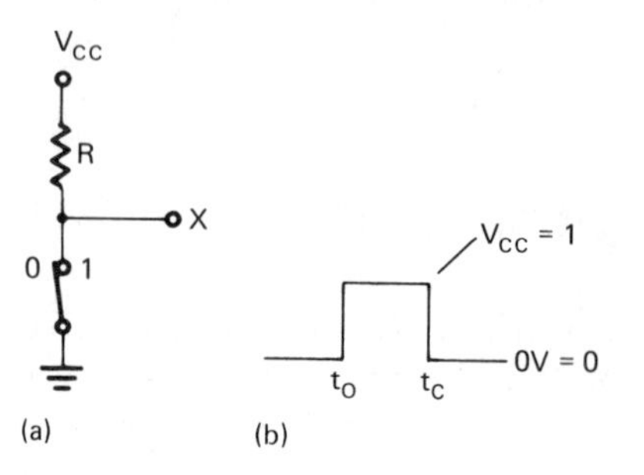

(b)

FIGURE 9-6

(a) Mechanical (inverting) switch. (b) Ideal output waveform under no load.

9-8(b). The 1-level output will be degraded by the loading effect, as was the case with the mechanical switch. Some additional reduction may result from the leakage current (I_{CBO}). This should, however, be a small amount. A major difference exists in the 0 level. It has been degraded upward by an amount equal to $V_{CE\ SAT}$. These two levels are an essential part of logic unit output specifications. The 0.4V 0 level and 2.4V 1 level in Figure 9-9 are sample levels taken from specifications for a particular type of TTL logic gate. The minimum 1 and maximum 0 output levels are marked on the trailing edge of the ideal rectangular pulse for comparison.

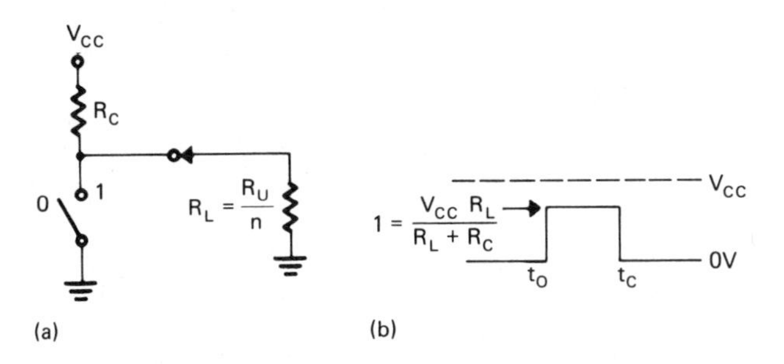

FIGURE 9-7
(a) Mechanical switch with n unit loads connected. (b) Output waveform showing reduction of 1 level due to loading.

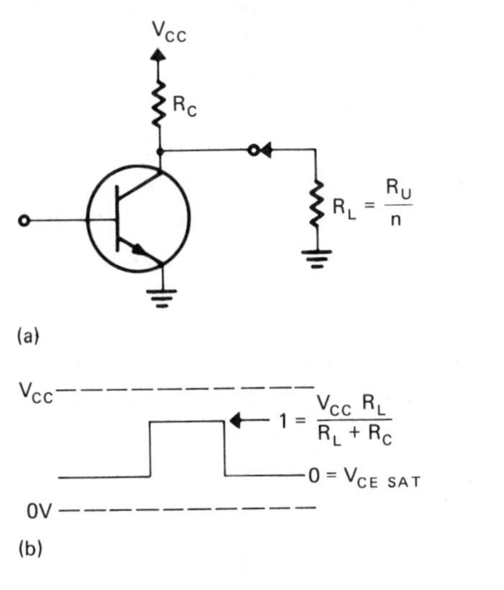

FIGURE 9-8
(a) Mechanical switch replaced by transistor. (b) Output waveform showing both 0 and 1 levels degraded slightly from the ideal.

9.6 Input-Level Parameters

The level to which a 1 output can fall is related to the minimum level an input will react to as a 1. If we look at the common emitter switch of Figure 9-10, we see that the input 1 voltage must be sufficient to produce an I_B that will saturate the transistor so that it will produce an output of $V_{CE\ SAT}$. The 0 level, on the other hand, should produce only a minute amount of base current; otherwise, it will excessively degrade the 1 level at the output. Using the same set of TTL gate specifications as in Figure 9-9, Figure 9-11 shows both input and output specifications marked on the ideal rectangular pulse. The output-level specification represents a wide-open window, into which the input specification will fit with room to spare at both top and bottom. In Figure 9-12, a hypothetical marginal pulse is applied to the input of a TTL NOR gate. If the gate is operating within specified limits, the inverted output will have at least the improved levels shown.

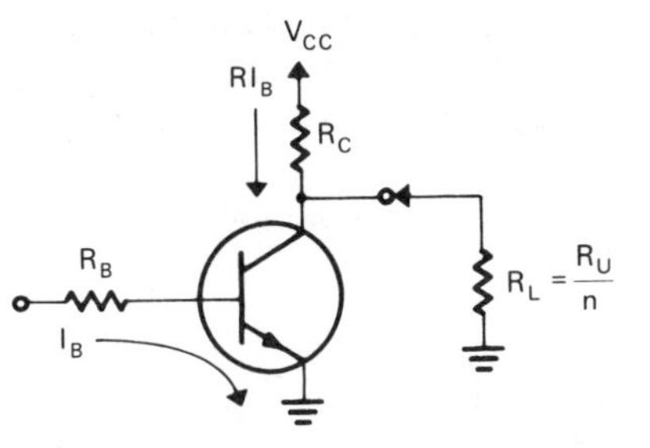

FIGURE 9-10
Common emitter switch with input requirements for maintaining 1 or 0 output levels.

FIGURE 9-9
Ideal rectangular pulse with output-level specifications for a particular type of TTL logic unit operating with a 5.0V V_{CC}.

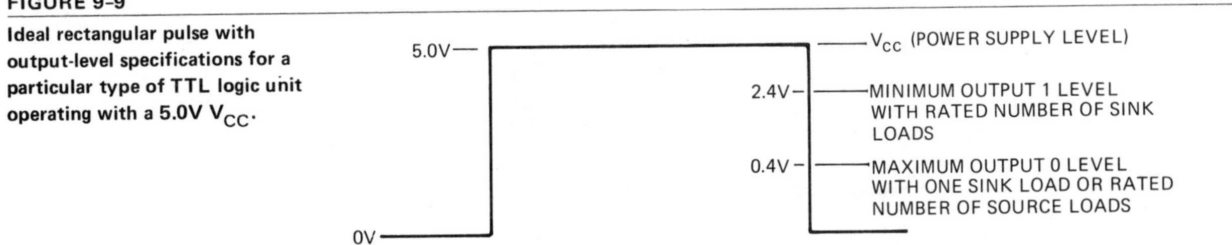

FIGURE 9-11

Ideal rectangular pulse with input-level specifications marked on leading edge and output-level specifications marked on trailing edge.

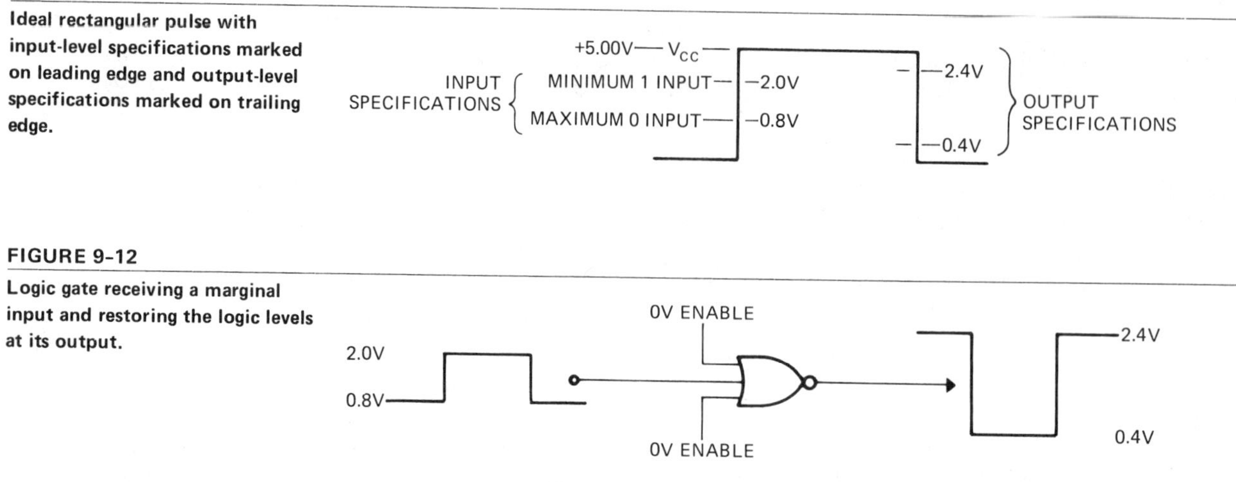

FIGURE 9-12

Logic gate receiving a marginal input and restoring the logic levels at its output.

9.7 Noise Immunity

The difference between the maximum 0 output and the maximum 0 input—an important quality characteristic of a logic circuit—is a measure of the degree to which the circuit will react to random noise spikes at the 0 level. The difference between the minimum 1 output and the minimum 1 input is a measure of the degree to which the circuit will react to random noise spikes at the 1 level. In any digital device, both ground and voltage lines will contain some quantity of random noise voltages.

Figure 9-13(a) shows the ideal switching gate with its clean 1 and 0 levels. Figure 9-13(b) shows the actual signal and noise combined. The noise voltages originate from many sources, some natural, some man-made, some internal to the machine, some external to the machine. They are normally voltage variations of very low level, but if we consider the variety of sources and their random nature, occasionally they may together produce a noise spike high enough to cause a 0 to switch to 1 or low enough to cause a 1 to switch to 0. The parity check system described in Section 8.6 is of some help in detecting such errors. It is, unfortunately, only practical to protect by this method certain long or parallel sets of lines that are particularly vulnerable to noise. Most of the protection must come from that narrow voltage difference between the input and output switching levels. The 2.4V minimum 1 output minus the 2.0V minimum 1 input gives a 0.4V 1-level noise immunity for our sample circuits. The 0.8V maximum 0 input minus the 0.4V maximum 0 output indicates a 0.4V 0-level noise immunity.

FIGURE 9-13

(a) Ideal output pulse showing noise immunity levels and switching threshold. (b) Actual output pulse showing noise spikes.

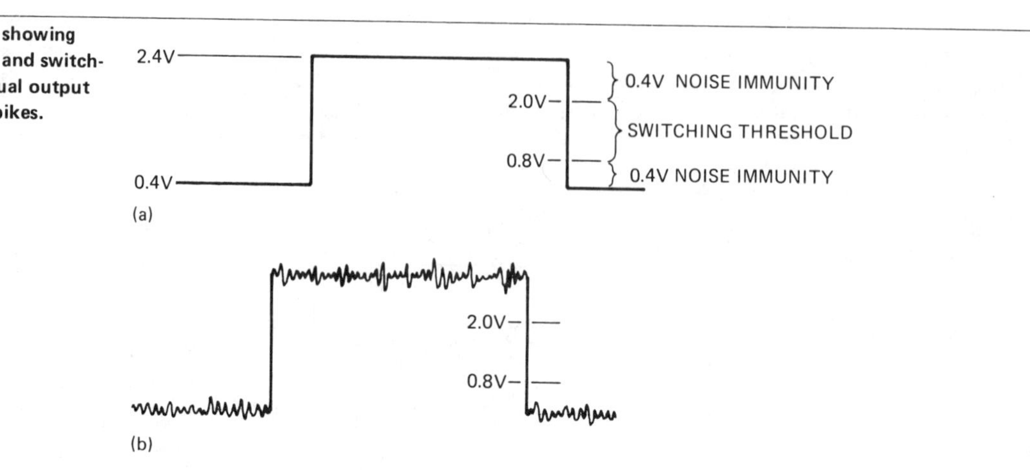

9.8 Switching Time Parameters (Rise Time, Fall Time, Propagation Delay Time)

If the logic circuits are to be used at low or intermediate speeds and do not have capacitor inputs, the specifications we have discussed thus far may be our only concern. But, with clock-over-1 megacycle or with input-to-logic circuits of the capacitor-resistor type shown in Figure 9-14, we must concern ourselves with time specifications on the output-voltage waveform. Figure 9-14(a) shows a sharp rectangular pulse applied to a typical capacitor input. After the RC differentiator reaches the full 1 level, the waveform can operate the circuit. Logic devices with RC inputs must have signals with fast leading or trailing edges. If the input pulse is not rectangular and the leading and trailing edges rise and fall gradually, as in Figure 9-14(b), the differentiated waveforms may not reach the full 1 level. In this case, the circuit will not operate.

Poor leading and trailing edges of a digital signal can adversely affect even those circuits not having capacitor inputs. Figure 9-15(a) shows a typical rectangular pulse with the maximum 0 and minimum 1 inputs marked on its edges. Somewhere between these two levels is the switching threshold, the point at which the circuit switches rapidly from 1 to 0 or vice versa. As the signal passes through this threshold, a noise spike of sufficient amplitude could cause double-switching, resulting in an extra pulse at the output, as in Figure 9-15(b).

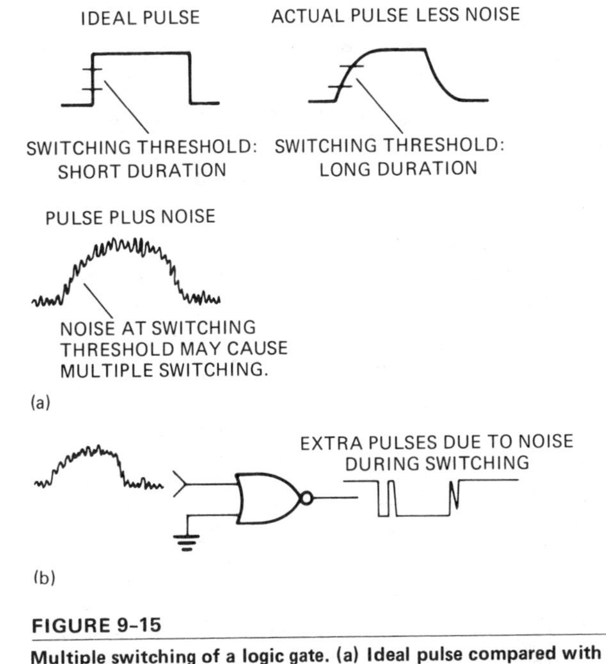

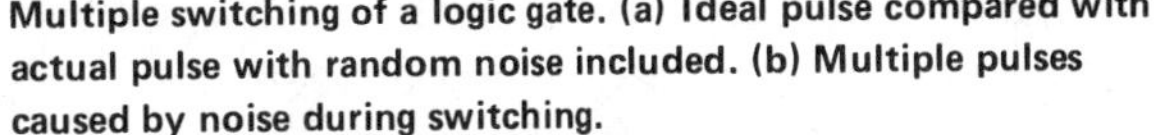

FIGURE 9-15

Multiple switching of a logic gate. (a) Ideal pulse compared with actual pulse with random noise included. (b) Multiple pulses caused by noise during switching.

FIGURE 9-14

(a) Sharp rectangular pulse applied to typical capacitor input. (b) Input pulse with poor leading and trailing edges applied to capacitor input.

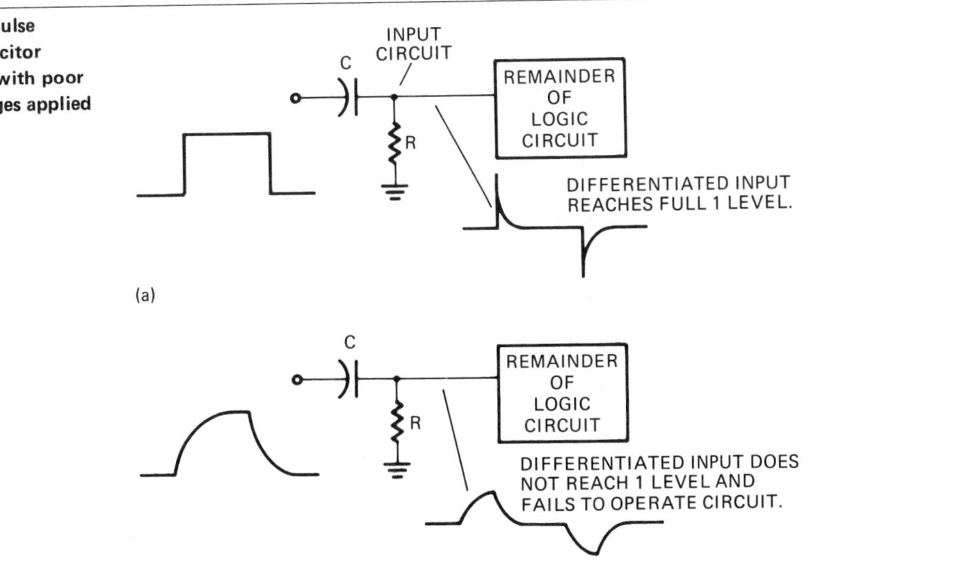

The operation of some digital circuits (to be discussed later) can definitely be disrupted by these extra pulses. The more rapidly the input signal passes through the switching threshold, the less likely it is that noise pulses will occur. For this reason, the rise time of the leading edge and the fall time of the trailing edge are parameters that concern us. Because of the indefinite or transient nature of the rise from or fall to an exact 0 and the possibility of overshoot and other indefinite characteristics at the 1 level, the rise and fall times are usually measured between the 10% and 90% levels of the waveform. Figure 9-16 shows these measurements.

Another timing problem is the delay in a digital signal as it passes through a logic circuit, sometimes referred to as propagation delay. Figure 9-17 shows a voltage-versus-time graph of an input and output pulse of the OR gate. The t_D measurement may be measured (usually in nanoseconds) as the time between the 50% level of the input leading edge and the 50% level of the output leading edge. It may also be measured between the trailing edges of the input and output waveforms. The delay measured at the leading edge will often be referred to as turn-on delay. The delay measured at the trailing edge is called turn-off delay. The delay times restrict the speed of gate operation and place an upper limit on the clock frequency that can be used. A 10-megacycle clock frequency has a 100-nanosecond period, and it would be difficult to use gates with delay times of over 50 nanoseconds. Today, gates are available with turn-on and turn-off delays of 3 nanoseconds. Such gates make possible the use of 20-megahertz and higher clock frequencies.

9.9 Power Drain

The amount of power consumed by the integrated circuit is another important consideration in selecting a type of circuit to be used for a logic system. Individual gate power drain for TTL may be determined from the 0-level and 1-level supply current specifications, $I_{CC}(1)$ and $I_{CC}(0)$. There is a wide variation between these, and for a given number of gates used in the usual system, only the maximum possible current drain can be computed with certainty.

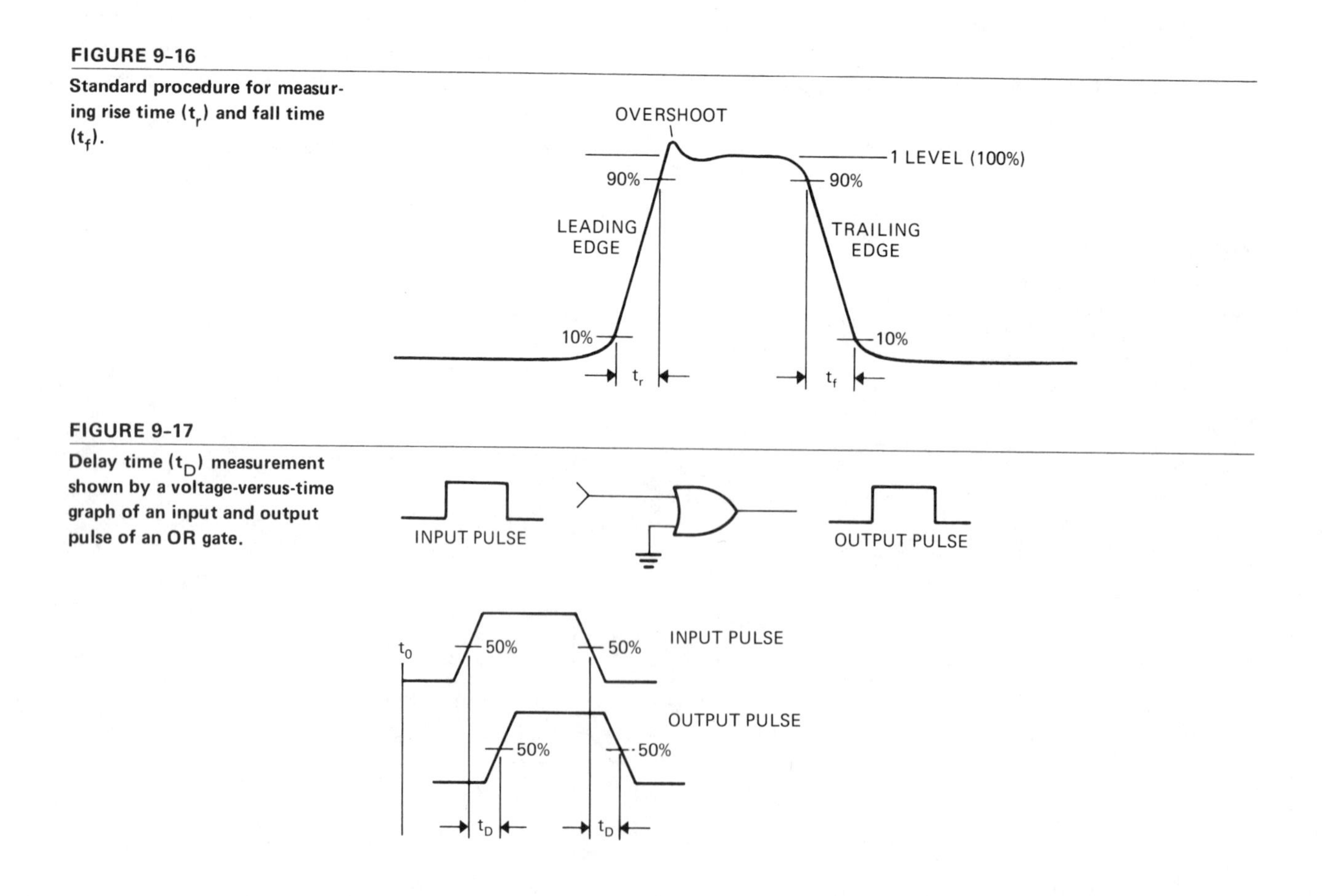

FIGURE 9-16

Standard procedure for measuring rise time (t_r) and fall time (t_f).

FIGURE 9-17

Delay time (t_D) measurement shown by a voltage-versus-time graph of an input and output pulse of an OR gate.

■ EXAMPLE 9-2

From the typical logic unit specifications of Figure 9-1, determine the maximum possible current drain by all eight NAND gates used in a particular logic system.

Solution: For this gate, the 0 state causes the highest current drain, $I_{CC}(0)$. Therefore, maximum possible current drain occurs whenever all eight NAND gates are in the 0 state.

$$\text{Maximum current drain} = 8 \times I_{CC}(0)\ \text{MAX} = 8 \times 4.8\ \text{mA} = 38.4\ \text{mA}$$

Depending on output and loading, the individual TTL gate may require as much as 25 milliwatts per gate. This is in contrast to a CMOS gate, which draws as low as 10 nanowatts per gate under typical load conditions.

Determining power drain by adding up the power drain of individual gates is no longer a major part of the power drain computation. The bulk of modern digital circuitry is contained in LSI (large-scale integrated circuit) or MSI (medium-scale integrated circuit) form. Again, examination of two like circuits (four-bit counter) in TTL- and CMOS-type integrated circuits finds TTL with a power drain as high as 250 milliwatts, in comparison to a CMOS power drain of 0.1 milliwatt.

9.10 Speed versus Power

Comparing the power drain specification of TTL and CMOS circuits as we did in Section 9.9 gives the impression that CMOS circuits are superior because of a many-fold lower power drain. Considering the importance of reducing power cost and heat dissipation, one might predict that in the future large digital machines will be built primarily from CMOS-type circuits to the exclusion of TTL and other bipolar circuits. At present, however, TTL circuits can be produced with propagation delay times about one-fifth those of CMOS circuits. This means that high-speed digital machines will probably be built with separate high-speed and low-speed sections—the former for bipolar circuits, the latter for MOS circuits. Future developments in the "state of the art" may, of course, alter this trend.

Users of modern, general-purpose computers often state that the cost of using computers is inversely proportional to the clock frequency; that is, the higher the clock frequency, the more work the computer can do, and the final cost per unit of work is reduced. This fact has stimulated development of higher-speed digital circuits. Looking again at the turn-off delay time specification for a TTL gate (Figure 9-1), we find a range of 5 to 20 nanoseconds. Primary clock frequencies of 25 megahertz might safely be used with these circuits. This compares with 3 to 5 megahertz for present-day CMOS circuits.

9.11 Special Circuits for High Speed

As discussed in Chapter 5, the 1 and 0 of a transistor switch are obtained from the operating states of saturation and cutoff. During saturation, a large number of charge carriers are stored in the base region. The time required to move these charge carriers—known as storage time—is a limiting factor in reducing the propagation delay time of a TTL gate. One solution is use of nonsaturated logic. Emitter coupled logic (ECL) is a family of logic circuits designed to use this method to obtain high speed.

Figure 9-18 shows an ECL NOR gate. A voltage designated V_{BB} is applied to the base of Q_4. This turns on Q_4 to conduct current through R_E to the extent that a voltage nearly equal, V_{BB}, appears on the emitters of the three input transistors. This establishes V_{BB} as the minimum 1 level. Until one or more of the input transistors receives a 1 level on its base lead, no collector current will flow through resistor R_1. A small base current, however, will flow through R_1 and the base emitter lead of Q_5. This drops very little voltage across R_1 and, the circuit of Q_5 being emitter follower, a 1 level substantially higher than V_{BB} will appear at the output. If an input 1 level occurs on the base of one or more input transistors, the collector current through R_1 drops the base voltage of Q_5 almost to the level of V_{BB}. Because of the base-to-emitter voltage drop (V_{BE}), the output 0 level on the emitter of Q_5 falls below V_{BB}. The circuit functions with a reasonable degree of noise immunity because the maximum output 0 level is

0.5V or more below V_{BB}, while the minimum input 1 level is 0.5V or more above V_{BB}. As all transistors in the gate have emitter resistors, there is no saturation, and propagation times as low as 2 nanoseconds are obtained. Unfortunately, ECL circuits are more expensive to produce and require several times more power than TTL gates. At present, there are few MSI or LSI logic circuits available in ECL.

Another approach to higher speed is use of Schottky clamped TTL circuits. During saturation of a transistor switch, as shown in Figure 9-19, the voltage V_{CE} is lower than the voltage V_{BE}. This means that the base-to-collector junction has a slight forward bias not quite equal to the barrier potential of a silicon diode. A Schottky barrier diode (SBD)—symbol shown in Figure 9-20(a)—has a barrier potential much

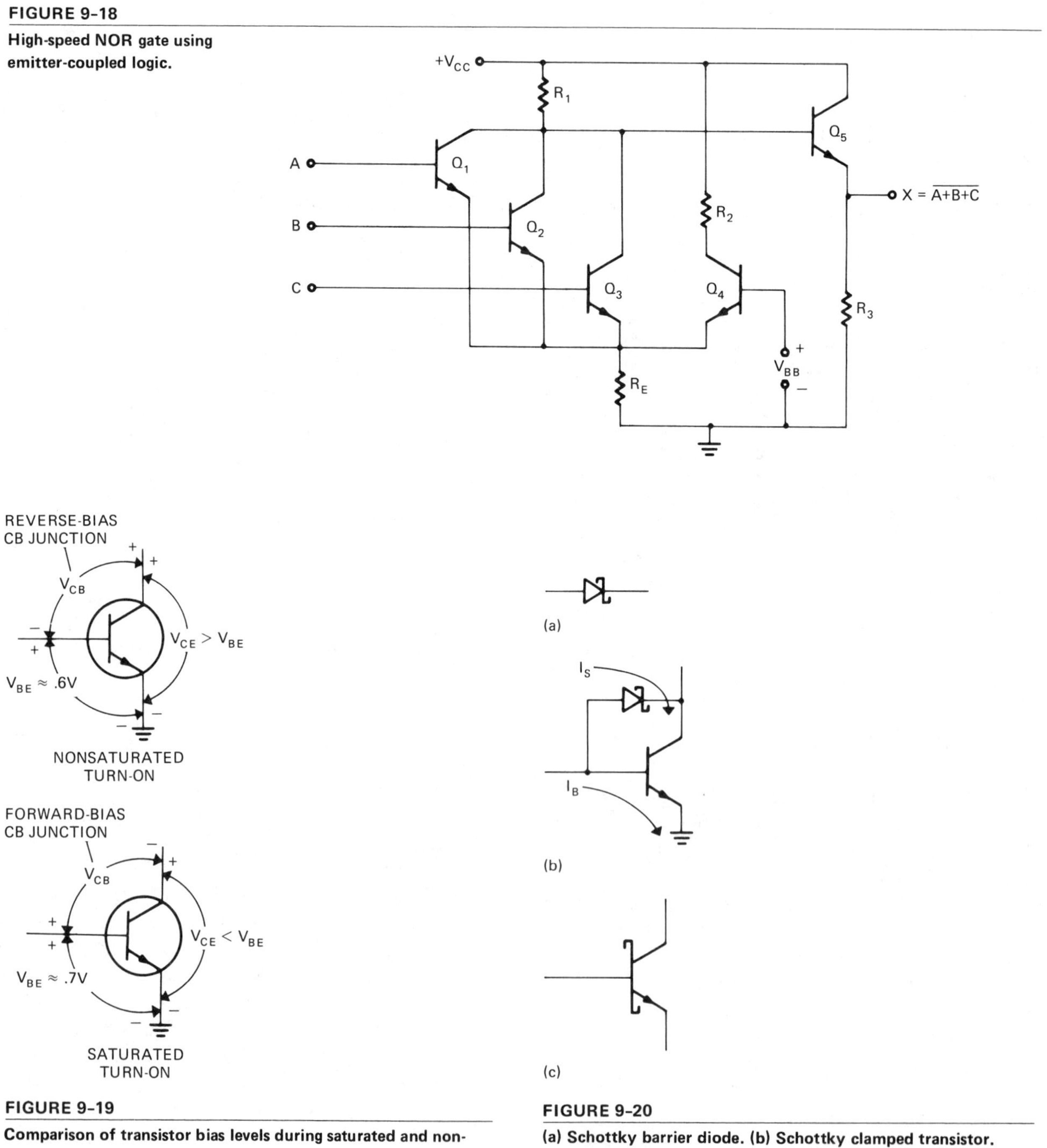

FIGURE 9-18
High-speed NOR gate using emitter-coupled logic.

FIGURE 9-19
Comparison of transistor bias levels during saturated and nonsaturated turn-on.

FIGURE 9-20
(a) Schottky barrier diode. (b) Schottky clamped transistor. (c) Symbol for Schottky clamped transistor.

lower than that of a silicon diode. The SBD is formed between base and collector of the integrated transistor, as in Figure 9-20(b). As the transistor approaches saturation, the collector voltage drops slightly below the level of the base voltage, foward-biasing the SBD. At this point, any excess base current passes through the forward-biased Schottky diode. This clamps the transistor at the threshold of saturation and reduces the build-up of charge carriers, which cause storage time delay. Figure 9-20(c) shows the symbol for a Schottky clamped transistor.

Figure 9-21(a) shows a standard TTL four-input NAND gate. Figure 9-21(b) shows a Schottky gate. The propagation time of the Schottky gate is about half that of the standard TTL gate. The power drain of the Schottky, however, is almost double that of standard TTL gates. A major advantage of the Schottky clamped TTL logic is that it is available in many of the varied MSI and LSI circuits developed for TTL logic.

FIGURE 9-21

(a) Standard TTL four-input NAND gate. (b) Schottky TTL four-input NAND gate.

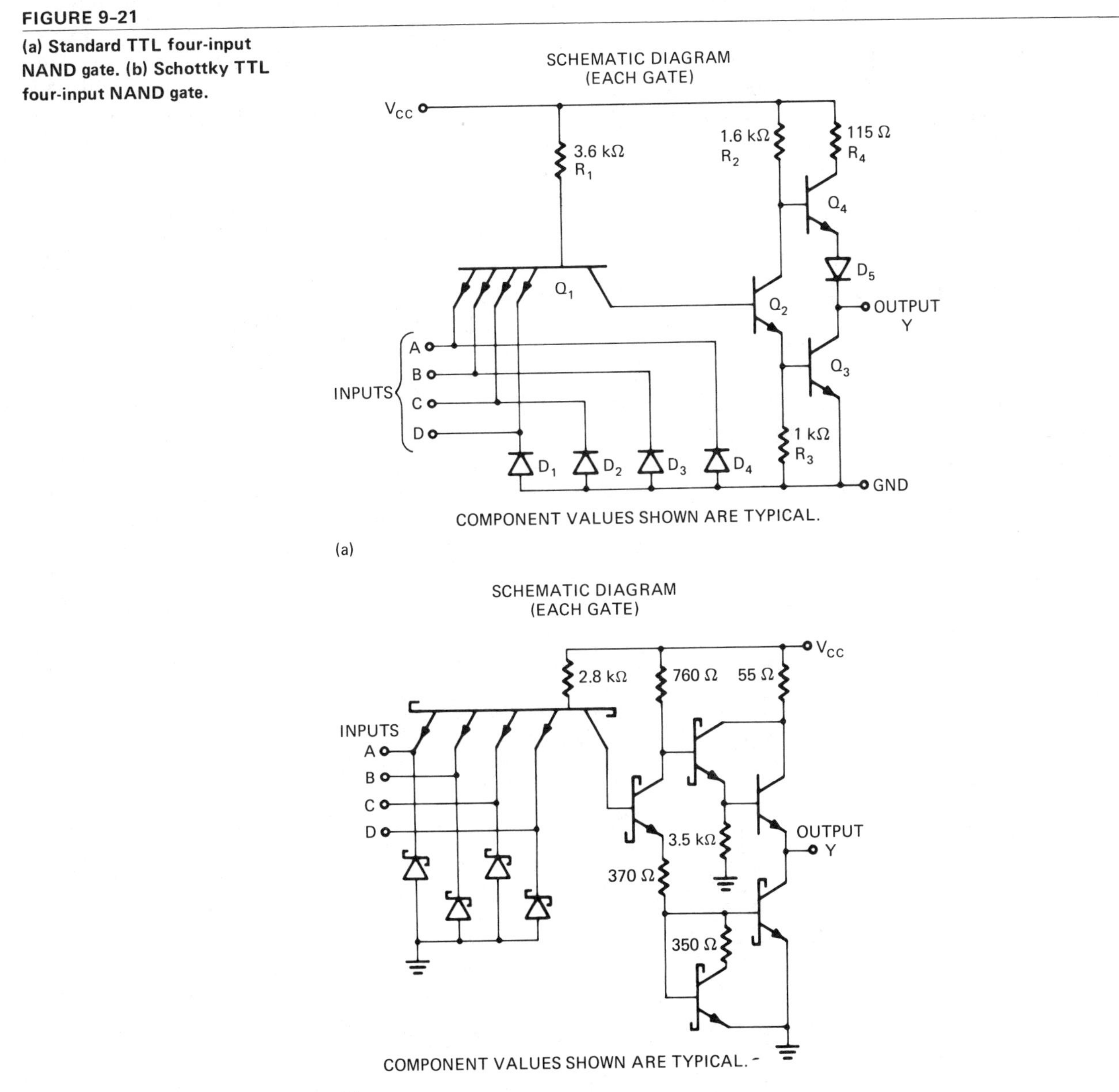

9.12 Integrated Circuit Technologies

Present-day integrated circuit technologies can be classified in two categories—bipolar and unipolar. In the bipolar class, there are a number of divisions: TTL, I^2L, and ECL. TTL is the dominant technology for small- and medium-scale integrated circuits. I^2L is more likely to be applied to large-scale or custom circuits. ECL is used where very high speed is required.

The unipolar, or MOS, technologies include PMOS, NMOS, and CMOS. These are, in general, lower speed and consume less power than equivalent-scale bipolar technologies. PMOS is becoming less popular because of the need for both negative and positive power supplies. NMOS technologies are used for memories and other large-scale circuits, including custom circuits. They are generally used with supporting TTL (SSI MSI) circuits and thus are often designed to have TTL-compatible input and output voltage levels.

CMOS is unique among the MOS technologies in that there is a complete series of small- and medium-scale logic circuits available in CMOS (4000 series). These devices are slower than equivalent TTL circuits but have a very low quiescent power consumption. They can operate at a much wider range of supply voltages (3.5V to 20V) than TTL (4.5V to 5.5V).

Because of the wide popularity of the 7400 TTL families and the 4000 series CMOS family, a detailed discussion of the specifications of these two families will be provided here.

In Figure 9-1 and in the preceding paragraphs, the digital parameters have been described in terms of binary levels 1 and 0. Figure 9-22 compares these high-low parameters with the 1-0 parameters given in Figure 9-1. The high-low parameters have been more commonly used in recent years.

9.13 TTL Specifications

9.13.1 TTL DC Parameters

To date, there are about eight speed/power versions of the 7400 series TTL integrated circuits. The three most popular are 74XX, 74LSXX, and 74SXX. The XX can be replaced with a number from 00 on up, depending on the MSI circuit. The 7400, 74LS00, and 74S00 are all quad two-input NAND gates. They differ from each other primarily by speed and power specifications. By logic function and pin configuration, they are identical. The number of available MSI circuits is increasing each year. The octal latch circuit 74LS374 has been available for a few years now. It might be more accurate to generalize the number as 74XXX, and, in the near future, a fourth X might be needed.

All speed/power versions of TTL circuits specify a V_{IL} of 0.8V and a V_{IH} of 2V. These mean simply that the circuit is guaranteed to respond to any input of 0.8V or lower as a binary 0 and to any input of 2V or higher as binary 1. The current drawn by these inputs under typical drive conditions is shown in Figure 9-23. Note that all three versions specify current not at the guaranteed 1 and 0 levels but at a worst-case drive condition. Generally speaking, the loading effect of a Schottky input is equivalent to five low-power Schottky inputs. A standard 74XX input is the equivalent of four 74LS inputs.

FIGURE 9-22

Comparison of DC parameters expressed in 1-0 and high-low symbols.

1-0 PARAMETERS		HIGH-LOW PARAMETERS	
DESCRIPTION	SYMBOL	DESCRIPTION	SYMBOL
1 INPUT VOLTAGE	$V_{IN}(1)$	HIGH-LEVEL INPUT VOLTAGE	V_{IH}
0 INPUT VOLTAGE	$V_{IN}(0)$	LOW-LEVEL INPUT VOLTAGE	V_{IL}
1 INPUT CURRENT	$I_{IN}(1)$	HIGH-LEVEL INPUT CURRENT	I_{IH}
0 INPUT CURRENT	$I_{IN}(0)$	LOW-LEVEL INPUT CURRENT	I_{IL}
1 OUTPUT VOLTAGE	$V_{OUT}(1)$	HIGH-LEVEL OUTPUT VOLTAGE	V_{OH}
0 OUTPUT VOLTAGE	$V_{OUT}(0)$	LOW-LEVEL OUTPUT VOLTAGE	V_{OL}
1 OUTPUT CURRENT	$I_{OUT}(1)$	HIGH-LEVEL OUTPUT CURRENT	I_{OH}
0 OUTPUT CURRENT	$I_{OUT}(0)$	LOW-LEVEL OUTPUT CURRENT	I_{OL}

At the output of the TTL circuit, currents are specified at the guaranteed output levels, as given in Figure 9-23. As Figure 9-24(a) shows, the I_{OL} is the maximum current the lower transistor of the totem pole can sink to ground and still maintain a $V_{CE\ SAT}$ of less than 0.4V and/or 0.5V in the case of Schottky. During a high-level output (V_{OH}), as in Figure 9-24(b), the bottom transistor of the totem pole output is turned off while the top transistor is operating in emitter follower configuration.

A sink load, such as a resistor to ground, would tend to reduce the V_{OH} level. The TTL input, however, is essentially a source load. The same would be true of a pull-up resistor connected between the output and V_{CC}. If the source load pulls the voltage to a level higher than the voltage present on the base of the top transistor, then the output is turned off and the V_{OH} level is determined not by the driving circuit but by the source load. If TTL inputs or similar source loads are the only loads connected to the output of a TTL circuit, then little concern is necessary over the likelihood of violating the (V_{OH}, I_{OH}) specification. In any usual TTL design, primary concern is given to the (V_{OL}, I_{OL}) specification. If resistors to ground or similar sink loads are connected, then the possibility of drawing an I_{OH} in excess of the specification must be considered. Although the output may sustain without damage a current well in excess of I_{OH}, there is no guarantee that it will maintain the correct V_{OH} while the overload condition exists. The output, in fact, has a short-circuit current rating specified for only one input at a time and a duration of 1 second.

It is not uncommon for 74, 74LS, and 74S circuits to be used in the same system. When this occurs, there is no interfacing problem since 1 and 0 levels are the same for all three versions. Unit load calculations, however, are complicated by the fact that three different input current levels are involved along with three different output current drive levels. It is normally convenient to add the rated I_{IL} MAX levels or the inputs connected to each output and compare it with the drive level, I_{OL} MAX, of that output.

Figure 9-25 illustrates the conventions used for current flow polarities on DC parameters. A conventional current flow (+ to -) into a circuit is considered positive. A current flow out of the circuit is considered negative. At the output of a TTL circuit, I_{OL} is a positive current, and the current I_{OH} is negative. At the input to the TTL circuit, the polarities are reversed. I_{IL} is negative, while I_{IH} is positive.

INPUT PARAMETERS

I_{IH} @ V_I = 2.4V		
74XX	74LSXX	74SXX
40 μA	20 μA	50 μA

I_{IL} @ V_I = 0.4V		
74XX	74LSXX	74SXX
-1.6 mA	-0.4 mA	-2 mA

OUTPUT PARAMETERS

I_{OH} @ V_O = 2.7V		
74XX	74LSXX	74SXX
-800 μA	-400 μA	-1000 μA

I_{OL} @ V_O = 0.4V		
74XX	74LSXX	
16 mA	4 mA	

I_{OL} @ V_O = 0.5V		
	74LSXX	74SXX
	8 mA	20 mA

FIGURE 9-23

DC voltage parameters of the 7400 series TTL circuits.

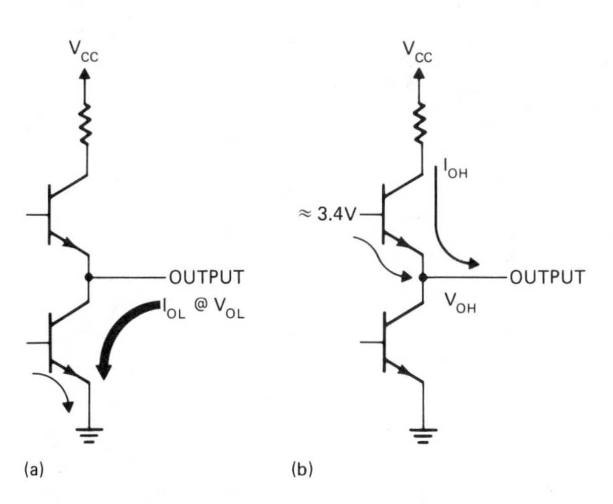

FIGURE 9-24

Output currents of the TTL circuits. (a) V_{OL} is the voltage across the bottom transistor when it is turned on and sinking a current of I_{OL}. (b) V_{OH} is the voltage between output and ground when the upper transistor is turned on and the bottom transistor is turned off. I_{OH} is, by convention, a negative current.

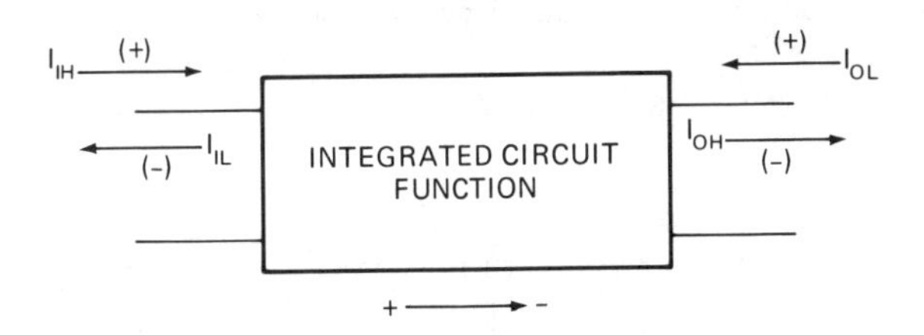

FIGURE 9-25

Diagram showing conventions for current flow polarities. A current flow (+ to -) entering the circuit is considered +; a current flow out of the circuit is considered –. Thus, TTL inputs are + for I_{IH} and - for I_{IL}; the outputs are + for I_{OL} and - for I_{OH}.

9.13.2 Operating Conditions and Power Supply Current

All of the foregoing TTL specifications are valid over the operating free air temperature range for 7400 series devices, 0° C to 70° C. They also apply to a power supply voltage range of 4.75V to 5.25V. The military version of TTL circuits, 54XX, including 54LSXX and 54SXX, is specified for a free air temperature range of -50° C to 125° C and a power supply range of 4.5V to 5.5V.

All circuits are given a maximum supply current rating (I_{CC}). This rating is different for each device type. Figure 9-26 compares the typical and maximum ratings for (a) a simple SSI device and (b) a complex MSI device. These figures are used to estimate power supply and air cooling requirements and to assure that power leads on circuit cards will carry the needed current without fusing. Whether to use typical or maximum values in the calculations depends on reliability level requirements.

9.13.3 Absolute Maximum Ratings

The absolute maximum ratings are specified for nonoperating stresses that may be applied to the circuit either accidentally or during a period when normal circuit operation is not expected. These ratings guarantee only that the circuit will return to normal specified operation after the stress is removed. These values are given in Figure 9-27.

I_{CC} @ V_{CC} = 5.25V

QUAD TWO-INPUT NAND	7400		74LS00		74S00		
	TYP	MAX	TYP	MAX	TYP	MAX	UNIT
ALL OUTPUTS HI	4	8	0.8	1.6	10	16	mA
ALL OUTPUTS LO	12	22	2.4	4.4	20	36	mA

(a)

HEX "D" FLIP-FLOP	74174		74LS174		74S174		
	TYP	MAX	TYP	MAX	TYP	MAX	UNITS
	45	65	16	26	90	144	mA

(b)

FIGURE 9-26

Typical and maximum supply currents. (a) For TTL (SSI) quad two-input NAND gate. (b) For TTL (MSI) hex "D" flip-flop.

9.13.4 TTL AC Parameters

There are only a few parameters for a simple gate circuit. For large-scale circuits, AC parameters may be quite numerous and often require a timing diagram to adequately describe them. For the quad two-input NAND gate, propagation delay is measured as t_{PLH} at the rising edge of the output and as t_{PHL} on the falling edge. As Figure 9-28 shows, the measurement is taken between the 1.3V or 1.5V level on the edge of the input signal and the 1.3V or 1.5V level on the resulting edge at the output. Figure 9-29 compares the propagation delay times of the three TTL families. The table shows 74S00 to be the fastest of the three families, but this exists at the expense of much higher supply current than the 74LS00.

9.13.5 Other 7400 Series Families

Two additional families of 74 series devices are the 74H and 74L families, which are still available from several vendors. The 74H is high-speed, but not as high as the 74S, in spite of the fact that it consumes more power than the 74S. There is no strong argument for the continued use of the 74H series. The 74L, on the other hand, consumes only half the power of an equivalent 74LS device and might be used

FAMILY	74XX	74LSXX	74SXX	UNIT
SUPPLY VOLTAGE	7	7	7	V
INPUT VOLTAGE	5.5	7	5.5	V
OPERATING TEMP.	0 TO 70			°C
STORAGE TEMP.	-65 TO 150			°C

FIGURE 9-27

Absolute maximum ratings for TTL circuitry.

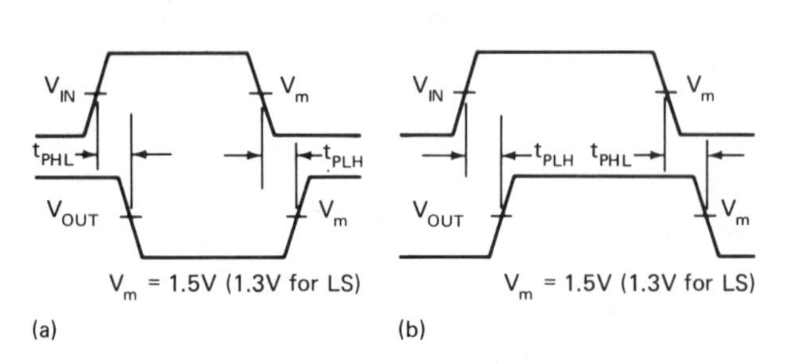

FIGURE 9-28

Input and output waveforms showing the measurement points for t_{PHL} and t_{PLH}. (a) Inverting function. (b) Noninverting function.

ideally where low power and a speed slower than LS, but faster than CMOS, are required. Figure 9-30 compares the speed/power specifications of these TTL devices with equivalent 74S and 74LS devices.

For the past several years, manufacturers of 74 family devices have been developing new and improved speed/power versions of the 7400 series. It appears to date that three of these versions have received market acceptance and will be used in an increasing number of future applications.

The 74ALS (advanced low-power Schottky), first introduced by Texas Instruments, has a speed that is about half again as fast as equivalent 74LS devices. In addition to this increase in speed, power dissipation is improved to about half that of equivalent 74LS devices.

Where speed is needed, the 74AS (advanced Schottky) series, also developed by TI, provides operating speeds about twice that of equivalent 74S devices. The power dissipation of the 74AS is about the same as equivalent 74S devices. Figure 9-31 shows graphically the speed/power spectrum occupied by identical sets of MSI circuits in each version of the 74 family.

A third new device in the 7400 family is the 74F (F for FAST, or Fairchild Advanced Schottky Technology) devices. These have operating speed as fast or faster than the 74S but with a power dissipation about one-quarter that of the 74S devices. Figure 9-32 compares the three common versions of 7400 devices with the three new speed/power versions. The SSI and MSI devices of the new families will be available with identical logic function and pin-outs as offered for the old families. The 1 and 0 voltage

TTL AC PARAMETERS

QUAD TWO-INPUT NAND	7400		74LS00		74S00		
	TYP	MAX	TYP	MAX	TYP	MAX	UNIT
t_{PLH}	11	22	9	15	3	4.5	ns
t_{PHL}	7	15	10	15	3	5	ns

FIGURE 9-29

Propagation delay times of TTL families.

TWO-INPUT GATE	74LS TYP	74S TYP	74L TYP	74H TYP	UNITS
t_{PLH}	9	3	35	5.9	ns
mW/GATE	2	19	1	22.5	mW

FIGURE 9-30

Speed/power comparison of older 74 series TTL devices.

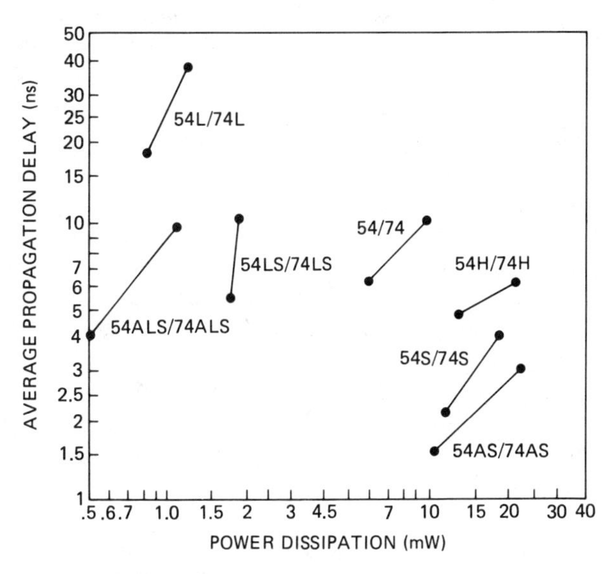

FIGURE 9-31

Speed/power spectrum for a set of SSI/MSI devices of Texas Instruments series 54/74 TTL family.

FIGURE 9-32

Speed/power comparisons for two-input NAND gates in advanced versions of TTL recently developed by Fairchild and Texas Instruments.

	ESTABLISHED TTL FAMILIES			NEW TTL FAMILIES[1]			UNITS
FAMILY	TTL	TTL LS	TTL S	TTL ALS	TTL FAST	TTL AS	
t_{PLH} TYP[2]	12	9	3.5	5	3.4	1.5	ns
t_{PHL} TYP[2]	8	10	3.5	5	2.9	1.5	ns
DC POWER DISSIPATION[3]	10	2	20	1	5	20	mW
I_{OL}	16	8	20	8	16	20	mA
I_{OL}	0.4	0.4	1.0	0.4	0.4	2	mA
I_{IL}	1.6	0.4	2.0	0.2	1.6	1	mA
I_{IH}	40	20	50	20	40	20	μA

[1] The new TTL families have guaranteed input switching levels of V_{IL} = 0.8V minimum and V_{IH} = 2V maximum. Output voltages V_{OH} and V_{OL} for the currents specified are also compatible between the new and old families.

[2] Typical values for NAND gate.

[3] Typical for NAND gate circuit and 50% duty cycle.

levels are compatible between them, and they can function in the same circuits without interfacing problems. A 7400 and a 74ALS00 will both be quad two input NAND gates in 14-pin DIP package with the same pin-outs. The 20-pin 74AS373 will be an octal latch circuit with pin-out identical to the older 74LS373. Cost being equal, the better speed/power ratios of the newer devices make them more advantageous than the older devices.

9.13.6 TTL Output Drivers

The usual output circuit for a TTL device is the totem pole, as shown in Figure 9-24. This output is ideal where a single output is driving a logic line to which only TTL inputs are connected. Two standard totem pole outputs should not be connected together on the same line because part of the time one driver would be pulling the line low while the other driver is pulling the line high. Where two or more drivers must be connected to the same line, special output circuits are required. One of these special output devices is the open-collector output. The 7401 is a quad NAND gate in which each of the gates is constructed with an open collector instead of the usual totem pole output. The circuit for this gate is shown in Figure 9-33. Open-collector outputs are found in other circuits, some of which are large- and medium-scale integrated circuits. They are particularly useful in memory circuits, where many outputs must be connected to the same input. The open-collector output must be connected to an external pull-up resistor, as Figure 9-33 shows, and, if a number of outputs must be connected to the same input, they can all be connected to the same resistor. For this to work properly, only one circuit can be active at a time and the inactive circuits must have their outputs turned off (in the 1 state). The terminal at the bottom of the pull-up resistor will become 0 if the active output is 0. If the active output is 1, then all outputs connected to the pull-up resistor are turned off, and the transistors draw no current through the resistor. This imposes a high-level voltage on the inputs connected to the line. The direct connection of two or more open-collector outputs to the same pull-up resistor is called a *wired OR circuit*.

The pull-up resistor, shown in Figure 9-33, normally connects to V_{CC}, and the resistor value is selected to limit current through the output transistor to I_{OL} MAX. This can be done by using a resistor value larger than $(V_{CC} - V_{OL})/I_{OL}$. Occasionally, it is desirable to have TTL devices drive circuits that use voltages in excess of V_{CC} or that sink currents in excess of normal I_{OL} levels. There are a number of open-collector buffers that can be used to provide these functions. Figure 9-34 shows V_{OH} and I_{OL} levels for these 7400 series TTL open-collector buffers. The maximum voltage, V_{OH}, that can be applied between output (collector) and ground for a standard 7400 series open-collector device is 5.5V. The maximum current, I_{OL}, that can be drawn through the output transistor for a standard 7400 series open-collector device is 16 mA. The open-collector output transistor of the devices listed in the table in Figure 9-34 is specially designed to withstand higher-than-standard voltage or higher-than-standard current or both.

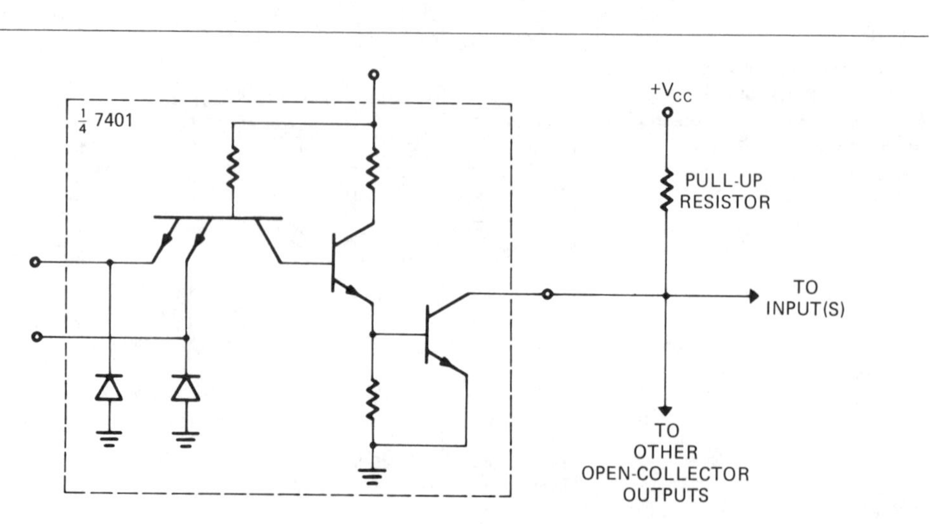

FIGURE 9-33

TTL 7401 two-input NAND gate with open-collector output connected to an external pull-up resistor.

The typical totem pole TTL output circuit, as shown in Figure 9-35(a), functions by having either the upper or lower transistor turned on at any given time. To produce a logic 1 or high output, Q_1 is turned on and Q_2 is turned off. To produce a logic 0 or low output, Q_1 is turned off and Q_2 is turned on. As compared to a pull-up of the open-collector circuit in Figure 9-33, the transistor Q_1 is an active pull-up and has a high resistance during a logic 0 out and a low resistance during a logic 1 output. This provides the best of both worlds for both high-speed and

DEVICE NUMBER	7406	7416	7426	7433	7438
CIRCUIT LOGIC	HEX INVERTER	HEX INVERTER	QUAD TWO-INPUT NAND	QUAD TWO-INPUT NOR	QUAD TWO-INPUT NAND
V_{OH} MAX (MAXIMUM VOLTAGE AT TOP OF PULL-UP RESISTOR)	30V	15V	15V	5.5V*	5.5V*
I_{OL} MAX (MAXIMUM CURRENT THROUGH OUTPUT TRANSISTOR)	40 mA	40 mA	16 mA*	48 mA	48 mA
V_{OL} MAX WITH I_{OL} MAX	0.7V	0.7V	0.4V	0.4V	0.4V

*Standard level

FIGURE 9-34

7400 series TTL open-collector buffers having either voltage or current ratings higher than the standard values for 7400 series.

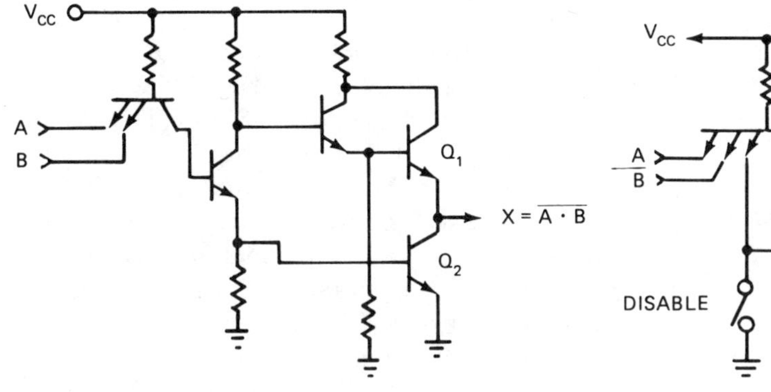

(a)

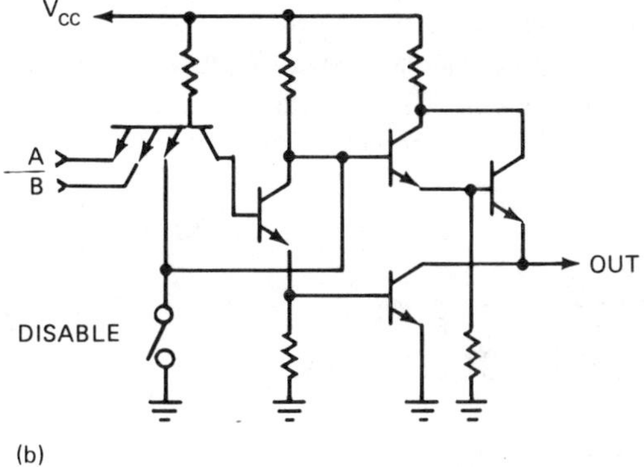

(b)

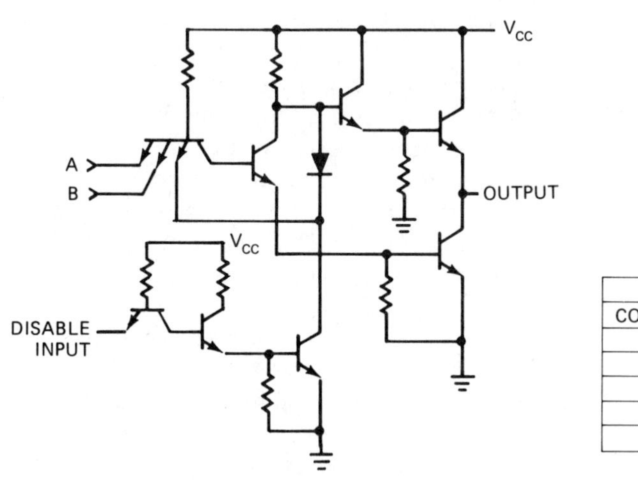

(c)

INPUT			OUTPUT
CONTROL	A	B	X
H	0	0	1
H	0	1	1
H	1	0	1
H	1	1	0
L	X	X	HIGH Z

X = DON'T CARE STATE

(d)

FIGURE 9-35

(a) TTL two-input NAND gate. (b) TTL NAND gate with manual switch equivalent of three-state output disable. (c) Two-input NAND gate with three-state output. (d) Truth table.

minimum power dissipation, but as previously mentioned, totem pole outputs cannot be connected on the same bus. With a minor change to the TTL circuit, however, a third state can be created during which both transistors of the totem pole are turned off, as shown in Figure 9-35(b). This third state produces a high impedance output.

The *three-state*, or *tri-state*,* bus allows a large number of outputs to be connected to a single bus line as long as all but one of these outputs is in the high impedance state (disabled) at any given time. Figure 9-35(c) shows the modification to a TTL circuit used to produce a three-state output. Note that an enable/disable control must be held in the enable state when this driver is outputting to the bus. It must be switched to the disable state when the bus is relinquished to another driver.

The number of three-state drivers that may be connected to a bus is limited by the operating frequency. Each disabled driver adds from 3 to 8 pF to the overall bus capacitance. At low switching frequencies, over a hundred three-state drivers may be connected to a single bus. The number of input circuits on that same bus is limited by the sink or source current rating of the weakest driver on the bus.

9.14 CMOS Specifications

9.14.1 Introduction

CMOS integrated circuits, if correctly employed, provide several major advantages over equivalent bipolar circuits. Two of these advantages are low quiescent power dissipation and excellent noise immunity. The primary limitation in the selection of CMOS digital circuits is a significantly lower switching speed than TTL circuits.

Many new MSI functions and many improvements to circuit parameters have come about since the introduction of CMOS circuits in 1969. A growing lack of standardization has caused major CMOS producers to agree upon a set of specifications called B-series specifications. In general, B-series devices have buffered outputs that allow higher output sink and source currents. However, devices without buffered outputs that can meet the B-series specifications may also be labeled as B-series. In addition to the drive current increase, the supply voltage limit of B-series devices was increased to 18V. The advantages of the standard B-series devices are obtained at the cost of only a minor reduction in switching speeds compared to that available in A-series devices. Figure 9-36 shows the electrical characteristics for the industry B-series specifications. Unfortunately, the switching characteristics have not yet been standardized.

*Tri-state® is a registered trademark of National Semiconductor Corporation.

9.14.2 DC Voltage Parameters

The process of defining DC voltage parameters at the inputs and outputs of CMOS circuits cannot be easily compared with practices used for bipolar circuits. A typical input for bipolar gates is defined as a sink or source load, and the fan-out is determined from the maximum sink or source current that the typical output can drive. Input resistance is very high for CMOS circuits. Therefore, fan-out is limited by the capacitive loading.

Each CMOS input is connected to a pair of insulated gates by way of a protection network, as shown in Figure 9-37(a). This results in a resistive load so high that DC loading is insignificant, but the capacitance formed by the pair of insulated gates is typically about 5 pF (7.5 pF maximum). The normal load of a CMOS circuit is the capacitance of the fan-out. As shown in Figure 9-37(b), the driver in the 1 state charges the capacitance of the input circuit to the level of V_{DD}. This capacitance is charged when current flows from V_{DD} through both the p-channel transistor and the input protection network. As the input capacitance approaches full charge, the current from the driver approaches 0. The current flowing from the driver is on the order of nA as long as the logic state is not changing. As shown in Figure 9-37(c), the driver in the 0 state provides a low resistance path through the n-channel transistor to V_{SS} (normally ground), through which the input capacitance is discharged. The power drain occurs only during the charging cycle; no power is drawn from V_{DD} during discharge. This input capacitance limits the operating speed of CMOS circuits to about 7 megahertz for a fan-out of 1. An increase in fan-out reduces the operating frequency. Fan-out is therefore limited more by problems of timing and pulse width than by problems of sustaining current flow to and from the load.

PARAMETER	TEMP. RANGE	V_{PP} (V DC)	CONDITIONS	LIMITS T_{LOW}* MIN	T_{LOW}* MAX	+25° C MIN	+25° C TYP	+25° C MAX	T_{HIGH}* MIN	T_{HIGH}* MAX	UNITS
I_{DD} (Quiescent Device Current)	Mil	5	$V_{IN} = V_{SS}$ or V_{PP}	–	0.25	–	–	0.25	–	7.5	μA DC
		10		–	0.5	–	–	0.5	–	15	
		15		–	1.0	–	–	1.0	–	30	
Gates		5	All valid input combinations	–	1.0	–	–	1.0	–	7.5	μA DC
		10		–	2.0	–	–	2.0	–	15	
		15		–	4.0	–	–	4.0	–	30	
	Mil	5	$V_{IN} = V_{SS}$ or V_{PP}	–	1.0	–	–	1.0	–	30	μA DC
		10		–	2.0	–	–	2.0	–	60	
		15		–	4.0	–	–	4.0	–	120	
Buffers, Flip-Flops	Comm	5	All valid input combinations	–	4.0	–	–	4.0	–	30	μA DC
		10		–	8.0	–	–	8.0	–	60	
		15		–	16.0	–	–	16.0	–	120	
	Mil	5	$V_{IN} = V_{SS}$ or V_{DD}	–	5	–	–	5	–	150	μA DC
		10		–	10	–	–	10	–	300	
		15		–	20	–	–	20	–	600	
MSI	Comm	5	All valid input combinations	–	20	–	–	20	–	150	μA DC
		10		–	40	–	–	40	–	300	
		15		–	80	–	–	80	–	600	
V_{OL} (Low-Level Output Voltage)	All	5	$V_{IN} = V_{SS}$ or V_{DD}	–	0.05	–	–	0.05	–	0.05	V DC
		10		–	0.05	–	–	0.05	–	0.05	
		15	$\lvert I_O \rvert \leqslant 1\ \mu A$	–	0.05	–	–	0.05	–	0.05	
V_{OH} (High-Level Output Voltage)	All	5	$V_{IN} = V_{SS} = V_{DD}$	4.95	–	4.95	–	–	4.95	–	V DC
		10		9.95	–	9.95	–	–	9.95	–	
		15	$\lvert I_O \rvert \leqslant 1\ \mu A$	14.95	–	14.95	–	–	14.95	–	
V_{IL} (Input Low Voltage)	All	5	V_O = 0.5V or 4.5V	–	1.5	–	–	1.5	–	1.5	V DC
		10	1.0V or 9.0V	–	3.0	–	–	3.0	–	3.0	
		15	1.5V or 13.5V; $\lvert I_O \rvert \leqslant 1\ \mu A$	–	4.0	–	–	4.0	–	4.0	
V_{IH} (Input High Voltage)	All	5	V_O = 0.5V or 4.5V	3.5	–	3.5	–	–	3.5	–	V DC
		10	1.0V or 9.0V	7.0	–	7.0	–	–	7.0	–	
		15	1.5V or 13.5V; $\lvert I_O \rvert \leqslant 1\ \mu A$	11.0	–	11.0	–	–	11.0	–	
I_{OL} (Output Low or Sink Current)	Mil	5	V_O = 0.4V, V_{IN} = 0 or 5V	0.64	–	0.51	–	–	0.36	–	mA DC
		10	V_O = 0.5V, V_{IN} = 0 or 10V	1.6	–	1.3	–	–	0.9	–	
		15	V_O = 1.5V, V_{IN} = 0 or 15V	4.2	–	3.4	–	–	2.4	–	
	Comm	5	V_O = 0.4V, V_{IN} = 0 or 5V	0.52	–	0.44	–	–	0.36	–	mA DC
		10	V_O = 0.5V, V_{IN} = 0 or 10V	1.3	–	1.1	–	–	0.9	–	
		15	V_O = 1.5V, V_{IN} = 0 or 15V	3.6	–	3.0	–	–	2.4	–	
I_{OH} (Output High or Source Current)	Mil	5	V_O = 4.6V, V_{IN} = 0 or 5V	-0.25	–	0.2	–	–	-0.14	–	mA DC
		10	V_O = 9.5V, V_{IN} = 0 or 10V	-0.62	–	-0.5	–	–	-0.35	–	
		15	V_O = 13.5V, V_{IN} = 0 or 15V	-1.8	–	-1.5	–	–	-1.1	–	
	Comm	5	V_O = 4.6V, V_{IN} = 0 or 5V	-0.2	–	-0.16	–	–	-0.12	–	mA DC
		10	V_O = 9.5V, V_{IN} = 0 or 10V	-0.5	–	-0.4	–	–	-0.3	–	
		15	V_O = 13.5V, V_{IN} = 0 or 15V	1.4	–	-1.2	–	–	-1.0	–	
I_{IN} (Input Current)	Mil	15	V_{IN} = 0 or 15V	–	±0.1	–	–	±0.1	–	±1.0	μA DC
	Comm	15	V_{IN} = 0 or 15V	–	±0.3	–	–	0.3	–	±1.0	μA DC
C_{IN} (Input Capacitance per Unit Load)	All	–	Any input	–	–	–	–	7.5	–	–	

*T_{LOW} = -55° C for military temperature range devices, -40° C for commercial temperature range devices.
*T_{HIGH} = +125° C for military temperature range devices, +85° C for commercial temperature range devices.

FIGURE 9-36

Electrical characteristics for B-series CMOS.

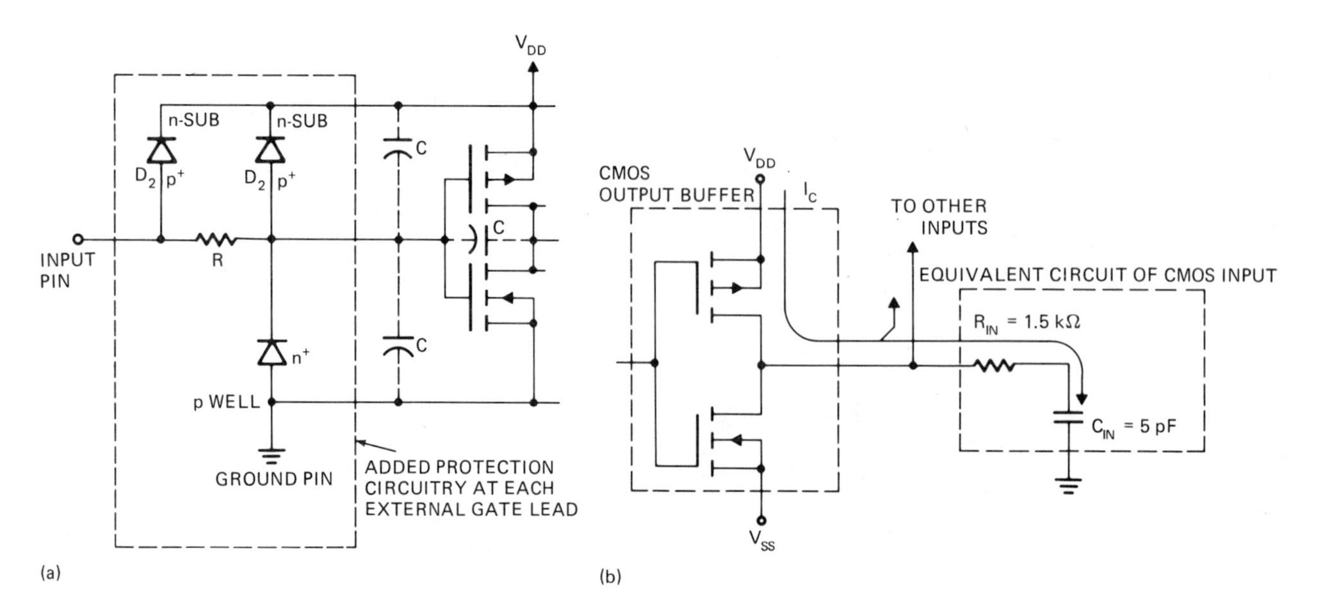

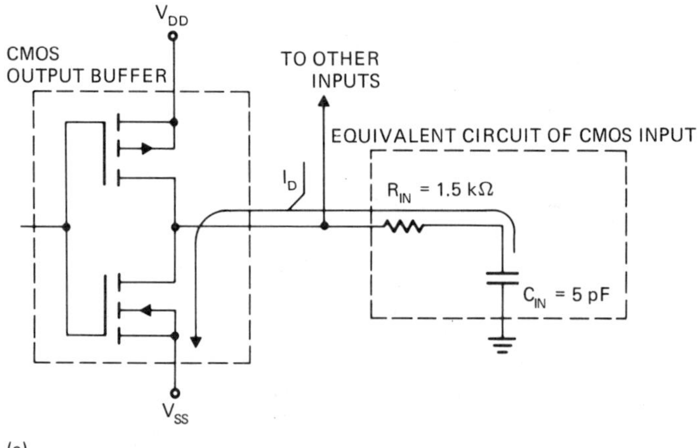

FIGURE 9-37

(a) Gate-oxide protection circuit used in CMOS integrated circuits. (b) CMOS IC switching to high state. (c) CMOS IC switching to low state.

The input voltage parameters—minimum input high level (V_{IH}) and maximum input low level (V_{IL})—are the 1 and 0 input levels that guarantee an output 1 level no lower than 90% of V_{DD} and an output 0 level no higher than 10% of V_{DD}. These parameters are provided in the table in Figure 9-38.

In part (a) of Figure 9-39, a rectangular pulse is shown with ideal digital levels $V_{IL} = 0V$ and $V_{IH} = V_{DD}$. If the ideal pulse is applied to the input of a CMOS gate (enabled-noninverting), the output levels, V_{OL} and V_{OH}, will be almost the same as the input levels, V_{IL} and V_{IH}. However, if for some reason the input levels are degraded to the levels specified for V_{IL} MAX and V_{IH} MIN, the output signal will deviate from the ideal ($V_{IL} = 0V$ and $V_{IH} = V_{DD}$) by not more than 10% of V_{DD}.

9.14.3 Noise Immunity

Noise Margin (V_N). V_{NL} is a measure of the degree to which a CMOS circuit reacts to random noise at logic level 0. This can be determined, as shown in part (b) of Figure 9-39, by the difference between V_{IL} and $V_{OL} = 10\% \times V_{DD}$. The level increases with V_{DD}: 1.0V at $V_{DD} = 5V$ and 2.5V at $V_{DD} = 15V$. This higher noise margin is

FIGURE 9-38

Voltage parameters of CMOS circuits.

SUPPLY VOLTAGE	INPUT LIMITS		OUTPUT LIMITS		NOISE MARGIN	
$V_{DD} - V_{SS}$	V_{IL} MAX	V_{IH} MIN	V_{OL} MAX	V_{OH} MIN	V_{NL}	V_{NH}
5V	1.5V	3.5V	0.5V	4.5V	1.0V	1.0V
10V	3.0V	7.0V	1.0V	9.0V	2.0V	2.0V
15V	4.0V	11.0V	1.5V	13.5V	2.5V	2.5V

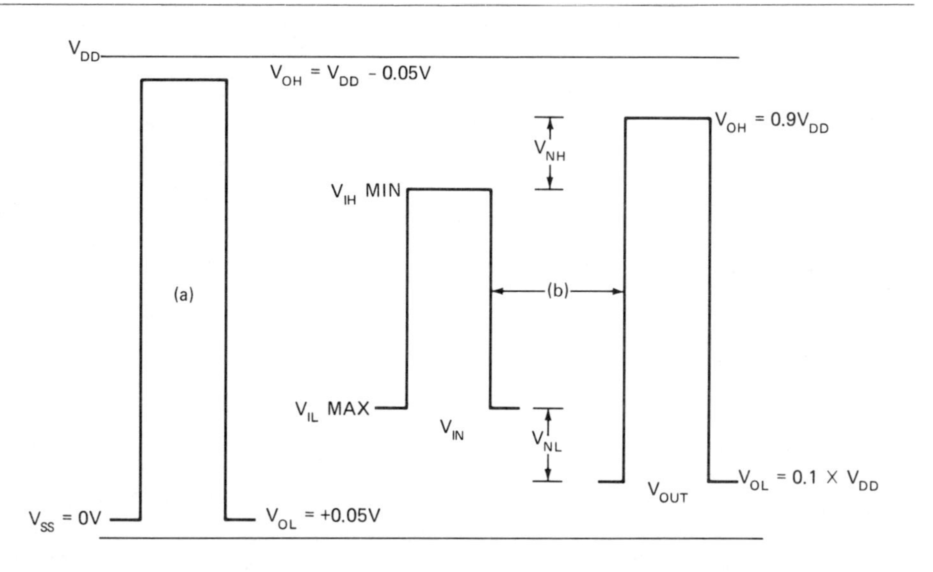

FIGURE 9-39

Digital pulses at B-series CMOS voltage levels.

one of several advantages when using the higher supply voltage level. (The disadvantages are a higher device failure rate and higher power dissipation.)

V_{NH} is a measure of the degree to which a CMOS circuit reacts to random noise at logic level 1. This can be determined, as shown in part (b) of Figure 9-39, by the difference between V_{IH} and V_{OH} = 90% of V_{DD}. The level is again 1V at V_{DD} = 5V and 2.5V at V_{DD} = 15V. Even at V_{DD} = 5V, the noise margins for CMOS circuits are twice as wide as for TTL circuits and become even wider as the supply voltage is increased.

Driver and Fan-out Impedance. The noise margin is not the only factor determining the noise immunity of CMOS circuits. Impedance of both the driver and its fan-out practically determine the amount of noise energy transferred from a noise generator to the circuit input. An understanding of this may be aided by the equivalent noise generator circuit of Figure 9-40. The driver impedance loads down high impedance noise generators. The typical input circuit presents a 3 dB cutoff at 21.2 megahertz to the noise spectrum.

CMOS circuits are helped by the fact that the CMOS driver has a fairly low output resistance, while the combined generator and coupling impedance of most noise generators, particularly those external to the circuit, are very high. Conversely, impedance of an open CMOS input is extremely high and very susceptible to noise operation.

The typical CMOS input has a high-frequency cutoff much lower than that of TTL circuits. The result of this is that noise pulses narrower than 50 nanoseconds are highly attenuated. This is shown in the graph of Figure 9-41(b).

Since this frequency response also limits the operating speed of CMOS systems, noise generated within the system may have sufficient pulse width to cause noise errors. Three types of internally generated noise must be considered: crosstalk between adjacent signal lines, power line noise, and ground noise.

Crosstalk. Crosstalk is noise caused by unwanted coupling between adjacent signal lines. This is seldom a problem in CMOS circuits unless a number of parallel signal lines switch simultaneously while one or more line(s) in the group

FIGURE 9-40

Equivalent noise generator circuit of a CMOS gate with a single fan-out.

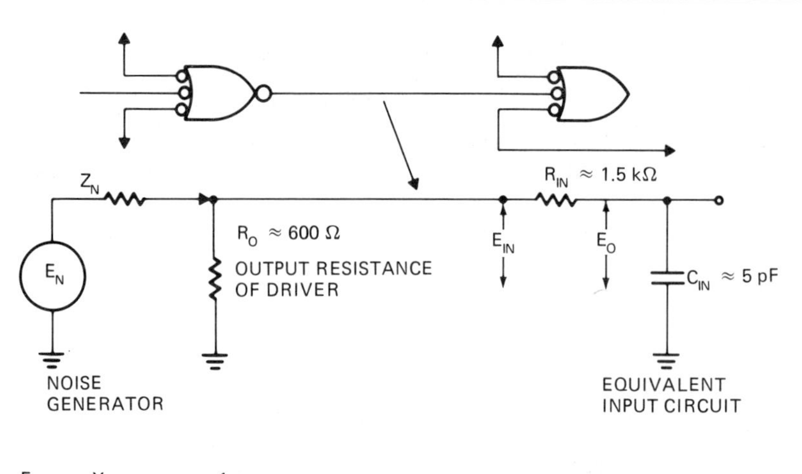

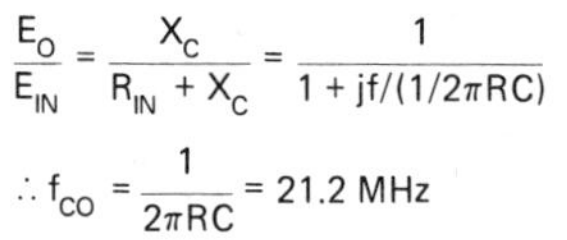

$$\frac{E_O}{E_{IN}} = \frac{X_C}{R_{IN} + X_C} = \frac{1}{1 + jf/(1/2\pi RC)}$$

$$\therefore f_{CO} = \frac{1}{2\pi RC} = 21.2\ \text{MHz}$$

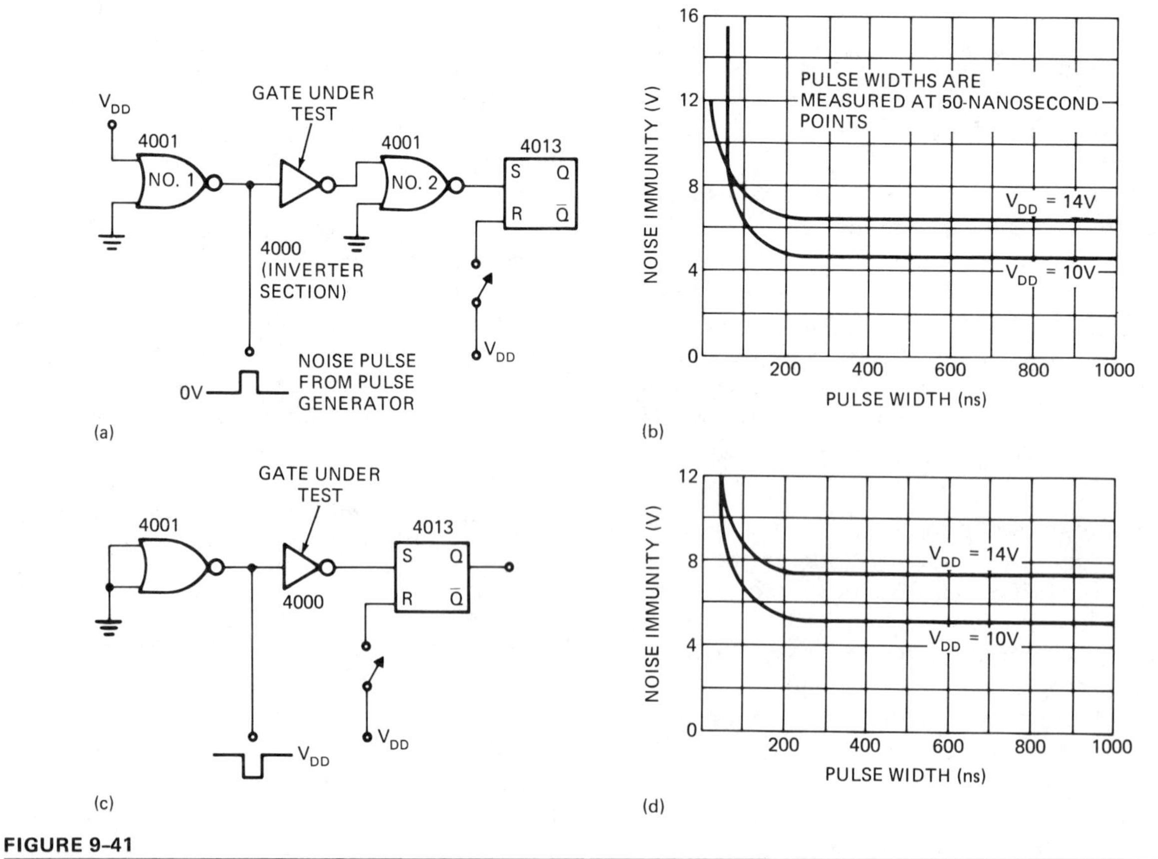

FIGURE 9-41

Noise immunity test.

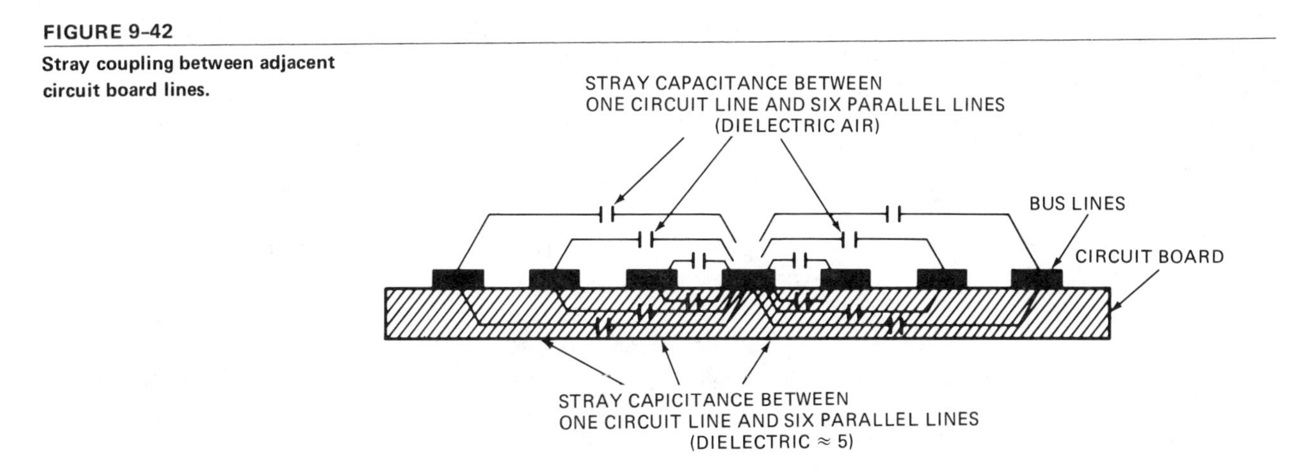

FIGURE 9-42

Stray coupling between adjacent circuit board lines.

remain(s) inactive. Figure 9-42 shows stray coupling between adjacent circuit board lines. The amplitude of the voltages that are coupled tends to add, increasing the likelihood that the inactive line will reach the switching level. The probability of a noise error due to crosstalk can be reduced by minimizing either the length or width of the parallel signal lines, increasing the distance between lines, or, where possible, delaying the clock pulse used to strobe the levels into registers at the receiving end of the parallel lines so as to sufficiently avoid storing the noise voltage.

Power Line Noise. In order to cause an error, negative noise pulses on power lines must have amplitude and pulse width approximately equal to those needed on signal lines. Considering the extremely high capacity of power lines, it is unlikely that this amount of energy would be coupled to power lines from external noise generators. IR drops from simultaneous switching of a large number of drivers could, however, cause a noise pulse of significant size. Proper placement of decoupling capacitors on the CMOS circuit board and use of double-clad circuit boards, with ground and power lines etched on opposite sides to form a distributed capacitance, as shown in Figure 9-43, is also helpful. The power line can be etched above the ground line to produce a distributed capacitance that will help filter noise from the power line.

Ground Line Noise. Similar to power line noise, ground line noise results from the IR drop occurring when a large number of devices switch simultaneously. However, the ground line cannot be protected by decoupling capacitors. CMOS circuit cards should not share the same ground or return line as high-current systems—in particular, those containing relays, solenoid, and motors.

Additive Effect of Noise. Due to the high noise margins of CMOS circuits, the probability of a noise voltage from any one of the previously mentioned sources reaching the switching threshold seems remote. However, noise voltages from two or more sources all too often combine to cause periodic errors. As this type of failure is extremely difficult to isolate, it is important that all reasonable measures be taken to minimize noise levels.

9.14.4 CMOS Power Requirements

Maximum power requirements for a given CMOS system are more difficult to predict, with any degree of accuracy, than are those for an equivalent bipolar system. The number of parameters altering the power drain and the range of variations are much greater than for bipolar systems.

In order to predict the worst case of total power dissipation for each circuit in a given CMOS system, it is necessary to determine the maximum switching frequency (f) and the load

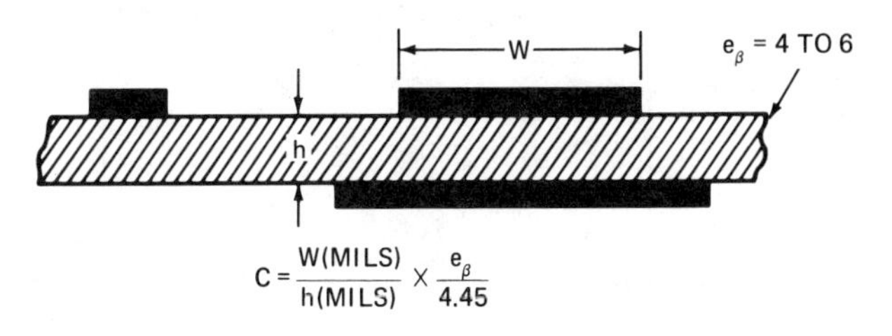

$$C = \frac{W(\text{MILS})}{h(\text{MILS})} \times \frac{e_\beta}{4.45}$$

FIGURE 9-43

Use of printed circuit board capacitance to filter power line noise.

capacitance (C_L). C_L is determined in most applications by multiplying the fan-out by the C_{IN} MAX of 7.5 pF. Where convenient, P_T for a given package can be determined from a power-versus-frequency graph, similar to that shown in Figure 9-44 for the CD4042B. This graph shows the typical frequency dependence of power dissipation, which, for this device at V_{DD} = 15V, increases from 20 μW to 200 mW as the clock frequency increases from 1 kHz to 10 MHz. Unfortunately, the lines of this graph are given for C_L = 15 pF or C_L = 50 pF, and it is assumed that all outputs are loaded with the same load capacitance. Normally, however, a circuit with numerous outputs may have some outputs that are not loaded at all, and the rest may have a wide variation in fan-out. Averaging and interpolation could be used to obtain an estimate, but that would be inconvenient. It is usually more accurate and convenient to estimate power on an output basis rather than by circuit. This is more readily understood if three mechanisms of power dissipation in a CMOS circuit are examined.

There are three distinct sources of current flow: *quiescent current* (I_{DD}), *through current* (I_{TC}), and *load current* (I_L). The total current is the result of all three current flows. Quiescent current (I_{DD}) is the only current given in the specification because it is independent of fan-out or the signals being processed. I_{DD} is the result of the leakage current through reverse-biased junctions within the circuit. This current may be as low as 50 pA for simple gate circuits and as high as 20 μA for more complicated MSI functions. Total quiescent power ($V_{DD} \cdot I_{DD}$) can be obtained by adding up the I_{DD} specified for each circuit in the system. However, when the operating frequency exceeds 100 kilohertz, this portion of the total power becomes insignificant.

Through current (I_{TC}) is the result of both the p- and n-channel transistors being partially turned on during the switching of a CMOS device, both from high to low and low to high. This current occurs on both leading and trailing edges of a switching waveform, as shown in Figure 9-45.

The power dissipated due to through current (P_{TC}) is proportional to rise and fall time ($t_r + t_f$), switching frequency (f), and V_{DD}^3, as we see from the following formula:

$$P_{TC} \propto f(t_r + t_f)V_{DD}^3$$

The through current portion of the power drain is not significant for rise and fall times of less than a few hundred nanoseconds. Even if waveforms with transition times longer than 200 nanoseconds are transmitted to the inputs of the CMOS system, it would affect the power drain of the interface circuits only. The interface circuits would speed up the transition times to a point where the through currents would be

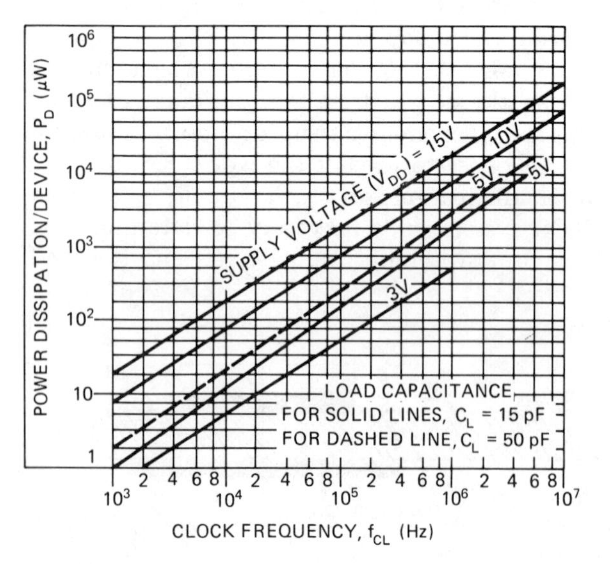

FIGURE 9-44

Total power dissipation versus clock frequency of the CMOS 4042B operating with V_{DD} at 3, 5, 10, and 15V.

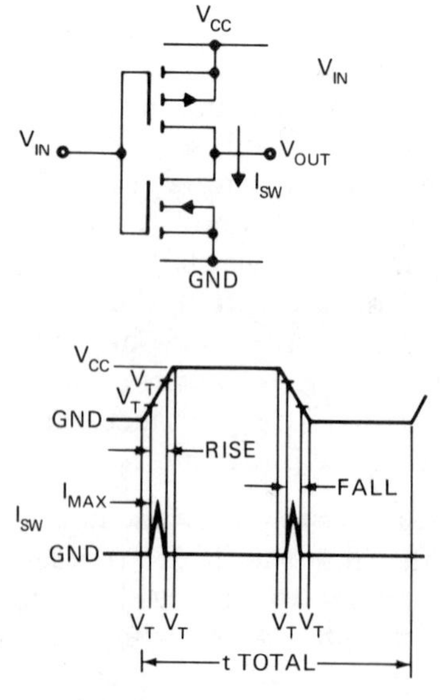

FIGURE 9-45

Through current, a component of CMOS power drain.

insignificant throughout the remainder of the system. Output rise and fall times increase in situations where large fan-outs are involved. This does not affect the through current of the driving circuit, but does increase the through current for all the circuits being driven. If power consumption is critical, it may prove useful to divide the fan-out between two or more buffers. Through current need not be considered except for these special circumstances.

The third component of power dissipation, the load current (I_L), results from the charging and discharging of the load capacitance, C_L. This component of the power dissipation increases linearly with the load capacitance, the switching frequency, and the square of the supply voltage ($P_L = C_L \cdot f \cdot V_{DD}^2$).

At the operating frequency of 10 kHz, this component of the power drain exceeds the quiescent power by an order of magnitude and becomes the only component worth considering in

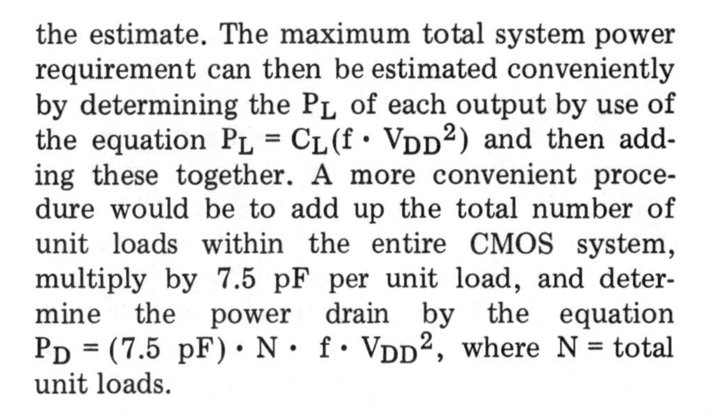
the estimate. The maximum total system power requirement can then be estimated conveniently by determining the P_L of each output by use of the equation $P_L = C_L(f \cdot V_{DD}^2)$ and then adding these together. A more convenient procedure would be to add up the total number of unit loads within the entire CMOS system, multiply by 7.5 pF per unit load, and determine the power drain by the equation $P_D = (7.5\ \text{pF}) \cdot N \cdot f \cdot V_{DD}^2$, where N = total unit loads.

9.14.5 CMOS Output Circuits

CMOS circuits are available with three types of output drivers: double buffer, transmission gate, and three-state output.

Double Buffer Output. Formation of basic NAND and NOR functions, as shown in Figure 9-46(a) and (b), requires either sink or source

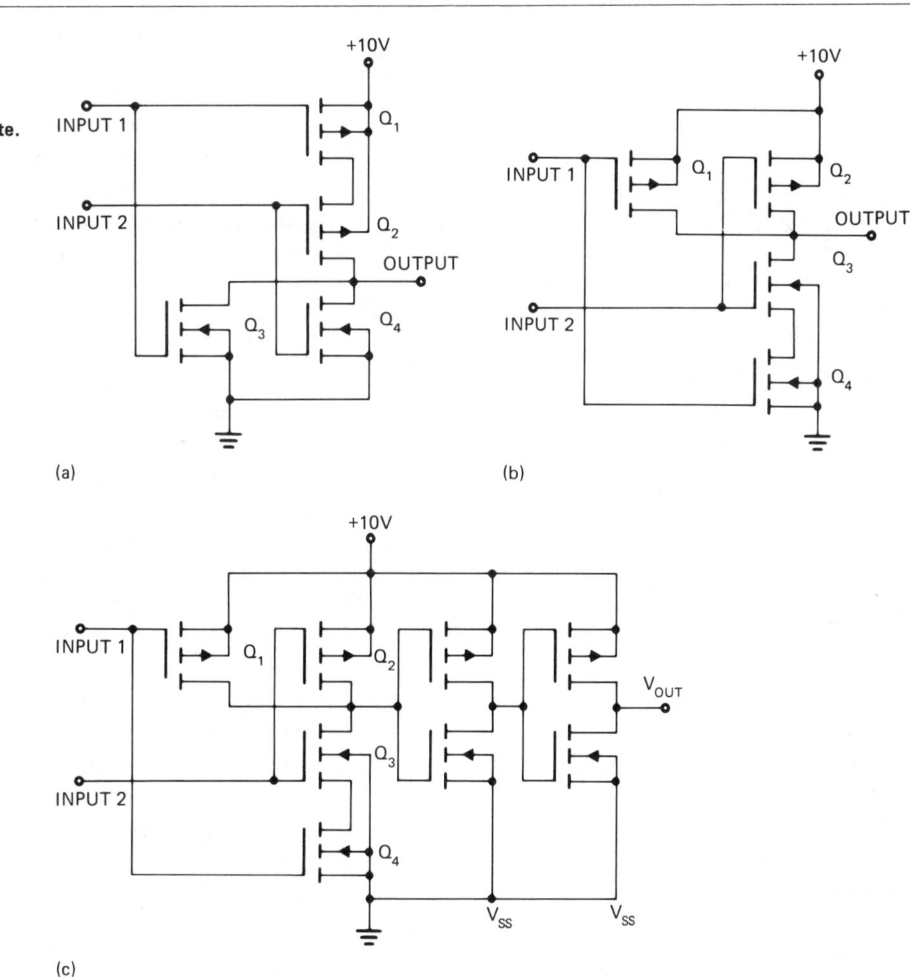

FIGURE 9-46
A-series outputs compared to double-buffered B-series. (a) A-series two-input NOR gate. (b) A-series two-input NAND gate. (c) B-series two-input NAND gate.

current to be drawn through parallel sets of transistors while the other current is drawn through an equal number of series transistors. The number of transistors in series and parallel increases with the number of inputs per gate. This results in a wide range of sink and source impedance for A-series CMOS devices. B-series CMOS devices have double-buffered outputs, as shown in Figure 9-46(c). This provides for a balanced sink and source impedance. (For clarity, input protection network is not shown.)

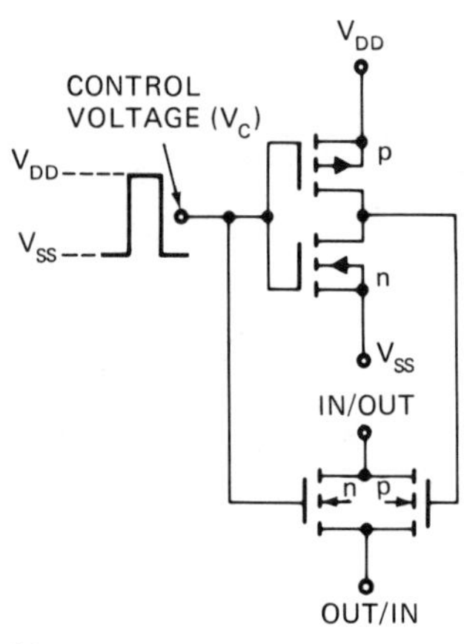

(a)

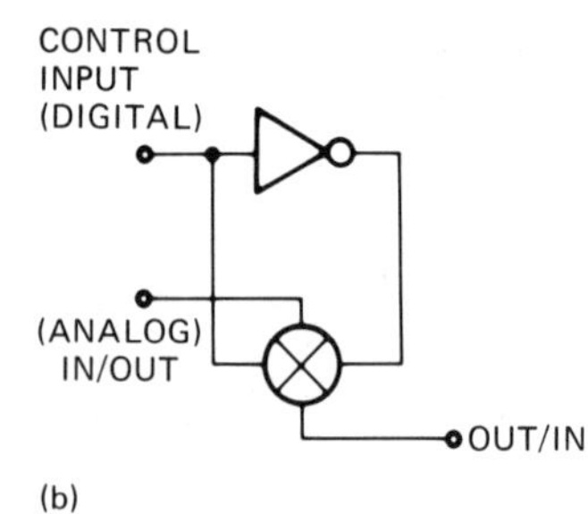

(b)

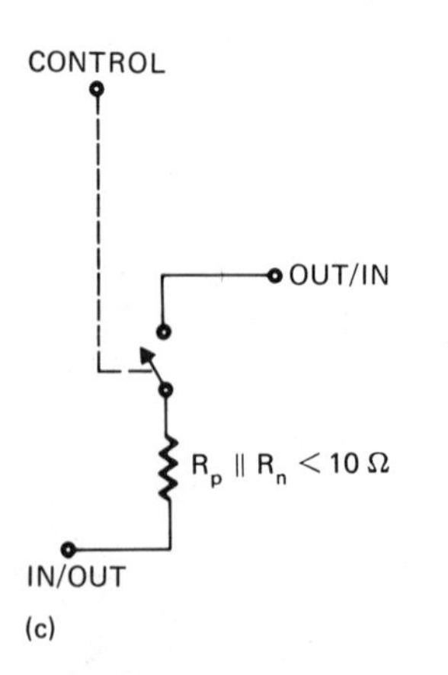

(c)

FIGURE 9-47

Transmission gate. (a) Schematic diagram. (b) Logic diagram. (c) Equivalent circuit.

Transmission Gate Output. The transmission gate is a digitally controlled gate, as shown in Figure 9-47. It is composed of both p- and n-channel transistors connected in parallel.

Complemented digital levels are applied to the gate inputs, causing both devices to be off or on at the same time. This provides a reasonably uniform "on" resistance to current flow in either direction; unlike most digital switches, the input and output terminals are interchangeable. If the inputs are digital levels and the outputs are applied to digital circuits, these devices operate as entirely digital. The inputs, however, may be analog voltages. The analog voltage excursions are limited by power supply levels V_{DD} and V_{SS}.

A few circuits, such as 4051, 4052, and 4053, have a third power pin, V_{EE}. The digital levels controlling the gates in these circuits remain V_{DD} and V_{SS}, but the analog excursion of signals passing through the gates is limited by V_{DD} and V_{EE}. V_{EE} can be more negative than V_{SS}. With V_{SS} at signal ground and V_{EE} negative, the analog signals passing through the gate may have both positive and negative voltage excursions. Transmission gates are typically used for multiplexing, demultiplexing, and three-state bus applications. A typical multiplex operation is shown in Figure 9-48. As the counter operates, it periodically and sequentially switches one of the eight inputs through a transmission gate onto the single output.

A problem that occurs with the multiplex operation is that the turn-off delay of the transmission gate being removed from the output is always longer than the turn-on delay of the transmission gate being enabled. This leaves a short time period during which two drivers are simultaneously applied to the outputs. With one driver in the low state and the other in the high state, a low impedance path exists between V_{DD} and V_{SS}, as shown in Figure 9-49.

Where digital levels are involved, this current drain will last no more than 750 nanoseconds. Special precautions may not be necessary except to ensure that adequate decoupling capacitors are located on the circuit board close to the driving circuits and that consideration is given to the increased power generated.

If the power drain proves excessive, the disable input can be used to turn off all transmission gates long enough to allow the decoding circuits to switch to the new channel. The disable signal can then be removed, allowing the single

FIGURE 9-48
CMOS circuit 4051 connected as an eight-channel to one-channel multiplexer.

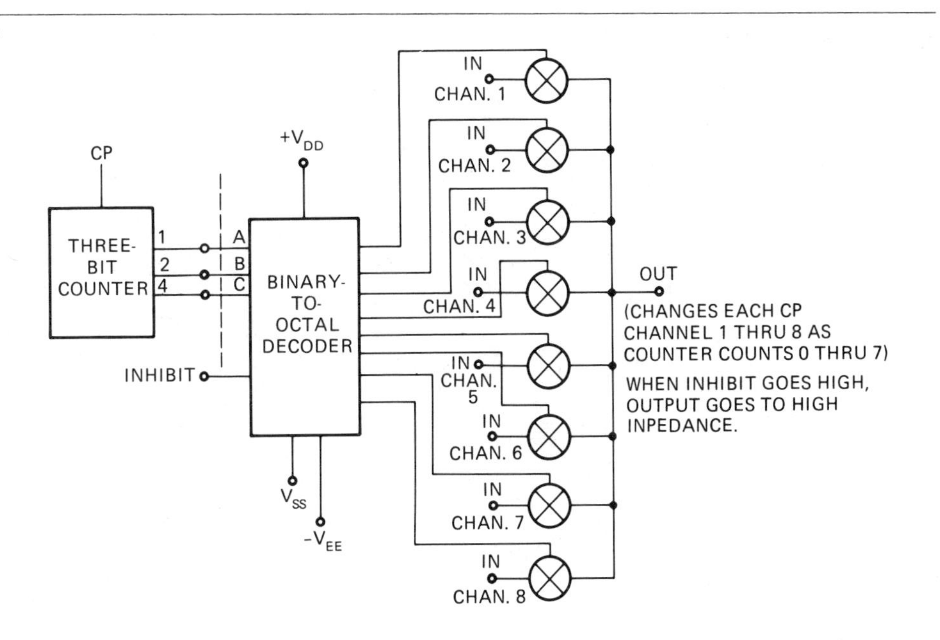

driver to take control of the line. If the disable signal is to remain on the multiplexer for more than 100 milliseconds, another precaution must be taken.

As mentioned previously, inputs to CMOS circuits must not be open-circuited. If all drivers are switched off the bus, a momentarily high impedance approximately equivalent to an open circuit exists. A solution to this is a pull-up or pull-down resistor. An ideal resistor value for most applications is approximately 100 kilohms.

The demultiplex operation, shown in Figure 9-50, uses a single driver circuit that is sequentially switched through transmission gates to one of numerous outputs. Since only one of the inputs is controlled by a driver at any given time, the remaining inputs are in the high-impedance state. This problem must be solved by using pull-up or pull-down resistors, whichever are appropriate.

Three-State Output. CMOS circuits with standard buffer outputs should not be connected in wired OR configurations. The three-state output circuit should be used for all applications that require the tying together of outputs on a common bus. Figure 9-51 shows several methods of implementing the three-state output buffer. Operation of the three-state bus differs from the open collector in that a disable signal must be applied to every output except the output controlling the bus. Under this condition, both the p- and n-channels of the disabled circuits are turned off. The single-enabled driver determines the logic state of the bus line.

The leakage current of a disabled driver is so low that it has little effect on the DC loading of the enabled driver. The capacitance of a disabled driver may vary from 8 pF to 15 pF and

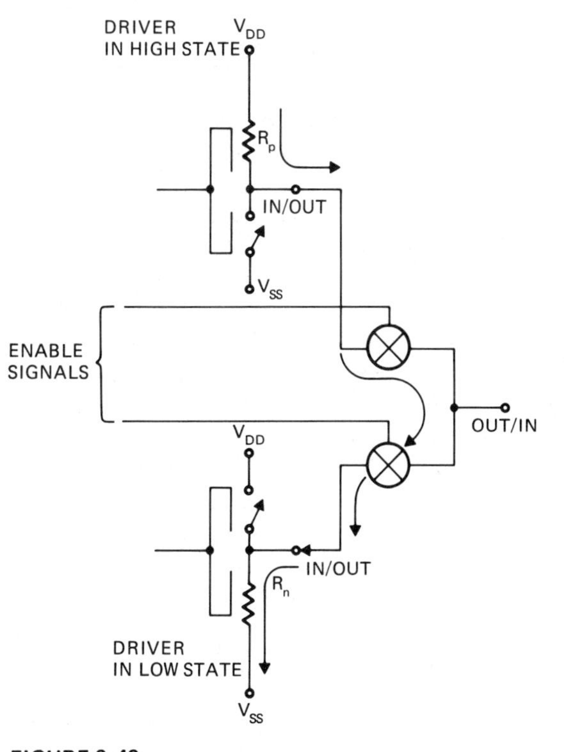

FIGURE 9-49
Two drivers on line simultaneously during multiplexing.

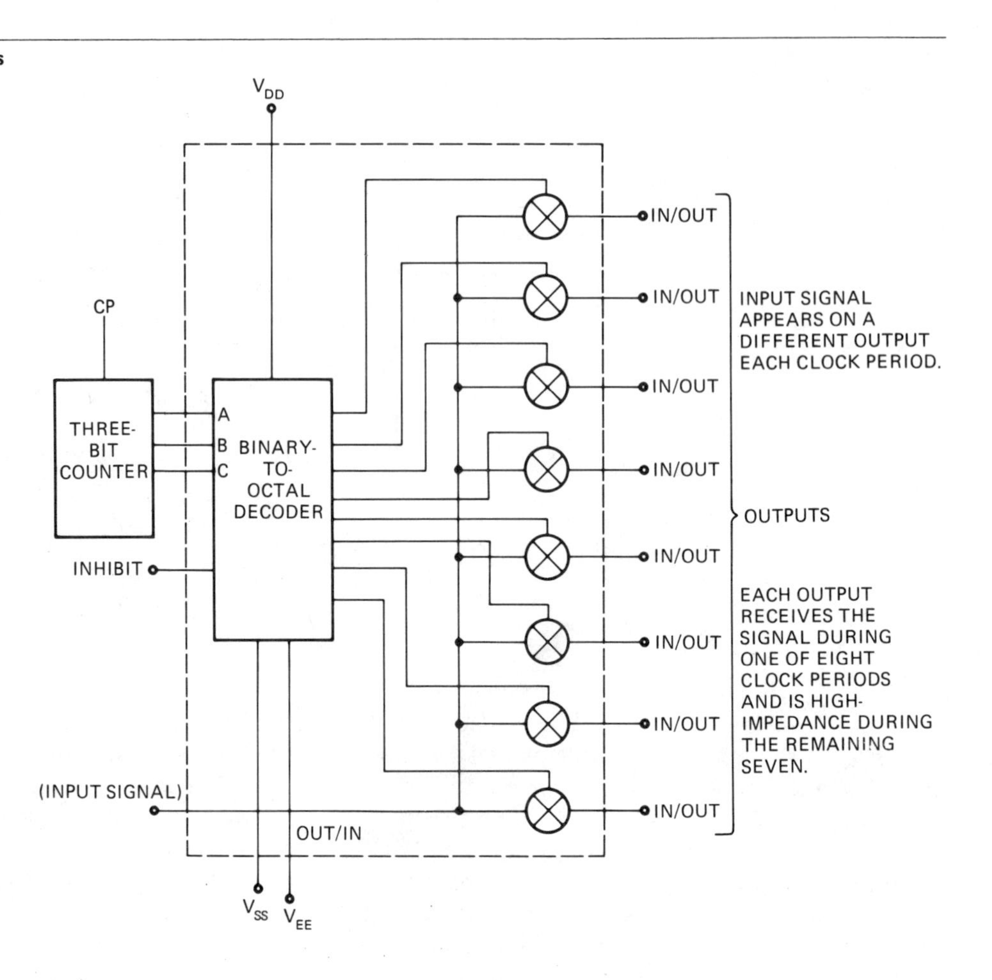

FIGURE 9-50
CMOS circuit 4051 connected as one-channel to eight-channel demultiplexer.

will have an effect on the rise and fall time, propagation delay, and AC power drain. For the sake of convenience, each disabled driver may be counted as two unit-capacitive loads.

Timing for a three-state bus is more troublesome than for an equivalent open-collector bus. There is no chance with an open-collector bus that the bus will be open-circuited (high impedance) or that the drivers will momentarily oppose each other when control of the bus transfers from one driver to another. Both of these possibilities must be guarded against with the three-state bus. If the driver coming on line is enabled before the driver going off the line is inhibited, the two drivers will be on the line simultaneously. With one driver in the low state and the other high, a low-resistance path exists between V_{DD} and V_{SS} through the two drivers. On the other hand, if the driver coming off the line goes to high impedance before the driver coming on the line leaves the high-impedance state, the inputs on the bus line are open-circuited (high impedance), leaving them susceptible to oscillation or noise triggering.

Applications that leave all drivers on a bus line in the high-impedance state for a period in excess of 100 milliseconds require a pull-up resistor (to V_{DD}) or a pull-down resistor (to V_{SS}) on each bus line. This avoids the difficulties experienced with open-circuited inputs but may have an adverse effect on rise and fall time and will increase power consumption.

9.14.6 Switching Frequency

The graph in Figure 9-41 indicates that CMOS circuits switch upon reception of rectangular pulses of 50 nanoseconds or wider. This seems to indicate that a switching speed of up to 10 megahertz could be used. Although circuits could be tested and screened for operation at this switching speed, both the advantage and the reliability of using CMOS at continuous switching rates at this high a frequency are question-

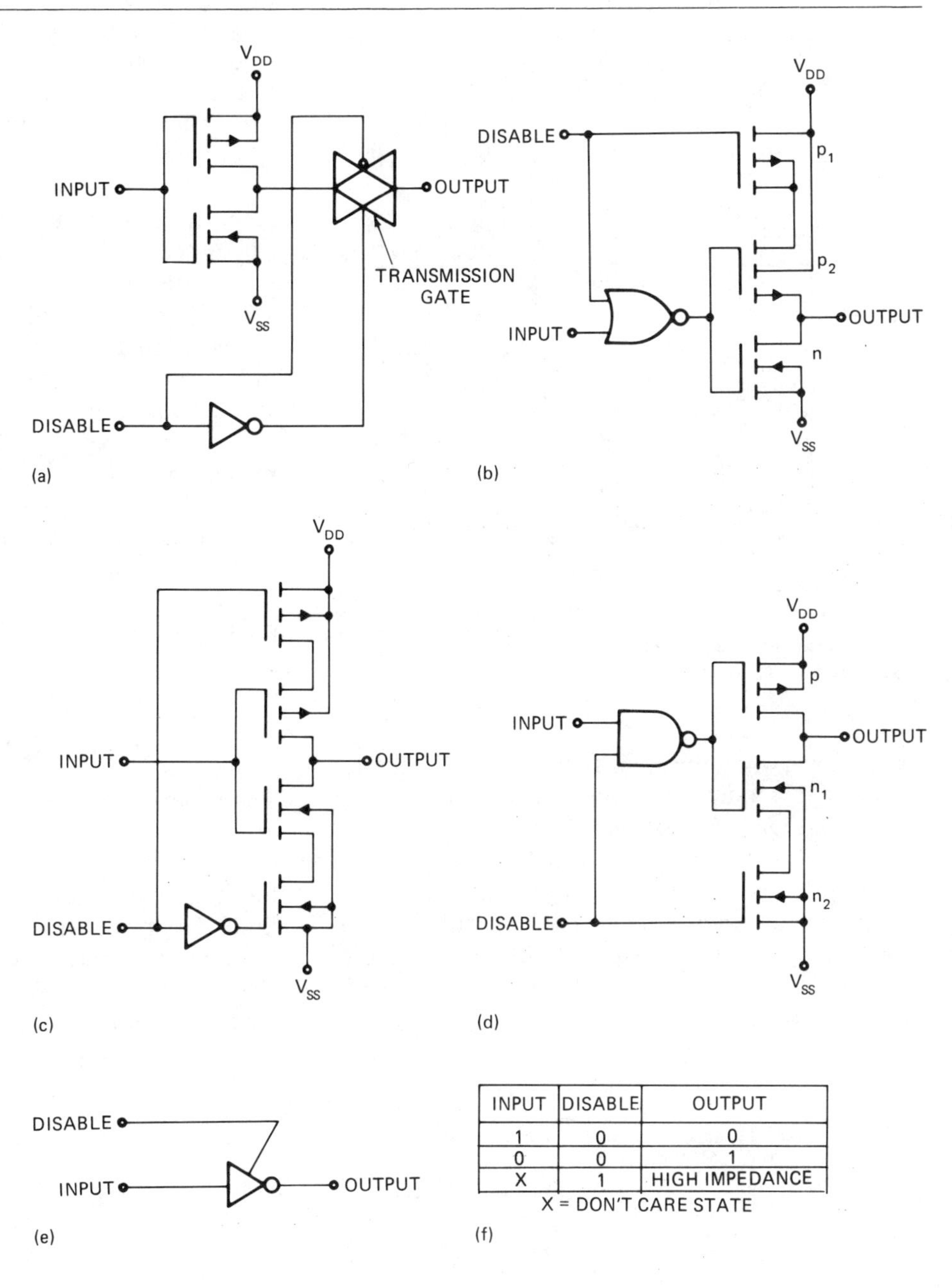

INPUT	DISABLE	OUTPUT
1	0	0
0	0	1
X	1	HIGH IMPEDANCE

X = DON'T CARE STATE

FIGURE 9-51

Methods used to produce a three-state output. (a) Using a transmission gate. (b) through (d) Three-state by turning off both a p- and n-transistor. (e) Logic symbol for three-state buffer. (f) Truth table.

able. The graph of Figure 9-52 shows that, at frequencies above 2 megahertz, a CMOS gate with a 50 pF load and a V_{DD} greater than 10V dissipates more power than even standard TTL circuits. If the fan-out is reduced to a single-unit load (7.5 pF), only then is the power dissipation of a 10V CMOS gate below that of TTL circuits. The lines in Figure 9-52 indicate that power dissipation of CMOS circuits at 5V remains below that of TTL to 3 megahertz. However, due to the higher driver impedance of CMOS circuits at 5V, not all standard B-series gates will operate at that high a frequency. Operation at 5V also reduces the noise immunity advantage of CMOS over TTL. Conversely, if circuits are processing signals at a high frequency rate for only a small percentage of the total time that the system power is turned on, then the advan-

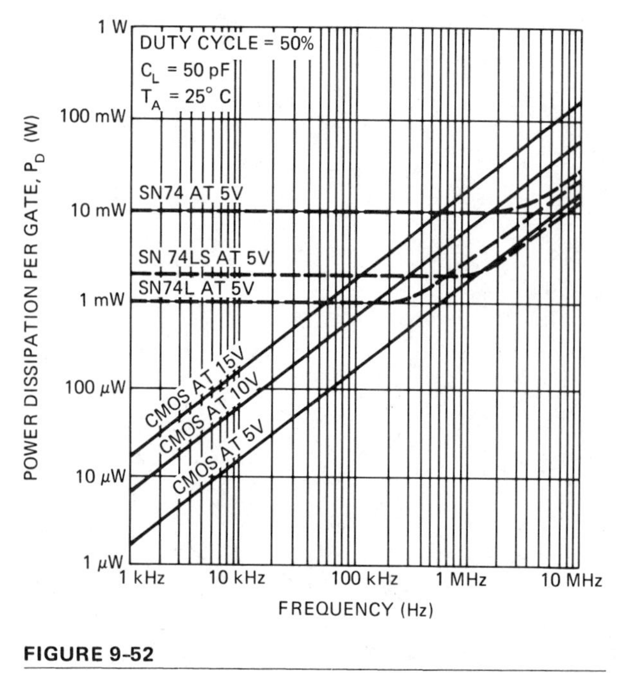

FIGURE 9-52

Texas Instruments graph of power dissipation per gate versus frequency for TTL and CMOS.

tage is overwhelmingly in favor of CMOS due to its very low quiescent power as compared to a continued high dissipation rate by TTL.

Whether operating at a high or low switching rate, there is a limit to how narrow a pulse width can be used. Factors limiting pulse width are found mainly between the driver circuit and the input. An equivalent circuit of these parameters (worst case at 10V V_{DD}) is shown in Figure 9-53. A minimum reliable pulse width may be estimated by either time constant or frequency response calculations applied to this circuit.

The minimum pulse width applies to both positive and negative-going pulses. The value for n should be the maximum fan-out at any point along a signal path for which the pulse width is calculated. As shown in Figure 9-54, the minimum pulse width varies from 65 nanoseconds at 7.5 pF load to 158 nanoseconds at 52.5 pF. Using a square wave, this corresponds to maximum switching frequencies of 7.8 megahertz and 3.16 megahertz, respectively.

In addition to the problem of minimum pulse width, tighter limits for switching frequency may be needed due to race problems caused by propagation delays. A number of CMOS MSI circuits call for clock speeds that are more restrictive of operating frequency and pulse widths.

9.14.7 Input Circuit Protection

The functional elements of a CMOS input circuit are the connections to the gates of p- and n-channel transistors. The dielectric of the gate input is a thin layer of silicon dioxide 1000 to 2000 angstroms thick with a breakdown voltage of approximately 100V to 120V. Electrostatic charges generated during normal handling can result in voltages exceeding this level. For this

FIGURE 9-53

Equivalent circuit of CMOS driver with n-input fan-out.

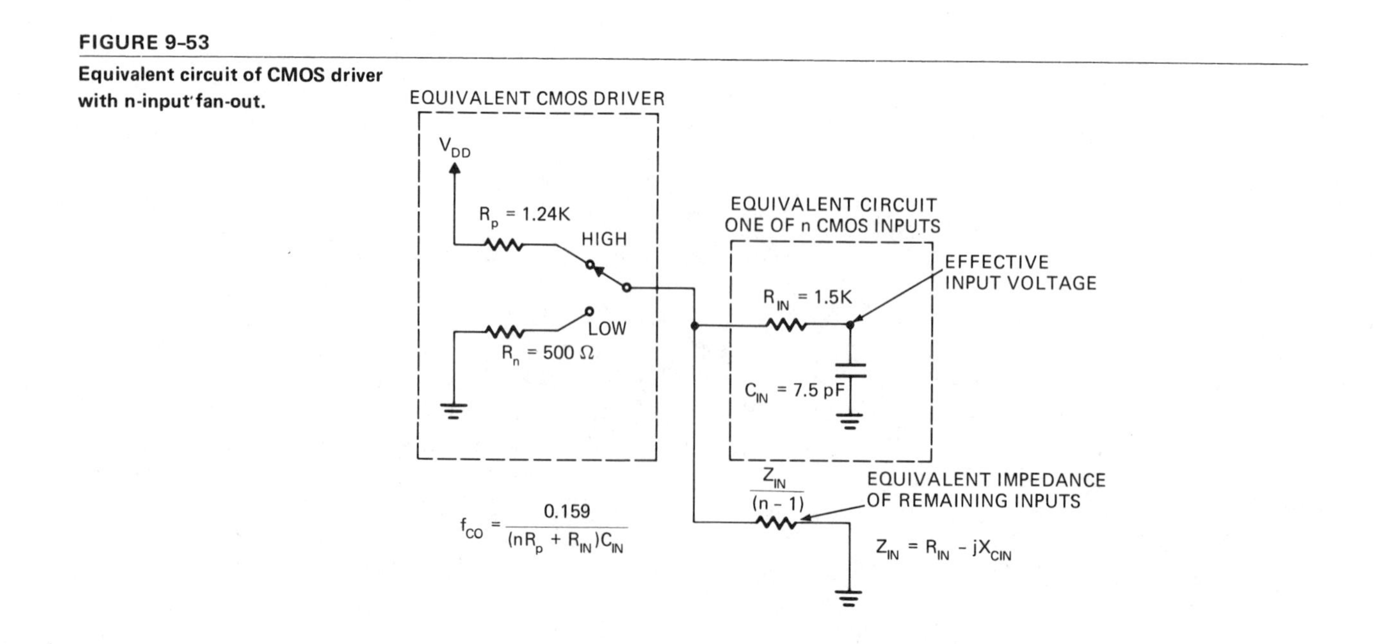

FIGURE 9-54
Minimum pulse width and switching frequency versus unit loads for CMOS at V_{DD} = 12V.

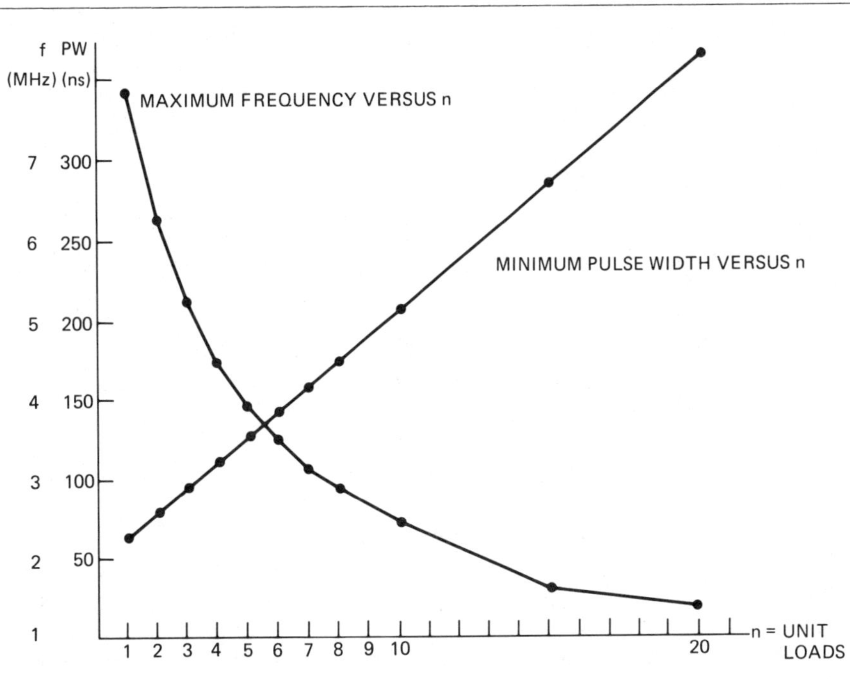

reason, each input circuit is preceded by a diode-resistor protection network, as shown in Figure 9-37. Both the forward and reverse characteristics of the diodes are useful in protecting the gate oxides during handling. This protection is not totally adequate, however, and the following precautions must be taken in handling CMOS devices.

General

- Use a conductive, grounded work surface. Keep operators at ground potential (use conductive wrist bands and a 1 megohm resistor to ground).
- Do not use nylon smocks.
- Repack devices in conductive or antistatic containers. Keep devices at a common potential.
- Use conductive or antistatic envelopes for storing and shipping devices. Never use untreated plastic.

Cleaning

- Use static-neutralizing ion blower when manually cleaning with brushes.
- Ground all automatic equipment.
- Ground cleaning baskets.

Assembly

- Insert CMOS devices last to avoid overhandling.
- Use conductive handling trays.
- Use conductive material between edge connections.
- Ground all automatic insertion equipment.
- Ground solder machines and metallic parts of conveyor systems.
- Ground soldering irons.

Testing

- Use grounded metallic fixtures where possible.
- Use static-neutralizing ion air blower when using automatic handlers.
- Use conductive handling trays.
- Do not insert or remove boards with power turned on.

Unused Inputs. Unused inputs to CMOS circuits should not be left open. The actual voltage level or logic state of a CMOS input that is left open cannot be predicted. It may drift into the transition zone and remain partially turned on, or it may disrupt circuit operation completely, cause excessive power drain, make the circuit vulnerable to noise, or cause intermittent operation. Inputs that are not used should be tied to the power rails—either V_{DD} or V_{SS}—whichever enables the circuit.

Power Sequencing. If an input to a CMOS circuit is driven by a low-impedance generator, the supply voltage (V_{DD}) must not be removed prior to removing the input voltage. With V_{DD} removed, the distributed diode (D_2) is no longer reverse-biased and, as Figure 9-55 shows, an excessive current may be drawn through D_2 to the V_{DD} line and ground. If more than one power supply is used in a CMOS system, those supplying high-current drivers should be turned on last when sequencing power on and should be turned off first when sequencing power off.

Removal of printed wiring boards without turning off system power may also subject the input circuits to the damaging current described in Figure 9-55. A momentary disconnect of the V_{DD} pin prior to the input pins coming free of the connector could produce the same damaging results. Insertion of input protection resistors between potential high-current drivers and the CMOS input may remove the necessity for power sequencing. The presence of such resistors will place a tighter limit on the maximum switching frequency that can be processed through the circuits being protected.

Design of Circuit Boards for Safe Handling. Ideally, circuit boards containing CMOS should be designed so that no leads from the edge connectors run directly to CMOS inputs. Where possible, line receivers or TTL devices should be used to interface at the inputs. Where this is not possible and CMOS inputs must run directly to the board connector, a series resistance of 10 kilohms or greater should be used to protect these inputs from damage due to static charge during handling of the boards. The limitation that this added resistance places on switching speed must again be considered.

9.14.8 Absolute Maximum Ratings

The absolute maximum ratings for CMOS B-series devices are provided in Figure 9-56. These ratings must not be exceeded, even momentarily. Precautions must therefore be taken against accidents during testing and troubleshooting of both circuit cards and systems.

Power Dissipation per Package. In light of the low quiescent power dissipation typical of a CMOS integrated circuit, the 300 milliwatts per-package power dissipation seems very safe for most applications. However, as Figure 9-52 indicates, at V_{DD} of 10V or 15V, it is possible for even simple gate circuits to draw more than 300 milliwatts per package if the switching frequency is high enough.

Section 9.14.4 describes, in detail, the

FIGURE 9-55
Current through input protection diode with V_{DD} turned off.

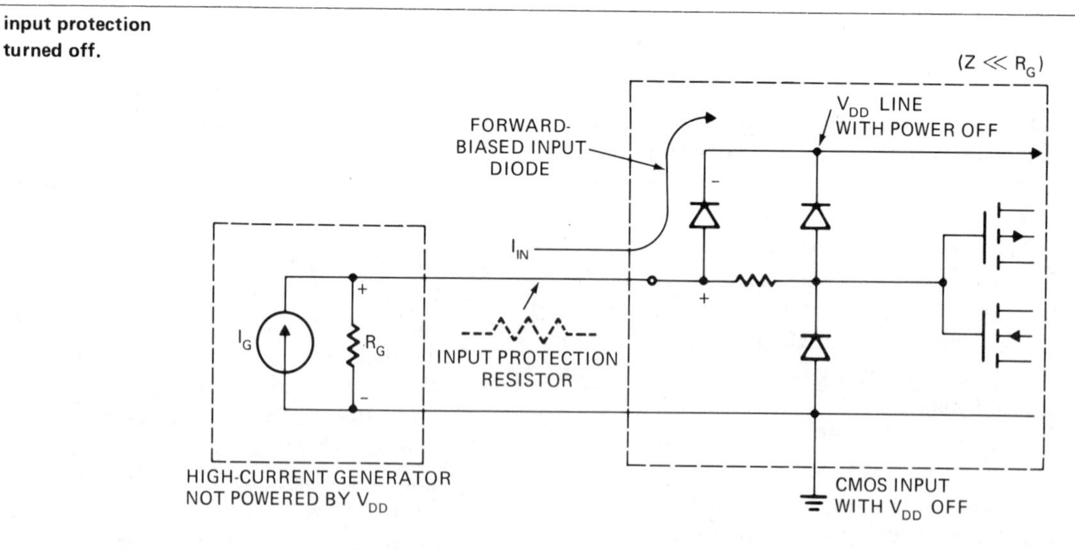

FIGURE 9-56
SCL4000B series CMOS specifications.

ABSOLUTE MAXIMUM RATINGS	SYMBOL	LIMITS	UNITS
DC SUPPLY VOLTAGE	V_{DD}	-0.5 TO +18	V DC
INPUT VOLTAGE	V_{IN}	-0.5 TO V_{DD} +0.5	V DC
DC INPUT CURRENT (ANY ONE INPUT)	I_{IN}	±10	mA DC
POWER DISSIPATION	P_T	300	mW
STORAGE TEMPERATURE RANGE	T_S	-65 TO +150	0_C

three elements of power drain occurring in a CMOS circuit. Similar principles are considered when controlling or determining the amount of this power that is dissipated in a given package. The quiescent power is insignificant. The through current, which is difficult to estimate, can be minimized by maintaining rapid rise and fall times in the switching waveforms. The remaining element of power drain caused by the load current is by far the most significant. It is proportional to the load capacitance (C_L), the switching frequency (f), and V_{DD}^2. The formula $P_L = C_L \cdot f \cdot V_{DD}^2$ can be used. It must be remembered, however, that on the average, 60% of this power is dissipated in the input circuits of the load devices and only 40% is dissipated in the driver.

DC Input Current. The input voltage and current ratings have a special significance when we are dealing with an interface to voltage levels exceeding V_{DD} in the positive direction or V_{SS} in the negative direction. The V_{IN} can exceed the supply voltages by more than the 0.5V of the maximum rating provided the V_{IN} is applied through a resistor R_{IN} that will limit the input current to 10 mA or less. If this is done, the ±0.5V limit will not be exceeded. The resistance of R_{IN} can be computed as shown in Figure 9-57. This method is not suitable for input signals which must be both high-voltage and high-speed because of the time constant existing between R_{IN} and 7 pF input capacitance.

9.14.9 Temperature Variations

Carrier mobility decreases as the temperature of a CMOS device increases. This results in an increase in the channel resistance. Both sink and source driver resistances increase. This increases the time constants at every point in the circuit, thus affecting such parameters as transition time, propagation delay, minimum pulse width, and, consequently, maximum switching frequency. The resistances change at the rate of +0.3% per °C.

9.15 Interfacing CMOS with TTL

9.15.1 TTL-to-CMOS

CMOS systems operating at 5V V_{DD} require a pull-up resistor to accomplish the interfacing between TTL and CMOS, as shown in Figure 9-58. The totem pole output connected to the CMOS input presents no special problem at the low output level (V_{OL}). The bottom transistor of the totem pole is turned on and discharges the input capacitance of the CMOS to the TTL V_{OL} level of 0.4V MAX. This compares with the CMOS V_{IL} of 1.5V MAX. The noise margin at the 0 level is 1.1V, and the overall noise immunity is the same or better than a CMOS-to-

FIGURE 9-57

Computation of resistance of R_{IN}. If a CMOS input is to be driven by a voltage more negative than V_{SS} or more positive than V_{DD}, a series resistance can be used to limit the current to ±10 mA.

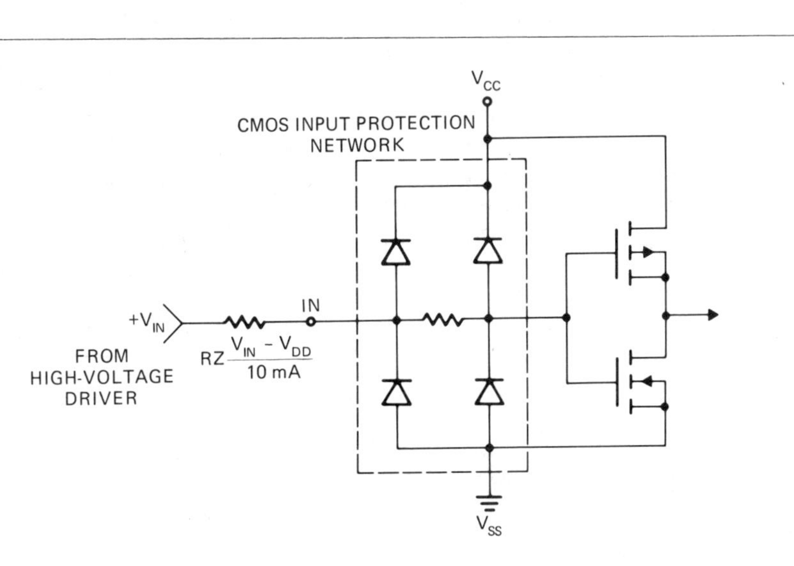

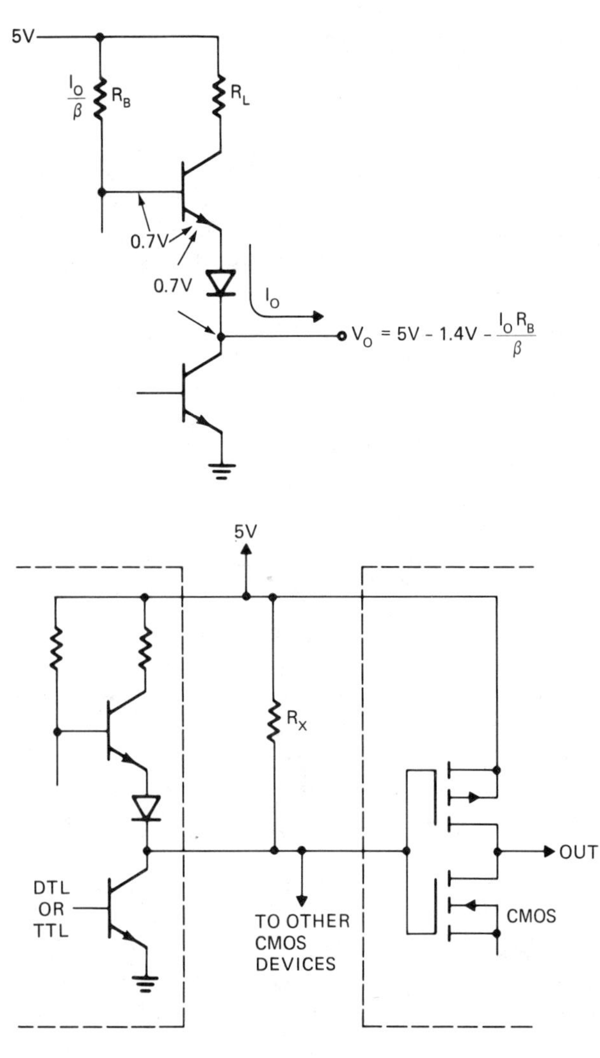

FIGURE 9-58

Interfacing standard TTL output to CMOS at V_{DD} = 5V.

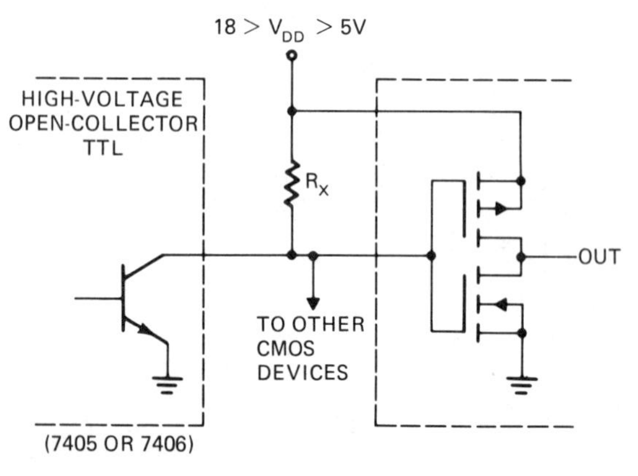

FIGURE 9-59

Interfacing TTL output to CMOS at V_{DD} > 5V. (See Figure 9-34.)

CMOS connection. A TTL totem pole output does not exceed 3.6V at the high output level. This, compared with a V_{IN} MIN CMOS input of 3.5V, allows only 0.1V noise margin. Pull-up resistor R_X is used to raise the V_{IH} to a level near V_{DD}. The minimum value of the pull-up resistor can be determined from the TTL parameters by the formula $R_X = (V_{CC} - V_{OL})/I_{OL}$.

Special interface circuits are needed for CMOS systems operating with V_{DD} greater than 5V. TTL open-collector circuits with a V_{OH} MAX rating as high or higher than V_{DD} may be used. The 7406 hex inverter or the 7405 hex buffer/driver are both open-collector drivers with V_{OH} MAX ratings of 30V. These devices can be used with pull-up resistors, as shown in Figure 9-59.

A minimum-value pull-up resistor will result in an interface with the least propagation delay (t_{PLH}). This, however, is at the cost of power dissipated in both driver and resistor. A higher-value pull-up resistor may be used to reduce the power drain, but the t_{PLH} will increase as shown in Figure 9-60. The graphs illustrate the trade-off between speed and power. The power figure shown is the power dissipated in both the bipolar driver and the pull-up resistor.

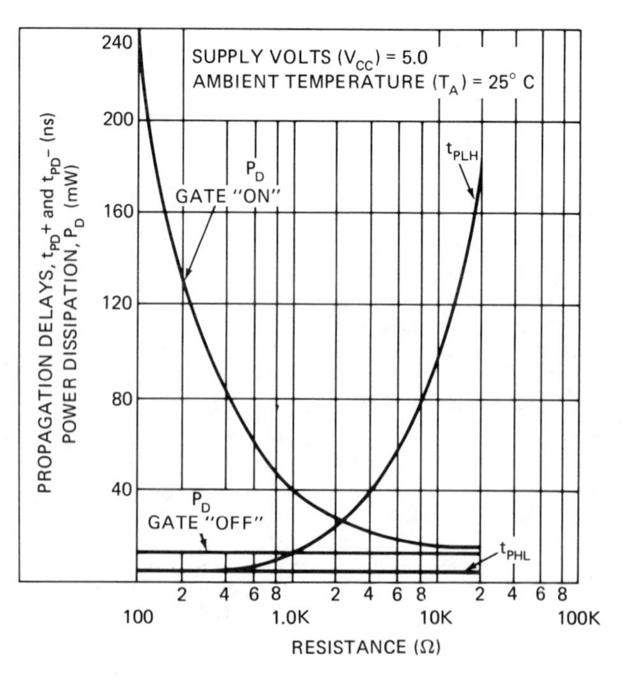

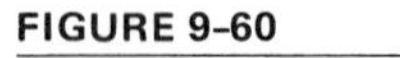

FIGURE 9-60

Speed/power relationships in selecting pull-up resistor value.

9.15.2 CMOS-to-TTL

The TTL input circuit is a source load requiring a current (I_{IL}) to pull down the voltage at the input to 0.8V or less. The I_{IL} for a Schottky low-power TTL is -0.36 mA. The B-series CMOS buffer has been designed to sink a single low-power Schottky TTL load. CMOS systems with V_{DD} of 5V interface directly to a single low-power Schottky TTL device and provide noise margins of 2.5V at level 1 and 0.4V at level 0. Although a 0.4V level 0 noise margin is the same as a TTL-to-TTL connection, the noise immunity is lessened due to the higher impedance of the CMOS driver. With a V_{DD} greater than 7V, a direct CMOS-to-low-power-Schottky connection cannot be used without exceeding the absolute maximum input voltage rating of the TTL circuit.

With a V_{DD} greater than 7V, interfacing from CMOS to TTL can be accomplished using the 4050B hex buffer or the 4049 hex inverting buffer. These buffers have inputs that have been modified to accept input levels (V_{IH}) greater than the supply voltage. They may be driven by CMOS circuits having V_{DD} greater than 5V, in spite of the fact that they are powered by the 5V TTL supply voltage. This results in a conversion to the TTL level with the capability of driving two TTL inputs (I_{OL} = 3.2 mA), as shown in Figure 9-61. The lower impedance of these buffers results in a noise immunity better than that of a standard B-series output-to-low-power-Schottky interface.

FIGURE 9-61

CMOS hex buffers used to interface CMOS to TTL (V_{DD} > 5V). Note the absence of clamp diodes to the power line such as exist on a standard input protection network.

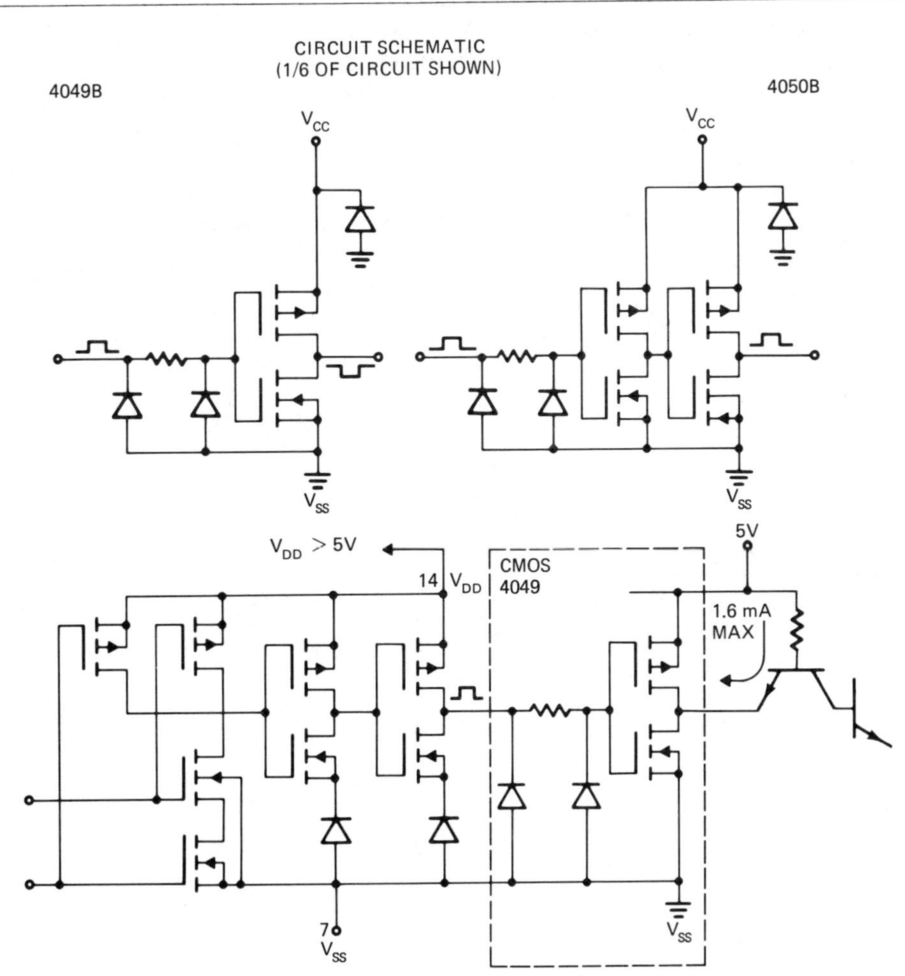

9.16 74C and 74HC Series Devices

The 4000 series CMOS devices, in most cases, have different logic and pin-out than the 7400 TTL devices. A special series of CMOS devices, numbered 74CXXX, was developed by National Semiconductor Corporation. These 74C devices have logic and pin-out identical to 74LS devices. Like 4000 series, 74C series devices can drive only one 74LS load. The speed of 74C devices is not much faster than 4000 series devices.

An improvement to the 74C series, a new version numbered 74HCXXX, is higher speed than 4000 series and is for most circuits close to the speed of 74LS circuits. The result is that most logic implemented by 74LS circuits can be implemented by 74HC. These applications will then have the advantage of the lower power requirements typical of CMOS.

As Figure 9-62 shows, in addition to the speed improvement, the 74HC devices have higher output drive current of 4 mA. This gives them capability of driving the equivalent of ten 74LS or two 74S loads. The single inconvenience remaining to interfacing CMOS with TTL using 74HC circuits is the V_{IH} level of 3.5V. TTL devices have guaranteed V_{OH} outputs of only 2.7V or lower. Therefore, a pull-up resistor is needed when driving 74HC circuits with TTL. The 74HCT series has addressed this problem with a V_{IH} level of 2.0V.

9.17 Summary

Logic circuits are not uniform devices delivering precise voltage levels for logic 1 or 0. The power supply voltages V_{CC} or V_{DD} are usually presented as the ideal 1 level, while a 0V ground is the ideal 0 level. Unfortunately, logic circuits are imperfect, and 1-level outputs are found to be significantly lower than power supply voltage, while 0 levels are significantly higher than ground level. The minimum 1 output and the maximum 0 levels will vary with the load that is connected to the output.

There are two types of loads that may be applied to a logic circuit—the *sink load* and the *source load.* The sink load can be represented by a resistor between the output and ground. The sink load reduces the 1 level of a logic circuit output but tends to improve the 0 level. The source load can be represented by a resistor between the output and the power supply ($+V_{CC}$). The source load raises the 0 level of a logic circuit output but tends to improve the 1 level.

A typical input circuit within a given logic system is usually defined as a *unit load.* In most systems of logic units, the unit load is either sink load or source load, but some logic systems have both, and both a unit sink load and a unit source load must be defined. The inputs of RTL circuits are *sink loads*; the inputs of TTL circuits are *source loads.* Once the typical load is defined, then all logic circuit inputs are rated according to the number of unit loads, while outputs are

FIGURE 9-62

Comparison of old and new CMOS families with 74LS circuits.

	TTL	CMOS (OLD)		CMOS (NEW)	
SERIES	74LS	4000	74C	74HC	74HCT
PROPAGATION DELAY[1]	10 ns	105 ns	35 ns	10 ns	10 ns
DC POWER DISSIPATION[2]	2 mW	FROM 1 μW @ 1 kHZ TO 3 mW @ 2 MHZ		FROM 1 μW @ 1 kHZ TO 3 mW @ 2 MHZ	
I_{OL}	8 mA	440 μA	360 μA	4 mA	4 mA
V_{OH}	2.7V	4.6V	4.6V	4.2V	4.2V
V_{IH}	2.0V	3.5V	3.5V	3.5V	2.0V
V_{IL}	0.8V	1.5V	1.5V	1.0V	0.8V
POWER SUPPLY RANGE	4.75V TO 5.25V	3.0V TO 15.0V		3.0V TO 6.0V	

[1]Typical for NAND gate.
[2]Power dissipated by a single NAND gate at 50% duty cycle.

rated according to the number of unit loads they can drive and still maintain an output within the specified limits. Another name for the output rating is fan-out. As given in the specifications of Figure 9-1, a typical TTL circuit rating is a fan-out of 1 to 10 circuits, meaning 1 to 10 unit loads.

The output-level parameters of a logic circuit are normally specified under full unit load test conditions. Figure 9-9 shows an ideal rectangular pulse output of a TTL gate with the minimum 1 and maximum 0 levels marked on the trailing edge. The reduced 1 level and raised 0 level do not disturb the functioning of the gate as long as the inputs connected to it will recognize them as 1 and 0. Input-level parameters for a logic gate must fall safely within the output limits. Figure 9-11 shows an ideal pulse with the specified output limits marked on the trailing edge and specified input limits marked on the leading edge.

In most logic systems, the logic circuits can restore marginal input levels. As Figure 9-12 shows, a marginal input pulse will be restored to improved levels at the output of the gate.

The difference between the maximum 0 output and the maximum 0 input is known as the 0-level *noise immunity*. The difference between the minimum 1 output and the minimum 1 input is called the 1-level noise immunity. Figure 9-13(a) shows the noise immunity levels marked on a marginal output pulse. Figure 9-13(b) shows how the noise immunity margins protect the circuit from accidental switching due to noise.

In drawing timing diagrams, we draw pulses and other waveforms as if the level changes between 1 and 0 occur instantaneously. We must on occasion face the fact that it requires time for a change from 0 to 1 level. This time measurement is called *rise time* (t_r). It also requires time for a change from 1 to 0, which is called *fall time* (t_f). If rise time or fall time is not rapid enough, it may lead to multiple switching of a logic gate, as in Figure 9-15.

Because of uncertainties at the 1 and 0 levels of a pulse, rise time and fall time are generally measured between the 10% and 90% levels of the pulse. The width of a pulse is usually measured between the 50% levels.

As a pulse passes through a logic gate, it is subject to some delay. If measured between the 50% levels of the leading edges of input and output pulse, it is called *turn-on delay*. Measured between 50% levels of the trailing edges, it is called *turn-off delay*. A general term for the larger of the two of these is *propagation delay*. This is shown in Figure 9-17.

Power drain is an important characteristic of logic circuits. A digital system requires so many logic elements that a minor difference in current drain per gate has a major effect on the total power supply requirements. It can be said in general that high-speed circuits require more power than low-speed circuits.

The highest practical speed for a standard TTL circuit is governed by a 5-nanosecond delay time. *Emitter coupled logic* (ECL) can provide gates with 2-nanosecond delays but requires more power than TTL. Figure 9-18 shows a typical ECL NOR gate. A method of improving the speed of a TTL logic gate is construction of gates with Schottky diodes between base and collector to prevent deep saturation, as Figure 9-20 shows. This technique reduces TTL propagation time to 3 nanoseconds but again at the expense of increased power consumption. Figure 9-31 compares the propagation time and power consumption of TTL devices.

The number of inputs a logic gate can have is known as fan-in. Integrated circuit gates are available with as many as eight inputs. If the number of inputs that must be connected to a point exceeds the fan-in of the available logic gates, then AND gates can be used to expand either AND or NAND gates, and OR gates can be used to expand OR and NOR gates. Special expandable gates are also available with gate expanders to provide very large fan-ins.

The dominant technology for small-scale integration (SSI) and medium-scale integration (MSI) is *transistor-transistor logic* (TTL). For low power, complementary metal-oxide semiconductor (CMOS) is usually used. The most popular TTL devices are 74XX, 74LSXX, and 74SXX. These devices respond to any input of 0.8V or lower as a binary 0, and any input of 2V or higher as a binary 1. The table in Figure 9-22 shows the current drawn by these inputs. Figure 9-24 shows the path of current for I_{OL} and I_{OH} along with voltages V_{OL} and V_{OH}. When working with TTL family unit, load calculations can simplify many interface problems. The tables in Figures 9-26 and 9-27 compare input and output parameters along with maximum ratings for the different TTL families. Figure 9-28 shows how the AC parameter of propagation delay is measured. The tables in Figures 9-30 and 9-32 compare the available TTL families. TTL devices can be purchased with either the standard totem pole output, open-collector output, or three-state output.

Low power consumption and excellent noise immunity are the two primary reasons for using CMOS devices. The table in Figure 9-36 shows the electrical characteristics for the B series. Unlike TTL gates, CMOS input resistance is very high, and therefore fan-out is limited by the capacitive loading. Figures 9-37(a)-(c) show input and output conditions for CMOS devices. Figure 9-38 illustrates input and output voltage levels. A description of noise margin, drive capability, and high frequency cutoff is given in Section 9.14.3 and Figure 9-40. Unwanted coupling between adjacent signal lines is a type of noise called *crosstalk*. In most CMOS applications, crosstalk is not a problem. Proper printed circuit board layout, decoupling capacitors, and not sharing the same ground line as high power devices should eliminate power supply and ground line noise problems. As described in Section 9.14.6, power requirements for CMOS systems are more difficult to predict than those for TTL systems.

CMOS devices are available with three types of outputs: double buffer, transmission gate, and three-state. Figures 9-46, 9-47, and 9-51 illustrate each type of output. Figure 9-49 shows a problem that occurs with the transmission gate output in a multiplexed system. That is, both gates are on for a short period of time. Precautions as described in Section 9.14.6 may be necessary to ensure proper circuit operation.

Although data sheets may show that a CMOS circuit will follow a 50-nanosecond pulse, the CMOS circuit is probably not designed for reliable performance at this high frequency. As the supply voltage and frequency of a CMOS gate increase, the CMOS gate's power dissipation may exceed that of a comparable TTL gate. Factors limiting pulse width are found mainly between the driver circuit and the input circuit.

The input circuit of a CMOS is protected for most handling. However, precautions must be taken. Section 9.14.8 divides the precautions into the following categories: general, cleaning, assembly, testing, unused inputs, power sequencing, and design of circuit boards. The table in Figure 9-56 gives the absolute ratings for CMOS B series.

At some point in building digital systems, CMOS and TTL circuits must be interfaced. Both TTL-to-CMOS and CMOS-to-TTL interfacing are covered in Section 9.15.

Glossary

Sink Load: A load that draws energy from the source or generator. In a positive logic system, a sink load reduces the 1 level and improves the 0-level output.

Source Load: A load that adds energy to the source or circuit it is loading. In a positive logic system, a source load raises the 0 level and improves the 1-level output.

Unit Load: A load that is representative of the most common input circuit in a logic system and that may be defined in terms of current draw or resistance. Other inputs are rated in number of unit loads. An output is rated according to the number of unit loads it can drive. The output rating is often referred to as fan-out.

Leading Edge: The voltage change between 1 and 0 that occurs first in time on a pulse or similar digital waveform. See Figure 9-16.

Trailing Edge: The voltage change between 1 and 0 that occurs second in time on a pulse or similar digital waveform. See Figure 9-16.

Rise Time (t_r): A pulse parameter—the time required for a voltage change to occur from 0 to 1 level. Generally measured between the 10% and 90% levels. See Figure 9-16.

Fall Time (t_f): A pulse parameter—the time required for a voltage change from 1 to 0 level. Generally measured between the 90% and 10% levels. See Figure 9-16.

Noise Margin: A measure of the degree to which a circuit reacts to random noise. The value is labeled V_{NL} at logic 0 and V_{HL} at logic 1.

Crosstalk: Noise caused by unwanted coupling between adjacent signal lines.

Power Line Noise: Noise that has been coupled into the power line.

Ground Line Noise: Noise that has been generated in the ground line due to a large number of power devices being switched on or off simultaneously.

Quiescent Current: The supply current drawn by a CMOS circuit when it is not being switched. It is composed of the leakage current through the p- and n-channel transistors when they are in the off state. It is normally an extremely small portion of the total operating current.

Through Current: I_{TC} for CMOS circuit. The result of both the p- and n-channel transistors being partially turned on during the switching from high-to-low and low-to-high.

Transmission Gate Output: A CMOS digitally controlled gate composed of both p- and n-channel transistors connected in parallel. See Figure 9-47. The transmission gate is bidirectional.

Three-State Output: A digital gate with an output that can produce a logic 0 or a logic 1 or go to a high impedance state. See Figure 9-51.

Absolute Maximum Ratings: The maximum voltage or current that can be applied to the terminals of an integrated circuit without causing permanent damage to the device. It is not a guarantee that the device will operate at these levels.

Turn-On Delay: The delay time between the leading edge of an input pulse and the leading edge of an output pulse. Generally measured between the 50% levels. See Figure 9-17.

Turn-Off Delay: The delay time between the trailing edge of an input pulse and the trailing edge of an output pulse. Generally measured between the 50% levels. See Figure 9-17.

Propagation Delay: The amount of time delay that a signal experiences as it passes through one or more logic circuits.

Emitter Coupled Logic (ECL): Logic circuits that accomplish higher speed than TTL circuits because the circuit does not saturate when it turns on. See Figure 9-18.

Schottky Barrier Diode: A diode with a barrier potential lower than that of silicon or germanium.

Schottky TTL Circuit: TTL circuits in which Schottky diodes are formed between base and collector. The Schottky diode becomes forward-biased as the circuit attempts to saturate. By avoiding deep saturation, the circuit accomplishes higher speed than standard TTL circuits. See Figures 9-19 through 9-21.

Fan-In: The number of inputs a logic circuit can have in a given logic system.

Expander Node: A special input of an expandable logic gate to which gate expanders can be connected.

Gate Expander: A logic circuit used to increase the fan-in of an expandable gate.

Open Collector: A TTL circuit with a single transistor output. The output lead is the collector of the transistor, and connection to V_{CC} must be applied through the external load. See Figure 9-33.

Wired OR: The shorting together of open-collector output circuits so that a single active circuit can be selected to control the output level. See Figure 9-33.

Questions

1. A logic gate has an input resistance of 500 ohms. Draw a schematic of this resistor as a sink load. Draw a schematic of this resistor as a source load.
2. A logic circuit that can receive 160 mA at its maximum 0 output level of 0.4V is rated at 10 unit loads. What type of load is this?
3. What characteristic of the transistor causes the 0 level to rise above 0V?
4. Why are long parallel sets of lines likely to cause noise problems?
5. A logic circuit has a turn-on delay time of 8 nanoseconds and a turn-off delay time of 13 nanoseconds. What will happen to the 20-nanosecond width of a pulse as it passes through the circuit?
6. A 50-nanosecond-wide reset pulse passes through 12 logic gates having a propagation delay of 15 nanoseconds each. Draw a timing diagram comparing the input and output pulses (use 50 nanoseconds/inch).
7. Is it necessary for all sections of a digital machine to operate at the same speed?
8. What advantage is gained by using high-speed TTL and lower-speed CMOS circuits in the same digital system?
9. Can a NAND gate be used to expand the fan-in of another NAND gate?
10. What logic gate fan-ins can be expanded by OR gates?
11. What logic gate fan-ins can be expanded by AND gates?
12. Explain how the Schottky barrier diode can be used to increase the speed of a TTL switch.
13. Why are ECL logic circuits higher-speed than TTL circuits?

14. It is not usual for logic circuit outputs to be connected. Explain why the open-collector wired OR circuit is an exception to this.
15. Eight open-collector output circuits are connected with a pull-up resistor to form a wired OR. Only one output at a time is active. What state should the inactive outputs be in?
16. What are two advantages of CMOS circuits compared to bipolar circuits?
17. What parameter levels were established by the B-series standard?
18. What is the typical input and maximum capacitance of a CMOS gate?
19. What relationship exists between fan-out and operating speed of a CMOS system?
20. Define noise margin.
21. List three methods that can be used to reduce the chance of noise errors in a CMOS system.
22. Define crosstalk.
23. What are the guaranteed 1 and 0 switching voltages for the inputs to 74HCT CMOS circuits?
24. How is power line noise reduced?
25. How is ground line noise reduced?
26. Does switching frequency affect power for CMOS circuits?
27. What are the three parts that make up the total current for a CMOS gate?
28. What is the name given to the current that is the result of both the p- and n-channel transistors being partially turned on during the switching of a CMOS device?
29. What are the guaranteed 1 and 0 switching levels for the input voltages to TTL circuits?
30. List the three types of output drivers available on CMOS circuits.
31. What type of CMOS output has both p- and n-channel transistors connected in parallel?
32. What are the guaranteed 1 and 0 switching voltages for the inputs to 4000 series CMOS?
33. How many 74S circuits can a standard 74LS circuit drive and still maintain the guaranteed V_{OL} of 0.5V?
34. List at least six precautions that should be taken to protect CMOS inputs.
35. Can unused inputs to CMOS circuits be left open?
36. To drive a 4000B input with a 7400 series device, what added component must be used? (V_{DD} = 5V; V_{SS} = GND)
37. To interface a 4000B series driver with a TTL input, what 4000 series devices might be used? (V_{DD} = 10V; V_{SS} = GND)
38. Which TTL buffer could be used to drive a 24V, 10 mA relay from a TTL control circuit?
39. What advantage occurs from using higher-than-minimum value of pull-up resistor for a particular open-collector driver?
40. What advantage occurs from using the minimum value of pull-up resistor for an open-collector driver?
41. A battery operated system using minimum pulse widths of 1 millisecond would ideally use what family of circuits?
42. A battery operated system using minimum pulse widths of 20 nanoseconds would use what family of circuits?
43. To obtain the highest possible noise immunity, what family of circuits would be used?
44. How many 74LS loads can a B-series CMOS circuit drive? How many CMOS circuits can a B-series CMOS circuit drive? How many 74S circuits can a B-series CMOS circuit drive?

Problems

9-1 Determine whether the gates in the following figures are sink or source loads:

Figure 6–8 Figure 6–15
Figure 7–14 Figure 7–3

9-2 Using the TTL specifications of Figure 9-1, label the pulses in Figure 9-63 with the marginal input and output levels.

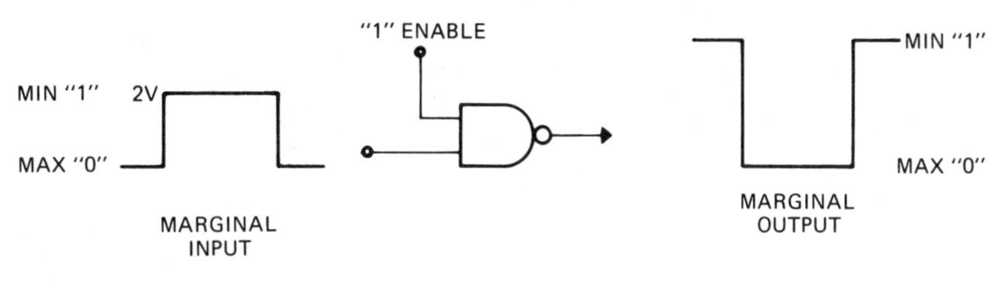

FIGURE 9-63. Problem 9-2

9-3 From the specifications of Figure 9-1, determine the 1- and 0-level noise immunity provided by these TTL logic circuits.

9-4 Using the specifications of Figure 9-1, label the trapezoidal pulses of Figure 9-64 with the worst-case (maximum) time parameters for those gates.

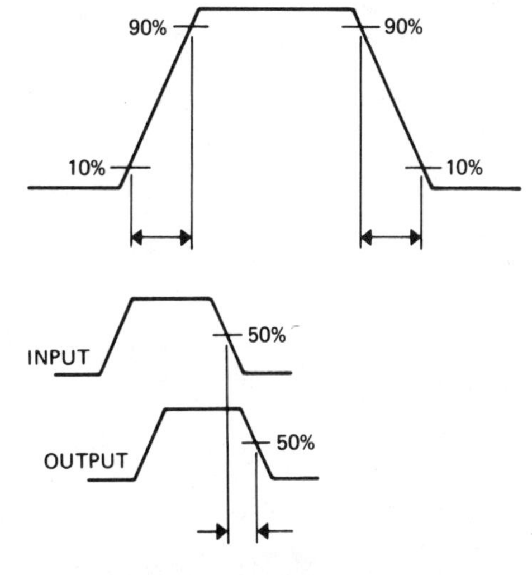

FIGURE 9-64. Problem 9-4

9-5 Using Figure 9-37(b), calculate the time constant of a CMOS input.

9-6 Calculate the noise margins V_{NL} and V_{NH} for a CMOS gate. (V_{DD} = 10 V)

9-7 Using Figure 9-44, determine the total power dissipation for a 4042B gate operating at a switching frequency of 100 kHz. (V_{DD} = 10 V; C_L = 50 pF)

9-8 Calculate the power dissipation due to load current for the following conditions: C_L = 50 pF, f = 100 kHz, and V_{DD} = 10V.

9-9 Draw a TTL-to-CMOS interface circuit and calculate the value of the pull-up resistor. Use typical values for the TTL gate given in Figure 9-1.

CHAPTER

Parallel Adder Circuits

Objectives

Upon completion of this chapter, you will be able to:

- Assemble three or more simple logic gates to form a half adder.
- Assemble two half adders and an OR gate to form a full adder.
- Interconnect integrated circuit two-bit and four-adder circuits to form large-scale adders.
- Connect additional logic to a four-bit binary adder to produce a binary-coded decimal (BCD) adder.
- Compare high-speed four-fit adders that contain internal carry generation with the interconnection of full adders that contain external carry generation.
- Compare propagation delays and power dissipation of CMOS adder circuits with TTL adder circuits.

10.1 Introduction

The adder is the main functional element of the digital computer. There are few services that a computer provides that do not in some way involve the adder circuit. Other smaller digital devices, such as data correctors and adding and tabulating machines, are all likely to use digital adders. Here we will explain the *parallel adder.* As Figure 10-1(a) shows, the parallel adder receives its input numbers with a separate lead for each binary bit. Its output, or sum, is also by a separate lead for each binary bit.

Adders may also be designed to work in serial form. As Figure 10-1(b) shows, the input and output numbers are a single line with each binary place appearing at its own time interval. Since understanding the serial adder requires a knowledge of register circuits, which are explained in Chapter 14, we will postpone discussion of the serial adder until Chapter 15.

Because each binary bit is handled by a separate set of logic circuits, the parallel adder requires more circuit elements per bit capacity. The serial adder, however, requires a separate time interval for each bit that must pass through the adder. Therefore, the parallel adder is much faster, and, except for a short propagation time required for the carry to ripple down through the adder, the answer is available the instant the numbers arrive at the inputs. It is often referred to as a *ripple carry adder.*

Other parallel adder circuits do not use the ripple carry method. Instead, they use a method known as *look-ahead carry generation*, which is the fastest method of generating a carry.

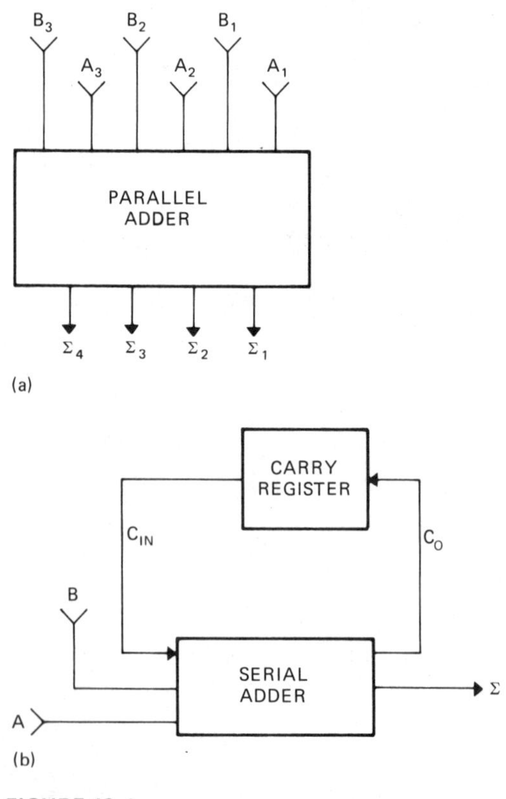

FIGURE 10-1

(a) Three-bit parallel adder. (b) Serial adder.

10.2 Binary Addition

If A and B are two three-bit binary numbers, we can add them and obtain the binary number Σ, so that $A + B = \Sigma$. (The Greek letter Σ is equivalent to the symbol S, which we used for sums in Chapter 3.) For the sake of explanation, we will express the three binary places of A, $2^2\ 2^1\ 2^0$, as $A_3\ A_2\ A_1$. Likewise, the three binary places of B can be expressed as $B_3\ B_2\ B_1$. Then:

$$\begin{array}{c} A \\ +B \\ \hline \Sigma \end{array} = \begin{array}{cccc} & A_3 & A_2 & A_1 \\ & B_3 & B_2 & B_1 \\ \hline \Sigma_4 & \Sigma_3 & \Sigma_2 & \Sigma_1 \end{array}$$

(Each subscript letter represents a digit 1 or 0 multiplied by its binary place value.)

An adder of three-bit capacity must handle A and B input numbers from 000 to 111 (0 to 7). It adds them as we discussed in Chapter 3 and must accommodate the sums at the output from 0000 to 1110 (0 to 14). Figure 10-2 shows the addition of the binary equivalent of 7 + 5. The letter C is used to indicate the carry—${}^nC_O$ for carry-out from column n, ${}^nC_{IN}$ for carry-in. A carry-out from column n is identical to the carry-in of column (n + 1):

$${}^nC_O = {}^{n+1}C_{IN}$$

To develop the logic system that can automatically yield these results, we follow the procedure for producing truth tables that was used in our explanation of logic gates in Chapter 6. We will develop a truth table for binary addition and try to find a combination of logic gates that will have that necessary truth table. The truth table of the three-input logic gate shows the single output (1 or 0) for each of the eight possible input conditions. A three-bit digital adder would necessarily have six inputs and four outputs, requiring us to show 64 (2^6) possible

FIGURE 10-2
Addition of the binary equivalent of 7 + 5.

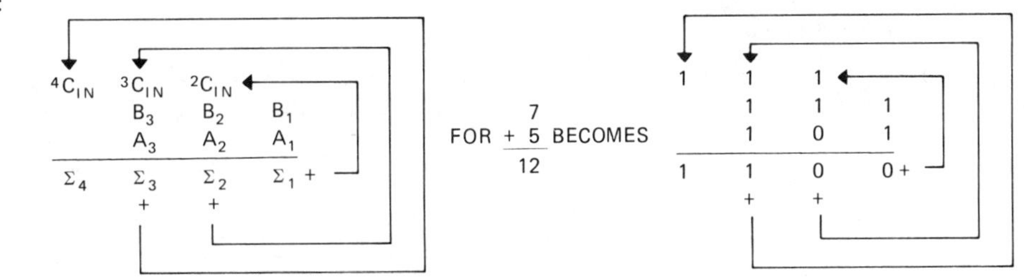

input conditions and the necessary output conditions occurring with each. A truth table like this does not help to simplify the design of the adder. Instead, we can form a truth table for each bit of the adder, from the least significant bit (LSB) through the second and third, or most significant, bit (MSB). We find these tables relatively simple and, fortunately, the second- and third-digit tables are identical.

INPUTS		OUTPUTS	
A_1	B_1	Σ_1	C_O
0	0	0	0
0	1	1	0
1	0	1	0
1	1	0	1

(a)

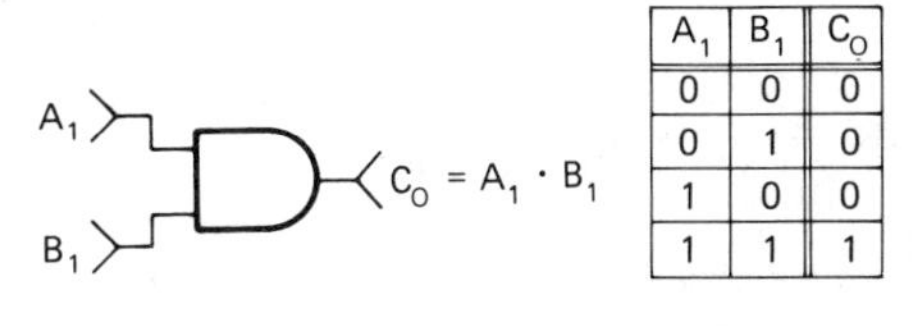

A_1	B_1	C_O
0	0	0
0	1	0
1	0	0
1	1	1

(b)

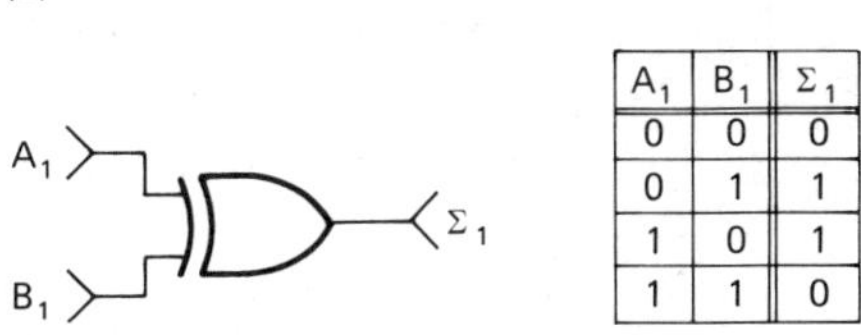

A_1	B_1	Σ_1
0	0	0
0	1	1
1	0	1
1	1	0

(c)

FIGURE 10-3
(a) Truth table of half adder function with two outputs, sum and carry. (b) Two-input AND gate with a truth table identical to the LSB carry function. (c) Exclusive OR gate with a truth table identical to the LSB sum function.

10.3 Half Adder

In the LSB addition of $A_1 + B_1 = \Sigma_1 + C_O$, the four possible combinations are as follows:

$$\begin{array}{r} 0 \\ +0 \\ \hline 0 \end{array} \text{ (carry 0)} \qquad \begin{array}{r} 1 \\ +0 \\ \hline 1 \end{array} \text{ (carry 0)}$$

$$\begin{array}{r} 0 \\ +1 \\ \hline 1 \end{array} \text{ (carry 0)} \qquad \begin{array}{r} 1 \\ +1 \\ \hline 0 \end{array} \text{ (carry 1)}$$

The truth table, therefore, has two inputs and two outputs. If we ignore the sum output, the carry output by itself has the truth table of a two-input AND gate. Only when the inputs are both 1 do we have a carry 1.

When we look at the sum output by itself, it has the truth table of the exclusive OR gate. Only when there is an odd set of 1s at the input is there a 1 at the output. It is obvious that an AND gate in parallel with an exclusive OR gate will provide the necessary logic function for LSB addition. This is developed in Figures 10-3(a) through 10-3(c) and 10-4.

Section 8.3 shows the numerous logic combinations that can be used to form the exclusive OR gate. One of these already contains an AND gate. With this exclusive OR logic, it is necessary only to take an extra lead off the AND gate output to provide the carry, as Figure 10-5 shows.

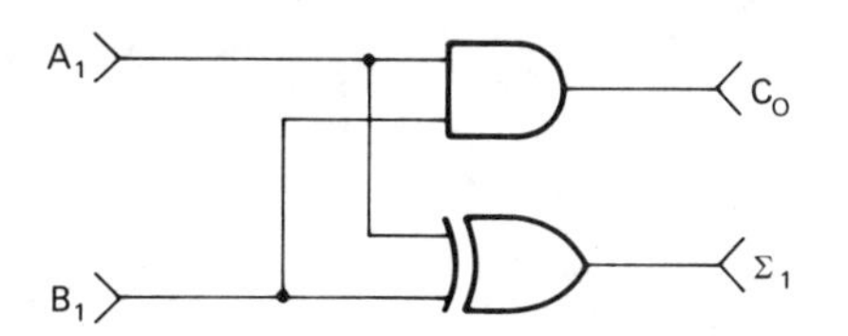

FIGURE 10-4
Half adder.

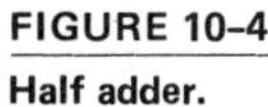

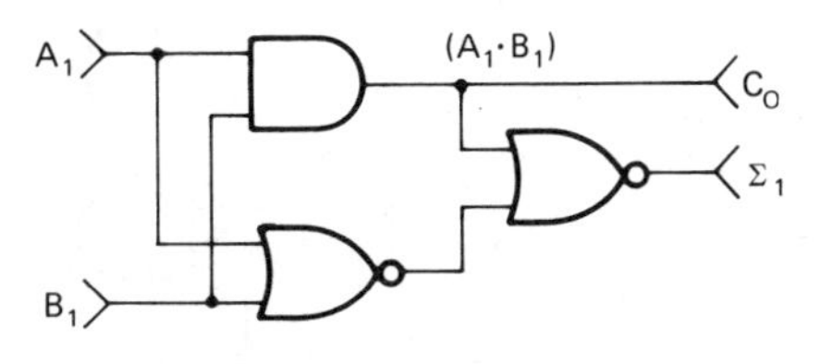

FIGURE 10-5
Half adder using AND and NOR gates.

FIGURE 10-6

Half adder carry-out connected to carry-in of next significant bit adder.

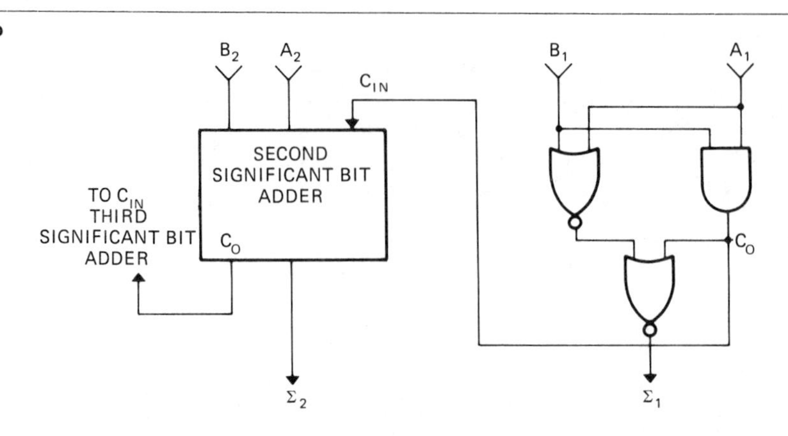

The sum output of the *half adder* provides the LSB or 2^0 bit of the three-bit adder. The carry-out becomes the carry-in for the second significant bit. Figure 10-6 shows this connection.

INPUTS			OUTPUTS	
C_{IN}	A_2	B_2	Σ_2	C_O
0	0	0	0	0
0	1	0	1	0
0	0	1	1	0
1	0	0	1	0
1	1	0	0	1
1	0	1	0	1
0	1	1	0	1
1	1	1	1	1

C_{IN}
A_2
B_2
$\Sigma_2 + C_O$

FIGURE 10-7

Full adder truth table.

10.4 Full Adder

The truth table for the second significant bit adder has three inputs and therefore eight possible combinations, as Figure 10-7 shows. It will be noted from the truth table that a 1 occurs in the sum only when there is an odd number of 1s on the inputs. Section 8.6 describes the parity generator, which uses exclusive OR

FIGURE 10-8

Two exclusive OR gates connected to provide the sum function of a full adder.

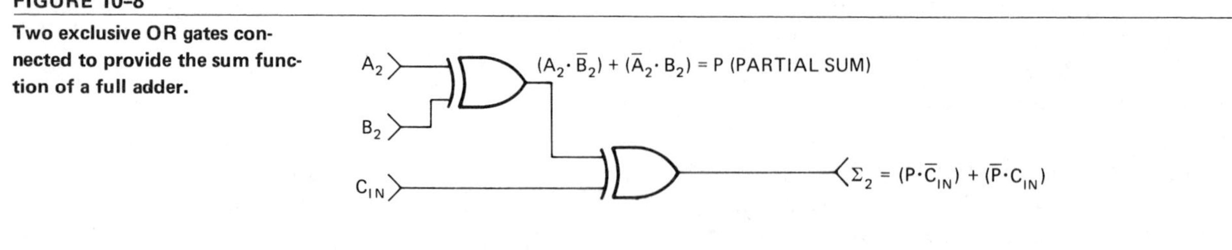

FIGURE 10-9

One of several logic circuits used to produce the carry function.

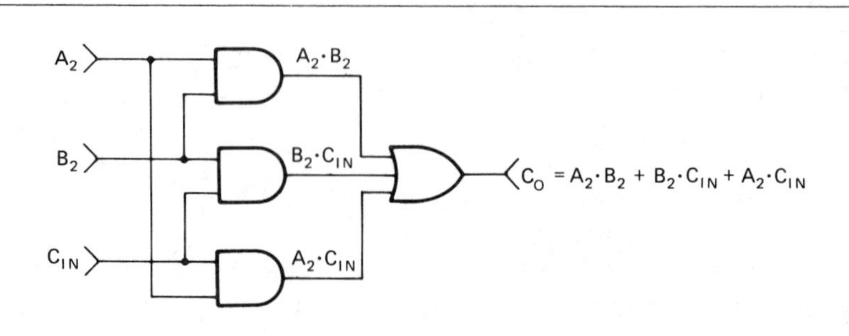

FIGURE 10-10

Full adder, composed of two half adders and an OR gate.

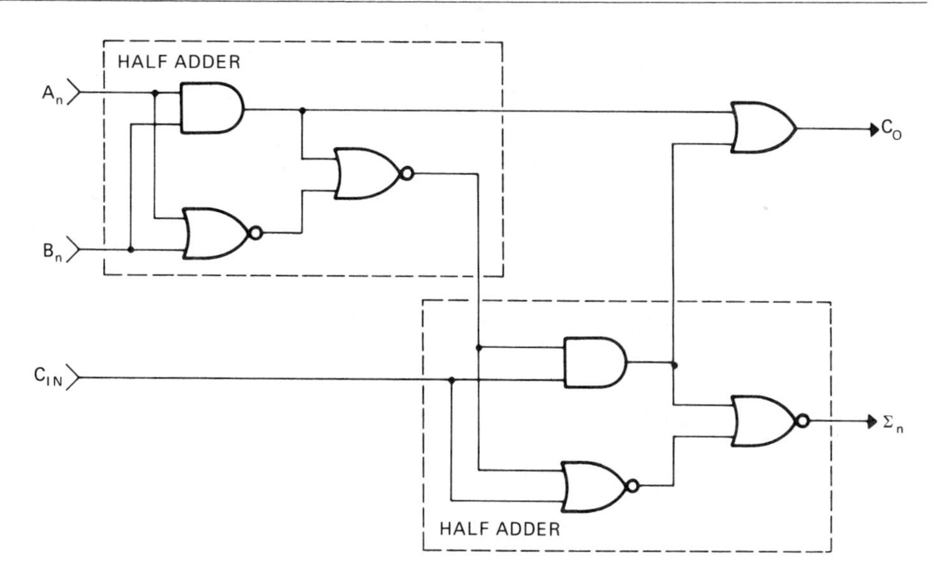

gates to determine if any number of inputs has an odd or even number of 1s. For the sum output, we need only a three-input parity generator using two exclusive OR gates, as Figure 10-8 shows.

Examination of the carry output shows a carry 1 if there are two or more 1s on the inputs. We could produce this function by using AND gates for each paired combination of inputs and connecting the AND gate outputs to a three-input OR gate, as in Figure 10-9.

In developing the half adder, we discovered that of the many circuit combinations that can produce an exclusive OR, selection of a convenient one might provide economies between the sum and carry logic. In fact, Figure 10-10 shows that with AND NOR logic a two-input OR gate added to the sum logic will effectively provide the carry logic.

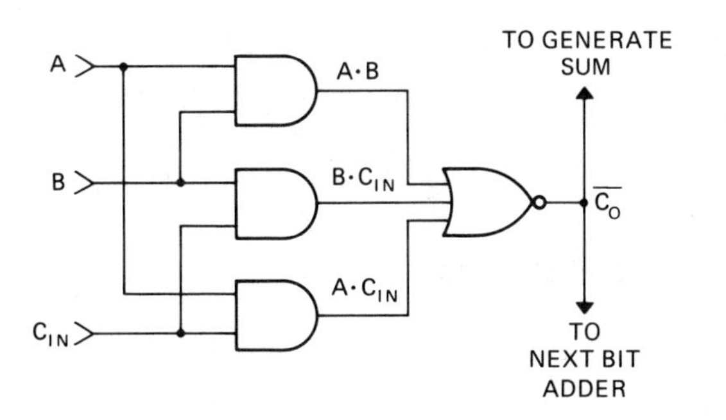

FIGURE 10-11

Logic circuit used in the TTL full adder to develop an inverted carry ($\overline{C_O}$).

The logic circuits for the third significant and even higher bits in a larger adder use the same logic as for the second significant bit. This logic, which can be formed from two half adders and an OR gate, is called a *full adder*.

The TTL integrated circuit full adder, available in either two- or four-bit size, uses the circuit of Figure 10-11 to generate an inverted carry ($\overline{C_O}$). This circuit is identical to that of Figure 10-9 except that the inversion results from use of a NOR gate output. The carry-not is then used in generating the sum output, as Figure 10-12 shows.

As the full adder truth table in Figure 10-12(b) indicates, if only one input is high, a $\overline{C_O}$ and a sum output are generated. These three conditions can be detected by ANDing the $\overline{C_O}$ with the three inputs. The remaining condition, calling for a sum of 1, occurs when all three inputs are high. The bottom AND gate in Figure 10-12(a) detects this.

The TTL full adder circuitry thus far explained produces a correct sum but an inverted carry. The inverted carry is passed on to the next full adder stage in the MSI chip.

As can be seen in the logic diagram and pinout in Figure 10-13(a) and 10-13(b), the second full adder is identical to the first except that it uses all three inputs in inverted or complement form. A rearrangement of the full adder truth

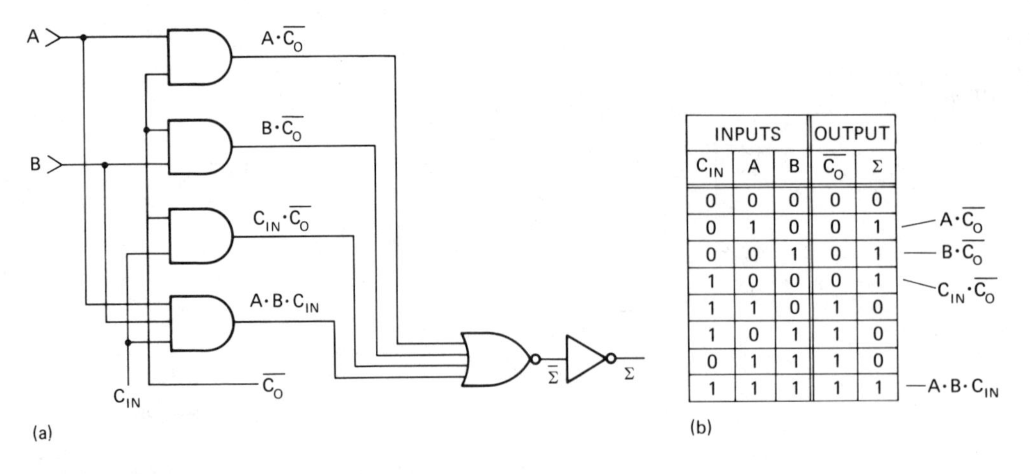

INPUTS			OUTPUT		
C_{IN}	A	B	$\overline{C_O}$	Σ	
0	0	0	0	0	
0	1	0	0	1	$A \cdot \overline{C_O}$
0	0	1	0	1	$B \cdot \overline{C_O}$
1	0	0	0	1	$C_{IN} \cdot \overline{C_O}$
1	1	0	1	0	
1	0	1	1	0	
0	1	1	1	0	
1	1	1	1	1	$A \cdot B \cdot C_{IN}$

FIGURE 10-12

(a) Logic circuit used in conjunction with an inverted carry to produce the sum output. (b) Truth table showing four AND conditions that produce a sum output.

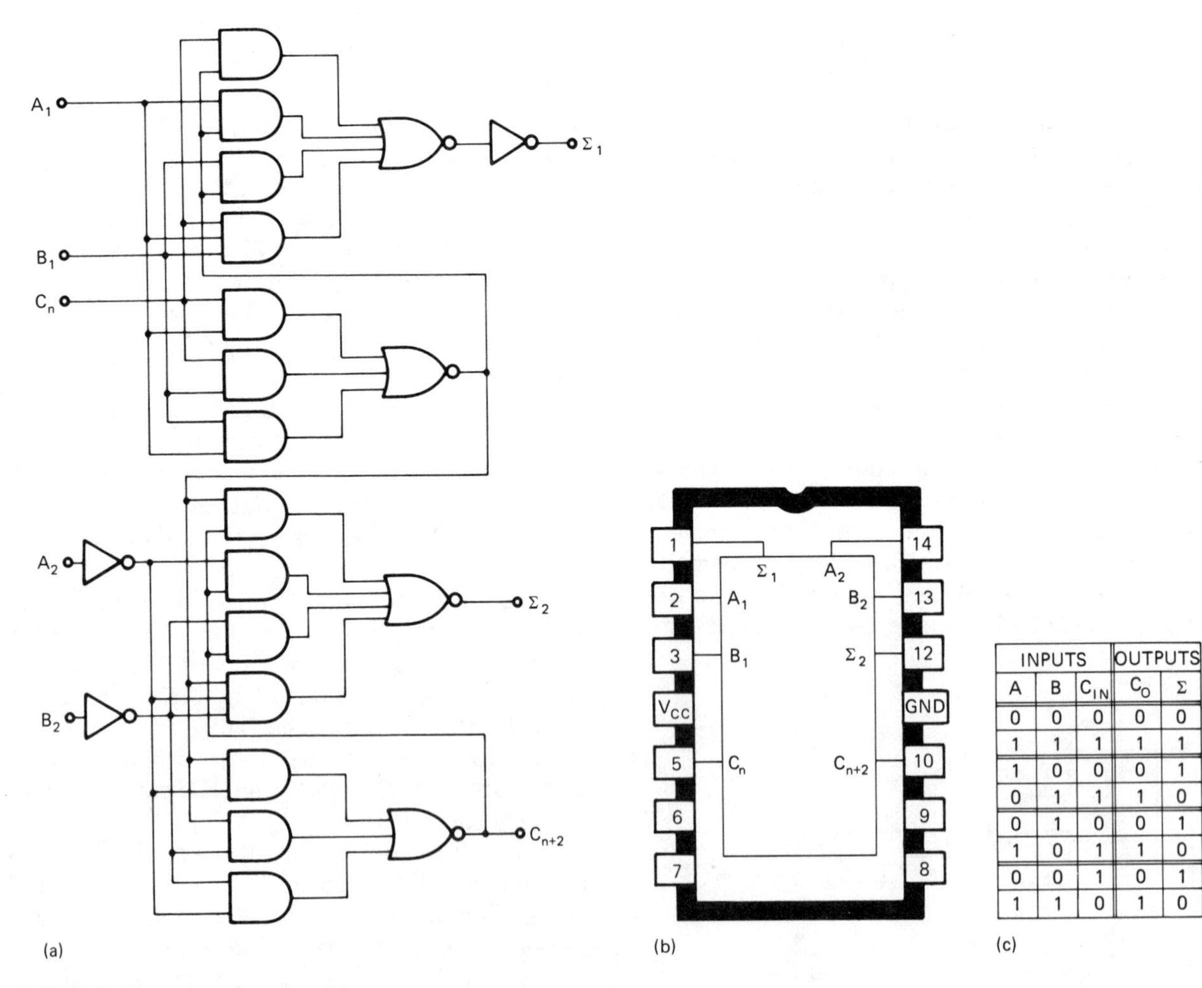

INPUTS			OUTPUTS	
A	B	C_{IN}	C_O	Σ
0	0	0	0	0
1	1	1	1	1
1	0	0	0	1
0	1	1	1	0
0	1	0	0	1
1	0	1	1	0
0	0	1	0	1
1	1	0	1	0

FIGURE 10-13

(a) Sprague TTL/MSI 5482 two-bit full adder circuit. Second digit uses inverted inputs; second inversion at the output makes them correct. (b) Pin-out. (c) Full adder truth table with inputs grouped in complementary pairs. When inputs are complemented, outputs are complemented also.

table, as in Figure 10-13(c), shows that complementary inputs have complementary outputs. This means that the second carry-out, for which there is an output pin, is not inverted. Elimination of the output inverter cancels the sum inversion.

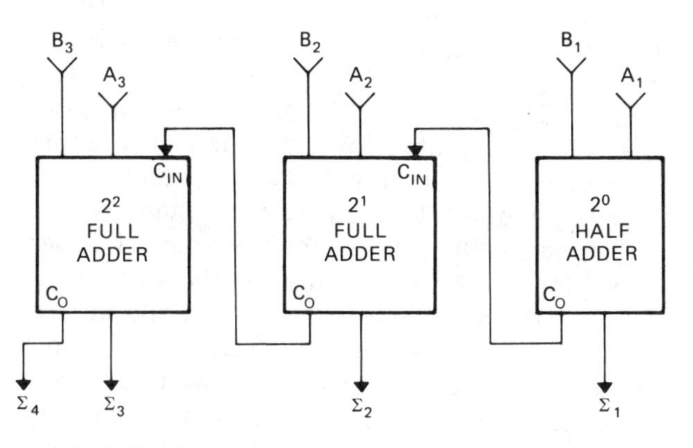

FIGURE 10-14
Block diagram of a three-bit digital adder.

10.5 Multibit Adders

Figure 10-14 shows a three-bit digital adder in block diagram form. This adder can add binary numbers from 000 to 111, or 0 to 7, with answers running from 0000 to 1111, or 0 to 14. We can double the adder's numeral capacity by merely connecting another full adder.

The MSI TTL two-bit full adders can also be connected to form an adder of more bits. When this is done, the LSB is a full adder where only a half adder is needed. This can be handled by merely grounding the first carry input.

■ EXAMPLE 10-1

Using two 14-pin integrated circuit two-bit adders like those of Figure 10-13, draw in the connections needed to produce a four-bit adder.

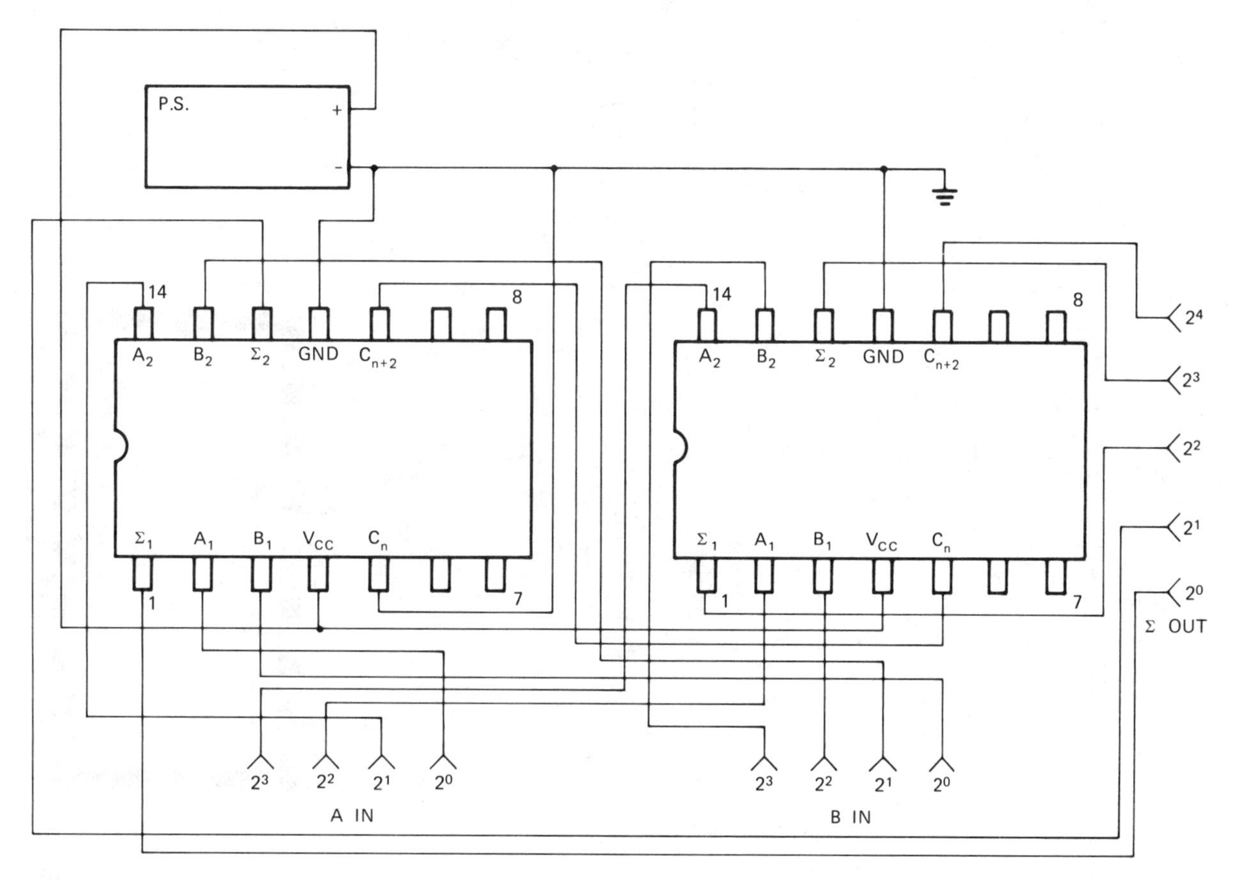

FIGURE 10-15
Solution for Example 10-1.

Solution: See Figure 10-15, which shows a pair of integrated circuit two-bit full adders interconnected to form a four-bit adder. A 16-pin integrated circuit has enough pin-out for a four-bit adder. Figure 10-16(a) and 10-16(b) shows the logic diagram and pin-out of a TTL/MSI 7483 four-bit binary full adder. The circuit is equal to a pair of two-bit adders with the carry between them internally connected.

The assembly of large adder circuits by interconnecting full adders, as shown in Figure 10-14, leaves a serious limitation in the time required to produce not only the final carry but also the carries to the higher significant digits. The process can be speeded up substantially by providing each bit of the adder with an individual carry-in generator. The carry generator has connected to it each of the lower significant bit inputs. This type of circuit is shown in the 74LS283 of Figure 10-17(a). Note how the carry generator portion of each bit becomes progressively more complex between the least significant bit and the final carry generator (C_4). This is due to the increasing number of inputs influencing the generation of the higher carries. Also, in comparison to the 7482 and 7483, the 74LS283 has the preferred power pin-out of V_{CC} and GND on the end pins of the DIP circuit rather than the middle, as shown in Figure 10-17(b). Figure 10-17(c) gives the propagation delays for the 74LS283. Note that the same delay time is quoted for generation of sum bits, regardless of whether it is S_0 or S_3, and the final carry C_4 requires less time than the sum.

The CMOS 4008 is another four-bit adder. The block diagram of this circuit is shown in Figure 10-18(a). This circuit has a special section for providing the final carry at a higher speed than would be possible by waiting for the final carry to ripple through the individual carry bits. This circuit feature is often called a carry look-ahead.

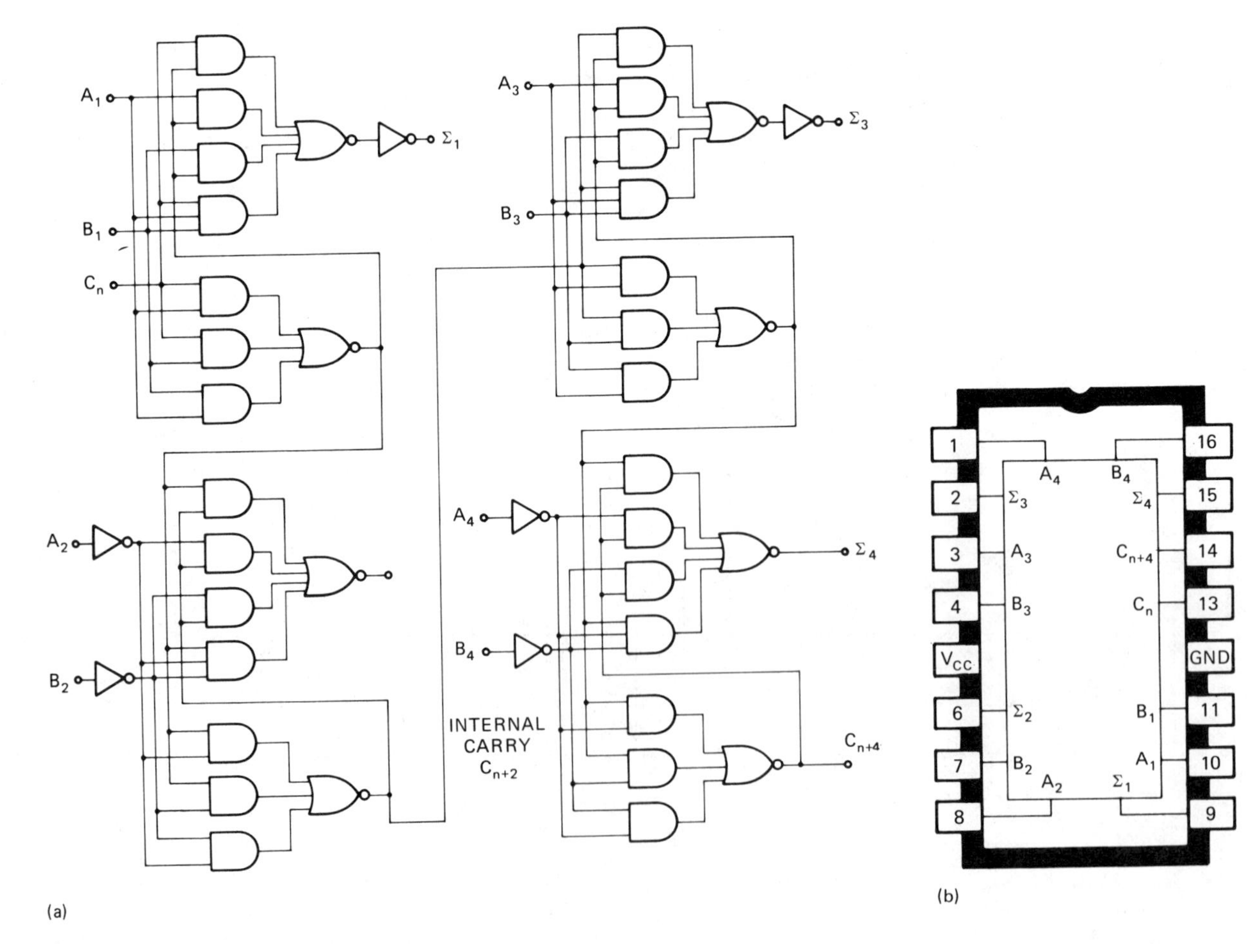

FIGURE 10-16

(a) Sprague 5483/7483 four-bit binary full adders. (b) Pin-out.

FIGURE 10–17

Fairchild 74LS283 high-speed four-bit binary adder. A separate carry generator is used for each bit of the adder and for the final carry. (a) Logic diagram. (b) Logic symbol and pin-out. (c) Propagation delay times.

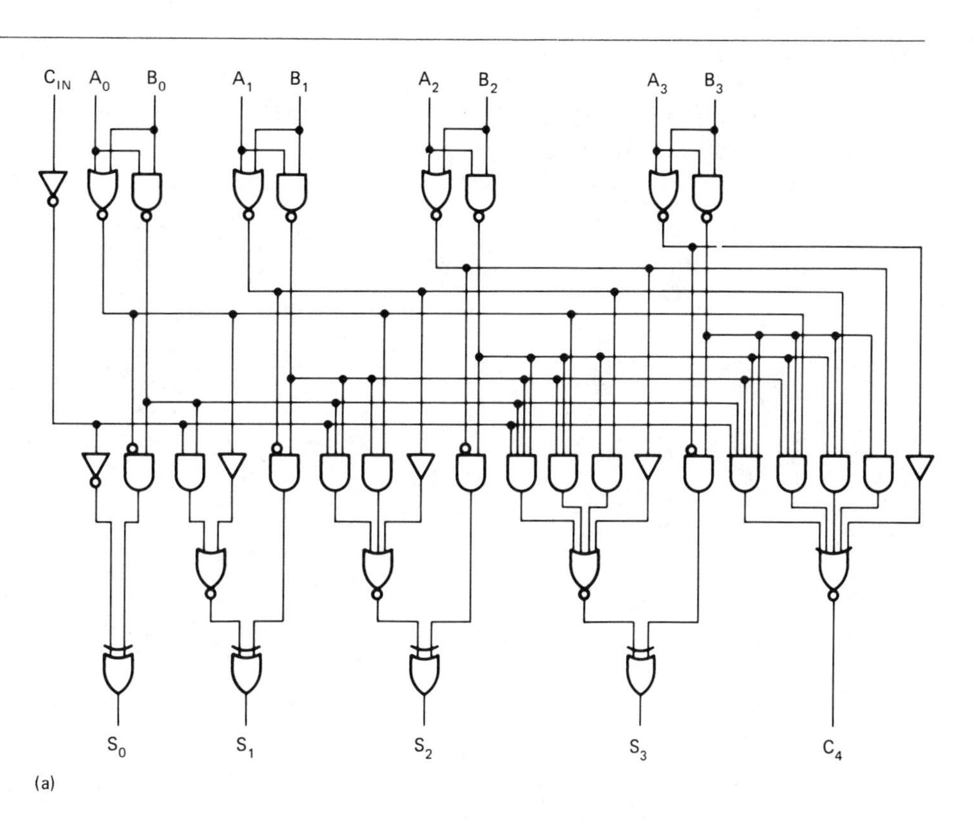

(a)

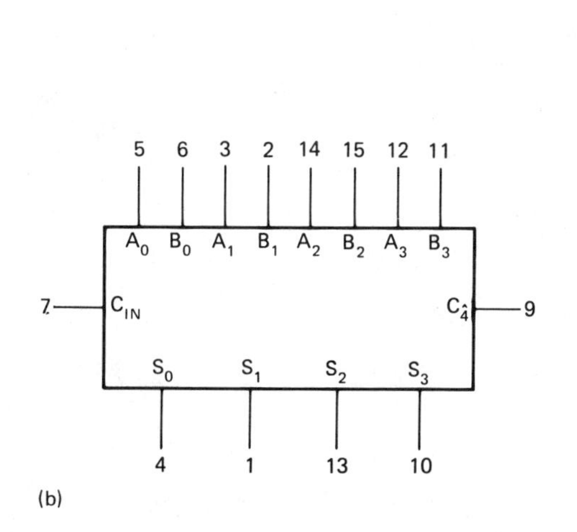

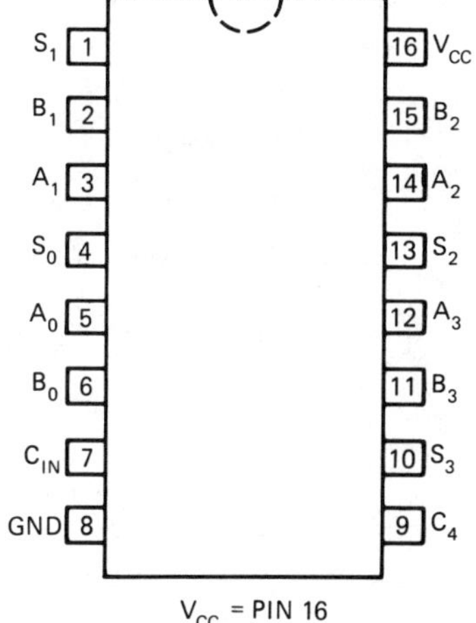

(b)

PARAMETER	MIN	MAX
PROPAGATION DELAY TIME (ns) C_{IN} TO S_n	–	24
A_n OR B_n TO S_n	–	24
C_{IN} TO C_4	–	17
A_n OR B_n TO C_4	–	17

(c)

Figure 10–18(b) shows the propagation delay time occurring for sum input to sum output and carry output. Note that these have, at 5V V_{DD}, maximum delays of 620, 760, and 800 nanoseconds. With these delays, it does not seem very useful to have a carry look-ahead of 360 nanoseconds maximum. The advantage occurs, however, when a number of these circuits are connected in parallel to form a large adder. As the final carry is always the limiting factor, reducing the carry-in to carry-out time has a marked effect. The final carry time for an adder

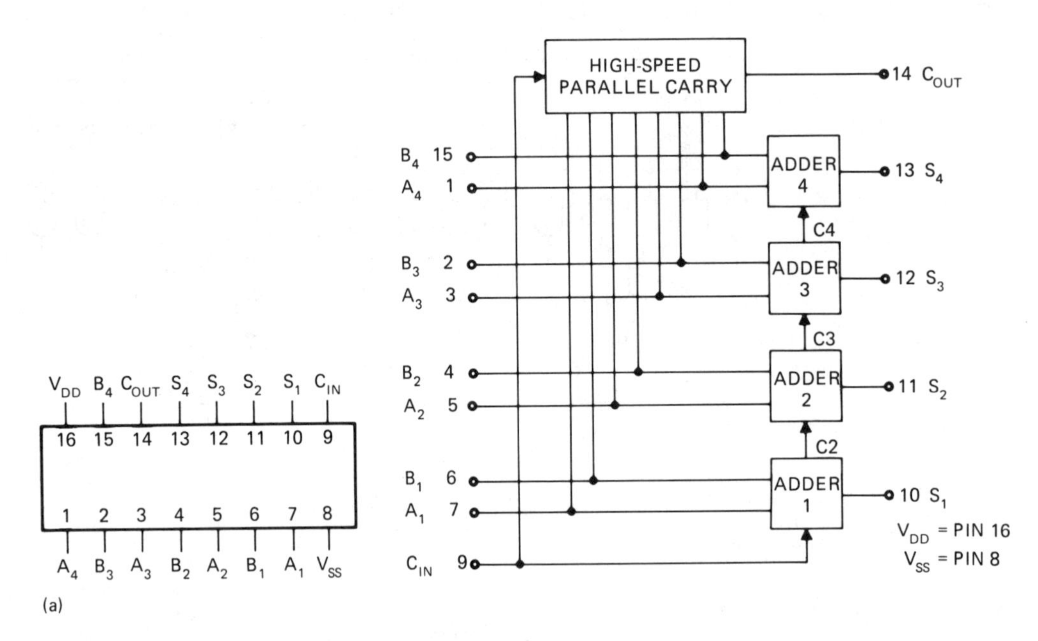

DYNAMIC CHARACTERISTICS (C_L = 50 pF, T_A = 25° C)

PARAMETER		V_{DD} (V DC)	MIN	TYP	MAX
PROPAGATION DELAY TIME (ns)					
SUM-IN TO SUM-OUT	t_{PLH}, t_{PHL}	5	–	400	800
		10	–	160	320
		15	–	115	230
SUM-IN TO CARRY-OUT	t_{PLH}, t_{PHL}	5	–	310	620
		10	–	140	280
		15	–	110	220
CARRY-IN TO SUM-OUT	t_{PLH}, t_{PHL}	5	–	380	760
		10	–	150	300
		15	–	115	230
CARRY-IN TO CARRY-OUT	t_{PLH}, t_{PHL}	5	–	180	360
		10	–	75	150
		15	–	55	110
OUTPUT TRANSITION TIME (ns)	t_{TLH}, t_{THL}	5	–	100	200
		10	–	50	100
		15	–	40	80

(b)

FIGURE 10–18

CMOS SCL4008B four-bit binary adder. The parallel carry circuit (carry look-ahead) speeds up the final carry so that the ripple carry time of the adder will be reduced when the circuits are cascaded to make adders of 8-, 12-, 16-, or larger bit size. (a) Block diagram. (b) Propagation delay times.

of n 4008 devices can be computed as sum-in to carry-out time plus n (carry-in to carry-out time). For a 16-bit adder at 5V V_{DD}, the maximum carry delay would be

$$620 + (4 \times 360)\ \text{ns} = 2.06\ \mu\text{s}$$

Figure 10–19 compares the operating speeds of four-bit adders in CMOS and the three main TTL families. The CMOS 4008 expends less than a microwatt when not in use, but as add cycle times increase, the power expended becomes significant but rarely approaches that of the 74 series adders. Note that higher speed in each case requires a sacrifice of higher power dissipation.

10.6 Binary-Coded Decimal Adder

The straight binary adder is not suitable for binary-coded decimal. As many general-purpose computers handle and store data in binary-coded decimal form, the binary-coded decimal adder is an important logic system. Figure 10-20 is a block diagram of a BCD adder of three decimal places.

Note that each decimal section of the BCD adder has four binary input lines for each A and B input and four binary output lines. Although four binary bits have a capacity of 0 through 15, in BCD they will never go higher than 9. If a normal four-bit binary adder is used for each digit of this addition, the outputs will be correct for sums up to 9. But, as Figure 10-21(a) shows, binary numbers and BCD numbers differ at sums of 10 and higher. At 10, we must generate a decimal carry with a value of 10 (or 2^0 times the next power of 10). At the same time, the four binary sum lines must be changed to 0. This can be accomplished by adding 6 and ignoring the binary carry, which would occur for a binary sum equal to 16. As Figure 10-21(b) shows, binary-coded decimal addition can be accomplished by using a normal four-bit binary adder for each decimal digit, decoding 10 and higher from the binary sum outputs with a logic circuit

FIGURE 10–19

Speed/power comparisons of the CMOS 4008 and three versions of the 74283.

DELAY PARAMETERS (@ C_L = 50 pF)	CMOS			TTL			UNITS
	5V	10V	15V	283	LS283	S283	
SUM-IN TO SUM-OUT	800	320	230	24	24	18	ns
SUM-IN TO CARRY-OUT	620	280	220	16	17	12	ns
CARRY-IN TO SUM-OUT	760	300	230	21	24	18	ns
CARRY-IN TO CARRY-OUT	360	150	110	16	22	11	ns
POWER DISSIPATION	SEE SECTION 9.14.5.			550	220	800	mW

FIGURE 10–20

Binary-coded decimal adder.

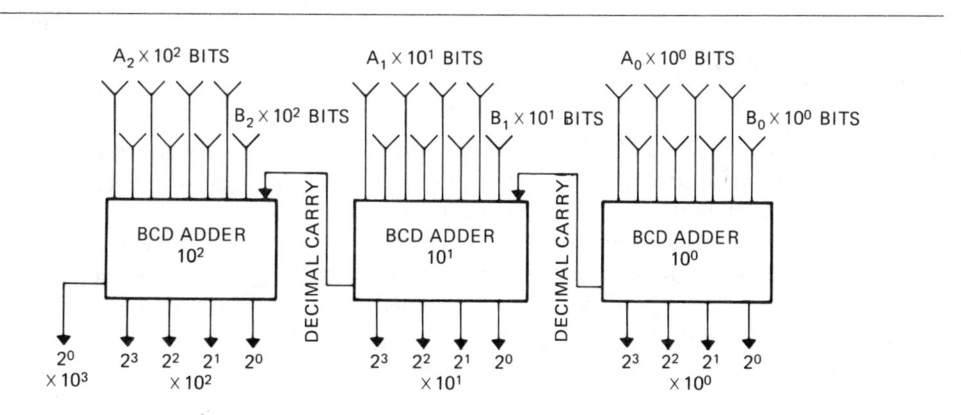

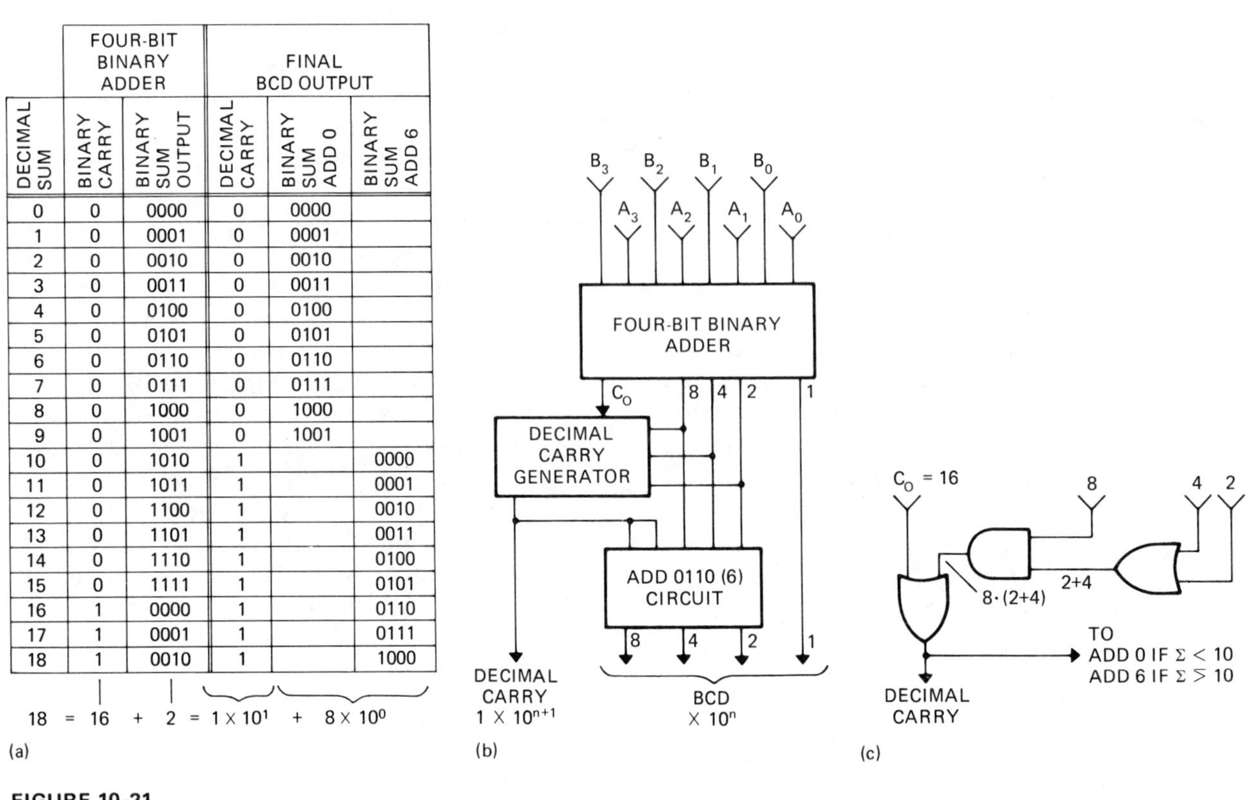

	FOUR-BIT BINARY ADDER		FINAL BCD OUTPUT		
DECIMAL SUM	BINARY CARRY	BINARY SUM OUTPUT	DECIMAL CARRY	BINARY SUM ADD 0	BINARY SUM ADD 6
0	0	0000	0	0000	
1	0	0001	0	0001	
2	0	0010	0	0010	
3	0	0011	0	0011	
4	0	0100	0	0100	
5	0	0101	0	0101	
6	0	0110	0	0110	
7	0	0111	0	0111	
8	0	1000	0	1000	
9	0	1001	0	1001	
10	0	1010	1		0000
11	0	1011	1		0001
12	0	1100	1		0010
13	0	1101	1		0011
14	0	1110	1		0100
15	0	1111	1		0101
16	1	0000	1		0110
17	1	0001	1		0111
18	1	0010	1		1000

$18 = 16 + 2 = 1 \times 10^1 + 8 \times 10^0$

(a)

FIGURE 10-21

(a) Comparative outputs of a four-bit binary adder and a binary-coded decimal adder. (b) Block diagram of a four-bit BCD adder. (c) Logic diagram of the decimal carry generator.

similar to Figure 10-21(c), and using additional adder circuits, as in Figure 10-22, to add 6 to the binary sum lines for sums of 10 or higher. Figure 10-21(a) confirms the validity of this logic system for all possible sums of BCD addition.

This solution for BCD addition has the obvious disadvantage of the increased circuitry required. Another solution for BCD arithmetic is to convert the input BCD numbers to binary, add them in a binary adder, and convert the answers back to BCD.

10.7 Summary

There are two fundamental types of adder circuits—the *serial adder* and the *parallel adder.* The parallel adder is discussed in this chapter. It is more expensive but of higher speed than the serial adder. The fundamental circuits are the same for both serial and parallel. The basic circuit used is the *half adder.* It has the capacity to add the least significant bits (LSBs) and produce a sum plus a carry bit. The circuit has two inputs (A and B) and two outputs (Σ_1 and C_O). The truth table for LSB addition, shown in Section 10.3, indicates that a two-input AND gate will handle the carry-out while an exclusive OR gate will provide the sum. This is shown in Figures 10-3 through 10-4. Figure 10-5 shows a convenient connection of three simple logic gates that produces a half adder. The carry-out from the LSB addition forms the carry-in to the second significant bit. This requires the second significant bit adder to have three inputs—A_1, B_1, and carry-in. Again, there are two outputs—Σ_2 and carry-out. The adder circuit handling three-input bits is called a *full adder.* The truth table for the full adder is shown in Figure 10-7. The Σ_2 output is that of a three-bit parity generator in that an odd number of 1 bits on the inputs produces a sum output. Two exclusive OR gates connected as shown in Figure 10-8 can provide this. A connection of three AND gates to an OR gate, as in Figure 10-9, is a most direct method of producing the carry. It is found, however, that if two half adders and an OR gate are connected, as shown in Figure

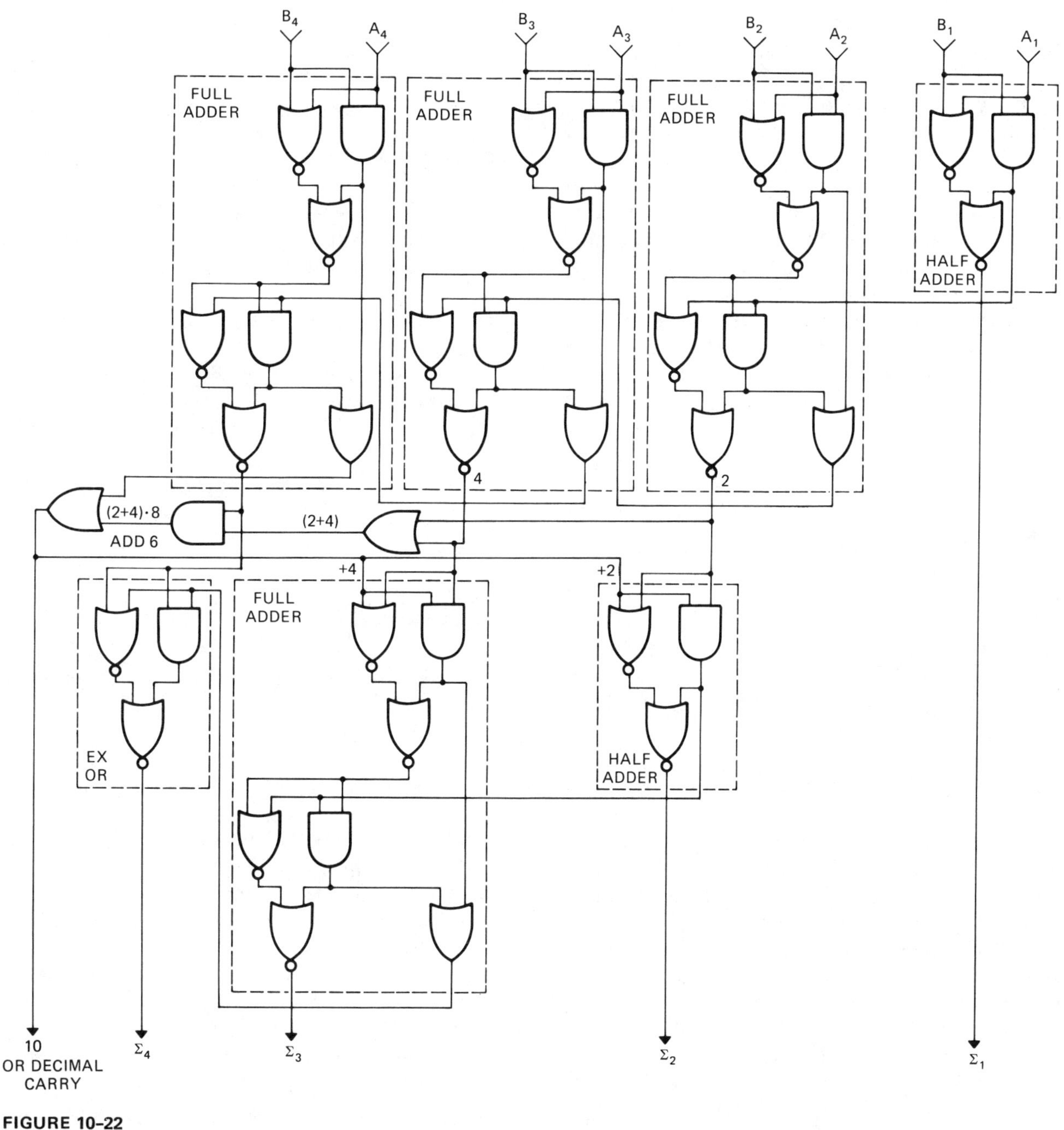

FIGURE 10-22

Binary-coded decimal adder for LSD (10^0 digit). For higher power of ten adders, the 2^0 bit must be a full adder to accommodate the carry-in.

10-10, the combined sum and carry logic can be produced with the minimum number of gates.

A two-bit full adder is available in 14-pin integrated circuit, as shown in Figure 10-13. The circuit uses the direct connection of AND gates explained in Figure 10-9, but is has a NOR gate instead of an OR gate, as shown in Figure 10-11. The result is an inverted carry, C_O. The C_O is used in conjunction with AND gates to form the sum logic, as Figure 10-12 shows. For the second bit of the two-bit adder, the inputs to the AND gates are inverted, but, as the truth table of Figure 10-13 shows, complemented inputs have complemented outputs. This means that the second carry-out will not be inverted. The absence of an inverter at the Σ_2 output eliminates inversion there.

A four-bit full adder is available in 16-pin

TTL integrated circuit. The logic circuits of the first and second bits of this adder are identical to that of the two-bit adder. The third and fourth bits also are identical in logic to the two-bit adder. They differ, however, in output pin connections.

Large adders can be assembled by connecting the carry-out of each bit to the carry-in of the next significant bit, with the final carry forming the MSB of the sum. Figure 10-14 shows this in block diagram form. In connecting the integrated circuit adders, only the final carry on each chip, C_{n+2} in two-bit adders or C_{n+4} in four-bit adders, need be connected to the C_n of the next higher circuit. The LSB C_n is grounded or 0; the highest-value carry forms the MSB of the sum. Figure 10-15 shows this for a pair of two-bit full adders.

A limitation of the ripple carry adder (Figure 10-14) is the time required to produce the final carry. A more complex, but high-speed, full-adder circuit with internal carry look-ahead is shown in Figure 10-17.

Many applications, such as portable equipment, require low power. In these situations, designers often use CMOS devices. The 4008 is a CMOS four-bit adder. However, the CMOS devices have longer propagation delays than TTL devices. Thus, the trade-off is power versus speed.

In many digital machines, it is more convenient to add numbers in binary-coded decimal instead of straight binary. The basic building blocks of half and full adders are used to form the BCD adder. Figure 10-21(a) compares the sums of a four-bit binary adder and a BCD single-digit adder. For sums that are 9 or lower, these adders operate identically. For sums of 10 and higher, they differ. At 10, the BCD adder must produce a carry and at the same time take the four lower digits to 0. The right-most column of the table shows the BCD digits to be attainable by adding 6 (0110) to the four-bit binary sum. Figure 10-21(b) shows the BCD adder to have three functional units—a four-bit binary adder, a circuit to decode sums of 10 and higher, and an adder to add 6 when the sum is 10 or higher. Figure 10-21(c) shows the decode-10 circuit. Identical circuits are used for each digit of the BCD adder.

Glossary

Half Adder: Logic circuit with the capacity to perform LSB addition. Figures 10-4 and 10-5 show two possible logic circuits for a half adder.

Full Adder: Logic circuit with capacity to perform addition of the second and higher significant bits of binary addition. Figure 10-10 is the logic symbol of a full adder.

MSI: Medium-scale integrated circuit.

Carry Look-Ahead Generator: A digital circuit that produces the final output carry by testing each input to the adder.

Questions

1. Draw the block diagram of a six-bit parallel adder.
2. Draw the truth table for LSB addition.
3. Draw the logic diagram of a half adder using AND and NOR gates.
4. Draw the truth table of a full adder.
5. The final carry of a seven-bit adder is used for what purpose?
6. Explain the difference between the first and second bit of the 5482 two-bit full adder.
7. What must be done with the carry-in (C_n) if a four-bit TTL full adder is used in the circuit of the LSB of a BCD adder?
8. Explain how you would use only three bits of a four-bit TTL adder.
9. The output of the decode-10-and-higher circuit has what two functions in the BCD adder?
10. Why is the LSB of the sum not involved in the add-6 circuit of a BCD adder?
11. For the 74LS283 adder, does the look-ahead carry generator circuit also test the input carry line, C_{IN}?

Problems

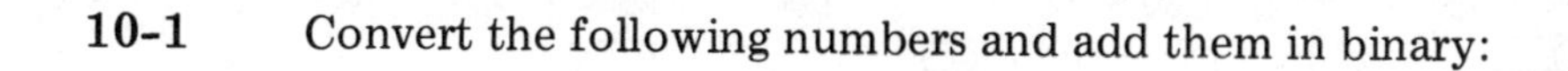

10-1 Convert the following numbers and add them in binary:

(a) 22 + 41 (b) 92 + 74 (c) 36 + 51 (d) 84 + 76

10-2 Figure 10-23 shows a two-bit adder, TTL 5482, and a four-bit adder circuit, TTL 5483. Connect the two to provide a six-bit binary adder.

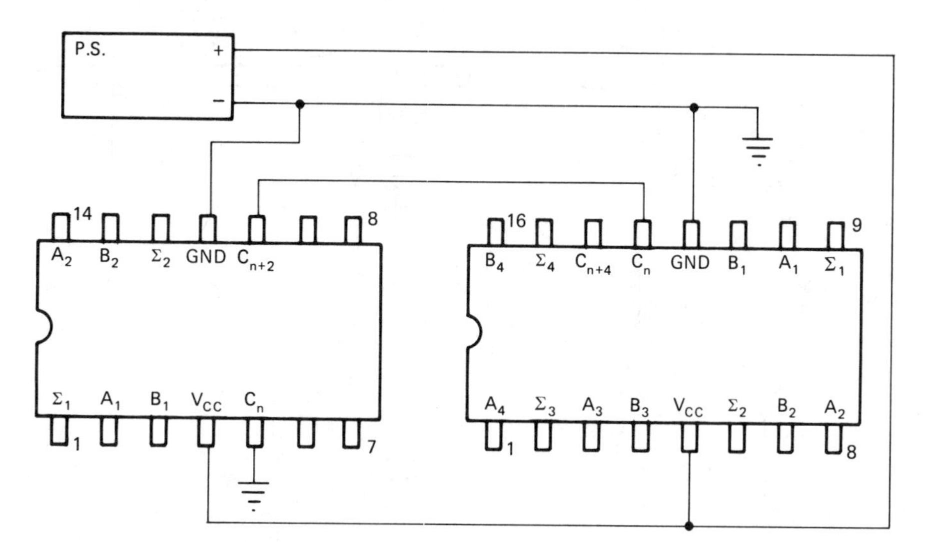

FIGURE 10-23. Problem 10-2

10-3 Convert the following decimal numbers to binary form and add them:

(a) 15 + 27 (b) 35 + 24 (c) 98 + 106 (d) 78 + 122

10-4 Convert the decimal numbers listed in Problem 10-3 to binary-coded decimal form and add them.

10-5 If only two-input NOR gates are available in your logic system, what connection of NOR gates could produce the half adder carry function?

10-6 Using only two-input NAND gates, produce a half adder carry function.

10-7 Draw the logic diagram of a half adder composed of all NOR gates (two or three inputs).

10-8 Develop the full adder by using two AND gates, two OR gates, and three NAND gates instead of using the method in Figure 10-10.

10-9 Draw the logic diagram of a full adder composed of all NAND gates.

10-10 At the inputs of the adder in Figure 10-24, the A number is the binary equivalent of 7 and the B number is the binary equivalent of 5. Label the 1 or 0 output state of each gate for these input numbers.

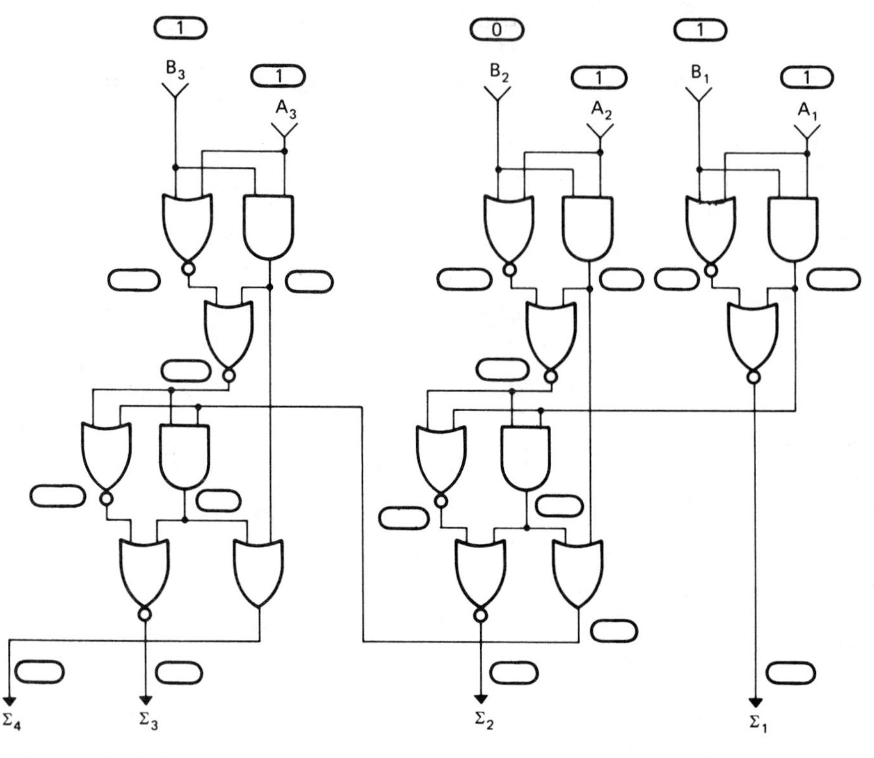

FIGURE 10-24. Problem 10-10

10-11 The BCD adder of Figure 10-25 has the A input of BCD 17, and the B input of BCD 16. Label the 1 or 0 output state at the points indicated.

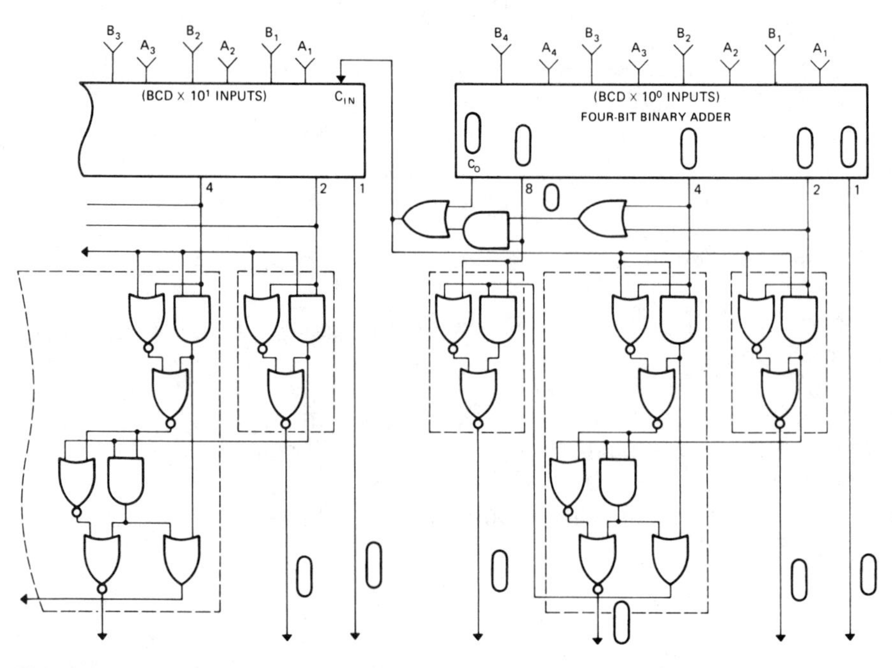

FIGURE 10-25. Problem 10-11

10-12 The circuits of Figure 10-26 are the TTL/SSI 5410 three-input NAND gates and the TTL/SSI 5405 hex inverter. Connect these circuits to form the decode-10-or-higher (carry-decode) circuits for a BCD adder. Use only one inverter.

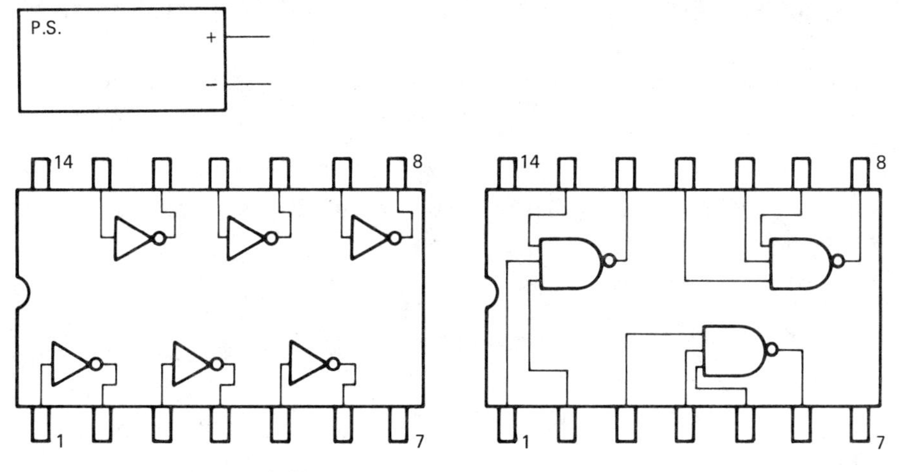

FIGURE 10-26. Problem 10-12

10-13 The circuits of Figure 10-27 are the TTL 7482 two-bit full adder and the 7486 quad exclusive OR gate. Draw in the connections needed to produce the add-6 circuit of the BCD adder.

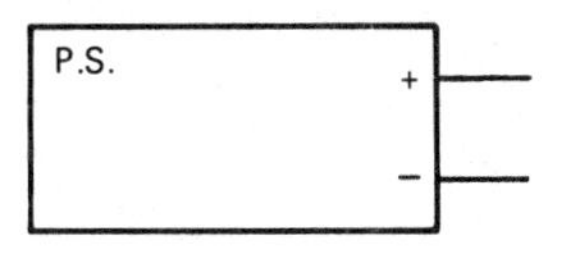

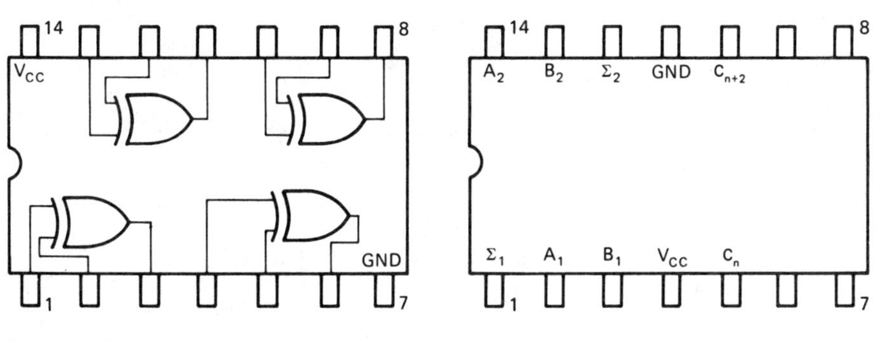

FIGURE 10-27. Problem 10-13

10-14 To produce a four-digit BCD adder, how many of each of the following circuits would be needed?

TTL/MSI	*TTL/SSI*
Four-bit full adder 7483	Quad exclusive OR 7486
Two-bit full adder 7482	Hex inverters 7405
	Quad two-input NAND 7400
	Triple three-input NAND 7410

10-15 A 16-bit adder is to be constructed using four 74S283 devices. What maximum time must be allowed for a complete add cycle? (No carry-in to the LSB. See Figure 10-19.)

10-16 A 16-bit adder is to be constructed using four 4008B CMOS devices. What maximum time must be allowed for a complete add cycle? (V_{DD} = 10V; C_L = 50 pF)

Boolean Algebra (Problems 10–17 and 10–18)

10-17 Prove by the theorems of Boolean algebra that the logic circuit of Figure 10-9 produces an identical carry output to that produced in Problem 10-8.

10-18 Prove that the output of Figure 10-9 is identical to the carry output of Figure 10-11.

CHAPTER

Parallel Subtraction

Objectives

Upon completion of this chapter, you will be able to:

- Connect logic gates to form a half subtractor.
- Connect two half subtractors and an OR gate to form a full subtractor.
- Assemble half and full subtractors to form a multibit subtractor.
- Use a subtractor circuit as a comparator for a three-way decision.

11.1 Introduction

Usually the logic described in Chapter 10 can also be used for subtraction, with some added circuitry. Adder-subtractor circuits, however, become more and more complex as we strive for an arithmetic unit with the highest degree of speed and versatility, and at some point it is worth considering using separate addition and subtraction circuits. For this reason, we will begin this discussion by developing logic circuitry for subtraction. Later chapters will explain the modification of adder circuits for doing both addition and subtraction in the same circuit.

11.2 Binary Subtraction

If A and B are two three-bit binary numbers, with A > B, we can subtract B from A and obtain the remainder R, a binary number: A - B = R. Let us again represent the three binary places 2^2 2^1 2^0 of A as A_3 A_2 A_1. Likewise, the number B as B_3 B_2 B_1. The letter b is used to indicate a borrow from the next higher place. The 1 that is borrowed from a higher power has a value of 2 in the column for which we borrow it. Therefore 0 - 1 results in a 1 borrow 1, as shown in Figure 11-1.

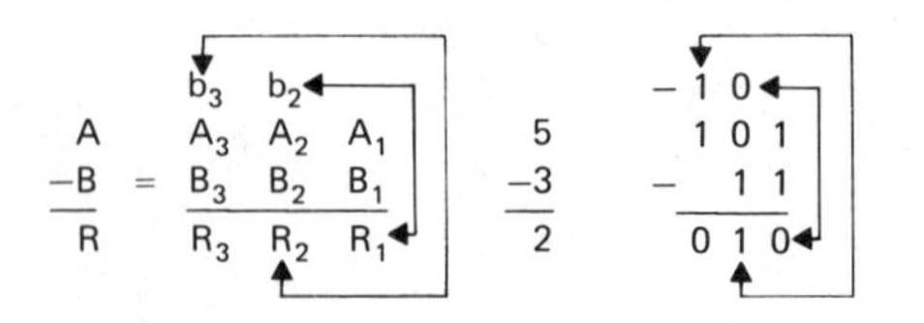

FIGURE 11-1
Subtraction of the binary equivalent of 5 - 3.

11.3 Half Subtractor

As in the case of the adder, let us develop a truth table for each individual bit of the subtraction, starting with the LSB $A_1 - B_1 = R_1 - b_0$. The four possibilities are as follows:

$$\begin{array}{cccc} 0 & 1 & 0 & 1 \\ \underline{-0} & \underline{-0} & \underline{-1} & \underline{-1} \\ 0 & 1 & 1 \text{ (borrow 1)} & 0 \end{array}$$

The truth table, therefore, will have two inputs and two outputs. If we ignore the borrow output, the truth table for the remainder is that of the exclusive OR. The borrow function is $b = \bar{A} \cdot B$. We could obtain this with a simple AND gate if the $\bar{A}$ function were already available. Combining the exclusive OR with the AND gate, as in the circuit of Figure 11-2, would accomplish the subtraction of the LSB bit.

If the complement of A_1 is not available, the same circuits can be connected, as Figure 11-3 shows. We can prove the validity of this as follows:

$$b_0 = B_1 \cdot R_1 = B_1 \cdot [(A_1 \cdot \bar{B}_1) + (\bar{A}_1 \cdot B_1)]$$
$$= (A_1 \cdot \bar{B}_1 \cdot B_1) + (\bar{A}_1 \cdot B_1 \cdot B_1)$$

Since $\bar{B}_1 \cdot B_1 = 0$, $B_1 \cdot B_1 = B_1$, and $A_1 \cdot 0 = 0$, therefore,

$$b_0 = (A_1 \cdot 0) + (\bar{A}_1 \cdot B_1) = (\bar{A}_1 \cdot B_1)$$

TRUTH TABLE

A_1	B_1	R_1	b_0
0	0	0	0
1	0	1	0
0	1	1	1
1	1	0	0

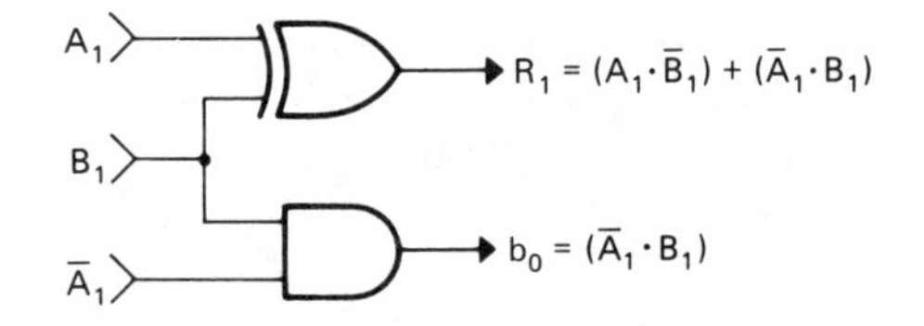

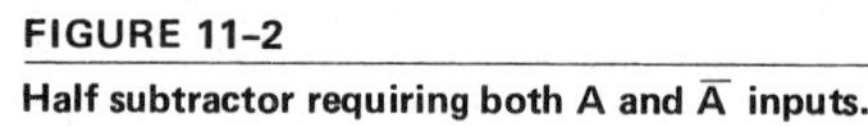

FIGURE 11-2
Half subtractor requiring both A and $\bar{A}$ inputs.

TRUTH TABLE

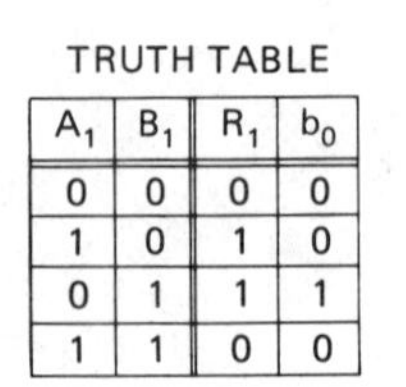

A_1	B_1	R_1	b_0
0	0	0	0
1	0	1	0
0	1	1	1
1	1	0	0

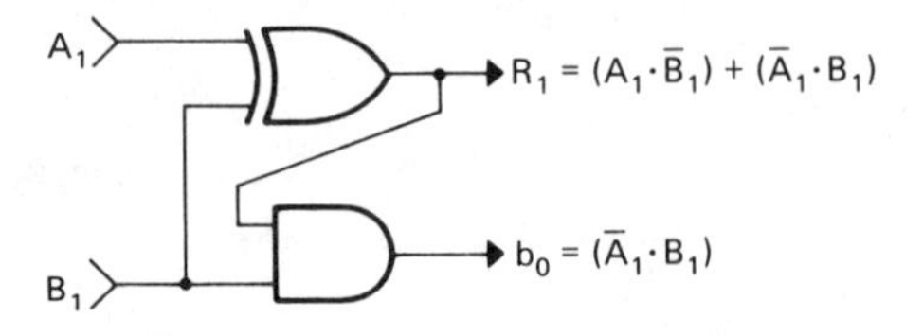

FIGURE 11-3
Half subtractor requiring only A and B inputs.

11.4 Full Subtractor

For subtraction of the second significant bit, the truth table has three inputs and two outputs, as in Figure 11-4. If we ignore the borrow column for the moment, the remainder column has a 1 only when an odd number of 1s appears on the inputs. This again is equivalent to a three-bit parity generator, or it can be accomplished with two exclusive OR gates, as in Figure 11-5.

Figure 11-6 shows the full subtractor truth table with the remainder column removed. We find four lines that produce a borrow-out, b_0. They produce Boolean functions and the identities shown. This borrow function could be produced with a variety of logic combinations, but in Chapter 10 we found that two half adders and an OR gate produce the most economical full adder. Fortunately, a similar economy occurs in combining two half subtractors and an OR gate to produce a full subtractor. Figure 11-7 shows this combination. The Boolean functions indicate the validity of the outputs. Figure 11-7 is therefore a full subtractor in its simplest form. The logic for a third significant digit and even higher digits in a larger subtractor would use the same full subtractor logic.

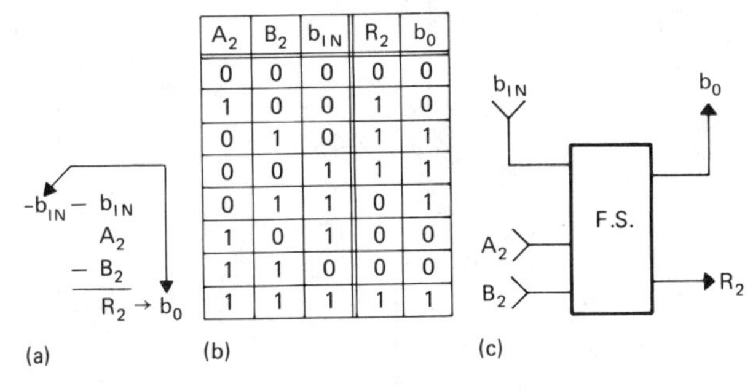

A_2	B_2	b_{IN}	R_2	b_0
0	0	0	0	0
1	0	0	1	0
0	1	0	1	1
0	0	1	1	1
0	1	1	0	1
1	0	1	0	0
1	1	0	0	0
1	1	1	1	1

FIGURE 11-4

(a) Full subtractor. (b) Truth table. (c) Block diagram.

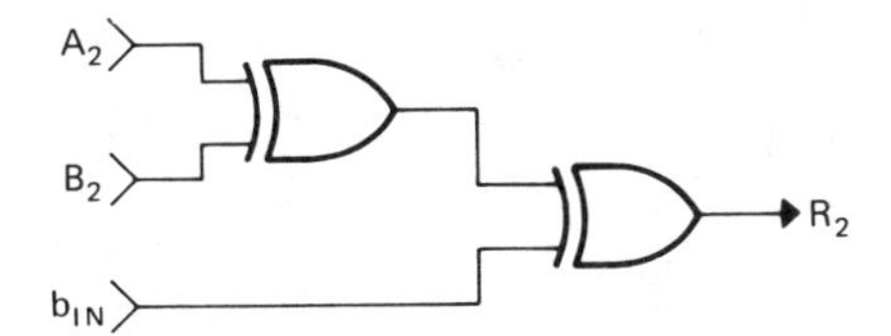

FIGURE 11-5

Remainder function obtained from a three-bit parity generator or two exclusive OR gates.

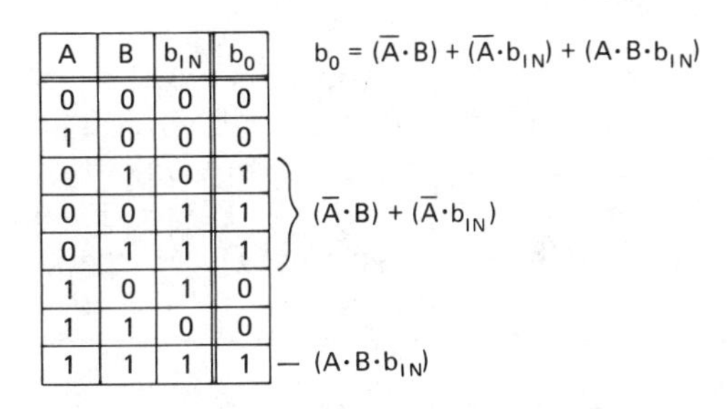

A	B	b_{IN}	b_0
0	0	0	0
1	0	0	0
0	1	0	1
0	0	1	1
0	1	1	1
1	0	1	0
1	1	0	0
1	1	1	1

FIGURE 11-6

Full subtractor borrow function truth table.

■ **EXAMPLE 11-1**

A full subtractor can be assembled using 14-pin integrated circuits—one TTL/SSI 7486 quad exclusive OR gate and one 7400 quad NAND gate. Show the connections needed.

FIGURE 11-7

Full subtractor logic diagram.

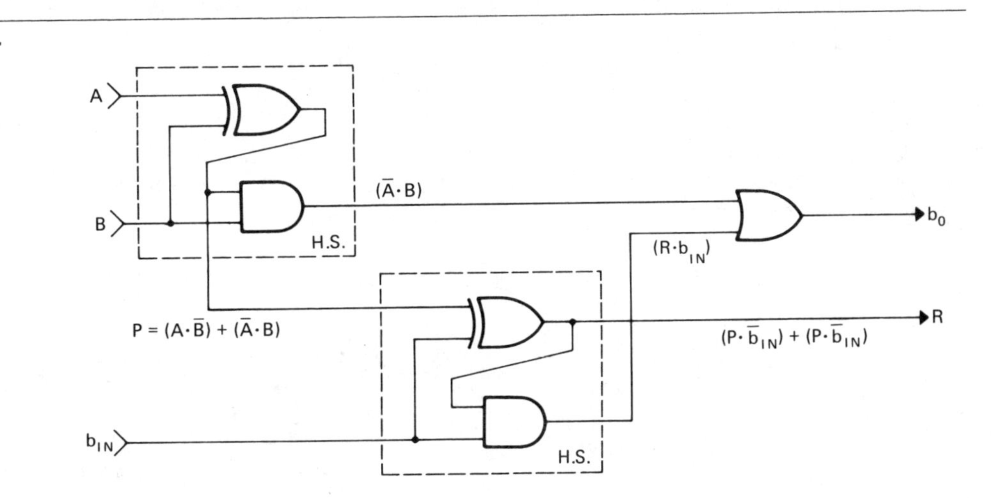

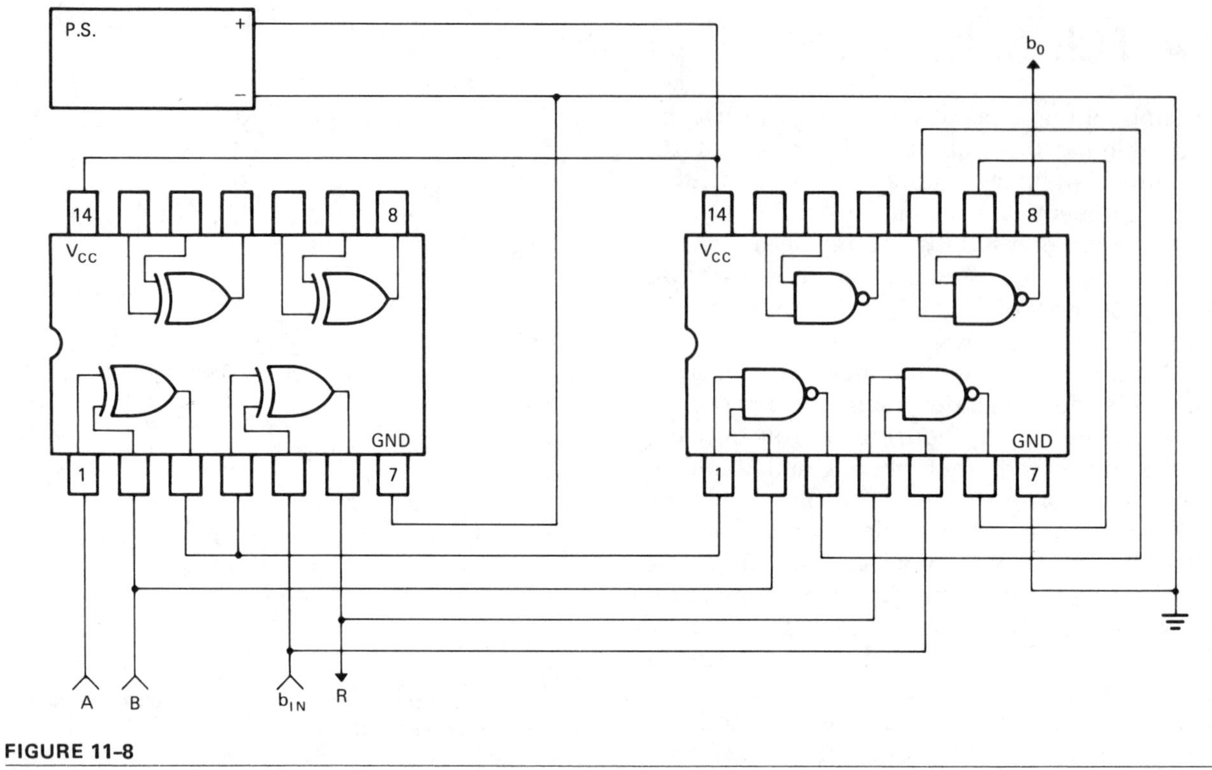

FIGURE 11-8
Solution for Example 11-1.

Solution: See Figure 11-8, which shows the circuit connection for a full subtractor. (*Note:* The borrow circuit of Figure 11-8 is the equivalent of the borrow circuit in Figure 11-7. Figure 11-9 shows them to be equivalent by De Morgan's theorem.)

11.5 Multibit Parallel Subtractor

A three-bit parallel subtractor would appear as shown in Figure 11-10. The size of the subtractor could be increased by connecting additional full subtractors. The outputs from this subtractor would be valid only for A > B. Those gates in dotted lines would not be needed in the MSB full subtractor, for if A > B, there would never be a borrow generated by the MSB.

The circuit has limited application. It could be used where a correction factor, B, which is always negative, must be subtracted from a larger number, A. If this need occurred in a circuit that otherwise did not require an arithmetic unit, the subtractor just described would offer the most economical solution.

FIGURE 11-9
Three-NAND-gate connection, the De Morgan equivalent to the full subtractor borrow function.

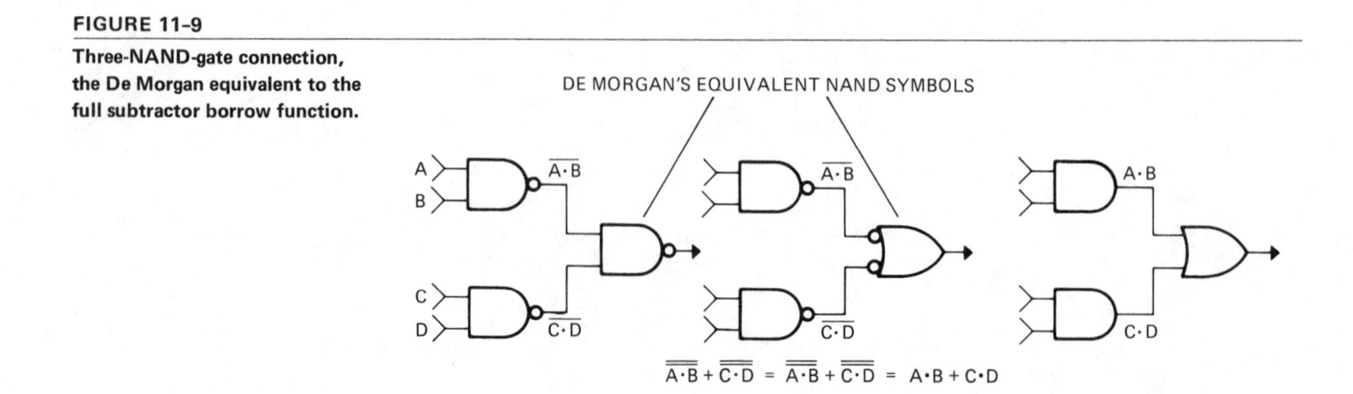

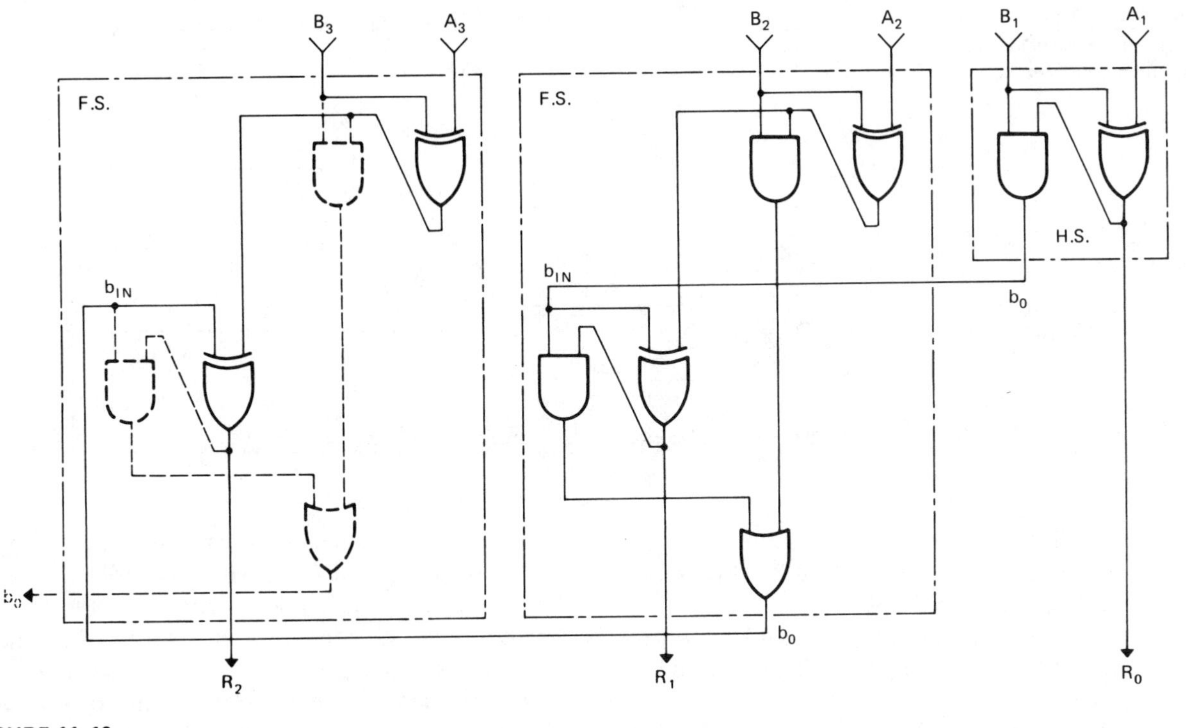

FIGURE 11-10

Three-bit parallel subtractor logic diagram.

11.6 Parallel Comparator

In Chapter 1, we stressed the technological advance resulting from the ability of electronic machines to make decisions and choose a course of action based on their decisions. Figure 11-11 shows several flow diagram symbols for machine decisions. The simplest machine decision is to compare two numbers in a comparator circuit, as explained in Section 8.7. Figure 11-11(a) shows such a decision made on the basis of $A = B$ and the two courses of action the machine might take. A two-way decision of this type can be made by comparing two numbers in the digital comparator of Figure 8-23. The digital subtractor can be used to make the three-way decisions of Figure 11-11(b). Figure 11-11(c) shows an application of such a three-way decision leading to different courses of action in the computation of payroll. The digital subtractor can be connected to provide three-way comparison, as in Figure 11-12.

If we use the subtractor without the restriction $A > B$, the three-way decision will be detected, as shown in Figure 11-12, as follows:

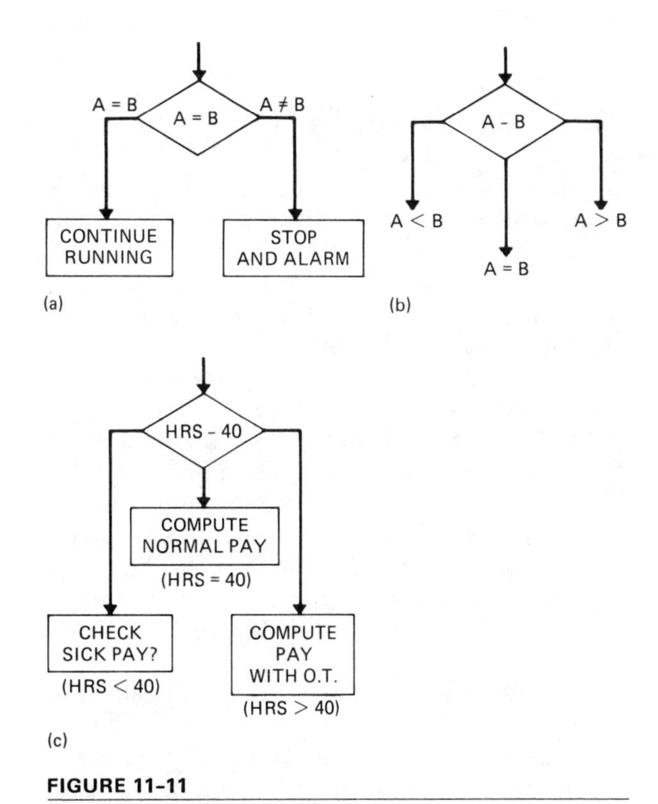

FIGURE 11-11

Flow diagram symbols of comparator functions. (a) Two-way decision made by comparator of Figure 8-23. (b) Three-way decision made by subtraction. (c) Typical application of subtraction-type comparison.

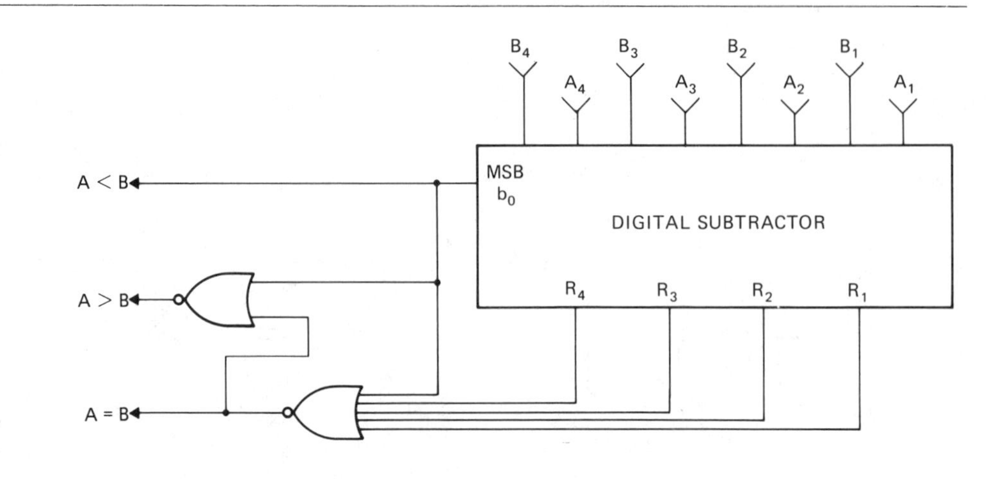

FIGURE 11-12
Logic circuit for digital comparison by subtractor.

1. If A = B, all the remainder output lines and the MSB borrow line will be 0. This can be detected with a NOR gate.
2. If A > B, one or more of the remainder lines will be high. The MSB borrow line will be 0.
3. If A < B, the MSB borrow line will be high. The remainder output will be incorrect, but if the only function is to compare, this will not matter.

11.7 Summary

In the majority of digital machines, subtraction is performed by ones or twos complement using circuits like those described in Chapter 10. There are, however, some circumstances where it is better to use circuits that are designed to subtract without complementing.

If we consider the truth table for binary subtraction of the LSB, there are only two inputs, A and B. The outputs are the remainder, R, and the borrow-out, b_0. The LSB truth table in Section 11.3 shows the remainder column to be an exclusive OR function. The borrow column is an AND function: $b_0 = (\overline{A_1} \cdot B_1)$. The logic circuit used for LSB subtraction is known as a *half subtractor.* Figure 11-3 shows this logic.

Subtraction of the more significant bits requires three inputs and two outputs. The circuit used for this is called a *full subtractor.* Figure 11-4 shows the full subtractor truth table. The logic can be accomplished by using two half subtractors and an OR gate, as in Figure 11-7.

Multibit subtractors can be made using a half subtractor for the LSB and a full subtractor for each higher bit. The borrow-out of each subtractor is connected to the borrow-in of the next higher bit. Figure 11-10 shows a three-bit parallel subtractor. This subtractor has no provision for handling negative remainders. Therefore, A must always be greater than or equal to B. This being the case, no borrow output will occur on the MSB and this bit can be simplified by using only the exclusive OR gates.

In Chapter 8, we discussed the parallel comparator circuit, with which two digital numbers can be compared to determine whether they are the same or different. This circuit is capable of a two-way decision, as shown by the flow diagram of Figure 11-11(a). If the MSB carry circuit is left in the parallel subtractor, it can be used to obtain a three-way comparison, as in Figure 11-11(b). Figure 11-11(c) shows a typical decision made by such a circuit. Figure 11-12 shows the logic circuit of a parallel subtractor used for comparison of A and B. The subtractor is wired for A - B. If A > B, there will be a remainder and no MSB borrow. Both inputs to the A > B NOR gate will be 0, producing a 1 on the A > B output. If A < B, there will be an MSB borrow. The 1 on this line will turn off the other two outputs. If A = B, there will be all 0 levels from the subtractor output, resulting in a 1 from the A = B NOR gate.

Glossary

Half Subtractor: Logic circuit used to provide binary subtraction of the LSB.

Full Subtractor: Logic circuit used to provide binary subtraction of bits higher than the LSB.

Questions

1. Draw the truth table and logic symbol of a half subtractor.
2. How does the half adder logic circuit differ from the half subtractor?
3. Why are subtractor circuits less common than adder circuits?
4. Draw the truth table and logic circuit of a full subtractor.
5. Can the subtractor circuit of Figure 11-10 work equally well for $A < B$ and $A > B$?
6. Why is it possible for the MSB of the subtractor to work without a borrow circuit?
7. Draw the flow diagram symbols and explain the difference between the comparator of Figure 8-23 and that of Figure 11-12.

Problems

11-1 Convert the following decimal numbers and subtract in binary:

$$\begin{array}{r} 23 \\ -12 \\ \hline \end{array} \qquad \begin{array}{r} 132 \\ -\ 69 \\ \hline \end{array} \qquad \begin{array}{r} 69 \\ -45 \\ \hline \end{array} \qquad \begin{array}{r} 72 \\ -38 \\ \hline \end{array}$$

11-2 Subtract the following binary numbers, convert to decimal, and check the results:

$$\begin{array}{r} 110110 \\ -101001 \\ \hline \end{array} \qquad \begin{array}{r} 101010 \\ -\ 11101 \\ \hline \end{array} \qquad \begin{array}{r} 1101101 \\ -1001111 \\ \hline \end{array}$$

11-3 How many of the following TTL circuits are needed to produce a five-bit subtractor?

(a) Quad exclusive OR (TTL/SSI 7486)
(b) Dual AND OR invert (TTL/SSI 7451)
(c) Hex inverter (TTL/SSI 7404)

11-4 Draw the block diagram of the circuit described in Problem 11-3.

11-5 Draw the logic diagram of the five-bit subtractor described in Problem 11-3 using logic symbols for the first, second, and last bit and block symbols for the third and fourth bits.

CHAPTER

Storage Register Elements

Objectives

Upon completion of this chapter, you will be able to:

- Identify the set-reset register and show how it stores parallel numbers.
- Draw the logic symbol and truth table of a set-reset register.
- Construct a set-reset flip-flop by crossing two NOR gates.
- Construct a set-reset flip-flop by crossing two NAND gates.
- Use toggle flip-flops to complement numbers in storage.
- Draw the logic symbol and truth table of a master-slave flip-flop.
- Use a toggle flip-flop in a serial parity check system.
- Steer numbers into storage by using the steer function of J-K flip-flops.
- Assemble strobe gates to strobe numbers into set-reset storage resisters.
- Use the integrated circuit bistable latch for large-scale storage.
- Use an integrated circuit J-K flip-flop.
- Draw timing diagrams for a J-K flip-flop.
- Compare the D latch and D flip-flop with the J-K flip-flop.
- Draw timing diagrams for the D flip-flop.

12.1 Introduction

As signals pass rapidly through a digital system, they must often be held in a particular location long enough for other signals to arrive or for certain operations to be performed. We have seen this in the case of the arithmetic circuits, in which two numbers, A and B, must be held at the adder inputs long enough for carry operations to occur. This holding or storing function is performed by register circuits. The register circuit is a device that can receive a temporary level such as a pulse and hold that level until instructed to change it. More complicated register circuits can toggle or convert numbers to complements. They are also used to shift numbers right or left in binary place and to convert numbers from serial to parallel and vice versa.

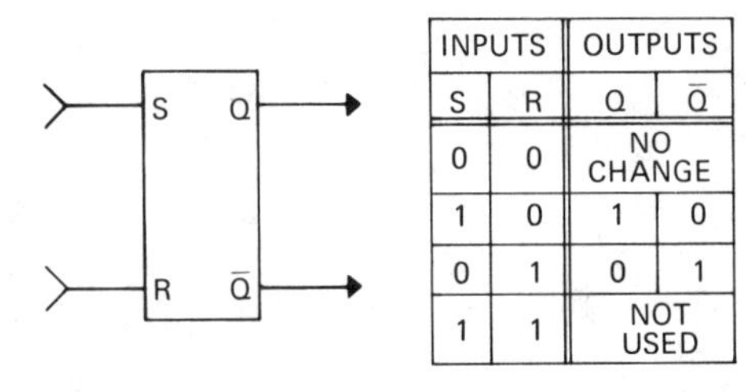

INPUTS		OUTPUTS	
S	R	Q	$\overline{Q}$
0	0	NO CHANGE	
1	0	1	0
0	1	0	1
1	1	NOT USED	

FIGURE 12-1
Logic symbol and truth table for a set-reset flip-flop.

12.2 Set-Reset Flip-Flop

The simplest form of storage register is the set-reset flip-flop. Figure 12-1 shows the logic symbol and truth table for this. The top line of the truth table is the resting state of the flip-flop. With 0 logic level on both inputs, S and R, one of the outputs must have a 1 on it and the other a 0. The outputs are labeled Q and $\overline{Q}$.

When a flip-flop is set, it may usually be considered as having a 1 stored in it. The flip-flop can be set by putting a pulse or temporary 1 level on the set input while keeping the reset input at 0. The outputs can be reversed by putting a temporary 1 or pulse on the reset input while the set input remains at 0. This places the flip-flop in the reset state, which produces a 0 on the Q output and a 1 on the $\overline{Q}$ output. Having a 1 on both inputs at the same time should not occur in normal operation. If a double 1 input does occur, the result will depend on the construction of the flip-flop, for there are numerous methods of constructing such a flip-flop, and the results of a double 1 input would depend on that construction.

Registers may be used to store binary numbers. Figure 12-2 shows a three-bit register storing the binary number 101.

Setting a number into registers usually occurs as follows: A reset pulse at time 0 resets all flip-flops to the 0 state, to remove the number previously stored therein. On the leading

FIGURE 12-2
Three-bit register being reset at time 0 and loaded with the binary number 101 at time 1.

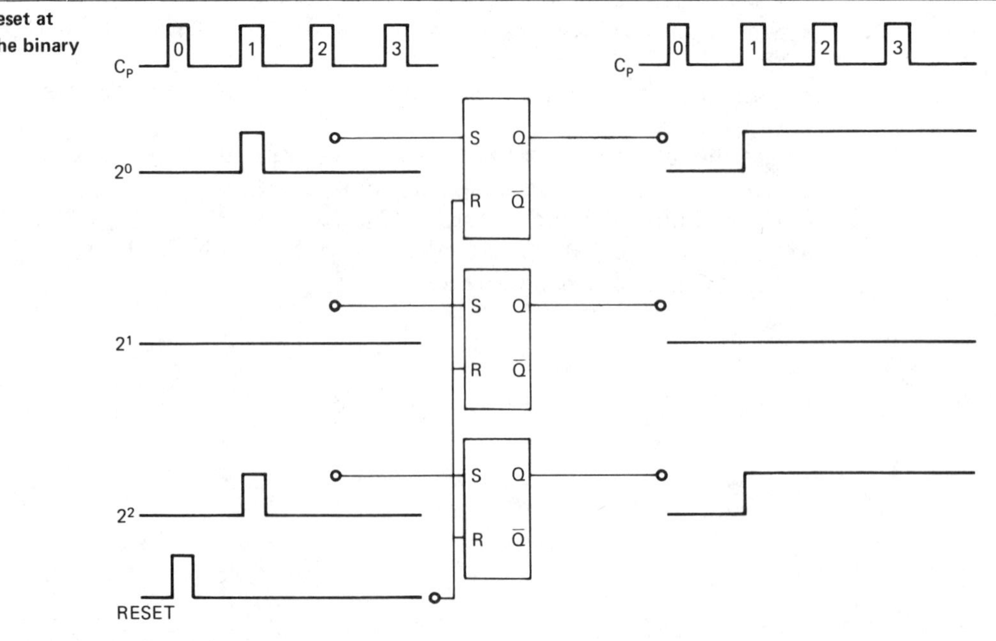

edge of the reset pulse, all flip-flop Q outputs will be at 0. One clock time later, a pulse will occur on the set inputs of only those flip-flops that are to store 1s. As shown for the three-bit register of Figure 12-2, to store a 101, only the 2^0 and 2^2 set lines receive pulses. The 2^1 line remains at 0. Note that the Q output levels are all 0 at reset but change to the value 101 at time 1. The output lines differ from the input lines in that they are in level form and will remain 101 until the next reset pulse occurs. With the exception of the number 111, the number stored cannot be changed without first resetting.

The terms *preset* and *clear* are often used in place of *set* and *reset*. Figure 12-3 shows a symbol and truth table using these designations.

The most economical method of providing a set-reset flip-flop is to cross two NOR gates, as in Figure 12-4. If any single input is in the 1 state, it produces a 0 out of its NOR gate. That 0 is transmitted to the second NOR gate. With both inputs at 0, the second gate will produce a 1 output that will hold the first gate in the 0 state even after the original 1 level is removed from its input. If the crossed NOR flip-flop is in the set state, it can easily be changed to the reset state by holding the set input at 0 and putting a pulse or temporary 1 level on the reset input. The pulse on the reset holds the 1 output (top NOR gate) at 0. This results in two 0s into the bottom NOR gate, resulting in a 1 output that, in turn, is applied to the top NOR gate, permanently holding its output at 0.

The crossing of NAND gates can also produce a set-reset flip-flop, as Figure 12-5 shows. As shown by the truth table, however, the resting state requires two 1s on the input. As designated, the effect of set and reset 1 inputs is also reversed. A temporary 0 level or negative-going pulse must be used to set or reset this flip-flop. A pair of input inverters convert this to the same function as the crossed NOR with the exception of the not-used state. This is shown in Figure 12-6.

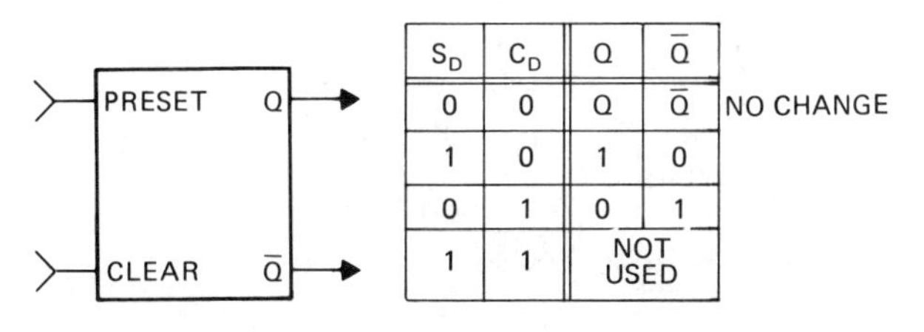

S_D	C_D	Q	$\bar{Q}$	
0	0	Q	$\bar{Q}$	NO CHANGE
1	0	1	0	
0	1	0	1	
1	1	NOT USED		

FIGURE 12-3
Alternative symbol and truth table for a set-reset flip-flop.

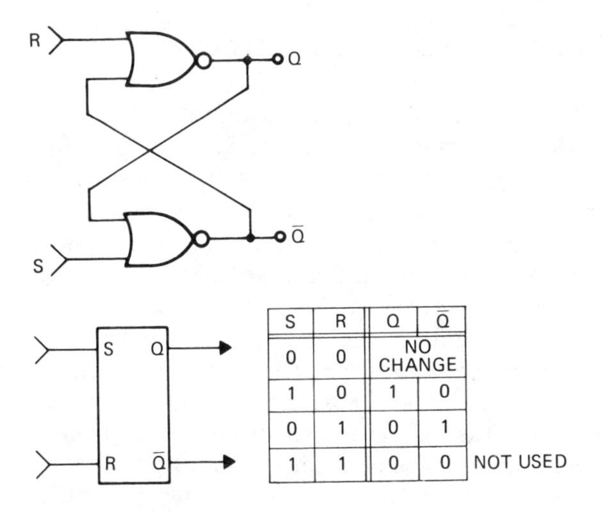

S	R	Q	$\bar{Q}$	
0	0	NO CHANGE		
1	0	1	0	
0	1	0	1	
1	1	0	0	NOT USED

FIGURE 12-4
Logic symbols and truth table for crossed NOR flip-flop.

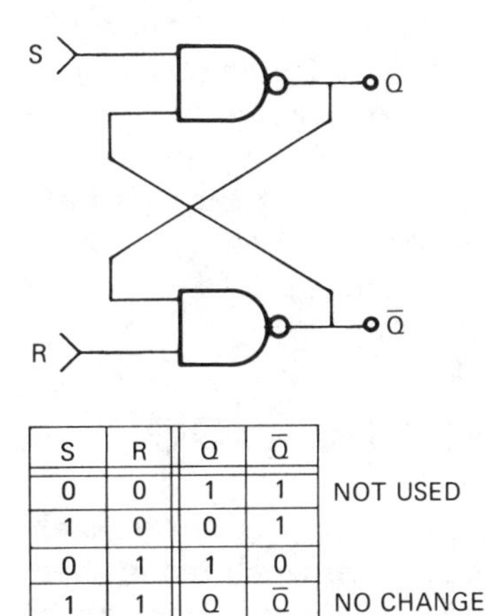

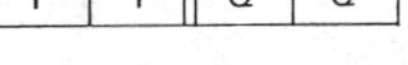

S	R	Q	$\bar{Q}$	
0	0	1	1	NOT USED
1	0	0	1	
0	1	1	0	
1	1	Q	$\bar{Q}$	NO CHANGE

FIGURE 12-5
Logic symbol and truth table for crossed NAND flip-flop.

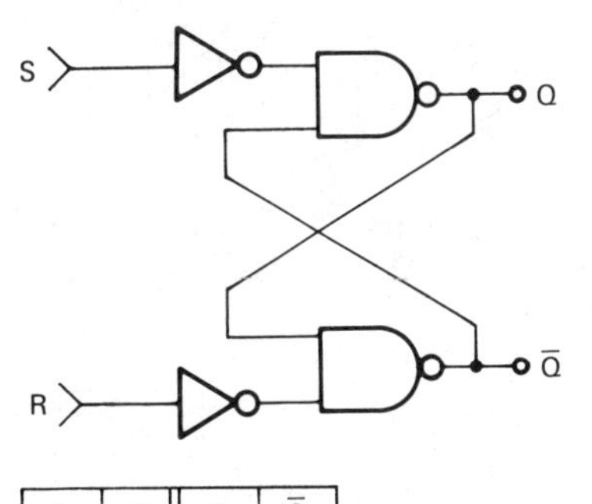

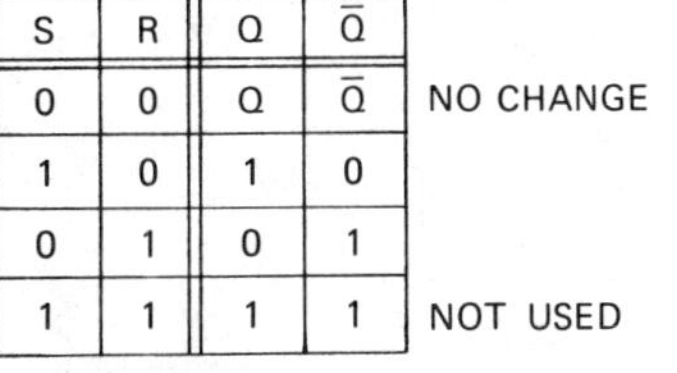

S	R	Q	$\bar{Q}$	
0	0	Q	$\bar{Q}$	NO CHANGE
1	0	1	0	
0	1	0	1	
1	1	1	1	NOT USED

FIGURE 12-6
Logic symbol and truth table for crossed NAND flip-flop with input inverters.

12.3 Strobe Gates

Often it is desirable to store information in registers from lines that are initially unstable or changing. To avoid having the registers set prematurely to some initial 1 levels, we do not allow the registers to see the data until they have settled down to their final value. Such would be the case if we were to store the sum output of a parallel adder. The inputs do not arrive at precisely the same time. The carries must ripple through the adder, causing the sum lines to vary between 1 and 0 several times before the answer is complete. If we connect the sum lines directly to the set inputs of the register, they may store 1s during this ripple time even though the output finally settles to 0. Figure 12-7 shows a solution to this. The sum lines are connected to gates that are strobed by a pulse occurring after the adder has settled down. The inhibited gates hold the set inputs at 0 until the strobe pulse occurs. The strobe pulse will pass through only those AND gates that are enabled by 1 levels from their respective sum output lines.

12.4 Trigger Flip-Flop

In arithmetic circuits, there is a need for registers that can complement the number stored in them when instructed to do so. This function can be provided by a modification of the set-reset flip-flop. Figure 12-8(a) shows the symbol for a flip-flop with this triggering capability.

Thus far we have seen that a pulse on a *set* input will result in no change if the flip-flop is already set. A pulse on a *reset* input will result in no change if the flip-flop is already reset. A pulse on the trigger input, however, will, regardless of which state the flip-flop is in, change it to the other state. Although there are some applications for a trigger flip-flop without set and reset capabilities, these inputs are an inexpensive addition and can be tied to 0 or ground when not used. Figure 12-8(b) and (c) show more likely symbols to be found in actual manufacture. The

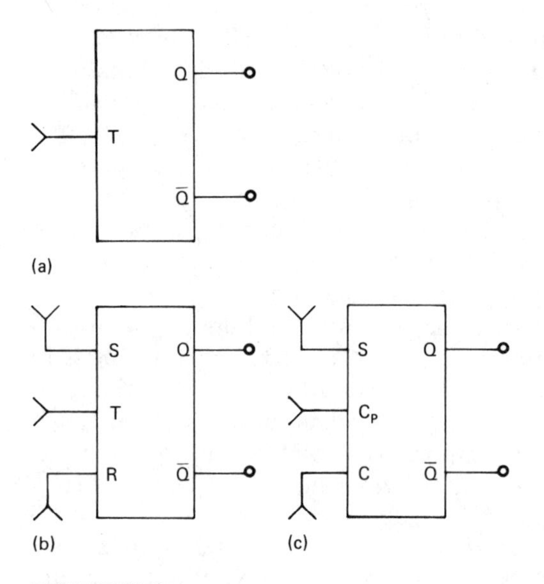

FIGURE 12-8

Logic symbols. (a) Toggle (trigger) flip-flop. (b) RST (reset-set and toggle) flip-flop. (c) Alternative symbol using C (clear) for reset, CP (clock) for toggle, and Q $\bar{Q}$ for outputs.

FIGURE 12-7

Adder outputs strobed into set-reset registers.

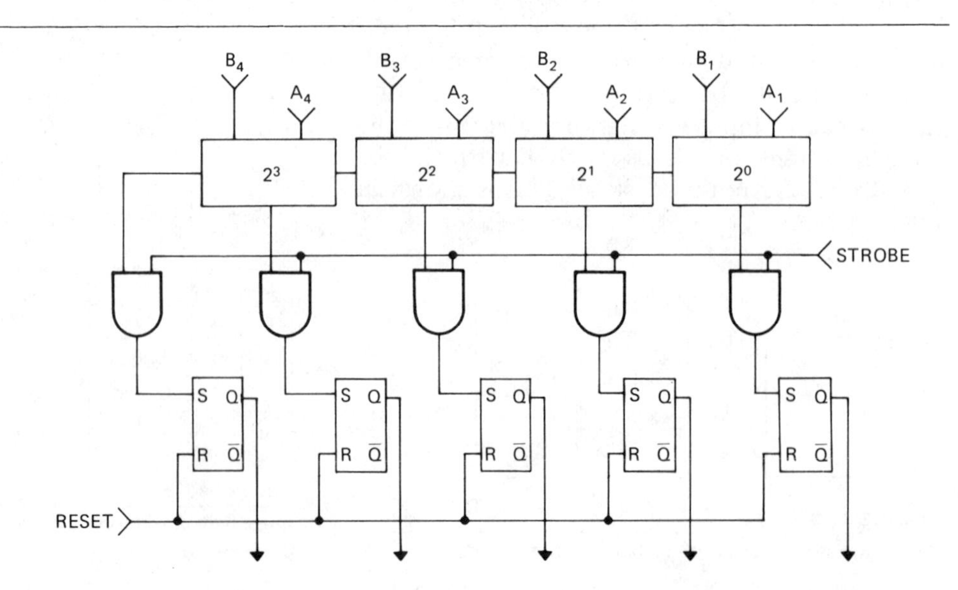

trigger function can be added by using capacitors, resistors, and diodes. It was such an easy matter to produce a set-reset flip-flop by crossing NOR gates or NAND gates that one might assume it to be simple to make this same circuit toggle. Unfortunately, there are several pitfalls to this assumption, which we shall describe here.

If we start with the crossed NOR, which has only a set and reset input, the trigger must be directed to the set input if the register is reset. It must be directed to the reset input if the flip-flop is set. It would seem that an AND gate on each input with the outputs crossed over to enable opposite inputs, as in Figure 12-9, would work; but this circuit is unreliable because of a *race problem.* A race problem is a condition in which a logic function, X, removes the conditions that are needed for its own existence. In order for a trigger pulse to pass through the AND gate and appear at the set input, the flip-flop must be in reset until the trigger has completed its job of setting the flip-flop. As the trigger appears, it tries to set the flip-flop, but the setting of the flip-flop may inhibit the AND gate, removing the trigger before it can complete the job.

Race problems of this type are a common pitfall in logic circuit design. Where they exist, they lead to unstable operation. The circuit of Figure 12-9 needs addition of some form of memory or delay for the crossed-over outputs. This could be provided by resistor-capacitor storage. Prior to the use of integrated circuits, toggle flip-flops composed of circuits with RC storage were commonly used. These were classified as AC flip-flops. Now that integrated circuits are so inexpensive, the use the AC flip-flops is a rarity.

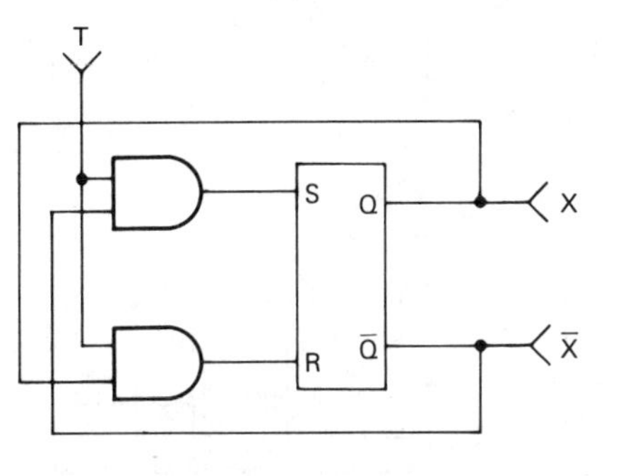

FIGURE 12-9

Race problem, resulting from the attempt to develop a toggle flip-flop by cross-coupling the outputs.

12.5 Master-Slave Flip-Flop

Need for capacitors is a serious disadvantage for integrated circuit fabrication of a flip-flop. For this reason, AC flip-flops are not suitable as integrated circuit toggle flip-flops. To provide a reliable means of toggling a crossed NOR or crossed NAND flip-flop, and at the same time avoid use of capacitors, a second flip-flop is used for storage. In this method, we use two flip-flops—one a master, the other the slave—as shown in Figure 12-10(a) and (b).

When the clock line is low, both master and slave are in the same state. Let us start with the set state and the trigger line low. Figure 12-11 shows the logic levels that occur at the inputs and outputs of the control gates before, during, and after the clock pulse. As Figure 12-11(a) shows, the reset AND gate of the master is enabled by the output 1 level of Q. The set AND gate is inhibited by the 0 from $\overline{Q}$. The output of both gates remains 0, however, until the clock line goes high. As Figure 12-11(b) shows, when the clock line goes high, the master flip-flop resets. Nothing happens to the slave, however, even though its reset gate is now enabled, because the inverted clock is at 0, inhibiting the AND gates to the slave. On the trailing edge of the clock pulse, the $\overline{C}_P$ goes high; and, as the logic levels in Figure 12-11(c) show, the slave will reset. On the next pulse, the same functions will occur, except that the set AND gates will be enabled and the outputs will return to the set state. The capabilities of this flip-flop can be further extended by adding the clear and preset inputs to the master NOR gates, as shown by the dotted lines in Figure 12-10. They make it possible to set or reset the flip-flop without waiting for a clock pulse.

An important difference between the clock or trigger operation and the set and clear operations is that using the clock input causes changes to occur on the trailing edge of the clock pulse. In use of the set or clear inputs, changes occur on the leading edge of an input pulse. If the flip-flop is already set, a pulse on the set input will have no effect. Likewise, if the flip-flop is already reset, a pulse on the reset input will have no effect. A pulse on the trigger or clock input, however, will—regardless of what state the flip-flop is in—change it to the other state.

FIGURE 12-10

(a) The master-slave flip-flop using crossed NOR gates as master and slave. (b) Master-slave flip-flop in block symbol.

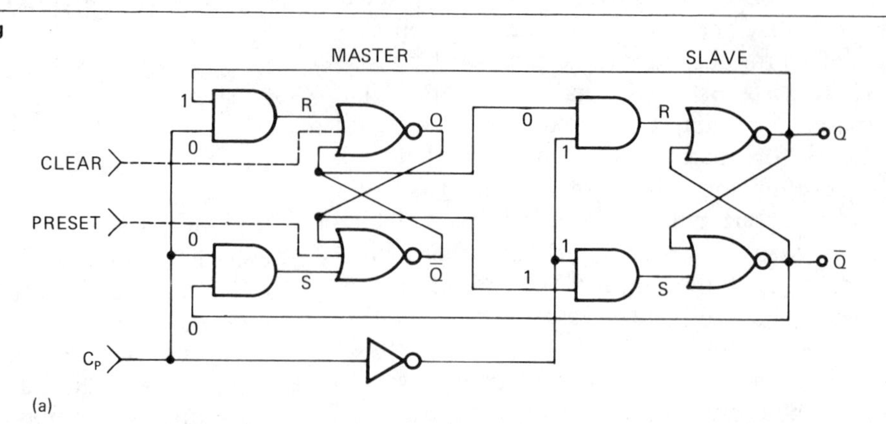

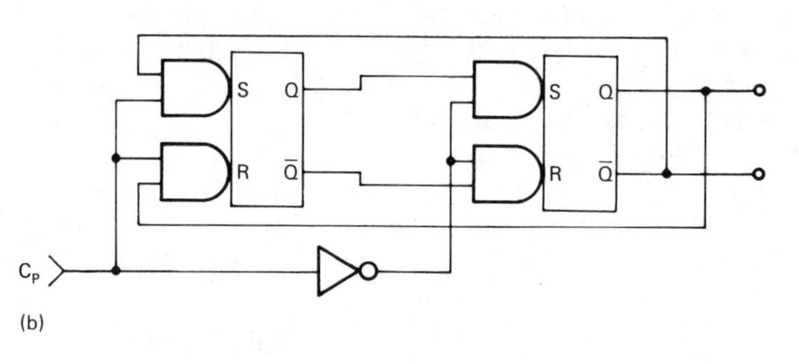

FIGURE 12-11

Master-slave flip-flop toggling from set to reset state. (a) Logic levels of set state before clock pulse. (b) Logic levels when clock pulse is high (master flip-flop resets). (c) Logic levels with clock pulse low and flip-flop in reset state (reset passed from master to slave on trailing edge of clock pulse).

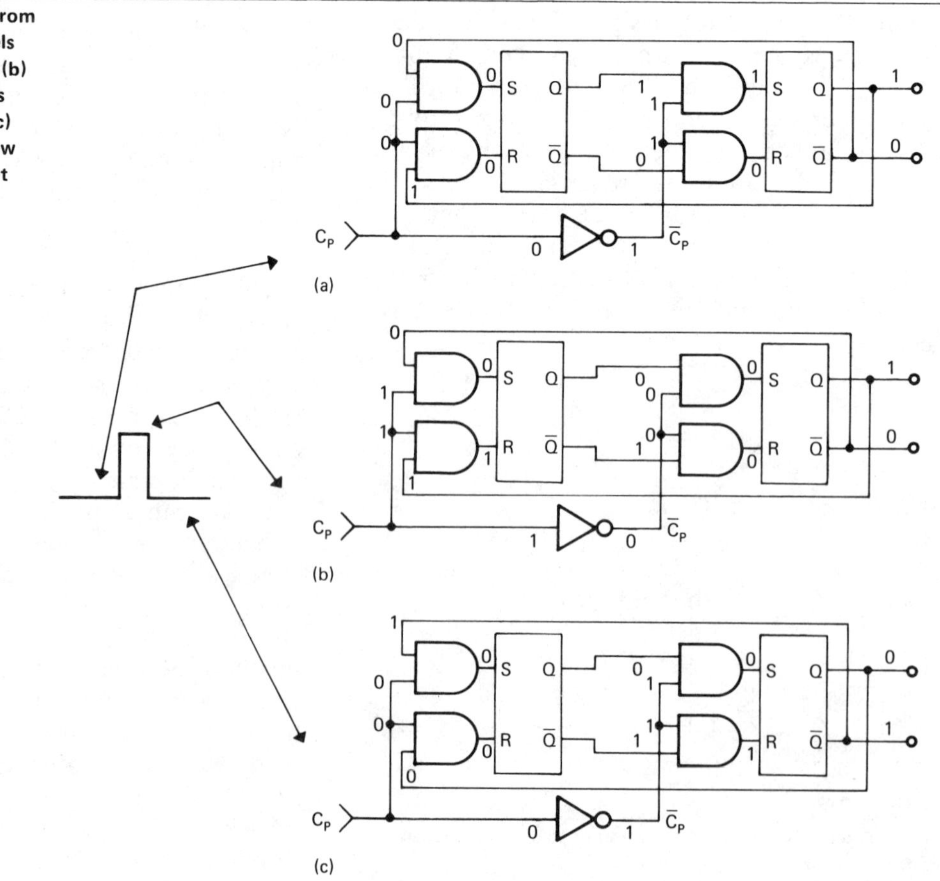

■ EXAMPLE 12-1

The flip-flop of Figure 12-12(a) has its clear input connected to ground (logic 0). The set and toggle input waveforms are shown in the timing diagram. Draw the resulting Q $\overline{Q}$ waveforms.

Solution: See Figure 12-12(b).

12.6 Serial Parity

In Section 8.6, we discussed a system of parity checks to detect loss or addition of a pulse to a digital word being transmitted in parallel. It is possible to generate and check parity in serial as well as parallel. Figure 12-13 is a logic system

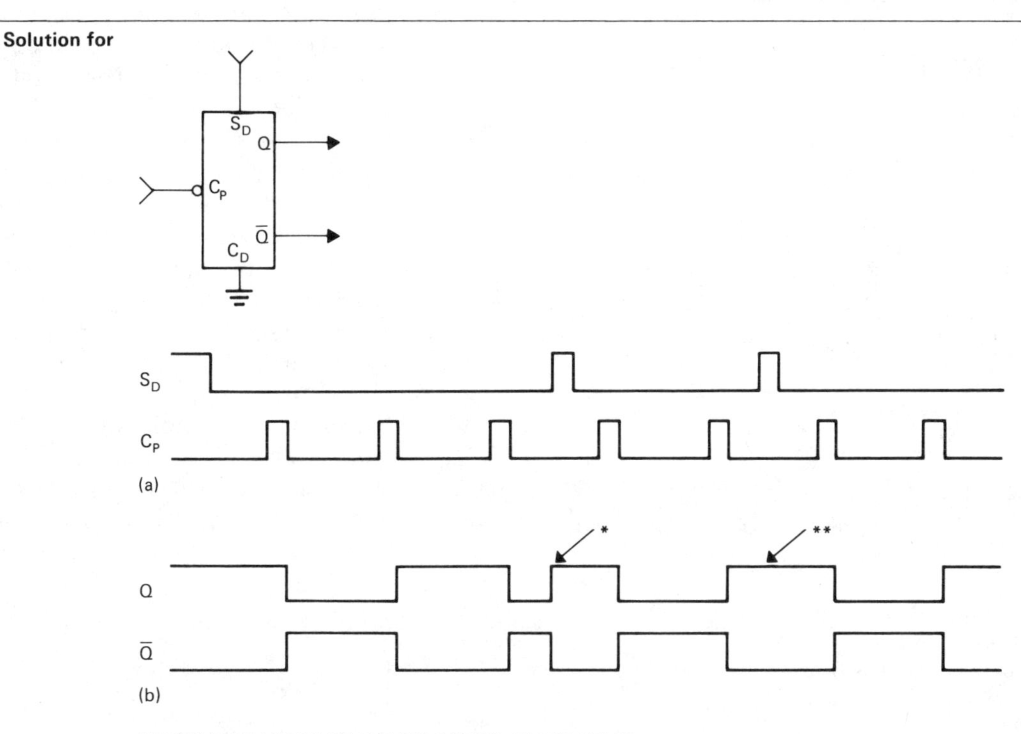

FIGURE 12-12
(a) Example 12-1. (b) Solution for Example 12-1.

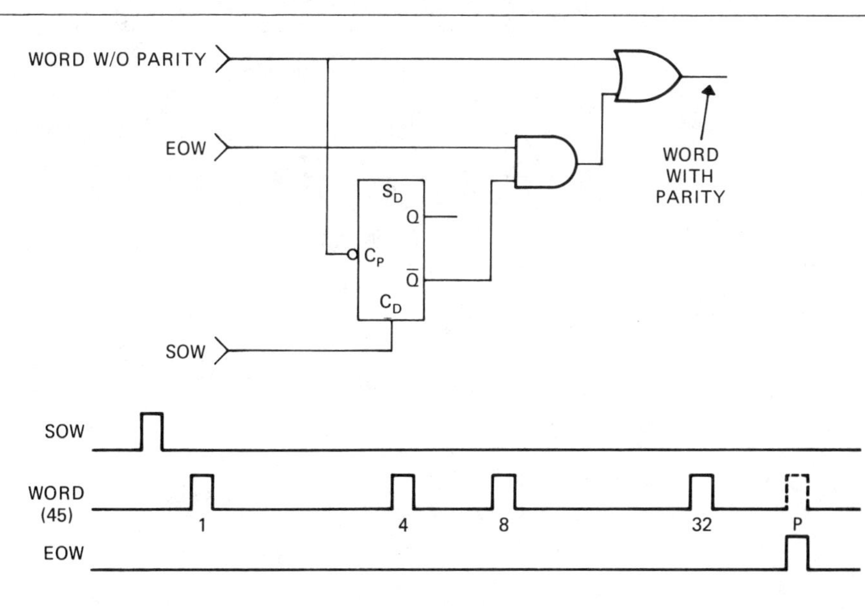

FIGURE 12-13
Logic system used to add serial parity pulse to a word or number.

used to add an odd parity bit to a serial word. The waveforms are for a six-bit number. The start-of-word (SOW) pulse clears the parity register. Each one-bit pulse on the word line toggles the register. If the number of toggle pulses are even, the flip-flop will be in the reset state at the end-of-word (EOW) time and will enable the EOW pulse through the AND gate. If the number of toggle pulses are odd, the flip-flop will be in the 1 state at the EOW time, and the EOW pulse will be inhibited from passing through the AND gate.

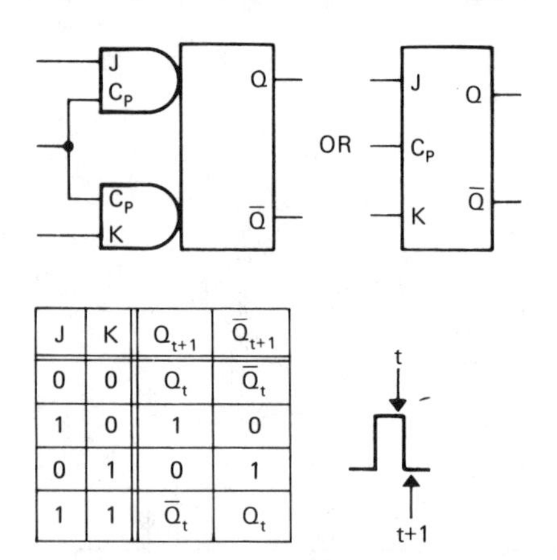

J	K	Q_{t+1}	$\bar{Q}_{t+1}$
0	0	Q_t	$\bar{Q}_t$
1	0	1	0
0	1	0	1
1	1	$\bar{Q}_t$	Q_t

FIGURE 12-14

J-K flip-flop and truth table. Changes occur on the trailing edge of the clock pulse.

12.7 Steerable Flip-Flop (J-K)

The triggered flip-flop of Figure 12-8 can be used to toggle or change state only as a result of a trigger or clock pulse on its input. The output cannot be directed to the 1 state, changing state, if need be, or remaining unchanged if already in the 1 state. Nor can it be directed to the 0 state, changing, if need be, and remaining the same if already in the 0 state. By using three-input AND gates to control the master flip-flop, these very useful functions can be added. These two inputs are referred to as *steer inputs*, in that they steer the registers to either 1 or 0. No actual change in state occurs, however, until the trailing edge of a clock pulse. Figure 12-14 shows the logic symbol and truth table of the J-K flip-flop. The truth table indicates that the flip-flop will ignore the clock pulse if J and K are both 0. If J is 1 and K is 0, the flip-flop will be in the 1 state after the trailing edge of the clock pulse. If J is

FIGURE 12-15

(a) Master-slave flip-flop with J, K, preset, and clear inputs added. (b) Simplified logic symbol for J-K flip-flop.

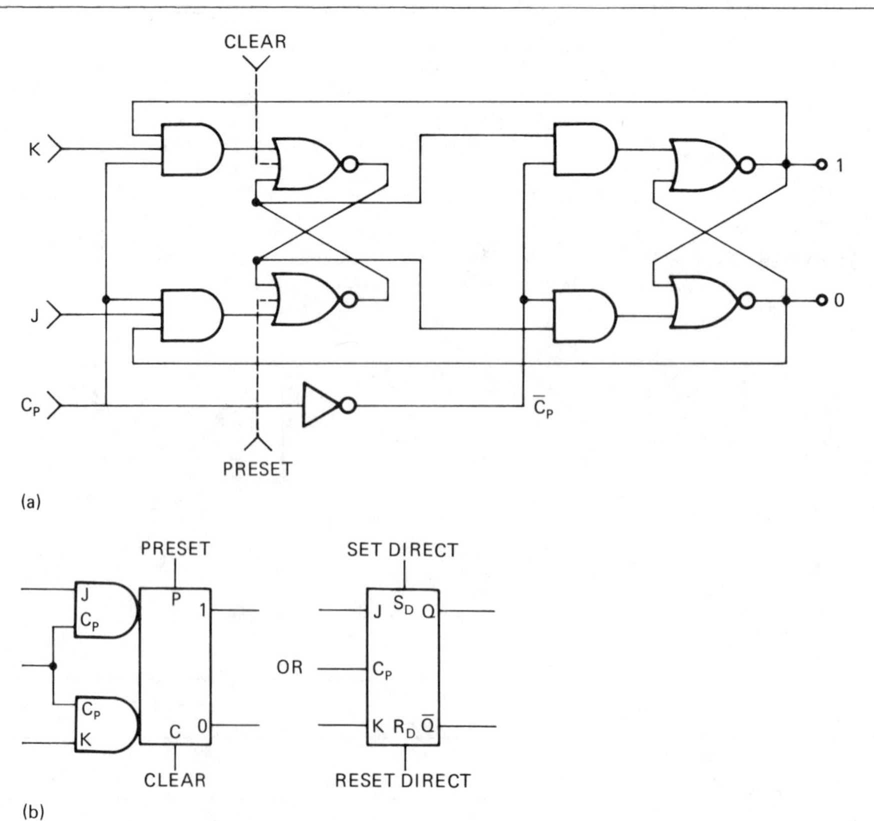

0 and K is 1, it will be in the 0 state after the trailing edge of the clock pulse.

Figure 12-15 shows the J-K functions added to the master-slave flip-flop by merely using three-input AND gates on the master input. If both J and K inputs are at logic 0 level, the master AND gates are both inhibited, and no change can pass through the flip-flop regardless of changes in the C_P and $\overline{C}_P$. If both J and K are at logic 1 level, the flip-flop will function as described in Figure 12-11 and toggle on the trailing edge of each clock pulse. If the J input is 1 and the K input is 0, the set gate is enabled and the reset is inhibited. If the K input is 1 and the J input is 0, the reset gate is enabled and the set is inhibited.

The capabilities of this flip-flop can be further expanded by including the DC clear and preset inputs to the master NOR gates, as shown by the dotted lines in Figure 12-15. These inputs are equivalent to the set and reset inputs of the set-reset flip-flop. They make it possible to set or reset the flip-flop without waiting for a clock pulse. Operation of the flip-flop with these DC inputs is called *asynchronous operation*, as compared to operation with the clock pulse, called *synchronous operation*. An important difference between these is that in synchronous operation all changes occur on the trailing edge of the clock pulse, while in asynchronous operation changes occur on the leading edge of the DC input signals.

Figure 12-16 shows the symbol and truth table of the J-K flip-flop with asynchronous inputs. This unit has every possible register capability. It can be cleared or preset independent of the clock pulse by using the DC inputs. It can be toggled or it can be steered to the 1 or 0 state by the clock pulse. As Figure 12-16(b) shows, the symbol for the flip-flop may be drawn with outputs to the right or left, depending on drafting convenience.

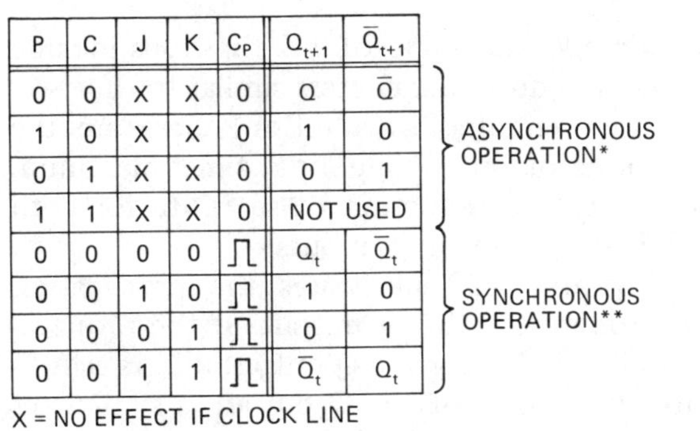

P	C	J	K	C_P	Q_{t+1}	$\overline{Q}_{t+1}$	
0	0	X	X	0	Q	$\overline{Q}$	ASYNCHRONOUS OPERATION*
1	0	X	X	0	1	0	
0	1	X	X	0	0	1	
1	1	X	X	0	NOT USED		
0	0	0	0	⎍	Q_t	$\overline{Q}_t$	SYNCHRONOUS OPERATION**
0	0	1	0	⎍	1	0	
0	0	0	1	⎍	0	1	
0	0	1	1	⎍	$\overline{Q}_t$	Q_t	

X = NO EFFECT IF CLOCK LINE HELD AT 0

 = CLOCK PULSE

* CHANGES OCCUR ON LEADING EDGE OF INPUT

** CHANGES OCCUR ON TRAILING EDGE OF CLOCK PULSE

(a)

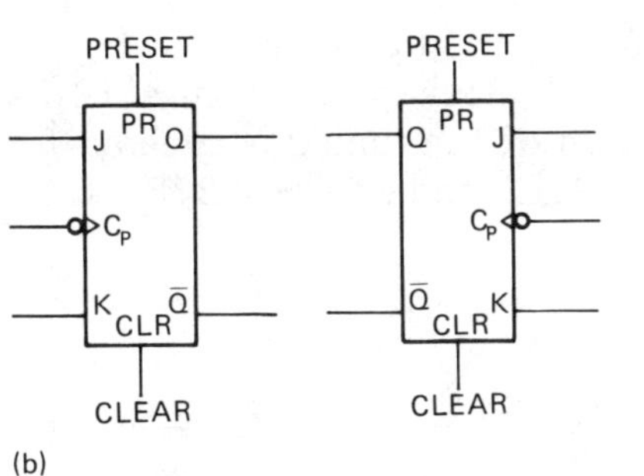

(b)

FIGURE 12-16

(a) Truth table of both synchronous and asynchronous operation of the J-K flip-flop. (b) Logic symbol for J-K.

12.8 Strobing Data into the J-K Flip-flop

Strobe gates are not needed to strobe data into the J-K flip-flop. The data lines are connected to the J input in conjunction with an inverter connected between J and K, as in Figure 12-17.

FIGURE 12-17

J-K flip-flops with an inverter between J and K.

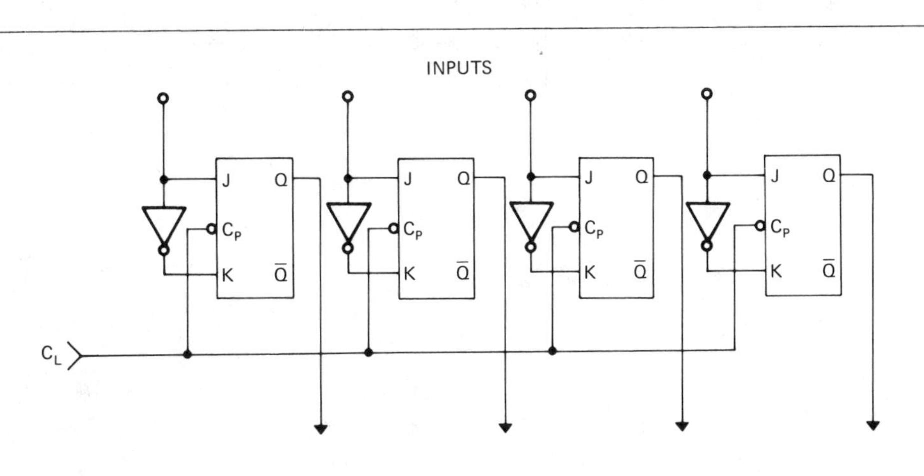

A 1 on the input (J) will result in a 0 on K, and, as the truth table in Figure 12-14 indicates, this will steer the flip-flop to the 1 state. A 0 on the input (J) will result in a 1 on K, steering the flip-flop to the 0 state. The levels on the input may vary between 1 and 0, preceding the clock pulse without changing the registers. The clock pulse should occur after the inputs have settled down. The input levels will enter the register on the trailing edge of the clock pulse.

12.9 Integrated Circuit J-K Flip-Flops

A wide selection of integrated circuit J-K flip-flops is available in both TTL and CMOS standard DIP packages. Typical of these is the 74LS112 dual J-K flip-flop, shown in Figure 12-18. The 74LS112 is much like the 7476 except that it has the preferred power pin-out. It differs from earlier versions in that the power pins are on the ends of the package rather than in the middle. Dual J-K flip-flops offered in 16-pin packages generally have both preset and clear inputs. Those offered in 14-pin packages have clear inputs only. Another circuit convenience offered in the 74LS109 is a dual J-K flip-flop that can be applied in a circuit like that shown in Figure 12-17 without the use of inverters.

The CMOS circuit 4027 is a dual J-K flip-flop. This circuit, shown in Figure 12-19, has no inverted inputs and, as the truth table indicates, it changes state on the rising edge of the clock pulse input. The set and reset are active high inputs.

12.10 Timing Parameters of J-K Flip-Flops

The timing parameters of J-K flip-flops include propagation delay parameters similar to those on logic gate circuits. In addition, there are the parameters of setup time, hold time, maximum clock rate, and minimum pulse width for both clock and asynchronous inputs.

Figure 12-20 illustrates the propagation delay timing for the Q output. If the propagation delay timing for the $\overline{Q}$ output differs significantly from that of the Q output, a similar set may need to be quoted for the $\overline{Q}$ output. Figure 12-21 illustrates the setup and hold times.

The J-K inputs must be stable for a period of time preceding the active edge of the clock pulse. In addition, the data at the inputs must remain stable for a period of time following the active edge of the clock pulse. Figure 12-22 gives a comparison of the timing parameters for the 74LS112, 74S112, and CMOS 4027B.

FIGURE 12-18

Logic symbol and truth table for the 54LS/74LS112 dual J-K negative edge-triggered flip-flop.

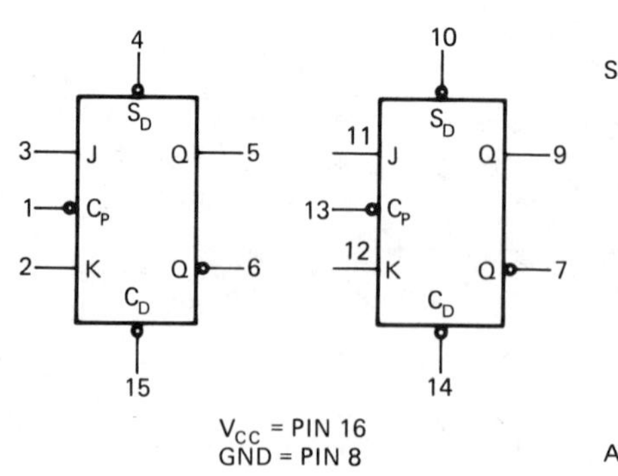

SYNCHRONOUS OPERATION

INPUTS @ t_n		OUTPUT @ t_{n+1}
J	K	Q
L	L	Q_n
L	H	L
H	L	H
H	H	$\overline{Q}_n$

ASYNCHRONOUS OPERATION:
LOW INPUT TO $\overline{S}_D$ SETS Q TO HIGH LEVEL.
LOW INPUT TO $\overline{C}_D$ SETS Q TO LOW LEVEL.
CLEAR AND SET ARE INDEPENDENT OF CLOCK.
SIMULTANEOUS LOW ON $\overline{C}_D$ AND $\overline{S}_D$ MAKES BOTH Q AND $\overline{Q}$ HIGH.

t_n = BIT TIME BEFORE CLOCK PULSE
t_{n+1} = BIT TIME AFTER CLOCK PULSE
H = HIGH VOLTAGE LEVEL
L = LOW VOLTAGE LEVEL

12.11 Bistable Latch (D Latch)

The integrated circuit TTL/MSI 7475 quadruple bistable latches can be used in applications like the set-reset flip-flops in Figure 12-7. This circuit has several advantages over the set-reset

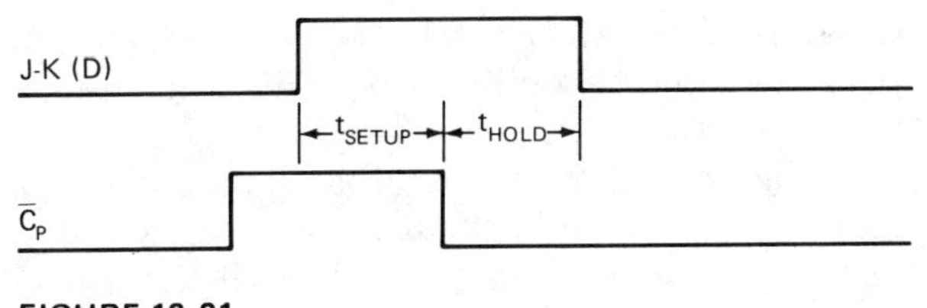

FIGURE 12-21

Setup and hold time parameters for a J-K flip-flop.

FIGURE 12-19

Block diagram and truth table for the 4027B dual J-K flip-flop.

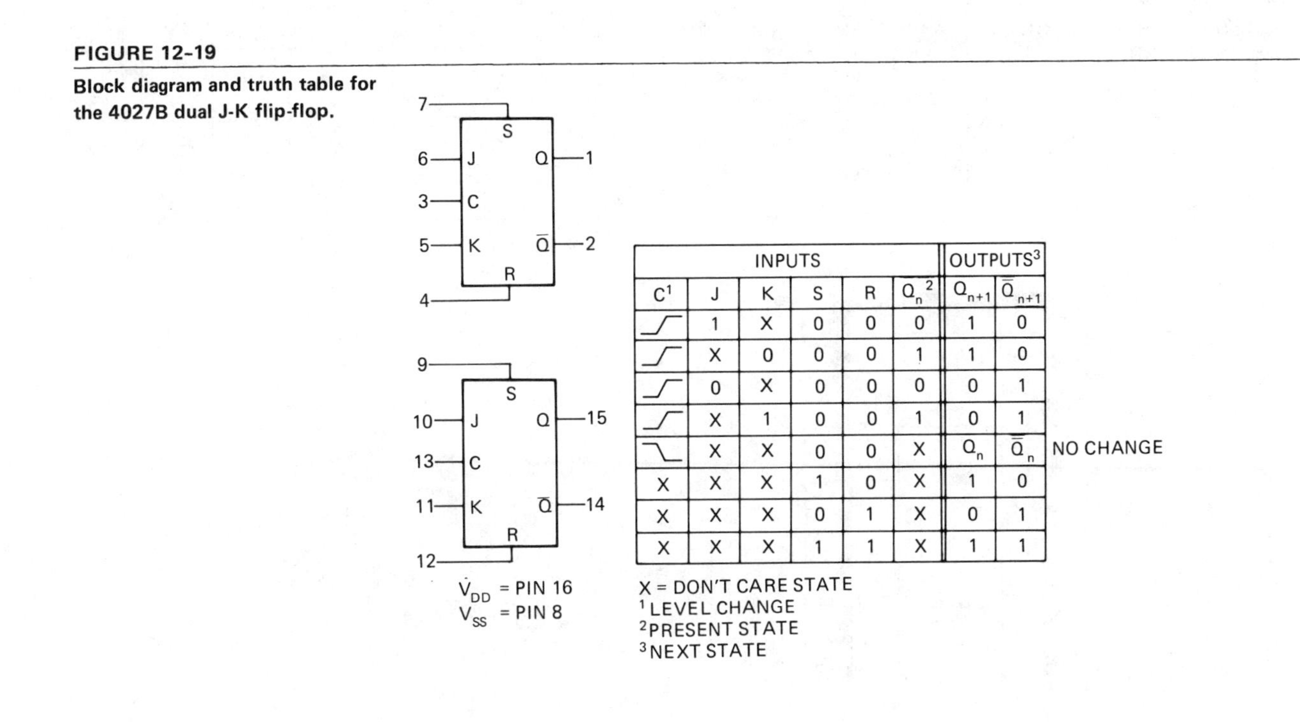

INPUTS						OUTPUTS[3]	
C[1]	J	K	S	R	Q_n[2]	Q_{n+1}	$\overline{Q}_{n+1}$
↑	1	X	0	0	0	1	0
↑	X	0	0	0	1	1	0
↑	0	X	0	0	0	0	1
↑	X	1	0	0	1	0	1
↓	X	X	0	0	X	Q_n	$\overline{Q}_n$ NO CHANGE
X	X	X	1	0	X	1	0
X	X	X	0	1	X	0	1
X	X	X	1	1	X	1	1

X = DON'T CARE STATE
[1] LEVEL CHANGE
[2] PRESENT STATE
[3] NEXT STATE

FIGURE 12-20

Propagation delay timing for the Q output. (a) Clock pulse (synchronous operation). (b) Synchronous operation.

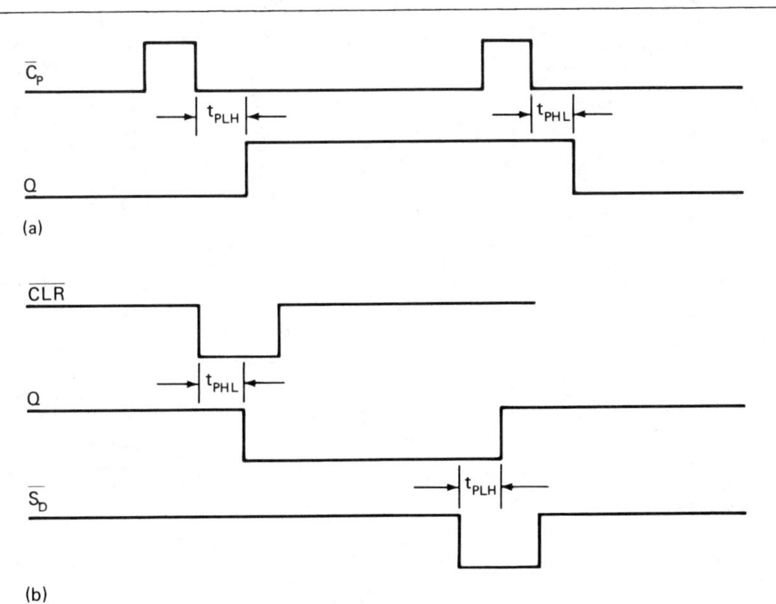

flip-flop. The latch has a D input, which can steer the circuit to either a 1 or a 0 state, and it is not necessary to reset before loading.

Figure 12-23(a) is the manufacturer's drawing of one such circuit. Figure 12-23(b) is a rearrangement of that circuit with the single-input AND gate left out. This figure shows the circuit to be a crossed NOR with strobe gates on both set and reset inputs. Because of the inverter, the D input will enable either the set or reset input. When the clock line goes high, a 0 on D will cause a reset and a 1 on D will cause a set.

FIGURE 12-22

Comparison of timing parameters for the high-speed, high-power, TTL J-K flip-flops (74S112 and 74LS112) and the low-speed, low-power CMOS 4027B.

PARAMETER	74S112	74LS112	4027B 5V	4027B 10V	4027B 15V	UNIT
MAXIMUM CLOCK FREQUENCY	80	30	1.5	4.5	6.5	MHz
MAXIMUM PROPAGATION DELAY						
C_P TO Q OR $\overline{Q}$	7	16	350	150	100	ns
SET TO Q OR $\overline{Q}$	7	16	350	150	100	ns
MINIMUM SETUP TIME	7	20	140	50	35	ns
MINIMUM HOLD TIME	0	0	140	50	35	ns
MINIMUM CLOCK PULSE WIDTH	6	20	330	110	75	ns
MINIMUM SET OR RESET PULSE WIDTH	8	15	250	100	70	ns

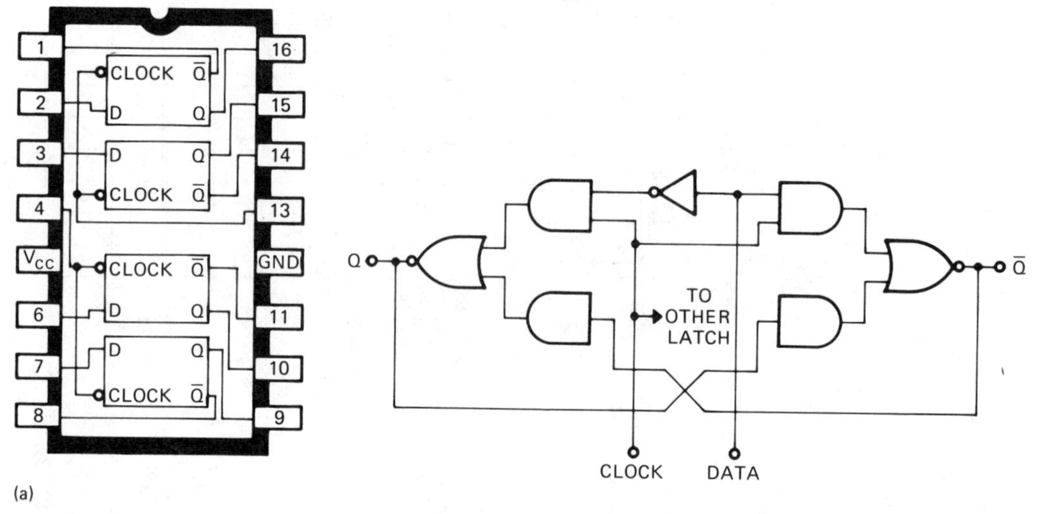

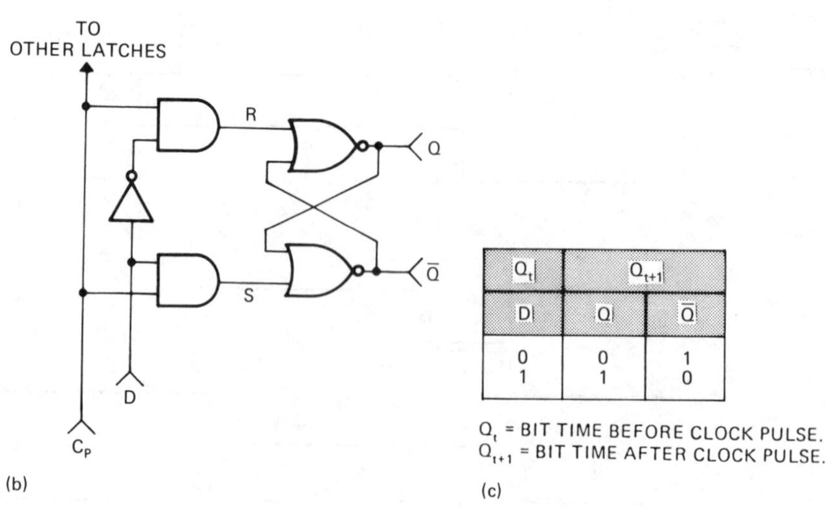

FIGURE 12-23

Sprague TTL/MSI 7475 integrated circuit quadruple bistable latch. (a) Pin-out and logic diagrams. (b) Latch circuit drawn in crossed NOR configuration (single-input AND gates removed). (c) Truth table.

In fact, as long as the clock line is high, the Q output will be whatever level is on the D input. When the clock line goes low, the Q output will stabilize to whatever level was on D before transition of the pulse from high to low. Unlike the master-slave flip-flop, however, change of state will occur any time the clock line is high. The inversion circle on the clock input does not indicate a need to invert the clock line. It indicates that a decision for the resting state of the flip-flop is based on the falling edge of the clock pulse. Note that the 74LS375 is a logically identical device, but power is applied through pins 8 and 16 instead of pins 5 and 12.

12.12 D Flip-Flop

For some applications, the D latch of Figure 12-23 is more convenient to use than the J-K flip-flop. The D latch, however, is *transparent* when the clock line is high (changes on D propagate immediately to Q). The D latch cannot toggle. However, the D flip-flop has some of the advantages of the D latch, but it is not transparent (Q changes on edge of clock pulse only) and it can be wired to toggle by merely connecting $\overline{Q}$ to D. It is slightly less versatile than the J-K flip-flop, but as an integrated circuit, it is more economical to produce.

The D flip-flop shown in Figure 12-24(a) is, for many applications, easier to use than the J-K flip-flop. Like the D latch, whichever level is applied to the D input when the clock input is low will be transferred to the Q output when the clock goes high. While the clock remains high, changes on the D input have no effect on the Q output. The flip-flop is steered when the clock line is low, and the change occurs when the clock line goes high. Unlike some J-K flip-flops, the change occurs on the leading edge of the clock pulse.

Figure 12-24(b) shows the logic method used to produce the D flip-flop. Any 0 on a NAND gate input results in a 1 out. Therefore, when the clock line is at 0, 1 levels are being passed to both output NAND gates, as shown in Figure 12-25(b), providing a no-change state. The D input level can in no way affect the output. When the clock line goes high, the level passed to the output NAND gates will depend on the D input. If D is high, the bottom line to the output NAND gate will be 1, the top line 0. This will provide a set state (1 on Q, 0 on $\overline{Q}$). If D is low, the bottom line to the output NAND gates will be 0 and the top line 1. This will provide a reset state (0 on Q, 1 on $\overline{Q}$). If D changes when the clock line is high, Q will not change. Unfortunately, $\overline{Q}$ does change with the clock line high when D changes from 1 to 0. This will produce a 1 output on both Q and $\overline{Q}$. For this reason, a valid $\overline{Q}$ is often produced by connecting an inverter to Q. Note that direct set and reset inputs are also provided.

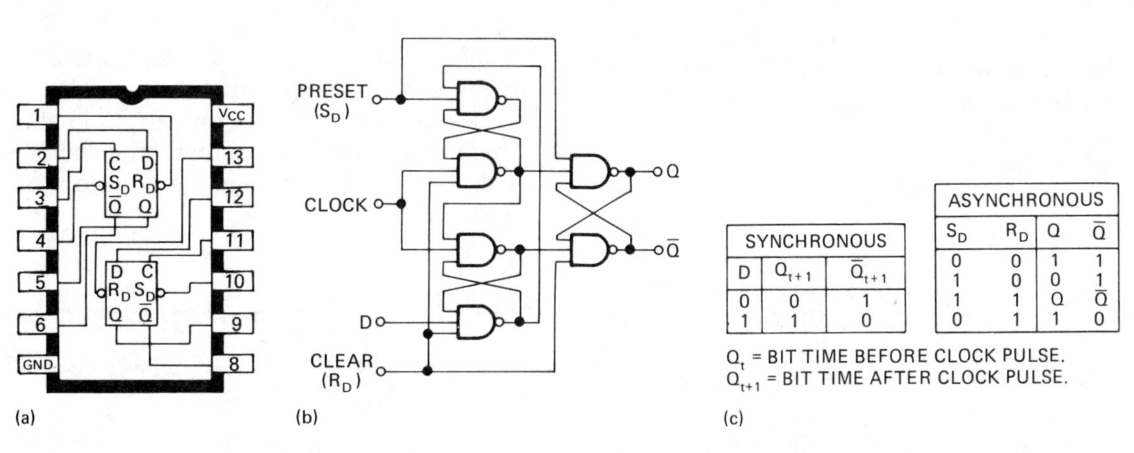

FIGURE 12-24

Sprague TTL/SSI 7474 D-type edge-triggered flip-flop with preset and reset inputs. (a) Pin-out diagram. (b) Truth tables. (c) Functional block diagram.

FIGURE 12-25

D flip-flop (S_D and R_D inputs excluded). (a) Truth table of synchronous operation. (b) D flip-flop in no-change state when clock line is low. (c) Flip-flop (with 1 level on the D input) in set state as clock line goes high. (d) Flip-flop (with 0 level on the D input) in reset state as clock line goes high.

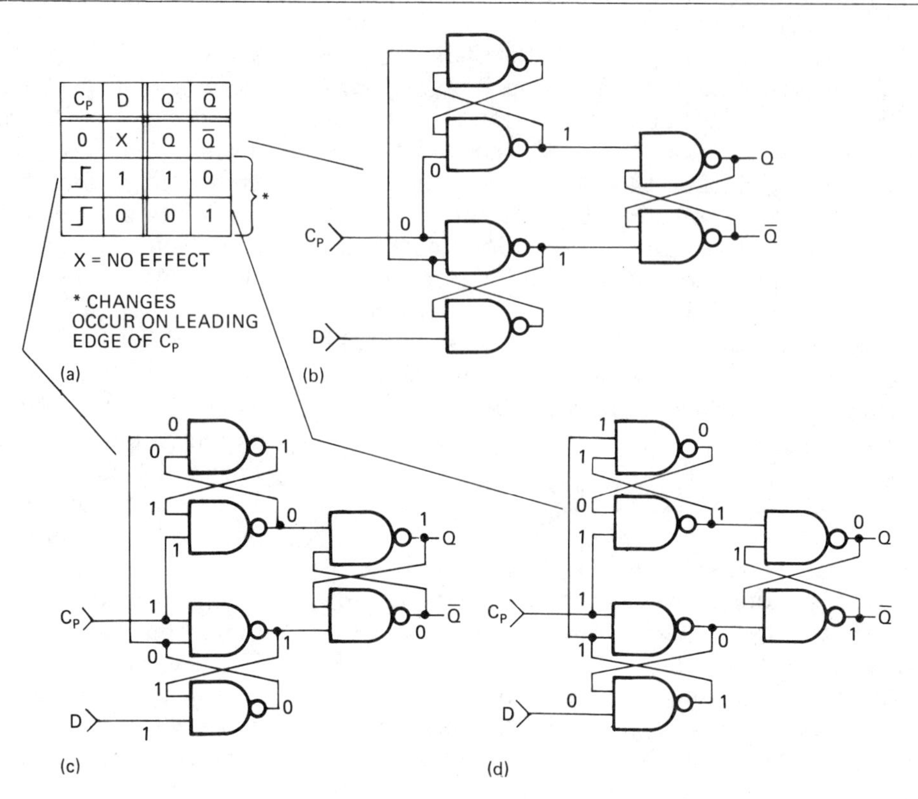

C_P	D	Q	$\overline{Q}$
0	X	Q	$\overline{Q}$
↑	1	1	0
↑	0	0	1

12.13 Integrated Circuit D Flip-Flops

In addition to the 7474 dual D flip-flop pictured in Figure 12-24, there is a CMOS version of the dual D flip-flop, the 4013B. Functionally, it differs from the 7474 by the fact that set and reset inputs are active high. D flip-flops are also available in quad, hex, and octal IC packages. Figure 12-26 shows the logic diagram and symbol for the quad D flip-flop.

Note that reset and clock functions to all flip-flops are internally connected, which is not the case for the dual D flip-flop. Both Q and $\overline{Q}$ outputs, however, are brought out. The quad D flip-flop is available as a 74175, 74LS175, 74S175, and as the CMOS 40175 (4175 from some manufacturers). The pin-out is identical for both the TTL and CMOS versions, but the usual speed/power trade-offs exist.

Using the same 16-pin DIP package as the 74175, a hex D flip-flop is available. The 16-pin package has insufficient pins to provide both Q and $\overline{Q}$ outputs, so only the Q outputs are made available, as Figure 12-27 shows. The hex D flip-flop also is available with identical pin-out as a TTL or CMOS device. The TTL versions are the 74174, 74LS174, and 74S174, and the CMOS version is the 40174 (or 4174).

Recently, an octal version of the D flip-flop has become available—the 74LS374 or 74S374. The logic diagram for this circuit is shown in Figure 12-28. It has the added feature of tri-state buffers. When the output enable, $\overline{OE}$, is high, the outputs are in high impedance. When $\overline{OE}$ is low, the true state of the flip-flops appears on the outputs.

12.14 Timing Parameters of D Flip-Flops

The D flip-flop timing parameters are like those of the J-K flip-flop, with D inputs replacing

FIGURE 12-26

Quad D flip-flops: 74175, 74LS175, 74S175, and 40175 (or 4175). (a) Logic diagram. (b) Logic symbol.

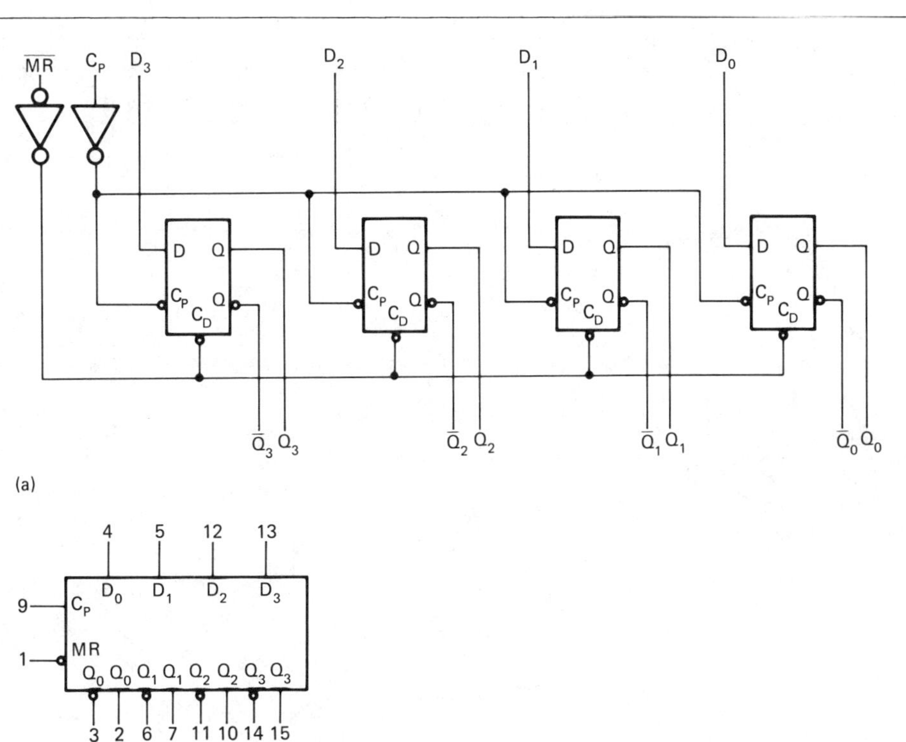

(a)

4 5 12 13
D_0 D_1 D_2 D_3
9 C_P
1 MR
Q_0 $\overline{Q}_0$ Q_1 $\overline{Q}_1$ Q_2 $\overline{Q}_2$ Q_3 $\overline{Q}_3$
3 2 6 7 11 10 14 15
V_{CC} = PIN 16
GND = PIN 8

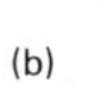

(b)

FIGURE 12-27

Hex D flip-flops: 74174, 74LS174, 74S174, and 40174 (or 4174). (a) Logic diagram. (b) Logic symbol.

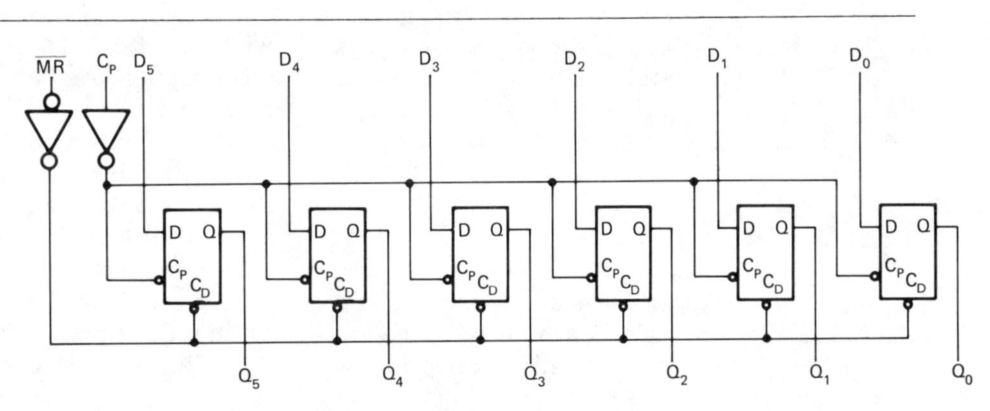

(a)

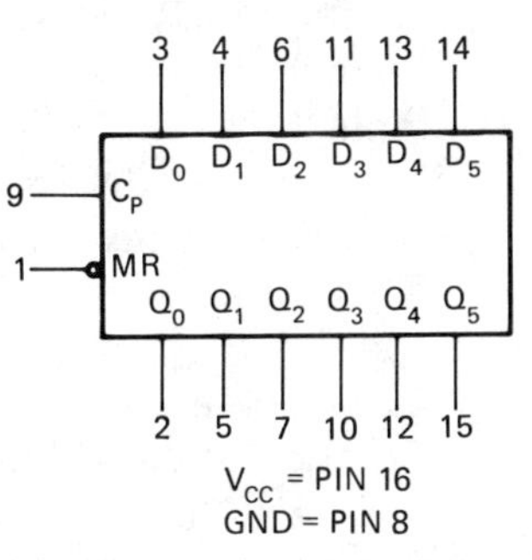

(b)

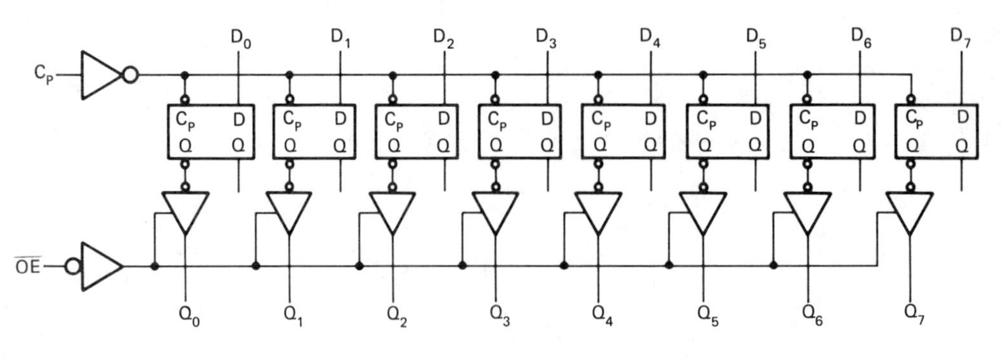

FIGURE 12–28

Logic diagram for the 20-pin octal D flip-flops with three-state buffers, 74LS374 or 74S374.

FIGURE 12–29

Comparison of timing parameters for J-K and D flip-flops of equivalent technology (LS).

PARAMETER	74LS112 J-K	74LS74 D	UNIT	
MAXIMUM CLOCK FREQUENCY	30	30	MHz	
MAXIMUM PROPAGATION DELAY				
C_P TO Q OR $\overline{Q}$	16/24	25/35	ns	t_{PLH}/t_{PHL}
SET TO Q OR $\overline{Q}$	16/24	15/30	ns	t_{PLH}/t_{PHL}
MINIMUM SETUP TIME	20	10/20	ns	D_{LOW}/D_{HIGH}
MINIMUM HOLD TIME	0	5	ns	
MINIMUM CLOCK PULSE WIDTH	20	18	ns	
MINIMUM SET OR RESET PULSE WIDTH	15	15	ns	

those of J-K where setup and hold times are involved. A partial comparison of these parameters is given in Figure 12-29 for the 74LS112 and the 74LS74. The timing differs only slightly in most respects. The maximum clock frequencies are the same. The clock to Q propagation delays of the D flip-flop are a bit slower than those of the J-K. The J-K flip-flop requires no hold time, while the D flip-flop requires 5 nanoseconds.

12.15 Summary

As signals pass rapidly through a digital system, they must often be held in a particular location long enough for other signals to arrive or for certain operations to be performed. *Register circuits* perform this holding or storing function. Each binary bit of a digital number must be stored in a separate register element. The simplest form of register element is a *set-reset flip-flop.*

Figure 12-1 shows the logic symbol and truth table of the set-reset flip-flop. The flip-flop has two states—the set and reset states. In the set state, a logic 1 exists on the Q output and a logic 0 voltage on the $\overline{Q}$ output. In the reset state, a logic 0 will be on the Q output and a logic 1 on the $\overline{Q}$ output. As the truth table indicates, when both inputs are 0, the flip-flop will maintain the state it is in. If a 1 level is applied to the set input only, the flip-flop will go to the set state. To return the register to the reset state, a 1 is applied to the reset input.

Figure 12-2 shows three set-reset flip-flops wired to store a three-bit digital number. The reset inputs are tied together, and before the number to be stored arrives, a reset pulse takes all three registers to 0. Later the number arrives as parallel bits applied to the set inputs. Although numbers arrive at the inputs as pulses, at the outputs they are in level form. When a new number is to be stored, it is again preceded by a reset pulse. The terms *preset* and *clear* are often used in place of *set* and *reset.*

The simplest method of making a set-reset flip-flop is to cross-couple two NOR gates, as shown in Figure 12-4. In systems that do not employ NOR gates, crossed NAND gates may be used. Figure 12-5 shows a logic circuit and truth table of the crossed NAND. As the truth table indicates, it differs from the crossed NOR in that the resting state is 1-level on both inputs. This difference can be corrected by using inverters on the inputs, as in Figure 12-6. This provides operation identical to crossed NOR except for the not-used state.

Another function available with some register elements is the ability to *toggle* or change state on application of a clock or trigger pulse. Figure 12-8 shows several logic symbols for a toggle flip-flop. In most cases, the set-reset capability is included with the ability to toggle. A register with all three capabilities will set if a 1-level pulse is applied to the set input. If it is already set, it will remain in that state. When a 1-level pulse is applied to the reset input, the flip-flop will go to the reset state. If it is already reset, it will remain in that state. A pulse applied to the clock or trigger input, however, will cause the flip-flop to change state from set to reset or reset to set. Regardless of which state it was in before receiving the clock pulse, it will change to the other state.

The most versatile of the storage flip-flops is the *J-K flip-flop*, shown in Figure 12-14. The J-K inputs to the flip-flop may be called *steer inputs*, in that they steer the flip-flop to a particular state, but the change (if any) does not occur until after a pulse is applied to the clock input. As the truth table of Figure 12-14 shows, when both J and K inputs are 0, no change results from application of a clock pulse. Any number of pulses may be applied to the clock input without causing change, in the state of the flip-flop. If a 1 is on J and a 0 is on K, the flip-flop goes to the set state on the trailing edge of the clock pulse. If it is already set, it will remain in the set state. If a 0 is on J and a 1 is on K, the flip-flop goes to the reset state on the trailing edge of the clock pulse. If both J and K inputs are 1, the flip-flop will toggle, changing state on the trailing edge of each clock pulse.

The J-K flip-flop will normally have the preset and clear inputs along with J-K and C_P inputs. The preset and clear operate exactly like the set-reset inputs. These inputs operate with the clock line held low or at 0. They are known as *asynchronous* operations and conform to the top of the truth table of Figure 12-16. The state of J and K inputs have no effect if the clock line is held at 0. The bottom half of the truth table is *synchronous* operation; preset and clear must remain at 0. On synchronous operation, all changes occur on the trailing edge of the clock pulse. On asynchronous operation, changes occur on the rising edge of the P or C inputs.

Very often, we want to enter data into registers from levels that are initially unstable or changing. To avoid having the flip-flops set prematurely to some initial 1 level, we do not allow the register to see the data until it has settled down to its final value. This can be accomplished with the set-reset register by connecting the data through AND gates, as in Figure 12-17. A second input to each AND gate is connected to a strobe line. After the data settle to their final levels, a pulse appears on the strobe line. The strobe pulse will pass through only those gates which are enabled by 1 levels. If J-K flip-flops are used, the *strobe gates* are not needed. Each unstable input is applied to J with an inverter between J and K, as shown in Figure 12-17. In this case, a clock pulse is applied to the registers after the input lines have settled down.

Another flip-flop widely used in TTL circuits is the *bistable latch*, or *D latch.* Figure 12-23 shows the logic circuit and truth table of a bistable latch. There are four to each integrated circuit.

As long as the clock input of a D latch remains at 0, the level on the D input has no effect. The output will not change even if the level applied to the D input changes. In this respect, it is like the synchronous operation of the J-K flip-flop. If a 1 is on the D input, the flip-flop goes to the set state on the leading edge of the clock pulse. If a 0 is on the D input, the flip-flop goes to the reset state on the leading edge of the clock pulse. To strobe or clock data into these registers, neither AND gates nor inverters are needed. The reason for this becomes apparent from the logic drawings of Figure 12-23. The D latch is a crossed NOR flip-flop with strobe gates and an inverter already connected to the inputs.

Glossary

Register: A logic circuit used to store digital information. It consists of one or more register elements or binaries that can be placed in a recognizable 1 or 0 state. The size of the register is measured in bits. Storing a word of 10 bits requires a register of 10 bits or more.

Flip-Flop: A bistable multivibrator; a logic circuit with two stable states, which can be designated 1 and 0. The state of the flip-flop can be recognized by the 1 or 0 voltage level on its outputs. There are usually complementary outputs, often labeled 1 and 0 or Q and $\overline{Q}$. Depending on the type of flip-flop, there are one or more inputs used to control the state of the flip-flop. See Figure 12-1.

Set: The state of a flip-flop in which a 1 level appears on the 1 output or a 1 level appears on the Q output. Usually represents a binary 1 stored in the register element. The input (S) used to place the flip-flop in the set or 1 state.

Preset: Sometimes used in place of *set* to designate an input (P) used to place the flip-flop in the 1 state. See Figure 12-3.

Reset: The state of a flip-flop in which a 0 level appears on the 1 output or a 0 level appears on the Q output. The input (R) used to place the flip-flop in the reset state. The state of a register in which all its bits have been returned to 0.

Clear: Sometimes used in place of *reset* to designate an input (C) used to place the flip-flop in the 0 state. See Figure 12-3.

Q and $\overline{Q}$: Designations given to the output leads of a flip-flop indicating that the outputs are complements. Replaces older designations 1 and 0. See Figure 12-3.

Crossed NOR: A flip-flop composed of two NOR gates with the outputs of each crossed over to an input of the other. Free inputs are designated *set* and *reset.* See Figure 12-4.

Crossed NAND: A flip-flop composed of two NAND gates with the outputs of each crossed over to an input of the other. See Figure 12-5.

Toggle: A change of state of a flip-flop without regard to its initial or final state.

Trigger (T): The input to a flip-flop that will cause a toggle or change of state each time it receives a pulse. Sometimes designated *clock input* (C_P). Also name given to a narrow pulse used to change the state of a flip-flop. See Figure 12-8.

Race Problem: An unstable condition caused when a signal removes a condition needed for its own existence—for example, when a flip-flop must be reset to generate a pulse and that pulse is applied to set the flip-flop. See Figure 12-9.

Master-Slave Flip-Flop: A dual flip-flop used to provide a toggle flip-flop without a race problem. In older discrete circuits, capacitors were used. In integrated circuits, two flip-flops are used. The first changes on the leading edge of the clock pulse, the second on the trailing edge. The first, or input, flip-flop is called the *master.* The second, or output, flip-flop is called the *slave.* See Figure 12-10.

Steer Input: An input that does not cause a change of state in a flip-flop but directs the flip-flop to a given state. The change does not occur until the flip-flop receives a pulse on the clock or trigger input.

J-K Flip-Flop: Flip-flop with J-K inputs. The changes that occur to this flip-flop when a clock pulse is applied depend on the levels of the J-K inputs. If both are 0, no change occurs. If both are 1, the register toggles. A 1 on J and a 0 on K steer it to the set state. A 0 on J and a 1 on K steer it to the reset state. See Figure 12-15.

D Flip-Flop: A flip-flop with a single steer input, the D input. On the leading edge of a positive-going pulse applied to the clock input, the state of the flip-flop goes to the level applied to the D input. No change occurs to the state of the flip-flop while the clock line is stable or during a negative transition of the clock input.

Bistable Latch (D Latch): A flip-flop with a D input. When the clock or enable input of the latch is low, the flip-flop will go to the level applied to the D input. If the D input changes while the clock is low, the state of the flip-flop will change with the level on the D input. The period during which the clock input is low is called the *transparent mode.* When the clock or enable goes high, the state of the flip-flop no longer changes with the level on the D input. The level on the D input just prior to the positive-going edge of the clock input is latched into the flip-flop for as long as the clock input remains high.

Synchronous Operation: Operation of a flip-flop in which changes occur on the leading or trailing edge of a pulse applied to the clock (C_P) input.

Asynchronous Operation: Operation of a flip-flop with the clock line kept low and changes affected by the DC set and reset inputs or the preset and clear inputs.

Transparent Operation: The mode of bistable or D latch operation during which changes occurring on the D input are propagated immediately to the state of the latch and to the Q output.

Strobe Gate: A logic gate used to inhibit passage of a level on a line that is initially unstable.

When the line stabilizes to its correct level, a strobe pulse is applied to the gates. If AND gates are used, the strobe pulse will pass through those gates which stabilize to 1 and set a flip-flop. The strobe pulse will not pass through those gates which stabilize to 0, leaving those flip-flops reset. See Figure 12-7.

Setup Time: The minimum amount of time required for the input data to be stable before the triggering edge of the clock pulse.

Hold Time: The minimum amount of time required for the input data to remain stable after the triggering edge of the clock pulse. The input data can change after the hold time specification is met.

Questions

1. What is the function of a register circuit?
2. Draw the logic symbol and truth table of a set-reset flip-flop.
3. If no reset pulse were received by the register in Figure 12-2 and the number 101(5) were followed by a 011(3), what would then be stored in the register?
4. Draw the logic diagram of a crossed NOR flip-flop and explain its operation.
5. How does operation of the crossed NAND differ from that of the crossed NOR?
6. How can a crossed NAND circuit be modified to operate like a crossed NOR?
7. Explain the difference between setting-resetting and toggling a flip-flop.
8. When a flip-flop is set, what level will appear on the Q output? Is the value stored in that register bit a 1 or a 0?
9. When a flip-flop is clear, what level will be on the Q output? What level will be on the $\overline{Q}$ output? Is the value stored in that register bit a 1 or a 0?
10. Explain the difference between synchronous and asynchronous operation of a J-K flip-flop. Which inputs are involved in each?
11. Using three-input bus lines—clock pulse, 1 level, and 0 level—draw a J-K flip-flop wired to toggle.
12. If the outputs of an adder circuit are to be stored in a register composed of crossed NOR flip-flops, what additional circuitry is needed between the sum outputs and the register flip-flops?
13. If J-K flip-flops are used in the register of Question 12, what added circuits are needed?
14. If bistable latches are used in the register of Question 12, what added circuits are needed?
15. Draw a simplified version of a bistable latch showing the crossed NOR configuration.
16. In Figure 12-28, how are the output lines affected if (a) $\overline{OE}$ is low and (b) $\overline{OE}$ is high?
17. What differences exist between the operation of a D latch and a D flip-flop?
18. List at least five versions of the quad D flip-flop in both TTL and CMOS. Are their pin-outs identical? What are the 74 series numbers for the octal D flip-flop?
19. The 74175 provides four D flip-flops per 16-pin package. What sacrifice was made in producing the 74174 with six D flip-flops in the same size package?
20. If a flip-flop must have a steer, a toggle, and a no response to clock pulse functions, which is the most likely choice: D flip-flop, D latch, or J-K flip-flop.

Problems

12-1 The circuit of Figure 12-30 is the CMOS SCL4001A quad NOR gate. Connect to form two crossed NOR set-reset flip-flops.

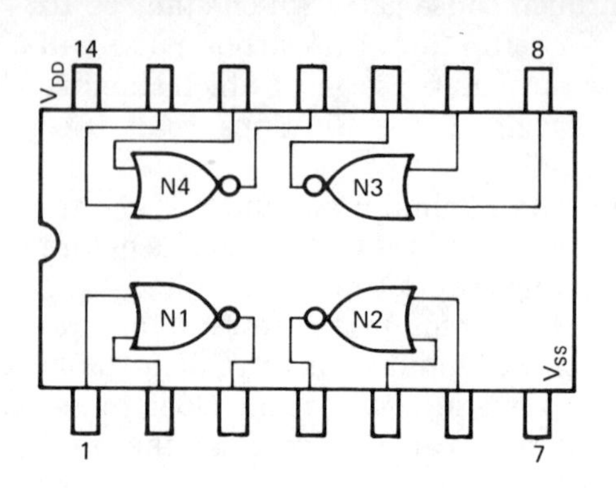

FIGURE 12-30. Problem 12-1

12-2 Draw the logic diagrams of the two flip-flops formed in Problem 12-1 using the NOR gate symbols. Identify each gate by letter designation and label each pin number.

12-3 Draw the logic symbols of the flip-flops formed in Problem 12-1. Label input and output pins.

12-4 The circuit of Figure 12-31 is a CMOS SCL4011A quad two-input NAND gate. Connect to form a crossed NAND flip-flop of the type shown in Figure 12-6.

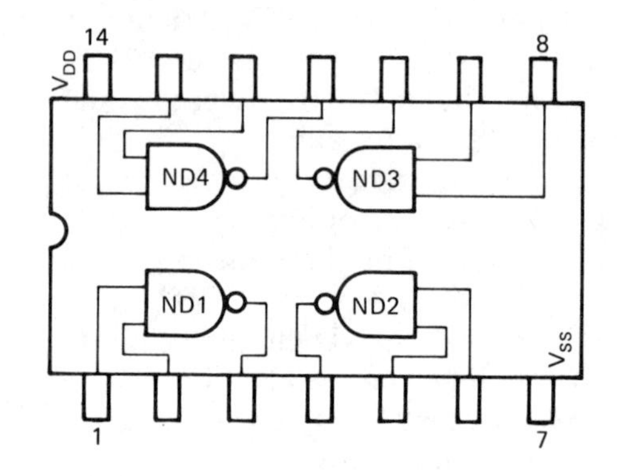

FIGURE 12-31. Problem 12-4

12-5 Draw the logic diagrams of the flip-flops formed in Problem 12-4 using the NAND and inverter logic symbols. Identify each gate by letter designation and label each pin number.

12-6 Draw the logic (block) symbols of the flip-flops formed in Problem 12-4. Label input and output pins.

12-7 The flip-flop of Figure 12-32 has its set input connected to ground (logic 0). The timing diagram of Figure 12-32 shows the clear and toggle input waveforms. Draw the resulting Q $\overline{Q}$ waveforms.

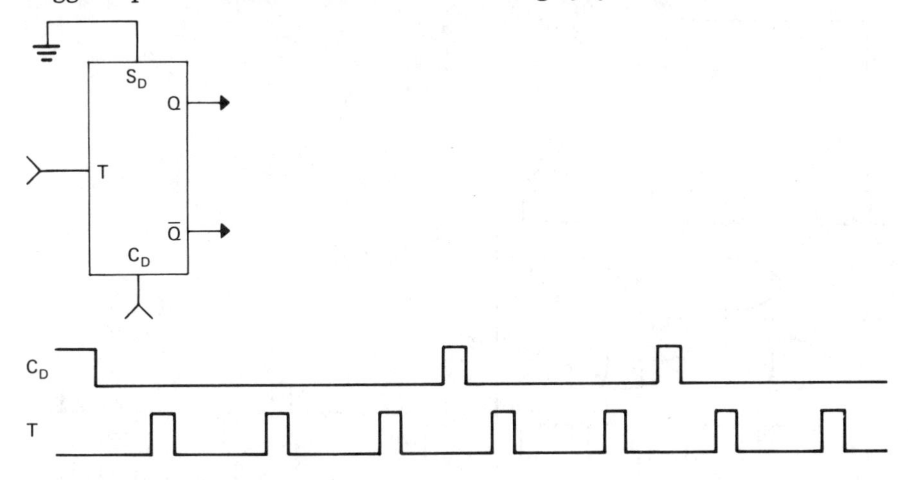

FIGURE 12-32. Problem 12-7

12-8 The IBM card shown in Figure 12-33(a) is read in 12 parallel output lines, each 80 bits long. Each of the 12 rows will appear as parallel output lines containing a serial pulse train. A pulse occurs for every punch. An eighty-first-bit position will receive an odd parity pulse according to the logic of Figure 12-13. The output waveforms are shown in Figure 12-33(b). Draw in the eighty-first-bit odd parity pulse.

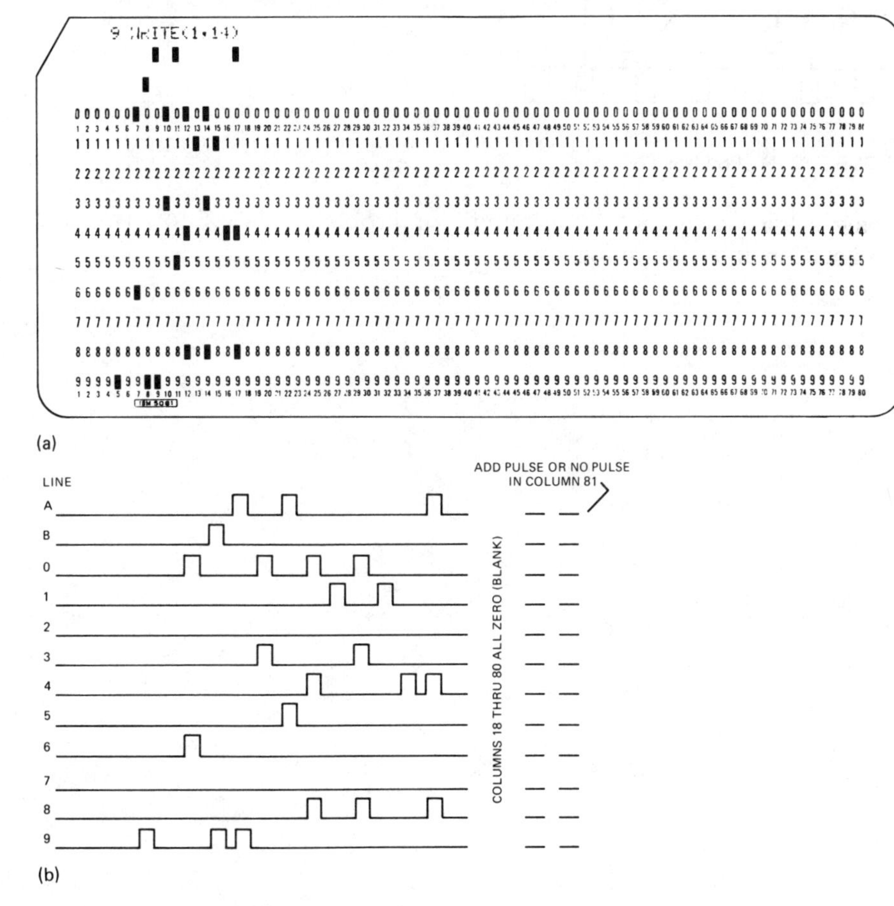

FIGURE 12-33. Problem 12-8

12-9 The circuits in Figure 12-34 are TTL/SSI 7402 quad two-input NOR gates and a TTL/SSI 7408 quad two-input AND gate. Draw in the connections needed to form a four-bit strobed register.

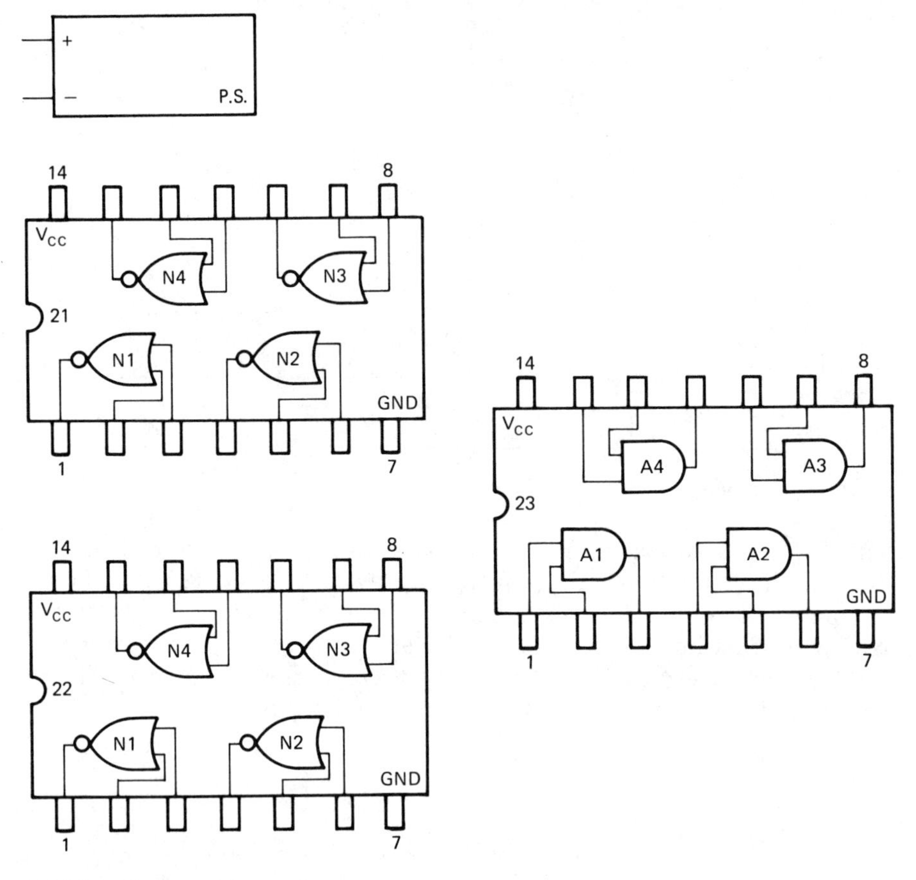

FIGURE 12-34. Problem 12-9

12-10 Draw a logic diagram of the circuit of Problem 12-9. Show the gate designation and pin numbers.

12-11 The circuits of Figure 12-35 are CMOS 4027 dual J-K flip-flops and 4001 dual-input NOR gates. Connect to form a four-bit register like that of Figure 12-17.

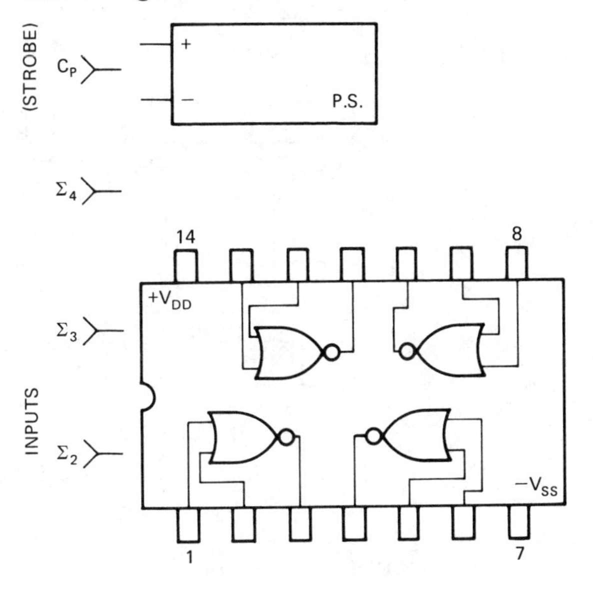

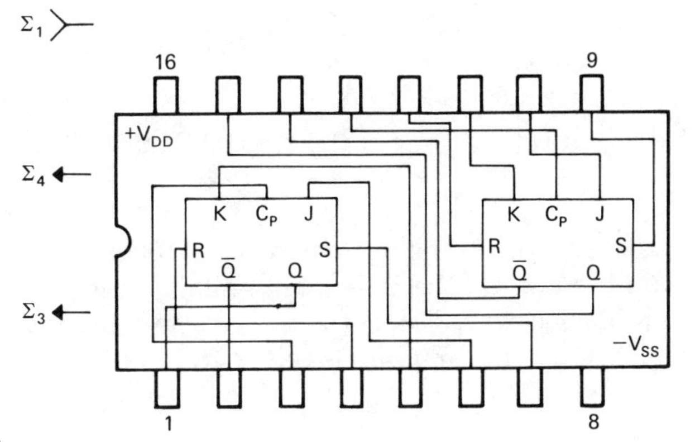

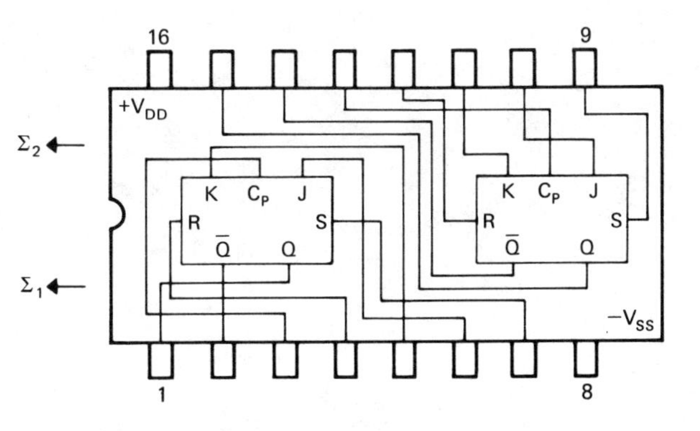

FIGURE 12-35. Problem 12-11

12-12 The circuit of Figure 12-36 is the TTL/MSI 7475 quadruple bistable latch. Connect to inputs and outputs to form a four-bit register equivalent to that of Figure 12-17.

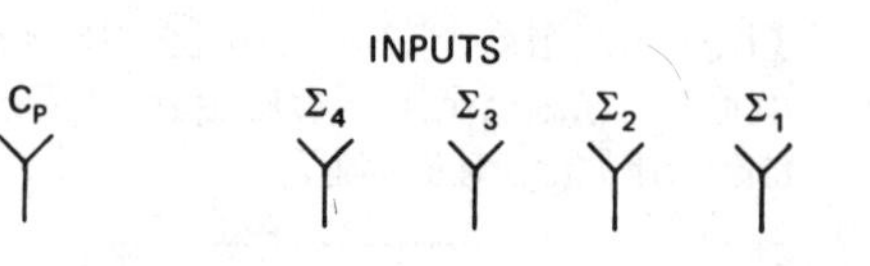

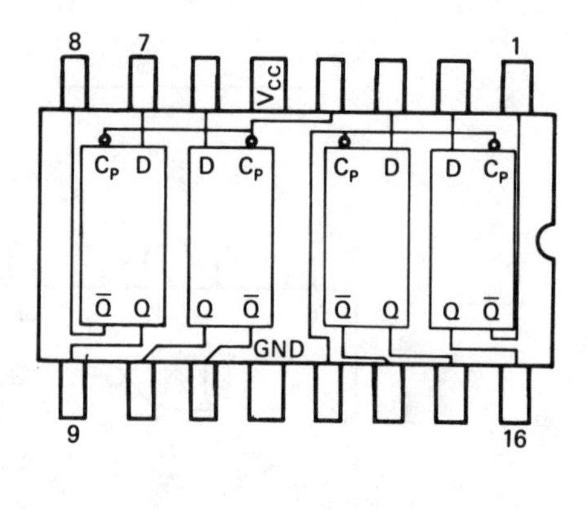

FIGURE 12-36. Problem 12-12

12-13 Draw the interconnections needed to produce a toggle flip-flop from a D flip-flop.

12-14 Draw the logic needed to convert the hex D flip-flop of Figure 12-27 to a hex toggle flip-flop.

12-15 A J-K flip-flop is needed to operate in a circuit with a 10 MHz clock. The supply voltage used is 5V. Of the integrated circuits described in this chapter, select the device with the correct speed and the lowest power requirements. What are the minimum setup and hold times of the device selected?

CHAPTER

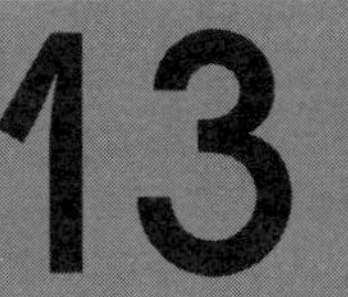

Shift Registers

Objectives

Upon completion of this chapter, you will be able to:

- Assemble J-K flip-flops to form a shift register that can receive digital data in serial form and read out when needed in either parallel or serial form.
- Assemble D flip-flops to form a shift register that can receive digital data in parallel form and read out when needed in either serial or parallel form.
- Convert digital data from parallel to serial and serial to parallel by using shift registers.
- Connect a TTL/MSI universal shift register to provide serial-to-parallel or parallel-to-serial conversion.
- Assemble a register with the capacity to shift both left and right.
- Use shift registers to assist in arithmetic operations.
- Compare a MOS static register with a MOS dynamic register.

13.1 Introduction

The J-K flip-flop introduced in Chapter 12 has every possible register function. It can be reset, preset, toggled, and steered to the 1 or 0 level. In Chapter 12, we loaded data into the registers in parallel and read the output in parallel.

In this chapter, we will show that a register can be loaded in serial from one side and read out in serial from the opposite side. It can also be loaded in serial and read out in parallel or loaded in parallel and read out in serial. Figure 13-1 shows these four possible functions, three of which require that numbers be shifted to the left or right between flip-flops.

The function of loading a register in parallel and reading its output in parallel was explained in Section 12.8 and shown in Figure 12-17. At this point, we will consider a new function, that of shifting binary numbers or levels through the individual flip-flops of a register. Although it is usual to consider the shift as occurring from left to right, the logic can very simply be rewired so that the shift is from right to left, and with more complex logic, a register can be designed for shift in either direction. Let us look again at the J-K input functions of the J-K flip-flop. These two inputs are referred to as steer inputs because they steer the flip-flop to either 1 or 0. No actual change in state occurs, however, until the trailing edge of a clock pulse. Figure 13-2 shows the logic symbol and truth table of the J-K flip-flop. The truth table indicates that the flip-flop will ignore the clock pulse if J and K are both 0. If J is 1 and K is 0, the flip-flop will be in the 1 state after the trailing edge of the clock pulse. If J is 0 and K is 1, it will be in the 0 state after the trailing edge of the clock pulse.

Note the circle or bubble on the clock input of the symbol in Figure 13-2. The bubble indicates operation on the trailing edge of the clock pulse. If the bubble was left out, it would indicate operation on the leading edge of the clock pulse.

Shift registers are an integral part of any computer. One of their applications through the computer, in conjunction with an adder circuit, is to perform multiplication. Section 13.5 describes such a circuit.

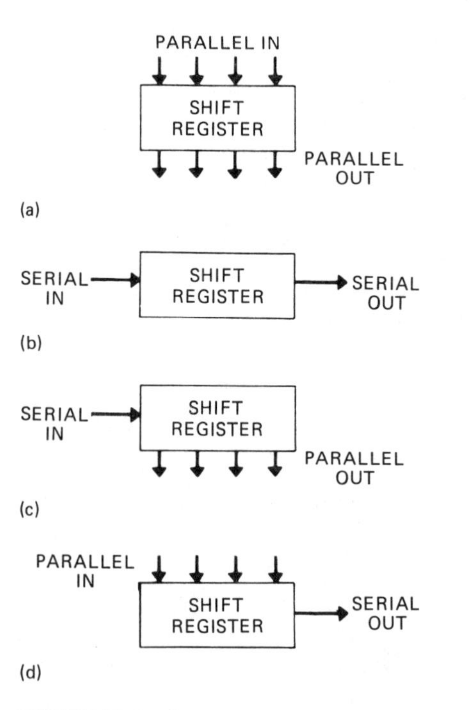

FIGURE 13-1

Four methods of loading and reading out a shift register. (a) Parallel in and parallel out. (b) Serial in and serial out. (c) Serial in and parallel out. (d) Parallel in and serial out.

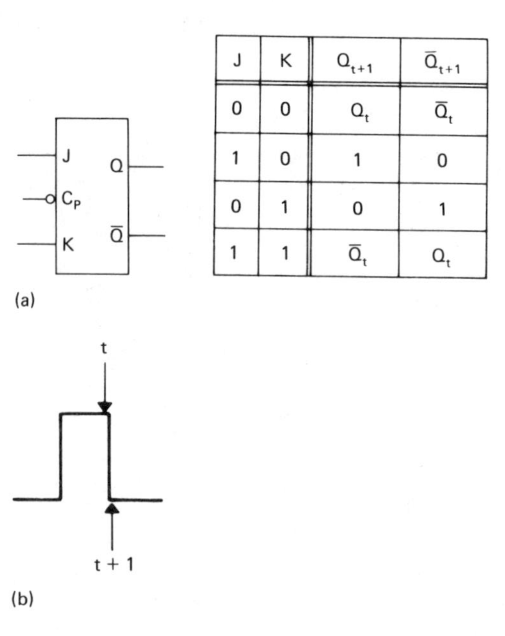

J	K	Q_{t+1}	$\bar{Q}_{t+1}$
0	0	Q_t	$\bar{Q}_t$
1	0	1	0
0	1	0	1
1	1	$\bar{Q}_t$	Q_t

FIGURE 13-2

(a) Logic symbol and truth table of J-K flip-flop. (b) Changes occurring on trailing edge of clock pulse.

13.2 Shift Registers

Shifting digital numbers or levels through a set of J-K flip-flops requires interconnections such as those shown in Figure 13-3(a). The Q of the flip-flop on the left connects to the J of the flip-flop on the right, while the $\overline{Q}$ connects to the K input. This means that, prior to the trailing edge of the clock pulse, each flip-flop (with the exception of the leftmost one) is being steered to the state of the flip-flop on its left. After the trailing edge of the clock pulse, the logic state (1 or 0) existing in flip-flop A will be the state in flip-flop B. The state that existed in flip-flop B will be the state in flip-flop C. Flip-flop A will clock to 0 because of the fixed logic level on its inputs.

Consider the initial state of the flip-flops, as shown in the simplified diagram of Figure 13-3(b). After the first clock pulse, the 1 level will shift to flip-flop B. After each clock pulse, the 1 level will move another bit to the right. After the third clock pulse, all bits of the shift register will be 0 and the 1 bit will be lost unless it is maintained elsewhere in the digital system.

The fact that a flip-flop is being steered by a preceding flip-flop, which may itself change state on the trailing edge of the same clock pulse, may appear to be a race problem; but a careful look at Figure 12-11 shows that the effect of the steer on the master is completed when the clock pulse goes high. On the trailing edge of the clock pulse, it merely passes the change to the slave or output half of the flip-flop.

For a trigger flip-flop to operate in a shift register, its propagation delay (clock to Q) must exceed its hold time specification. Note that the 74LS112 J-K flip-flop described in Chapter 12 has a propagation delay (clock to Q and $\overline{Q}$) of 15 nanoseconds and, being a master-slave flip-flop, its hold time is 0 nanoseconds. The steer conditions will, therefore, remain stable long enough for the shift to take place. Examination of the AC specifications of the 74LS74 D flip-flop shows a hold time of 5 nanoseconds, while the propagation delay (clock to Q) is more than 10 nanoseconds, indicating that it also will provide stable shift register operation.

The D latch, such as the 7475 shown in Figure 12-23, should not be used for shift register operation because changes in the level on the D input will propagate to Q as long as the clock pulse is high. Also, depending on the width of the clock pulse, levels may shift through more than one bit of the register during each clock pulse.

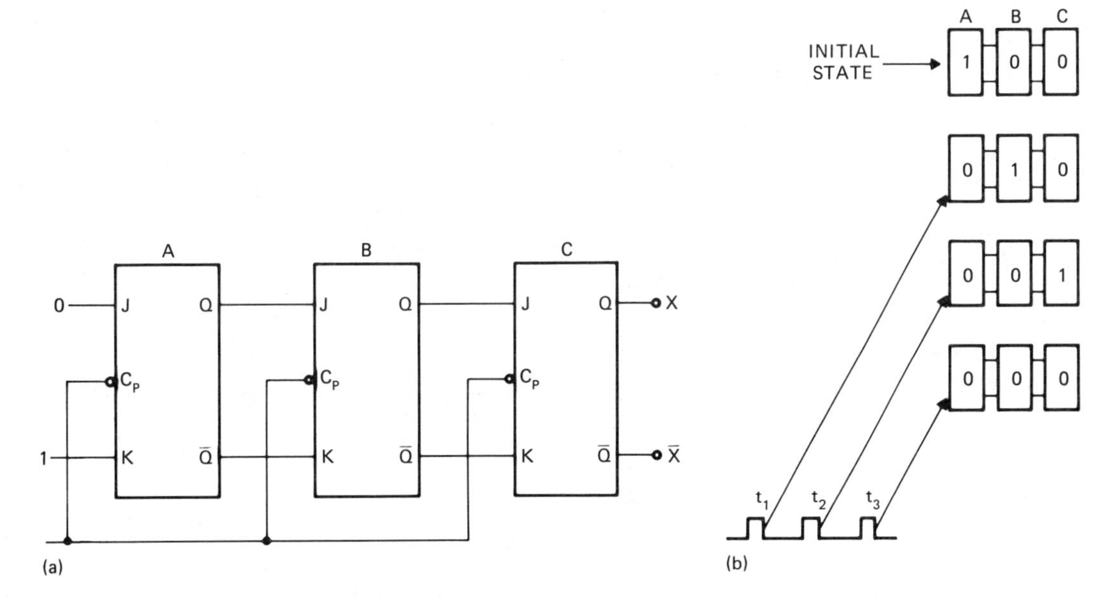

FIGURE 13-3

(a) J-K flip-flops connected to form a three-bit shift register. Digital levels stored in this register will shift to the right. (b) Initial state of the flip-flops. The high level in register A will shift to the right, one bit for each clock pulse, with the changes occurring on the falling edge of the clock pulse.

Figure 13-4(a) shows four D flip-flops connected to form a shift register. In addition to the fact that no connection is needed from $\overline{Q}$, the register of Figure 13-4 differs from the J-K register of Figure 13-3 in that changes occur on the leading edge of the clock pulse.

The shift registers described thus far are serial-in-serial-out and, in this form, have several applications. One of the applications is the delay of serial signals. Figure 13-4(b) shows a waveform for 0110, the binary equivalent of 6_{10}. If this signal is applied to the input of a four-bit shift register, such as the D register in Figure 13-4(a), the output will be delayed by four clock periods. Note, however, that the signal must have synchronization with the clock. There must be one clock pulse for each bit time of the signal. At the output of the shift register, changes will occur after the rising edge of the clock pulse. If the same signal had been applied to the J-K flip-flop, an inverted signal would have been needed for the K input, and changes at the output would occur after the falling edge of the clock pulse.

The shift register can also be used for storage of serial signals by merely turning off the clock pulses at the time the first bit of the signal reaches the rightmost register. When the signal is needed, the clock line is enabled and the signal appears at the output of the rightmost register.

Figure 13-4(c) shows the waveforms that would occur if the serial number equivalent to 6_{10} were to be stored for a period of seven clock pulses. Note that a clock inhibit signal must be generated and used to inhibit the clock pulses to the register for the period during which the number is to be stored.

FIGURE 13-4

(a) D flip-flops connected to form a four-bit shift register. (b) Waveforms for a digital level train 0110 being shifted through the four-bit register. The output at X is delayed by four clock pulses. (c) Waveforms showing that clock inhibit signal goes low after four clock pulses, permitting the number 0110 to be stored in the register. Output appears at X after the inhibit signal goes high seven clock pulses later.

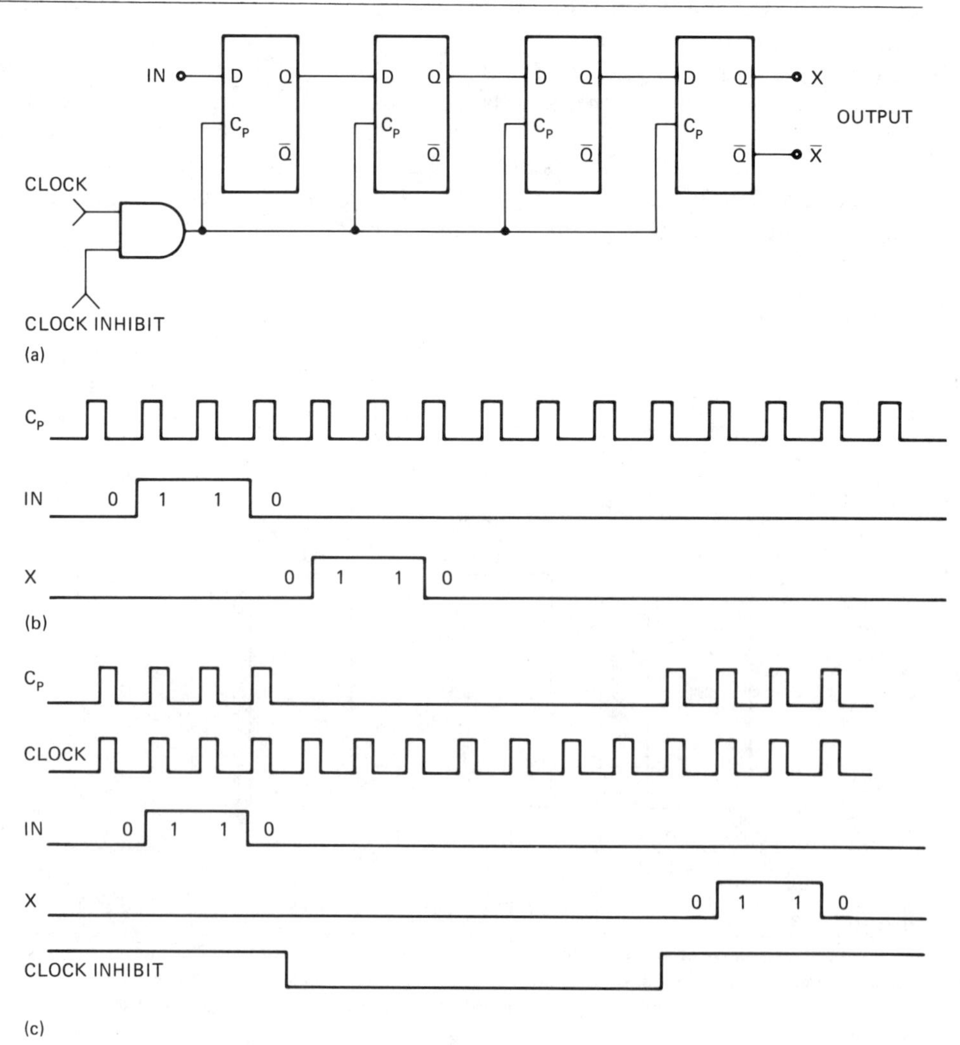

13.3 Serial-to-Parallel Conversion

In a parallel adder, like that of Figure 12-7, the inputs are applied to the register by setting them in parallel. The outputs are obtained by taking parallel leads from the 1 outputs of the sum register. Those registers are loaded in parallel and read out in parallel. But often a number must be received in serial and transmitted in parallel. This conversion is accomplished by connecting and loading the register as in Figure 13-5(a). The process begins before the first clock pulse, when the reset pulse resets all four flip-flops to 0. As shown by the waveforms in Figure 13-5(b), the serial number

FIGURE 13-5

(a) Register circuit connected for serial-to-parallel conversion. (b) Waveforms showing input arriving at A in level-train form, complement applied at B, clock line inhibited after LSB reaches 2^0 register, and parallel outputs available from Q outputs of the flip-flop.

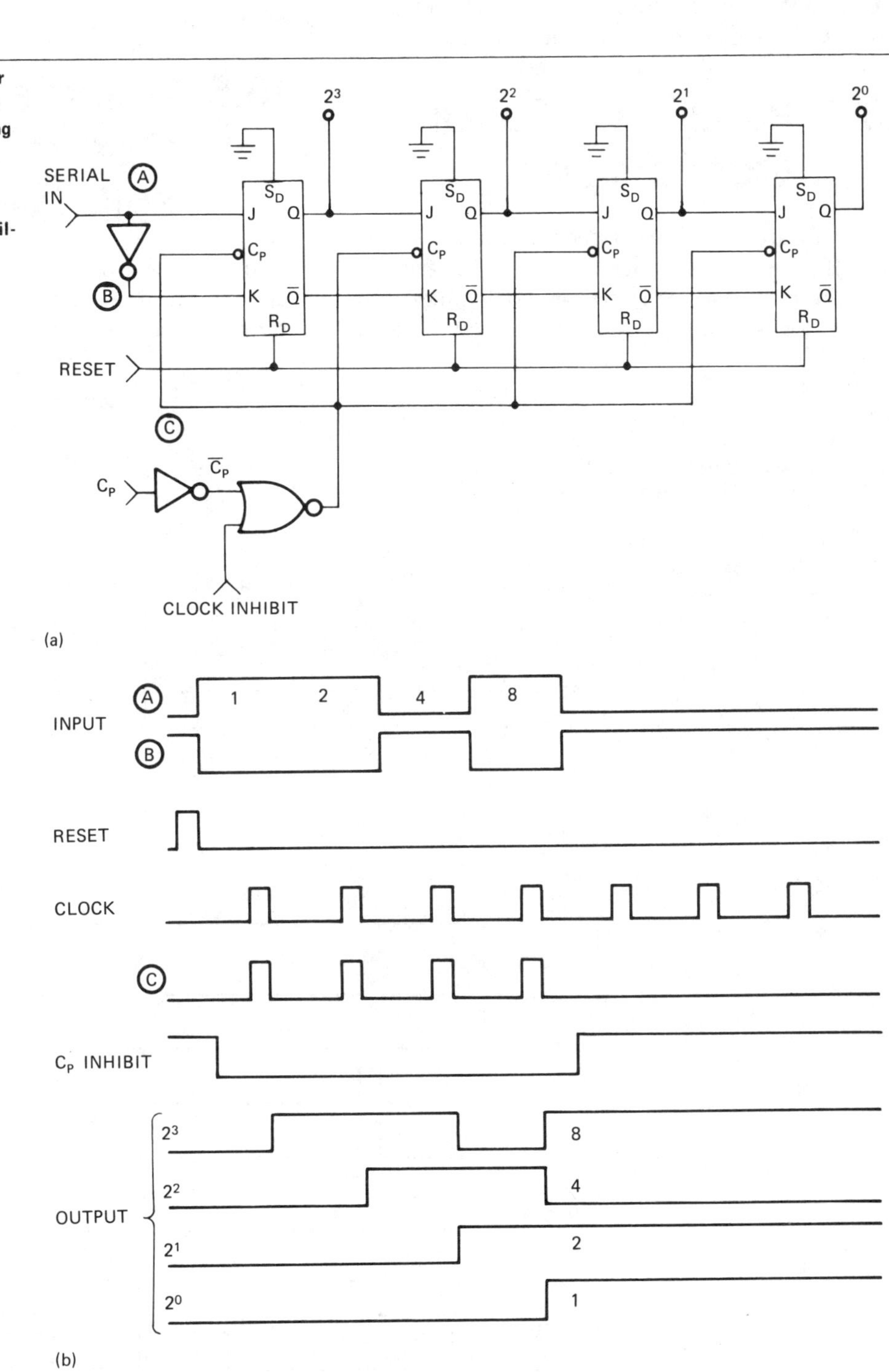

arrives at point A as a level-train LSB. To steer numbers into the MSB register, the complement of A must be applied to B. This is accomplished with the inverter. If the complement is already available, the inverter will not be needed. The clock is applied through an inverter and the NOR gate so that the positive-going inhibit signal can inhibit the clock pulses after the fourth one. By that time, the LSB has reached the LSB flip-flop, and the entire number is in storage and available as a parallel output.

13.4 Parallel-to-Serial Conversion

Numbers are often processed in parallel for the sake of speed but are then converted to serial for the sake of economy in transmitting them. This conversion is provided by flip-flops connected as shown in Figure 13-6. The operation begins with all registers being set to 0 by application of the reset pulse to the clear inputs. Then pulses representing the parallel number (in this case) are applied to the preset inputs, no pulse representing a 0. The clock pulse then shifts the number to the right and out through the LSB flip-flop. The output, shown by the waveform X and $\overline{X}$, is a level train representing the number 11_{10}.

13.5 Arithmetic Operation with Shift Registers

Shift registers provide an important function in arithmetic operations. In addition to storing the terms to be applied to the inputs of the adder,

FIGURE 13-6

(a) Register circuit connected for parallel-to-serial conversion. (b) Waveforms representing the number 11_{10} (1011). LSB begins on leading edge of preset input and changes to next higher bit (power of 2) on trailing edge of each clock pulse. After the last 1-level bit is shifted out, all flip-flops are at 0.

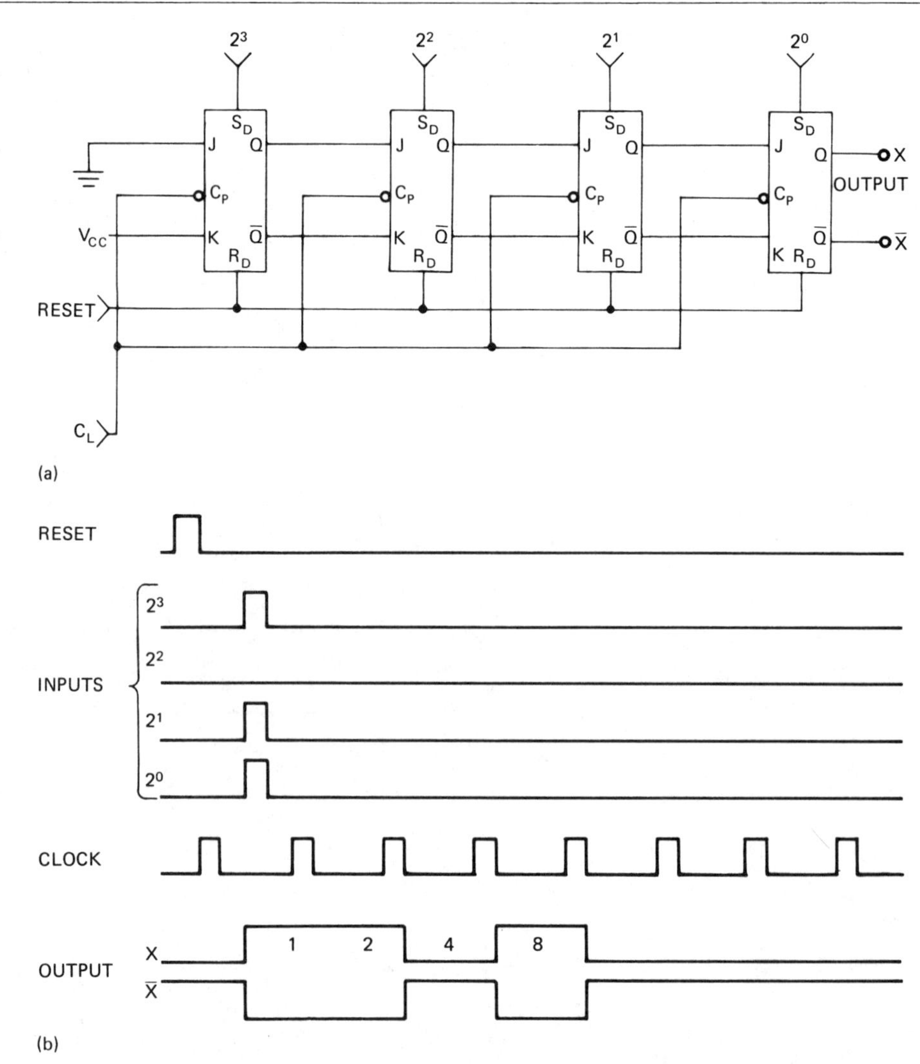

registers may be used to shift terms either to the right or to the left. As we recall from Chapter 3, binary multiplication is a matter of shifting the multiplicand to the left in binary place and either adding or not adding after each shift, depending on the state of successive bits in the multiplier. To multiply two binary numbers, X and Y, requires an X register that shifts to the right and a Y register that shifts to the left. Assuming that both are four-bit numbers, a four-bit register like the one shown in Figure 13-6 will serve for X. A larger register of at least seven bits is required for Y, and it must be wired for a shift left. Figure 13-7(b) shows the Y

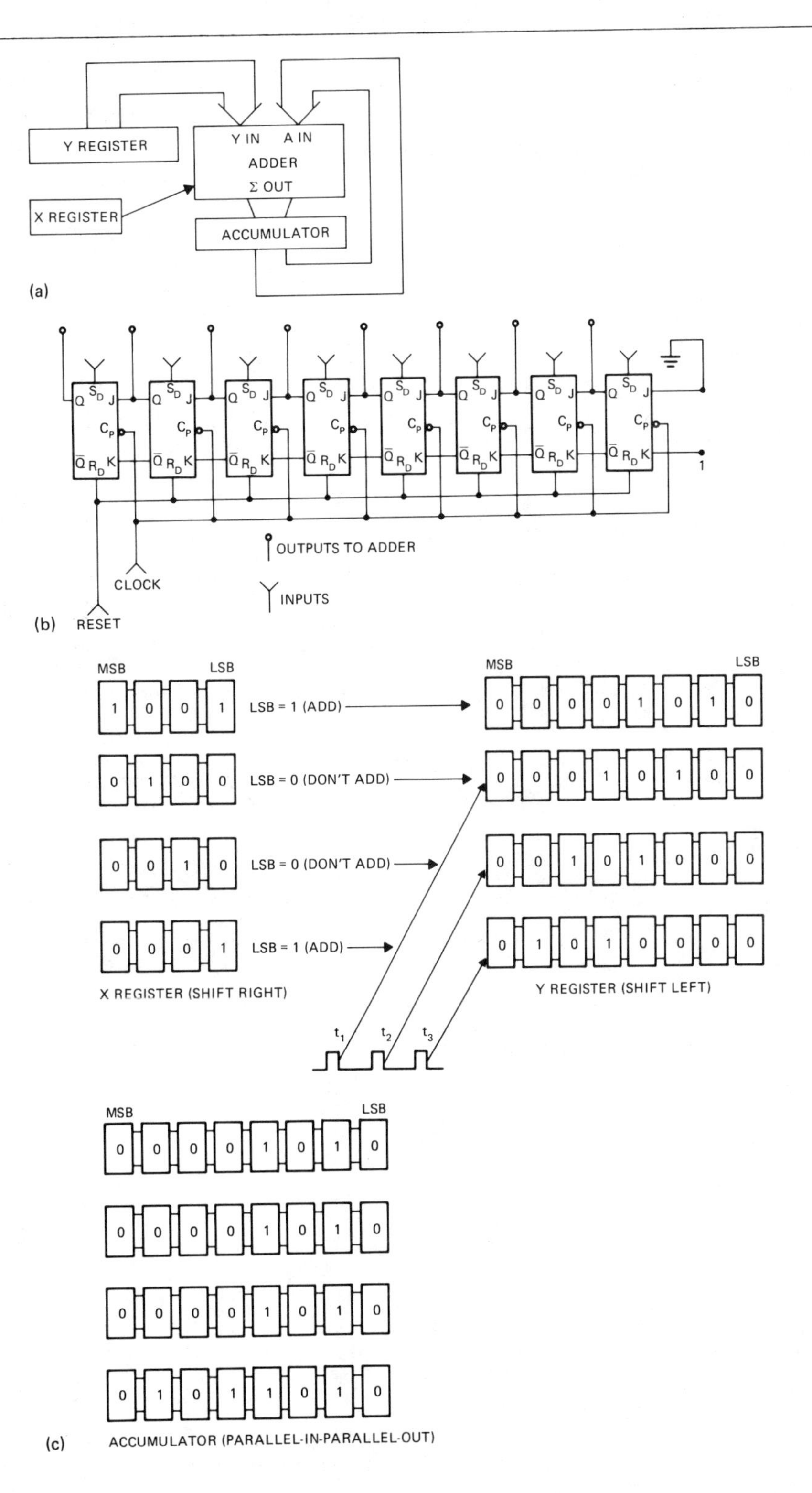

FIGURE 13-7

Binary multiplication using shift registers and an adder. (a) Multiplier block diagram. (b) Y register wired for shift left. (c) X register, Y register, and accumulator in initial state and after each clock pulse.

register. The adder is connected as shown in Figure 13-7(a). The accumulator, which stores the results of each arithmetic operation, is switched back to one adder input, while the Y register output is connected to the other adder input.

Prior to each shift, the adder control circuitry examines the LSB output of the X register and initiates an add cycle only when the LSB is a binary 1. Figure 13-7(c) shows a simplified diagram of the X register, Y register, and accumulator as they appear initially and after each clock pulse. The state of the registers is given for X = 1001 and Y = 1010. After the third clock pulse, the accumulator will contain 01011010 (9 × 10 = 90).

13.6 Four-Bit Universal Shift Register (MSI)

Figure 13-8 shows the Fairchild 9300 four-bit universal shift register in medium-scale integrated circuit (MSI) form. The lead labels indicate that this is a four-bit shift register. It can be employed for either of the three shift register applications just discussed. The device is

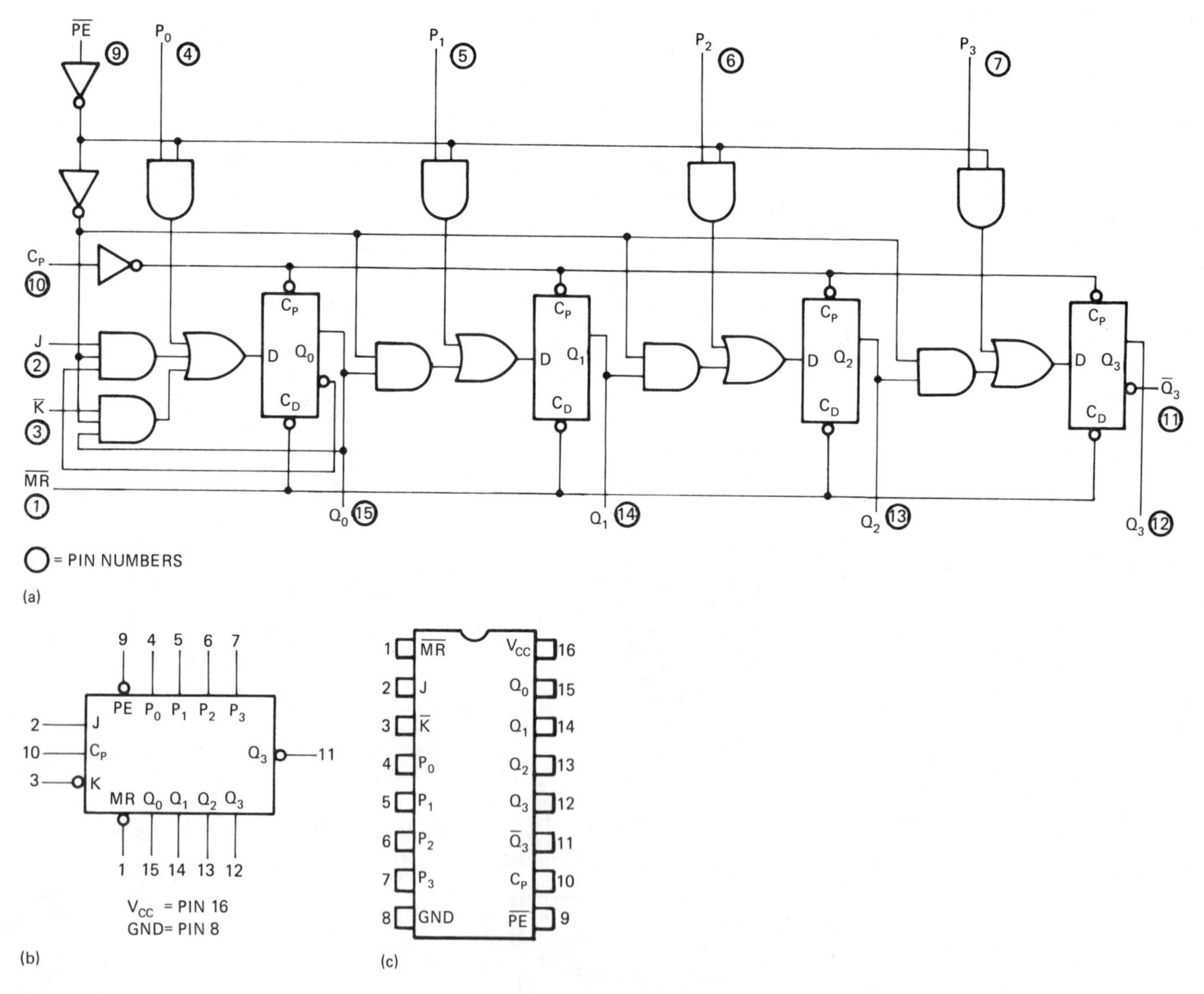

FIGURE 13-8

Fairchild 9300 four-bit universal shift register in MSI form. (a) Logic diagram. (b) Logic symbol with 16 pins labeled by input and output functions. (c) Connection diagram of DIP (top view).

a 16-pin IC with just enough output leads for these applications.

The logic diagram shows the construction to be of D flip-flops, but the inverter on the clock line and logic gates at the D inputs cause it to function more like a J-K flip-flop register. The parallel inputs are not direct inputs. The $\overline{\text{PE}}$ ($\overline{\text{parallel enable}}$) must go low and a clock pulse must occur in order to load the register in parallel. The register will shift right on the trailing edge of each clock pulse when the $\overline{\text{parallel enable}}$ is high.

Figure 13-9 shows the connection needed for a four-bit serial-to-parallel conversion. The waveforms for its operation are identical to those of Figure 13-5(b). The J and K shorted together form the input. The inversion circle on the K input tells us that an inverter is not needed between J and K. The preset inputs are not needed for serial-to-parallel conversion and are, therefore, grounded. With the exception of $\overline{Q}_3$, the complement outputs are not made available. Q_0 is the MSB output; Q_3, the LSB.

Figure 13-10(a) shows the connections needed to provide parallel-to-serial conversion. The waveforms for this operation, shown in Figure 13-10(b), differ slightly from those in Figure 13-6(b). To load the data in parallel, the $\overline{\text{parallel enable}}$ must go low for one or more clock pulses. After the $\overline{\text{parallel enable}}$ goes high, the data will shift to the right one bit on the trailing edge of each clock pulse. The waveforms for this operation are identical to those shown in Figure 13-6(b). P_0 through P_3 provide the input leads. The output is Q_3, and if the complement is needed, pin 11 provides an inverted Q_3, equivalent to $\overline{Q}_3$. In shifting to the right, the leftmost flip-flop is steered to 0. This would normally call for a 0 on J and a 1 on K, but the inversion symbol on the K input indicates that a 0 or ground on that input should serve as a 1 on K, and therefore both J and K are grounded.

13.7 MOS Registers

There are various forms of registers in MOS and CMOS integrated circuits. We can first classify them as *static* or *dynamic*. The dynamic register stores the 1 or 0 level in the very minute capacitance that is inherent to the gate-to-source junction, as Figure 13-11 shows. A 1-level charge will leak off this capacitance in a few milliseconds. It is, therefore, necessary to keep the data shifting through the register. Since the dynamic register is more widely used in memory applications, it will be discussed in detail in Chapter 23.

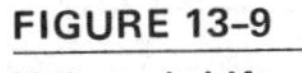
FIGURE 13-9

Universal shift register (MSI) connected for serial-to-parallel conversion.

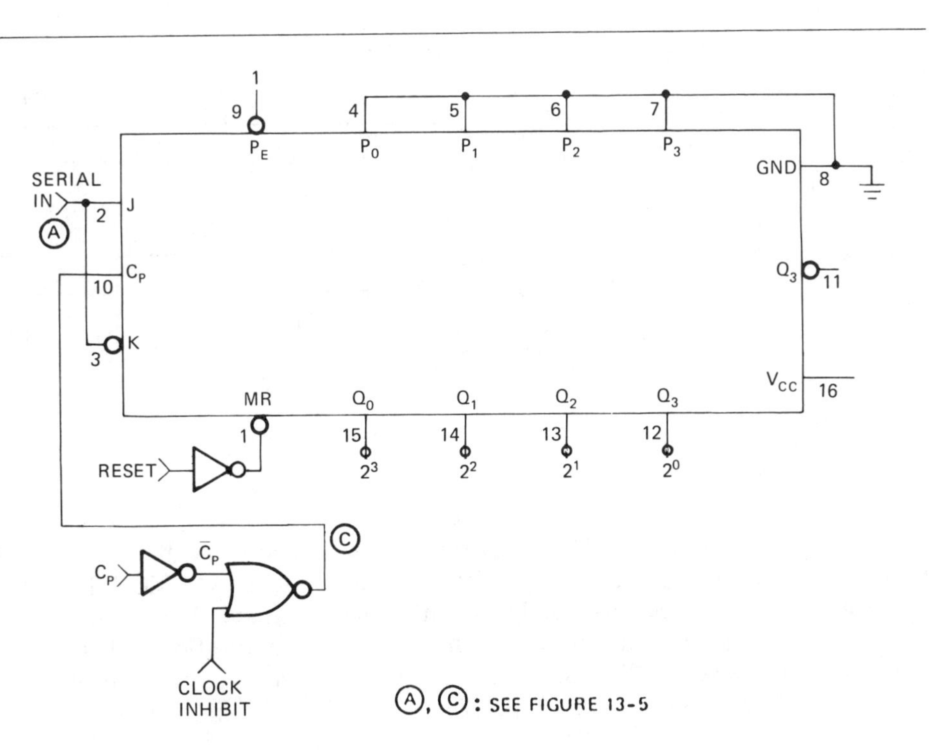

FIGURE 13-10

(a) Universal shift register (MSI) connected for parallel-to-serial conversion. (b) Voltage waveforms that would occur as a result of parallel-to-serial conversion of 1011 (11) using the four-bit universal shift register.

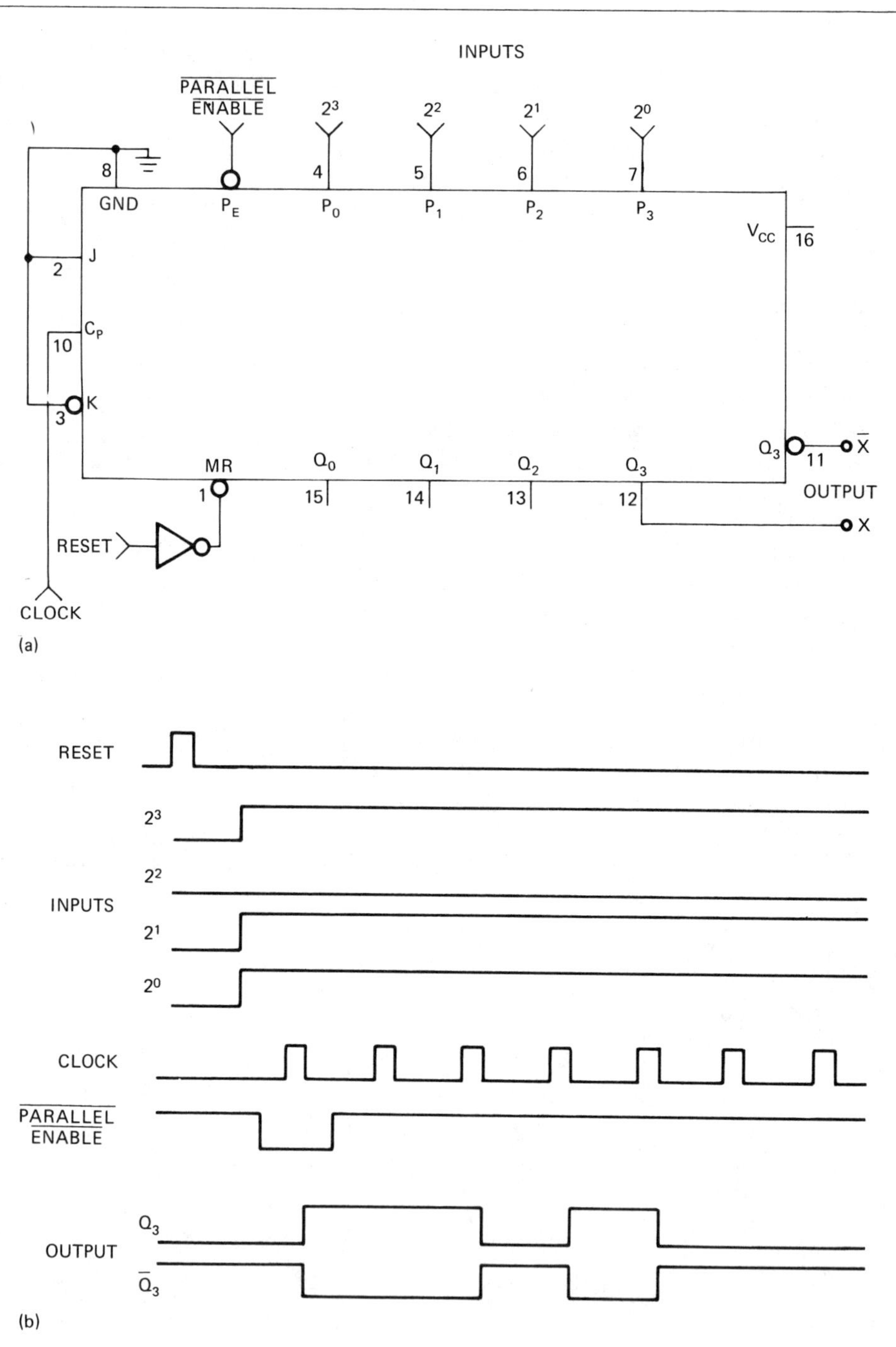

The static form of a MOS register uses a flip-flop and, unlike the dynamic register cell, it can hold a 1 or 0 level indefinitely, or until the power is removed from the circuit. Figure 13-12 is a typical MOS static flip-flop. From this type of cell, flip-flops similar to the bipolar J-K flip-flop can be produced. Figure 13-13 is the SCL 4013A CMOS D flip-flop. As the truth table shows, this flip-flop functions just like a bipolar D flip-flop. This makes it convenient for shift register application.

Numerous CMOS shift registers are avail-

FIGURE 13-11

MOS dynamic register, which stores logic levels in gate-to-source capacitance of MOSFET. Charge leaks off in a few milliseconds, requiring that data circulate through the register.

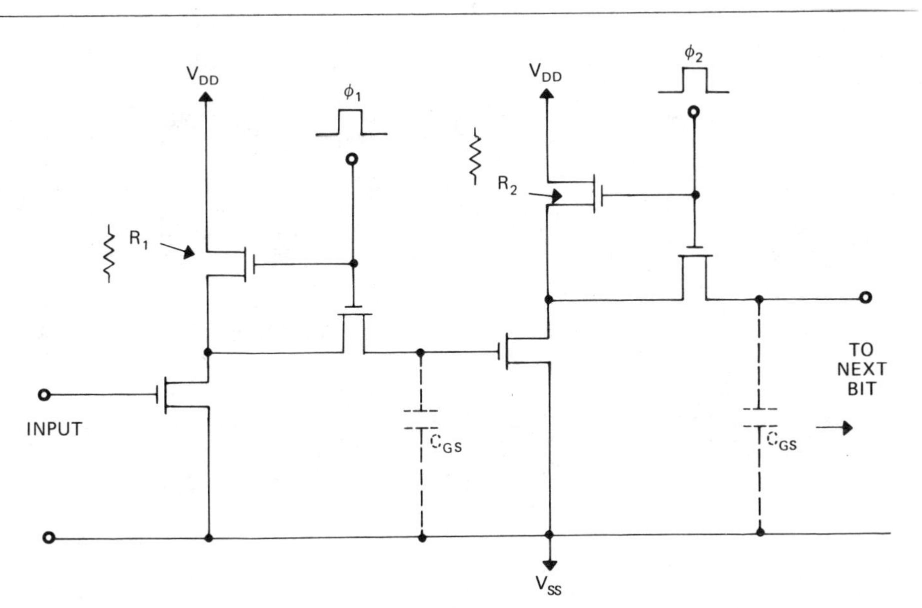

able in MSI form. Figure 13-14 is the SCL 4014B eight-stage static shift register. This integrated circuit can be used as a serial-in-serial-out register or as a parallel-to-serial conversion of six- to eight-bit capacity.

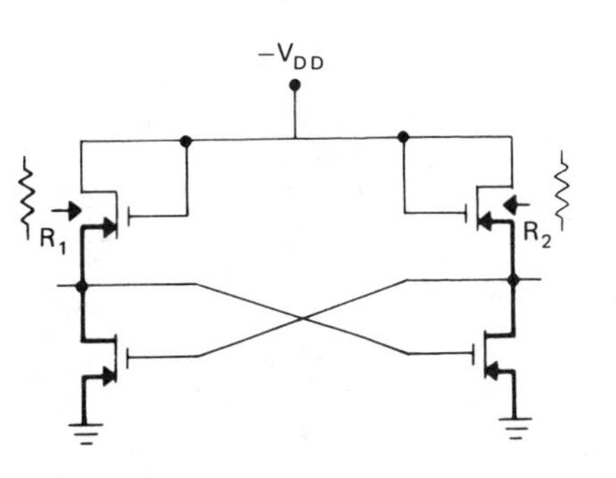

FIGURE 13-12

MOS static register, which functions like a bipolar flip-flop.

13.8 Right-Shift-Left-Shift Register

There is occasional need for a register that can shift the number stored in it to either left or

FIGURE 13-13

SCL 4013A CMOS D flip-flop. (a) Logic symbol. (b) Truth table.

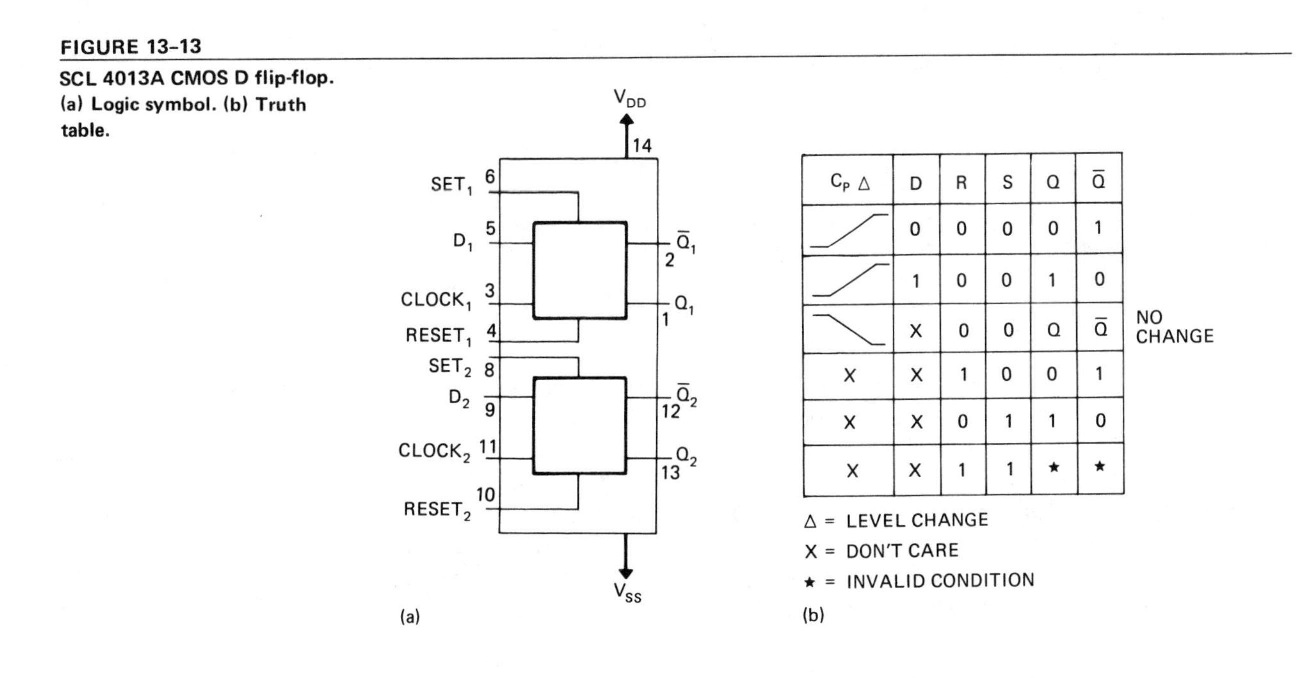

C_P Δ	D	R	S	Q	$\bar{Q}$
↑	0	0	0	0	1
↑	1	0	0	1	0
↓	X	0	0	Q	$\bar{Q}$ NO CHANGE
X	X	1	0	0	1
X	X	0	1	1	0
X	X	1	1	★	★

Δ = LEVEL CHANGE
X = DON'T CARE
★ = INVALID CONDITION

(b)

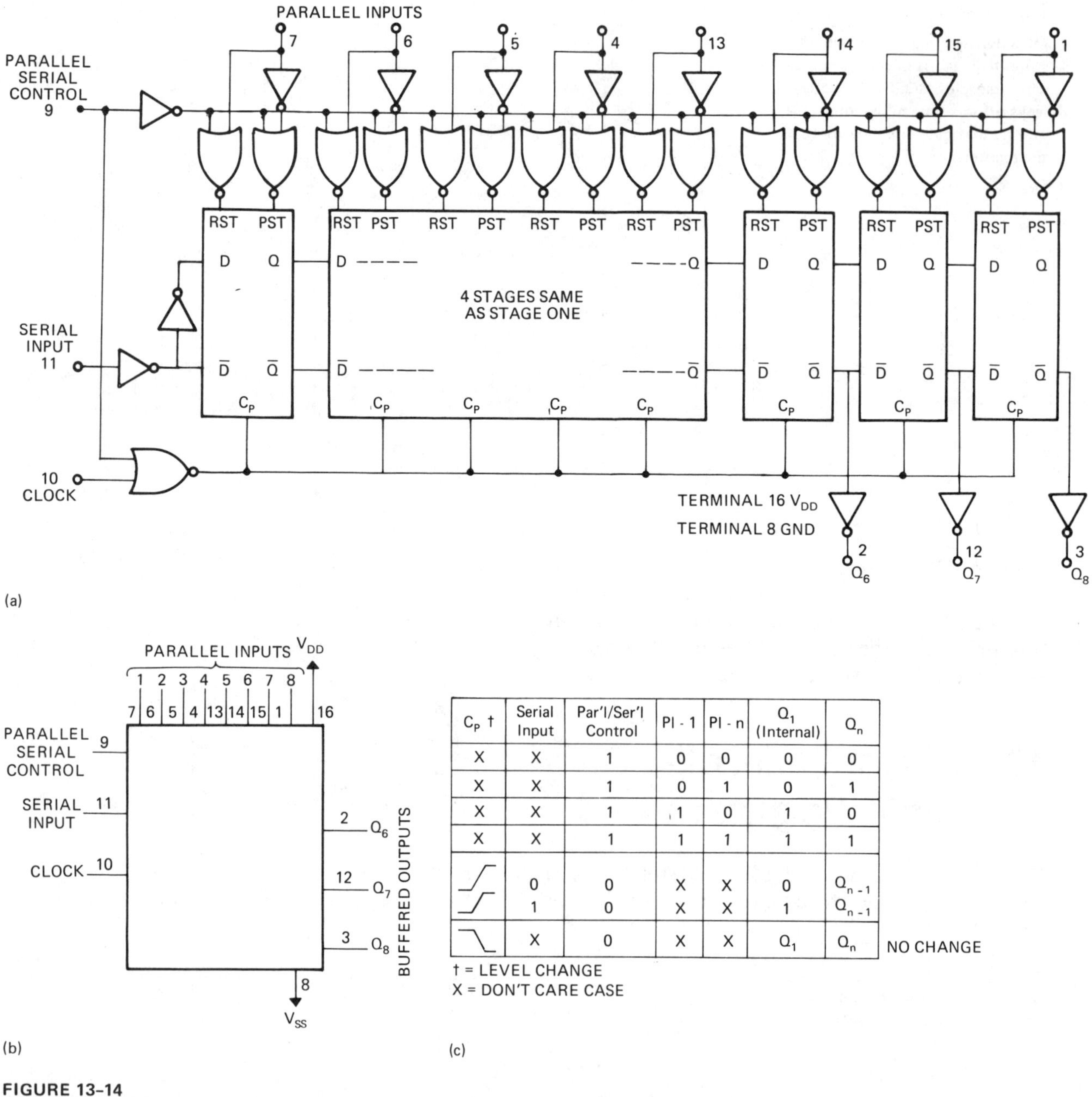

C_P ↑	Serial Input	Par'l/Ser'l Control	PI - 1	PI - n	Q_1 (Internal)	Q_n	
X	X	1	0	0	0	0	
X	X	1	0	1	0	1	
X	X	1	1	0	1	0	
X	X	1	1	1	1	1	
_/‾	0	0	X	X	0	Q_{n-1}	
_/‾	1	0	X	X	1	Q_{n-1}	
‾_	X	0	X	X	Q_1	Q_n	NO CHANGE

↑ = LEVEL CHANGE
X = DON'T CARE CASE

FIGURE 13-14

SCL 4014B CMOS eight-stage static shift register, useful for serial output register capacity of six to eight bits. (a) Logic diagram. (b) Logic symbol. (c) Truth table.

right. This can be accomplished in a register like the TTL/MSI 7495 shown in Figure 13-15. If shift to the right is desired, a 0 level is applied to the mode control, enabling the number 1 AND gates and inhibiting the number 2 AND gates. Under this condition each flip-flop is steered by the flip-flop on its left. The leftmost flip-flop is controlled by the serial input. In shifting to the right, CP_1 controls the register. The flip-flops in the register are steered like J-K flip-flops in that changes occur on the trailing edge of the clock pulse. A 1 level on R and 0 level on S steer the register to the 1 state.

If shift to the left is desired, a 1 level is applied to the mode control, enabling the number 2 AND gates and inhibiting the number 1

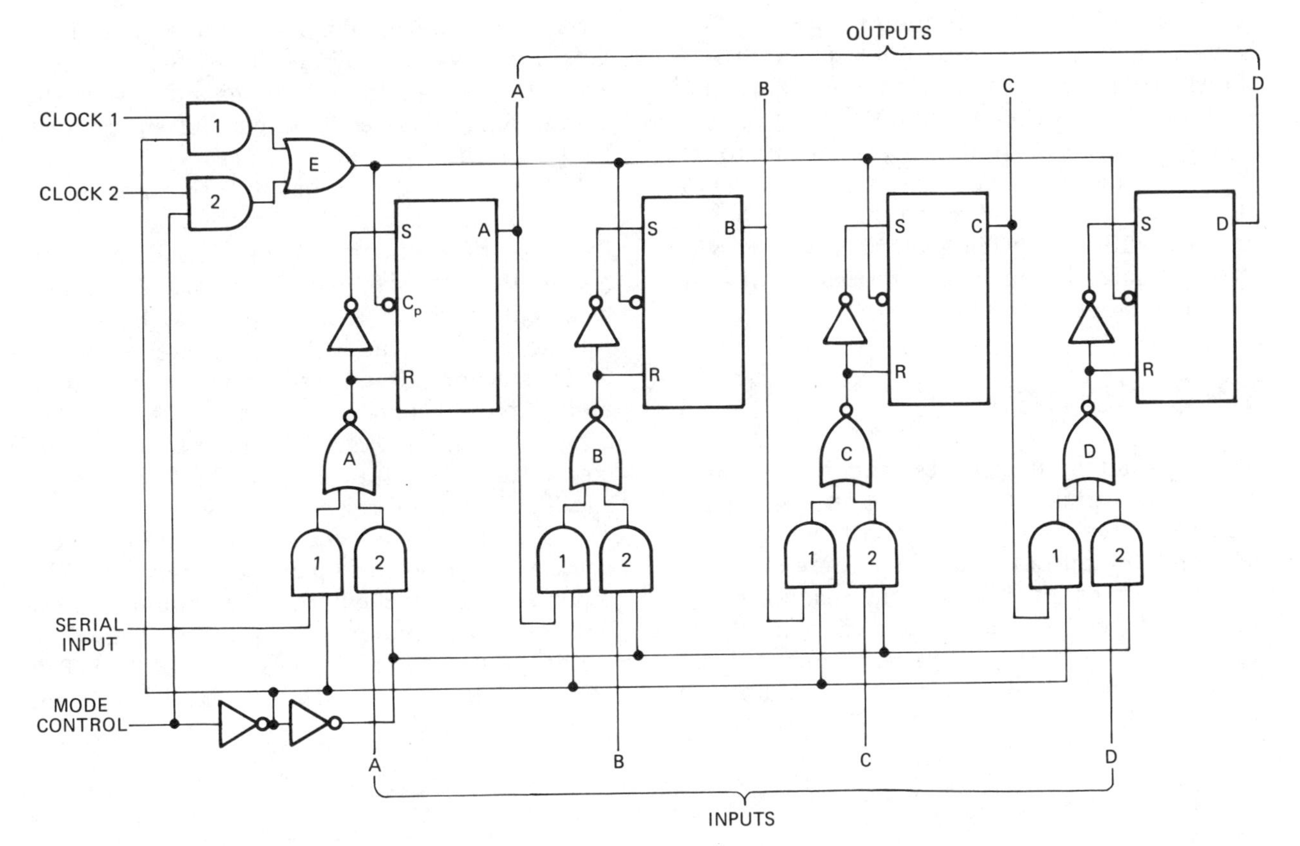

(a)

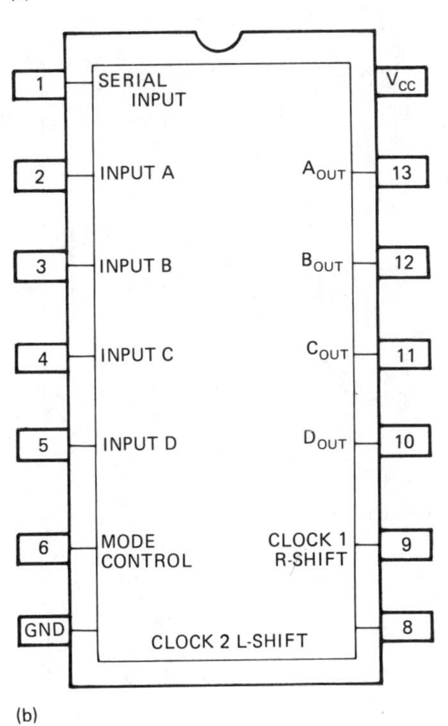

(b)

FIGURE 13–15
TTL/MSI 7495 right-shift-left-shift register. (a) Logic diagram. (b) Connection diagram.

AND gates. Under this condition, each flip-flop is steered by an external input. Figure 13-16 shows the external connections needed, so that B steers A, C steers B, and D steers C. This accomplishes a shift to the left. In shifting to the left, CP_2 controls the register. This versatile IC register can be used for all the register modes of Figure 13-1 with shifts left or right, and with external gates it can be automatically switched from one mode to another.

13.9 Summary

The capabilities of a register circuit go beyond storage of digital data in parallel. In Chapter 12, we loaded the data into the registers in parallel and read the output in parallel. There are three other methods of loading and reading out a register. A register may be loaded in serial and read out in serial, loaded in serial and read out in parallel, or loaded in parallel and read out in serial. The block diagrams of Figure 13-1 illustrate these four modes of register operation. The latter two of these produce a change in the digital data from serial-to-parallel or parallel-to-serial form.

To handle data in serial form, the register must be able to shift the data bits between the flip-flops or elements of the register. For this reason, the term *shift register* is used. To operate in a shift register, a flip-flop must have steer functions such as the J-K or D flip-flop possesses. Shift is accomplished by having the output of each flip-flop connected to steer the flip-flop adjacent to it. In Figure 13-5, this is done by connecting the output of each flip-flop to the J and K input of the adjacent flip-flop.

Figure 13-5(a) shows J-K flip-flops connected for a serial-to-parallel conversion. The input lead is attached to the J terminal of the leftmost flip-flop, and an inverter is used to apply the signal complement to the K lead. This will steer the left flip-flop to the level of the

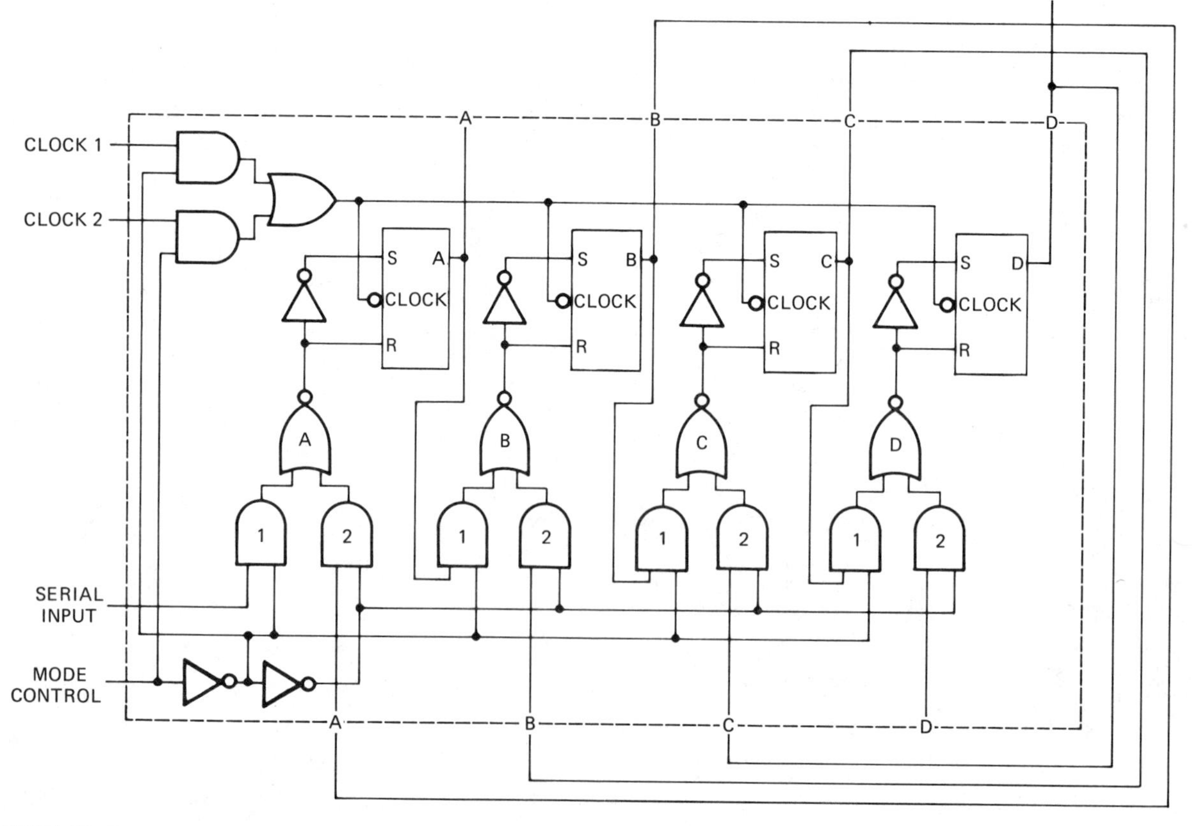

FIGURE 13-16

Right-shift-left-shift register with external wiring needed for shift left. A is serial output during shift left; D is serial output during shift right.

data line. Figure 13-5(b) shows the waveforms for 1011 (11) being shifted into the register. Note that the LSB level travels through the register shifting to the right one flip-flop for each clock pulse. Four clock pulses after reset, the LSB is stored in the rightmost flip-flop. The other data bits have followed in correct order behind the LSB, and all four bits are now positioned in their correct flip-flop. At this time, the clock line must be inhibited; otherwise, the data will continue to shift out through the right side of the register. Note that after the fourth clock pulse, the levels found at the Q outputs of the four flip-flops form a parallel number 1011.

Figure 13-6(a) shows J-K flip-flops connected for a parallel-to-serial conversion of the data. The parallel input leads are connected to the direct set inputs. The waveforms of Figure 13-6(b) are for a data word of 1011 (11) being set into the register shortly after a reset. The output is taken from the Q and $\overline{Q}$ of the rightmost flip-flop. On the leading edge of the input set pulses, the LSB is seen at the output. On the trailing edge of each clock pulse, the output changes to the level of the next higher bit. The leftmost flip-flop is steered to 0. The MSB is, therefore, followed through the register by zeros. For this reason, it is not necessary to inhibit the clock line after shift out of the MSB.

In addition to just storing data, shift registers are used with other digital circuitry to build more complex networks. Figure 13-7 shows how shift registers and an adder are interconnected to build a binary multiplication circuit. Binary multiplication is a matter of shifting the multiplicand to the left and either adding or not adding the partial result, depending on the multiplier bits.

A four-bit universal shift register is available in TTL/MSI form. Figure 13-8 shows the logic diagram and symbol of a Fairchild 9300 universal shift register, which can be wired for all four modes of operation shown in Figure 13-1. The logic diagram indicates a construction of D flip-flops, but added gates and inverters have been used to alter its functions to conform to those of J-K flip-flops. For this reason, changes occur on the trailing edge of the clock pulse.

Figure 13-9 shows a universal shift register wired to produce serial-to-parallel conversion. The waveform for this circuit will the the same as those shown in Figure 13-5(b). Note that an inverter is not needed between J and K. The $\overline{\text{parallel enable}}$ must have a 1 level applied to it. Figure 13-10 shows a universal shift register wired for parallel-to-serial conversion. It differs from the J-K register of Figure 13-6 in that the parallel inputs are not direct set inputs; they must be enabled by a 0 level on the $\overline{\text{parallel enable}}$. This applies them to take the D inputs of the flip-flops, and one clock pulse is needed to enter data bits into the register flip-flops. Shift of the register is inhibited until the $\overline{\text{parallel enable}}$ goes high. Figure 13-10(b) shows the waveforms needed to load the number 1011 (11) in parallel and shift it out in serial.

Both J-K and D-type flip-flops are available for register construction in MOS circuits. Figures 13-13 and 13-14 show typical CMOS integrated flip-flops. Figure 13-14 shows an eight-bit TTL/MSI shift register. It is designed primarily for serial output operation, as outputs are supplied for only the rightmost three flip-flops—permitting its use as a six-, seven-, or eight-bit register.

There are some applications for a register that can shift data bits both to the left and to the right (called right-shift-left-shift registers or shift-right-shift-left registers). This is accomplished by having the steer inputs to each flip-flop applied through a set of gates, as shown in the logic diagram of Figure 13-15. For shift left, the gates enable the steer to come from the flip-flop on the right. For shift right, the gates enable the steer to come from the flip-flop on the left. Figure 13-16 shows connection that will allow shift in both directions. A 1 on the mode control produces shift to the left; a 0 on the mode control produces shift to the right.

Glossary

Shift Register: A register that can shift the data bits stored in it. If it is wired for a shift from left to right, on application of a clock pulse, the level stored in each flip-flop will be transferred to the adjacent flip-flop on its right.

Parallel-to-Serial Converter: A register into which data are entered in parallel form and out of which the data are shifted in serial form, after application of a series of clock pulses. See Figure 13-6.

Serial-to-Parallel Converter: A register into which data are shifted in serial form and out of which the data are taken in parallel form. On completion of shift-in, the data are available from the register Q leads as a parallel data word. See Figure 13-5.

Questions

1. Describe four modes of storage register operation.
2. Why is a set-reset flip-flop not sufficient for shift register operation?
3. Can the D latch of Figure 12-23 be used for shift register application?
4. Can a J-K flip-flop provide shift left as well as shift right?
5. In a parallel-to-serial converter, what levels remain in the register after the data have been shifted out?
6. A serial-to-parallel converter has eight-bit capacity. How many clock pulses are needed to shift in the data?
7. An eight-bit parallel-to-serial converter is composed of two TTL/MSI universal shift registers. How many clock pulses are needed to load and shift out the data?
8. Draw the diagram of the register of Question 7.
9. Draw the timing diagram for the register of Question 7 if a data word of 10110011 is entered and read out with succeeding clock pulses.
10. Explain the need for the clock inhibit shown in Figure 13-5.
11. In Figure 13-9, could the $\overline{\text{clock inhibit}}$ gate be eliminated by using the $\overline{\text{parallel enable}}$?
12. If the register of Figure 13-5(a) were composed of D flip-flops, what difference would the output waveforms have in comparison to those of Figure 13-5(b)?
13. Explain the difference between a dynamic and a static MOS shift register.
14. Show how two CMOS 4014B eight-stage static shift registers can be connected to form a 15-bit serial-in-serial-out register.
15. Explain how a shift register can be connected to give both left shift and right shift. Draw the logic diagram of one set of gates needed to change the steer of one flip-flop in the register.
16. Draw the logic diagram of an eight-bit right-shift-left-shift register. Use two TTL/MSI 7495 circuits.

Problems

13-1 Draw an eight-bit shift right register using the TTL 7474 D flip-flop.

13-2 Expand the serial-to-parallel register of Figure 13-5 to six bits. Draw the logic diagram.

13-3 Expand the parallel-to-serial register of Figure 13-6 to six bits. Draw the logic diagram.

13-4 Draw the shift register of Figure 13-6 using the D flip-flops SCL 4013A shown in Figure 13-13. What changes, if any, must be made in the timing diagram of Figure 13-6?

13-5 Draw the shift register of Figure 13-5 using the SCL 4014B-type flip-flop shown in Figure 13-14. Correct the timing diagram of Figure 13-5 to account for any difference in the two registers.

13-6 Show the connections to wire the TTL/MSI 7495 shown in Figure 13-15 for serial input on the left, shift right, and parallel output.

13-7 The circuits of Figure 13-17 are two MSI 9300 universal shift registers. Draw the connections needed to form an eight-bit shift register capable of serial-to-parallel conversion. Label all inputs and outputs.

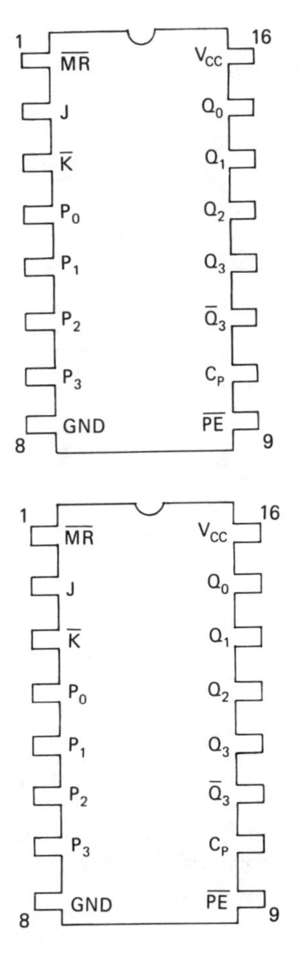

FIGURE 13-17. Problem 13-7

13-8 The voltage waveforms of Figure 13-18 are inputs to the register of Figure 13-9. Draw the resulting output waveforms.

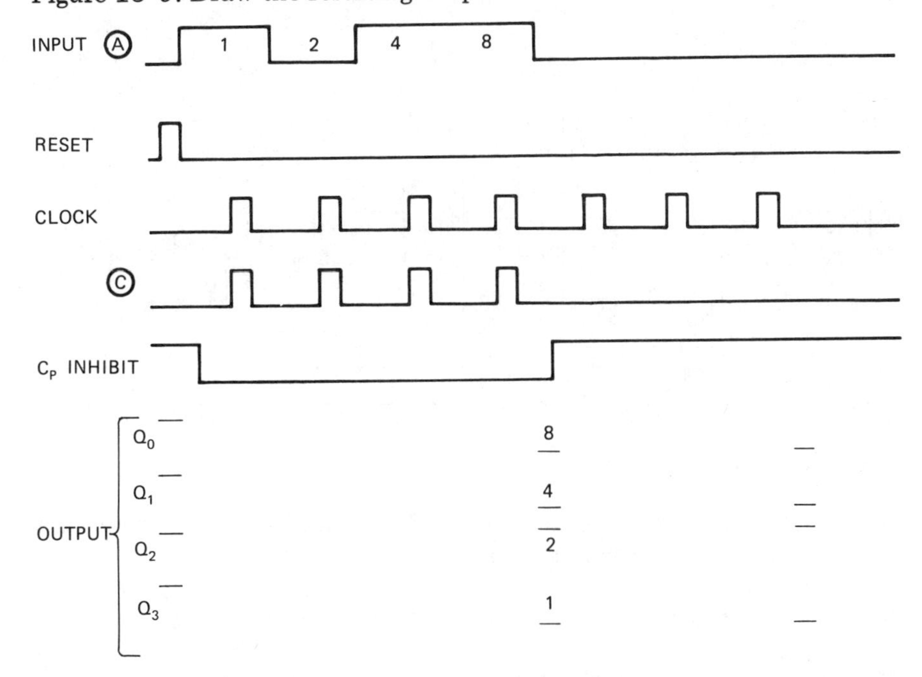

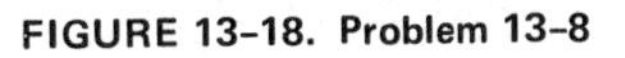
FIGURE 13-18. Problem 13-8

13-9 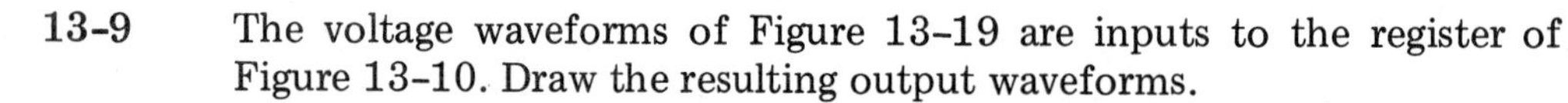
The voltage waveforms of Figure 13-19 are inputs to the register of Figure 13-10. Draw the resulting output waveforms.

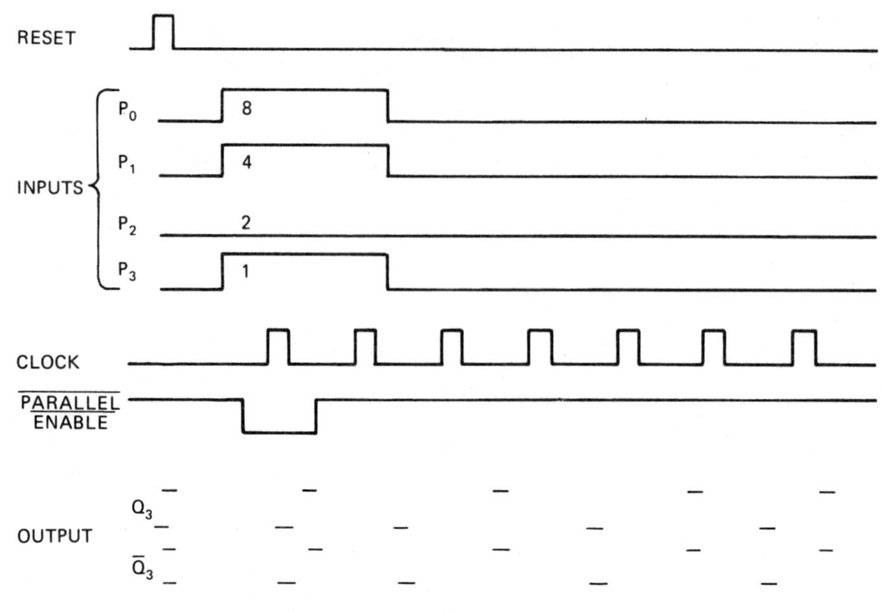

FIGURE 13-19. Problem 13-9

13-10 The 7473 dual J-K flip-flop has no preset inputs, and it appears that it cannot be used for parallel-to-serial conversion. Using two NAND gates per flip-flop, draw a four-bit register that will toggle preset in parallel, then shift out in serial.

13-11 The circuits in Figure 13-20 are D flip-flops TTL/SSI 7474. Draw in the connections needed to provide a four-bit serial-in-serial-out register.

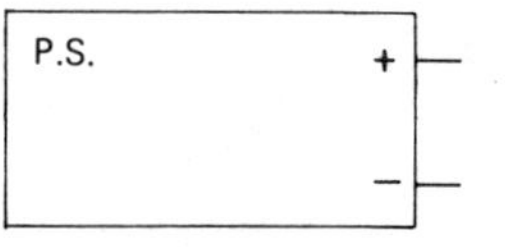

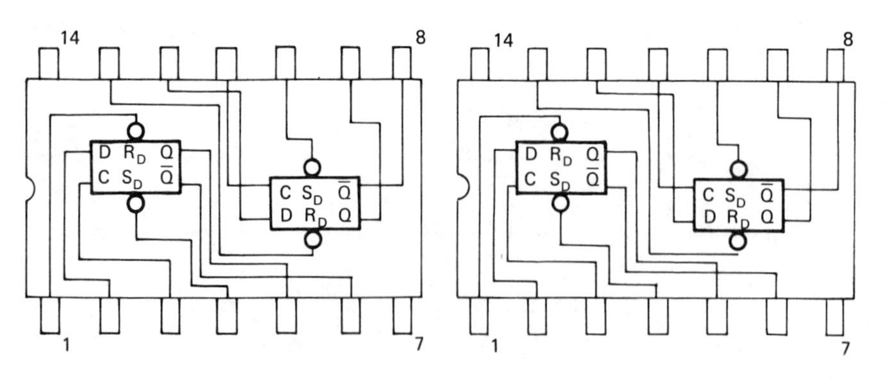

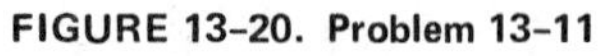
FIGURE 13-20. Problem 13-11

13-12 If the initial state in the shift register of Figure 13-3 is 110, what data will be stored in the register after two clock cycles?

13-13 Repeat Problem 13-12 if the J input is 1 and the K input is 0.

13-14 In Figure 13-4(a), consider that the $\overline{Q}$ output, instead of the Q output, is connected to the D input. What binary numbers will be in the register after four clock pulses? (Remember that a binary number stored in a flip-flop is the logic level at the Q output.)

13-15 Consider the initial state in the X register in Figure 13-7(c) to be 0111. Determine the data stored in the accumulator after each clock pulse.

CHAPTER

Counter Circuits

Objectives

Upon completion of this chapter, you will be able to:

- Assemble toggle flip-flops to form an asynchronous digital counter.
- Connect decoding circuitry to counter outputs.
- Assemble circuits that will stop the count at a given number.
- Wire the TTL divide-by-N ripple counter for counts of 0 through 15 or lower.
- Assemble one or more TTL ripple counters for counts higher than 0 through 15.
- Determine the upper frequency limit of a ripple counter.
- Assemble flip-flops to form a synchronous counter.
- Steer a synchronous counter back to 0 after the desired count has been reached.
- Assemble decimal counting units to form a digital counter.
- Measure pulse width and period with a digital counter.
- Use the D flip-flop to assemble a synchronous counter.
- Assemble both synchronous and asynchronous down counters.
- Use synchronous up/down IC counters.
- Compare timing parameters of CMOS counters with those of TTL counters.

14.1 Introduction

Operation of digital machines usually demands a circuit that can count electric pulses, time intervals, or events per unit time. Exact timing between machine functions can be accomplished by using a counter to count an exact number of clock pulses between the end of one function and the start of another. Most counters are made by the interconnection of toggle, D-type, or J-K flip-flops, discussed in Chapter 12.

Each flip-flop in a counter has two possible states, 1 and 0 (set and reset). The maximum number of possible combinations of 1 and 0 states for a counter is called the *modulus.* A counter composed of N flip-flops cannot have a modulus higher than 2^N. Additional logic circuits used with the counter can reduce the modulus to a value lower than 2^N. In the majority of applications, we must allow for a 0 state of the counter. Therefore, the maximum count of a counter composed of N flip-flops is $(2^N - 1)$, 1 less than its modulus. If the count of 0 must be included, a counter with three flip-flops can count no higher than 7. In this explanation, the counters described will be counting clock pulses. This is for convenience of description. In actual use, they may be counting other waveforms, either periodic or random.

14.2 Ripple Counter (Trigger Counter)

Two flip-flops connected as in Figure 14-1 can count from 0 through 3. The first flip-flop will change state on the trailing edge of each clock pulse. The second flip-flop will change state every time the first flip-flop changes from 1 to 0. Figure 14-2 shows the output waveforms of the two flip-flops in comparison to the clock pulse line. The AND gates are not an integral part of the counter itself, but act as decoder gates, similar to the binary-to-decimal decoder described in Section 6.10. The flip-flops themselves have weighted values of 1 and 2. The four states of the counter are indicated by the AND functions:

$$\bar{1} \cdot \bar{2} = 0$$

$$1 \cdot \bar{2} = 1$$

$$\bar{1} \cdot 2 = 2$$

$$1 \cdot 2 = 3$$

In the example of Section 6.9, the 1 and 2 input lines have weighted values, as we have given our two flip-flops the weighted values 1 and 2.

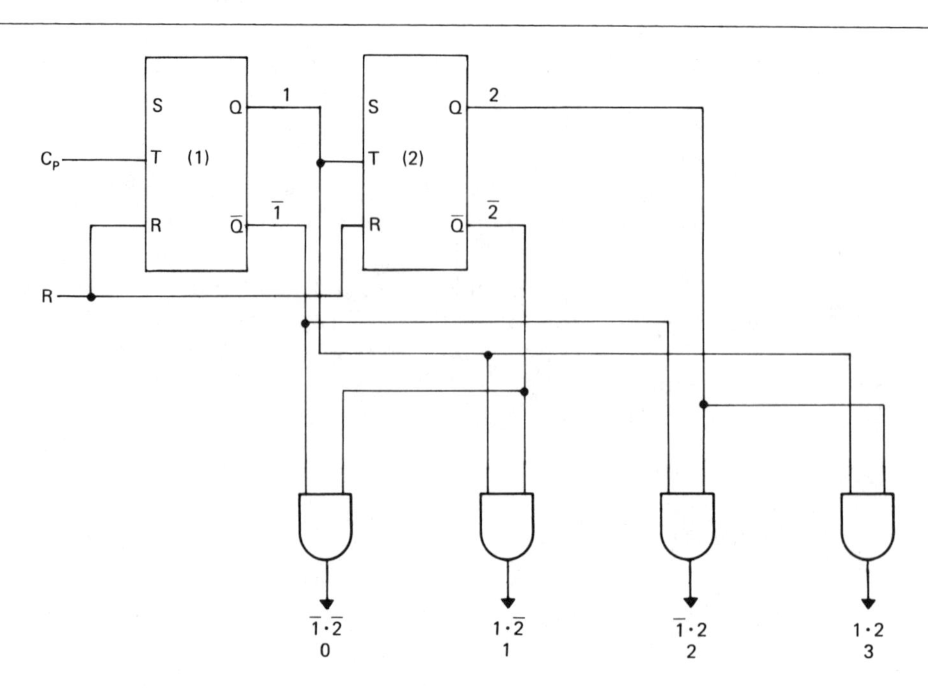

FIGURE 14–1

Two toggle flip-flops connected to form a divide-by-4 ripple counter. Dual-input AND gates decode a count of 0 through 3. The flip-flops are assigned weighted values of 1 and 2.

In Section 6.9, it was necessary to connect inverters on the lines to obtain the complements (not functions), but the $\overline{Q}$ outputs of the flip-flop provide the not function for us. As Figure 14-2 shows, if the clock pulses continue beyond number 3, the counter will continue to count. The 0 gate will go high every fourth pulse.

For the two-flip-flop counter, a table can be drawn comparing the states of the flip-flops with the count, as in Figure 14-3(a). The size of a ripple counter can be increased by merely adding another toggle flip-flop, as in Figure 14-3(b). This third flip-flop has a weighted value of 4, and, as the table shows, it can extend the count to 0 through 7. Decoding of this counter can be accomplished with one three-input AND gate equivalent to each line of the count table of Figure 14-3(c). Decoding with a single gate per count requires that the gates have a number of inputs equal to the number of flip-flops in the counter. NOR gates are equally convenient for decoding, as complementary outputs are available from each flip-flop. In using counters, it is not always necessary to decode all states. In many cases, we are interested only in the last count. When that count is decoded, the counter

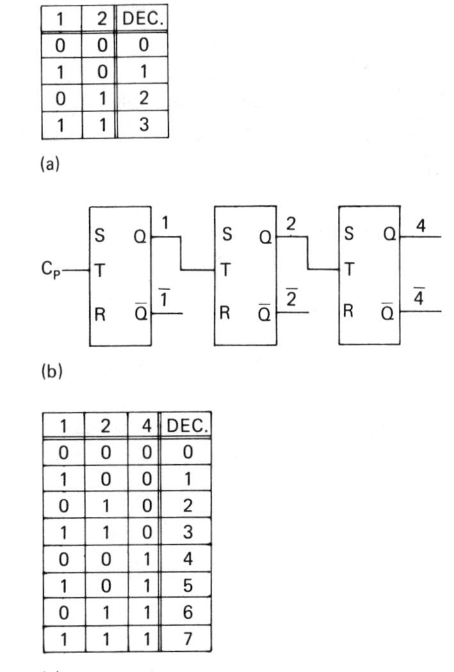

1	2	DEC.
0	0	0
1	0	1
0	1	2
1	1	3

(a)

(b)

1	2	4	DEC.
0	0	0	0
1	0	0	1
0	1	0	2
1	1	0	3
0	0	1	4
1	0	1	5
0	1	1	6
1	1	1	7

(c)

FIGURE 14-3

(a) Truth table for a two-flip-flop ripple counter. (b) Three-bit ripple counter. (c) Truth table of a three-bit ripple counter showing count of 0 through 7.

FIGURE 14-2

Output waveforms of the 1 and 2 flip-flops and the four decoder gate outputs. The 0 output occurs after every fourth clock pulse.

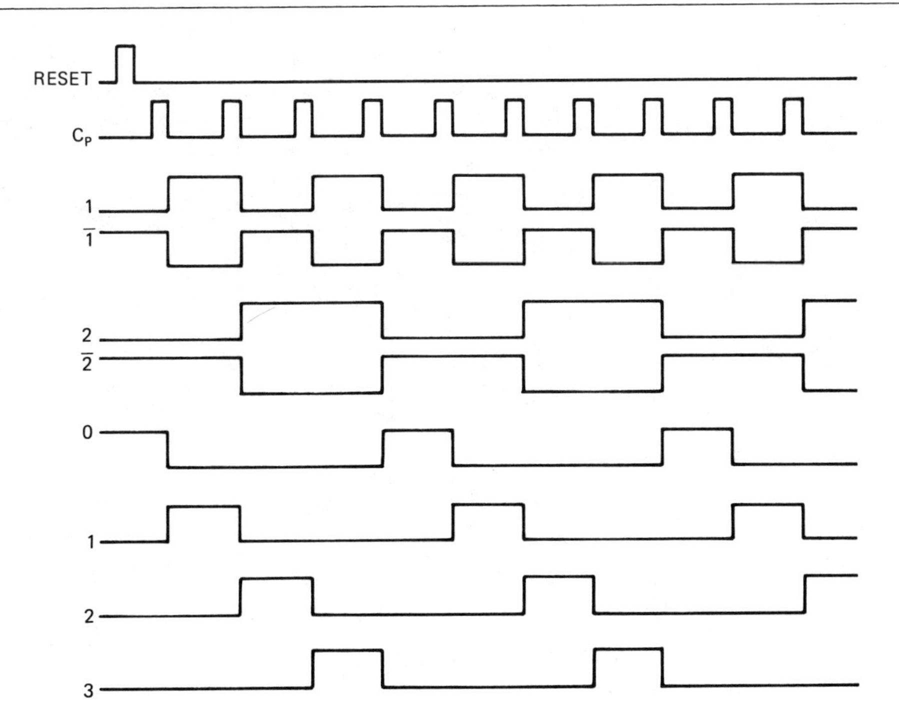

may be reset or inhibited until a new count is needed.

The ripple counters, as shown in Figures 14-1 and 14-3, can be expanded to give us counts of 0 to (2^N - 1), where N is the number of flip-flops in the counter. Although the number of states in counting is 2^N, one of these states is 0 and therefore the highest count is (2^N - 1). This provides counters with the maximum count of 3, 7, 15, 31, 63, 127, and higher (2^N - 1) values.

To provide counters that can count to levels between these (2^N - 1) values, it is practical in most applications to decode one count above the maximum and use this decoded output to reset the counter. This is a simple approach, but for some logic units, it might create a race problem in that the reset level is removed by resetting the counter. Figure 14-4(a) shows a ripple counter used to count to 5. Having three flip-flops, it is capable of a count of 0 through (2^3 - 1) or 0 to 7, but at the count of 6 it is reset. Therefore, the count of 6 lasts only long enough to reset the counter back to 0. Figure 14-4(b) shows the waveforms of the flip-flop and gate outputs. The decoded 6 shows a pulse of very short duration. The 1 flip-flop is in no way affected by the reset procedure. The 2 flip-flop, however, contains a spike at the beginning of each 0 count. The 4 flip-flop remains high for a short time into the 0 period. There are few applications for which these minor imperfections

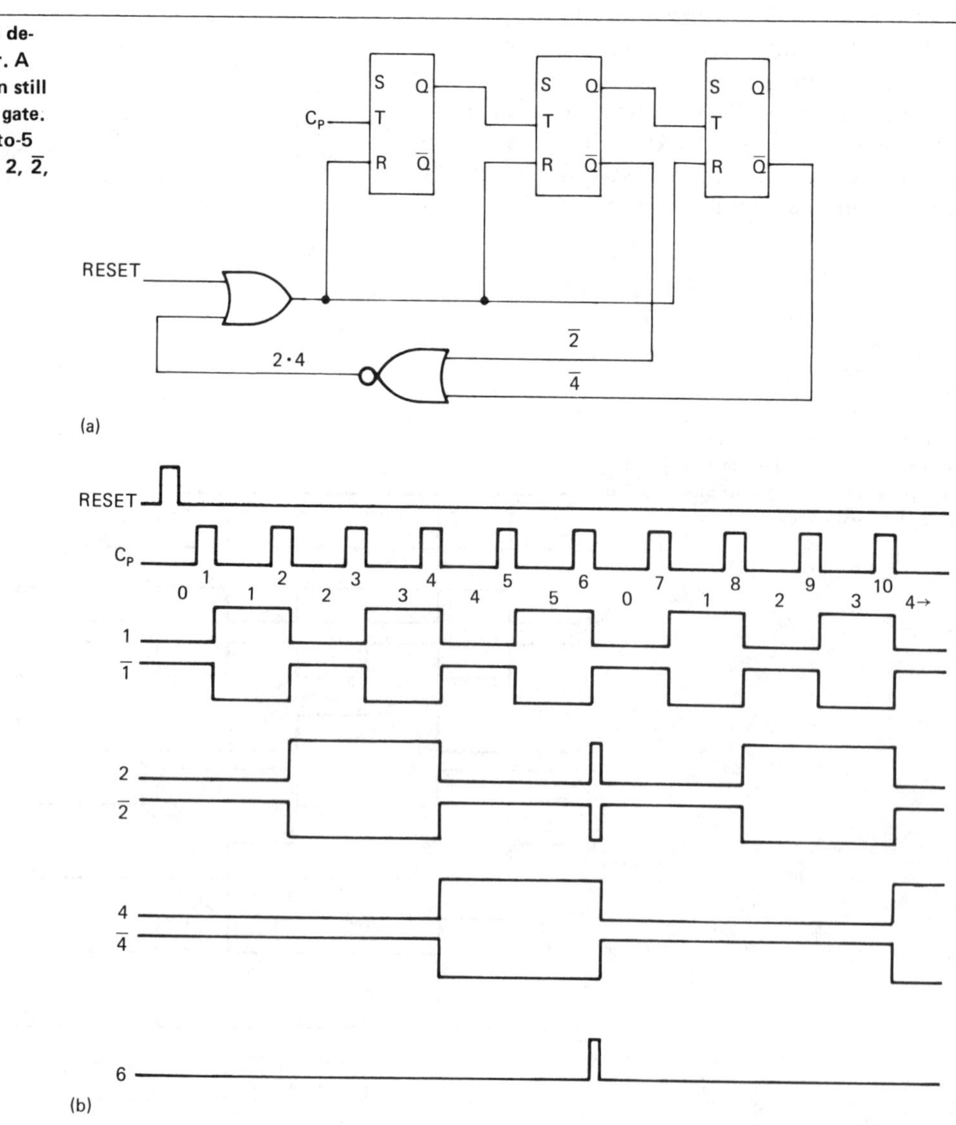

FIGURE 14-4
(a) Counting to 5 by using a decoded 6 to reset the counter. A master or power on reset can still be included by using an OR gate. (b) Output waveforms of 0-to-5 counter showing short pulse 2, $\overline{2}$, and decoded 6; 4 and $\overline{4}$ are slightly extended into the 0 period.

will create a difficulty. If the counter of Figure 14-4 need only count to 5 and then could wait for a reset generated elsewhere in the system, a decoded 5 could be used to inhibit further counts above 5, as in Figure 14-5.

14.3 Divide-by-N Ripple Counter

One of the many available MSI circuit ripple counters in the 14-pin four-bit binary counter 74LS293, shown in Figure 14-6. If the A_{OUT} is connected to the B_{IN}, the counter will divide by 16 or count through 0 to 15 states. The reset inputs show an inversion symbol telling us that a 0 is needed at the output of the NAND gate for a reset. This occurs only when both inputs are high. With the use of some external logic gates, the count can be changed to divide by any number lower than 16. Figure 14-7 shows a connection that provides a (divide-by-10) 0-through-9 count. This is accomplished by connecting the 8 and 2 outputs (D_{OUT} and B_{OUT}) to the reset NAND gate inputs. At the count of 10, two 1s at the NAND gate inputs reset the counter to 0. If a master reset is needed, OR gates may be used on the reset inputs. For a divide by N lower than 7, the first flip-flop can be excluded by connecting the clock input to B_{IN}.

Increasing the count beyond 15 can be accomplished by connecting two circuits in series. Figure 14-8 shows two four-bit binary ripple counters connected to obtain a divide by 256 (2^8), a count of 0 to 255.

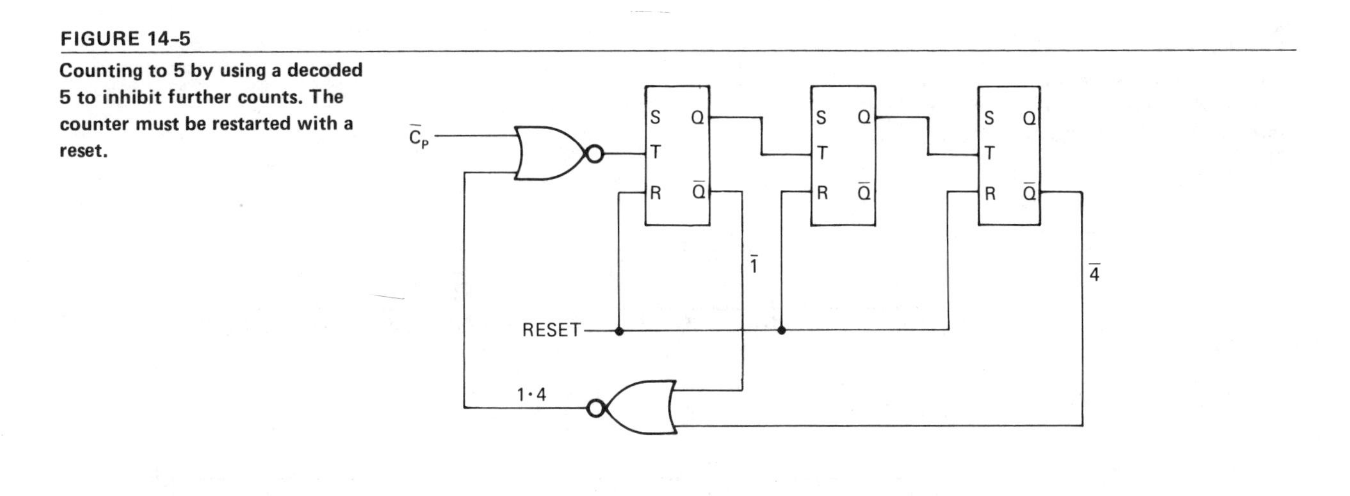

FIGURE 14-5
Counting to 5 by using a decoded 5 to inhibit further counts. The counter must be restarted with a reset.

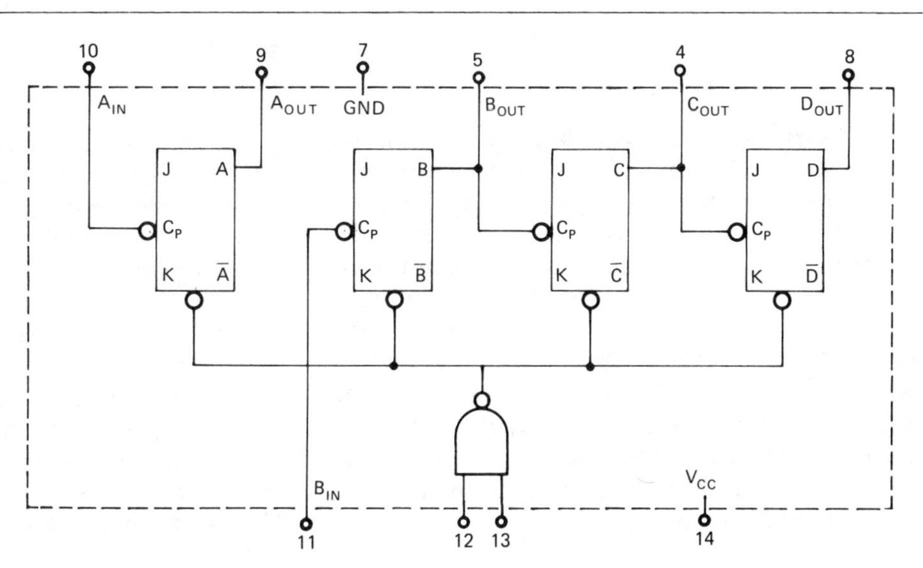

FIGURE 14-6
74LS293, a 14-pin MSI circuit divide-by-N ripple counter connected for a count of 0 through 15 or lower. The J and K leads of each flip-flop are understood to be connected for a toggle (both J and K high).

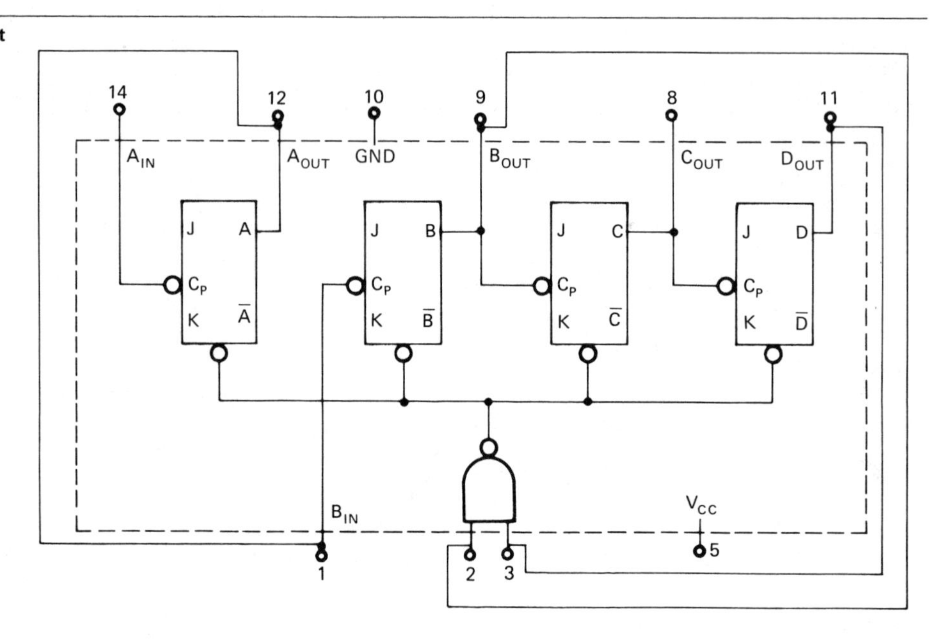

FIGURE 14-7

7493 divide-by-N integrated circuit ripple counter connected for a count of 0 through 9 by using the internal reset NAND gate to decode 10 for a reset. (*Note:* The 74LS293 and the 7493 are logically the same but differ by pinout and power ratings.)

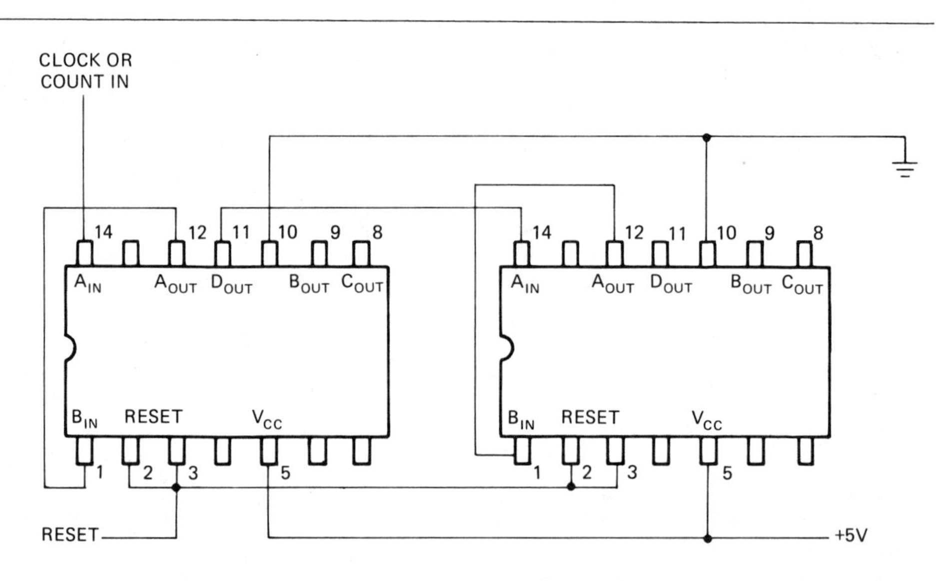

FIGURE 14-8

Two divide-by-N ripple counters connected for a count of 0 through (2^8 - 1), or 0 through 255.

A serious problem with large ripple counters is the ripple delay time. This is the time required for a change in state to ripple through from the input of the LSB to a change in state of the MSB output. This is determined by a summation of the typical propagation delay of a flip-flop times the number of flip-flops in the counter. By manufacturers' specifications, turn-on and turn-off delays of a TTL flip-flop are 75 nanoseconds typical, 135 nanoseconds maximum. As the count ripples through the eight flip-flops, a total of eight time delays have added up in the toggling of the final flip-flop.

Figure 14-9 shows a clock line of pulses 0.5 microsecond apart (2 MHz). With these pulses connected to the clock-in, the final flip-flop (128) would not change state until 0.6 microsecond after the trailing edge of the one hundred and twenty-eighth clock pulse. This is a delay of more than one clock period. The amount of delay in terms of clock periods becomes even greater at higher clock frequencies and for counters with more flip-flops. Delays of this magnitude might easily disrupt operation of the digital system. The solution to this problem is use of a synchronous counter.

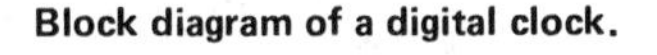

FIGURE 14-9

Timing diagram showing the effect of ripple delay time. The sum of the time delays in the toggling of eight successive flip-flops results in a count of 128 being more than a clock pulse late.

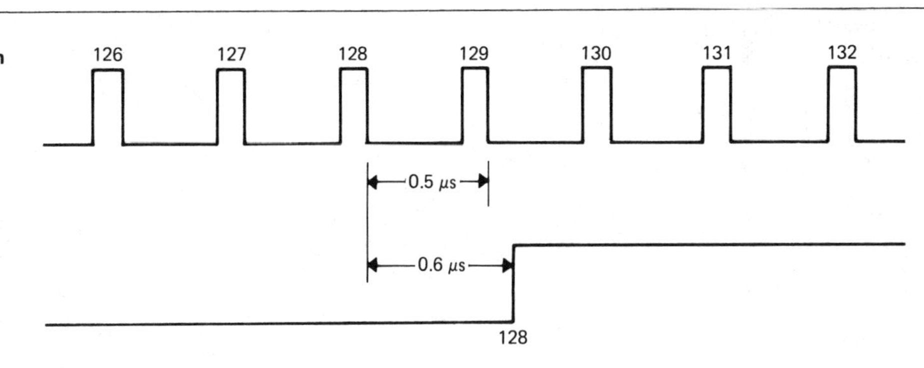

FIGURE 14-10

Block diagram of a digital clock.

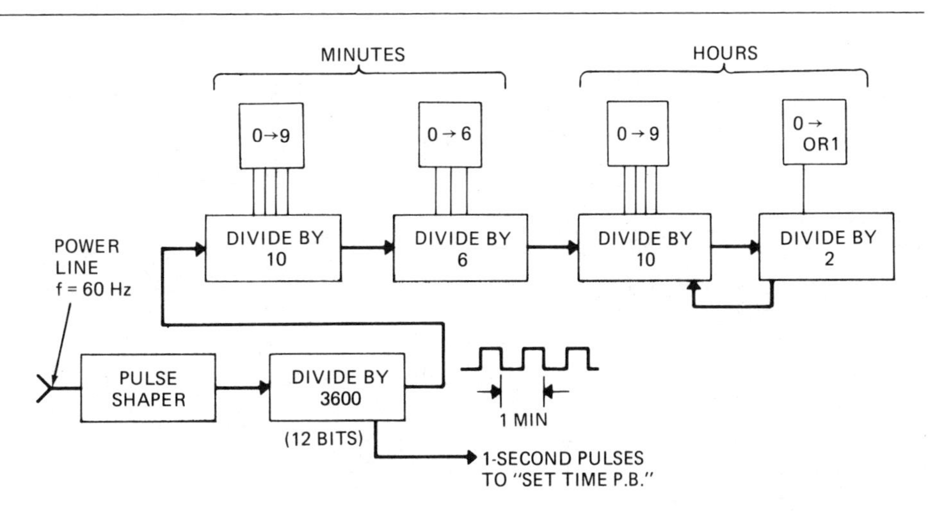

14.4 Digital Clock

A set of ripple counters arranged as in Figure 14-10 can provide a digital clock. A 12-bit section that resets at 3600 will divide the 60 cycles per second from the power line into one-minute pulses. A decade counter with a divide-by-6 counter provides the minute readout. Another decade and a single flip-flop to divide by 2 provides the hour readout.

There is an interesting problem with the hours section in that the LSD of the hour section must reset to 0 both at 10 o'clock and 12 o'clock. This can be handled by decoding 10 on the hours LSD section, using it to reset the LSD section only. This will trigger the single-bit MSD section to a 1, resulting in 10 o'clock. An AND function between the MSD 1 and the LSD $1 \cdot 2$ will generate a reset at 1300. This reset, however, should allow the LSD 1 to remain high, resetting only the LSD 2 and the MSD of the hours section. Another solution is use of a divide-by-12 counter for the hours section. This simplifies the reset problem but complicates the decoding.

Setting the time can be accomplished in several ways. Bringing a one-second gate out of the divide-by-3600 section and ORing it to the clock inputs of the minute and hour sections through separate push-buttons would be a simple solution.

FIGURE 14-11

Count table showing conditions that must exist before the toggling of each successive J-K flip-flop in the counter. These conditions, if decoded, can be used to steer the flip-flop to a toggle on the correct count pulse.

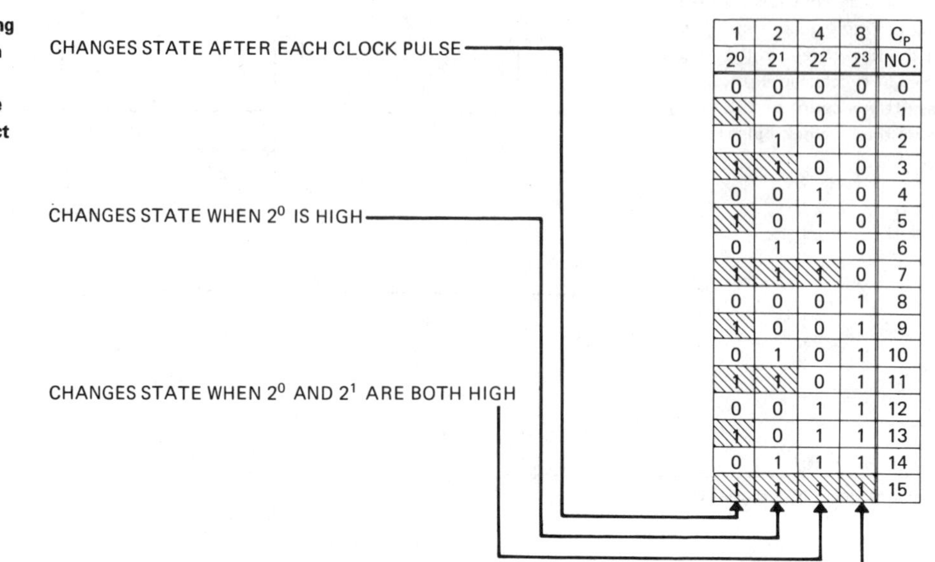

1 2^0	2 2^1	4 2^2	8 2^3	C_P NO.
0	0	0	0	0
1	0	0	0	1
0	1	0	0	2
1	1	0	0	3
0	0	1	0	4
1	0	1	0	5
0	1	1	0	6
1	1	1	0	7
0	0	0	1	8
1	0	0	1	9
0	1	0	1	10
1	1	0	1	11
0	0	1	1	12
1	0	1	1	13
0	1	1	1	14
1	1	1	1	15

14.5 Synchronous Counter

The delay difficulties of the ripple counter can be avoided if the J-K flip-flop is used and steered in a fashion to permit each flip-flop to be clocked by the same clock line. Figure 14-11 is a count table for a four-bit counter with each flip-flop having the weighted value shown at the top of the column. The first flip-flop is 1 for all odd numbers and 0 for all even numbers of the count. Therefore, as shown in the partial count table of Figure 14-12(b), it changes state on each clock pulse. This is a toggle and can be accomplished by merely placing a 1 level on both J and K of the first flip-flop, as shown in Figure 14-12(a).

The second or 2^1 flip-flop must toggle on the clock pulse after 2^0 is 1. For this we need only connect the 1 output of 2^0 flip-flop to both J and K of 2^1 flip-flop, as in Figure 14-13. During the count of 3, both 2^0 and 2^1 are in the 1 state. If these are decoded and used to steer the 2^2 flip-flop to toggle, it will toggle the 2^2 flip-flop to 1 on the fourth clock pulse, as Figure 14-14 shows.

The three flip-flops, as connected in Figure 14-14, will provide a count to 7. At 7, the 2^3 flip-flop should be steered to toggle. On the trailing edge of clock pulse 8, this flip-flop will toggle to the 1 state, giving a count of 8. Figure 14-15 shows the complete four-bit counter. There is no race problem in steering the J-K flip-flop, as the steer voltage has its effect starting

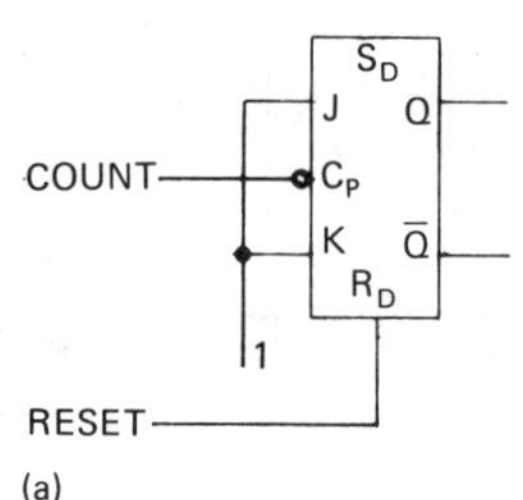

(a)

C_P NO.	2^0
0	0
1	1
2	0
3	1
4	0
5	1
6	0
7	1

▨ CONDITION THAT SHOULD STEER THE 2^1 FLIP-FLOP TO TOGGLE ON THE NEXT CLOCK PULSE

(b)

FIGURE 14-12

(a) First flip-flop of a synchronous counter set to toggle by a fixed 1 level on the J-K. The 1 output can be used to steer the next flip-flop. (b) Count table.

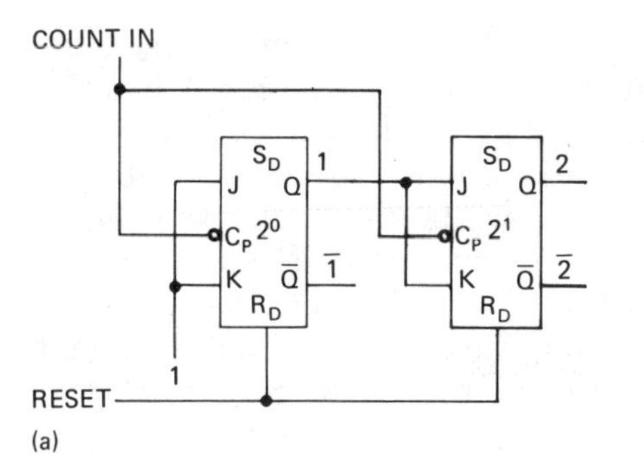

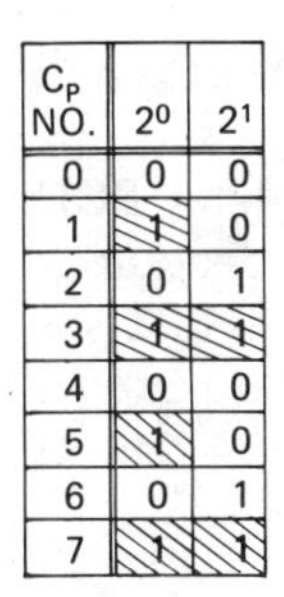

C_P NO.	2^0	2^1
0	0	0
1	1	0
2	0	1
3	1	1
4	0	0
5	1	0
6	0	1
7	1	1

CONDITION THAT SHOULD STEER THE NEXT HIGHER FLIP-FLOP TO TOGGLE ON THE NEXT CLOCK PULSE

(b)

FIGURE 14–13

(a) First two flip-flops of a synchronous counter. A double 1 can be decoded to provide a steer to toggle the third register on pulses 4 and 8. (b) Count table.

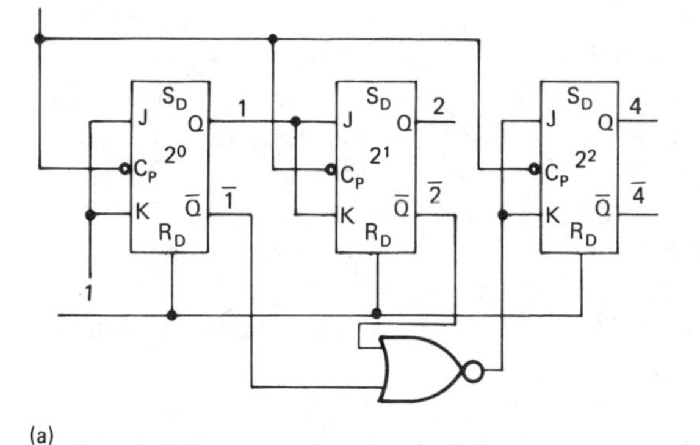

(a)

C_P NO.	2^0	2^1	2^2
0	0	0	0
1	1	0	0
2	0	1	0
3	1	1	0
4	0	0	1
5	1	0	1
6	0	1	1
7	1	1	1

CONDITION THAT MUST STEER THE NEXT HIGHER FLIP-FLOP TO TOGGLE ON THE NEXT CLOCK PULSE

(b)

FIGURE 14–14

(a) First three registers of a synchronous counter, showing NOR gate used to decode and steer the third register. (b) Count table.

with the leading edge of the clock pulse, even though the change itself does not occur until the trailing edge. As a binary counter, this counter has the advantage of having all flip-flops change on the trailing edge of the clock pulse, and there is no accumulation of delays between the LSB and the MSB of the counter, regardless of how many flip-flops are in the counter.

14.6 Binary-Coded Decimal Counter

The techniques just outlined can be used to develop counters of any size. Adding a flip-flop and the necessary gates to steer it can expand

FIGURE 14–15

Complete four-bit synchronous counter. The fourth flip-flop is steered to toggle by an AND function of 1 • 2 • 4.

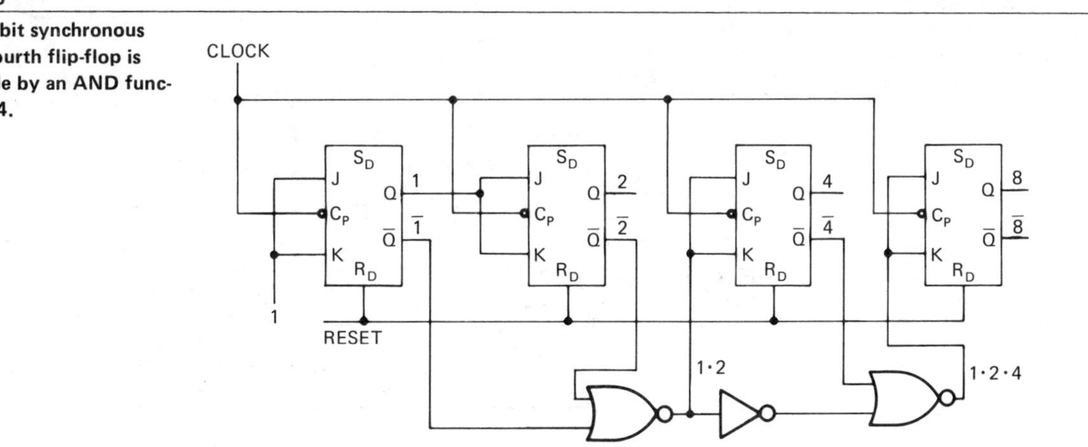

the count in increments of $(2^N - 1)$, where N is the number of flip-flops in the counter.

Unlike the ripple counter, however, the synchronous counter can be steered back to 0 at the end of any count without using the reset and without danger of a race problem. This can be seen with the BCD counter, which must have four flip-flops, but, instead of cycling back to 0 at $(2^4 - 1)$ or 15, it must cycle back to 0 after the count of 9.

This counter, with a modulus of 10, is often called a *decade counter.* The term *decimal counting unit* (DCU) is also used to describe a counter that counts 0 to 9. Figure 14-16 shows the binary counter of Figure 14-15 except that one NOR and one OR gate have been added. Neither of these gates has changed the operation of the binary counter. They do, however, make it possible for us to inhibit the toggling of the 2^1

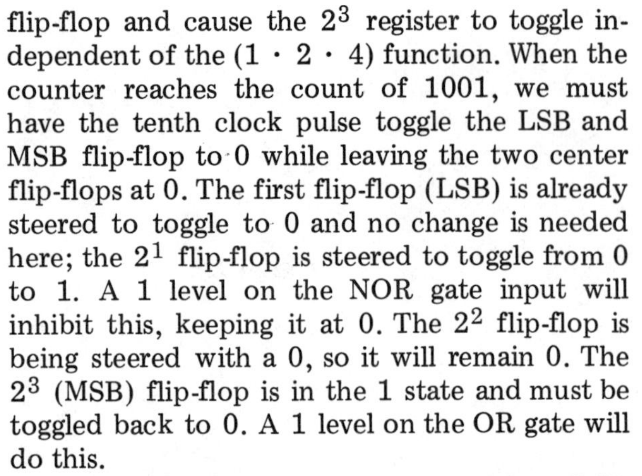

flip-flop and cause the 2^3 register to toggle independent of the $(1 \cdot 2 \cdot 4)$ function. When the counter reaches the count of 1001, we must have the tenth clock pulse toggle the LSB and MSB flip-flop to 0 while leaving the two center flip-flops at 0. The first flip-flop (LSB) is already steered to toggle to 0 and no change is needed here; the 2^1 flip-flop is steered to toggle from 0 to 1. A 1 level on the NOR gate input will inhibit this, keeping it at 0. The 2^2 flip-flop is being steered with a 0, so it will remain 0. The 2^3 (MSB) flip-flop is in the 1 state and must be toggled back to 0. A 1 level on the OR gate will do this.

Figure 14-17 shows the complete BCD counter. A single two-input gate can decode 9. This 1 level, which occurs during 1001, is applied to the added NOR and OR gate inputs that we left hanging loose in Figure 14-16. If

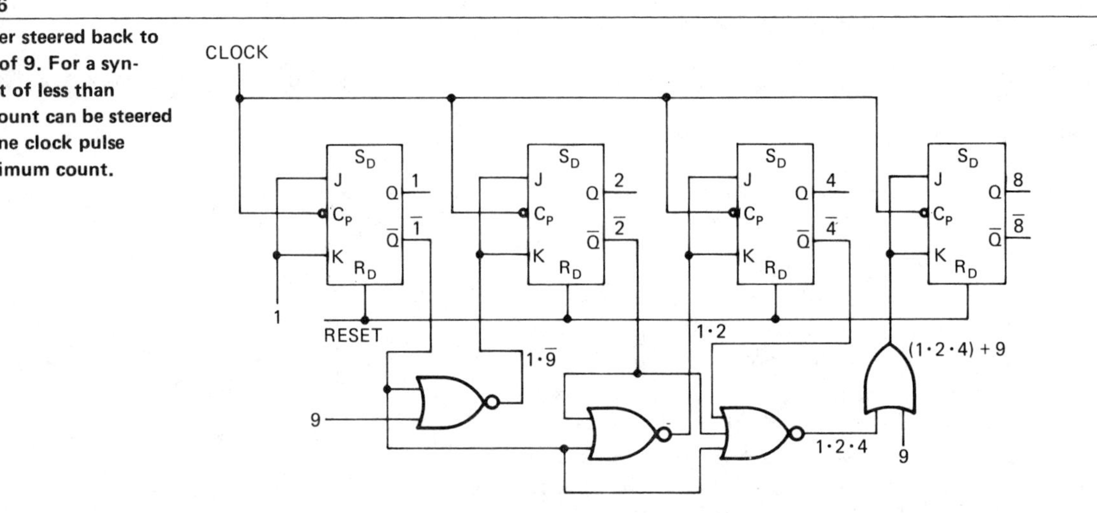

FIGURE 14-16

Four-bit counter steered back to 0 at the count of 9. For a synchronous count of less than $(2^N - 1)$, the count can be steered back to 0 on one clock pulse above the maximum count.

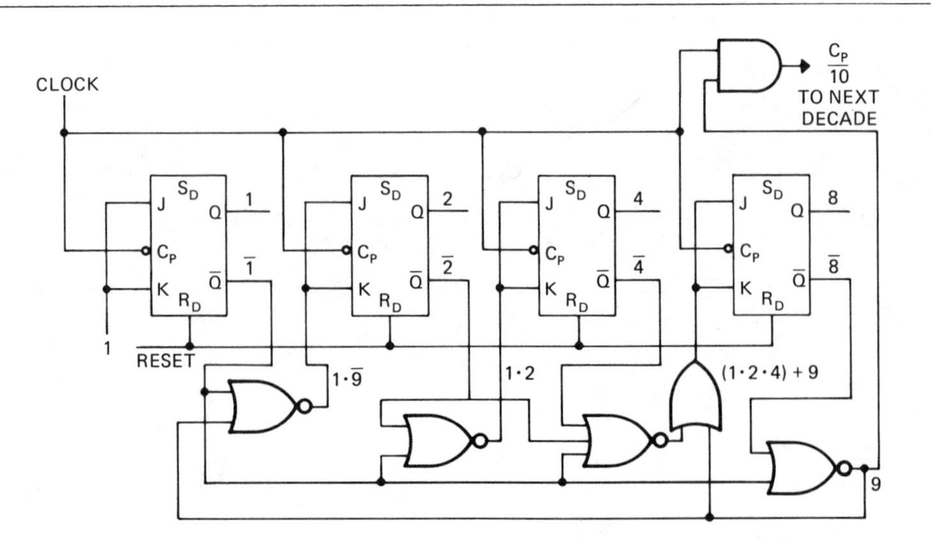

FIGURE 14-17

A synchronous (BCD) binary-coded decimal counter. The decoded 9 inhibits the 2 bit from toggling to 1 and causes the 8 bit to toggle to 0, resulting in a return to 0 after each count of 9.

the counter of Figure 14-17 is one decade in a multidigit decimal counter, the decode 9 may also be used to enable every tenth clock pulse to the clock line of the next decade.

14.7 D Flip-Flop Counter

Counters can be made from any form of flip-flop that can be wired to toggle (change state on receipt of a clock pulse). The J-K flip-flop is steered to toggle on receipt of a clock pulse by having both J and K inputs at the high logic level. The D flip-flop can be steered to toggle by merely connecting the $\overline{Q}$ output to D. Figure 14-18 shows a three-bit counter composed of D flip-flops. This counter will function like the three-bit counter of Figure 14-3 except that the change in count will occur on the leading edge of each clock pulse.

Figure 14-19 shows a synchronous version

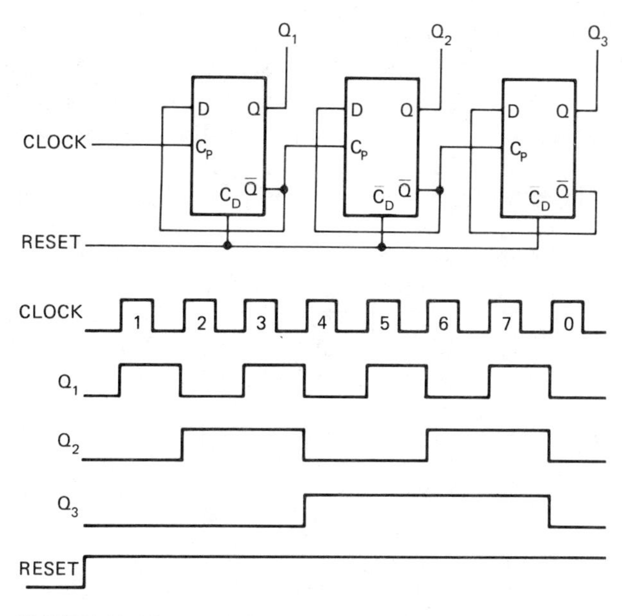

FIGURE 14-18

Three-bit asynchronous counter composed of D flip-flops. The D flip-flop is wired to toggle by interconnecting D and $\overline{Q}$. $\overline{Q}$ must be used for clock between the flip-flops.

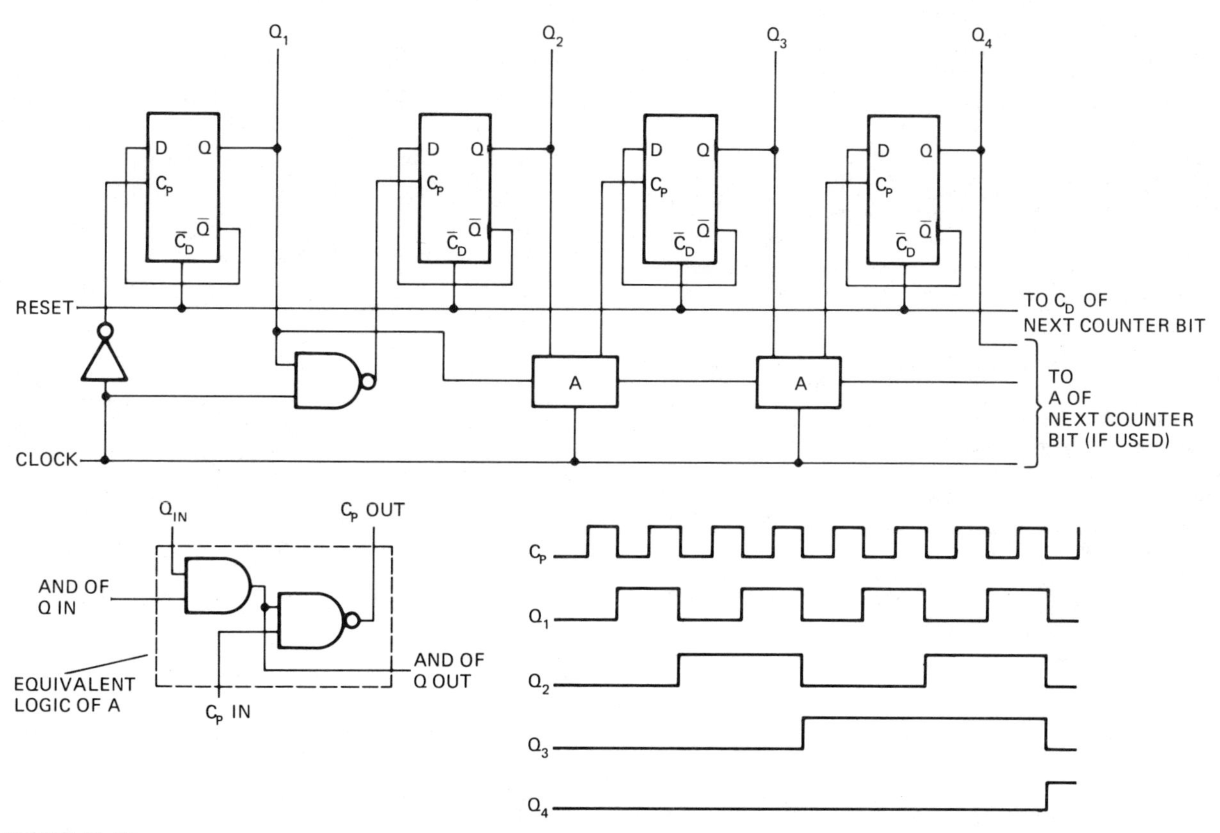

FIGURE 14-19

D flip-flop synchronous counter. An AND function of the Q outputs of all preceding counter bits is used to gate the clock pulse to the D flip-flop C_P inputs. The synchronous D flip-flop counter triggers on the trailing edge of the clock pulse.

of the D flip-flop counter. Its function is similar to the J-K counter of Figure 14-14 except that the clock pulse is gated to the flip-flop by the steer condition. When all lower-bit flip-flops are in the 1 state, the AND gate in A enables the clock through the NAND gate, toggling the next higher flip-flop.

COUNT	Q_2	Q_1	Q_0	$\overline{Q}_2$	$\overline{Q}_1$	$\overline{Q}_0$
0	0	0	0	1	1	1
1	0	0	1	1	1	0
2	0	1	0	1	0	1
3	0	1	1	1	0	0
4	1	0	0	0	1	1
5	1	0	1	0	1	0
6	1	1	0	0	0	1
7	1	1	1	0	0	0

(Hatched cells:) CONDITION THAT WILL CAUSE THE NEXT HIGHER BIT TO TOGGLE DURING A COUNT DOWN (DECREMENTING COUNT)

FIGURE 14-20

Truth table of three-bit binary count compared with its complement. The $\overline{Q}$ outputs count down (decrement), while the Q outputs count up (increment). The steer condition for toggling the next higher bit for a down counter exists when all lower bits are in 0 state.

14.8 Down Counter

In a digital circuit where a variable count is needed, a down counter with presettable inputs is often the best solution. The counter is preset to the desired count and decremented with each clock pulse until a count of 0 is detected. This method is usually easier than counting upward to a variable count. Counting up to a variable count requires a separate decode for each count used. Presetting to the count and decrementing to 0 requires only a decode of 0 to detect the end of count.

The difference between an incrementing counter and a decrementing counter can be a simple matter of using $\overline{Q}$ instead of Q as an output. Figure 14-20 shows a truth table that compares the Q outputs with the $\overline{Q}$ outputs for a three-bit counter. The same circuit that provides an incrementing count from the Q outputs provides a decrementing count from the $\overline{Q}$ outputs.

Figure 14-21 shows four D flip-flops connected to provide a down counter that is presettable for any count between 0 and 15. Note that the count in true form is forced onto the set inputs after reset. This provides a true output from the $\overline{Q}$ leads. A simple four-input NOR gate can be used to detect 0 (end of count). The 0 detect is decoded from the $\overline{Q}$ leads. The validity of this is seen in the fact that a count from a binary number (X) down to all bit 0s is equal to a count from $\overline{X}$ up to all bit 1s. This method can be used equally well with both synchronous and asynchronous counters.

FIGURE 14-21

Four-bit D flip-flops wired to form a down counter. A count applied to the S_D inputs after reset will set the counter to the complement of the applied number. The $\overline{Q}$ leads will provide an output that will decrement from the applied input value down to 0, at which time a high will appear at the end of count output.

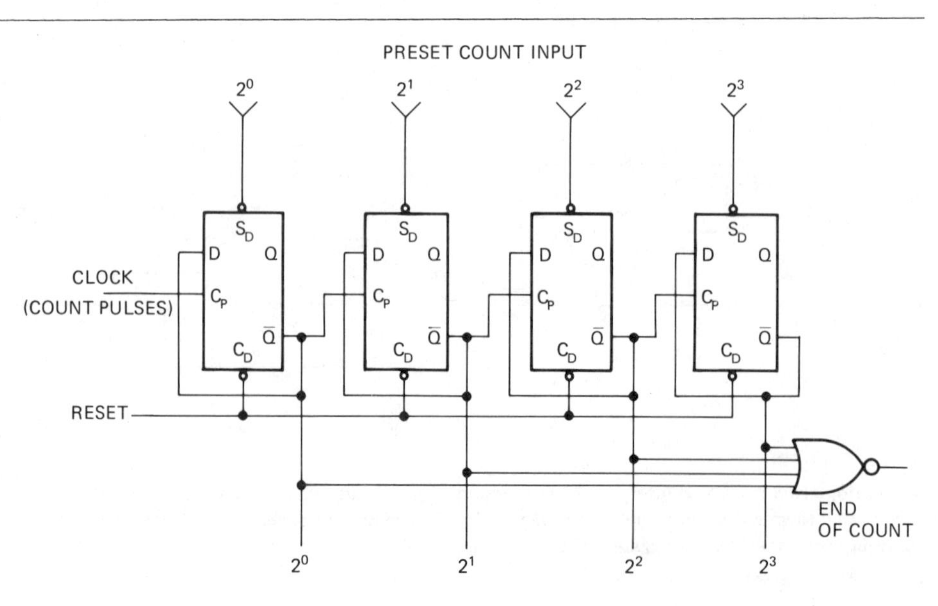

14.9 Up/Down Counter

Switching from an up count to a down count with a logic signal control can be accomplished more easily using a synchronous counter. Synchronous up/down counters are available in integrated circuit form. Figure 14-22 shows the 74LS169 synchronous up/down binary counter. As can be noted from the truth table in Figure 14-20, for both up and down counts, the LSB of the counter changes state with each count. The difference between up and down counts occurs in the conditions for which the higher

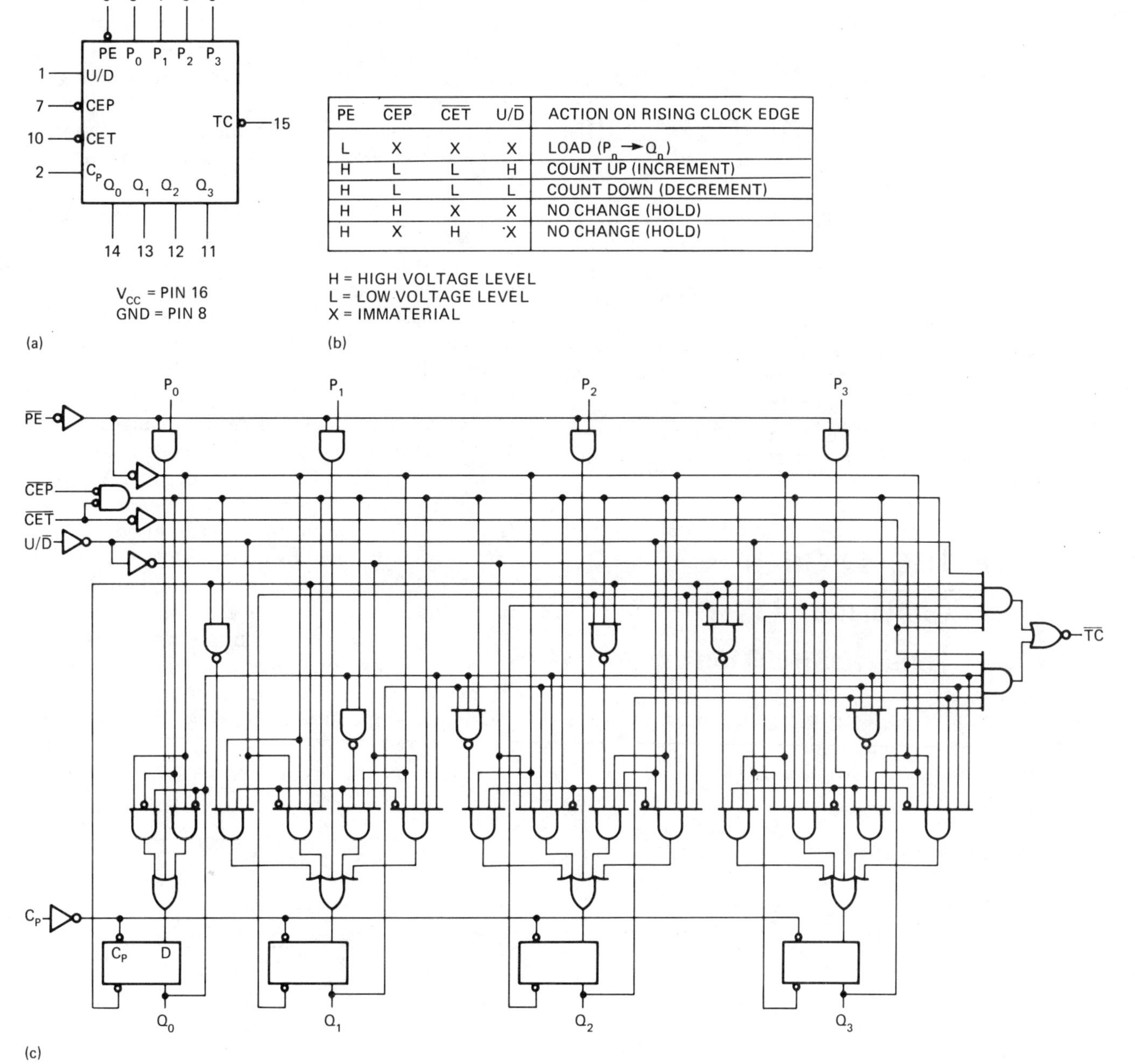

$\overline{PE}$	$\overline{CEP}$	$\overline{CET}$	$U/\overline{D}$	ACTION ON RISING CLOCK EDGE
L	X	X	X	LOAD ($P_n \rightarrow Q_n$)
H	L	L	H	COUNT UP (INCREMENT)
H	L	L	L	COUNT DOWN (DECREMENT)
H	H	X	X	NO CHANGE (HOLD)
H	X	H	X	NO CHANGE (HOLD)

FIGURE 14-22

74LS169 synchronous up/down binary counter. The D input of each flip-flop receives the output of an AND/OR select gate. For an up count, one AND gate of each higher-bit select gate causes the flip-flops to toggle only when all lower bits are 1. For a down count, one AND gate of each higher-bit select gate causes the flip-flops to toggle only when all lower bits are 0. (a) Logic symbol. (b) Mode select table. (c) Logic diagram.

bits change state. For an up count, the higher bits are steered to toggle only when all lower bits are in the 1 state. For a down count, the higher bits are steered to toggle only when all lower bits are in the 0 state. As shown by the logic in Figure 14-22, to switch from an up count to a down count, the AND/OR select logic switches the steer from 1 decoders to 0 decoders. A third gate in the AND/OR select logic switches the steer to synchronous preset inputs.

14.10 Timing Parameters of Counter Circuits

In addition to propagation delay parameters (clock edge to output delays), maximum clock frequency and minimum pulse widths are normally specified for counter circuits. The table in Figure 14-23 shows some typical timing parameters for counter circuits made in a variety of integrated circuit technologies. The CMOS counter has slower operating speed resulting in lower clock frequencies and requiring wider pulse widths. TTL counters are high-speed but consume more power.

14.11 The Digital Counter as a Test Instrument

The digital counter is a valuable item of electronic test equipment used in both digital and communication electronics. It can be employed to measure the frequency of sinewaves and rectangular and other waveforms. It can measure electrical events per unit time regardless of whether they are periodic or irregular in occurrence. As Figure 14-24 shows, the input is first converted to a digital rectangular form. The counter is reset and then a waveform exactly one second wide enables the conditioned f_x signal onto the clock line. Each decade of the counter is connected to a decimal readout. After a one-second-long count, further count is inhibited for a period of display time, which gives the operator sufficient time to observe the display. At the end of the display time, another reset and one-second-wide enable pulse provide another count.

The decade counters used to count the signal and provide the display are only about half the number of decades needed. The one-second-wide enable signal must be accurate to within a few microseconds. This kind of accuracy can be provided only by a crystal oscillator operating in a small thermostatically controlled oven to reduce the change in frequency, which would be caused by ambient temperature changes. Figure 14-25 shows the circuitry needed to count down the output of a one-megahertz crystal oscillator and provide a one-second-wide pulse.

If decimal counting units are used, the output of each unit can be used to shift the decimal point of the readout. At the one-millisecond pulse width, the units are read out in kilohertz and the decimal point is shifted accordingly. At the one-microsecond position, the units change to megahertz. The decimal point shift is handled by simply controlling it from a second wafer on the range switch.

The digital counter can also be used to

FIGURE 14-23
Timing parameters of counters in several technologies.

TIMING PARAMETER	4020B		7493	74S162	UNITS
	5V	15V			
PROPAGATION DELAY CLOCK TO Q_1	400	160	10	10	ns
Q_n to Q_{n+1}	300	120	12	–	ns
MINIMUM CLOCK PULSE WIDTH	200	80	30	10	ns
MAXIMUM CLOCK FREQUENCY	2	5	16	40	MHz
PROPAGATION DELAY RESET	600	240	26	–	ns
MINIMUM RESET PULSE WIDTH	300	120	15	10	ns

measure time intervals, period, or pulse width. As Figure 14-26 shows, the display counter is switched from counting an unknown to counting the one-megahertz crystal oscillator output. This makes each count equal to one microsecond. The counter is gated by the input signal. For period measurement, the counter is gated to count between rising edge and rising edge of the input signal. For pulse width measurement, the counter is gated to count between leading edge and trailing edge.

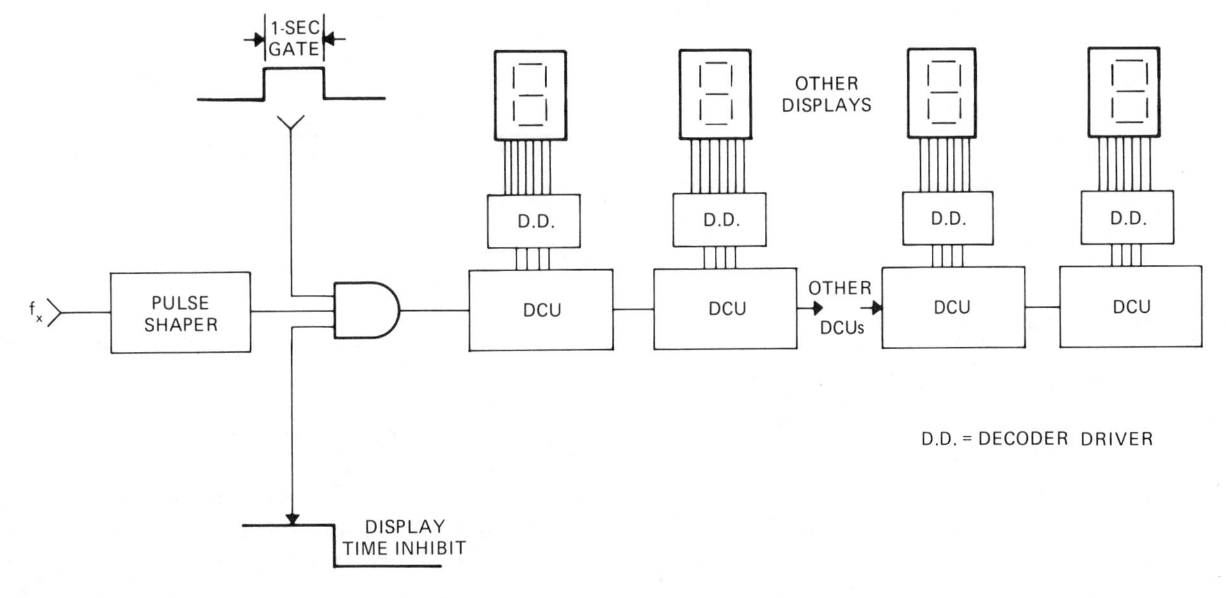

FIGURE 14-24

Block diagram of an f_x counting section of a digital counter showing inputs used to gate the counter off or on.

FIGURE 14-25

Block diagram of the control gate generating section of the digital counter, showing the development of precise one-millisecond and one-second gates to control the f_x display counter section.

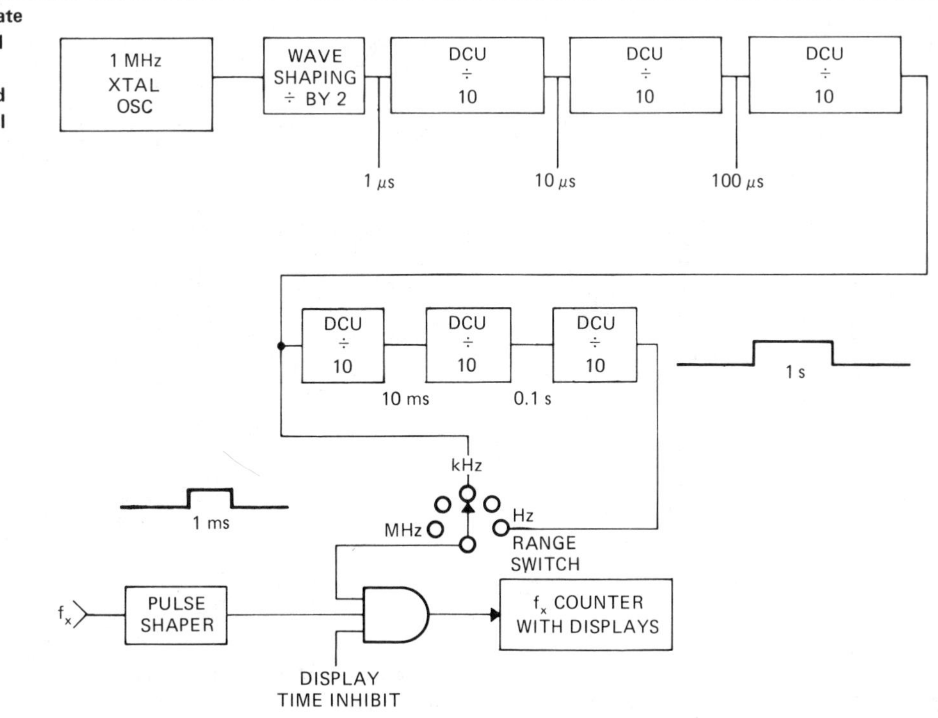

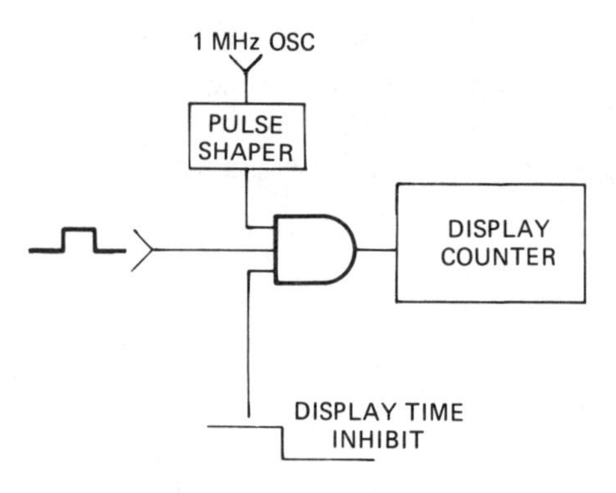

FIGURE 14-26

Block diagram of digital counter connected to measure pulse width by using the input pulse to gate one-megahertz reference oscillator into the display counter.

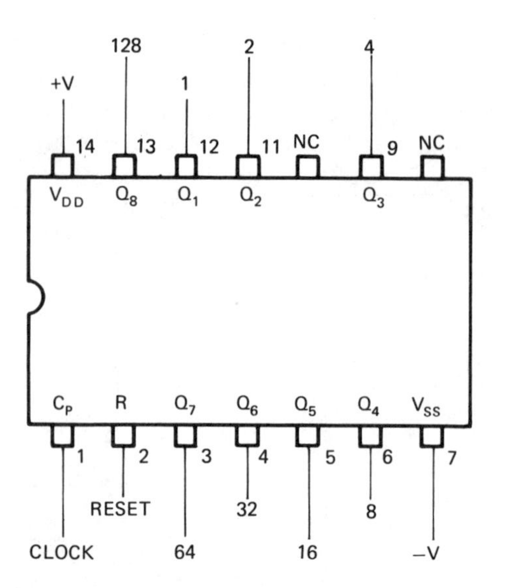

FIGURE 14-27

SCL4404A eight-bit synchronous counter in CMOS integrated circuit.

14.12 CMOS MSI Synchronous Counter

The Solid State Scientific SCL4404A is an eight-stage binary counter. The DIP 14-pin package has inputs and outputs as shown in Figure 14-27. The outputs Q_1 through Q_8 can be given the weighted values of 2^0 through 2^7 or 1 through 128.

Figure 14-28 shows the logic diagram of the counter. The flip-flops used are trigger flip-flops that change state on the rising edges of the clock input, but, as each trigger input is preceded by an inverter or inverting gate, the net effect is triggering on the trailing edge of the input clock pulse. The clock pulses are enabled through a NAND gate by a positive 1 level obtained by inverting the $\overline{Q}_1$. The second NAND gate is enabled by a $1 \cdot 2$; the third, by a $1 \cdot 2 \cdot 4$. These functions are the same as the outputs of the succession of NOR gates used to steer the J-K flip-flops of Figure 14-14.

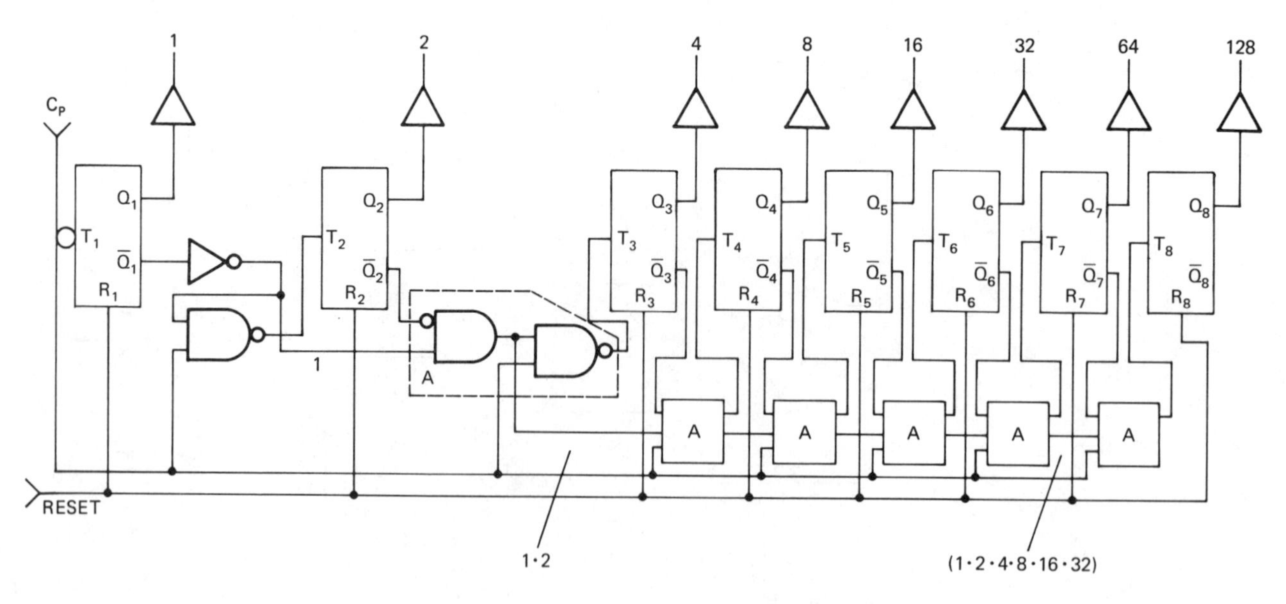

FIGURE 14-28

Logic diagram of the SCL4404A. Each A circuit is identical to the pair of gates in the dotted lines except that a term is added to the AND function input.

14.13 Summary

The digital counter is another highly useful circuit formed by the interconnection of flip-flops. The simplest form of counter is the *trigger* or *ripple counter*. The ripple counter can be formed by connecting the clock or other line to be counted to the trigger input of the first flip-flop in the counter. The output of the first flip-flop is then used to trigger the second flip-flop. The output of each flip-flop in the counter, except the MSB, is used to trigger the next-higher-bit flip-flop. Figure 14-1 shows a two-bit ripple counter capable of counting from 0 through 3. Each flip-flop in an N flip-flop counter is given a binary value from 2^0 through $2^{(N-1)}$. The flip-flops of a three-flip-flop counter would have weighted values of 1, 2, and 4, respectively. The state of the counter can be decoded into decimal by using the methods described in Section 6.9. Figure 14-2 shows the waveforms at the inputs and outputs of a set of decoding gates attached to a two-bit binary counter. The counter is counting the clock pulses. The outputs of gates 0 through 3 will go high during their respective clock periods in numerical order. After the count of 3, the 0 gate goes high and the count will repeat.

If a ripple counter is composed of N flip-flops, it will count to $(2^N - 1)$ before returning to 0 and cycling through the count again. This means that without added circuitry we would be limited to counters of 3, 7, 15, 31, 63, and higher $(2^N - 1)$ values. To obtain counts below the $(2^N - 1)$ value of a counter, a decode of one number higher than the maximum count desired can be used to reset the counter back to 0. This is shown in Figure 14-4 for a count of 5 from a three-bit counter. For this counter, $(2^N - 1)$ is $(2^3 - 1)$, so it would normally count from 0 through 7 and return to 0. A count of 5 is desired; therefore, the count of 6 is decoded and used to reset all flip-flops to 0.

The counters discussed thus far will reach their maximum count and return to 0 and cycle through the count continuously as long as the clock or trigger pulses continue. In some cases, we may want a counter to reach its maximum count and stop, remaining at that count until a reset occurs. This can be done by decoding the maximum count and using the decoded level to inhibit the counter trigger input. This has been done for the count of 5 in Figure 14-5.

An integrated circuit *divide-by-N ripple counter* is available as TTL/MSI 74LS293, as Figure 14-6 shows. It is a four-bit counter with its inverted resets tied to a two-input NAND gate. The double inversion results in an AND function. The C_P inputs are also inverted, resulting in a count that changes on the leading or positive-going edge of the clock pulse. The maximum count available is 0 through 15, but the first bit can be excluded for a 0 through 7. Other counts can be produced by decoding with the NAND gate reset input. Figure 14-7 shows a connection to provide 0 through 9. For counts higher than 0 through 15, two or more of these circuits can be connected together. Figure 14-8 shows two divide-by-N counters connected for a count of 0 through 255.

Only the first flip-flop in a ripple counter is operated by the clock or count pulse. Each succeeding flip-flop is operated by the preceding flip-flop output. For each flip-flop, there is a significant time delay between the changing edge of the input and the change that occurs on the output. These time delays add up between the LSB and the MSB of the counter. A counter of eight bits has about eight propagation delays between the clock input and a change that it will cause at the MSB flip-flop. When high-frequency clock pulses are being counted by a counter with a large number of bits, a cumulative delay in excess of a clock period may occur. A delay of 75 nanoseconds is typical for a TTL flip-flop. If the count rippled through eight flip-flops to a count of 128, and if the clock frequency being counted were 2 MHz or higher, there would be a delay in excess of one clock period. This is shown in the timing diagram of Figure 14-9. In fact, the count of 128 would not occur, as the one hundred and twenty-ninth pulse would set the LSB flip-flop before the setting of the eighth flip-flop. Delays of this magnitude could certainly cause difficulty.

The solution to the *ripple delay* in the asynchronous counter is found in the use of the *synchronous counter*. Each flip-flop in a synchronous counter is clocked by the same clock line. Thus, each flip-flop in the synchronous counter changes state at the same time, approximately one propagation delay after the clock pulse. Instead of the higher-magnitude flip-flops being triggered by the output of the preceding flip-flop, they are steered by the state of the preceding flip-flops and all are clocked by the same clock line. A four-bit synchronous counter is shown in Figure 14-15. Each flip-flop is steered to toggle only when every preceding flip-flop is in the 1 state. NOR gates are used to decode these conditions. The synchronous counter can also be used at values between 0 and (2^N - 1). A four-bit synchronous counter can count 0 to 15 but is often used in decades that count 0 to 9. This is accomplished by decoding 9 and using this level to affect the steer of the flip-flop, so that those at 0 remain 0 and those at 1 toggle to 0 on the count of 10. This is shown in Figures 14-16 and 14-17.

D flip-flops can also be used to assemble asynchronous and synchronous counters, as shown in Figures 14-18 and 14-19. In shift registers (Chapter 13), the D input was steered by the Q output of the preceding flip-flop. For counters, however, the D flip-flop is steered to toggle by connecting each $\overline{Q}$ output to its own D input.

Down counters can be used to signal the end of a timing event. The counter is preset to the desired binary number. On each clock pulse, the counter decrements by 1. A NOR gate can be used to detect when the counter has reached 0 (end of count), as shown in Figure 14-21. The end of count is easier to decode using a down counter than with an up counter. Up counters would require a different decoding circuit for each new binary value.

IC manufacturers have built a versatile *up/down synchronous counter*—the 74LS169, shown in Figure 14-22. A logic signal controls which direction of count will occur.

The table in Figure 14-23 compares the timing parameters of CMOS counters with TTL counters. Remember, there is a power/speed trade-off. TTL devices are faster but consume more power than a comparable CMOS device.

The digital counter is one of the most widely used items of electronic test equipment. To ensure an accurate count, it uses a one-megahertz crystal as a clock. When used to measure time, the one-megahertz pulses are counted directly. When used to count pulses, the one-megahertz clock is counted down to provide an accurate one-second-wide waveform. Two separate counter circuits are involved. One is used to divide down the clock frequency. The second is a display counter. The outputs of each decade of the display counter are connected to decoder-driver and display circuits used to read out the results of the count to the operator. The counter may operate in several modes. One counts the number of pulses, cycles, or events per unit time occurring on the line connected to the input. This is shown in Figure 14-24. The input is first converted to the correct binary levels. An internally generated one-second-wide gate allows the count to continue for only one second. A display time inhibit prevents another count from occurring until the operator has had time to read the display.

Figure 14-25 shows the block diagram of the sections, which divides the one-megahertz oscillator into control gate waveforms of one-millisecond to one-second width.

Another mode of operation is measurement of pulse width. In this case, the display counter section counts the number of one-megahertz clock pulses occurring between the leading and trailing edges of the input pulse. This is shown in Figure 14-26.

Glossary

Asynchronous Counter: A counter in which the clock or count pulses operate only the first flip-flop. The remaining flip-flops are operated by the output of the flip-flops preceding them (often called a *ripple counter*). See Figure 14-3.

Ripple Counter (see *Asynchronous Counter*).

Ripple Delay: The cumulative delay between the change in state of the pulses being counted and the change in the higher-significant-bit flip-flops of the counter.

Synchronous Counter: A counter designed so that it has no ripple delay. Each flip-flop above the LSB is steered to change state by a decoded condition of the preceding flip-flops. The actual change of state of all flip-flops is caused by the same clock pulse line. See Figure 14-15.

Modulus: The total number of states (conditions of 1- and 0-state flip-flops) available from a given counter. A counter of N flip-flops has a modulus of 2^N.

N: Used to represent any whole number (for this chapter, positive whole numbers only).

2^N: The modulus of a counter composed of N flip-flops. If a counter consists of four flip-flops, it can count through 2^4 or 16 states of 1 and 0 flip-flop combinations.

$2^{(N-1)}$: Values of the MSB flip-flop in a binary counter. In an N flip-flop counter, each flip-flop may be given a weighted value, from 2^0 for the LSB to $2^{(N-1)}$ for the MSB. Thus, a four-flip-flop counter has flip-flops with weighted values of 1, 2, 4, and 8.

(2^N - 1): Allowing for 0 state, the highest count available from an N flip-flop counter. Thus, a four-flip-flop counter can count no higher than 15.

DCU (Decimal Counting Unit): A four-bit counter with logic circuits needed for a count of 0 through 9 (sometimes called a *decade counter*). A four-digit decimal counter would have four DCUs. The LSD is triggered by the pulses being counted; each higher significant decade is triggered each time the preceding digit counts back from 9 to 0.

Down Counter: A digital counter loaded with a preset binary number. On each clock pulse, the count is decremented by 1.

Up/Down Counter: A counter that can be switched from an up count to a down count with a logic control signal.

Questions

1. If we include the count of 0, what maximum count is available from a counter composed of five flip-flops?
2. List the modes of operation that can be accomplished by the 74LS293 divide-by-N ripple counter without use of additional gates.
3. What causes ripple delay time in an asynchronous counter? Why is it not a problem with the synchronous counter?
4. If the digital clock of Figure 14-10 were constructed of ripple counters with flip-flops having average propagation delays of 60 nanoseconds, what would be the total ripple delay time in the changing of the minute readout? Would this cause any difficulty?
5. What advantage is there to using a divide-by-12 counter in the hour section instead of separate divide-by-10 and divide-by-2 sections in Figure 14-10?
6. If separate divide-by-10 and divide-by-2 counters were used for the hours readout in Figure 14-10, how would the resets need to be connected?
7. What method could be used to set the correct time on the digital clock?
8. What state must the lower significant bits of a synchronous counter be in before the MSB flip-flop is steered to toggle? Is it the same for a toggle to both 1 and 0 states?
9. For the synchronous BCD counter of Figure 14-17, the count of 9 is decoded to steer the counter back to 0 on the tenth clock pulse. How does this differ from an asynchronous BCD counter?
10. The digital counter used as an item of electronic test equipment has two separate counter sections. Describe both sections and their functions.
11. Why is a 1 MHz crystal oscillator used in the counter operation of Figure 14-25?
12. What is the display time inhibit?
13. Explain how the digital counter is used to measure pulse widths.
14. The CMOS 4404A eight-stage counter shown in Figure 14-28 consists of toggle flip-flops. Explain why this is a synchronous counter.
15. Will the counter of Figure 14-28 change state on the leading or trailing edge of the clock pulse?
16. What circuit elements are needed in synchronous counters and not in asynchronous counters?
17. For the 74LS169 up/down counter, what must be the logic state of the PE input (a) to load a binary number, (b) to count up, and (c) to count down?
18. What factor makes it possible to use the same logic circuits for a down count as for an up count?
19. When the circuit of Figure 14-21 is used to count down from 1010, what value is initially applied to the preset count inputs?
20. If 0110 is applied to the preset inputs of Figure 14-21, what is the value from the $\overline{Q}$ outputs? What is the value from the $\overline{Q}$ outputs after the first clock pulse?

Problems

14-1 Connect the two integrated circuits of Figure 14-29 to form a four-bit ripple counter.

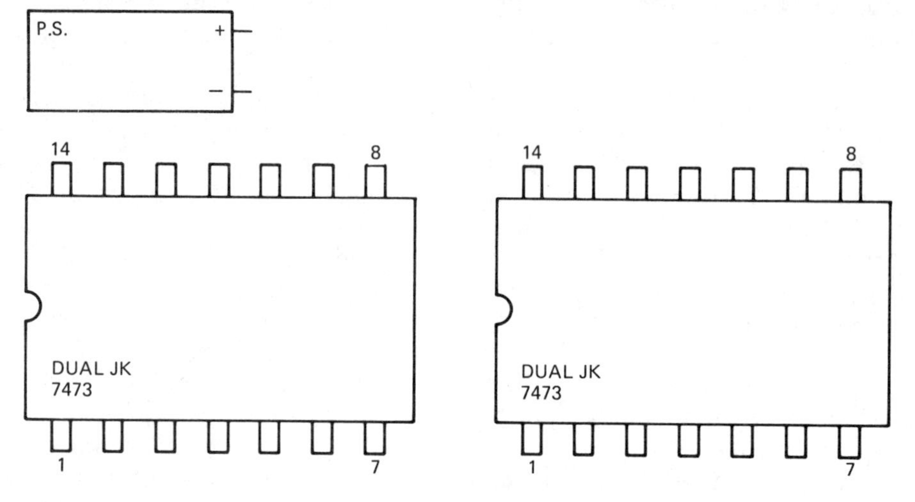

FIGURE 14-29. Problem 14-1

14-2 Show the connection of three TTL/MSI 74LS293 ripple counters (see Figure 14-6) to form a divide-by-3600 counter.

14-3 Connect the three integrated circuits of Figure 14-30 to form a synchronous counter like that of Figure 14-15.

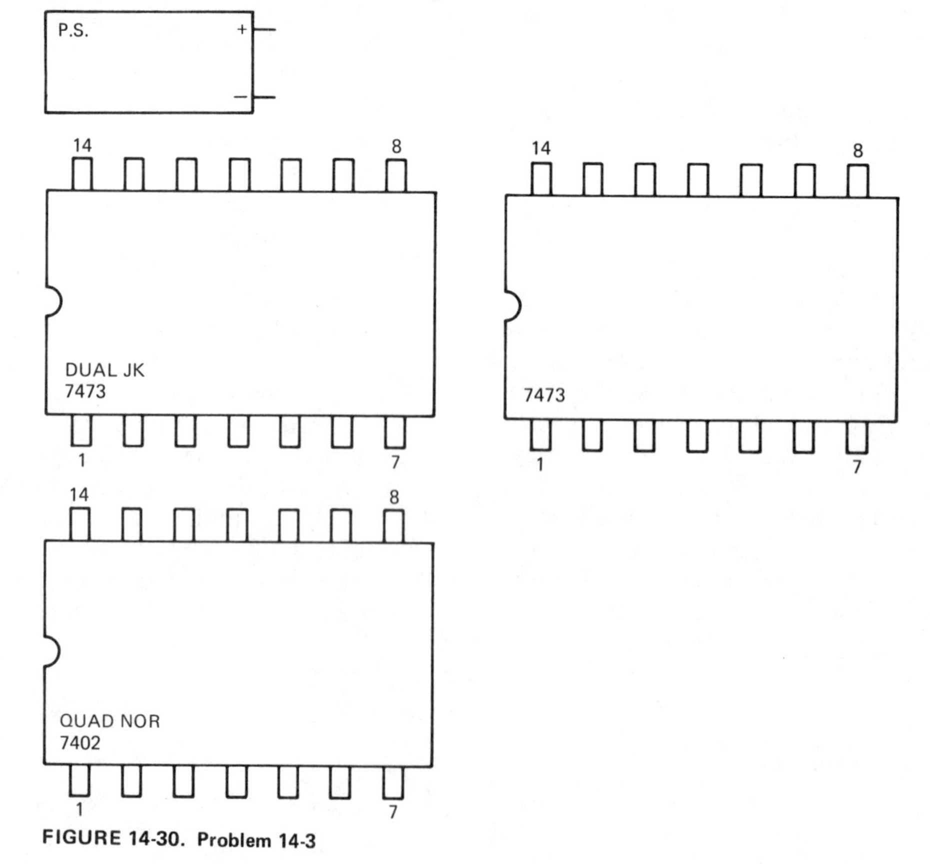

FIGURE 14-30. Problem 14-3

14-4 Draw a decoder for the 0–3 counter of Figure 14-1 using all NOR gates.

14-5 Draw the counter of Figure 14-3 using the J-K flip-flop.

14-6 Draw a four-bit ripple counter with the added gates needed for a count of 0 to 9.

14-7 Draw a three-bit ripple counter with the added gates needed to count to 6 and stop.

14-8 Draw the synchronous counter of Figure 14-15 using AND gates to develop the steer inputs.

14-9 Draw a three-bit asynchronous down counter. Draw the timing diagrams.

14-10 Draw the logic diagram of a four-bit down counter like that of Figure 14-21, but use 74112 J-K flip-flops shown in Figure 12-16.

14-11 If in Figure 14-18, Q_1 and Q_2 were used for clock between the flip-flops instead of $\overline{Q}$, what difference would occur in the output waveforms? Draw the output waveforms with reference to the clock input.

CHAPTER

Serial Adder Circuits

Objectives

Upon completion of this chapter, you will be able to:

- Assemble a single full adder and a flip-flop to form a serial adder.
- Connect a serial adder and the associated circuits needed to perform twos complement subtraction.
- Assemble flip-flops and gates to provide a serial comparator.
- Combine the serial comparator and serial adder to perform ones complement subtraction.
- Perform BCD addition by digits in serial.

15.1 Introduction

The adder circuits discussed previously were parallel adders, for which each bit in the addition from LSB through MSB required its own full adder circuit with supporting gates and registers. The individual full adders were interconnected by the carry lines. The add function occurred during a single clock period, requiring only the delay needed for the carries to ripple through from LSB to MSB.

The *serial adder* has the advantage of needing only one full adder circuit with a carry store regardless of the number of bits in the addition. This is a substantial saving in circuits if the adder is to have a large number of bits. Unfortunately, the serial adder requires a clock period for each bit in the addition. This makes the serial adder too slow for use in high-speed, general-purpose computer applications. It may, however, be used in desk calculators, whose speed is limited by human reaction time anyway. Serial adders may also be used in small, low-priced, general-purpose computers and in many special-purpose computer applications.

Figure 15-1 shows the block diagram of the serial adder compared with that of a four-bit parallel adder. The serial adder requires fewer input/output leads—a substantial advantage to integrated circuit design, for which pin numbers are limited.

15.2 Serial Addition

Let us begin the discussion of serial addition with a review of serial numbers as they will appear at the input and output of the adder. The diagram in Figure 15-2 shows the associated waveforms that will occur at the inputs and outputs of the adder during a five-bit add cycle if the numbers 9 and 11 are added. Before the first clock pulse in the add cycle, the A and B inputs are both 1; the carry-in (C_{IN}) is 0. As the truth table in Figure 15-2 shows, this condition produces a 0 sum and a 1 carry-out. The carry-out steers the carry register to a 1, but the change does not occur until the clock pulse.

After the first add cycle clock pulse, the carry-in becomes 1 and the A and B inputs are on the second significant bit; A is 0, B and C_{IN} are 1. As the truth table indicates, the outputs are, again, 0 sum and 1 for carry-out.

After the second add cycle clock pulse, the carry-in is again 1, but both A and B are 0. This produces a sum of 1 and a carry-out of 0.

After the third add cycle pulse, the carry-in is at 0, but both A and B are 1, resulting in a 0 sum, carry 1.

After the fourth add cycle pulse, the A and B inputs are 0, but the final carry-in is 1. This generates the final sum bit. Stored now in the sum register is the number 10100_2 (20_{10}).

15.3 Serial Adder-Subtractor

Before discussing the more versatile but complex forms of serial adder-subtractors, let us look at a simplified case in which B is a small correction factor, either positive or negative, to be added to A, which is always larger than B.

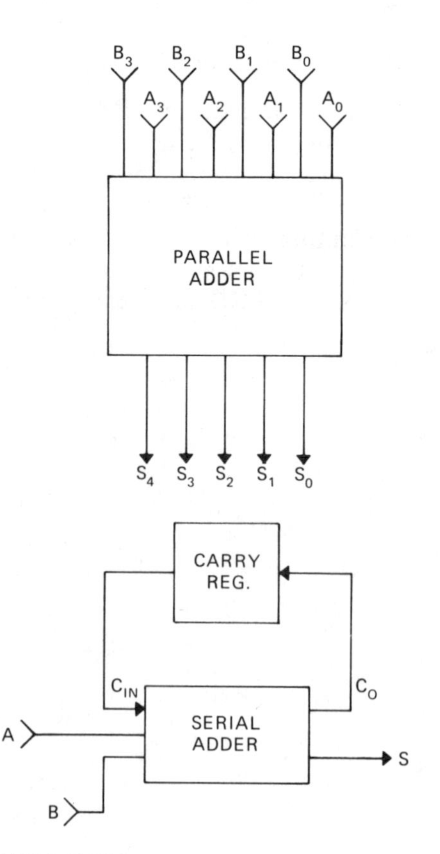

FIGURE 15-1

Block diagram comparison of parallel and serial adders, showing advantage of fewer lead-ins for the serial adder.

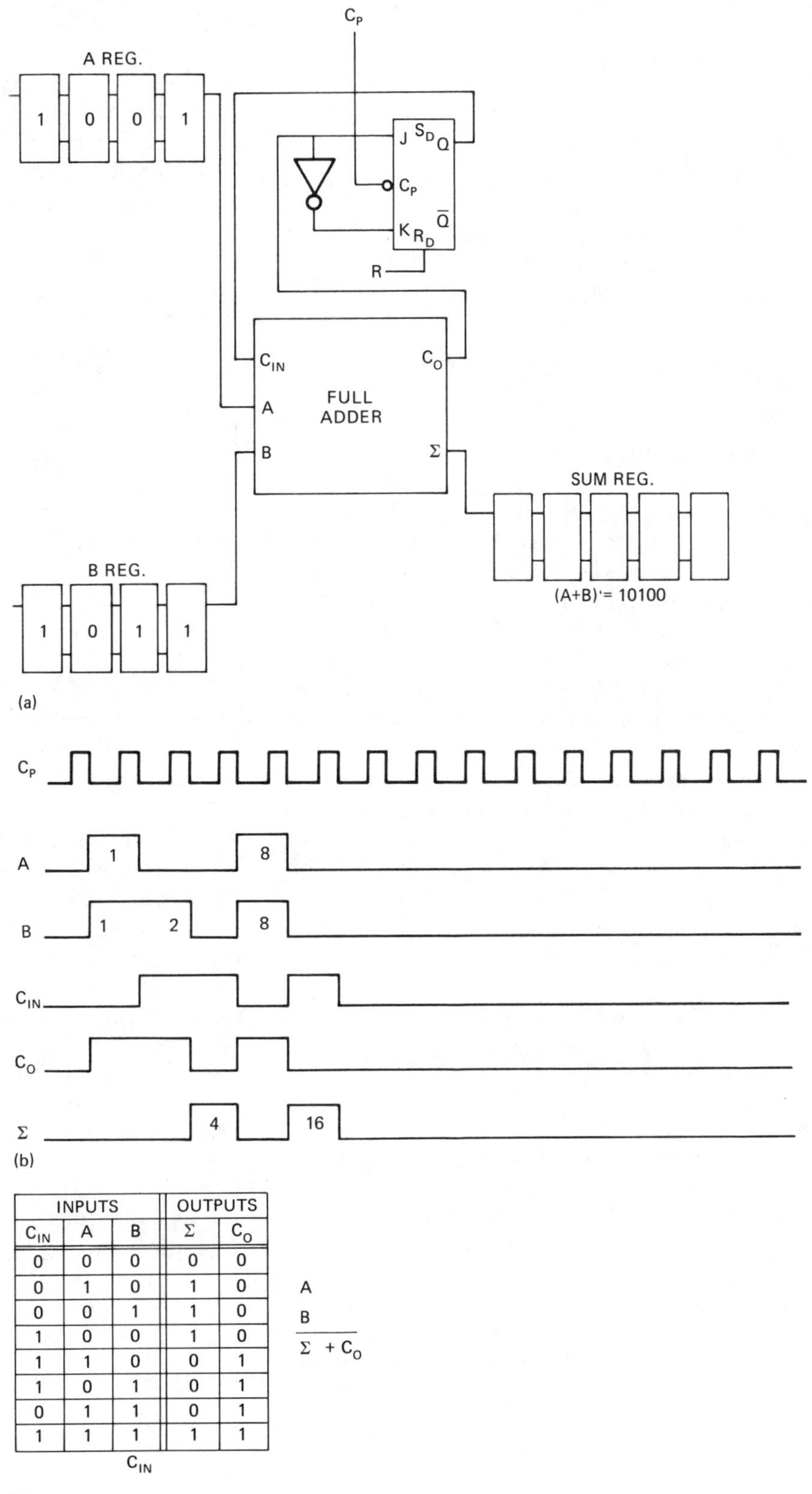

INPUTS			OUTPUTS	
C_{IN}	A	B	Σ	C_O
0	0	0	0	0
0	1	0	1	0
0	0	1	1	0
1	0	0	1	0
1	1	0	0	1
1	0	1	0	1
0	1	1	0	1
1	1	1	1	1
	C_{IN}			

FIGURE 15-2

(a) Block diagram of serial adder. (b) Associated waveforms at the inputs and outputs of the adder for addition of 9 + 11. Reset and clock signals are applied to all registers. (c) Truth table.

Assume:

$A = 101101_2\ (45_{10}) \qquad B = \pm 1010\ (10_{10})$

If B is positive, A + B = S.

$$\begin{array}{rr} \text{So,} & 101101 \\ & +\quad 1010 \\ \hline & 110111 = 55 \end{array}$$

If B is negative, the ones complement is used and A - B = R becomes $\bar{A}$ + B = S and the remainder R = $\bar{S}$. Therefore,

$$\begin{array}{c} \bar{A} \\ +B \\ \hline S \end{array} \quad \text{or} \quad \begin{array}{r} 010010 \\ +\quad 1010 \\ \hline 011100 \end{array}$$

So, R = $\bar{S}$ = 100011 = 35

To provide this operation, B must travel with a sign bit 0 for positive and 1 for negative. As shown in Figure 15-3(a), the sign of B can be used as a control input to an exclusive OR gate, the result being that A will be complemented when B is negative and passed through without change when B is positive. A second exclusive OR gate at the output will complement the sum when B is negative. Figure 15-3(b) shows the waveforms that will occur for A = 13, B = +9 on the left and for A = 13, B = -9 on the right.

15.4 Twos Complement Adder-Subtractor

The adder-subtractor of Section 15.3 is applicable only if we know A to be greater than B. For a more versatile situation, in which we are free to enter into the adder ±A and ±B, whose magnitudes are restricted only by the size of the adder registers, a ones or twos complement adder may be used. Such adder circuits display many variations, and the exact circuitry used may be dictated not only by the needs of addition and subtraction, but also by difficulties encountered in using the adder circuit for multiplication, division, and other arithmetic functions. We will simplify this discussion by restricting ourselves to only those functions needed for addition and subtraction. We will explain twos complement first because it is more widely used in serial adders.

As we recall from Chapter 2, a twos complement is a ones complement plus 1, but it is easier to form in a serial operation by beginning the complement after the first 1 bit has passed. The identity of this method with the ones complement plus 1 is shown as follows:

If A = 1101, then

$$\begin{array}{rr} \bar{A} + 1 = & 0010 \\ & +\quad 1 \\ \hline & 0011 \end{array}$$

Or, 1101 ⟶ 0011
(↑ begin complementing; ↑ let pass)

If B = 1100, then

$$\begin{array}{rr} \bar{B} + 1 = & 0011 \\ & +\quad 1 \\ \hline & 0100 \end{array}$$

Or, 1100 ⟶ 0100
(↑ begin complementing; ↑ let pass)

To subtract by twos complement, we need a circuit that can be energized to perform the twos complement of the negative number during a subtraction but that can be inhibited from functioning when the number is positive or during addition. The circuit of Figure 15-4 will twos-complement the B number only for (+A -B). The flip-flop is reset at start of the add cycle. If (+A -B) line is high, the J-K flip-flop is steered to 1 but cannot set until the first 1 level of number B enables a clock pulse to set it. From that point on, a 1 level on the exclusive OR will complement the input bits until the end carry control resets the flip-flop.

In review of subtraction by twos complement:

If	-A	-13	-1101	twos	+0011
	+B	+ 9	+1001		+1001
	R	- 4			1100

No end carry means the answer is a negative remainder in twos complement form. End carry

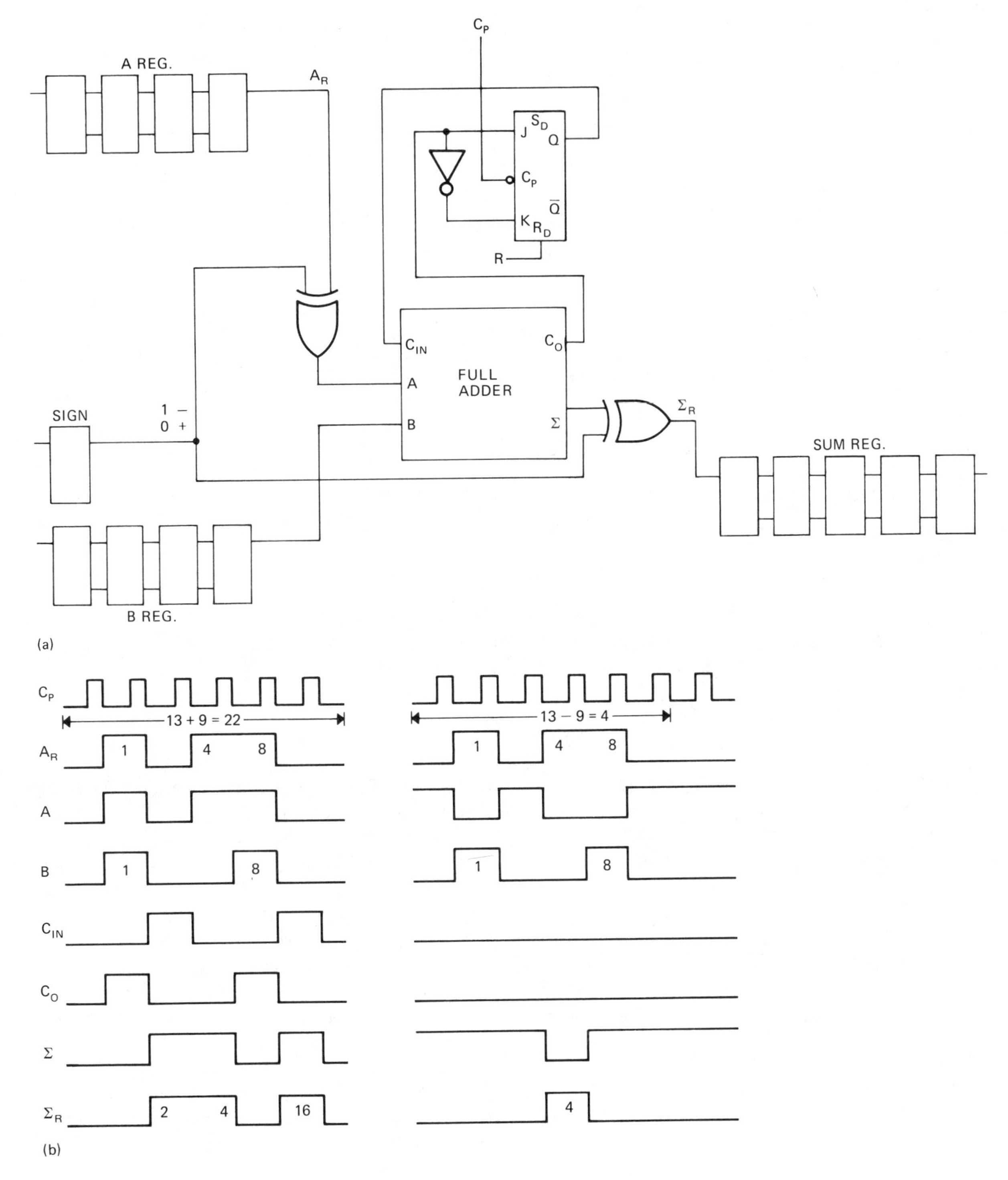

FIGURE 15-3

(a) Simple serial adder-subtractor, valid only for A > B. The number A will pass through the exclusive OR without being complemented if the sign lead is 0 (pos. B) but will be complemented if the sign lead is 1 (neg. B). (b) Waveforms that occur for 13 + 9 = 22 and for 13 - 9 = 22.

FIGURE 15-4

Circuit designed to twos complement the serial number.

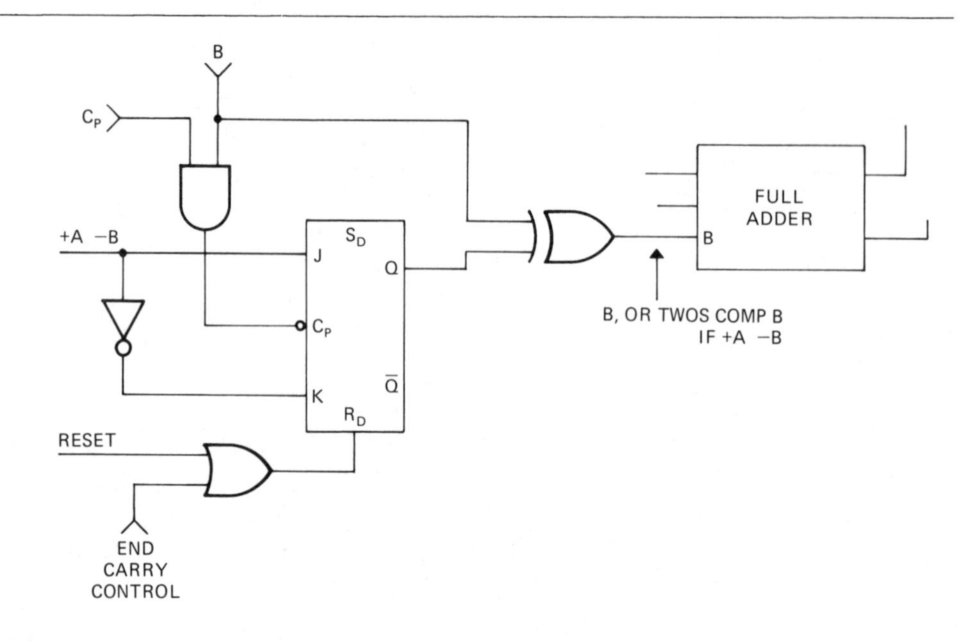

indicates a positive remainder in normal form if the end carry is dropped. For example,

-1100 = -0100 = -4
begin complementing after first 1 bit

If	+A	+13	+1101	1101
	-B	- 9	-1001	+0111
	R	4		(1)0100 = +4

end carry (drop)

The block diagram of Figure 15-5 shows how these twos complementers for A and B would be used at the inputs to the full adder. The twos complementer for the sum cannot be connected directly to the full adder output, as was the case with the adder-subtractor in Figure 15-3. During twos complement subtraction, the output is complemented only if it is a negative remainder. The circuit cannot recognize the negative sign of the remainder until it sees the absence of an end carry. By end carry time, it is too late to complement at the full adder output. This means the output must be stored, either by recirculating into the A register or in a separate sum register.

If we handle the number in magnitude plus sign bit form, as follows,

+ sign bit
+13 = 01101
-13 = 11101
- sign bit

the A and B registers must have an added flip-flop to handle the sign bit.

The sign function circuit must examine the sign bits and develop the correct control signals. This circuit can be much like the sign function circuit of the parallel adder except that flip-flops will be needed to store the sign level.

The adder-subtractor must operate in three phases—shift in, add and store, and shift out. Each phase is initiated by a control signal, as Figure 15-6 shows. The reset zeros all the registers at the beginning of an add cycle. The clock pulses after reset shift A and B number bits into their respective registers. This requires as many pulses as there are bits in the register. The last bit to enter is the sign bit. Then comes the check sign pulse, which will cause the sign function circuit to look at the signs and determine whether either of the inputs should be twos-complemented.

As soon as the last bits of A and B have passed through the full adder, the end carry control pulse occurs. It energizes a circuit that will examine the end carry and determine the sign of

FIGURE 15-5

Block diagram showing twos complementer circuits inserted in both A and B adder input lines. The sign function circuit examines the sign of the input numbers and controls the twos complementers accordingly.

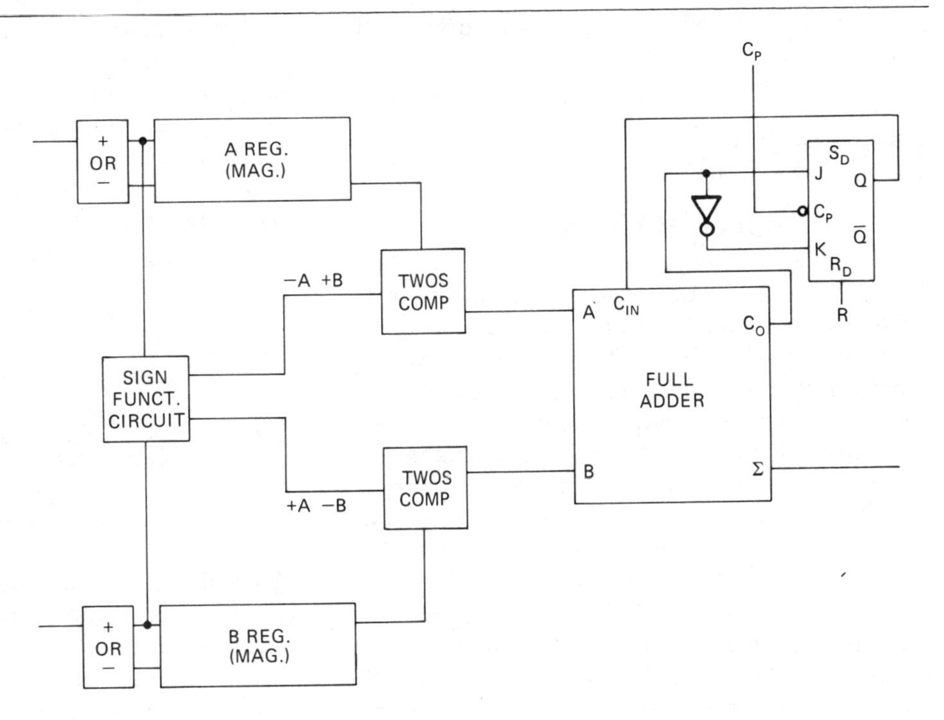

FIGURE 15-6

Control signals and their time relation to a 13 − 9 = +4 subtraction: *Reset*—begins add cycle, starting with shift in. *Check sign*—pulse marks beginning of add-and-store phase of cycle. *End carry control*—starts the shift out of the sum.

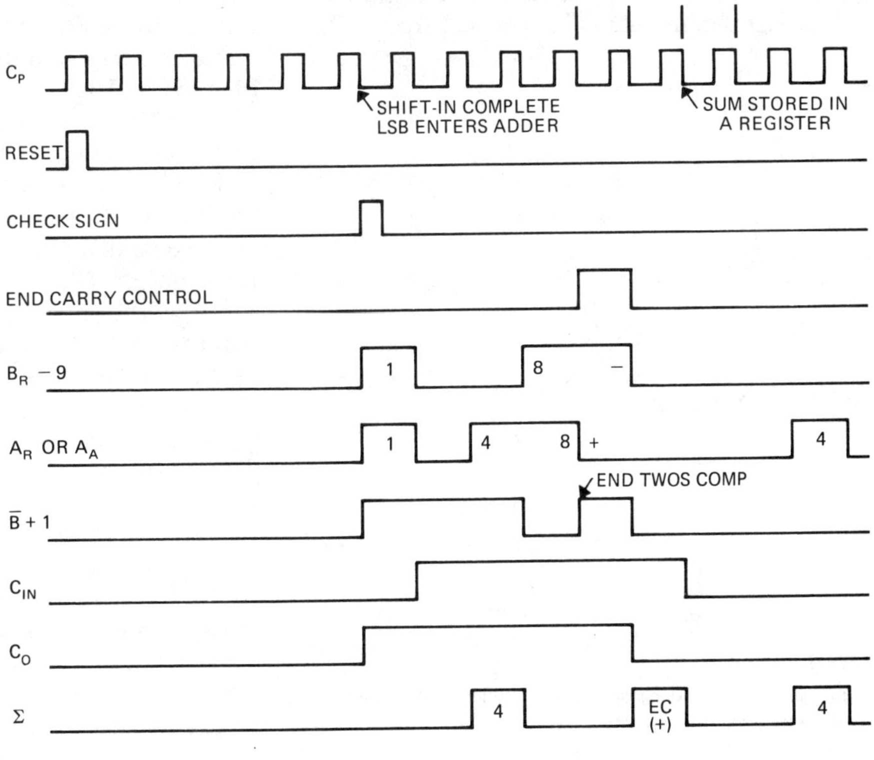

C_P
SHIFT-IN COMPLETE
LSB ENTERS ADDER
SUM STORED IN
A REGISTER
RESET
CHECK SIGN
END CARRY CONTROL
B_R −9 1 8 −
A_R OR A_A 1 4 8 + 4
END TWOS COMP
$\overline{B}$ + 1
C_{IN}
C_O
Σ 4 EC (+) 4

the output, and, if there is a negative remainder, cause it to be twos-complemented during shift out.

15.5 Serial Comparator

The serial comparator enables the machine to make three-way decisions, as in the flow diagram of Figure 15-7. There are many uses for this function. When thousands of information records are placed on a magnetic tape hundreds of feet long, the machine can find a particular record in a few seconds by comparing the record number with those that pass over the magnetic read head in numerical order. It can also be used to simplify the functioning of a serial adder.

Two numbers, A and B, whose bit times are synchronized can be compared in the circuit of Figure 15-8(a). The exclusive OR gate will enable the clock pulses through the AND gate only on those bits for which $A \neq B$. If $A = B$, the levels on J and K will not matter, for no change will occur, since a clock pulse is absent. If A is 1 and B is 0, the upper register will be clocked to 1 and the lower register to 0. If B is 1 and A is 0, the lower register will be clocked to 1 and the upper one to 0. The comparator will not, however, have made its final decision until the most significant bits have passed through it.

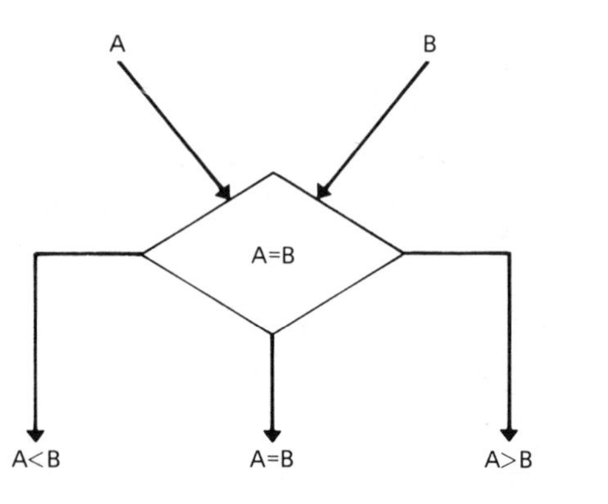

FIGURE 15-7

Flow diagram three-way decision symbol, a function that can be provided by a serial comparator.

Figure 15-8(b) shows the waveforms that occur if A = 0101 and B = 1001. During the LSB, both bits are 1 and the C_P line is inhibited. During the second bit, both A and B are 0, again inhibiting the C_P line. During the third bit, A is 1, B is 0, and a C_P pulse passes to the flip-flops, which are steered for $A > B$. The $A > B$ line goes high. The NOR gate output $A = B$ drops to 0. During the last digit, A is 0, B is 1—which also causes a C_P pulse to clock the registers, but this time they are steered $B > A$. The $A < B$ line goes high; the $A > B$ line goes low.

15.6 Serial Adder with Comparator

The simple complementing adder of Section 15.3 can be made more versatile by shifting the numbers through a serial comparator at the same time that they are being shifted into the adder registers. Thus, the adder will detect, as soon as the subtract cycle commences, which number is the larger. In every case of subtraction, we can complement the larger number going into the adder and subsequently complement each bit the moment it leaves the adder. By this method, the serial adder can operate in two-thirds the time required by the twos complement adder of Section 15.4. The comparator need have only one of the J-K flip-flops, making the combined circuitry simpler and faster than the twos complement adder.

In the block diagram of Figure 15-9, the magnitude bits of A and B are simultaneously shifted into their registers and through the comparator. When the LSBs of A and B have arrived at the output flip-flops of their registers, the sign function gates have determined whether there is to be addition or subtraction, while the comparator has determined which of the two numbers is the larger. If addition is called for, by like signs (--) or (++), the 0 on the subtract line prevents complementing at either inputs or outputs. If subtraction is called for, by difference of signs (+-) or (-+), the 1 on the subtract line causes the adder output (sum) to be complemented. It will also enable the pair of AND gates, so that the comparator outputs can cause the larger number to be complemented at the input of the adder. There is no need to recycle the sum for the purpose of end-around carry or twos-complementing.

15.7 Binary-Coded Decimal Serial Adder

In Chapter 10, we saw the binary-coded decimal parallel adder to be considerably more complicated than a binary parallel adder of equivalent bits. It is possible to obtain a saving of these complex circuits by serial operation. Reducing to a single full adder circuit would require recycling each four-bit decimal group several times; and to subtract, we would have to allow for recycling the remainder. The circuits needed to control this would be more complex than the twos complement adder described in Section 15.4.

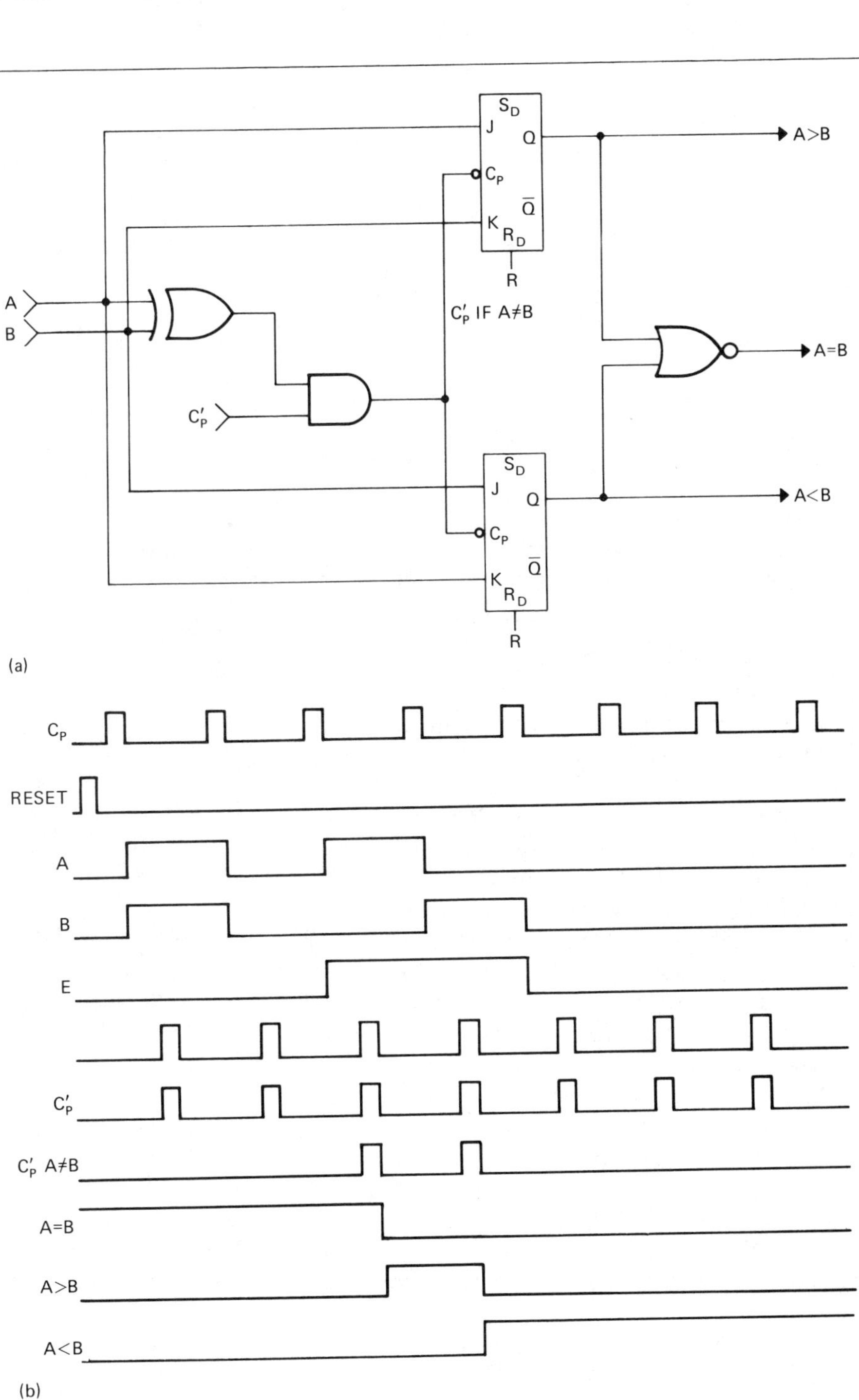

FIGURE 15-8

(a) Serial comparator circuit. When flip-flops are reset, two 0 inputs produce a 1 from the A = B NOR gate. If the serial numbers A and B are identical, the AND gate remains inhibited and both flip-flops remain reset: A = B. If A is 1 and B is 0, the A > B will set and the B < A reset. If B is 1 and A is 0, the B > A will set and the A < B reset. The final decision is not made until the MSB. (b) Waveforms that occur if A = 0101 and B = 1001.

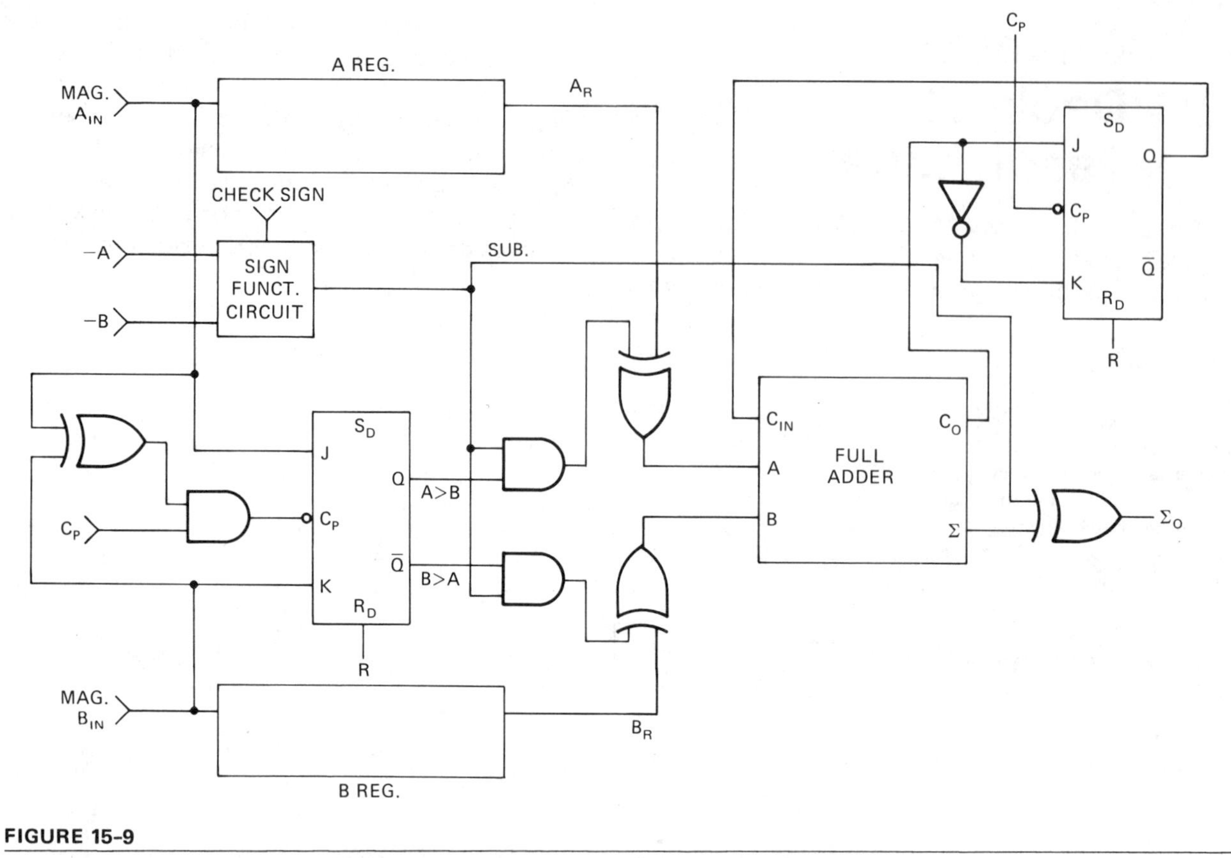

FIGURE 15-9

Ones complement adder, like that of Figure 15-3, except that the comparator determines the larger number and complements it and the sum during subtraction, removing the restriction of A > B.

An even more serious drawback would be the manifold increase in add cycle time. By using a single-digit parallel BCD adder of the type described in Section 10.6, we can reduce the add cycle time to a single clock pulse per bit. Figure 15-10 shows a block diagram of such a circuit for addition only. In this method, each pair of four-bit digits are added in parallel; yet they move to and from the adder in serial. This adder has input registers for eight decimal digits, or 32 bits, but only the lower four bits of the A and B registers are connected to the adder inputs. Only the higher four bits of the output or sum register are connected to the adder outputs. As the numbers are shifted through the A and B registers, the adder output is valid only when decimal digit 1 bits are in the LSB flip-flops of the entry registers. For this reason, the adder output and carry register are strobed only after every fourth clock pulse.

Figure 15-11 is the timing diagram for A digits 1356 and B digits 2769. The carry register must be clocked by every fourth clock pulse occurring after the LSBs are in correct position. The significant bits of the output (S) are exactly four bits (clock periods) delayed from those at points (A) and (B). The inputs to the adder are valid only when the LSBs of the decimal digits are stored in the LSB flip-flops of the entry registers. The outputs of the adder become valid a short time later.

The time required from valid input to valid output (T_r) is the time required for carries to ripple through the adder. It can be estimated by tracing through the shortest path from LSB in to MSB out and counting the number of gates the signal passes through. Multiply the number of gates by the propagation time per gate. The adder in Figure 10-14 has a maximum propagation path of 13 gates. Even at 10 nanoseconds per gate, this would be 130 nanoseconds. If the clock rate were 1 MHz, we would expect no difficulty in allowing half a clock period (500 nanoseconds) before strobing the output into

FIGURE 15-10

BCD serial adder for serial addition of digits in a parallel BCD adder circuit.

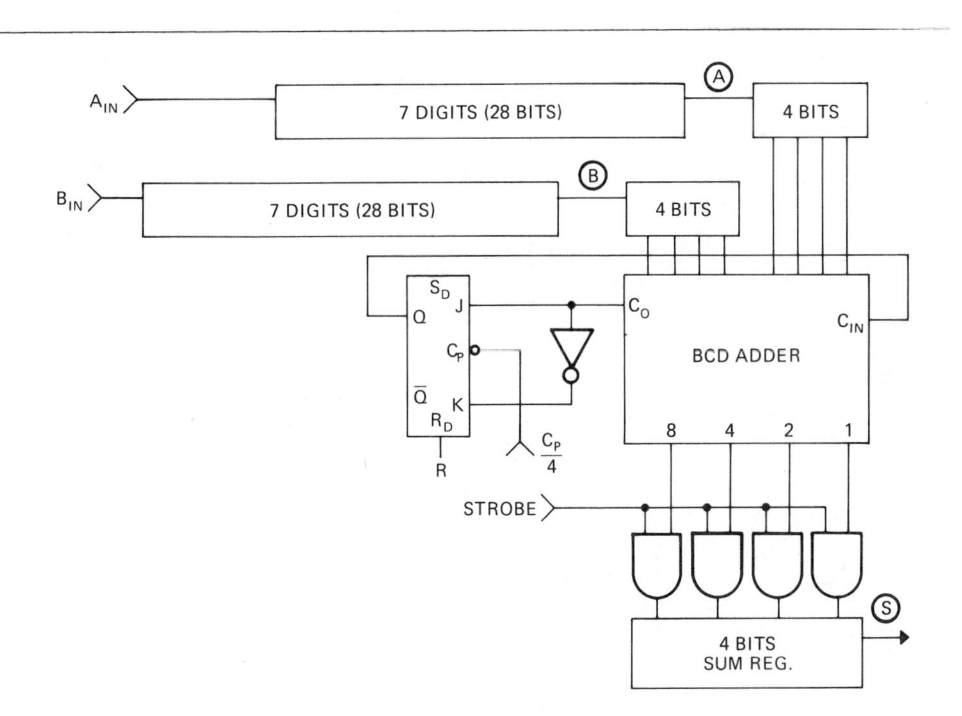

FIGURE 15-11

Timing diagram of BCD serial adder for A = 1356, B = 2769. T_r shows the critical times, between which the adder inputs become valid and output is strobed to the sum register.

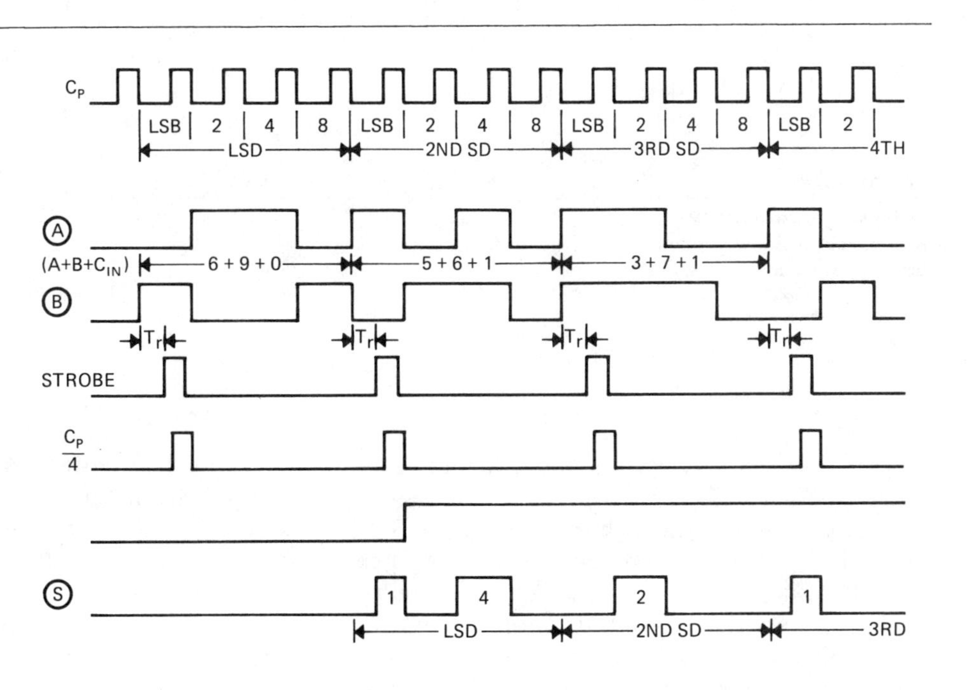

the registers. If the clock frequency were 10 MHz, half a clock period would be only 50 nanoseconds, and the strobe could occur before the sum bits from the adder were correct. When we consider that one remedy for the slowness of the serial adder, in terms of clock pulses per add cycle, is to increase the clock rate, we see the (T_r) ripple time as a critical limitation. An obvious solution is to skip one or more clock pulses after each set of four pulses; yet this would increase the add cycle time and possibly create a synchronization problem. A better method is to

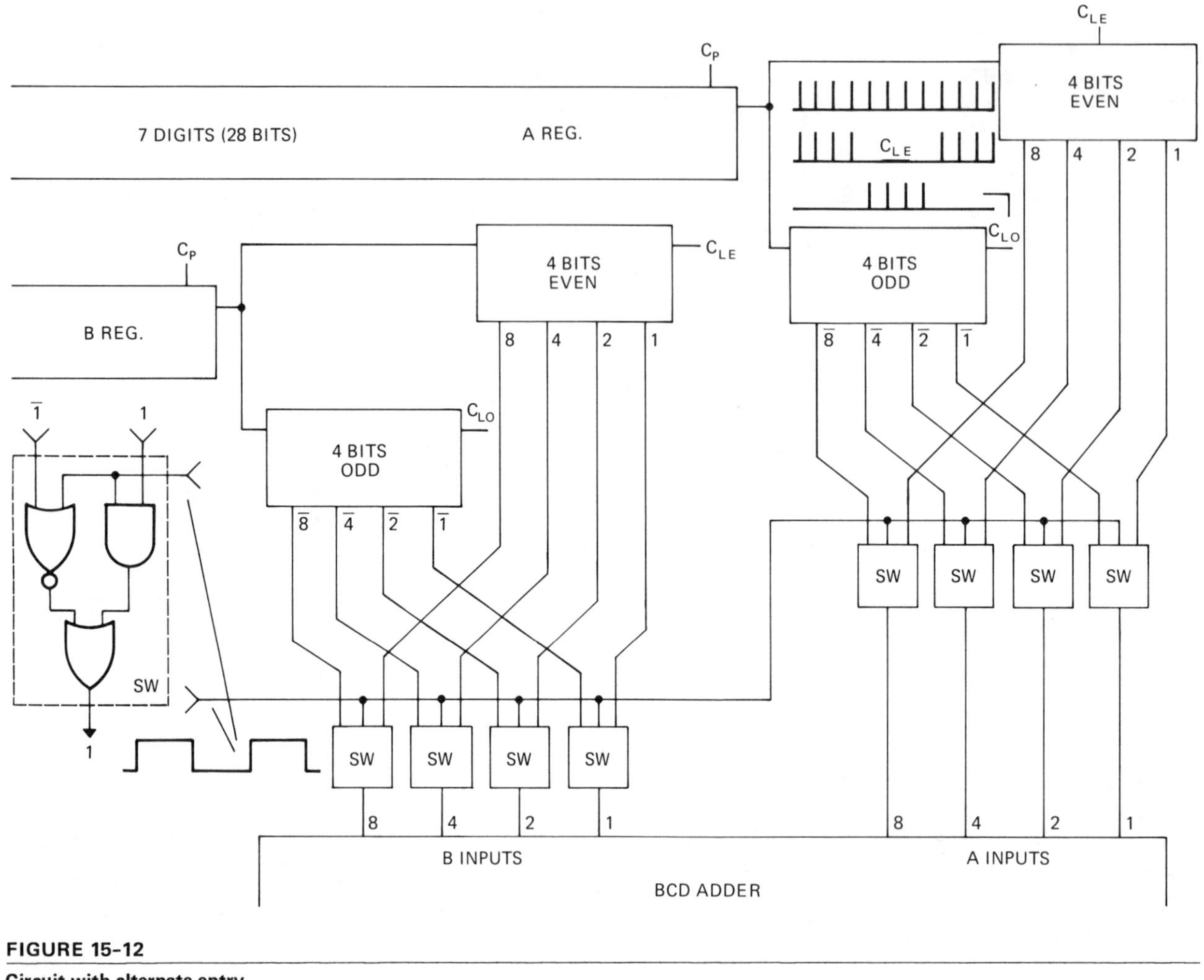

FIGURE 15-12

Circuit with alternate entry registers for the A and B inputs that is used to increase available ripple carry time (T_r).

use separate entry registers for odd and even digits, as Figure 15-12 shows.

In this circuit, the higher bits of A and B are clocked continuously through the input register, moving at a rate of one bit per clock pulse. These registers are clocked by a normal continuous clock line (C_L). At the end (last four bits) are pairs of registers in parallel. These are clocked by alternate bursts of four clock pulses (C_{LE} and C_{LO}). The clock pulses are timed so that even-digit (powers of 10) bits shift into one register and remain static while the next four odd-digit bits are shifting into the other registers. In this system, the adder is allowed almost four clock periods of ripple carry time. The outputs of the adder entry registers are alternately inhibited from the inputs of the adder during shift-in time and enabled during the static time. This requires that the switching voltage be high, enabling the even inputs through the switches during the time the odd clock pulses, C_{LO}, are shifting the odd-digit entry register. It must then go low, enabling the odd-digit inputs through the switch during the time the even clock pulses, C_{LE}, are shifting the even-digit entry register. The waveforms for operation of this adder are identical to those of Figure 15-11 except for the added clock and switching lines, shown in Figure 15-12.

15.8 Summary

The parallel adders described in Chapter 10 require a full adder circuit for each bit in the largest number to be processed. It is possible to operate the adder with a single full adder circuit plus one flip-flop regardless of the number of bits in the numbers being processed. To do this, we use a *serial adder*. Figure 15-1 is a block diagram comparison of a four-bit parallel adder with a serial adder. The serial adder accepts the numbers A and B in serial, one bit of each at a time, beginning with the LSB. The sum is not complete until one clock period after the MSB has passed through the adder. This requires one clock period for each bit of the numbers being processed plus one for the final carry.

Figure 15-2(a) shows a serial adder with its associated registers. Normally a shift register is required for the A and B numbers. A separate sum register is shown, but in many applications the sum is recirculated into the A register.

A single flip-flop is used to store the carry bit. The carry-out generated during any bit time is not used for a carry-in until the next significant bit time. The carry-out is used to steer the flip-flop, which does not change to that level until the clock pulse. The same clock pulse also shifts the next significant bit out of the A and B registers, applying them to the adder inputs. Figure 15-2(b) shows the waveforms that result from the serial numbers 1001 (9) and 1011 (11) being applied to the adder, producing a serial output of 10100 (20).

Modifying the adder of Figure 15-2 for subtraction is a simple matter if the larger of the numbers being subtracted can always be directed to the same register. In this case, an exclusive OR gate can be used between the register output and the full adder input and a second exclusive OR at the full adder sum output. Figure 15-3(a) shows this arrangement. If we can allow the condition of $A > B$, then for all cases of subtraction, complementing both A and the sum will result in the correct remainder. For the adder-subtractor to recognize the need for complementing, the numbers must travel with a sign bit. The sign bit may be a period in the serial number and therefore require a flip-flop at one end of the register, or it may occur as a level on a separate line. Figure 15-3(b) compares the waveforms that would result from the operations 13 + 9 and 13 - 9.

There are many applications of serial adders for which we cannot allow for $A > B$. In most cases, a more versatile adder is needed—one that will accept $\pm A$ and $\pm B$ whose magnitudes are restricted only by the size of the adder registers. In these applications, we must consider either ones or twos complement techniques. The ones complement method requires that we add the end-around carry to all positive remainders, forcing us to cycle the sum back through the adder and doubling an already time-consuming operation. For this reason, twos complement is often used. The twos complement is formed by waiting until the least significant 1 level has passed before beginning the complement. This method produces identical results to a ones complement plus 1. Figure 15-4 shows and explains a circuit that can be energized to provide this function. A twos complement adder requires one of these at both the A and B inputs. If a separate sum register is used, a third such circuit is needed at the output of that register. Figure 15-5 shows the location of the twos complementers at the inputs of the adder.

Another circuit that is useful in both adders and other applications is the *serial comparator*. The serial comparator can compare two serial numbers (A and B) one bit at a time and determine whether A is greater than B, equal to B, or less than B. Figure 15-7 shows a flow diagram symbol for this circuit. Figure 15-8(a) explains the logic diagram and its operation. Figure 15-8(b) shows the waveforms that would occur if $A = 0101$ and $B = 1001$ were passed through the comparator.

In our discussion of both ones and twos complement adders, we faced the problem that no action could be taken to complement the outputs during a subtraction until *end carry time.* This meant that the output of an N-bit adder-subtractor could not be put to use for N clock periods after it passed through the full adder. This problem is easily remedied by first passing the inputs through a serial comparator, which will identify the larger of the two, and if the signs indicate subtraction, it will complement the larger number at the input and the remainder at the output. This adds only a single clock period during shift-in and makes the sum available directly from the output exclusive OR gate.

Figure 15-9 shows an adder-comparator combination. Note that $A = B$ output is not needed, and therefore both a flip-flop and NOR gate have been eliminated from the comparator.

The sign function circuit enables both AND gates during subtraction. The comparator flip-flop output will provide a 1 level, causing the larger of the input numbers to be complemented. The output will be complemented in all cases of subtraction.

It would be difficult to do binary-coded decimal addition in a serial adder composed of a single full adder circuit. It can, however, be accomplished by using a single-digit BCD adder, which processes four bits or one digit at a time. Figure 15-10 shows a BCD serial adder of this type. It uses a BCD single-digit adder like that developed in Chapter 10. Only the last four bits of the A and B registers are connected to the adder inputs. After each fourth clock pulse, a strobe pulse sets the BCD output into the sum registers. This is shifted out serially during the next four clock pulses. The first of each four clock pulses is used to clock the decimal carry into the carry register in preparation for the next digit.

Figure 15-11 is a timing diagram for the addition 1356 + 2769. Note that the strobe and the $C_P/4$ must occur just before the trailing edge of the clock pulse that shifts the LSB of the next digit into the entry section of the register. The period of time, T_r, must be at least long enough to allow for the ripple carry time of the four-bit adder. As the clock pulse goes higher in frequency, there would not be enough time during one clock period to allow for ripple carry time and the strobe pulse. In this case, separate odd and even entry registers can be used. These are clocked with alternate sets of four clock pulses. Connections between the entry registers and the adder are made through switching gates, which are shown in Figure 15-12. The gates switch the odd registers to the adder inputs at the time the clock pulses are shifting the next even digit into its entry register. When the even-digit registers are loaded, they are switched to the adder input, while the alternate set of clock pulses shifts the next odd-digit bits into the odd entry registers. This procedure allows almost four clock periods for ripple carry time, increasing the speed at which the serial BCD adder can operate.

Glossary

Serial Adder: A digital circuit that adds two binary numbers one bit at a time. The circuit requires only one full adder and a flip-flop. The flip-flop is used to store the carry bit.

Serial Comparator: A digital circuit that can compare two serial numbers one bit at a time and determine if $A > B$, $A = B$, or $A < B$.

Questions

1. Compare the advantages and disadvantages of serial and parallel adders.
2. What must be added to a full adder circuit to form a serial adder?
3. Explain why the adder-subtractor of Figure 15-3 will be valid for subtraction only if $A > B$. Will addition be valid for $A < B$?
4. Why can't the output of the twos complement adder-subtractor of Figure 15-5 be used directly from the full adder sum output?
5. Explain the operation of the twos complementer circuit. How does it provide a ones complement plus 1 without using an adder?
6. In what way does the serial comparator of Figure 15-8 exceed in capabilities the parallel comparator of Chapter 8?
7. In the comparator circuit of Figure 15-8, why won't the flip-flop toggle when both A and B are 1 levels?
8. When the serial comparator is used to simplify the ones complement adder, as is done in Figure 15-9, why is only one flip-flop needed?
9. For the adder of Figure 15-9, if the numbers A and B are equal but one is negative, which would be complemented? What output would result?
10. What is the advantage of using odd and even entry registers in the BCD adder-subtractor of Figure 15-12?

Problems

15-1 With the timing waveforms of Figure 15-2 changed to A = 1010 and B = 0111, as shown in Figure 15-13, draw the resulting C_O, C_{IN}, and sum waveforms.

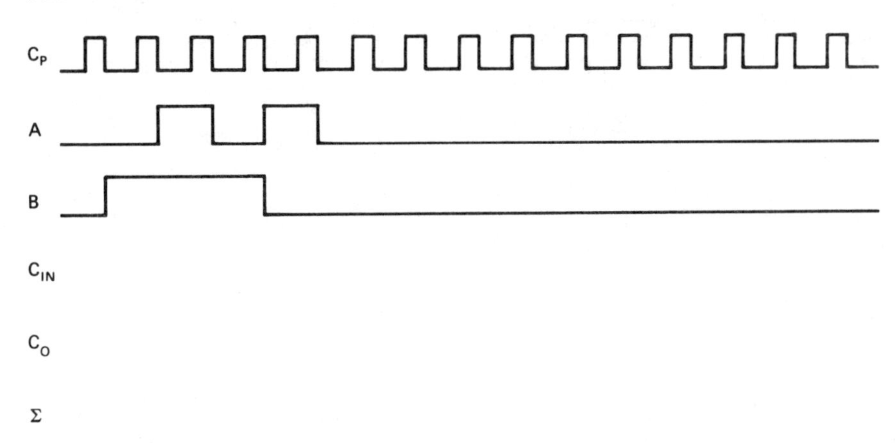

FIGURE 15-13. Problem 15-1

15-2 In place of the exclusive OR gates in Figure 15-3, the TTL/SSI 7450 AND OR invert gate can be used in conjunction with both Q and $\overline{Q}$ outputs of the input registers. Draw the circuit of Figure 15-3 using the AND OR invert gates (see Figure 8-10).

15-3 Subtract the following numbers by twos complement:

24	63	43	21
-63	-24	-21	-43

15-4 Using the necessary logic gates and J-K flip-flops to store the signs of A and B, draw the sign function logic needed for the circuit of Figure 15-5.

15-5 The inputs to the comparator of Figure 15-8 are changed to A = 1010, B = 0110. Correct the remaining waveforms of Figure 15-8(b) to correspond to these input numbers.

15-6 The timing diagram of Figure 15-14 shows inputs to the serial adder of Figure 15-9 for +5 -13. Complete the missing waveforms that will result from this subtraction.

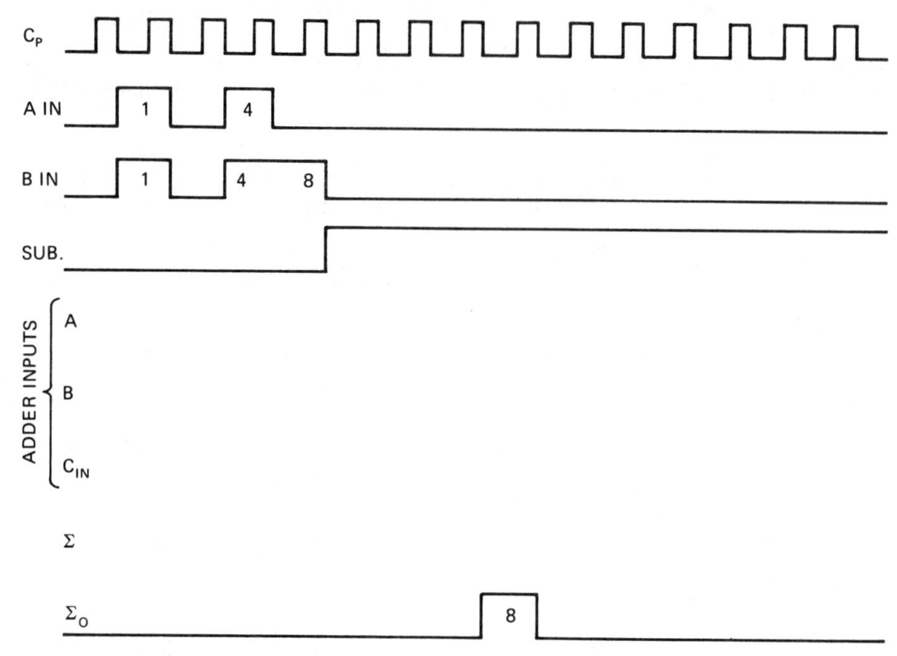

FIGURE 15-14. Problem 15-6

15-7 For a 22 - 9 = 13 operation in the adder of Figure 15-3(a), redraw the rightmost set of waveforms in Figure 15-3(b).

15-8 The four-bit serial adder of Figure 15-2 uses the 74LS112 for its input register, sum register, and carry flip-flop. The full adder uses the 7482 with A_2 and B_2 rounded (low). What is the maximum clock frequency that this serial adder can use? What is the add cycle time if the maximum clock frequency is used? (See timing parameters in the table of Figure 12-22. The maximum propagation delay for the 7482 is 42 nanoseconds.)

15-9 For a +A -B and magnitude of B = 1011, draw a set of waveforms for the operation of the twos complementer of Figure 15-4. Show clock pulse waveform and B shifted in on falling edge of clock pulse. Show also C_P and Q of the J-K flip-flop and the B input to the full adder.

CHAPTER

Digital Multiplexers and Demultiplexers

Objectives

Upon completion of this chapter, you will be able to:

- Assemble digital multiplexers using gate circuitry.
- Use a CMOS circuit for multiplexing either analog or digital signals.
- Study pulse code modulation systems.
- Use commercially available TTL and CMOS digital multiplexers.
- Assemble digital demultiplexers using gate circuitry.
- Learn how decoder glitches can be eliminated.
- Compare analog signals before multiplexing and after demultiplexing.
- Use commercially available TTL and CMOS demultiplexers.

16.1 Introduction

Computers often receive serial data from several different sources—keyboards, CRTs, sensors, other computers, and so forth. However, a computer can only handle one source of information at a time. Therefore, a digital circuit is needed to perform this selection process. It is called a *multiplexer* or *data selector.* A multiplexer has many inputs but only one output. Section 16.2 shows how multiplexers are built using the basic gates. Multiplexers are also used in analog applications. Manufacturers have designed some CMOS devices to multiplex either digital or analog signals.

A *demultiplexer* is the reverse of a multiplexer. A demultiplexer has one input and several outputs. A digital demultiplexer is built using the basic logic gates, as shown in Section 16.3. There is available a CMOS device, 4051, that can be used for either multiplexing or demultiplexing. TTL demultiplexers are generally available as decoders/demultiplexers. This chapter explains the operation and shows other uses for these devices.

16.2 Multiplexers

A digital multiplexer is a circuit that permits information from several channels to be switched onto a single channel. Each input uses the output channel at a different time. The time share of the output channel may be random and uneven between the inputs or may be shared periodically and equally.

Except for being high-speed and automatic, the digital multiplexer acts similarly to the mechanical rotary switch shown in Figure 16–1(a). As shown by the two-to-one multiplexer of Figure 16–1(b), there are two types of inputs required for a digital multiplexer but a single output. When select input S_0 goes high, it enables the signal from D_0 through the AND-OR circuit to the single output lead. When S_0 goes low and S_1 high, it enables the signal from D_1 onto the output.

The AND-OR function is easy to implement in TTL logic. The NAND gate is the cheapest logic gate to produce in TTL technology. As Figure 16–1(c) shows, if all elements of the AND-OR function are NAND gates, there is, by DeMorgan's theorem, the equivalent of AND-OR logic. The same is true if the AND-OR function is expanded either by number of AND gates or by number of inputs to each AND gate.

Figure 16–2(a) shows the AND-OR function expanded to provide a four-to-one multiplexer. Note that the number of AND gates has been increased to equal the number of inputs, but two-input AND gates are still being used. If encoded control signals are used, they can be decoded by the multiplexer circuit as shown in Figure 16–2(b). The decode of the select function is made easy by the fact that it requires only an inverter for each control input and an increase in the number of inputs to the AND gates. Encoded select inputs are generally called *address inputs.*

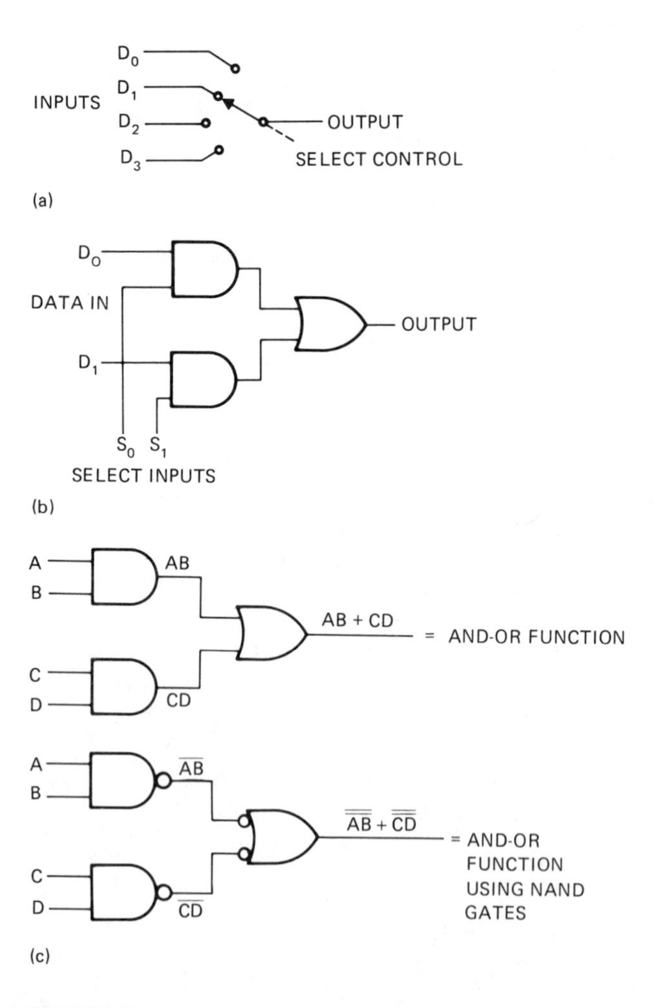

FIGURE 16–1

(a) Four-input rotary switch. Either of the four inputs can be selected by rotating the switch. (b) Two-input digital multiplexer or select circuit. Either D_0 or D_1 can be selected by supplying a high level to its select input. (c) AND-OR function of circuit (b) implemented in TTL logic using NAND gates.

Selection of inputs through the multiplexer can be random in that neither a specific order nor uniform time interval is allotted to each input. In the usual application, however, each input is sampled periodically. When this is the case, the address inputs can be generated using a binary counter. This is shown in Figure 16–3 for an eight-to-one multiplexer. Each input is

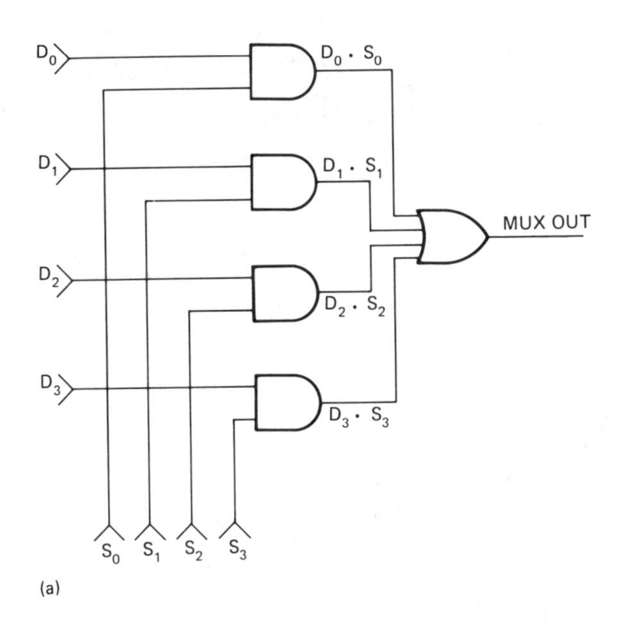

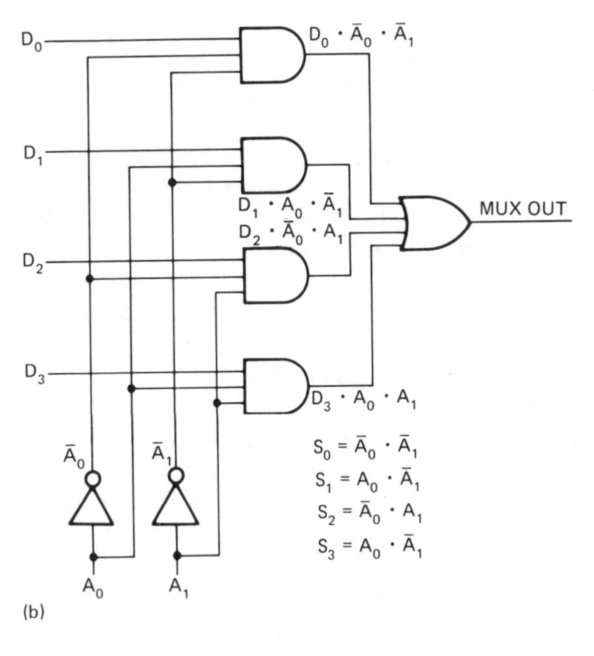

FIGURE 16–2

Four-input multiplexers. (a) Four select signals needed with two-input AND gates. (b) Expanding to three-input NAND gates for an encoded select input.

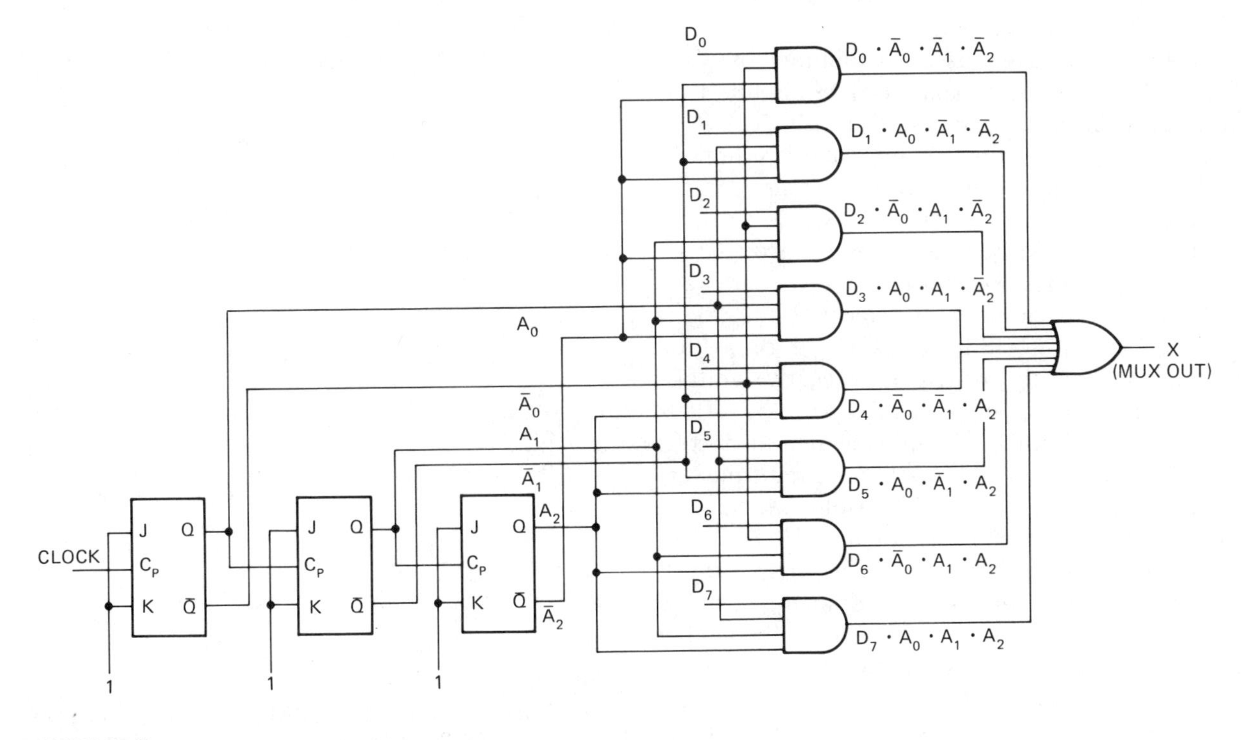

FIGURE 16–3

Eight-input multiplexer controlled by a binary counter such that each input appears at the output, X, for the duration of one clock pulse every eighth clock pulse.

gated onto the output for a period of one clock pulse and reappears on the output once every eighth clock pulse.

There are three types of input signals commonly multiplexed. They are digital, analog, and pulse code modulated (PCM) signals. The digital and PCM signals have the advantage of being compatible with the digital circuits of the multiplexer. They consist of 1 and 0 levels within the 1 and 0 input levels of the multiplexer. Analog signals may be multiplexed by using switches that, when turned on, can pass analog signal levels without altering them significantly. To be controllable by a digital system, the analog switches must turn off and on under control of digital 1 and 0 levels.

A CMOS circuit commonly used for multiplexing either analog or digital signals is the *transmission gate*. The transmission gate is composed of an N-channel FET connected in parallel with a P-channel FET, as shown in Figure 16–4. This provides a relatively uniform conduction of current in either direction. If an inverter is connected between the gates of the two FETs, then both will be turned on with a logic 1 or both will be turned off with a logic 0. That is, a high level on the control input turns the gate on (closes the switch). The resistance $R_P \cdot R_N/(R_P + R_N)$ is typically 250 ohms. A low on the control input turns the device off.

The logic levels used are defined by V_{DD} (1) and V_{SS} (0). For some CMOS circuits that contain transmission gates, the voltage excursions through the transmission gates are limited to V_{DD} and V_{SS}. A few CMOS transmission gate circuits have a third voltage pin, V_{EE}. For these circuits, analog voltage excursions through the transmission gates are limited to V_{DD} and V_{EE}. V_{EE} may be negative with respect to V_{SS}. This feature makes possible the passage of analog voltages with both positive and negative voltage excursions. Three such circuits are the 4051, single eight-channel multiplexer/demultiplexer; the 4052, differential four-channel multiplexer/demultiplexer; and the 4053, triple two-channel multiplexer/demultiplexer. Figure 16–5 shows the logic symbols of these three circuits.

Usually the return for the analog signal is digital ground (normally V_{SS}). Where the analog signals do not return to digital ground but require differential inputs, both positive and negative inputs to be multiplexed simultaneously, the 4052 is ideal. The 4052, however, can be used as a dual four-channel multiplexer where the signal return is to V_{SS}. Figure 16–6 shows segments of four analog signals being applied to a four-channel multiplexer and the resulting multiplexed waveform. Note that even though each signal shares the output for only one-fourth of the time, the frequency and amplitude characteristics of each signal are present in the output. For this to be true, the sampling rate (1/the sampling period) must exceed twice the frequency of the highest frequency signal com-

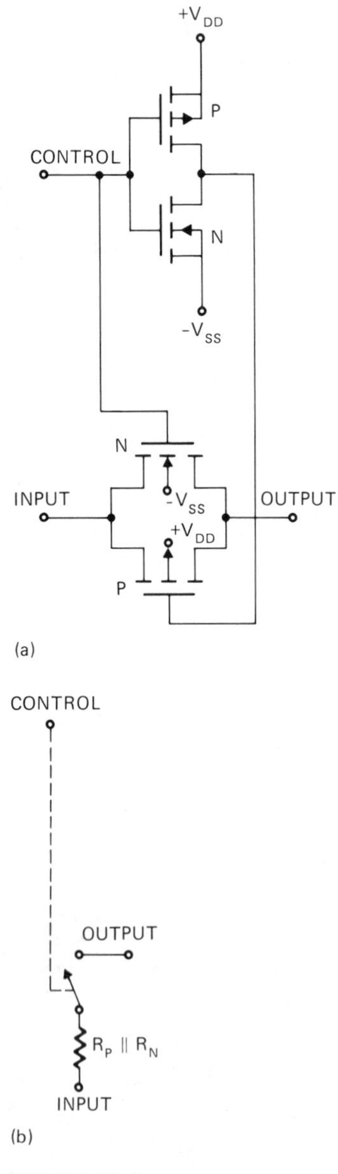

FIGURE 16–4

CMOS transmission gate. (a) CMOS bilateral switch. (b) Equivalent circuit. The P- and N-type transistors in parallel allow signal conduction in both directions when turned on.

FIGURE 16–5

CMOS multiplexers/demultiplexers. (a) CD4051B. (b) CD4052B. (c) CD4053B. Because the transmission gate conducts in both directions, the same devices can be used for either multiplexer of demultiplexer.

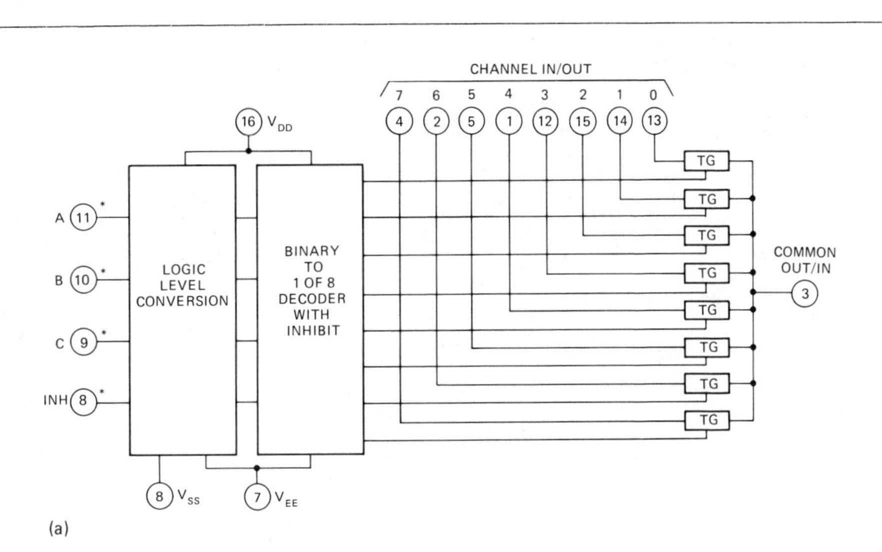

(a)

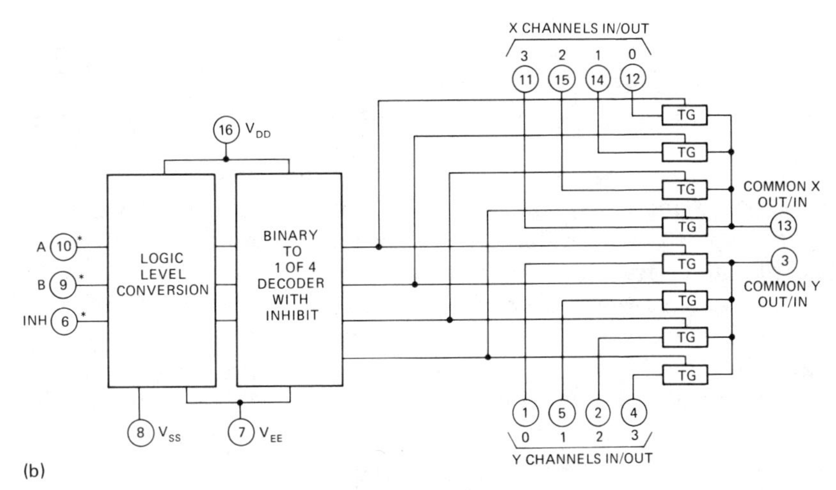

(b)

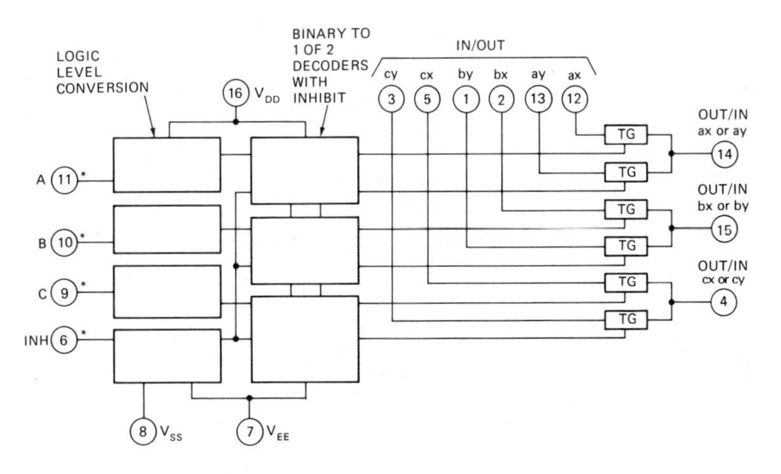

*ALL INPUTS PROTECTED BY STANDARD COS/MOS PROTECTION NETWORK

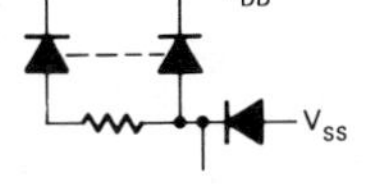

(c)

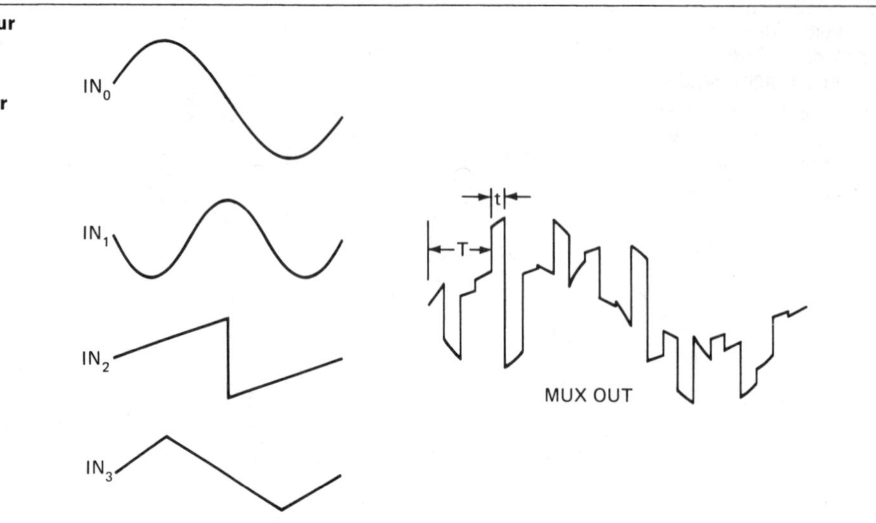

FIGURE 16-6

The results of multiplexing four analog signals. The MUX OUT has amplitude frequency and phase characteristics of all four input signals.

ponent applied to the inputs. Thus, each signal shares the output for a sampling time t. It is sampled at an interval T, where the sampling frequency, f, equals 1/T.

In the multiplexing of pulse code modulated signals, the analog voltage of each channel is sampled prior to its sampling period, and an analog-to-digital conversion is completed, thus loading the A-to-D register with a binary code representing the amplitude of the signal. During the sampling period, the A-to-D registers are read out serially onto the PCM highway. Figure 16-7 shows a typical PCM multiplexing system. The varied techniques for doing this are discussed in Chapter 21.

16.3 Integrated Circuit Digital Multiplexers

A wide variety of digital multiplexers are available in both TTL and CMOS MSI circuits. Figure 16-8 shows the 74LS157 quad two-input multiplexer. As the truth table indicates, when the $\overline{E}$ input is high, the outputs of all four channels are low and neither input is on the output bus. When $\overline{E}$ is low and select (S) is high, I_1 inputs are on the output, Z. When $\overline{E}$ is low and S low, the I_0 inputs are on the output, Z.

Another version of this IC is the 74LS257.

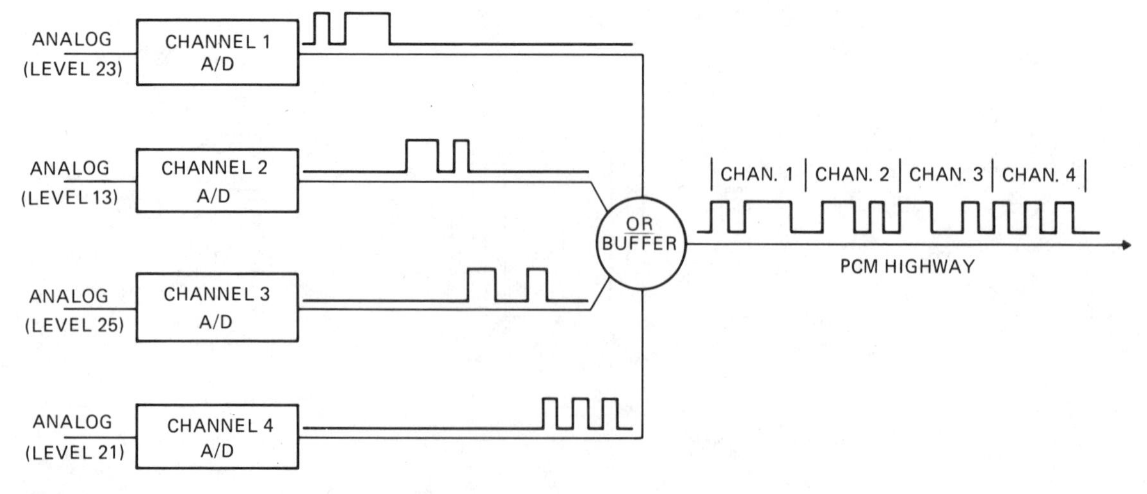

FIGURE 16-7

Four-channel PCM multiplexing system. Each channel uses an A/D converter that, just prior to the sampling time, stores the analog level and converts it to a proportional digital number. During its allotted time, it outputs that number to the PCM highway.

It is identical to the 74LS157 except that the outputs are three-state and go to high impedance when not enabled. The 74LS158 is identical to the 74LS157 except that outputs are complements of the inputs. Likewise, the 74LS258 is identical to the 74LS257 except for inversion at the outputs. The CMOS 4019 is similar except that select leads are separate and outputs are off when both select leads are low.

The 74LS153, shown in Figure 16–9, is a dual four-input multiplexer. It has a separate enable lead for each of the two multiplexers. Two select leads, S_0 and S_1, are decoded to select one of the four inputs, as shown in the truth table. The 74LS253 differs from the 74LS153 in that its outputs are three-state and go to high impedance when not enabled. The 74LS352 is identical to the 74LS153 except that the outputs of the 74LS352 are inverted. The 74LS353 is identical to the 74LS352 except that the outputs of the 74LS353 are three-state.

The 74LS151, shown in Figure 16–10(a), is an eight-input multiplexer with both inverted and noninverted outputs. The 74LS251 is identical to the 74LS151 except that both the inverted and noninverted outputs of the 74LS251 go to high impedance when not enabled. As Figure 16–10(b) shows, the 4512 CMOS provides an eight-input multiplexer with noninverted output. The output is three-state and goes to high impedance when not enabled.

FIGURE 16–8

(a) Quad two-input multiplexer 74LS157. (b) Truth table. The output Z will contain the level of the I_0 leads when select is low, and the level on the I_1 leads when select is high. When enable is high, all outputs are low.

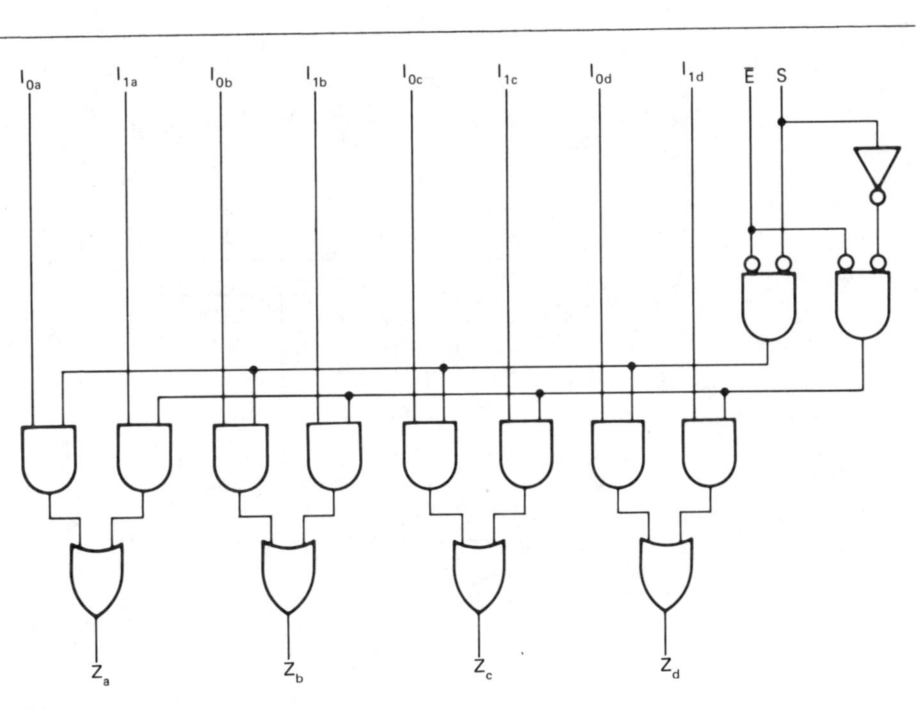

(a)

INPUTS				OUTPUT
$\bar{E}$	S	I_0	I_1	Z
H	X	X	X	L
L	H	X	L	L
L	H	X	H	H
L	L	L	X	L
L	L	H	X	H

H = HIGH VOLTAGE LEVEL
L = LOW VOLTAGE LEVEL
X = IMMATERIAL

(b)

FIGURE 16–9

(a) Dual four-input multiplexer 74LS153. (b) Truth table. The state of the two select inputs determines which of four input levels will appear on the output Z.

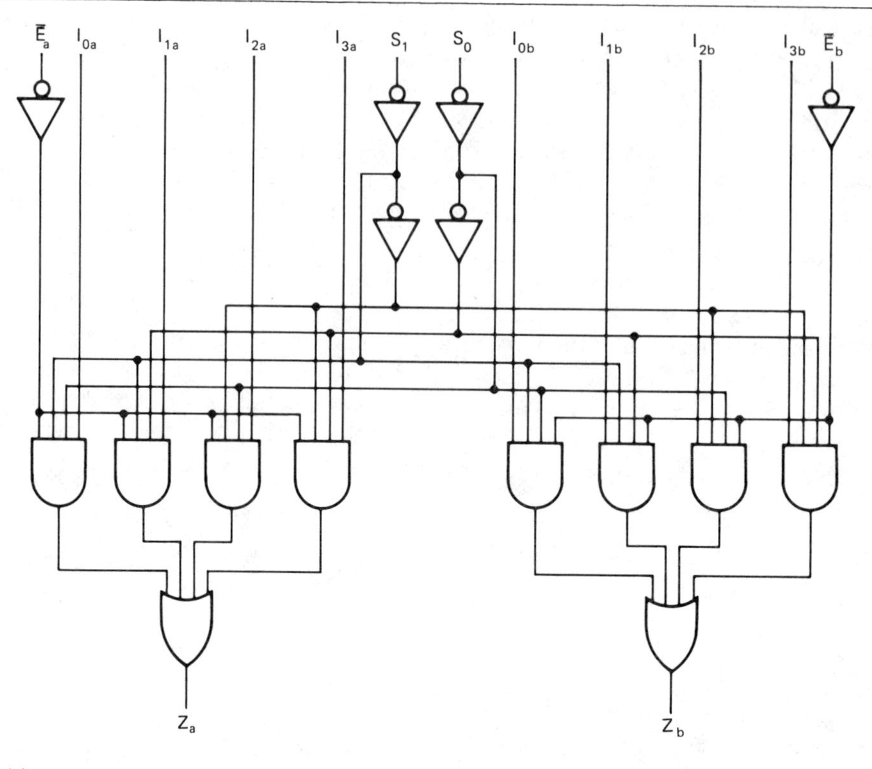

(a)

SELECT INPUTS			INPUTS (a or b)				OUTPUT
S_0	S_1	$\bar{E}$	I_0	I_1	I_2	I_3	Z
X	X	H	X	X	X	X	L
L	L	L	L	X	X	X	L
L	L	L	H	X	X	X	H
H	L	L	X	L	X	X	L
H	L	L	X	H	X	X	H
L	H	L	X	X	L	X	L
L	H	L	X	X	H	X	H
H	H	L	X	X	X	L	L
H	H	L	X	X	X	H	H

H = HIGH VOLTAGE LEVEL
L = LOW VOLTAGE LEVEL
X = IMMATERIAL

(b)

16.4 Demultiplexers

The rotary switch of Figure 16–1(a), drawn to represent a multiplexer, can just as easily represent a demultiplexer by interchanging the labels between inputs and output. Figure 16–11 shows a one-to-four demultiplexer circuit. Note that the decoding AND gates are the same as those used on the multiplexer of Figure 16–2(b) except that on the decoding AND gates a single input lead goes to each of the gates and the outputs remain separate (no OR gate). Integrated circuit demultiplexers usually double as decoder circuits because it is a simple matter of connecting the single data input to a high (1) level for the function to become a binary-to-decimal decode of the select inputs.

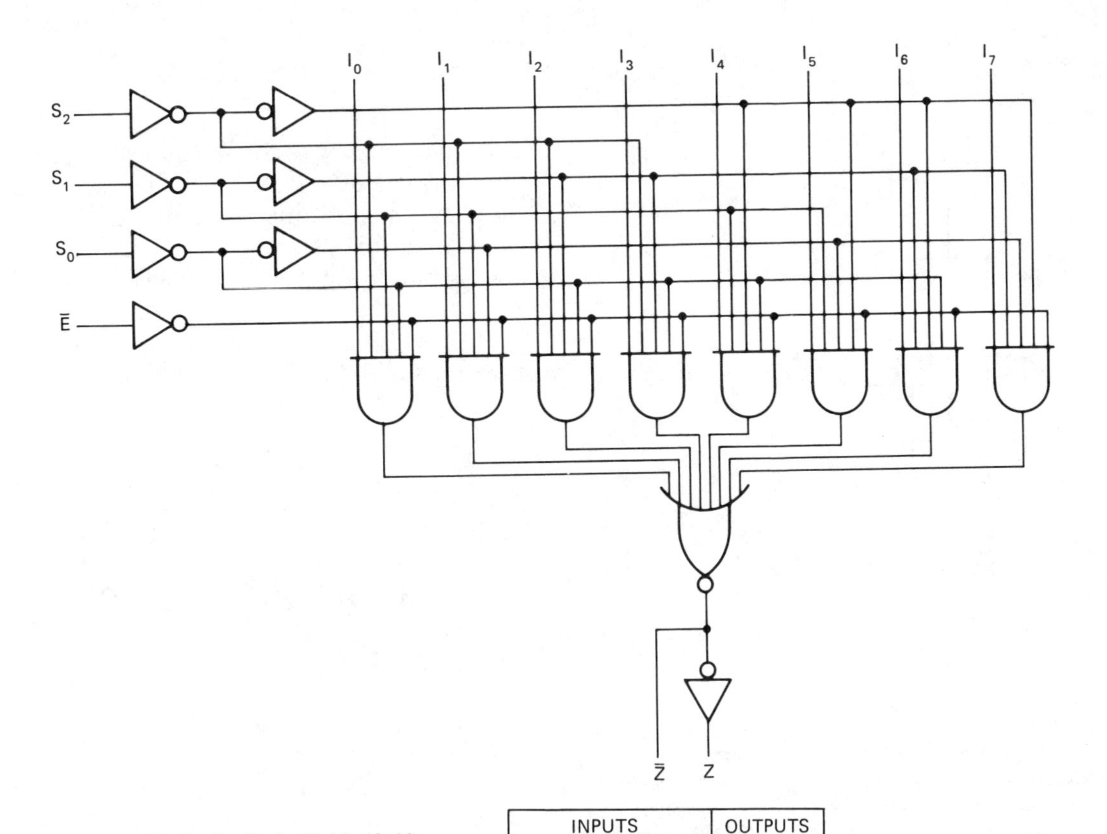

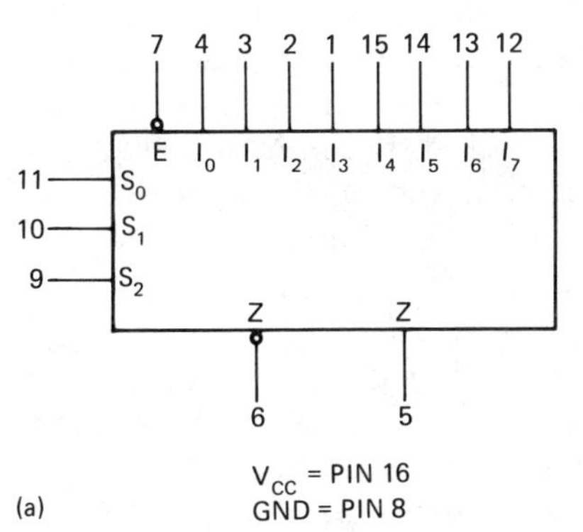

(a)

INPUTS				OUTPUTS	
$\bar{E}$	S_2	S_1	S_0	$\bar{Z}$	Z
H	X	X	X	H	L
L	L	L	L	$\bar{I}_0$	I_0
L	L	L	H	$\bar{I}_1$	I_1
L	L	H	L	$\bar{I}_2$	I_2
L	L	H	H	$\bar{I}_3$	I_3
L	H	L	L	$\bar{I}_4$	I_4
L	H	L	H	$\bar{I}_5$	I_5
L	H	H	L	$\bar{I}_6$	I_6
L	H	H	H	$\bar{I}_7$	I_7

H = HIGH VOLTAGE LEVEL
L = LOW VOLTAGE LEVEL

15 DISABLE Z 14
10 INHIBIT
11 A
12 B
13 C
1 X0
2 X1
3 X2
4 X3
5 X4
6 X5
7 X6
9 X7
V_{DD} = PIN 16
V_{SS} = PIN 8

(b)

FIGURE 16–10

(a) Logic diagram, logic symbol, and truth table for eight-input multiplexer 74LS151. (b) Block diagram of CMOS equivalent device 4512. A, B, and C are the select inputs.

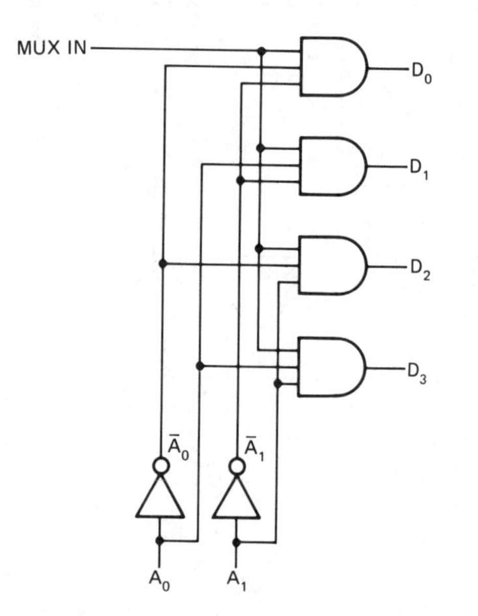

FIGURE 16-11

One-to-four demultiplexer. The demultiplexer circuit appears like a multiplexer circuit with the absence of an OR gate and a single data lead going to all AND gates. The demultiplexer can also serve as a decoder with MUX IN being used as an enable or strobe.

16.5 Decoder Glitches

Figure 16-12 shows a counter circuit connected to a decoder/demultiplexer circuit and the resulting outputs. The data input is fixed high, giving the circuit a decoder function. Although the waveforms of the counter show that all outputs from the counter bits change exactly on the trailing edge of the clock pulse, in actual fact, the counter bits do not change simultaneously due to minor differences in propagation delays within the circuit. In addition the inverters within the decoder circuit cause an uneven arrival of address bit level changes to the decoder AND gate inputs. The decoder gates will output for every combination of 1 and 0 levels they see even though it occurs only momentarily. This leads to a problem of *decoder glitches.* Glitches are unintended momentary outputs that occur during transition of the inputs. For some circuit functions, glitches cause no difficulties, but in many cases, they cause malfunctions particularly if the outputs are connected to clock, set, or reset lines. To avoid decoder glitches, an additional input is connected to each AND gate for a strobe. As Figure 16-13 shows, the strobe is a pulse with a leading edge occurring later than the input transition times and with a width narrow enough to inhibit the outputs during the following transition time.

16.6 Demultiplexed Signals

Some multiplexing and demultiplexing operations create no problems with regard to reclaiming the signal's original form. In many cases, the selected signal serves its complete function when it has control of multiplexing highway. In a periodic system where each signal is on the multiplexed bus periodically but for only a short interval each period, the demultiplexed signal may require additional processing to return it to its original form. Figure 16-14 compares the differences of signals before multiplexing and after demultiplexing for analog signals. Analog signals can be reclaimed to original form by charging a capacitor followed by an amplifier with high input impedance. The high impedance of the amplifier input prevents the capacitor from discharging when the demultiplexed output is turned off. Added improvement is attained with a low-pass filter following the amplifier. The low-pass filter removes the frequency components of the multiplex system that are normally much higher than the signal frequency.

Demultiplexing with an addressable latch is an ideal solution for digital signals. This system is shown in Figure 16-15. The multiplexed signal line is supplied to the D input. As the address input counts through the states 0 through 7, the enable clock $\overline{E}$ is applied to latches 0 through 7. The latch keeps the outputs constant during the period the channel is not addressed. Thus, instead of the signal being demultiplexed through the decoder gates, a delayed clock pulse is demultiplexed, producing a separate clock for each signal to be demultiplexed. The outputs of each latch will store the digital levels through the off periods.

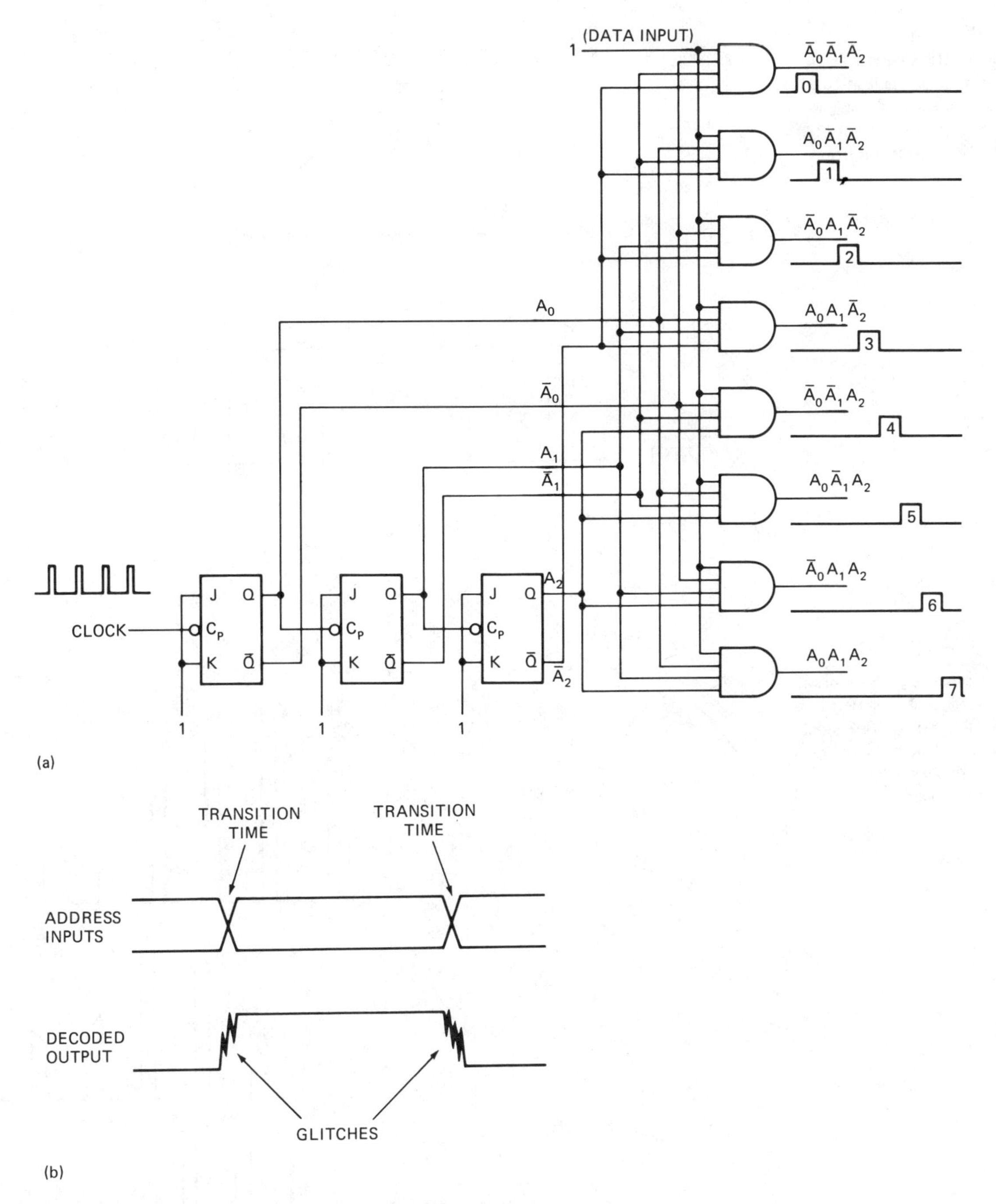

FIGURE 16–12

Three-bit counter/decoder circuit. (a) The input of the demultiplexer is held high, giving the circuit a decoder function. (b) Expanded look at the decoder output shows glitches at both leading and trailing edges caused by the uneven transition of the address inputs.

FIGURE 16–13

A strobe pulse that is used to remove glitches. Its leading edge occurs after the beginning address transition; its trailing edge, before the ending address transition.

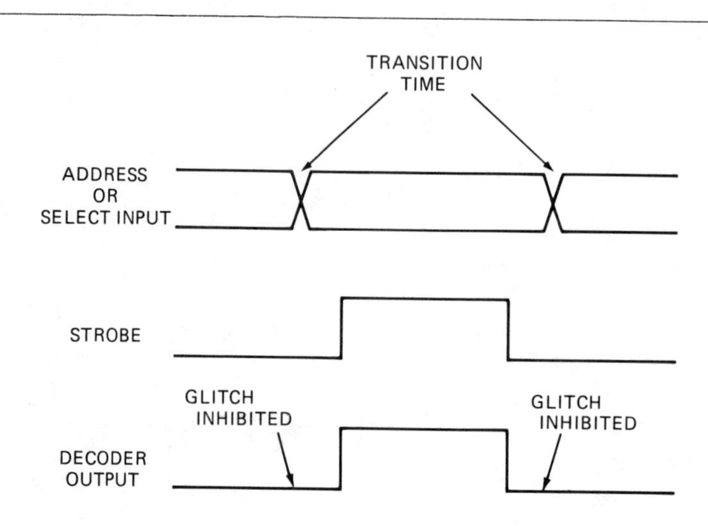

FIGURE 16–14

Comparison of signals before multiplexing and after demultiplexing. After demultiplexing, the analog signal is still only segments of the original and must be further processed.

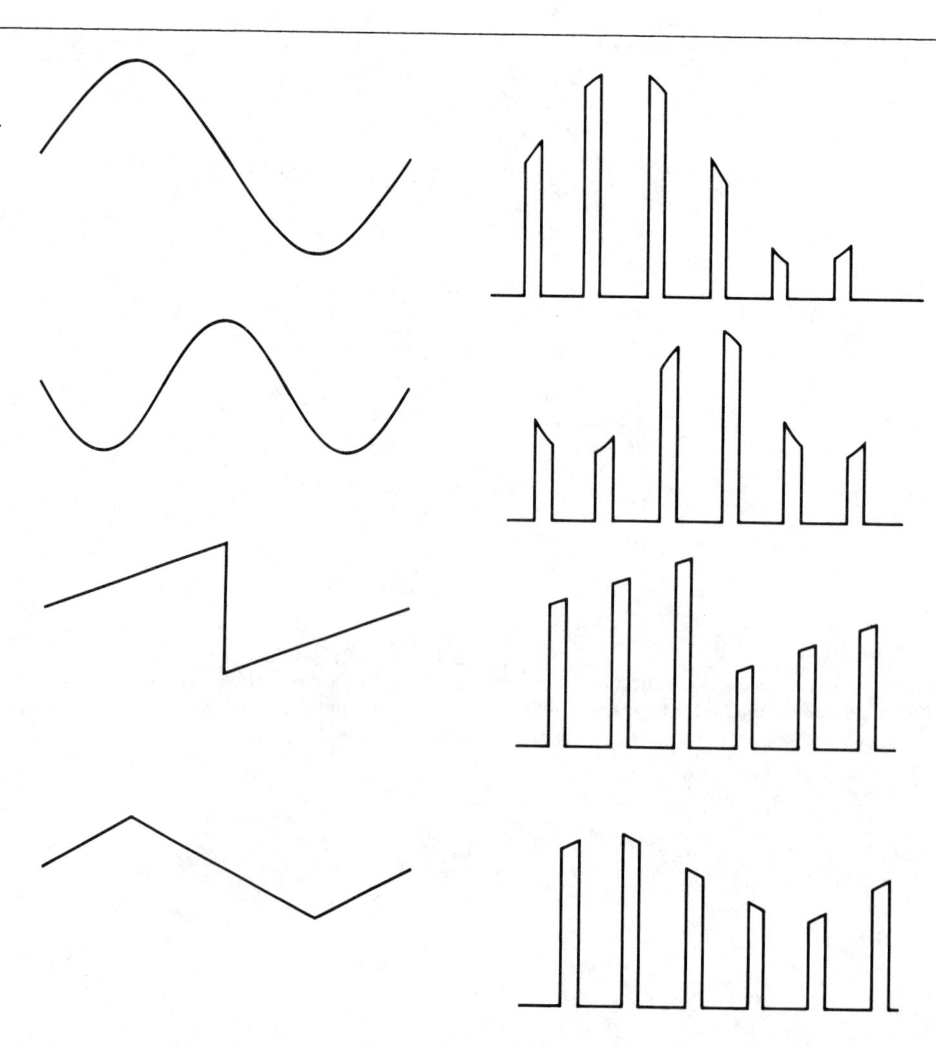

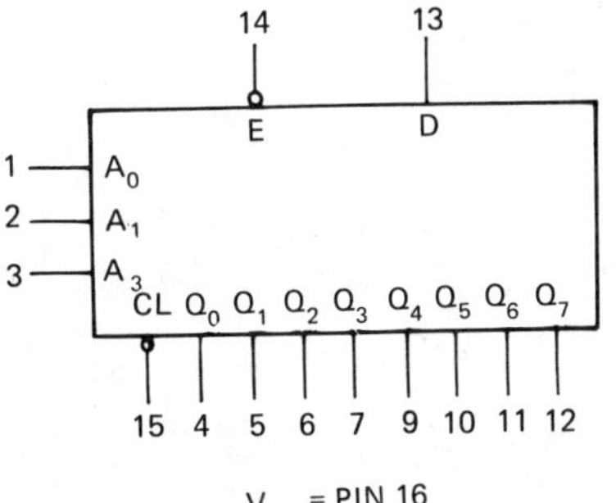

PIN NAMES	DESCRIPTION
A_0-A_2	ADDRESS INPUTS
D	DATA INPUT
$\overline{E}$	ENABLE INPUT (ACTIVE LOW)
$\overline{CL}$	CONDITIONAL CLEAR INPUT (ACTIVE LOW)
Q_0-Q_7	LATCH OUTPUTS

(a)

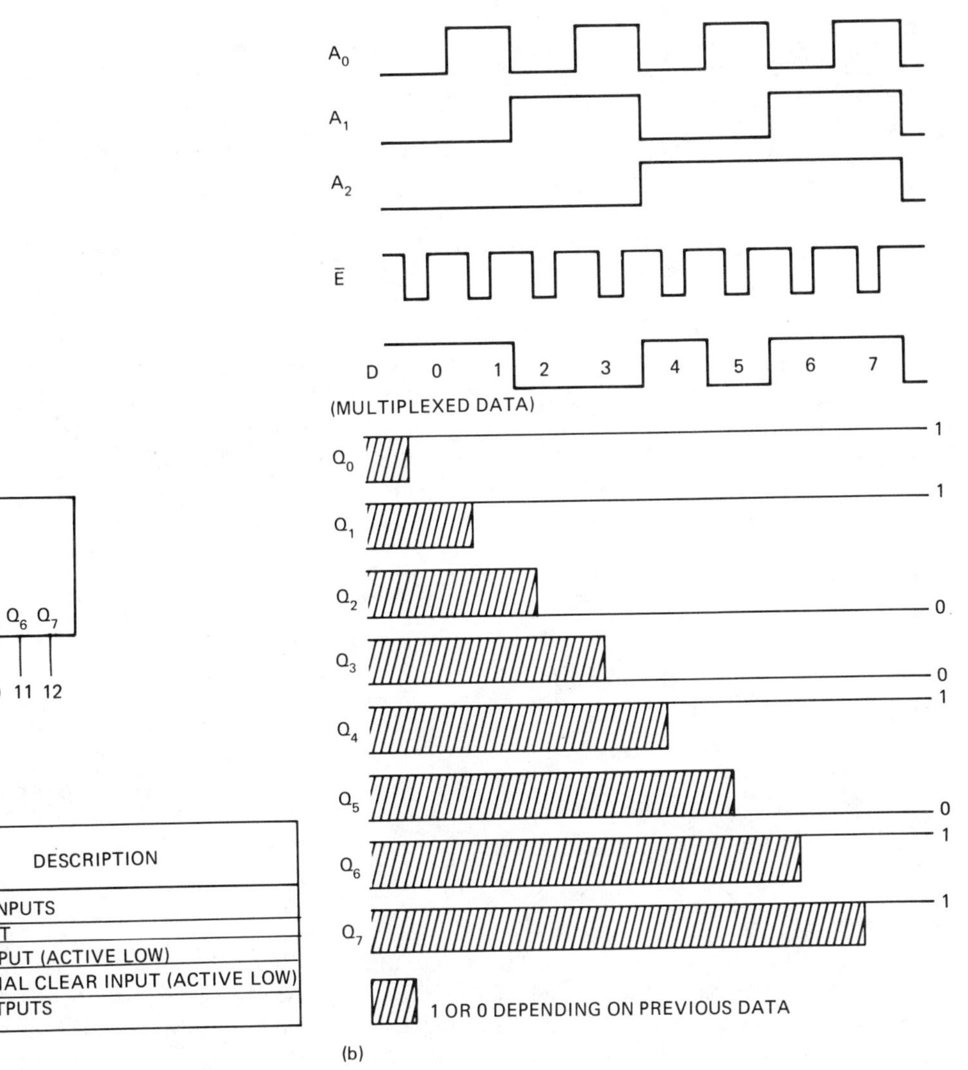

FIGURE 16–15

(a) Pin-out and description of the 74LS259 eight-bit addressable latch. (b) Typical waveforms of demultiplex with an addressable latch. With the unlatched demultiplex, the output channels 0 through 7 contain the correct data only during its respective channel period. With the addressable latch, the data level is stored until the level during the channel period changes.

16.7 Integrated Circuit Demultiplexers

In most cases, integrated circuit demultiplexers have alternate functions. CMOS versions are available as multiplexer/demultiplexer circuits in that they may be wired for either of the two applications. In TTL, they are typically decoders/ demultiplexers.

Figure 16-16 shows the 4051 multiplexer/ demultiplexer connected for a demultiplexing operation. With a multiplexed signal connected to the common input, the device can act as a demultiplexer. The clock pulse to the counter must be synchronized with the multiplexed signal. With $V_{EE} = V_{SS}$, the circuit is a digital demultiplexer, but if V_{EE} is negative with respect to V_{SS}, then the device can work as an analog demultiplexer. In this mode, the control signal will remain digital ($0 \approx V_{SS}$ and

FIGURE 16-16
CMOS 4051 connected as a demultiplexer.

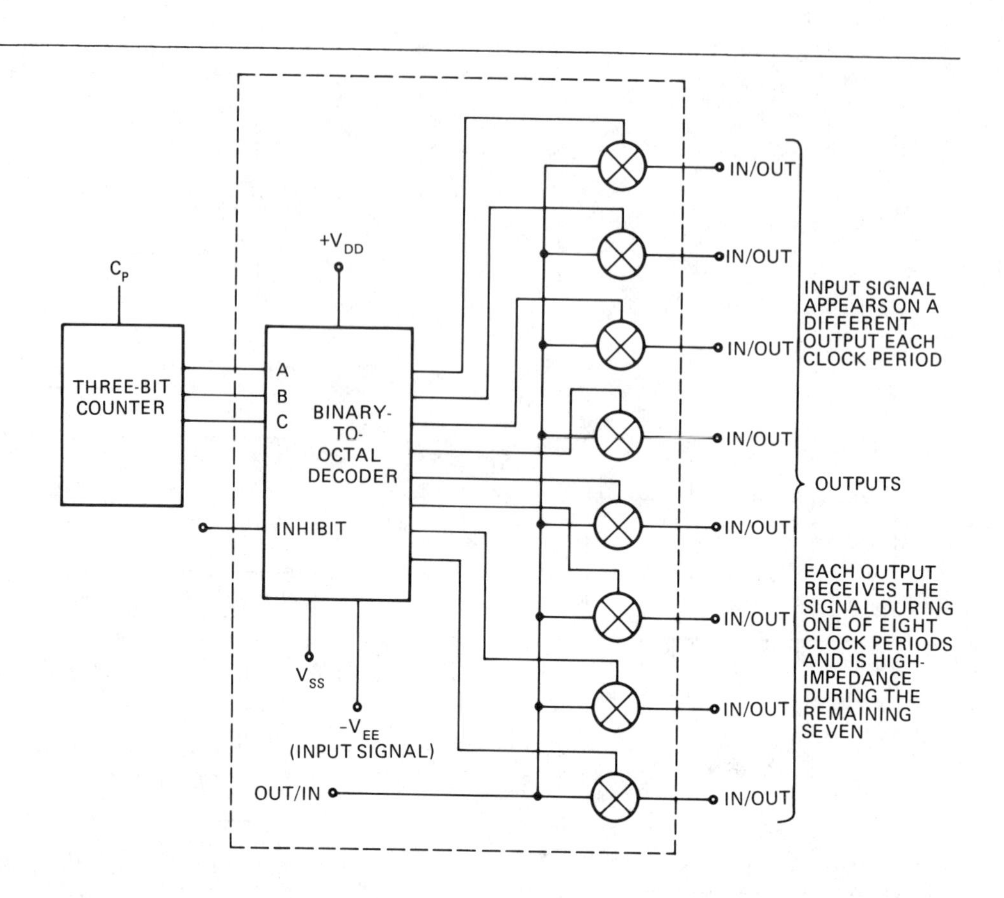

$1 \approx V_{DD}$) while the channel signal passing through the transmission gates may have excursions between the positive V_{DD} and negative V_{EE}. The inhibit in this case is being used as a strobe. During the off time, the demultiplexed outputs are at high impedance. These may be pulled up by connecting a resistor from outputs to V_{DD} or pulled down by connecting a resistor from output to ground. The resistor value may range from 10K to 100K, depending on the circuits loading the output. It is possible for the 4051 to work equally well as multiplexer or demultiplexer because the channels are composed of transmission gates that pass signals in either direction.

TTL demultiplexers are generally available as decoders/demultiplexers. Figure 16-17 shows the 74LS138 one-of-eight decoder/demultiplexer. If both $\overline{E}_1$ and $\overline{E}_2$ are low and E_3 is high, the device acts as an active low binary decoder. If $\overline{E}_2$ is held low and E_3 is held high, then $\overline{E}_1$ can be used as the demultiplexer input. The demultiplexing under this condition is straightforward except for the glitches that are likely during the channel changes. To avoid the glitches, a negative-going strobe may be applied to the $\overline{E}_2$ input or a positive-going strobe may be applied to the E_3 input. The off period of the strobe time will result in an inactive high level at the output.

A highly advantageous circuit that can be used as a demultiplexer is the 74LS259 eight-bit addressable latch. As Figure 16-18 shows, the common D input can be applied directly to the addressed output, producing an active high demultiplexing operation. By placing the device in addressable latch mode, the D input is applied to the input of the addressed latch. This stores the demultiplexed data between sampling intervals and comes closer to recreating the original set of waveforms that existed before the multiplex operation. When using the addressable latch mode, the transition of the address may cause loading of incorrect latches. This can be prevented by using a negative-going pulse at the $\overline{E}$ input. Latches will not load until the leading edge of that pulse. The simultaneous high level on $\overline{CL}$ and $\overline{E}$ during address transition will hold the device in memory mode until the address is stable.

FIGURE 16-17

One-of-eight decoder/demultiplexer 74LS138. (a) Logic diagram. (b) Logic symbol. (c) Truth table.

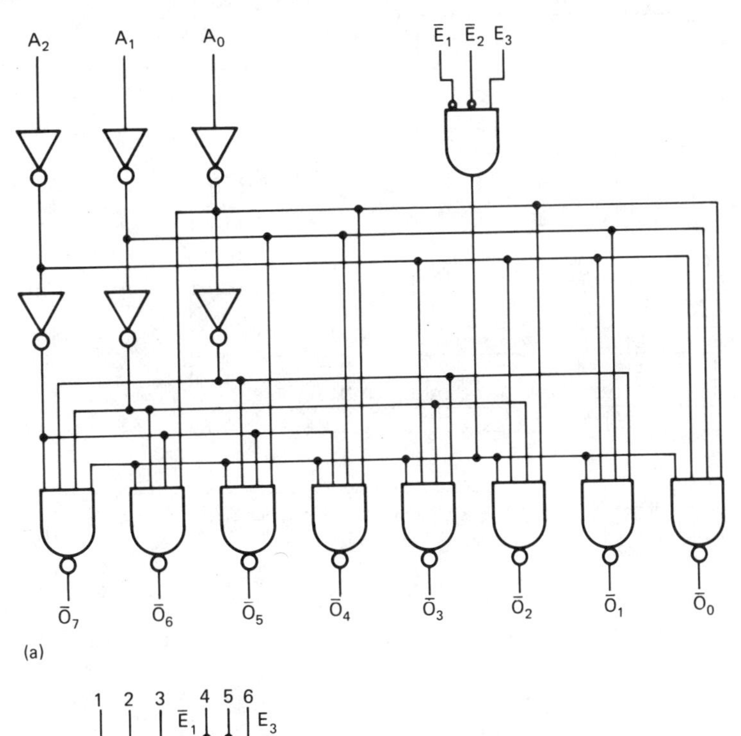

(a)

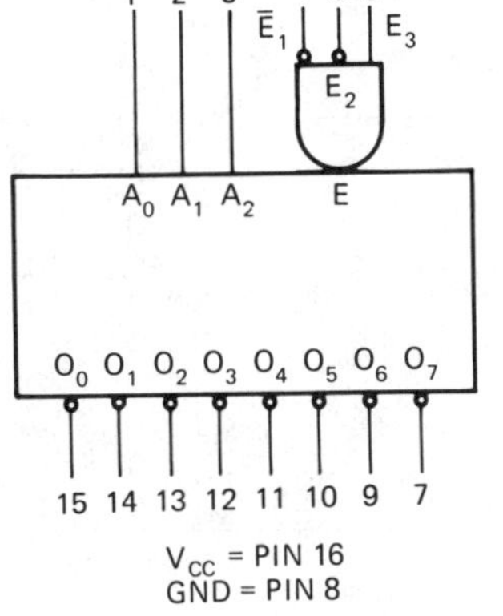

(b)

INPUTS						OUTPUTS							
$\bar{E}_1$	$\bar{E}_2$	E_3	A_0	A_1	A_2	$\bar{O}_0$	$\bar{O}_1$	$\bar{O}_2$	$\bar{O}_3$	$\bar{O}_4$	$\bar{O}_5$	$\bar{O}_6$	$\bar{O}_7$
H	X	X	X	X	X	H	H	H	H	H	H	H	H
X	H	X	X	X	X	H	H	H	H	H	H	H	H
X	X	L	X	X	X	H	H	H	H	H	H	H	H
L	L	H	L	L	L	L	H	H	H	H	H	H	H
L	L	H	H	L	L	H	L	H	H	H	H	H	H
L	L	H	L	H	L	H	H	L	H	H	H	H	H
L	L	H	H	H	L	H	H	H	L	H	H	H	H
L	L	H	L	L	H	H	H	H	H	L	H	H	H
L	L	H	H	L	H	H	H	H	H	H	L	H	H
L	L	H	L	H	H	H	H	H	H	H	H	L	H
L	L	H	H	H	H	H	H	H	H	H	H	H	L

H = HIGH VOLTAGE LEVEL
L = LOW VOLTAGE LEVEL
X = IMMATERIAL

(c)

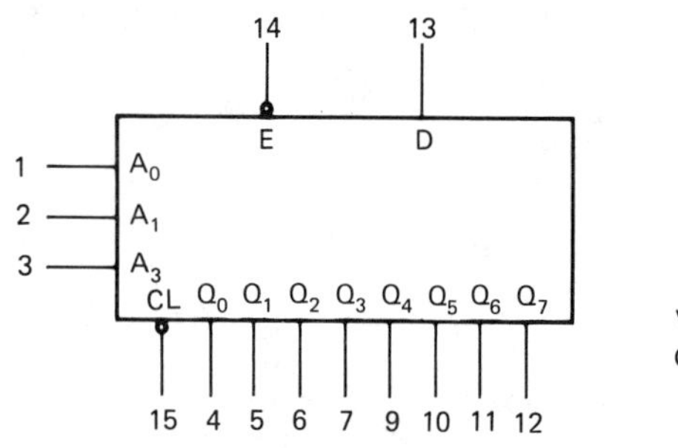

(a)

$\overline{E}$	$\overline{CL}$	MODE
L	H	ADDRESSABLE LATCH
H	H	MEMORY
L	L	ACTIVE HIGH EIGHT-CHANNEL DEMULTIPLEXER
H	L	CLEAR

(b)

INPUTS						OUTPUTS								
$\overline{CL}$	$\overline{E}$	D	A_0	A_1	A_2	Q_0	Q_1	Q_2	Q_3	Q_4	Q_5	Q_6	Q_7	MODE
L	H	X	X	X	X	L	L	L	L	L	L	L	L	CLEAR
L	L	L	L	L	L	L	L	L	L	L	L	L	L	DEMULTIPLEX
L	L	H	L	L	L	H	L	L	L	L	L	L	L	
L	L	L	H	L	L	L	L	L	L	L	L	L	L	
L	L	H	H	L	L	L	H	L	L	L	L	L	L	
.	.	.		.					.					
.	.	.		.					.					
.	.	.		.					.					
L	L	H	H	H	H	L	L	L	L	L	L	L	H	
H	H	X	X	X	X	Q_{t-1}	Q_{t-1}	Q_{t-1}	Q_{t-1}	Q_{t-1}	Q_{t-1}	Q_{t-1}	Q_{t-1}	MEMORY
H	I	I	L	L	L	L	Q_{t-1}	Q_{t-1}	Q_{t-1}	Q_{t-1}	Q_{t-1}	Q_{t-1}	Q_{t-1}	ADDRESSABLE
H	L	H	L	L	L	H	Q_{t-1}	Q_{t-1}	Q_{t-1}	Q_{t-1}	Q_{t-1}	Q_{t-1}	Q_{t-1}	LATCH
H	L	L	H	L	L	Q_{t-1}	L	Q_{t-1}	Q_{t-1}	Q_{t-1}	Q_{t-1}	Q_{t-1}	Q_{t-1}	
H	L	H	H	L	L	Q_{t-1}	H	Q_{t-1}	Q_{t-1}	Q_{t-1}	Q_{t-1}	Q_{t-1}	Q_{t-1}	
.	.	.		.				.						
.	.	.		.				.						
.	.	.		.				.						
H	L	L	H	H	H	Q_{t-1}	Q_{t-1}	Q_{t-1}	Q_{t-1}	Q_{t-1}	Q_{t-1}	Q_{t-1}	L	
H	L	H	H	H	H	Q_{t-1}	Q_{t-1}	Q_{t-1}	Q_{t-1}	Q_{t-1}	Q_{t-1}	Q_{t-1}	H	

Q_{t-1} = PREVIOUS OUTPUT STATE
H = HIGH VOLTAGE LEVEL
L = LOW VOLTAGE LEVEL
X = IMMATERIAL
Z = HIGH IMPEDANCE

(c)

FIGURE 16–18
Eight-bit addressable latch 74LS259. (a) Logic symbol. (b) Mode select table. (c) Truth table.

16.8 Summary

A *multiplexer* is used to route data from several sources to one output line. This time sharing of the output channel is an economical way of sending serial data. Each input is connected to the output for a particular time interval. The multiplexer's select lines choose which input is connected to the output. Figure 16–2 shows a four-input multiplexer with encoded select input lines. Often the select inputs are called *address inputs.* If a binary counter is connected to the address lines, each multiplexer input will be connected to the output channel for one clock pulse, as shown in Figure 16–3.

A CMOS multiplexer can be used for both digital and analog applications. Some CMOS devices can pass both digital and analog signals.

In the multiplexing of pulse code modulation, each analog voltage must be converted to a digital signal before being transmitted. Figure 16-7 shows a typical PCM multiplexing system.

A variety of digital multiplexers are available in both TTL and CMOS medium-scale integration circuits. Section 16.3 examines the differences among the most often used IC multiplexers.

A *demultiplexer* is used to separate input data onto one of several output channels. Figure 16-11 shows how a one-to-four demultiplexer circuit is assembled. TTL demultiplexers usually double as decoder circuits because the data input line can be tied high.

Glitches are a problem in most digital circuits. These unwanted signals can be eliminated from the output by using a strobe pulse. This pulse enables or inhibits the output gates. The inhibit mode is activated until the input data are stabilized, after which the strobe pulse enables the output gates.

The 4051 is a versatile CMOS device because it can be wired for either multiplexing or demultiplexing applications.

Glossary

Demultiplexer: A circuit that separates incoming multiplexed data to selected output lines.

Glitches: Unwanted pulses on the output that occur during transition of the inputs.

Multiplexer: A circuit that permits information from several input channels to be switched onto a single output channel.

Transmission Gate: A CMOS gate that provides a relatively uniform conduction of current in either direction. It can be used to transmit either analog or digital signals.

Pulse Code Modulation (PCM): The conversion of analog signals into a digital code. The value of the code is proportional to the amplitude of the analog signal.

Questions

1. What is a multiplexer? What other name is it likely to be given?
2. Which form of multiplexer can function as either analog or digital?
3. What name is often given to select signals that are encoded in binary?
4. What are the three types of input signals that are commonly multiplexed?
5. What simple function converts a TTL decoder-demultiplexer into a decoder?
6. What is the function of the binary counter in a multiplexing or demultiplexing operation?
7. An eight-to-one multiplexer, with all channels in use, would be addressed by a binary counter of what size?
8. What basic difference exists between analog and digital multiplexers?
9. What additional circuitry is needed for pulse code modulation of analog signals?
10. In what way does a transmission gate differ from the CMOS inverter?
11. How can decoder glitches be prevented?
12. What is the required timing for the leading and trailing edges of a strobe pulse?
13. On the 74LS138, can either $\bar{E}_1$, $\bar{E}_2$, or E_3 be used as a strobe input?
14. What is the reason for the voltage V_{EE} on a CMOS 4051 multiplexer/demultiplexer?
15. The CMOS 4052 multiplexer/demultiplexer has two transmission gates that are turned on with each address input. What special application does this facilitate?
16. What relationship must exist between the frequency components of the signal being multiplexed and the sampling rate?
17. What difference exists in the logic of the 4019 and the 74LS157?
18. Why is timing for a PCM multiplexer likely to be more critical than for an analog multiplexer?
19. If the rotary switch shown in Figure 16-1(a) were changed to represent a demultiplexer, what simple change would be made?
20. Why do demultiplexed analog signals usually require additional processing?
21. Why is the addressable latch ideal for demultiplexing digital signals?
22. In using an addressable latch, what signal is actually processed through the demultiplexer?

Problems

16-1 Draw an eight-to-one multiplexer with encoded select inputs.

16-2 Draw the four-to-one multiplexer of Figure 16-2(b) using all NAND gates.

16-3 Draw the timing diagram for the binary counter of Figure 16-3. On the diagram, show when each input data line is connected to the output.

16-4 What additional gate(s) is needed on the output of Figure 16-3 to obtain X and $\overline{X}$.

16-5 Draw a four-to-one demultiplexer using a CD4052B addressed from a two-bit counter. Have V_{SS}, V_{EE}, inhibit, and the Y channels grounded. Have the MUX OUT of Figure 16-11 applied to the X input. Draw the following waveforms: (a) clock pulses of period t, (b) MUX OUT (assume MUX OUT voltage excursions between 0.5V and 3.5V), (c) counter outputs/address inputs, and (d) demultiplexed X channel outputs.

16-6 If a 16-channel multiplexer is addressed by a four-bit counter clocked at a 5 kHz rate, how long does each input channel occupy the multiplexed bus and at what rate does it reappear?

16-7 A CMOS 4051 is to be used to multiplex an analog signal having voltage excursious between +3V and -3V (6V peak to peak). How would you bias V_{DD}, V_{SS}, and V_{EE}? The control signals are TTL levels.

16-8 With the input waveforms of Figure 16-6, except with IN_1 and IN_2 180° changed in phase, draw the MUX OUT.

CHAPTER

Clock and Timing Circuits

Objectives

Upon completion of this chapter, you will be able to:

- Select or construct an ideal clock circuit for a given application.
- Use a Schmitt trigger circuit to convert irregular waveforms to rectangular waveforms at the digital levels.
- Develop a system of clock and reset pulses by using a clock and cycle multivibrator and by using a digital counter.
- Develop a system of clock and delayed clock pulses.
- Use a one-shot multivibrator to change the length of a rectangular pulse and to delay a rectangular pulse.
- Use a modified one-shot multivibrator to accomplish pulse width modulation.
- Use the 555 timer as an astable multivibrator, and determine the frequency of oscillation.
- Use CMOS ICs as astable multivibrators, and determine the frequency of oscillation.
- Use crystal oscillators with integrated circuits.
- Calculate the component values of a crystal, and determine the frequency of oscillation for crystal oscillators.
- Use clock generator circuits.
- Use CMOS ICs as monostable multivibrators.
- Use the 555 timer as a monostable multivibrator.

17.1 Introduction

At the introduction of each new logic gate or circuit, we began with the truth table or other methods of analyzing it statically. We immediately followed that first analysis with a timing diagram analysis, for most complex digital circuits operate with some degree of timing, cycling, or sequencing. Timing and cycling are done by using either digital clock circuits, which provide a square wave at the logic levels, or a generator, which provides rectangular pulses at the logic levels. The timing waveforms may be generated by an astable multivibrator, a unijunction transistor, or a crystal oscillator, depending on the frequency and accuracy needed.

A number of circuits are used to build astable multivibrators, but one of the most widely used devices is the 555 timer, covered in Section 17.3. The frequency of oscillation depends on only two resistors and one capacitor. A CMOS multivibrator is available or can be assembled using CMOS gates. Crystal-controlled oscillators are commonly used in microprocessor systems. Section 17.7 shows how the frequency of oscillation is calculated.

17.2 Astable Multivibrator

The astable multivibrator, shown in block form in Figure 17-1, is one of the most popular clock circuits used at intermediate speeds. It can be made from two transistors or two FETs. Figure 17-2 shows a typical transistor astable multivibrator using a single power supply.

If, at the time power is turned on, Q_1 saturates, its collector voltage will be dropped across R_{C1}, placing 0V on C_1. This momentarily places 0V on the base of Q_2, turning it off. In a period of time that depends on C_1 R_{B2}, C_1 will charge to V_{CC}. At some point in this charging, Q_2 will saturate, making the X output 0 level and passing a 0 level through C_2 to the base of Q_1, turning it off. With no collector current flowing through R_{C1}, the $\overline{X}$ output will be at the 1 level.

FIGURE 17-1

Astable multivibrator, which produces a continuous rectangular wave between the logic levels and is widely used as a digital clock at intermediate speeds.

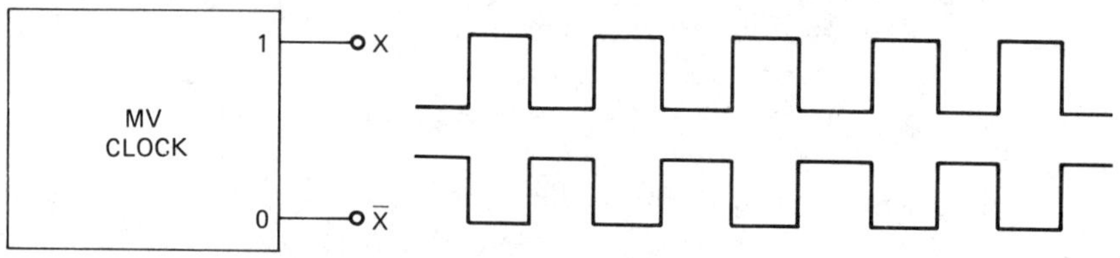

FIGURE 17-2

A discrete component astable multivibrator.

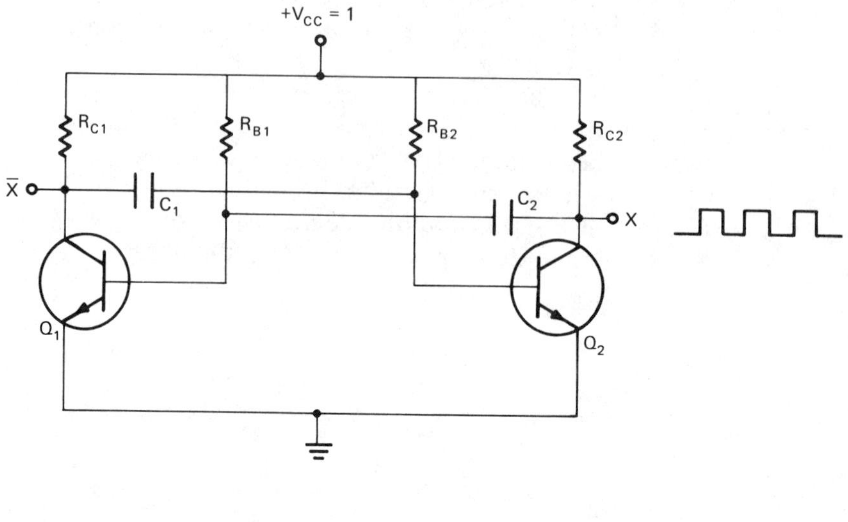

In a period of time that depends on $C_2 R_{B1}$, C_2 will charge to V_{CC}, turning Q_1 back on again, discharging C_2, and turning Q_2 off, starting the cycle over again. This cycling of off and on will continue until the supply voltage is removed. If the components are balanced so that $R_{C1} = R_{C2}$, $C_1 = C_2$, and $R_{B1} = R_{B2}$, then the rectangular waveform will be symmetrical, as in Figure 17-3. The frequency of this square wave is:

$$f = \frac{1}{T_1 + T_2}$$

It is not always desirable to have a symmetrical clock pulse. It may be better to have a longer off time, to allow for numerous events to occur between clock pulses. To accomplish this nonsymmetry, a calculated imbalance between $R_{B1} C_2$ and $R_{B2} C_1$ is used in the circuit.

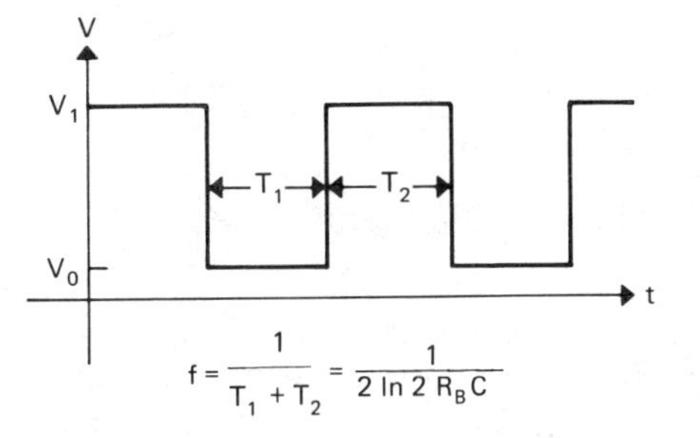

FIGURE 17-3

Frequency waveform of the astable multivibrator (assuming $R_B = R_{B1} = R_{B2}$ and $C = C_1 = C_2$).

17.3 Integrated Circuit Astable Multivibrator

A convenient device for producing an astable multivibrator at frequencies below 100 kHz is the 555 timer. This is available as an integrated circuit in eight-pin TO-5 metal can or eight-pin mini DIP packages, as shown in Figure 17-4. A 556 dual timer is available in a 14-pin package. Used as an astable multivibrator, this device is connected as shown in Figure 17-5(a). The rectangular output coming from pin 3 can be made TTL compatible by using a $V_{CC} = 5V$. The off time and the on time of the output can be controlled rather accurately. The on time equation is $T_{ON} = 0.693(R_A + R_B)C$. The off time equation is $T_{OFF} = 0.693(R_B)C$. This results in a frequency calculation of $f = 1/(T_{ON} + T_{OFF}) = 1.44/(R_A + 2R_B)C$. Figure 17-5(b) shows the variation of frequency with values of resistance and capacitance.

The 555 timer can operate with V_{CC} from 5V to 15V. The supply voltage changes within that range cause only minor changes in output frequency, so the foregoing equations for frequency and pulse width apply over the entire V_{CC} range. The output waveform amplitudes will vary with V_{CC} and are compatible for driving standard CMOS circuits operating at the same supply voltage as the 555 timer.

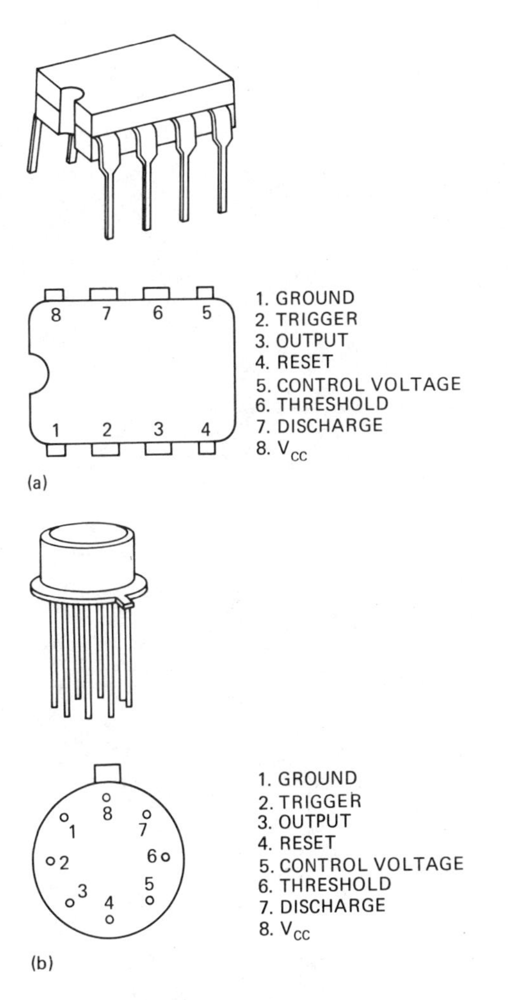

FIGURE 17-4

MC1555 timer (Motorola equivalent to the standard 555 timer). (a) Eight-pin mini DIP package. (b) Eight-pin TO-5 metal can package.

17.4 Astable Multivibrators from CMOS ICs

RC-type oscillators can be assembled from CMOS gates or inverters. The circuit shown in Figure 17-6 can be assembled from 4000A series inverters or inverting gates. The oscillator of Figure 17-6(a) produces an approximate frequency of 1/(2.2RC). The symmetry of the output is dependent on the switching level of the devices being used. If they happen to switch at exactly 1/2 V_{CC}, which is unlikely, the outputs will be symmetrical.

The resistor, R_S, shown in the circuit of Figure 17-6(b), is used to reduce variations of frequency with variations in supply voltage.

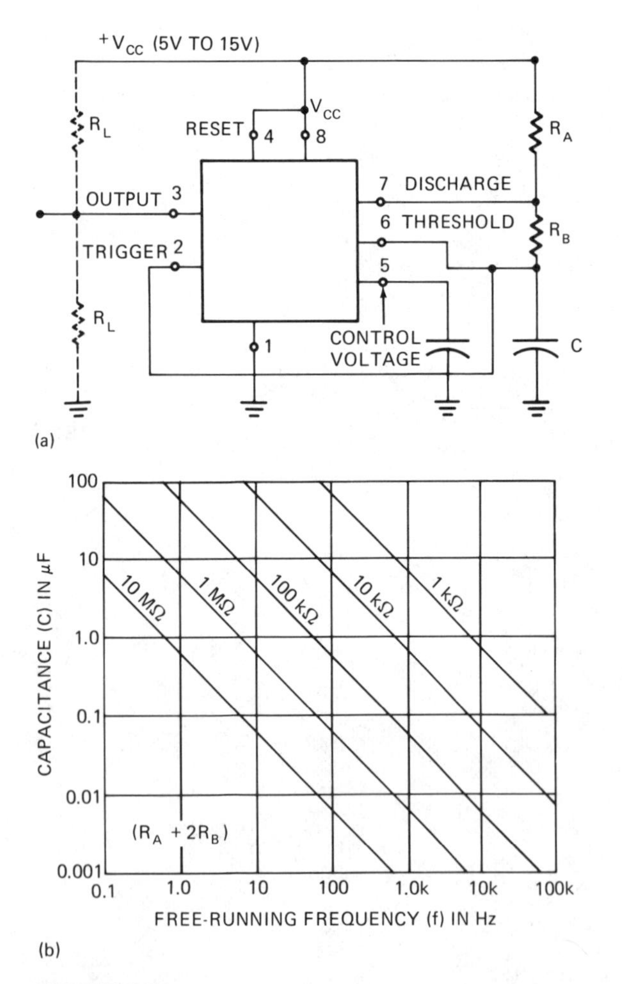

FIGURE 17-5

(a) MC1555 timer connected to provide an astable multivibrator. (b) Free-running frequency provided by selected values of R_A, R_B, and C. Each line is for a value of (R_A + $2R_B$).

R_S does not enter into the calculations for frequency, and the approximate equation remains 1/(2.2RC) for this circuit also.

When a precise frequency is required, the resistor R can be made a potentiometer. In fact, both symmetry and frequency can be made adjustable using the circuit shown in Figure 17-6(c). R_1 can be adjusted for symmetry. If symmetry is off in the opposite direction, the diode must be reversed. Due to the fact that the

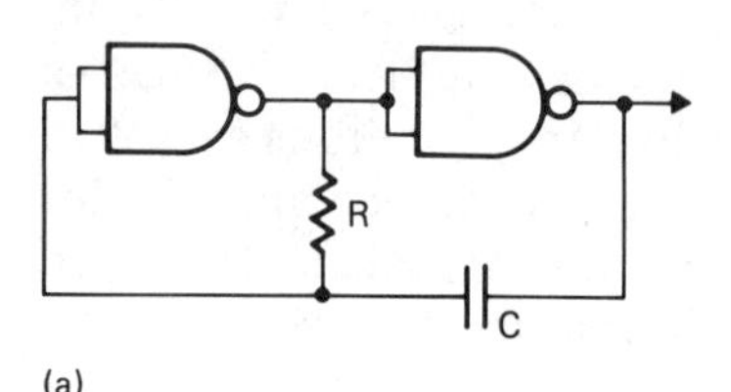

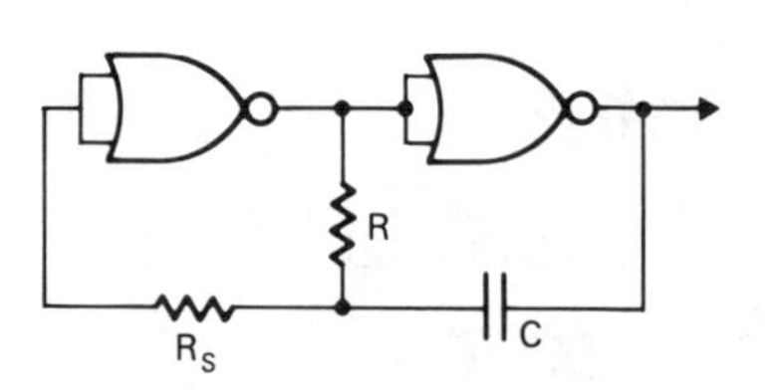

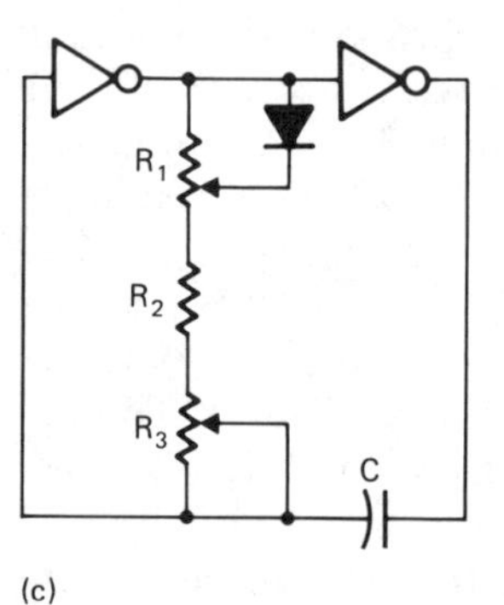

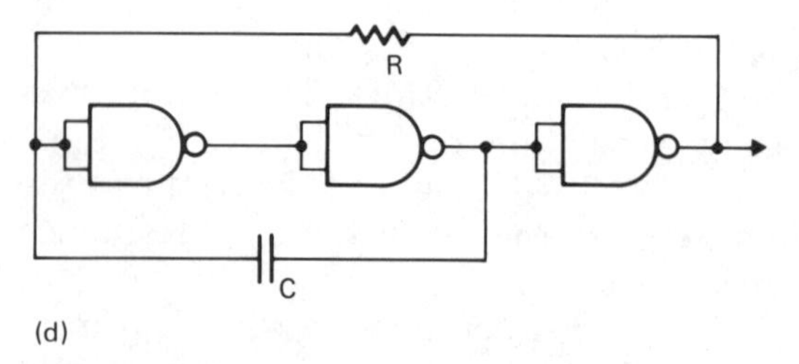

FIGURE 17-6

CMOS astable multivibrator. (a) Circuit with single resistor and capacitor to change frequency slightly with supply voltage changes. (b) Circuit with resistor R_S added to reduce change in frequency caused by supply voltage variations. (c) Diode and two potentiometers added to provide both symmetry and frequency control. (d) Three-gate circuit used to provide astable waveform that is free of ringing at the rising or falling edges.

symmetry adjustment will also change the frequency, R_3, the frequency adjustment, must be readjusted after each symmetry adjustment.

As mentioned earlier, these circuits work well with A series CMOS gates or unbuffered inverters such as the 4069UB. When B series gates are used, the oscillator outputs often show ringing on either leading or trailing edges of the waveforms. If B series CMOS gates must be used to make the RC oscillator, a three-gate device, connected as shown in Figure 17-6(d), can be used to provide an astable waveform that is free of ringing at the edges.

17.5 Unijunction Pulse Generator

A unijunction circuit can generate clock pulses with accuracy and stability comparable to those of the astable multivibrator. Its use is also limited to intermediate clock frequencies. Figure 17-7 shows a typical circuit for a UJT pulse generator. The unijunction symbol differs from the FET symbol in that the arrow of the emitter is slanted. The unijunction has a pair of terminals, base one and base two, labeled B_1 and B_2. For the N-channel unijunction shown in Figure 17-7, these are normally biased with a positive voltage of 5V or more on B_2, with B_1 on the common or returning to common through a low-value resistor. The emitter-to-B_1 junction is a very high-resistance junction and conducts very little current until a positive voltage, V_P, is applied between the emitter and B_1. This voltage, V_P, causes the junction to break down and conduct current like a low-value resistance. This high conductivity will continue until the emitter voltage drops below the level V_V, accounting for the waveform across the capacitor (e_c) shown in Figure 17-7. The capacitor will charge through resistor R until it reaches the level V_P. At that point, the emitter junction breaks down and discharges the capacitor through the low-resistance junction and the resistor R_L. When the capacitor discharges to V_V, the emitter B_1 junction returns to its high-resistance value, allowing the capacitor to charge up again through R until it reaches V_P, starting the cycle over.

The frequency of the UJT pulse generator is a function of a number of variables, one of which is the intrinsic standoff ratio, η. The values of η may vary from 0.4 to 0.8. The intrinsic standoff ratio and the V_{CC} determine the breakdown potential, V_P, of the unijunction, in that $V_P = \eta V_{CC}$. The frequency, however, can be determined as follows:

$$f = \frac{1}{RC \ln (1/1 - \eta)}$$

As η varies from 0.4 to 0.8, the $\ln (1/1 - \eta)$ varies from 0.513 to 1.61, but a middle value $\eta = 0.6$ produces $\ln (1/1 - \eta)$ close to 1 (actually 0.916). When we consider the fact that semiconductor manufacturers supply unijunctions with η specifications in a range of several tenths of a point anyway, an approximate formula of $f \cong 1/RC$ may be used in most practical cases.

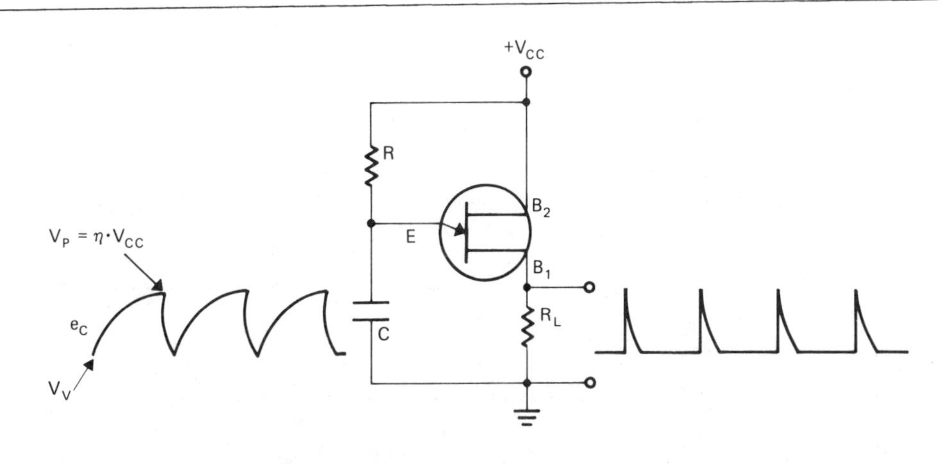

FIGURE 17-7
Typical unijunction pulse generator and the waveforms at the emitter and at the output, B_1.

■ EXAMPLE 17-1

A unijunction transistor has an intrinsic standoff ratio, η, of 0.6. If we use a capacitor of 0.02 μF and a V_{CC} of 5V, what size resistor, R, will produce a pulse rate, f, of 10,000 PPS?

Solution:

$$R = \frac{1}{fC \ln (1/1 - \eta)}$$

$$= \frac{1}{10^4 \text{ PPS} \times 2 \times 10^{-8} \text{ F} \times 0.916}$$

$$= \frac{1}{1.83 \times 10^{-4}} = 5.46\text{K ohms}$$

$$\text{Or,}\quad R \cong \frac{1}{fC} \cong \frac{1}{10^4 \text{ PPS} \times 2 \times 10^{-8} \text{ F}}$$

$$\cong \frac{1}{2 \times 10^{-4}} \cong 5\text{K ohms}$$

The output pulses do not have the ideal rectangular shape but can usually be shaped effectively by running them through some form of shaping network, as in Figure 17-8. The output pulses have a relatively fast leading edge, but the trailing edges are too slow for reliable operation of digital circuits. After the pulses have been passed through an inverter, the relatively fast rising edge has become an even faster falling edge, which can operate a toggle flip-flop, producing a square wave at half the unijunction frequency.

Another circuit having the particular function of converting nonrectangular waveforms to rectangular waveforms at the logic levels is the Schmitt trigger circuit. As Figure 17-9 shows, the unijunction pulse generator output can be shaped by applying it to a Schmitt trigger circuit.

17.6 Schmitt Trigger Circuit

Although the Schmitt trigger circuit is not in itself a timing circuit, it is often used with other timing generators whose outputs are not rectan-

FIGURE 17-8

Unijunction pulse generator output shaped by inverting. The fast leading edge becomes a fast falling edge, reliable for triggering a toggle flip-flop and producing a rectangular waveform at half the unijunction frequency.

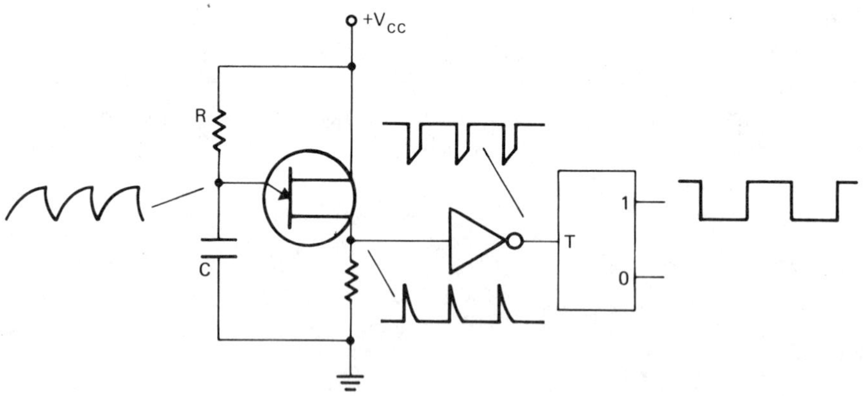

FIGURE 17-9

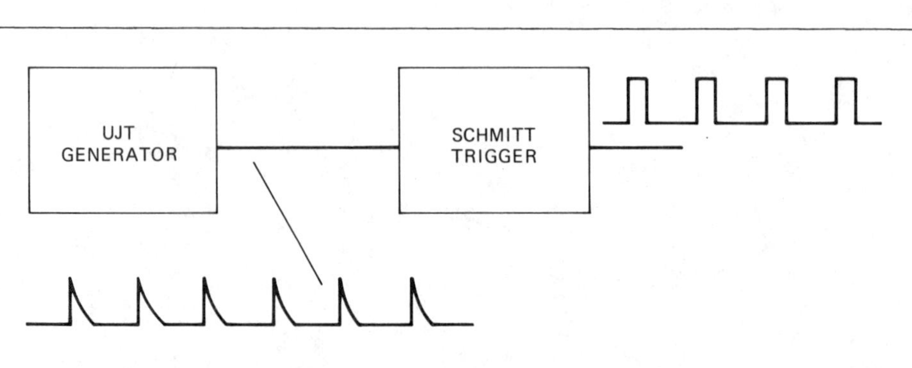

Schmitt trigger circuit, designed to speed up the leading and trailing edges of waveforms applied to its input and thus make the UJT pulses rectangular.

gular waveshapes between the logic levels. Figure 17-10 shows a typical Schmitt trigger circuit. There are two possible outputs from this circuit—the 1 level, which occurs when Q_2 is turned off (Q_1 on), and the 0 level, which occurs when Q_2 is turned on (Q_1 off). The 1 level will be approximately V_{CC}; the 0 level will be $V_0 = V_{EE} + V_{CE\ SAT}$. To be useful in a particular logic system, the V_0 must be safely below the maximum 0 input specification of the logic unit it is driving. Unfortunately, a Schmitt trigger with R_E returned to ground tends to have a logic 0 output too high to be compatible with integrated circuit logic units. It may, therefore, be necessary to return the emitter and R_2 resistors to a negative power source to translate the 0 level downward. The object of the Schmitt trigger is not only to convert the input into voltage transitions between the two logic levels but also to make the transitions between the 1 and 0 levels as rapid as possible. This rapid transition is accomplished by regenerative feedback through the resistor, R_E.

R_{C1} is made larger than R_{C2} by about two to one or higher. This ensures that the voltage V_E is about twice as high with Q_2 conducting than it is with Q_1 conducting.

If Q_2 is turned on, there will be a voltage, V_E, developed across R_E, helping to ensure that Q_1 remains turned off. To turn Q_1 on and have Q_2 turn off, the input on the base of Q_1 must rise to the level of $V_{BET} + V_E$. This level is referred to as the upper triggering level—the input level that results in the output switching to the 1 level. As Q_1 turns on, Q_2 turns off, reducing the voltage V_E, ($I_{E1} = \frac{1}{2} I_{E2}$), which helps to speed Q_1 to the on state. Q_1 will remain on and the output will remain 1 until the input signal falls to a voltage even lower than the Q_1 turn-on level. The turn-off level for Q_1 is lower than the turn-on level because of the reduced value of V_E. However, when Q_1 turns off, its turn-off is quickened because the Q_2 subsequently turns on, raising the level of V_E. The voltage V_E acts as a positive feedback, quickening the change of state, v, and reducing the rise and fall time of the rectangular waveform. Figure 17-11 shows the rectangular waveshape occurring at the Schmitt trigger output resulting from a sine wave applied at the input.

The Schmitt trigger is available in integrated

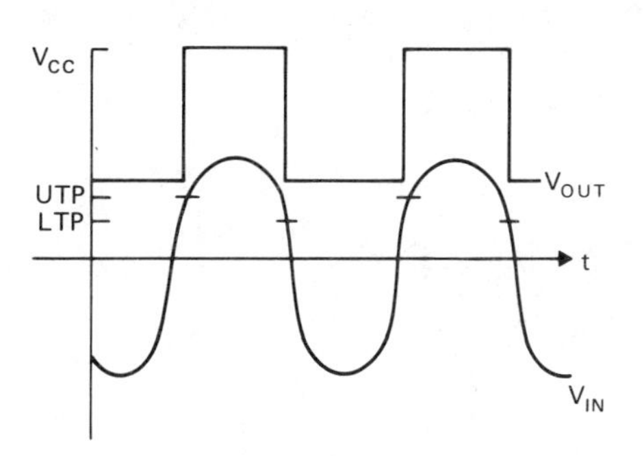

FIGURE 17-11

Width of output pulses determined by upper and lower trip points on input signal.

FIGURE 17-10

Schmitt trigger circuit, used to convert a sine wave to a rectangular wave.

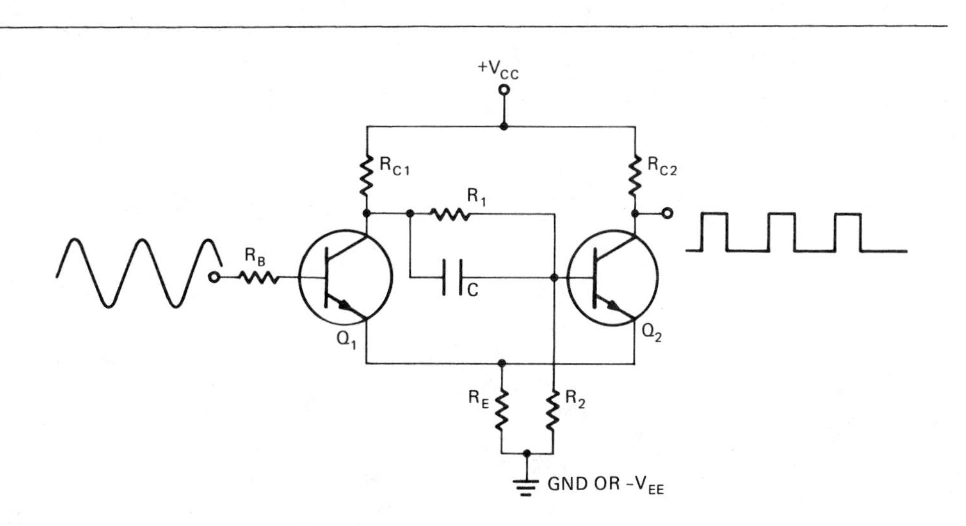

circuit form. Figure 17-12(a) is a Fairchild TTL/SSI 7413 dual NAND Schmitt trigger. Figure 17-12(b), the graph of output voltage versus input voltage, shows a hysteresis of about 0.8V. The typical turn-off delay is 18 nanoseconds, indicating an operating range in excess of 20 megahertz.

17.7 LC and Crystal Oscillators

Modern integrated circuits have reduced propagation times to the point that clock frequencies of 10 megahertz and higher have become common. At these frequencies, the multivibrator and UJT circuits are not practical because their frequency of operation depends on RC time constants and, at these frequencies, the RC values become too low for practical circuit functions. If a high degree of frequency, accuracy, and stability is not needed, then an LC oscillator can be used, but crystal oscillators have become so economical in past years that crystals have generally replaced the LC tuned circuit in this type of oscillator. Figure 17-13 shows a simplified diagram of the LC or crystal oscillator. It contains three essential elements—the amplifier, a feedback path, and the crystal or tuned circuit. The feedback must provide a 180° phase shift. The product of gain times the feedback ratio must be greater than 1.

Figure 17-14 shows two forms of crystal oscillators. The circuit of Figure 17-14(a) uses the fundamental crystal frequency. Crystals can be ground for fundamental oscillation at frequencies as high as 20 megahertz. Beyond this, the crystal becomes too thin and fragile. To obtain clock pulses above 20 megahertz, the oscillator circuit must operate on overtones or harmonics of the crystal frequency. This is true of the circuit of Figure 17-14(b). The output circuit is tuned to the third harmonic of the crystal frequency. Under these conditions, the oscillator output is very low in power and a buffer amplifier must be used to bring the sine wave up to the necessary voltage and power levels. Another method is to use a crystal oscillator of the fundamental type followed by a frequency multiplier circuit, as shown in Figure

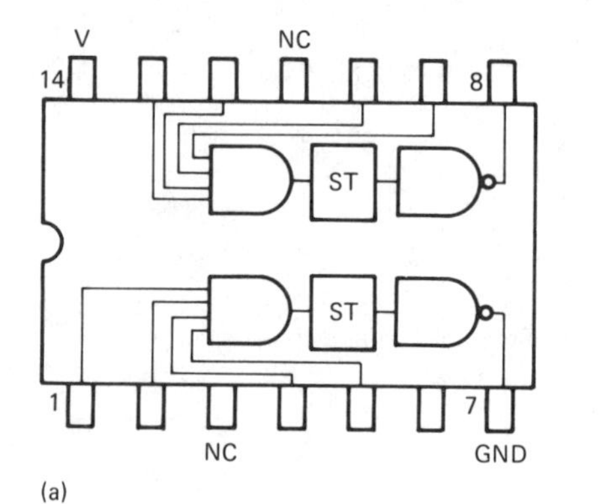

(a)

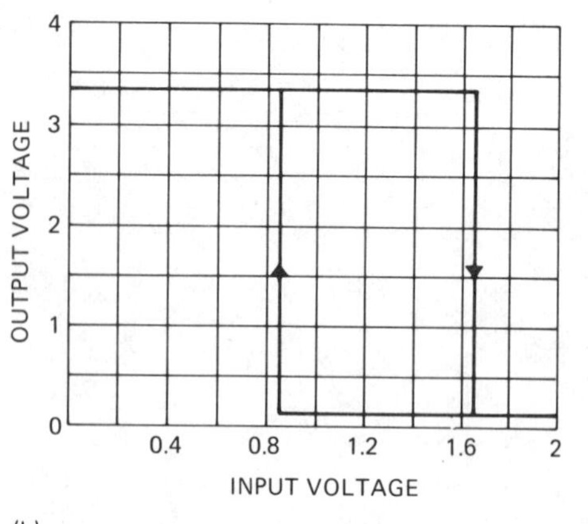

(b)

FIGURE 17-12

(a) Fairchild TTL/SSI 7431 dual NAND Schmitt trigger. (b) Output voltage versus input voltage graph, showing hysteresis and inversion caused by NAND gates.

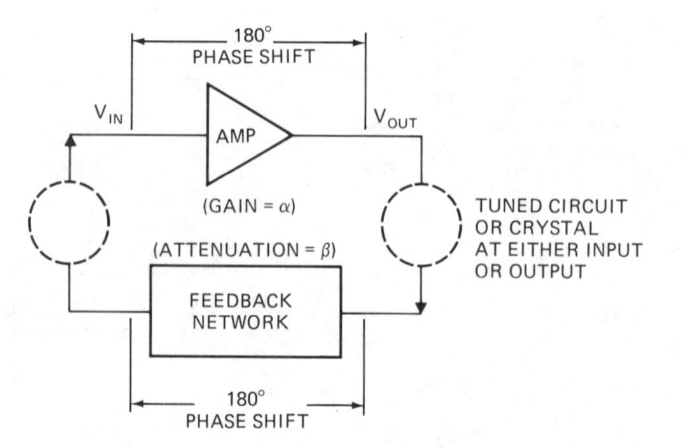

FIGURE 17-13

Simplified diagram of an LC or crystal oscillator showing essential elements—amplifier, feedback path, and crystal or tuned circuit.

17-15. The sine wave outputs of oscillators and subsequent frequency multipliers may be converted to logic-level waveforms by using overdriven amplifiers, diode clamping, and Schmitt trigger and other wave-shaping circuits.

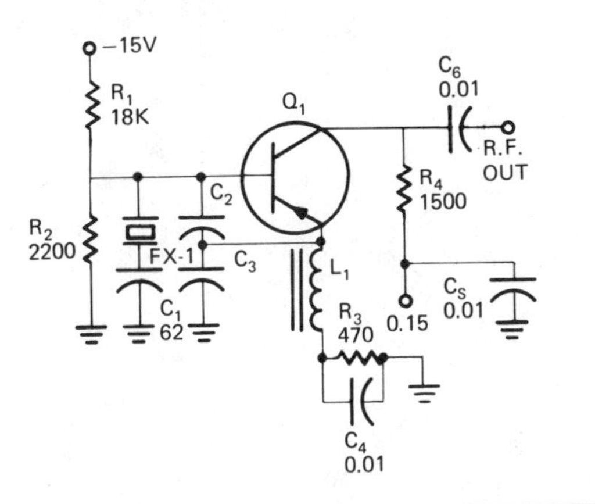

FREQUENCY	C_2	C_3	L_1
2,000–10,000 kHz	500 pF	500 pF	5 mH
10,000–20,000 kHz	150 pF	36 pF	11 μH

(a)

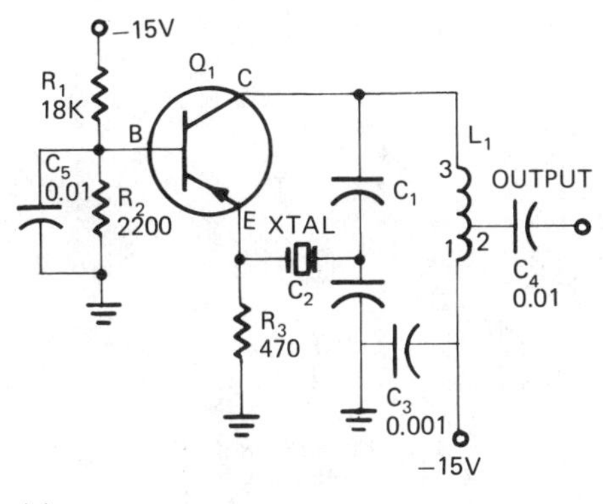

(b)

FIGURE 17-14

(a) Crystal oscillator using the fundamental crystal frequency for oscillations below 20 MHz. (b) Crystal oscillator using overtones or harmonics of the crystal frequency to obtain oscillation of 15 to 60 MHz.

17.8 Integrated Circuit Crystal Oscillators

Integrated circuits that require only the connection of a crystal and a load capacitor to form a crystal-controlled oscillator are commonly used with microprocessors. Figure 17-16 shows the 8224 clock circuit, which is designed to supply two nonoverlapping clock phases for the 8080 microprocessor. The oscillator operates at the crystal frequency or, if the proper tank circuit is connected at pin 13, it will operate at a multiple of the crystal frequency. The clock frequencies are one-ninth of the oscillator frequency. In addition to the two clock phases, the 8224 supplies correct timing for other 8080 microprocessor signals, such as reset ready and synchronization.

The CMOS inverter proves to be an effective component as the amplifier element in a crystal-controlled oscillator circuit. The normal operation of the p-n pair of the CMOS inverter is to switch between cutoff and saturation states. To function in a crystal oscillator, the complementary pair must be biased into the transition zone, making it sensitive to the vibrations of the crystal. This is accomplished by connecting a high-value resistor between the input and output as shown in Figure 17-17(a). This results in nearly equal input and output voltages that are midway between ground and V_{DD}. Figure 17-17(b) shows the transfer characteristics of the device. Note that the voltage gain around the bias point $\Delta V_O / \Delta V_I$ is relatively high. At V_{CC} = 5V, the line is steeper, indicating a higher gain at the lower voltage.

All the details of crystal oscillator design cannot be described here, but some information about the crystal and how it affects the frequency and feedback is essential at this point.

FIGURE 17-15

Frequency multiplier used with crystal oscillator for crystal control of clock frequencies in excess of 20 MHz. The sine wave outputs may be shaped by using an overdriven amplifier.

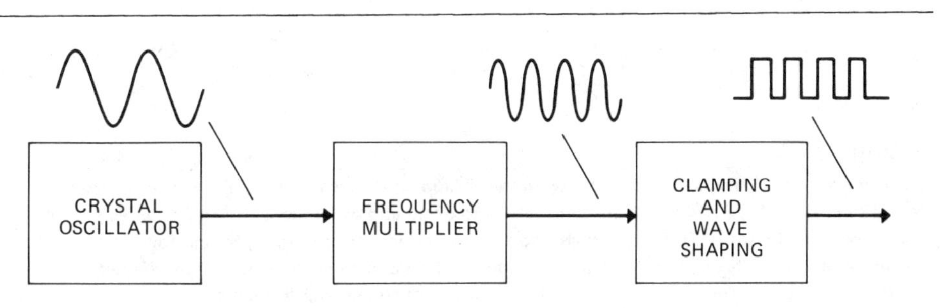

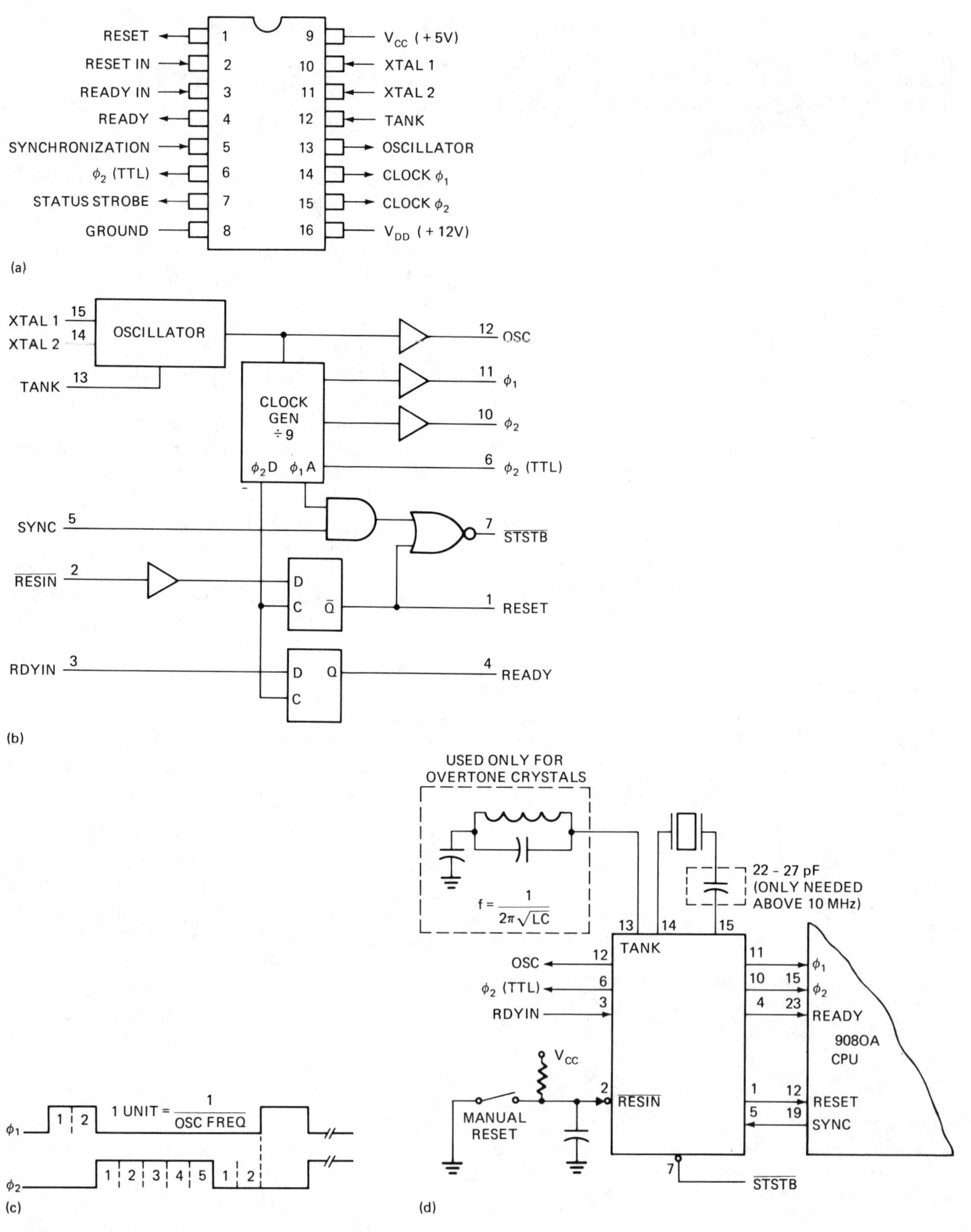

FIGURE 17-16

8224 clock circuit designed to provide two nonoverlapping clocks for the 8080 microprocessor. (a) 8224 pin-out. (b) 8224 block diagram. (c) Ratio of on and off times of ϕ_1 and ϕ_2 clocks measured in oscillator periods. ϕ_1 is high two periods and low for seven. ϕ_2 is high for five periods and low for four. (d) Connection diagram of the 8224 clock generator. Tank circuit is used only if oscillator is to operate at third harmonic of the crystal frequency.

The crystal is a finely ground piece of quartz mounted between two metallic plates. It functions like a high Q tuned circuit. Its symbol and equivalent circuit are shown in Figure 17-18. The crystal has a series resonant frequency of inductance (L_S), series capacitance (C_S), and series resistance (R_S). This branch is often called its motional arm. When the crystal is used as a series resonant circuit at frequency $f_S = 1/(2\pi L_S C_S)$, the parallel capacitance (C_0) is of little concern. The parallel capacitance (C_0) results from the capacitance between the plates and connecting leads. In conjunction with the motional arm, C_0 forms a parallel tuned circuit resonating at f_0, which is a slightly higher frequency than f_S. At the higher frequency, the inductive reactance in the series resonant branch dominates, and an equivalent parallel resonant circuit results, as shown in Figure 17-19. The series resonant branch of the circuit functions like an inductance, $L_0 < L_S$, which resonates with the stray capacitance, C_0, at a parallel resonant frequency $f_0 = 1/(2\pi\sqrt{L_0 C_0})$.

Parallel capacitance from the loading circuit has little effect on the series resonance of the crystal, but the loading capacitance does have significant effect on the parallel resonance. Therefore, crystal manufacturers normally grind crystals to provide parallel resonance at a specified frequency when loaded with a standard load capacitance. A standard load capacitance of 30 pF or 32 pF is normally used. When properly loaded, the crystal will form a parallel resonant circuit like the one shown in Figure 17-20. The series resonant branch again forms the inductive branch, while the parallel capacitance is the sum of $C_0 + C_L$. This parallel resonant frequency, $f_L = 1/2\pi\sqrt{L_0(C_0 + C_L)}$, is the frequency nor-

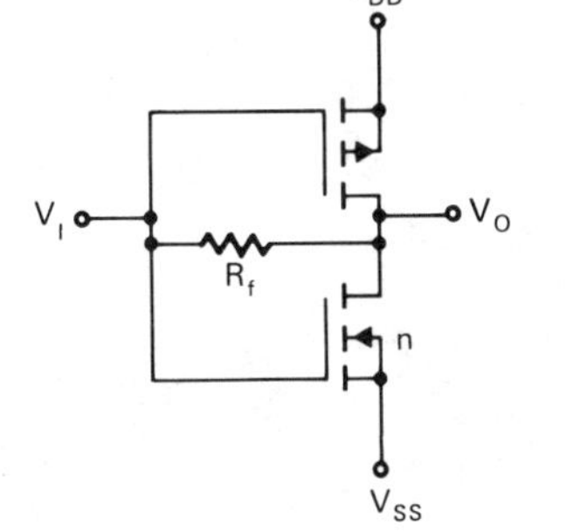

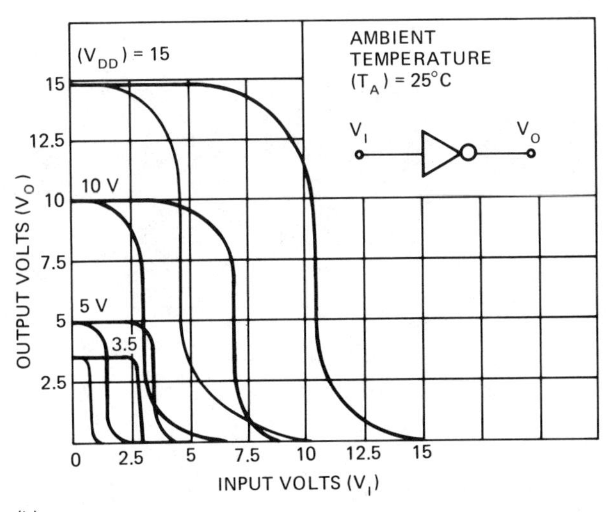

FIGURE 17-17

(a) CMOS inverter, biased in its transition zone by connecting a high-value resistor between input and output. (b) Graph of transfer characteristics of CMOS inverter.

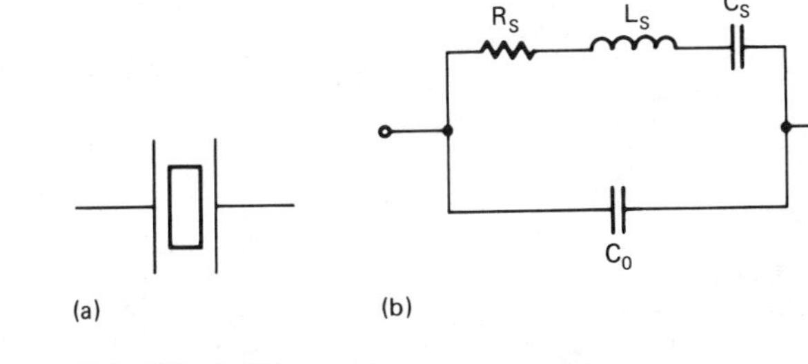

FIGURE 17-18

(a) Schematic symbol of crystal. (b) Equivalent circuit of crystal in terms of R, L, and C values.

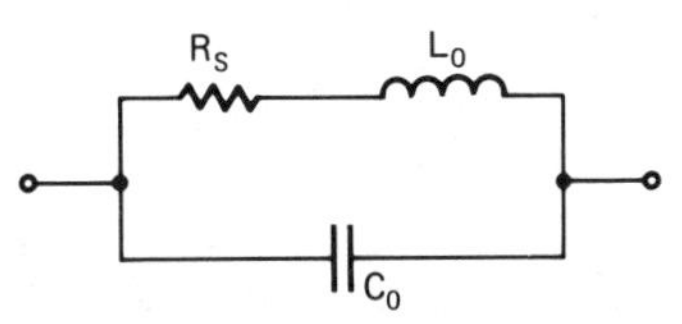

FIGURE 17-19

Equivalent circuit of a crystal with parallel resonance of $f_0 = 1/(2\pi\sqrt{L_0 C_0})$, assuming $X_{L_0} = X_{L_S} - X_{C_S}$.

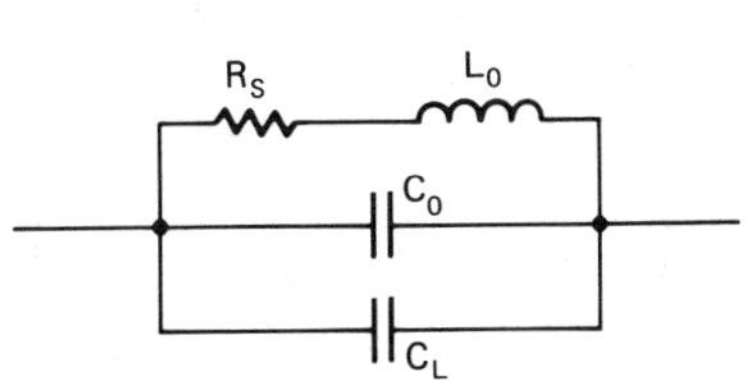

FIGURE 17-20

Equivalent circuit of a crystal with its parallel load capacitance. Use of the crystal must allow for the presence of capacitance in the external circuit parallel to the crystal. The parallel resonance is $f_L = 1/(2\pi\sqrt{L_0(C_0 + C_L)})$.

mally used in CMOS oscillator design. The circuit shown in Figure 17-21 is a CMOS oscillator using a crystal pi network to provide the tuned circuit, the feedback, and the 180° phase shift.

In selecting the correct component values for a particular oscillator frequency, some parameters of the crystal must be known either by measuring them or taking them from the manufacturer's data sheet. In our discussion thus far, we have shown the crystal to have numerous parameters, most of which are difficult to measure with any degree of accuracy. Because of the cost of these measurements, crystal manufacturers supply only a minimum of such information.

The four components of the crystal—R_S, L, C, and C_0—form an impedance which by complex arithmetic can be equated to a series circuit containing an equivalent reactance and resistance.

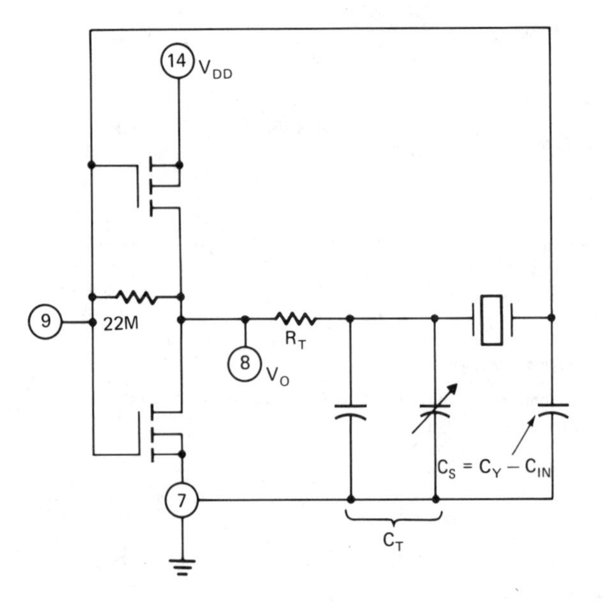

FIGURE 17-21
CMOS oscillator using a crystal pi network as a parallel tuned circuit.

The crystal is most often defined only in terms of this equivalent resistance and reactance at the specified frequency. As the quartz is ground to provide the parallel resonant frequency while loaded with capacitance C_L, the equivalent reactance must then be $X_e = 1/(2\pi f_X C_L)$. This is shown in the equivalent circuit of Figure 17-22.

Knowing the parameters f, R_e, and C_L, and assuming a feedback ratio of $\beta = 0.75$, we can determine the components of the circuit in Figure 17-21 as follows:

1. Equivalent reactance, X_e:

$$X_e = \frac{1}{2\pi f C_L}$$

2. Circuit gain, K_A:

$$K_A = \frac{4X_e + 1.07R_e}{X_e - 1.07R_e\beta}$$

3. Capacitor C_Y:

$$X_{CY} = \frac{X_e \beta K_A}{1 + \beta K_A}$$

$$C_Y = \frac{1}{2\pi f X_{CY}}$$

4. Resistor R_T:

$$R_T = \frac{K_A - 1}{R_e} \cdot \frac{X_e^2}{1 + \beta K_A}$$

5. Capacitor C_T:

$$C_T = \frac{C_L C_Y}{C_Y - C_L}$$

FIGURE 17-22
Four components of crystal tuned circuit equated to its equivalent series circuit, specified as $X_e = X_{C_L}$ (R_e is larger than R_S).

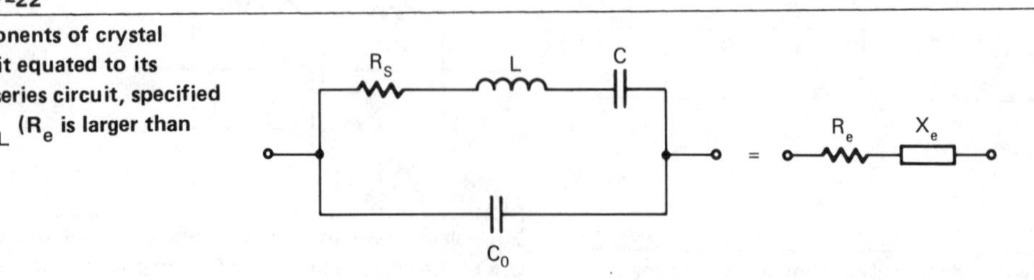

The capacitance C_Y includes the input capacitance of the inverter plus the lead capacitance. To compensate for these, about 10 pF is subtracted from the computed value to obtain the component value.

The maximum oscillator frequency for which a CMOS inverter can be used varies with the supply voltage. At $V_{DD} = 5V$, the upper limit is near 2 MHz. At $V_{DD} = 15V$, frequencies of 10 MHz are possible.

■ EXAMPLE 17-2

Determine the values of R_T, C_T, and C_Y for the CMOS oscillator of Figure 17-23 if a crystal is specified as $f = 280$ kHz, $R_e = 2.5K$, and $C_L = 32$ pF.

Solution:

$$X_e = \frac{0.159}{fC_L} = \frac{0.159}{280 \text{ kHz} \times 32 \text{ pF}} = 17.8K$$

$$K_A = 4\,\frac{X_e + 1.07R_e}{X_e - 1.07R_e\beta}$$

$$= \frac{17.8K + 1.07 \times 2.5K}{17.8K - 1.07 \times 2.5K \times 0.75} = 4.67$$

$$X_{CY} = \frac{X_e\beta K_A}{1 + \beta K_A} = \frac{71.2K \times 0.75 \times 4.67}{1 + 0.75 \times 4.67}$$

$$= 13.85K$$

$$C_Y = \frac{1}{2\pi f X_{CY}}$$

$$= \frac{0.159}{280 \times 103 \times 13.85 \times 103} = 41 \text{ pF}$$

$$R_T = \frac{K_A - 1}{R_e} \cdot \frac{X_e^2}{1 + \beta K_A}$$

$$= \frac{4.67 - 1}{2.5K} \cdot \frac{(17.8K)^2}{1 + 0.75 \times 4.67}$$

$$= 23K \quad \text{(Use standard value, 22K.)}$$

$$C_T = \frac{C_L C_Y}{C_Y - C_L} = \frac{32 \times 41}{41 - 32} = 146 \text{ pF}$$

As Figure 17-23 shows, the capacitor C_Y is reduced by the lead capacitance between the crystal and the inverter input in addition to the CMOS input capacitance of about 7 pF. The capacitor C_T is ideally supplied by a fixed and variable combination to allow for minor frequency and feedback adjustment.

Power dissipation ratings of crystals may range from 2 mW to 10 mW. In determining the amount of power dissipated by the crystal in the circuit of Figure 17-23, it is important to remember that the purely reactive components of a tuned circuit do not dissipate power. Only resistance dissipates power. Therefore, the power dissipated by the crystal is dependent on the resistance presented to the circuit by its parallel tuned resonance. This resistance (R_P) is higher than R_S or R_e and is equal to QX_e. As Q is equal to X_e/R_e, then $R_P = X_e^2/R_e$. The crystal presents a near open circuit to the DC current component. Therefore, only AC values need be considered. The highest possible rms value of AC voltage external to the CMOS inverter is equal to $0.707\ V_{CC}/2$. The total AC power dissipated is, therefore,

$$P_D = \frac{E^2}{R_D}$$

where:
P_D = total AC power
R_D = circuit resistance
E = rms value of AC voltage

Power in a series circuit divides in proportion with resistance; therefore, the power dissipated by the crystal is $P_X = P_D\ [R_P/(R_T + R_P)]$.

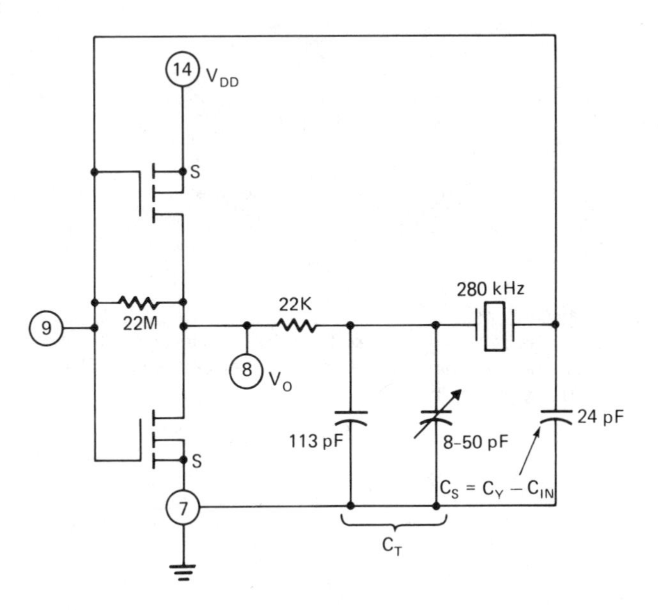

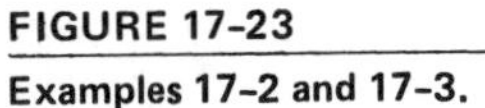
FIGURE 17-23
Examples 17-2 and 17-3.

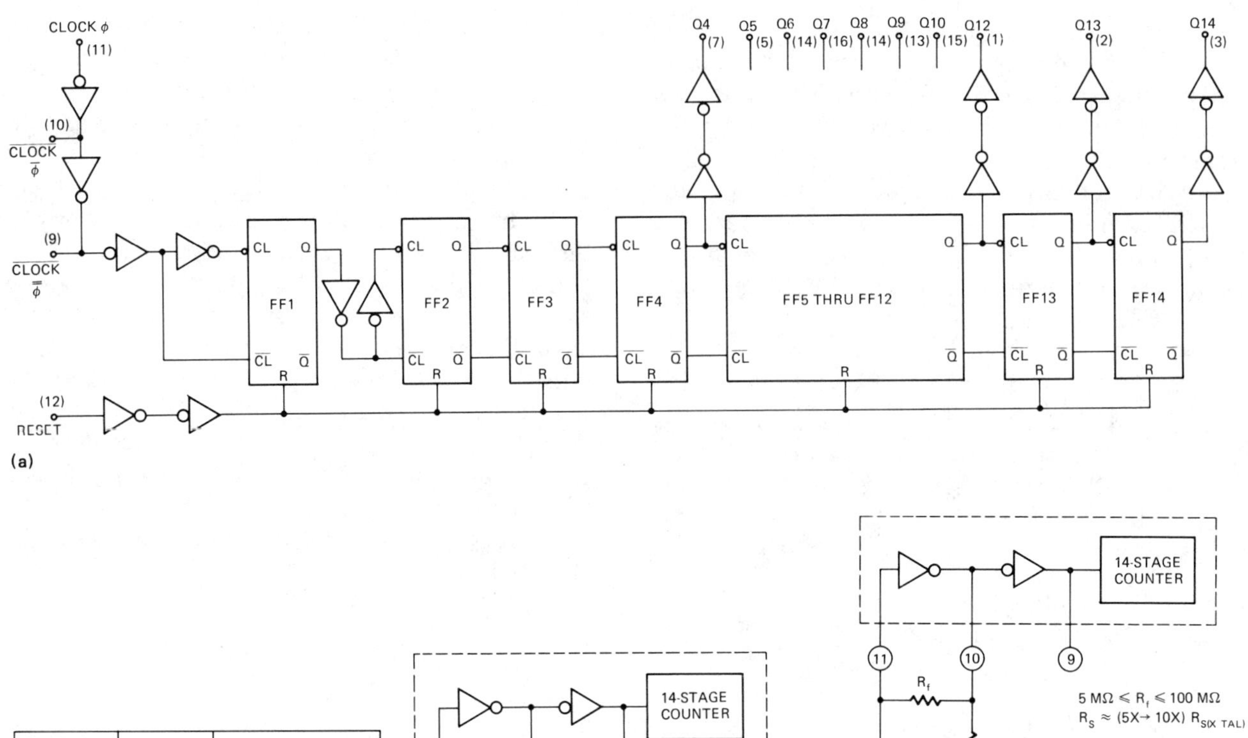

(a)

CLOCK	RESET	OUTPUT STATE
(rising edge)	0	NO CHANGE
(falling edge)	0	ADVANCE TO NEXT STATE
X	1	ALL OUTPUTS ARE LOW

X = DON'T CARE

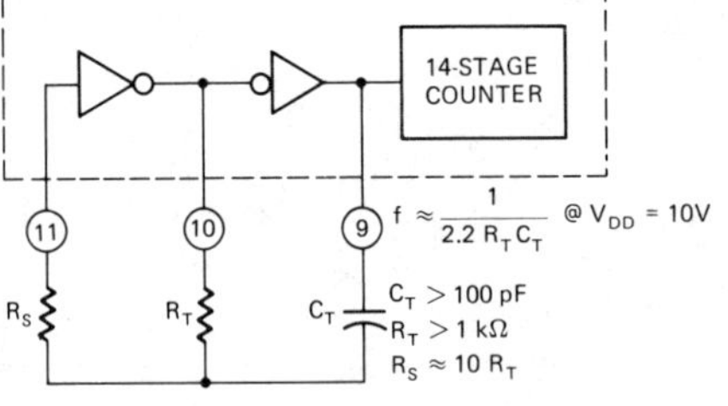

(b)

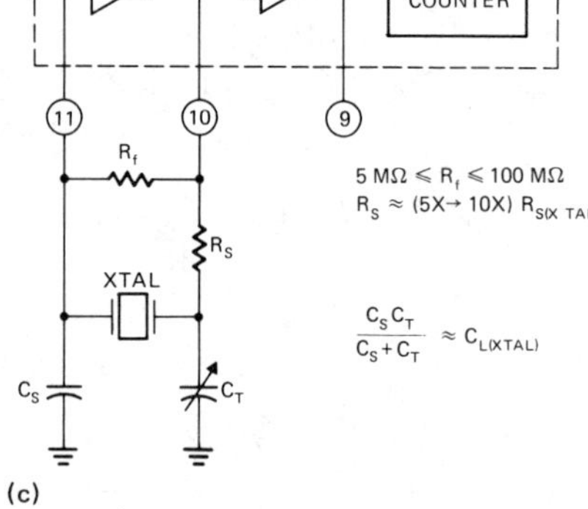

(c)

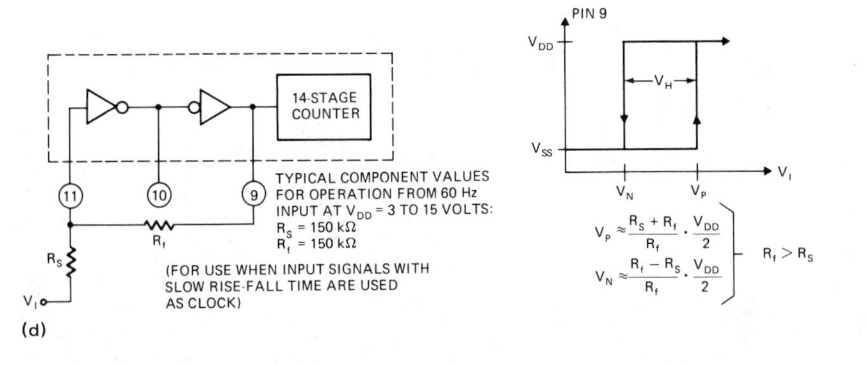

(d)

FIGURE 17–24

CMOS SCL4060 counter oscillator. (a) Logic diagram. (b) Inverters connected to provide an RC oscillator. (c) Inverters connected to provide a crystal oscillator. (d) Inverters connected to form a Schmitt trigger input.

■ **EXAMPLE 17-3**

Determine the power dissipated by the crystal in the circuit of Figure 17-23.

Solution:

$$P_D = \frac{E^2}{R_D} = \frac{3.12}{148.7K} = 21\ \mu W$$

$$E = 0.707\frac{V_{CC}}{2} = \frac{0.707 \cdot 5V}{2}$$

$$= 1.7675\ V_{rms}$$

$$R_D = R_T + R_P$$

$$R_P = \frac{X_e{}^2}{R_e} = \frac{(17.8K)^2}{2.5K} = 126.7K$$

$$P_X = \frac{R_P}{R_T + R_P} \cdot P_D = \frac{126.7K}{148.7K} \cdot 21\ \mu W$$

$$= 18\ \mu W$$

The CMOS circuit CD4060, shown in Figure 17-24, provides a very convenient clock generator. The device is a 14-stage binary counter with an input preceded by a set of inverters that may be used as either an astable (RC) multivibrator, when connected as shown in Figure 17-24(b), or as a crystal oscillator with components connected as shown in Figure 17-24(c). The resistor and capacitor values for astable operation can be determined by the equation $f = 1/(2.2RC)$, as described in Section 17.4. The crystal oscillator components can be determined by the methods described in this section.

Q outputs are available from stages 4 through 10 and 12 through 14. This provides a choice of frequencies $f_0/2^n$ for n from 4 through 10 and 12 through 14 (f_0 = the frequency of the astable or crystal oscillator).

17.9 Cycling

Thus far we have shown the type of circuits that produce continuous pulses. To stop here would be like making a clock with no face. The clock pulses themselves are of little use to us unless we can number them, just as we put numbers on the face of a clock. We must somehow designate pulse 0 and consecutively number each pulse thereafter until the necessary numbers have occurred, at which time we begin again with a 0 pulse. This selection of a 0 pulse and consecutive numbering of pulses in periodic sets is responsible for the cycles that occur in computers and other digital devices.

The general-purpose computer has at least four built-in operating cycles: the instruction cycle, the execution cycle, the memory cycle, and the arithmetic cycle. We have already seen the arithmetic cycle for a parallel adder. The other cycles will be discussed in later chapters.

A cycle is a set of operations that must occur in sequence. The same sequence of operations, possibly with modifications, occurs repeatedly. A first pulse isolated from the clock line is usually designated as a 0 or reset pulse, as Figure 17-25 shows. This 0 pulse will reset the A and B registers, canceling out numbers from the previous addition. One method of designat-

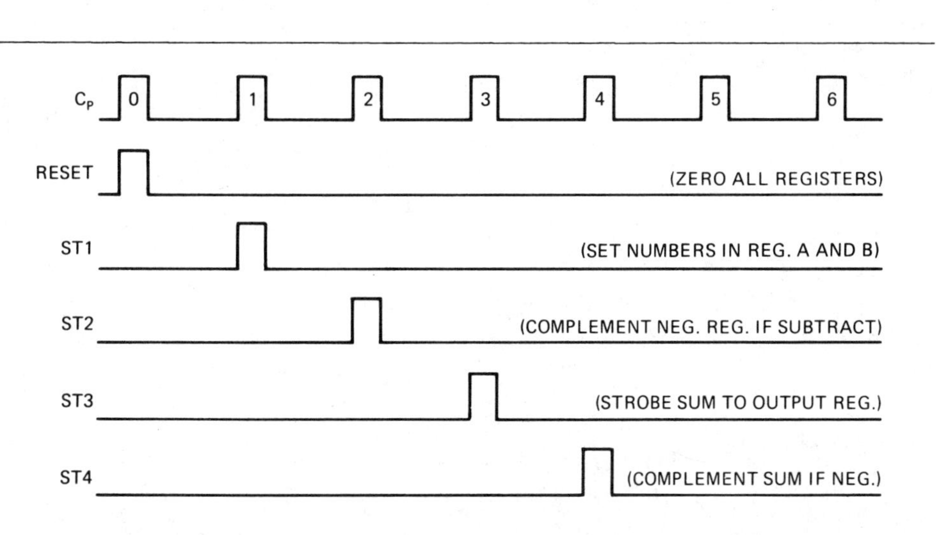

FIGURE 17-25
Reset and four strobe pulses generated for proper timing of the operations of the ones complement adder-subtractor. These pulses are repeated periodically to form an arithmetic cycle.

ing the 0 pulse is to use a second clock generator of lower frequency, as Figure 17-26 shows. As the timing diagram shows, the cycle generator need not be synchronized with the clock and may change state in the middle of a clock pulse. To prevent the possibility of a fragmented 0 pulse, the first clock pulse during or after the positive rise of the cycle generator output is used merely to set the first register. Setting the first register enables the gate to pass the next clock pulse—the 0 pulse. On the trailing edge of the 0 pulse, the second register will set and inhibit any further pulses from occurring until the cycle generator goes to its low state and back again to high. If the cycles are of short duration in terms of clock pulses, a continuously cycling counter can be used as a generator. The 0 state of the counter starts the cycle by generating a 0 pulse each time the counter returns to 0. Figure 17-27 shows a circuit used to generate

FIGURE 17-26

Second multivibrator of lower frequency used to establish a cycle of clock pulses. The second pulse occurring after the cycle multivibrator goes high is gated and designated as the 0 pulse.

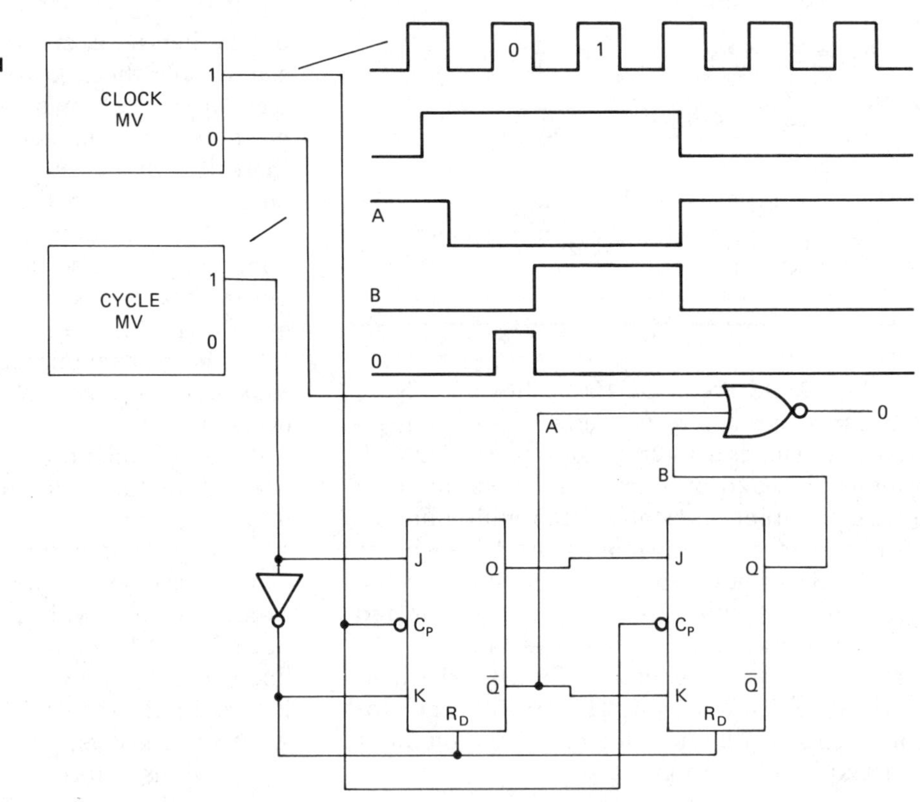

FIGURE 17-27

Three-bit counter and decode gate used to obtain a reset pulse for a cycle of eight pulses.

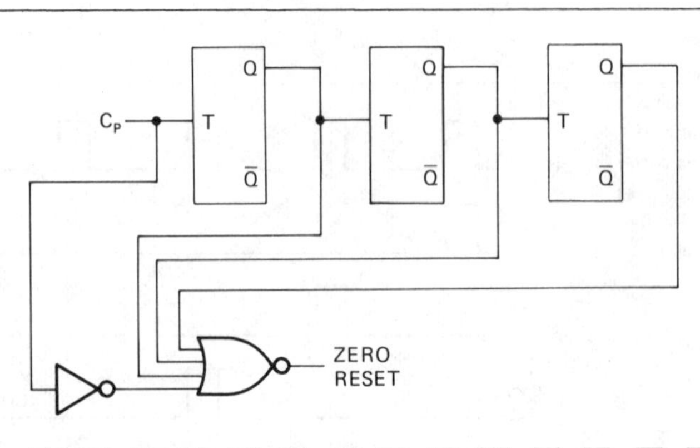

a cycle of eight clock pulses. The counter returns to the 0 state (all three flip-flops reset) every eighth clock pulse. During that state, the NOR gate is enabled to pass a 0 or reset pulse for the cycle.

17.10 Delayed Clock (Clock Phases)

We have used numerous examples of dual clocks that have identical frequency but differ in that on one clock line the pulses are delayed from those on the other line. Those may be referred to as a clock and a delayed clock. The delayed clock pulses occur midway between the clock pulses, as Figure 17-28 shows. The control signals that enable or inhibit passage of pulses through logic gates are themselves created by clock pulses. This results in their having leading or trailing edges that are coincident with the clock pulses. Operating logic gates with signals having coincident leading or trailing edges can produce erratic and unstable conditions. It is, therefore, desirable to enable or inhibit clock pulse signals using gates created from delayed clock pulses, and vice versa. The two clock phases needed to operate the dynamic shift registers of Chapter 13 present another use for dual clocks. Additional uses will be found in later chapters.

In this text, the two clock lines are designated C_P and C_P', or C_L and C_L', and occasionally as ϕ_1 and ϕ_2. Figure 17-29 shows one method of separating a single clock output into separate clock and delayed clock lines. The J-K flip-flop

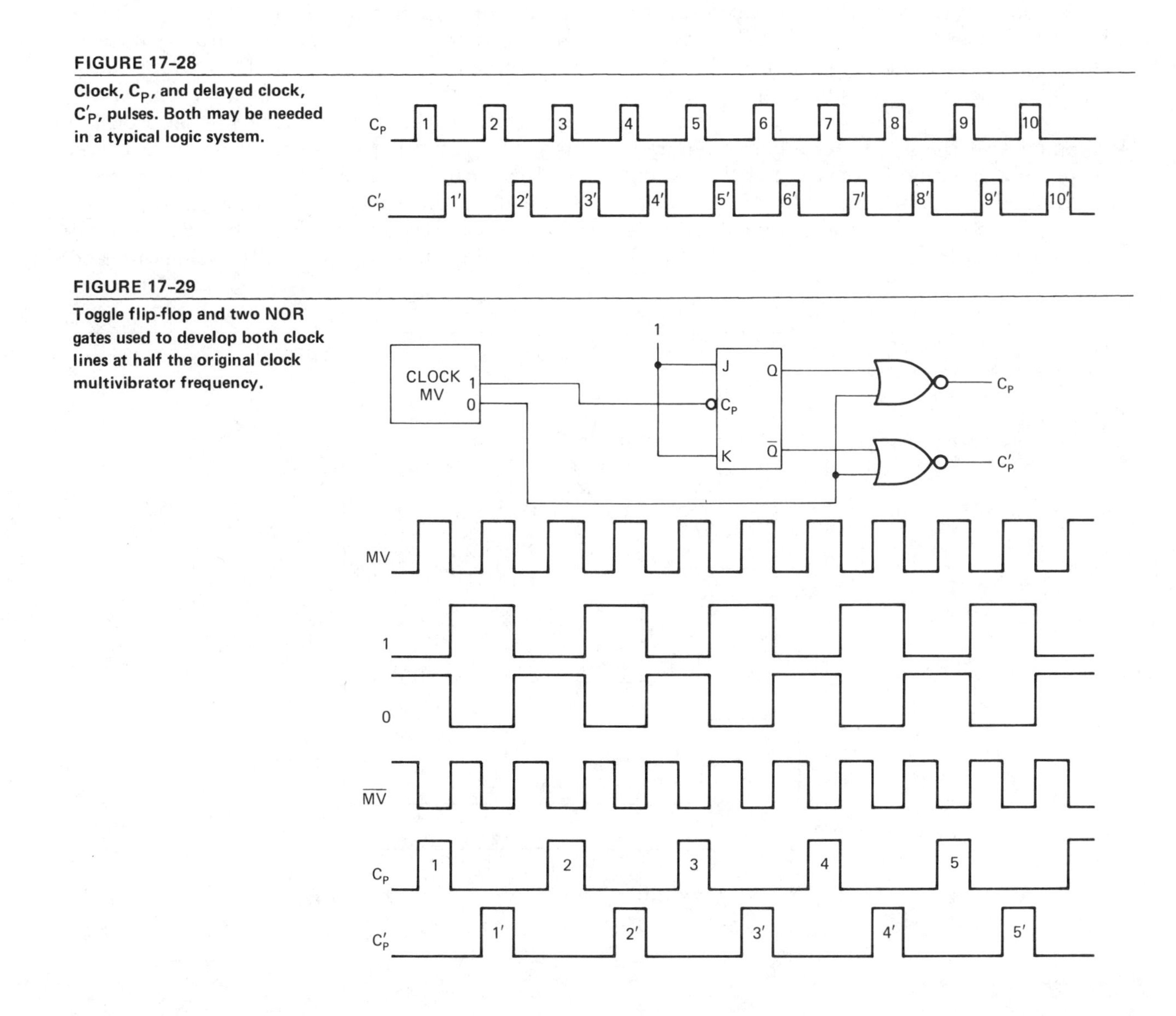

FIGURE 17-28
Clock, C_P, and delayed clock, C_P', pulses. Both may be needed in a typical logic system.

FIGURE 17-29
Toggle flip-flop and two NOR gates used to develop both clock lines at half the original clock multivibrator frequency.

with 1 level on both J and K will toggle or change state on each trailing edge of the multivibrator output. The 1 and 0 outputs are connected to C_P and C_P' NOR gates, causing one to be enabled while the other is inhibited. Each time the multivibrator line is low, it results in a high level out of the enabled gate. As the gates are enabled alternately, alternate pulses will switch between the C_P and C_P' lines.

17.11 Monostable (One-Shot) Multivibrator

17.11.1 Need for the Monostable Multivibrator

The one-shot multivibrator is a device used for delay and other forms of irregular timing. As Figure 17-30 shows, a pulse applied to a one-shot input results in an output pulse whose leading edge occurs at about the same time as the leading edge of the input pulse, but whose trailing edge occurs either before or after the trailing edge of the input pulse. The width of the output pulse depends on the size of a capacitor in the multivibrator circuit.

Figure 17-31(a) shows a typical use for the one-shot multivibrator. The top waveform in Figure 17-31(b) is the output of the demodulator of a three-channel pulse width modulation receiver. The higher-amplitude pulse is a synchronization pulse, which is separated from the rest of the signal by amplitude clipping and used to pulse three one-shots. The outputs of the one-shots are used to enable the AND gates only during the correct channel time for each of the three channels. Each channel will receive a pulse of varying width once every 16 microseconds, occurring only during the channel's sampling period.

17.11.2 Transistor Monostable Multivibrator

Figure 17-32 is a schematic diagram of a typical transistor one-shot multivibrator circuit. In the resting state, the base current through R keeps Q_2 turned on, providing a 0 output. When a positive-going input pulse occurs, Q_1 temporarily turns on, placing a temporary low level on the base of Q_2, turning it off and producing a 1 level at the output. The output will remain in the 1 state until the capacitor, C, charges to a level high enough to turn Q_2 back on. The time required to do this depends on the value of R and

FIGURE 17-30

A one-shot (monostable) multivibrator used to widen or narrow down the width of a pulse applied to its input. The width of the pulse at the output is proportional to the value of C.

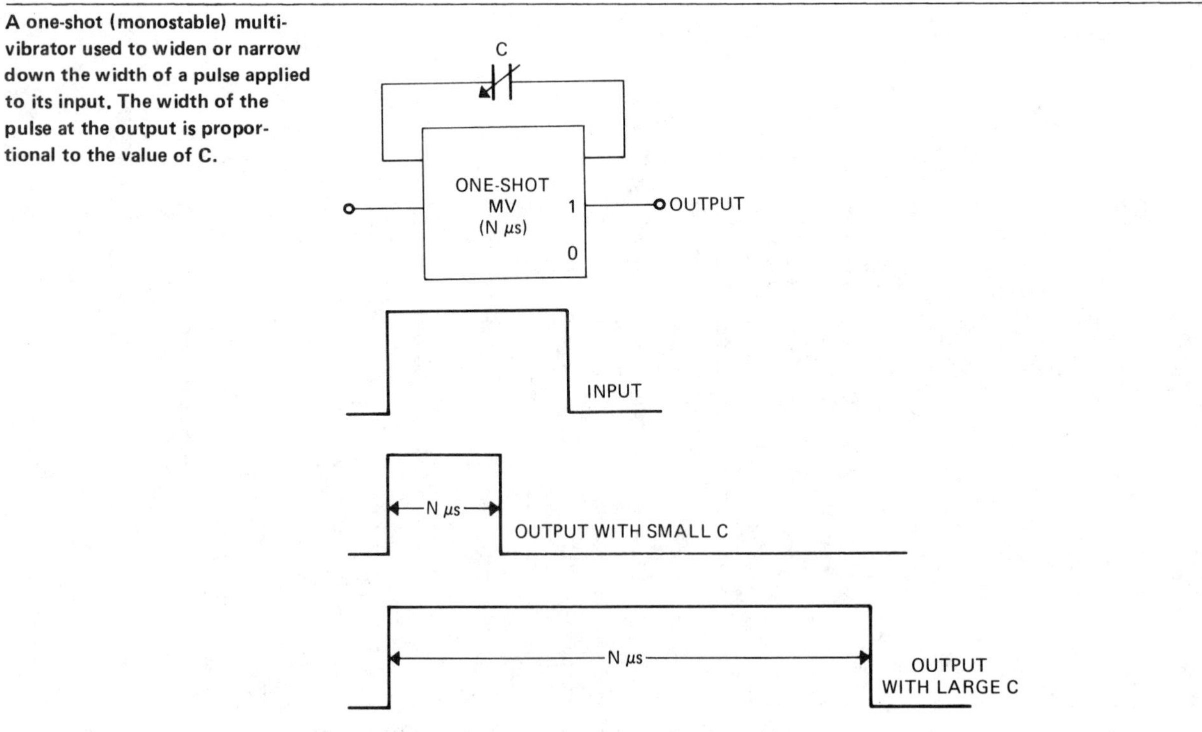

C and can be adjusted, within limits, for the required output width. The value of R must, of course, be low enough to allow saturation of Q_2; the value of C is limited by size and cost considerations.

17.11.3 Integrated Circuit One-Shot

Monostable multivibrators lacking C and R can be obtained in integrated circuit form. The user

FIGURE 17–31

(a) One-shot multivibrators used to separate a received three-channel pulse width modulated signal into its three separate channels. The sampling period for each channel is 5 microseconds and the sync is 1 microsecond wide. (b) Waveforms occurring in channel separation of pulse width modulation signals.

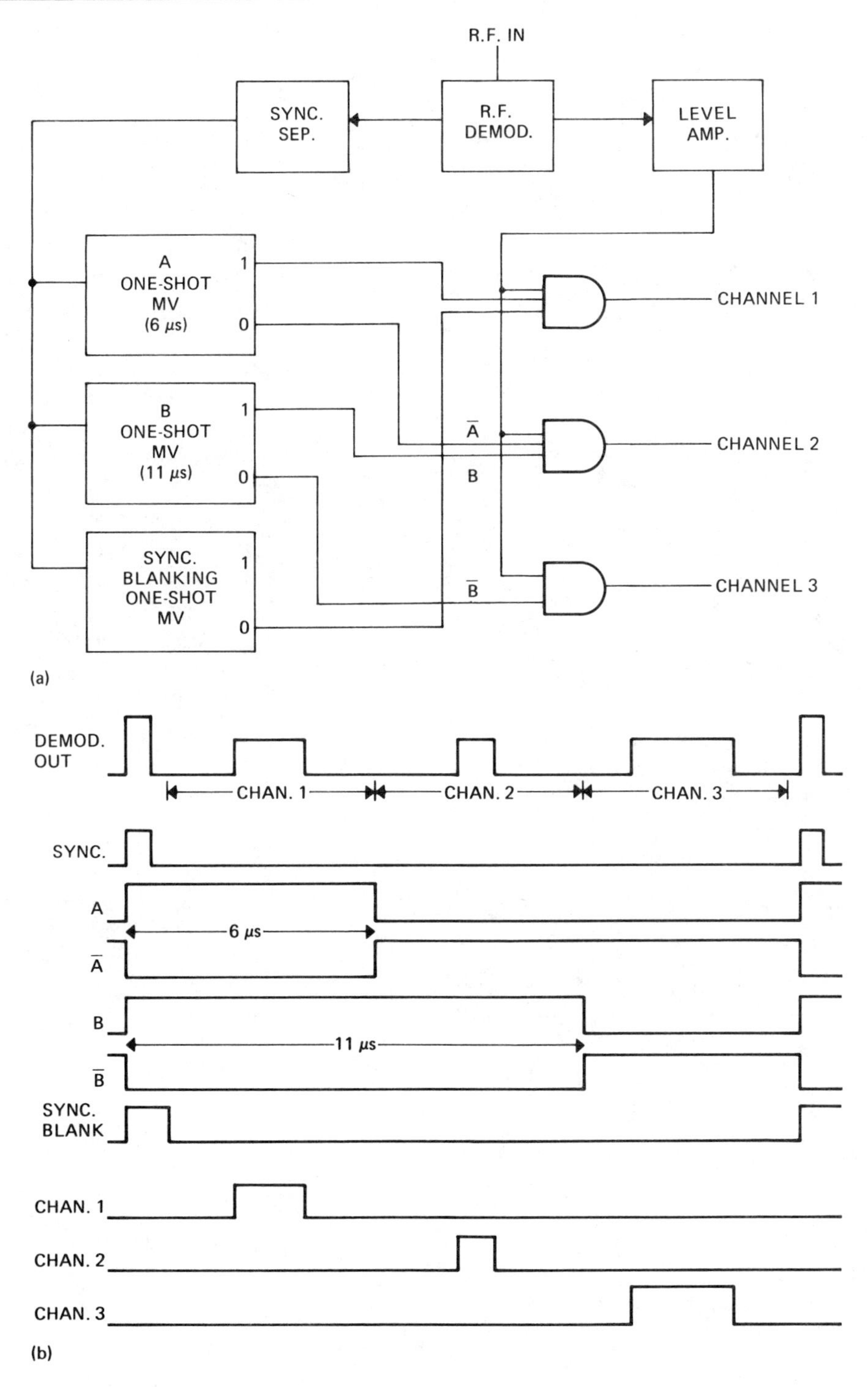

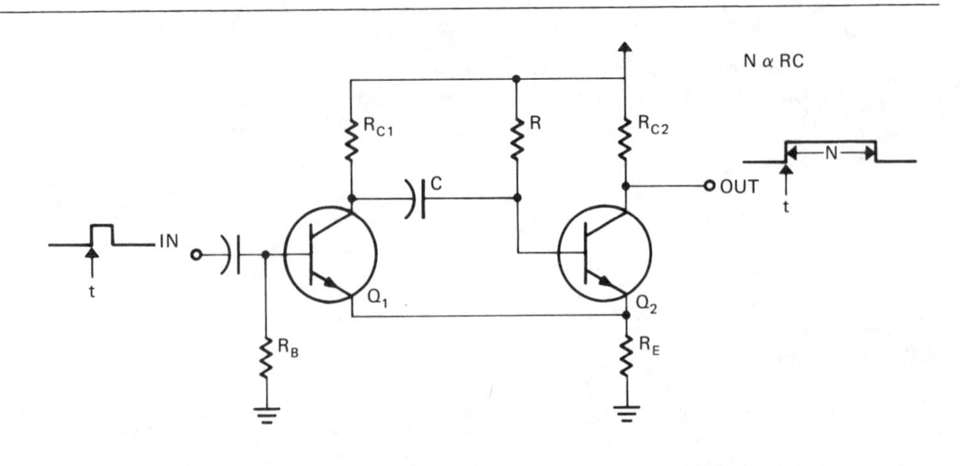

FIGURE 17–32
Discrete circuit one-shot multivibrator. Width of output pulse depends on size of R and C.

can obtain the needed delay or pulse width by connecting the appropriate R and C. The Fairchild 96L02 monostable, shown in Figure 17–33, has complementary outputs. It also has an inverting input, which causes the output to start on the falling edge of the input pulse instead of the rising edge. The clear input can be used to override or inhibit the output. There are two identical one-shots for every 16-pin integrated circuit.

Figure 17–33(a) shows a 1-microsecond input pulse and the difference in output resulting in its application to IN or $\overline{\text{IN}}$. The R and C were selected for a 3-microsecond output. Figure 17–33(b) is a graph of time delay or pulse width (t) versus capacitance C_x. For capacitors in excess of 1000 pF, the equation is as follows (C_x in pF, R_x in kΩ, t in ns):

$$t = 0.33 R_x C_x \left(1 + \frac{3.0}{R_x}\right)$$

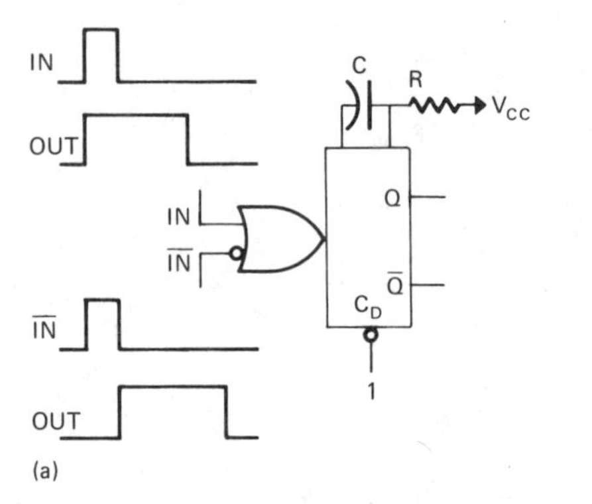

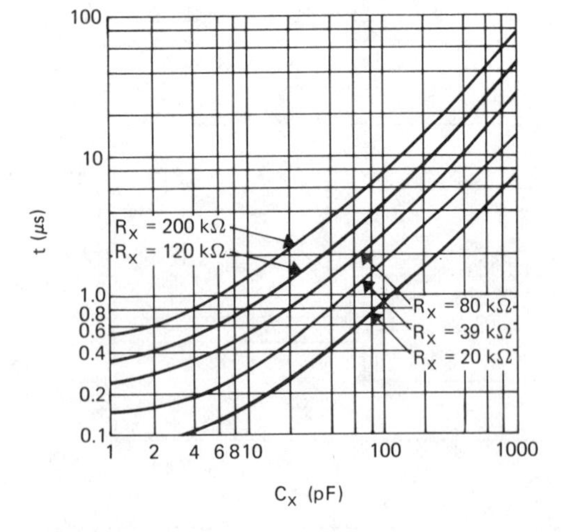

FIGURE 17–33
(a) Fairchild 96L02 integrated circuit one-shot. User determines width of output by connecting the necessary R and C externally. (b) Time delay/pulse width versus capacitance graph. ($C_X \leqslant$ 1000 pF).

■ EXAMPLE 17–4

If we use a 96L02 monostable multivibrator to obtain a pulse width of 10 milliseconds, with a capacitor of 0.01 μF, what size resistor do we need?

Solution:

$$t = 0.33 R_x C_x \left(1 + \frac{3.0}{R_x}\right)$$

$$R_x = \frac{t}{0.33\ C_x} - 3$$

$$= \frac{10 \times 10^6\ \text{ns}}{0.33 \times 104\ \text{pF}} - 3 = 3 \times 10^2 - 3$$

$$= 297\ \text{k}\Omega$$

The 96L02 will output the same pulse width regardless of whether the input pulse width exceeds or is shorter than the output pulse width. These conditions are shown in Figure 17-34 ① and ②. If, however, a new edge of pulse occurs while Q is high, the 96L02 will retrigger. This is shown in Figure 17-34 ③ and ④. It can, however, be connected for non-retriggerable operation by connecting either Q to IN or $\overline{Q}$ to $\overline{IN}$.

Figure 17-35 shows other integrated circuit one-shot multivibrators available in TTL and CMOS. The 74121 is a single with an internal 2K

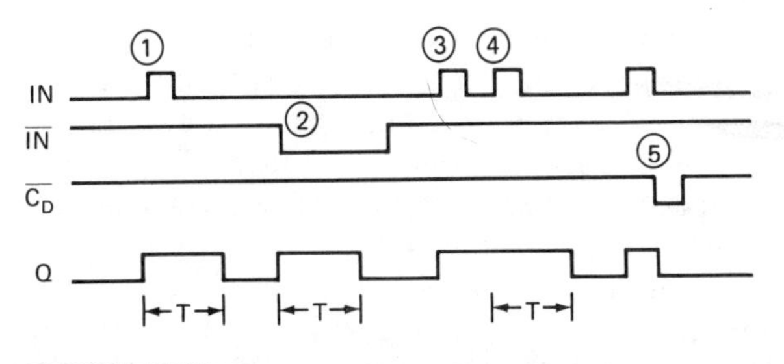

FIGURE 17-34

Timing diagram of 96L02 showing ① positive edge triggering, and trigger pulse width less than output pulse width, ② negative edge triggering and trigger pulse width greater than output pulse width, ③ and ④ effect of retriggering, and ⑤ operation of $\overline{C_D}$.

FIGURE 17-35

Other integrated circuit monostable multivibrators. (a) 74121. (b) 74122. (c) 74123. (d) 4538B.

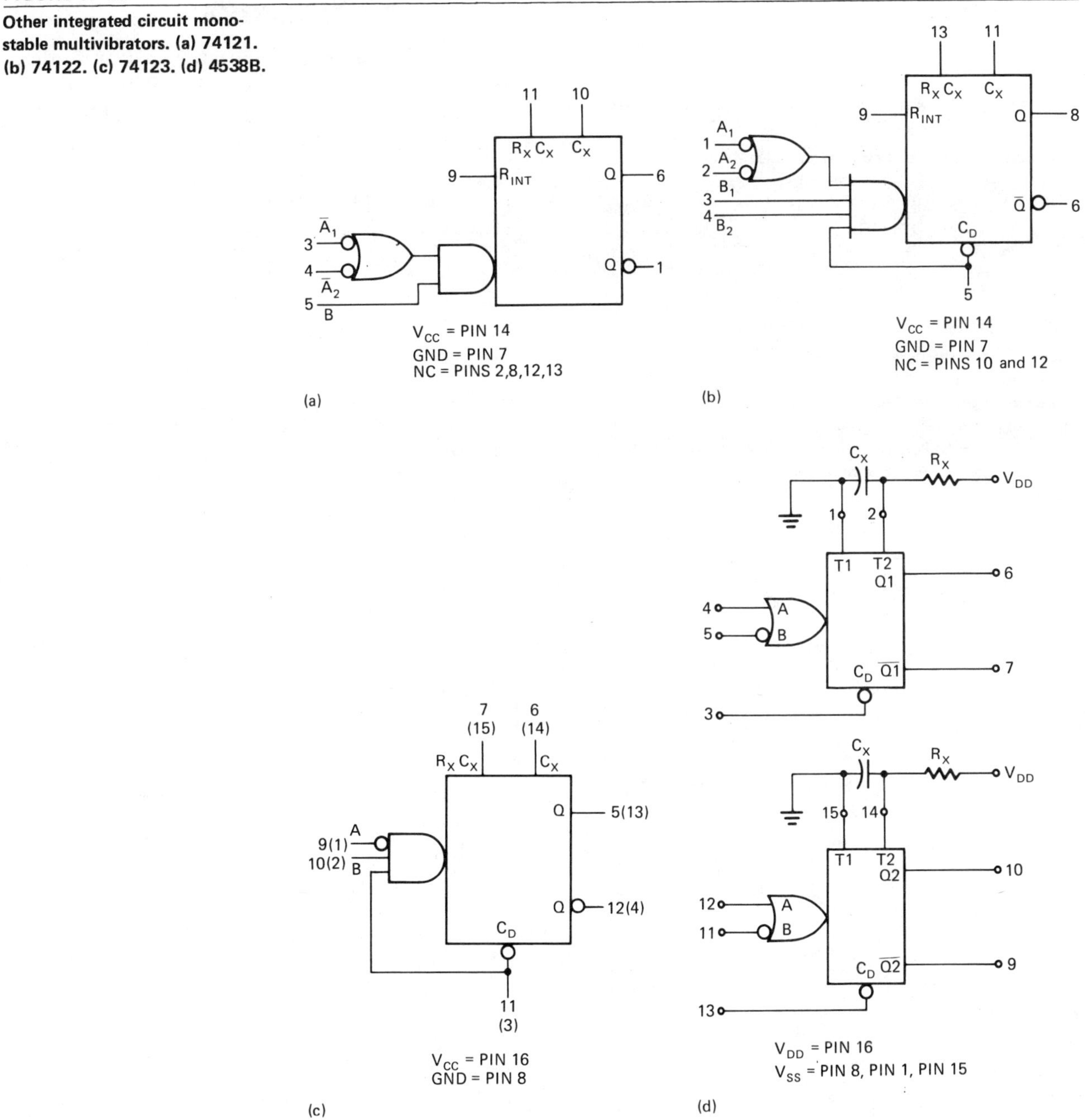

resistor used instead of an external resistor. A jumper between pin 9 (R_{INT}) and pin 14 employs the 2K resistor, and the capacitor is the only external component required. If pin 9 is left open, then an external resistor between pin 14 and pin 11 is required.

The 74122 has a 10K internal resistor for optional use. It is retriggerable and resettable. The 74123 is a dual, retriggerable, resettable, one-shot multivibrator. It is in a 16-pin package and has no internal timing resistor option.

The 4538 is a CMOS dual monostable multivibrator. It is retriggerable and resettable. It can be made nonretriggerable by connecting either Q to A or $\overline{Q}$ to $\overline{B}$. R_x and C_x are external components.

17.11.4 CMOS Gate Monostable Multivibrator

Figure 17-36(a) shows a pair of CMOS inverters connected to provide a monstable multivibrator. It is trailing-edge-triggered and provides output pulse widths that vary with the value of RC. It also varies with the triggering levels of the gates so that different ICs will provide different pulse widths for the same values of RC. Figure 17-36(a) is a negative-going edge-triggered monostable multivibrator. The pulse width varies with R_2C_2. Figure 17-36(b) is a positive-going edge-triggered monostable multivibrator. The multivibrator of 17-36(b) requires more components than the circuit of 17-36(a), but the pulse widths vary only slightly for triggering levels of the CMOS circuits used.

17.11.5 Pulse Delay by One-Shot Multivibrator

A single one-shot multivibrator as a pulse delay circuit does not actually delay the pulse itself but delays the trailing edge of the pulse and, in effect, stretches the pulse width. Two one-shot multivibrators can provide a delay and recreate the original pulse width at the output. If we require a pulse to be delayed, as in Figure 17-37(a), it can be accomplished as shown in Figure

FIGURE 17-36

Monostable multivibrators made from standard CMOS inverters or gates. (a) Negative-going edge-triggered monostable multivibrator. (b) Positive-going edge-triggered monostable multivibrator.

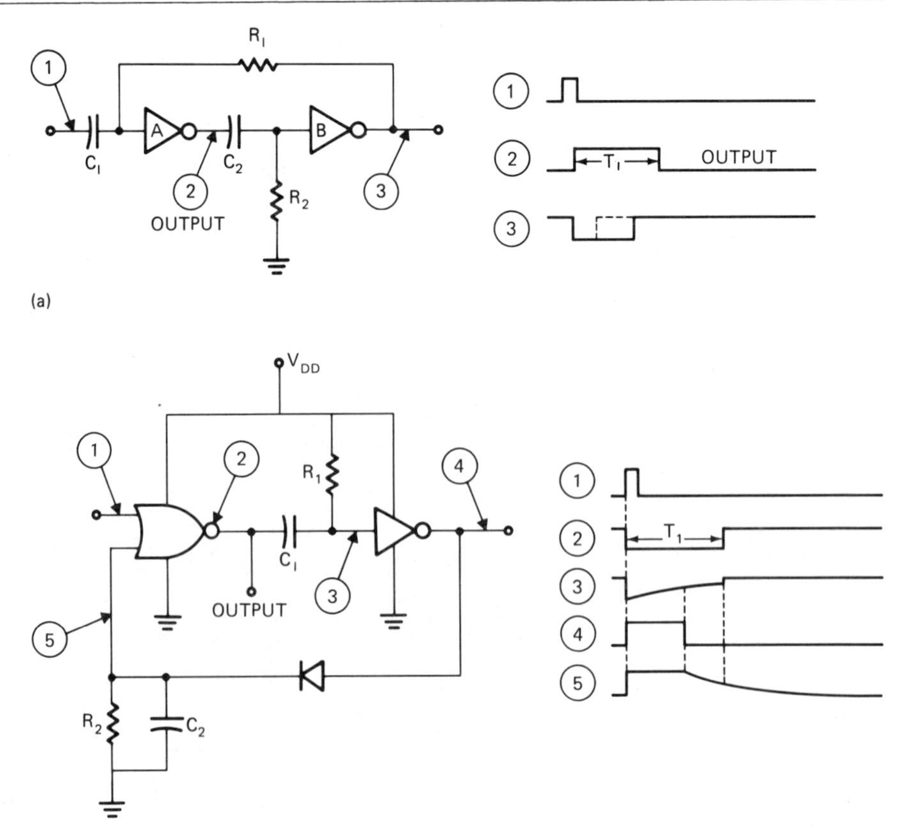

17-37(b). The first one-shot changes output on the leading edge of the input pulse. The $\overline{Q}$ output goes low and returns to the high level after 10 microseconds. That causes the Q output of the 2-microsecond one-shot to go high for 2 microseconds, recreating a 2-microsecond pulse delayed 10 microseconds from the original input pulse.

17.11.6 555 Timer as a Monostable Multivibrator

Figure 17-38 shows a 555 timer connected to provide a monostable multivibrator. The pulse width can be determined by the equation $T = 1.1R_AC$.

FIGURE 17-37

(a) A 2-microsecond pulse to be delayed 10 microseconds. (b) Pulse delay accomplished by using two one-shot multivibrators. (c) Timing diagram comparing $\overline{Q}_1$ with input and output waveforms.

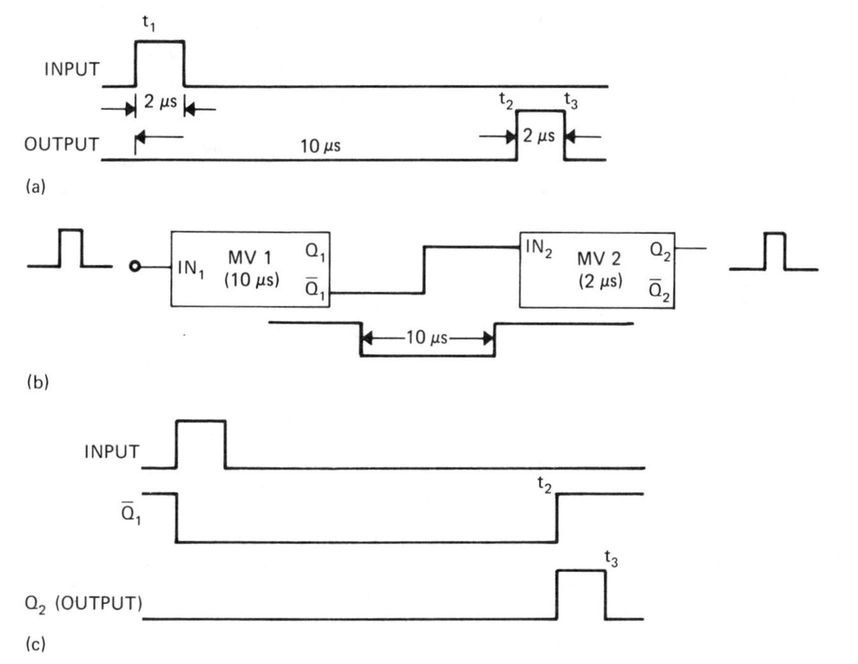

FIGURE 17-38

555 timer connected as a monostable multivibrator. The output pulse width is determined by $T = 1.1R_AC$.

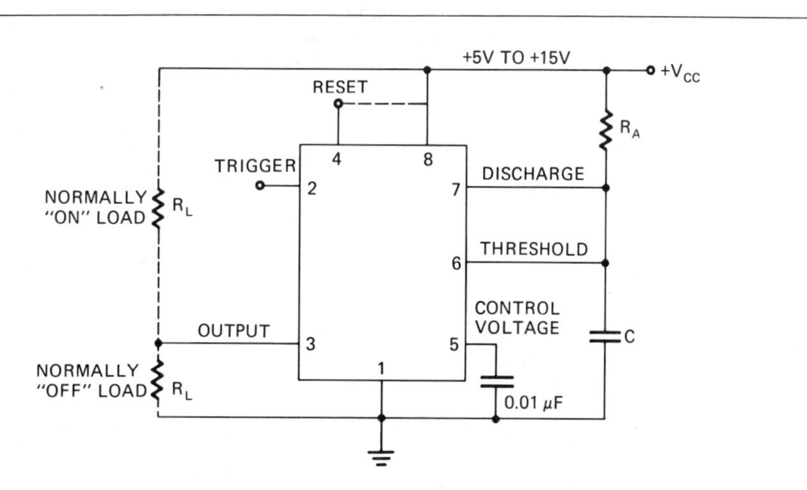

17.11.7 Variable Pulse Width Multivibrator

Various designs of the one-shot multivibrator can be used to give a variable pulse width under control of an external voltage. Figure 17-39 is a one-shot multivibrator used to provide pulse width modulation of a radio frequency transmitter. The one-shot is continuously pulsed by a clock pulse, but, as Figure 17-40 shows, the output pulse width will vary with the amplitude of the audio input signal. The output pulses are used to key the radio frequency in the transmitter. Pulse width modulation has the advantage of being more noise-free than standard AM modulation. Demodulation is only slightly more difficult and can be handled in digital fashion. It lends itself well to multiplexing, as described in Section 17.10.1.

The 555 timer is one of various integrated circuit approaches to a voltage-controlled pulse width from a monostable multivibrator. Figure 17-41 shows the 555 timer connected for pulse width modulation. The modulating voltage is applied between pin 5 (control voltage input) and ground.

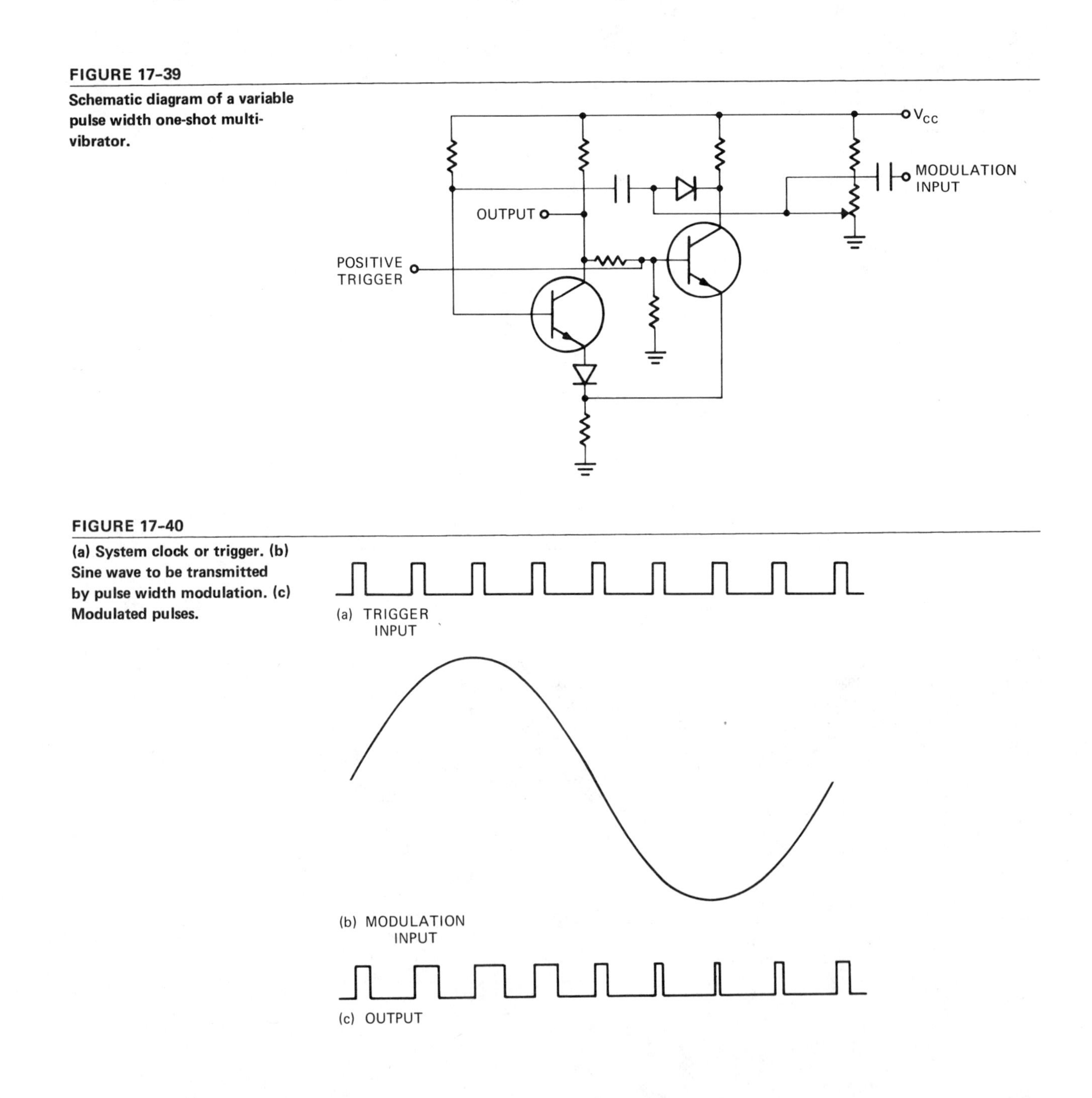

FIGURE 17-39
Schematic diagram of a variable pulse width one-shot multivibrator.

FIGURE 17-40
(a) System clock or trigger. (b) Sine wave to be transmitted by pulse width modulation. (c) Modulated pulses.

17.12 Summary

There are few uses of digital systems that do not require some form of timing, cycling, or sequencing. In most systems, there is a circuit that provides periodic pulses or square waves at the logic level. This circuit is called a *clock.*

An *astable multivibrator* is often used as a digital clock. Figure 17-1 shows the typical output waveform of the astable multivibrator. It is a rectangular waveform changing periodically between the logic levels. Figure 17-2 shows a typical circuit for a transistor astable multivibrator. Its operation differs from that of bistable multivibrators described in Chapter 12 in that there is no need for an external pulse on set, reset, or trigger inputs to obtain a change of state. The circuit will change state periodically at a frequency determined by the size of the resistors and capacitors in the circuit. If $R_{B1} = R_{B2}$ and $C_1 = C_2$, the output will be symmetrical in that off and on time will be equal, or T_1 will equal T_2 (see Figure 17-3).

For some applications, a nonsymmetrical waveform may be desired with off time longer than on time. In this case, the output will be in the form of a pulse rather than a square wave and will allow for more things to occur between clock pulses. This lack of symmetry can be accomplished by altering the balance between R_{B1} C_2 and R_{B2} C_1. The astable multivibrator is a reliable, accurate clock circuit and can be designed to provide frequencies from below 1000 Hz to over 1 MHz.

For frequencies of 100 kHz or lower, the *555 timer* or *unijunction pulse generator* are sometimes used. Although both are relatively stable circuits, the output waveforms are markedly different. The output waveform of the 555 timer is a rectangular wave that can be made TTL or CMOS compatible. The output waveform of the unijunction pulse generator generally requires some shaping in order to be sufficiently rectangular for digital circuits.

The 555 timer can operate from a voltage range of 5V to 15V. The circuit of Figure 17-5 produces a rectangular wave. The high, or on time, depends on R_A, R_B, and C.

Figure 17-6(a) shows how an astable multivibrator can be assembled using resistors, capacitors, and CMOS inverter gates. The frequency of oscillation for this network is given by 1/(2.2RC).

Figure 17-7 shows a schematic of a typical unijunction pulse generator. The active element in this circuit is the *unijunction transistor* (UJT). The symbol for a UJT is similar to a FET symbol except that the arrow points slightly downward instead of straight inward. The leads to the UJT are called emitter base 1 and base 2. With no voltage on the emitter, the device conducts no current, but as capacitor C charges through resistor R, it reaches a level of about $V_{CC}/2$ and the emitter B_1 junction breaks down and conducts. This causes a conduction between B_2 and B_1. It also produces a voltage drop across R_L. The conduction through the emitter base 1 lead quickly discharges the capacitor, taking the device out of conduction and returning the output to 0. The UJT will not conduct again until the capacitor has had time to charge back to the breakdown potential. The time required for the capacitor to charge to breakdown potential depends primarily on R and C, and these components can therefore be selected to provide the desired frequency.

Note that the output pulses from the B_1 lead have a fast leading edge but a very slow trailing edge. For this reason, the output is generally shaped by added digital circuits. Figure 17-8 shows an inverter used to convert the relatively fast leading edge of the unijunction pulse to an even faster trailing edge at the output of the inverter. The flip-flop that operates on the fast falling edge produces an ideal digital waveform at half the frequency of the unijunction pulse generator.

The *Schmitt trigger circuit* is ideally suited for shaping nonrectangular waveforms, making them compatible for use in a digital system. Figure 17-10 shows a Schmitt trigger circuit

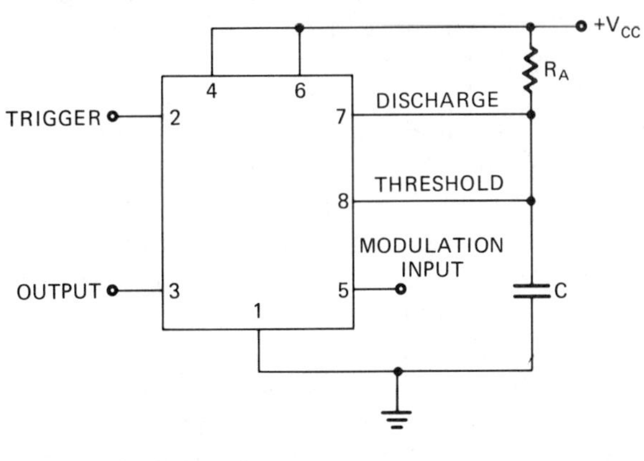

FIGURE 17-41

555 timer connected to provide a voltage-controlled pulse width.

receiving a sine wave at its input and converting it to a rectangular waveform at the output.

The Schmitt trigger is available in integrated circuit form. Figure 17-12 is a TTL 7413 integrated circuit dual Schmitt trigger. The input is through a four-input AND gate, giving additional control. The output is inverted. The graph of output voltage versus input voltage indicates that the signal to be converted may need to be attenuated or amplified to provide excursions between the trip points at 0.82V and 1.62V. This circuit can operate at frequencies in excess of 20 megahertz.

The existence of the high-frequency Schmitt trigger and other circuits that can shape sine waves into rectangular form suitable for operating digital circuits makes it possible to use sine wave oscillators, which can provide clock frequencies high enough to take advantage of the high-speed capabilities of present-day digital circuits. Accurate and stable crystal oscillators are available to 20 megahertz, and sine wave frequencies exceeding these can be obtained with *frequency-multiplying circuits.* Figure 17-15 shows a block diagram of the method used to obtain clock frequencies in excess of 20 megahertz.

Most microprocessor systems require a stable clock generating circuit. This requires the clock circuit to be driven by a *crystal* oscillator. Figure 17-16 shows such a network for the 8080 microprocessor.

The crystal functions like a high Q tuned circuit; its equivalent circuit is shown in Figure 17-18. In some microprocessor designs, the manufacturer recommends that the crystal be operated in its series mode. For other applications, its parallel mode is recommended. Most often only the crystal's equivalent resistance and reactance are defined at a specified frequency. The crystal's power dissipation depends on its internal AC resistance.

The CMOS 4060 counter oscillator is an easy-to-use clock generator with a 14-stage binary counter. It can be used as an astable multivibrator with an RC circuit or with a crystal oscillator.

A circuit that merely generates rectangular pulses is in itself as useless as a conventional clock without hands. The pulses must be organized into *cycles.* The first pulse in the cycle is generally designated as a 0 or reset pulse. Each pulse thereafter may be numbered or, in some cases, given a functional name. The periodic separation of one pulse onto a 0 or reset line can be accomplished by using an astable multivibrator of a lower frequency than the clock. Figure 17-26 shows a circuit that performs this function. Figure 17-27 shows another technique for performing this function. Here a counter is allowed to cycle continuously. Each time it reaches the 0 state, the decoded 0 is used to enable a clock pulse onto a reset line.

There are many occasions in digital systems when both a clock and a delayed clock are needed. Figure 17-29 shows a method by which clock pulses are alternately passed and inhibited onto separate clock lines, producing two clock lines that are identical except that the C_P pulses are spaced midway between like numbered pulses on the C_P' line.

The timing we have discussed so far is periodic in nature, but in some cases, we need a random form of timing. An event may need to start or be finished instantly, without waiting for a clock pulse. For this application, the *monostable*, or *one-shot*, *multivibrator* can be used. Figure 17-30 illustrates the one-shot multivibrator's capacity to lengthen or shorten a pulse. When a pulse is applied to the input, the leading edge of the output occurs at about the same instant as the leading edge of the input. The trailing edge of the output, however, may occur sooner or later than that of the input, depending on the size of the capacitor, C. This gives us the capacity to lengthen or shorten a pulse.

Figure 17-31 shows a communications application of the one-shot multivibrator. A three-channel pulse width modulated signal is separated into three separate lines with the aid of three one-shots and three AND gates. Figure 17-32 shows a discrete one-shot. The output pulse width can be adjusted by varying R and C. Figure 17-33 is an integrated circuit one-shot. R and C are connected externally to allow time adjustments. Inverted or complementary input and output leads increase its versatility.

The 96L02 is a versatile IC one-shot circuit. The width of the output pulse is independent of the width of the input triggering pulse, as shown in Figure 17-34. This circuit can also be wired for nonretriggerable operation by connecting Q to IN and $\overline{Q}$ to $\overline{IN}$. Other IC one-shot multivibrators are shown in Figure 17-35.

Figure 17-37(b) shows a pulse delay network composed of two one-shots. From the first one-shot, a delay of the leading edge is accomplished by using the $\overline{Q}$ output. The second one-shot merely reconstructs the pulse after the delay time by having RC components that provide a width equal to the original input pulse.

The 555 timer can also be operated as a one-shot multivibrator. A comparison of Figure 17-5(a) with Figure 17-38 shows that resistor R_B between pins 7 and 6 is not used in the monostable multivibrator. An application of the versatile 555 timer as a pulse width modulation circuit is shown in Figure 17-41.

Glossary

Astable Multivibrator: A logic circuit that automatically and periodically changes between logic 1 and logic 0, normally producing a square wave or rectangular pulse on its output terminals. See Figures 17-1 and 17-2.

Unijunction Transistor: A semiconductor device widely used in timing or pulse generator circuits. It has three terminals—emitter, base 1, and base 2. See Figure 17-7.

Schmitt Trigger: A circuit used to convert irregular waveforms to rectangular waveforms at the logic levels. See Figure 17-10.

Frequency Multiplier: A circuit used to multiply the frequency of a sine wave voltage. The input is tuned to the original frequency. The output is tuned to an exact multiple of the input frequency. See Figure 17-15.

Monostable Multivibrator (One-Shot Multivibrator): A logic circuit that goes high at its output on the positive-going edge of an input pulse and remains high for a period dependent on the selected value of an RC component. The circuit has the capacity to increase or decrease the width of a pulse applied to its input. See Figure 17-30.

Pulse Width Modulation: A method of applying intelligence to a carrier by varying the width of a radio frequency pulse in proportion to an audio frequency amplitude. See Figure 17-40.

Crystal Load Capacitance: The value of capacitance used by the crystal manufacturer to simulate the capacitance that will load the crystal when it is in the circuit. For crystals ground to a parallel resonant frequency, load capacitance of 30 pF or 32 pF is standard. The crystal is ground to resonate at a specified frequency when loaded with the load capacitance. The user of the crystal must then ensure that the total of stray capacitance plus lumped capacitance loading the crystal is equal to the load capacitance in order to obtain resonance at the specified frequency.

Questions

1. Name three circuits that might be used as digital clocks. Which has the highest frequency limit?
2. How can an astable multivibrator be made to produce rectangular pulses rather than a square wave? Does the pulse have any advantage over a square wave?
3. What type of output waveform is generated by a 555 timer operating as an astable multivibrator?
4. Can the 555 timer be used in both TTL and CMOS circuits? Explain your answer.
5. What is the equation for the frequency generated by the RC-type oscillator shown in Figure 17-6(a)?
6. How does the unijunction schematic symbol differ from the symbol for a junction FET?
7. What deficiencies may exist in the output of a unijunction pulse generator, and how can they be remedied?
8. What purpose does the Schmitt trigger serve?
9. Where in the counter described in Section 14.11 are Schmitt trigger circuits likely to be used?
10. Can a Schmitt trigger operate at frequencies in excess of 10 MHz?
11. What are the approximate trip points of the TTL/SSI 7413 Schmitt trigger?
12. Where stability and accuracy are important, what type oscillator is most often used?
13. What zone of operation must a CMOS inverter be operating in when it is used in a crystal-controlled oscillator circuit?
14. Draw the equivalent circuit of a crystal.
15. Which resonant frequency of a crystal is higher—series or parallel?
16. What name is used to describe the series branch of a crystal's circuit?
17. What will happen if a crystal is used in a circuit where the load capacitance differs from that used when the crystal was ground?
18. How are a crystal's parameters most often defined?
19. What is the power dissipation range of most crystals?

20. In using the cycle multivibrator in Figure 17-26, what prevents the selection of a fragmented pulse for the 0 reset? What prevents additional pulses from entering the 0 reset line?
21. How can a counter be used to select a periodic reset pulse?
22. Describe a method for obtaining separate 1 MHz clock and delayed clock pulse lines from a single 2 MHz square wave.
23. How does operation of a monostable multivibrator differ from that of a bistable multivibrator?
24. Which of the following cannot be accomplished with one or more monostable multivibrators: generating periodic pulses, lengthening a pulse, shortening a pulse, delaying a pulse?
25. Describe three of the four operations listed in Question 24.
26. How does the operation of a retriggerable monostable multivibrator differ from that of a nonretriggerable monostable multivibrator?
27. Describe use of a one-shot multivibrator in pulse width modulation.
28. Describe the functional differences between three types of multivibrators—astable, monostable, and bistable. What basic function do they have in common?

Problems

17-1 Determine what size resistor is needed for use with a 96L02 monostable multivibrator to obtain a pulse width of 1 millisecond if the capacitor to be used is 0.0047 μF.

17-2 Draw a block diagram and timing diagram of a circuit designed to delay a 500-microsecond pulse by 2 microseconds.

17-3 If we assume the intrinsic standoff ratio of a unijunction to be 0.6, what is the approximate frequency of a UJT pulse generator having R = 1.8K and C = 0.033 μF?

17-4 A 5 MHz crystal oscillator circuit is to be used as the generator for a 10 MHz clock. Describe the circuits needed to bring the output to correct frequency at TTL logic levels (the output to be a symmetrical square wave). Draw a block diagram and describe the output of each block with respect to frequency amplitude and waveshape.

17-5 Draw a four-bit counter and the added circuits needed for a cycle of 0 through 12 pulses.

17-6 The delay circuit of Figure 17-37 is being implemented with the two one-shots of a 96L02 integrated circuit. Using the graph of Figure 17-33, determine the approximate capacitance, C_x, needed for MV 1 and MV 2. Both one-shots have R_x = 120K ohms.

17-7 Draw a 555 timer connected as an astable multivibrator using values of R_A, R_B, and C to give a free-running frequency of 12 kHz.

17-8 Draw a CMOS astable multivibrator with values of R and C selected to provide a frequency near 20 kHz.

17-9 An 8224 must provide ϕ_1 and ϕ_2 frequencies of 1.5 MHz. If no tank circuit is to be used, what frequency crystal must be used?

17-10 Figure 17-42 shows a Fairchild LPTTL/monostable 96L02. Timing capacitors (C_x) connect between pins 1 and 2 and pins 14 and 15. Timing resistors (R_x) connect to pins 2 and 14. Draw the added components and interconnections needed to convert the integrated circuit of Figure 17-42 to the pulse delay system described in Problem 17-6.

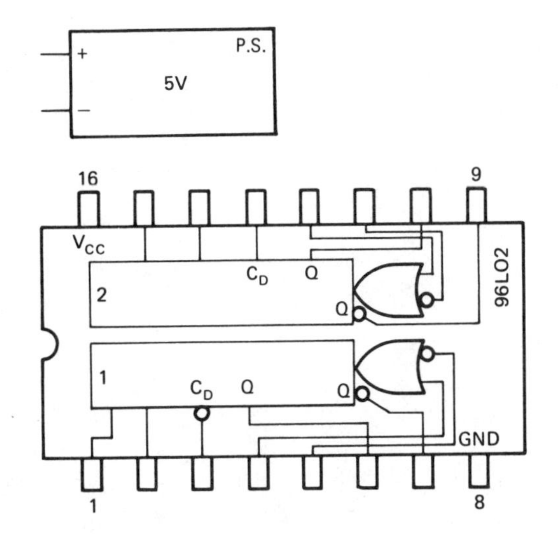

FIGURE 17-42. Problem 17-7

17-11 Draw a 555 timer connected to provide a monostable multivibrator with an output pulse width of 8 milliseconds. Use C = 10,000 pF. Determine R_A.

17-12 Using a crystal where f = 3.58 MHz, R_3 = 90Ω, and C_L = 30 pF, determine the component values of R_S, C_T, and C_S for the CMOS crystal oscillator of Figure 17-21.

17-13 Determine the power dissipated in the crystal in the circuit of Problem 17-8.

17-14 Draw a 96L02 monostable multivibrator connected for triggering on the falling edge of the input pulse and for nonretriggering operation.

17-15 Use the CMOS 4060 to obtain f_x, $f_x/2$, and $f_x/8$ for f_x = 64 kHz. Determine values C_T and R_T for the astable multivibrator tuning. Specify astable frequency and Q outputs to be used for each frequency.

17-16 Repeat Problem 17-15 for f_x = 80 Hz.

CHAPTER

Display

Objectives

Upon completion of this chapter, you will be able to:

- Construct a transistor lamp driver.
- Construct a transistor relay driver.
- Use integrated circuit arrays for large numbers of lamp or relay drivers.
- Assemble an in-line decimal readout.
- Provide the necessary coding and driving circuits for a seven-segment readout.
- Assemble decoder/drivers with correct lamp test and ripple blanking for multidigit seven-segment displays.
- Correctly assemble and power seven-segment readouts of incandescent, fluorescent, neon, and LED types.
- Use a seven-segment display for hexadecimal digits.
- Use a hexadecimal latch/decoder driver circuit.
- Provide multiplexing for display with large numbers of readouts.
- Provide failsafe circuits for multiplexed displays.
- Use LED dot matrices for hexadecimal display.

18.1 Introduction

For some digital devices, the end result of their function is a visual display. These displays are in most cases numerical, possibly including a decimal point and a few special characters, such as plus and minus signs. This is the case with the counter, discussed in Chapter 14. The digital clock and digital voltmeter are two other devices whose output is a digital readout.

On maintenance or operator panels of larger digital machines, an assortment of labeled lamps may light up to indicate to the operator the mode in which the machine is operating or the reason for a stop or delay in operation. In many cases, the state or content of important registers or counters is continuously displayed. This presents a need for special circuits to decode, power, drive, and in some cases multiplex the information to be read out.

In numerical control and other automated processing techniques, the end result of even very complex digital functions is a mechanical action requiring application of high-current AC power to operate motors or solenoids. This is generally done by energizing a relay. The relay coil is driven by a circuit very much like the lamp driver.

18.2 Lamps and Lamp Drivers

The incandescent lamp is the most widely used indicator. It consumes more electrical power per candlepower of light than the neon lamp, but it is available for voltages as low as 1.2V, while neon lamps require 50V to 100V. A reasonably visible indication requires 100 to 500 milliwatts of power, which exceeds the drive capability of most integrated circuits. They must, therefore, be operated by a lamp driver circuit, as Figure 18-1 shows. If the logic-level power supply is used to power the lamp, then the lamp voltage selected should be approximately V_{CC} - $V_{CE\ SAT}$. The lamp is normally lighted by the logic 1 level applied to the input of its driver. The 1-level input must produce an I_B of $2.5 I_L/\beta$. The 2.5 is to ensure saturation regardless of variations in beta of the transistor or variations in the input 1 level. The resistor, R_B, would be determined by $(V_{ON} - V_{BE})/I_B$.

■ **EXAMPLE 18-1**

A lamp is to be used to indicate a set condition of a register having a 3.6V logic 1 level. With a V_{CC} of 5V, design a single transistor driver. As shown in Figure 18-2, we have selected a number, 680 (5V at 60 mA). It will be operated slightly under voltage but without a noticeable loss of brilliance. The transistor to be used has a minimum β of 100.

Solution:

$$I_B = 2.5\frac{I_L}{\beta} = 2.5 \times \frac{60 \text{ mA}}{100} = 1.5 \text{ mA}$$

$$R_B = \frac{V_{ON} - V_{BE}}{I_B} = \frac{3.6\text{V} - 0.6\text{V}}{1.5 \text{ mA}} = 2\text{K}\Omega$$

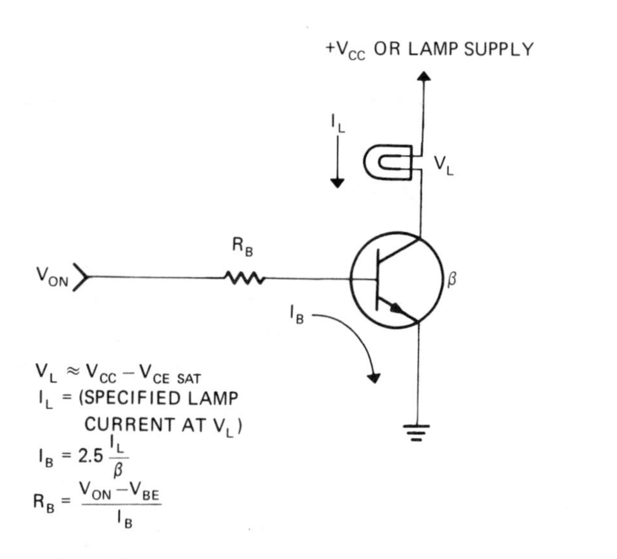

FIGURE 18-1
Transistor lamp driver.

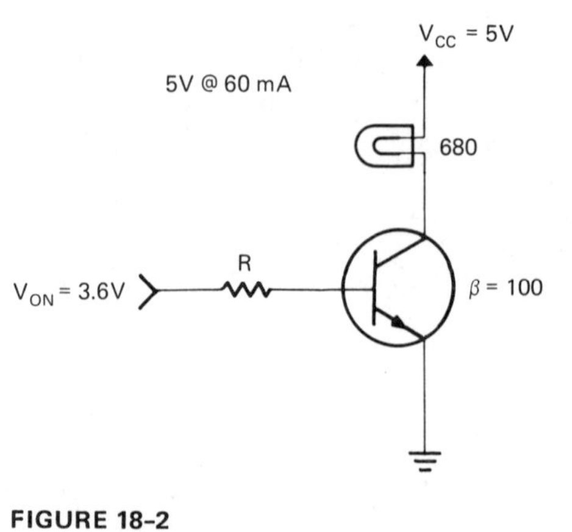

FIGURE 18-2
Example 18-1.

The 1.5 mA I_B of the circuit in Example 18-1 is within the drive capability of most TTL integrated circuit outputs. Ten or more unit loads at 150 μA is typical and this is enough to supply the 1.5 mA to the lamp driver. If low-power logic circuits are involved, a single transistor circuit of Figure 18-1 would not make a suitable driver. It might also be inadequate for a medium-power circuit if some of the unit loads were needed to drive logic circuits simultaneously with the lamp. A lower-current lamp is not a feasible solution, as 60 mA is already low for a lamp of reasonable size and brilliance. The solution may be to use a two-transistor driver. The Darlington pair circuit of Figure 18-3 is favored if the 1-level voltage is sufficiently high. The V_{BE} of the two transistors is in series. Because of this, the minimum voltage needed to operate the driver is 1.4V. If we assume both transistors to have the same minimum β, and if we substitute β^2 for β and $2V_{BE}$ for V_{BE}, the same design procedure may be used for this circuit as was used for the circuit of Figure 18-1.

■ EXAMPLE 18-2

As shown in Figure 18-4, a number, 680 (5V at 60 mA lamp), is to be used to display a set level stored in a register having a 3.6V 1 output level. With a V_{CC} of 5V, design a two-transistor driver. The transistors to be used have a minimum β of 100.

Solution:

$$I_B = 2.5\frac{I_L}{\beta^2} = \frac{150\text{ mA}}{10{,}000} = 15\ \mu\text{A}$$

$$R_B = \frac{V_{ON} - 2V_{BE}}{I_B} = \frac{3.6\text{V} - 1.4\text{V}}{15\ \mu\text{A}} = 147\text{K}\Omega$$

The I_B computed in Example 18-2 is less than a unit load for any type of logic circuit except, possibly, CMOS I_C. It therefore represents a minimal load to the logic circuit driving it. For some RTL logic, the minimum 1 levels may fall below 1V and therefore fail to drive the lamps, but few modern types of TTL- or MOS-type circuits present this problem.

The transistor peak collector current rating should exceed the lamp current by at least 2½ times. Although the lamp will limit the collector current to I_L once it begins to glow, the lamp's cold resistance is many times lower than its hot resistance. This results in an initial surge current that could destroy the transistor. If the transistor rating is marginal, a series resistor in conjunction with a lamp of lower voltage might provide a solution. This is shown in Figure 18-5.

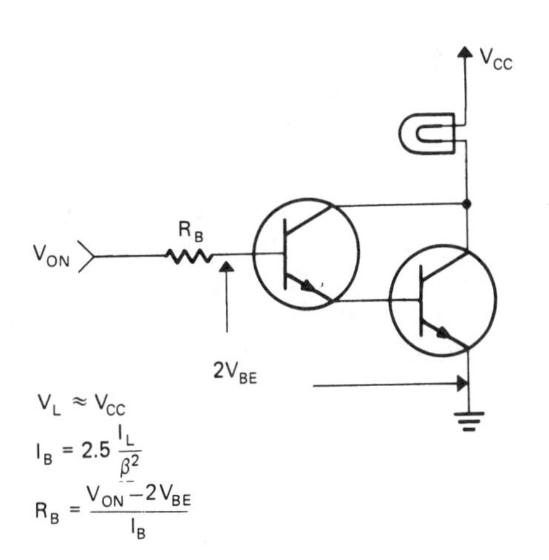

FIGURE 18-3
Use of Darlington pair as a lamp driver.

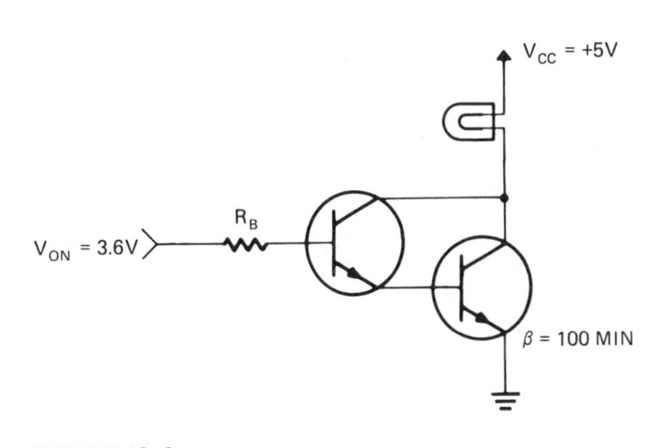

FIGURE 18-4
Example 18-2.

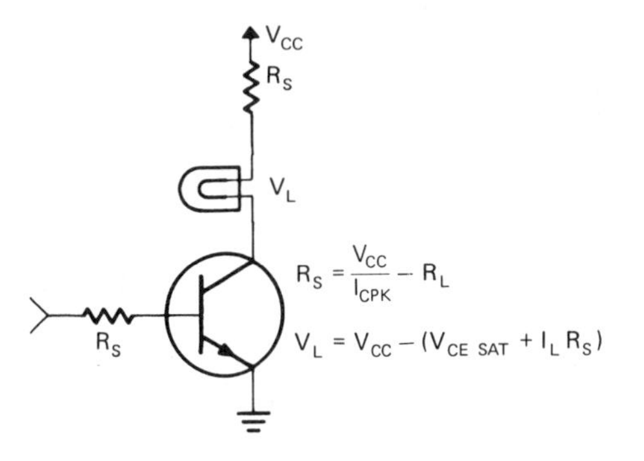

FIGURE 18-5
Resistor in series with lamp, resulting in reduced initial surge current.

If it is likely that many lamps will be turned on at one instant, the surge current might prove damaging to the power supply. The series resistor in the lamp driver might be used to alleviate this.

18.3 Relay Driver

Figure 18-6 shows a relay driver logic symbol, a single-transistor relay driver, and a Darlington pair relay driver. The design methods are essentially the same as those for the lamp driver. A reverse-biased diode is normally placed across the relay coil to short out the inductive voltage spike that occurs when the relay is turned off.

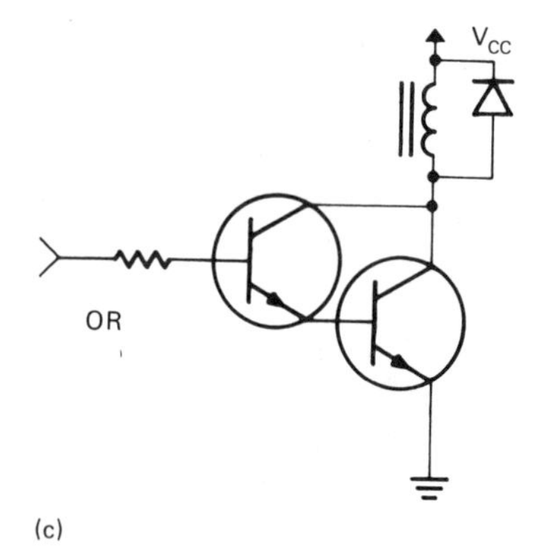

FIGURE 18-6
(a) Logic symbol of relay driver. (b) Single-transistor relay driver. (c) Darlington pair relay driver.

■ EXAMPLE 18-3

A relay with 12V, 100 mA coil is to be driven by a TTL logic circuit output. The logic circuit output is a 2.4V logic 1 level and can supply 1.2 mA. Using transistors with a minimum β of 80, design the necessary single-transistor relay driver. The relay supply is 12.5V, as shown in Figure 18-7.

Solution:

$$I_C = 100 \text{ mA}$$

$$I_B = 2.5\frac{I_C}{\beta} = \frac{250 \text{ mA}}{80} = 3.13 \text{ mA}$$

■ EXAMPLE 18-4

The 3.13 mA required input computed in Figure 18-7 exceeds the 1.2 mA rating of the logic circuit. Therefore, design a relay using two transistors. As shown in Figure 18-8, the 39 μA is low enough for the circuit to operate the driver.

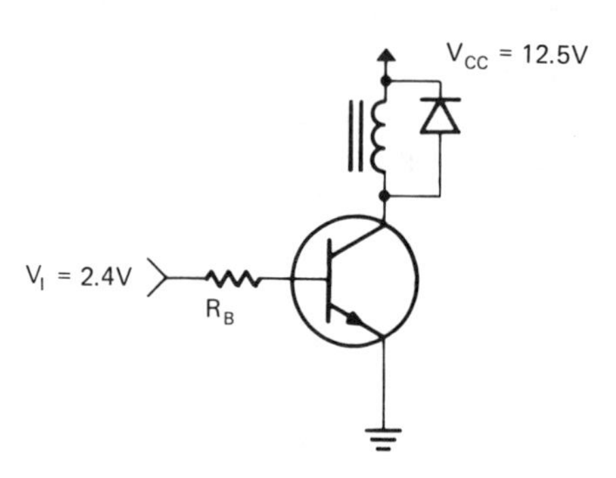

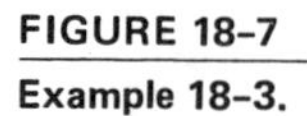

FIGURE 18-7
Example 18-3.

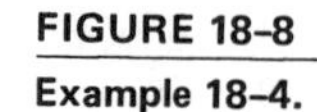

FIGURE 18-8
Example 18-4.

Solution:

$$I_B = 2.5\frac{I_C}{\beta^2} = \frac{250\text{ mA}}{6400} = 39\ \mu\text{A}$$

$$R_B \approx \frac{V_E - 2V_{BE}}{I_B} = \frac{2.4 - 1.4}{39\ \mu\text{A}}$$

$$= 25.6\text{K} \approx 27\text{K}$$

When many lamps are to be driven, it is economical to use monolithic transistor arrays, like those shown in Figure 18-9. These are available in 16-pin DIP form with a power dissipation of 750 mW for all seven devices and 300 mW for each transistor. This is adequate for driving seven relatively low-power lamps. Another device convenient for lamps of 250 mA or less is the quad

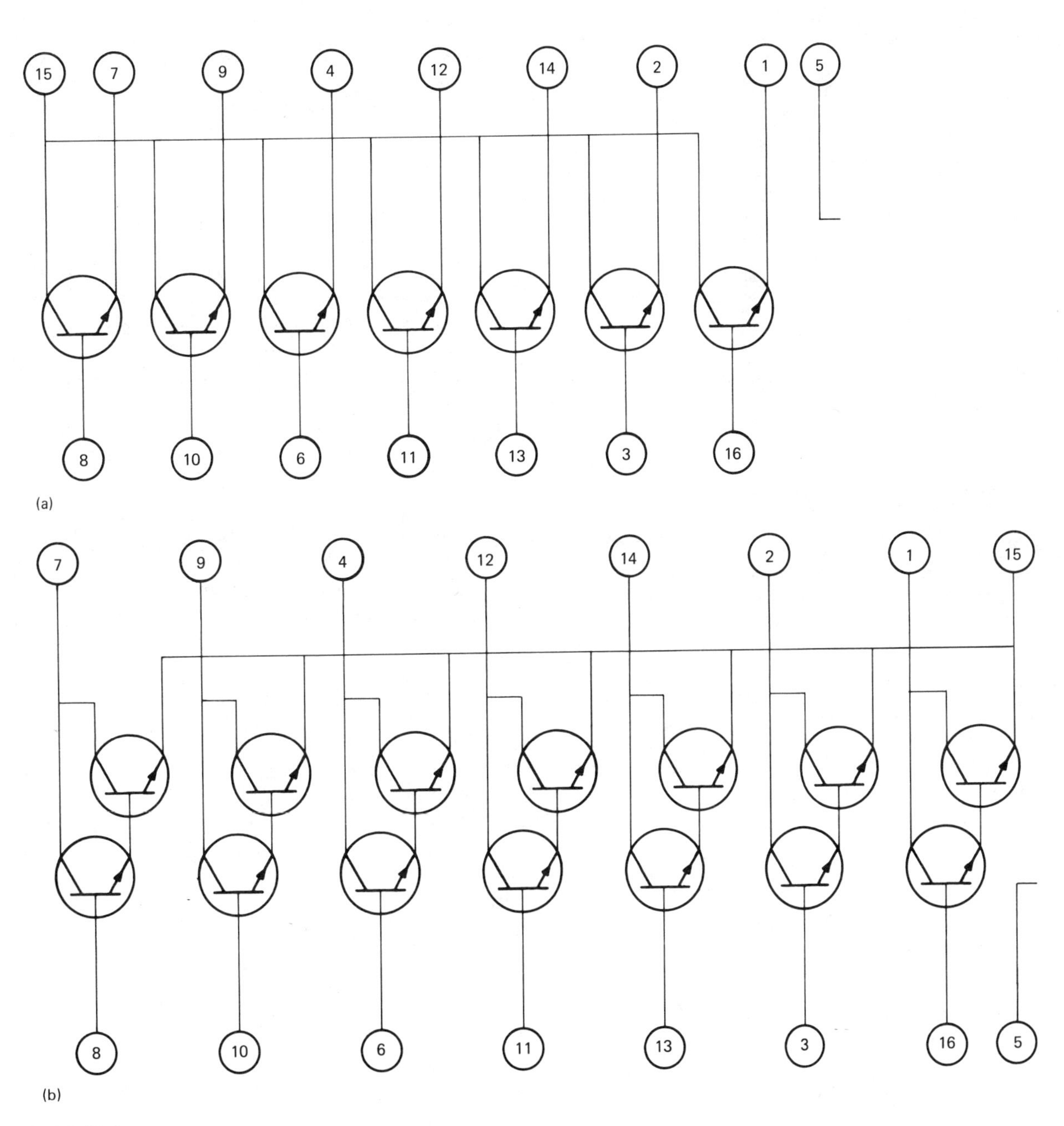

FIGURE 18-9

Seven-transistor arrays for use where many lamp drivers are needed. (a) Single 16-pin DIP form. (b) Darlington pair version.

power driver, shown in Figure 18-10, which can be operated by TTL or DTL logic units. A logic 0 to either NAND gate input will short the lamp to ground through the transistor. The NAND gate is also a convenience for lamp testing, as one input of each NAND gate can be tied to a lamp test line. When the lamp test line is grounded, all lamps will light. When the lamp test line is open, the drivers are enabled for control by the other NAND gate input.

The circuit of Figure 18-10 is called an AND power gate because the transistor provides a second inversion to the NAND gate, and if it is used with pull-up resistors for logic applications other than lamp drivers, the end result is an AND function. Other quad power gates are available with AND, OR, and NOR input gates.

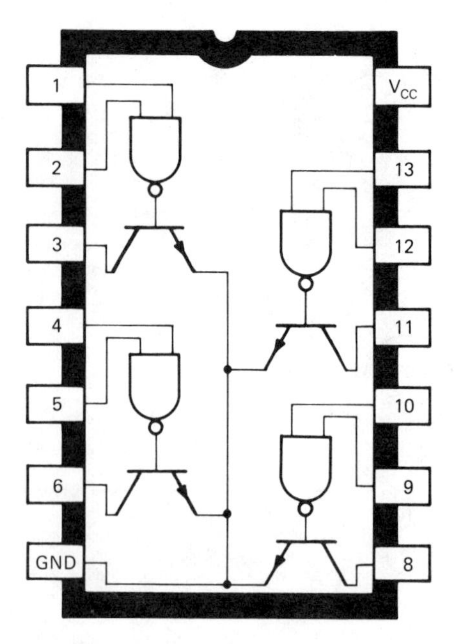

FIGURE 18-10

Sprague Electric quad AND power gate. The inputs are NAND gates. If they are used as lamp drivers, a logic 0 will turn the lamp on.

18.4 BCD or Binary Readouts

On maintenance or test panels, the readouts of small counters or registers may be direct and involve no decoding, as shown for the counter of Figure 18-11. In a simple case of this type involving experienced personnel, the operator, tester, or maintenance person will decode mentally by adding up the value of the lamps that are lighted. When many decimal counting units are involved or the readout is a register of many bits, mental decoding may be too difficult or too slow.

18.5 In-Line Readouts

A BCD decimal counting unit can be decoded and the numerals 0 through 9 read out in a line

FIGURE 18-11

Lamp and lamp drivers connected to read out the contents of a universal shift register storing BCD numbers.

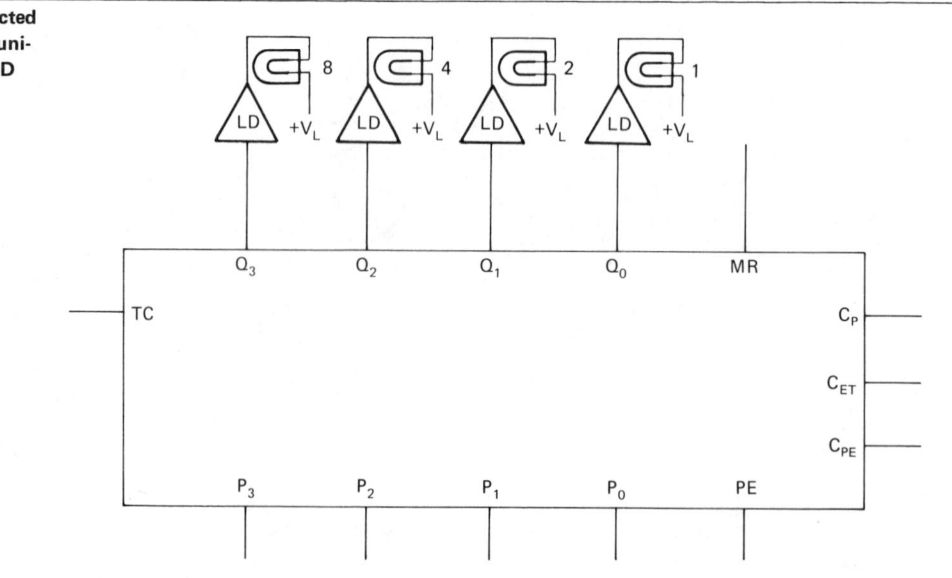

(usually vertical) of incandescent or neon lamps. This method was once widely used with vacuum tubes and, in some cases, semiconductor counters. Decoding could be accomplished with the AND gate decoder shown in Figure 6-33, but in this day of integrated circuits, a decoder like the TTL 7442, shown in Figure 18-12, might also be used. The outputs are inverted, providing a logic 0 to turn the lamp on; this would, therefore, be ideal for use with the NAND power gates of Figure 18-10.

18.6 Segmented Readouts

18.6.1 Description

Seven-segment readouts have, in recent years, become the most popular form of numerical readout. They are available in incandescent, fluorescent, neon, and LED (light-emitting diode) form. The segments are as shown in Figure 18-13 and are designated by the letters A through G. If all segments are lighted at one time, the numeral 8 results. The numerals 0 through 9 are as shown in Figure 18-14.

The numeral 1 can be displayed by lighting either B and C segments or E and F segments. This, unfortunately, leaves an unnaturally wide space on either side of a numeral 1. A nine-segmented readout is available that places the 1 in the center. The segmented readout is easier to read and more attractive, and it occupies less panel space than the vertical in-line readout of Figure 18-12. Decoding from BCD to seven-segment readout is more complicated than decoding from BCD to decimal. Let us assume a normal BCD-to-decimal decode and then decode from decimal to seven-segment. Decoding direct from BCD to seven segments would save very

FIGURE 18-12

BCD output decoded by a BCD-to-decimal decoder TTL/MSI 7442 before driving ten-lamp readout.

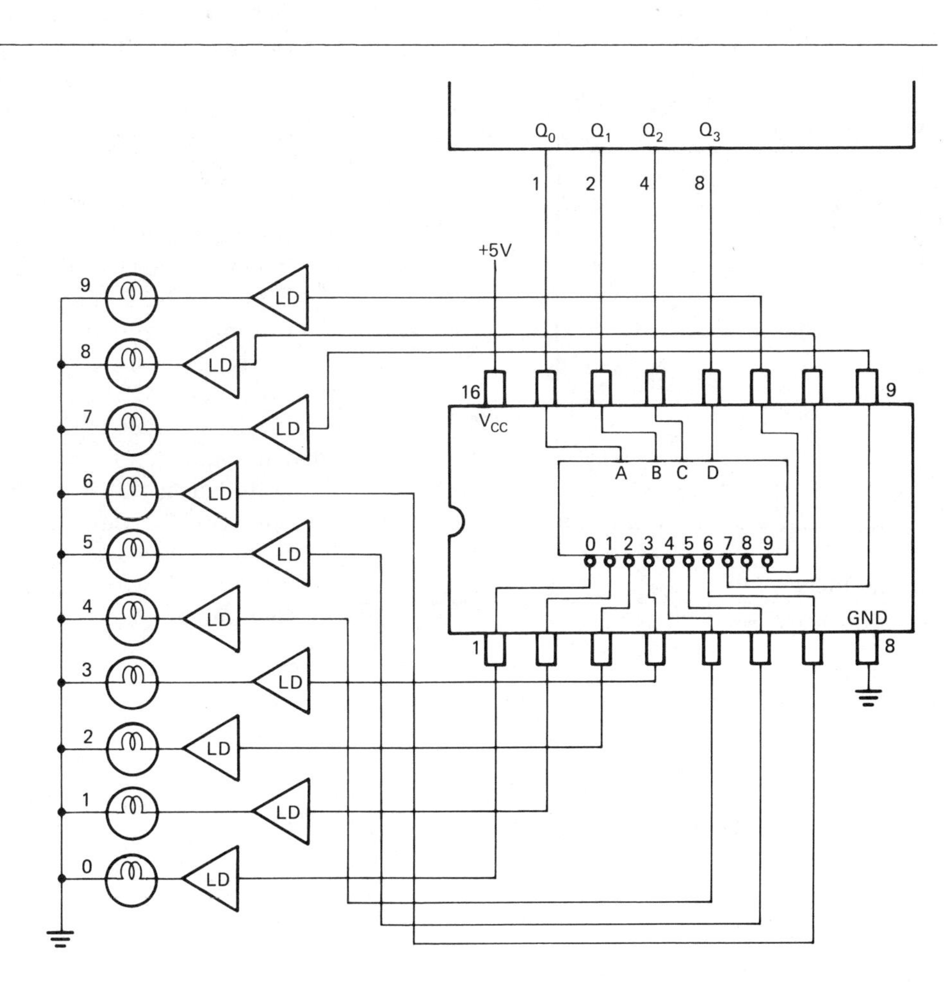

few circuits and would excessively complicate the explanation. Figure 18-15 shows a block diagram of the BCD decode drive and readout circuits.

The table of Figure 18-16 compares the decimal lines 0 through 9 with the seven segments A through G. Of the 63 blocks in the table, only 21 are OFF. This means it would be simpler to base our decode on the OFF conditions. The bottom line of the table is a NOR function of the OFF numbers for each segment. The NOR gates of Figure 18-17 would therefore be a suitable encoder for decimal to seven-segment.

If the seven-segment readouts were incandescent, each filament segment would require a lamp driver similar to those described in Section 18.2. Fortunately, decoder/drivers are available in integrated circuit form. Figure 18-18 shows a logic symbol for a seven-segment decoder/driver Fairchild 9317. The truth table for the segment outputs is the same as in Figure 18-15. The outputs, however, provide a ground (saturated transistor to ground) for the end of the lamp segment and turn it on. Three additional inputs are also included—the lamp test (LT), ripple blanking in (RBI), and ripple blanking out (RBO). The lamp test is used to test the seven segments for burnout. A 0-level input to the lamp test will turn on all seven segments regardless of the state of the four BCD input lines.

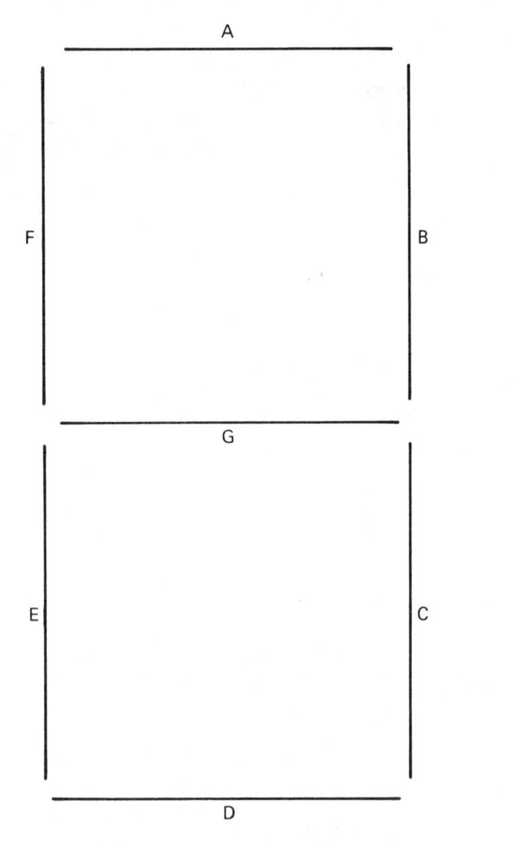

FIGURE 18-13

Seven-segment readout.

18.6.2 Automatic Blanking

To the left of the decimal point, the RBO and RBI inputs are used for automatic blanking of leading-edge zeros (those zeros to the left of the highest nonzero digit) or for automatic blanking of the trailing-edge zeros (those zeros to the right of the LSD to the right of the decimal point). The number 0070.0500 would be less confusing if displayed as 70.05. It is not practical to merely blank all zeros, as this would also remove zeros that may be needed between nonzero digits.

The unnecessary zeros can be blanked by proper interconnection of the RBI (ripple blanking in) and RBO (ripple blanking out) leads. Figure 18-19 shows the truth table of these inputs. As shown in the table, a 0 on the RBI input, combined with a BCD 0, turns off all segments and at the same time produces a 0 (active level) for RBO. This is the only condition that will blank (turn off all segments). For all other conditions, the segments will light in accordance with the BCD input. For all other conditions, an RBO 1 level occurs. This RBO 1 level is used to prevent lower significant-figure zeros from blanking.

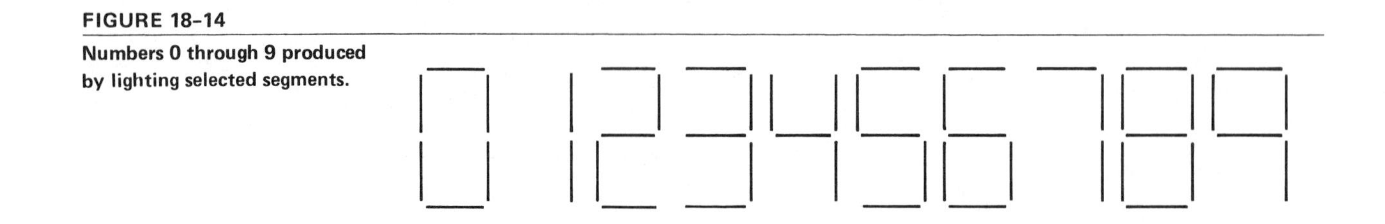

FIGURE 18-14

Numbers 0 through 9 produced by lighting selected segments.

FIGURE 18–15

Transition of BCD outputs to seven-segment readout levels A through G.

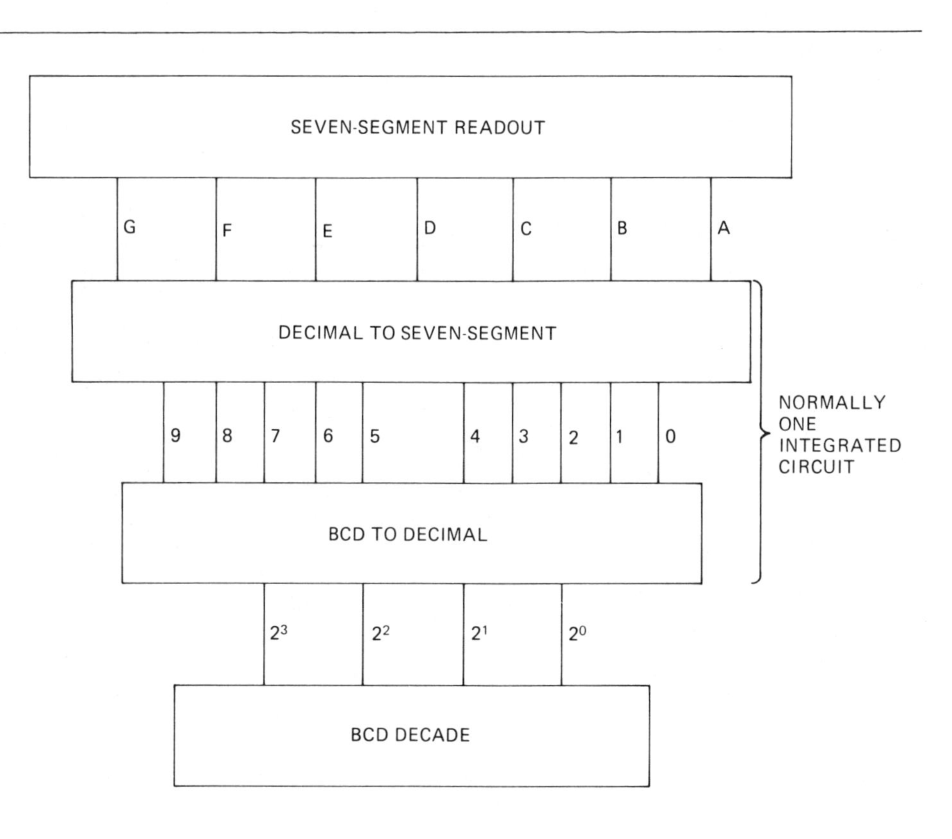

FIGURE 18–16

Table of active lines during seven-segment readout of decimal numbers 0 through 9 and NOR functions needed to provide the encoding.

	SEGMENTS LIGHTED							
DEC	A	B	C	D	E	F	G	READOUT
0	0	0	0	0	0	0		0
1					1	1		1 *
2	2	2		2	2		2	2
3	3	3	3	3			3	3
4		4	4			4	4	4
5	5		5	5		5	5	5
6	6		6	6	6	6	6	6
7	7	7	7					7
8	8	8	8	8	8	8	8	8
9	9	9	9	9		9	9	9
NOR FUNCTION PER SEG.	$\overline{1}\cdot\overline{4}$	$\overline{1}\cdot\overline{5}\cdot\overline{6}$	$\overline{1}\cdot\overline{2}$	$\overline{1}\cdot\overline{4}\cdot\overline{7}$	$\overline{3}\cdot\overline{4}\cdot\overline{5}\cdot\overline{7}\cdot\overline{9}$	$\overline{2}\cdot\overline{3}\cdot\overline{7}$	$\overline{0}\cdot\overline{1}\cdot\overline{7}$	

*OPTIONAL B·C

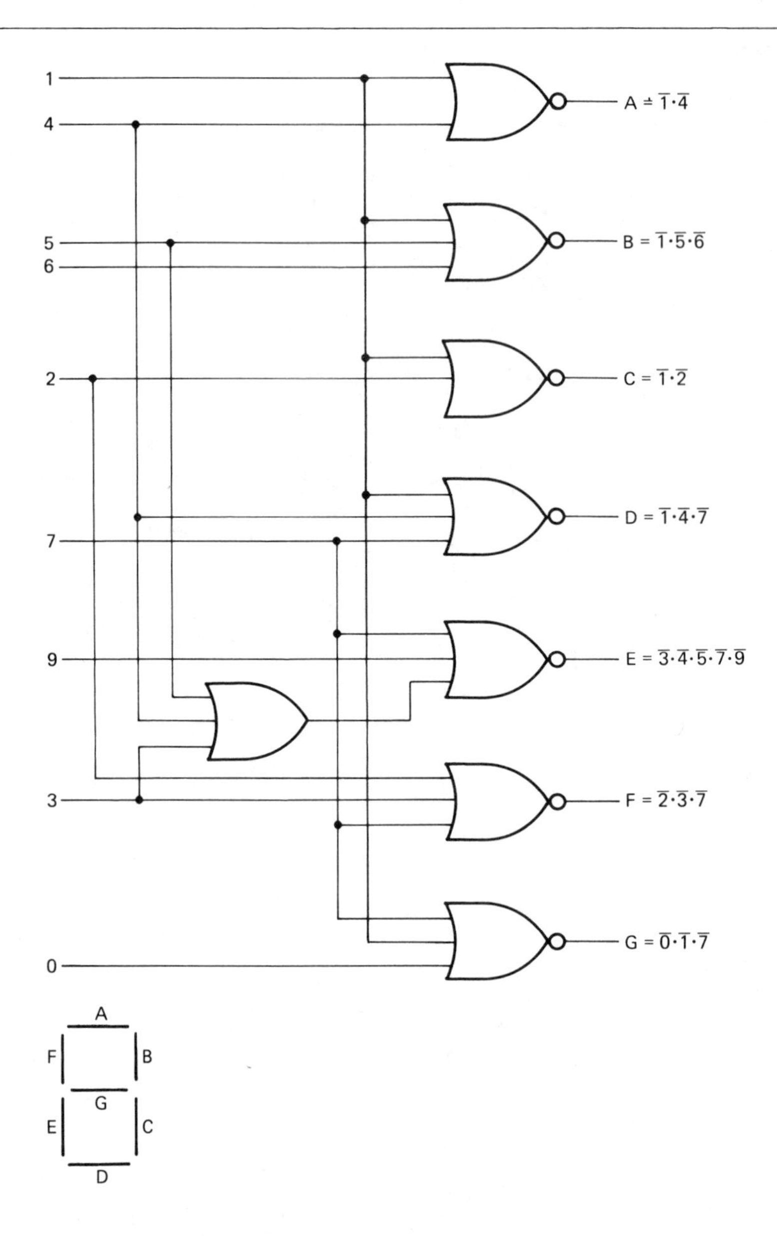

FIGURE 18–17

Decimal-to-seven-segment encoder.

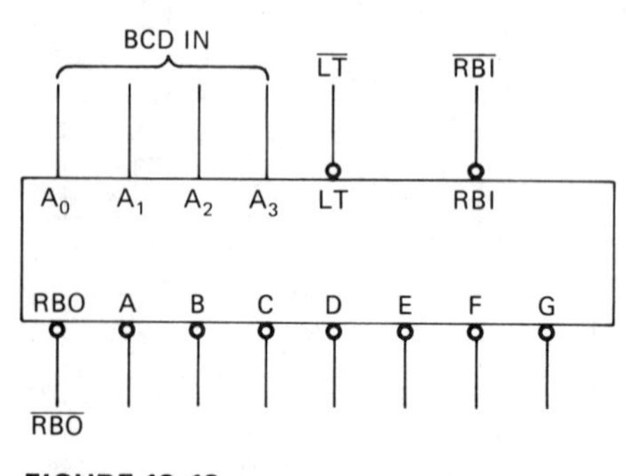

FIGURE 18–18

TTL/MSI Fairchild 9317 BCD-to-seven-segment decoder driver with lamp test and blanking functions.

The blanking connections are made as shown in Figure 18–20. The MSD ripple blanking input is grounded. A BCD 0 is enough to cause it to blank. If it blanks, it will also send an RBI 0 to enable the 10^2 digit to blank on a BCD 0. If it blanks, it sends an RBI 0 to enable the 10^1 digit to blank on BCD 0. In Figure 18–20, the 10^1 digit receives a BCD 7. It does not blank and therefore inhibits the 10^0 digit from blanking, despite the fact that it receives a BCD 0. To the right of the decimal point, the function differs only by the fact that the lower significant digits enable or inhibit blanking of the higher significant digits, as Figure 18–21 shows.

$\overline{RBI}$	BCD (DEC)	$\overline{RBO}$	A	B	C	D	E	F	G
0	0	0	OFF	OFF	OFF	OFF	OFF	OFF	OFF
0	1→9	1	ENABLE	ENABLE	ENABLE	ENABLE	ENABLE	ENABLE	ENABLE
1	0	1	ON	ON	ON	ON	ON	ON	OFF
1	1→9	1	ENABLE	ENABLE	ENABLE	ENABLE	ENABLE	ENABLE	ENABLE

FIGURE 18–19

Truth table of the ripple blanking functions, used to turn off unnecessary zeros in the readout.

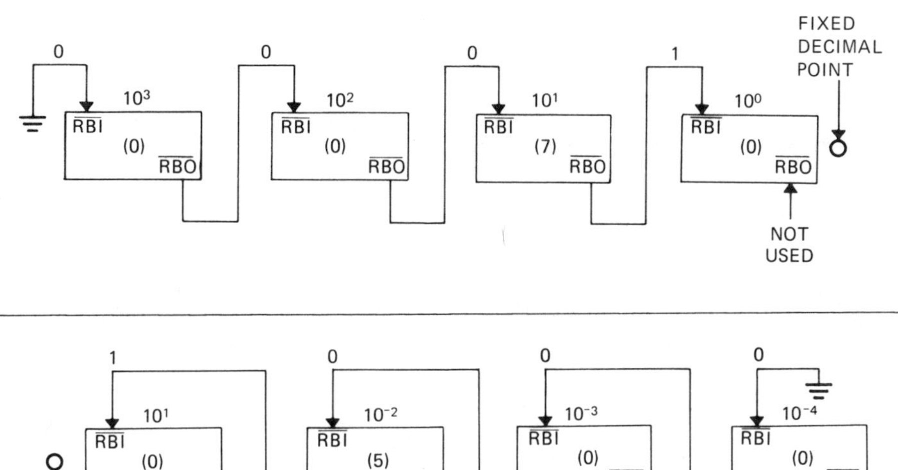

FIGURE 18–20

Ripple blanking connections to the left of the decimal point.

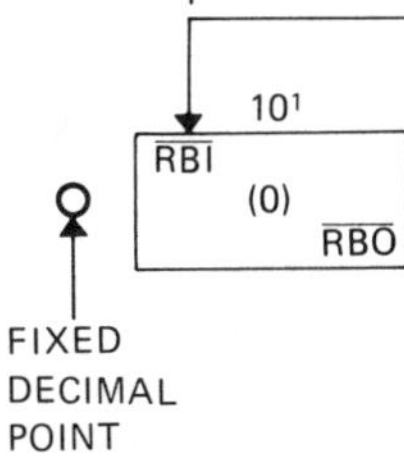

FIGURE 18–21

Ripple blanking connections to the right of the decimal point.

For microprocessors of 4-, 8-, 12-, and 16-bit sizes, it is common to use hexadecimal displays rather than BCD. The standard seven-segment display can be used for hexadecimal by using lowercase letters for the digits B and D which, in uppercase, would use the same segments as 8 and 0. Figure 18–22 shows the upper six digits used in seven-segment display of hexadecimal numbers. The seven-segment decoder/drivers developed for early use did not include hexadecimal capability, but decoders can be made from 32-by-8-bit PROMS (described in Chapter 24) or from some newly developed

HEXADECIMAL DIGIT	A	B	C	D	E	F
SEVEN-SEGMENT DISPLAY USED	A	b	C	d	E	F
REASON FOR LOWERCASE		8		0		
CAUSE OF CONFUSION		6				

FIGURE 18–22

Seven-segment display of hexadecimal digits A through F. Lowercase letters are displayed for B and D to avoid duplication with 8 and 0. The lowercase B is often confused with the number 6.

FIGURE 18–23

Motorola CMOS MC14495 hexadecimal latch/decoder driver. This circuit receives a four-bit hexadecimal input and will drive a common cathode seven-segment display. (a) Block diagram. (b) Alphanumeric display. (c) Truth table.

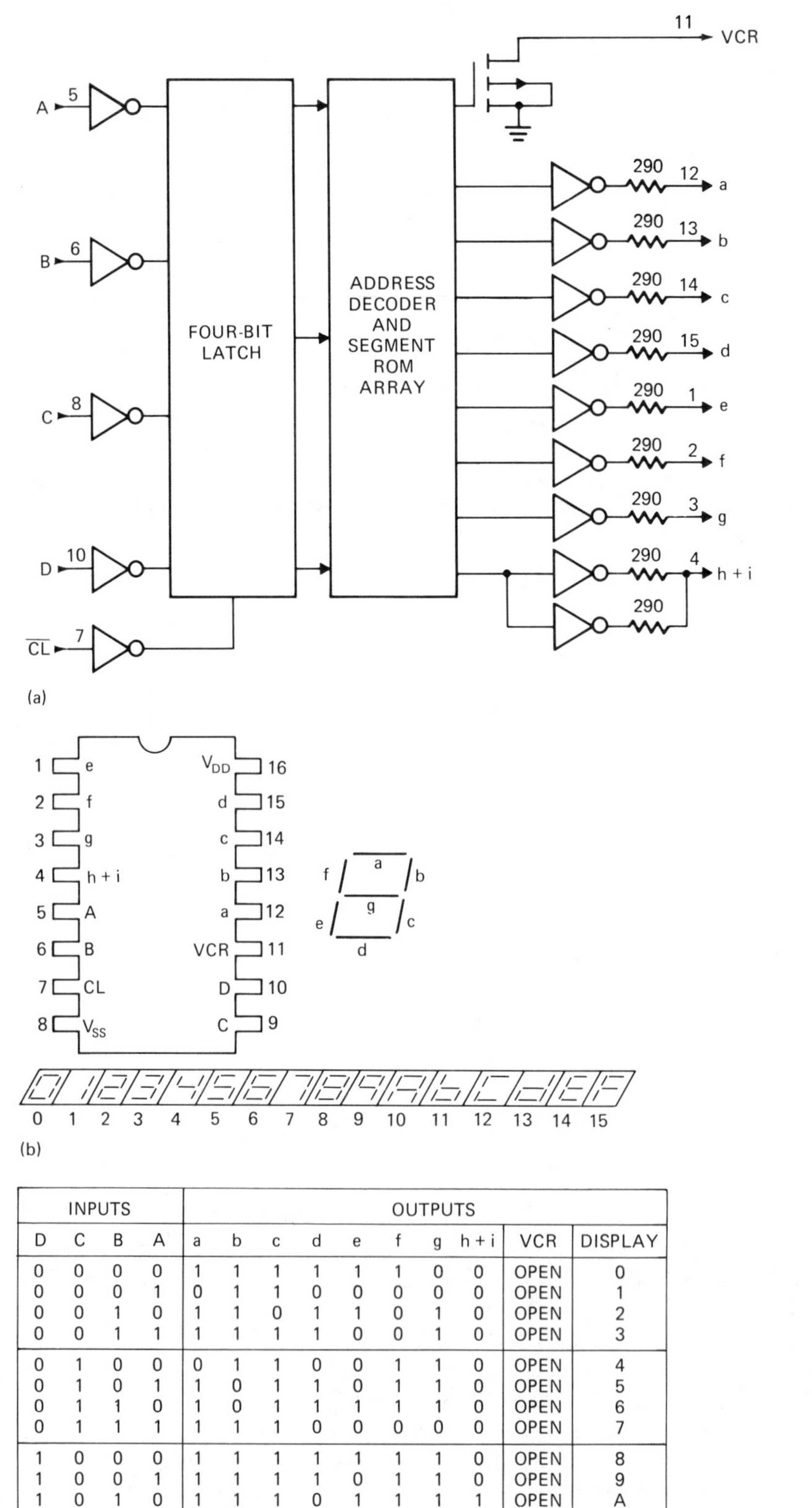

INPUTS				OUTPUTS									
D	C	B	A	a	b	c	d	e	f	g	h + i	VCR	DISPLAY
0	0	0	0	1	1	1	1	1	1	0	0	OPEN	0
0	0	0	1	0	1	1	0	0	0	0	0	OPEN	1
0	0	1	0	1	1	0	1	1	0	1	0	OPEN	2
0	0	1	1	1	1	1	1	0	0	1	0	OPEN	3
0	1	0	0	0	1	1	0	0	1	1	0	OPEN	4
0	1	0	1	1	0	1	1	0	1	1	0	OPEN	5
0	1	1	0	1	0	1	1	1	1	1	0	OPEN	6
0	1	1	1	1	1	1	0	0	0	0	0	OPEN	7
1	0	0	0	1	1	1	1	1	1	1	0	OPEN	8
1	0	0	1	1	1	1	1	0	1	1	0	OPEN	9
1	0	1	0	1	1	1	0	1	1	1	1	OPEN	A
1	0	1	1	0	0	1	1	1	1	1	1	OPEN	b
1	1	0	0	1	0	0	1	1	1	0	1	OPEN	C
1	1	0	1	0	1	1	1	1	0	1	1	OPEN	d
1	1	1	0	1	0	0	1	1	1	1	1	OPEN	E
1	1	1	1	1	0	0	0	1	1	1	1	0	F

(c)

devices like the Motorola CMOS MC14495, shown in Figure 18-23.

The MC14495 is a latch/decoder driver. It can latch the input by using a negative-going clock, or the $\overline{CL}$ can be held low, allowing the display to respond to all input changes. The decoder produces the 16-figure alphanumeric display shown in Figure 18-23. The segment drivers are designed for common cathode displays and contain internal resistors, limiting the diode current to 7 to 10 mA per segment. The device is CMOS and may require pull-up resistors on the input leads if driven by TTL-compatible devices.

18.7 Types of Segmented Readouts

18.7.1 Incandescent

An incandescent lamp is one in which a thin metal filament is heated white-hot by running an electric current through it. Seven separate filament wires are used to supply the seven-segmented readout. As Figure 18-24 shows, each filament has one end connected to the common. The seven opposite ends form connections A through G.

18.7.2 Fluorescent

The fluorescent lamp is similar to a vacuum-tube diode. The plate or anode is coated with a phosphor that glows when bombarded by the electrons that are attracted to it. In the seven-segment fluorescent display, a single filament serves all seven segments, but each segment is a separate anode that can be turned off and on independently by application or removal of the positive anode potential. Figure 18-25 shows typical wiring of the seven-segment fluorescent display. The control grid is included only for those applications requiring multiplexing. A negative potential applied to the control grid turns off all segments, regardless of application of anode potential.

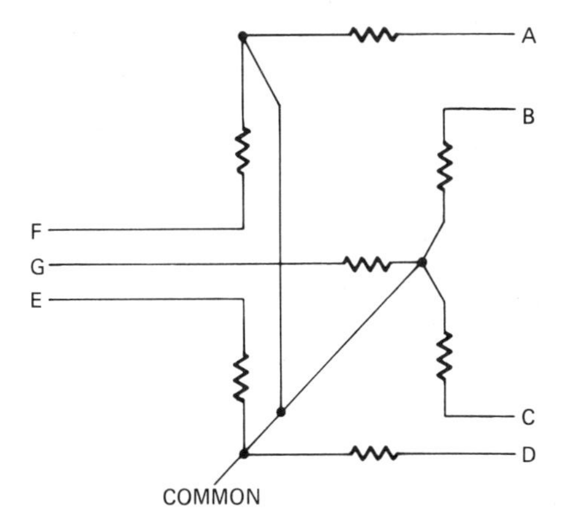

FIGURE 18-24

Connection of incandescent lamp filaments for a seven-segment readout.

18.7.3 Neon and Other Gas Tubes

Cold cathode diode gas tubes will conduct electricity once the voltage between anode and cathode reaches the ionization potential. This conduction causes an illumination or glow in the vicinity of the cathode. The NIXIE* tube, once widely used in digital equipment, produces the digits 0 through 9 by having ten separate cathodes bent to form the figures 0 through 9. Figure 18-26 shows a schematic of a ten-segment NIXIE tube. The decoder for this display is a simple

*NIXIE® is a registered trademark of Burroughs Corporation.

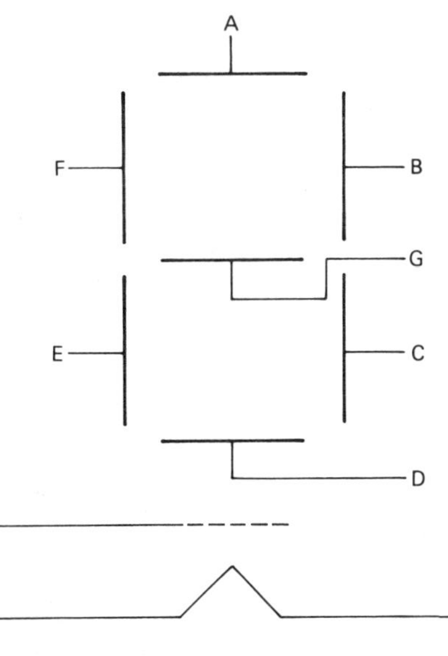

FIGURE 18-25

Seven-segment fluorescent readout, with a common cathode but a separate anode for each of the seven segments. Control grid is available to turn all segments off with a negative control signal.

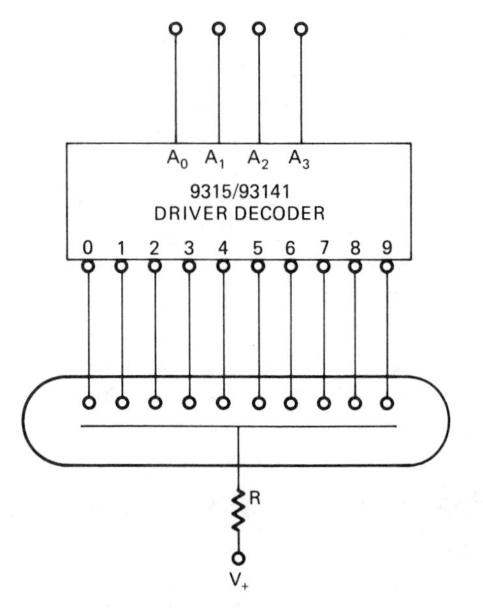

FIGURE 18–26

Fairchild ten-cathode NIXIE tube with driver decoder.

BCD-to-decimal decoder. The driver transistors used must be high-voltage breakdown and low leakage. The neon or gas tube is also used to form seven-segment readouts by constructing it with a single anode and seven separate cathodes to form the seven sgements A through G.

18.7.4 Light-Emitting Diode (LED)

The light-emitting diode is a transparent semiconductor diode using gallium phosphide or gallium arcinide phosphide to produce light emission. It appears electrically like a normal diode with respect to its VI curve, as Figure 18–27(a) shows. There is little current until the barrier potential is exceeded, and suddenly the current rapidly rises for voltages beyond V_D. Light emission begins and increases linearly at low values of diode current. Light saturation occurs within safe-operating current levels, as shown in the current/CP curve of Figure 18–27(b). To prevent device failure, current should be limited at this level. The drivers for the LED segments should therefore be a current source. The seven-segment LED can be made as common-cathode or common-anode devices, as Figure 18–28 shows, and can be selected for compatibility with the driving circuits used.

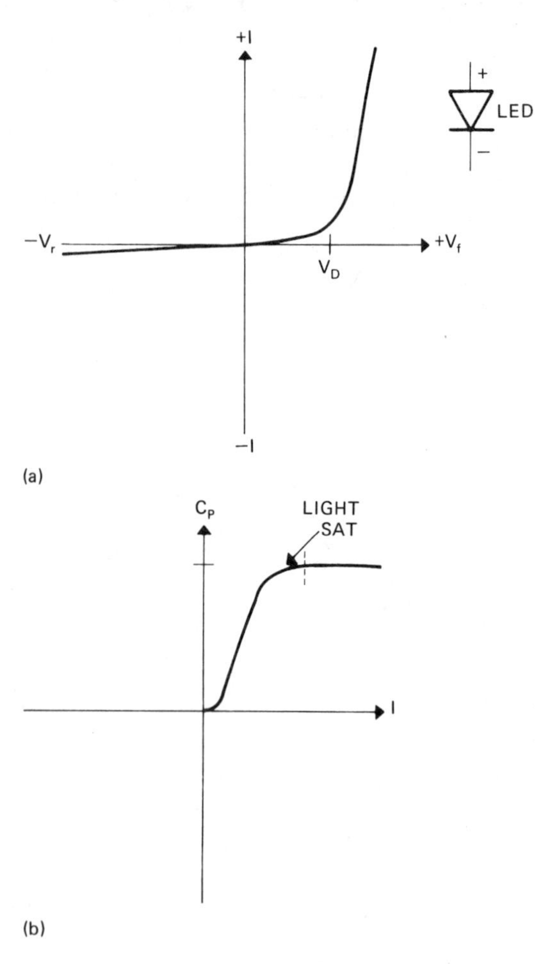

FIGURE 18–27

(a) VI curve of a light-emitting diode similar to that of a signal diode. (b) Current-versus-candle-power curve, showing light saturation, the point at which the current should be limited.

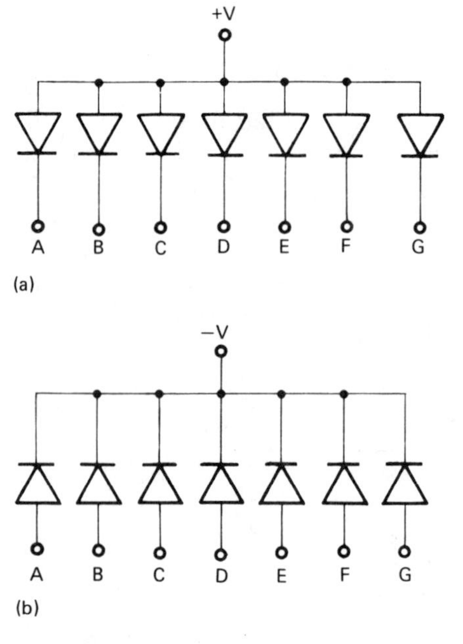

FIGURE 18–28

(a) LEDs connected for seven-segment readout with common anode. (b) LEDs connected for seven-segment readout with common cathode.

18.8 Display Multiplexing

A display that is not multiplexed requires the logic and circuits of Figure 18-29—including a separated decoder/driver—for each decimal digit in the readout. If the decoder/drivers are in the circuit card holder chassis, each digit requires at least seven leads from chassis to readout panel. If the decoder/driver is mounted as an integral part of the readout, at least, four leads per digit are required between chassis and panel. If there are many digits in the display, both the number of circuits and the number of leads to panel can be reduced by multiplexing. In a multiplexed system, the individual leads A through G are connected in parallel and only one set of leads is needed between the panel and the circuit card containing a single decoder/driver circuit. The BCD inputs to the single decoder/driver are switched between the decimal counting units or register, while the corresponding digits are enabled by the scan decoder. Figure 18-30 shows this system.

The scan decoder ensures that the correct digit is enabled to turn on, while the multiplexing unit ensures that the correct segments representing the data in the DCU light up. If the control counter has eight states, each indicator will be on only one-eighth of the time. The human eye retains a light stimulus for about 1/20 of a second. If the indicators are pulsed to full brilliance at a rate higher than 1 kHz, the eye will detect no flicker.

When we compare the eight-unit multiplexed system with an unmultiplexed system, there is a saving of seven decoder/driver circuits at the expense of adding three new circuits, a counter, a scan decoder, and a multiplexer. If the system does not have a suitable clock to operate the counter, a multivibrator will be needed also. Of the four units added, only the multiplexer is as complex as a decoder/driver. The scan counter could be a simple ripple counter, as in Figure 14-6. The scan decoder for eight digits is a simple binary-to-octal decode combined

FIGURE 18-29
Decimal counting unit with decoder/driver.

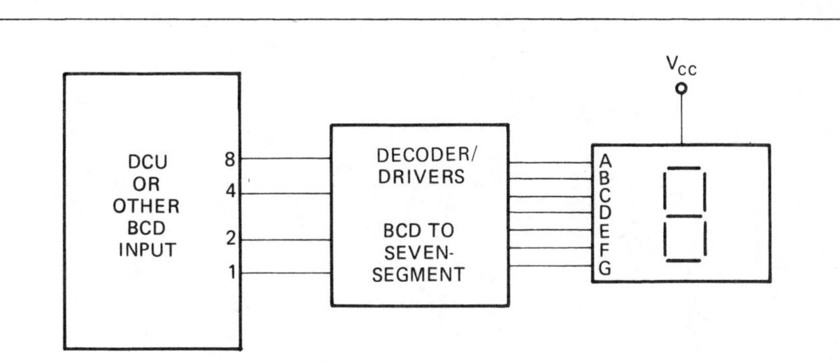

FIGURE 18-30
Fairchild eight-digit multiplexed display system.

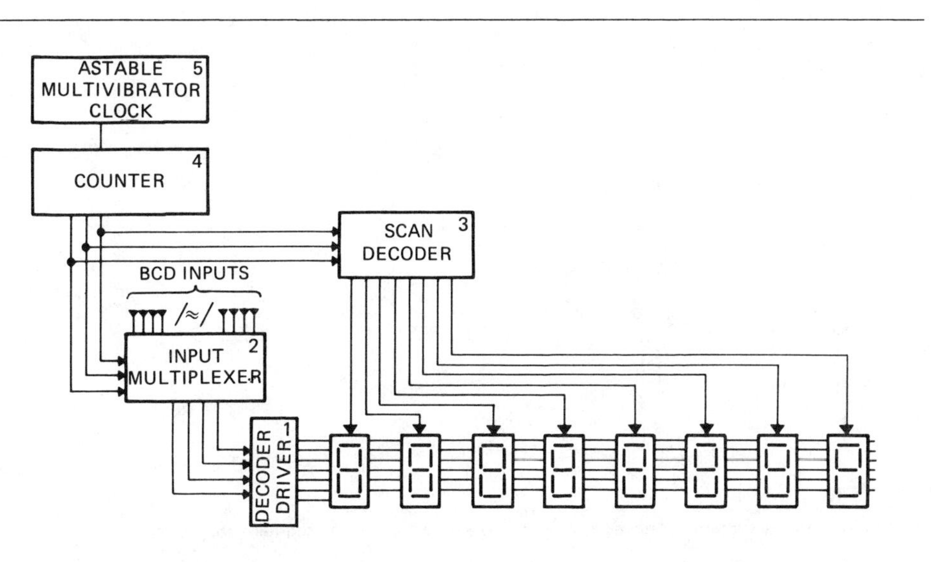

with the necessary drivers for the type of display. The drivers in Figure 18-31 are for LEDs with common anode.

Figure 18-32 shows typical logic for an eight-digit multiplexer. The input lines from the counter are decoded to octal. Each of the eight sets of BCD inputs is connected to a separate set of four AND gates. The outputs of the AND gates are connected to the inputs of a single set of eight-input OR gates. As the count continuously cycles 0 through 7, a different set of AND gates is enabled with each count. The output OR gates will agree with the LSD input during the count of 0 and will change to a higher digit with each count as a different set of AND gates is enabled with each count. At the count of 7, the MSD bits will be on the OR gates and the cycle will start over. The scan decoder must, of course, be synchronized to drive a particular power-of-ten display during the same period as it appears at the output of the multiplexer.

18.9 Power Averaging

A 5V, 40 mA lamp must receive 200 mW of power to light at normal brilliance. In an eight-digit multiplexed display, the power is applied one-eighth of the time and is turned off seven-eighths of the time. To maintain brightness, we must apply eight times as much power during ON time. From the power formula P = EI, we might assume that increasing the voltage by eight times would be required, but a voltage increase also brings a current increase. We can better use the equation $P = V^2/R$, knowing that R is fixed by the filament temperature at normal brilliance of:

$$\frac{V}{I} = \frac{5V}{0.04A} = 125 \text{ ohms}$$

If we were to use the rated lamp voltage of 5V, the average power would be $P_{AV} = P_{ON}/N = V^2/NR$, with N the number of digits in the display. In an eight-digit display:

$$P_{AV} = \frac{(5V)^2}{8 \times 125 \text{ ohms}} = \frac{25V^2}{1000 \text{ ohms}} = 25 \text{ mW}$$

But if, instead of 5V, we use $\sqrt{N} \times 5V$, the results are as follows:

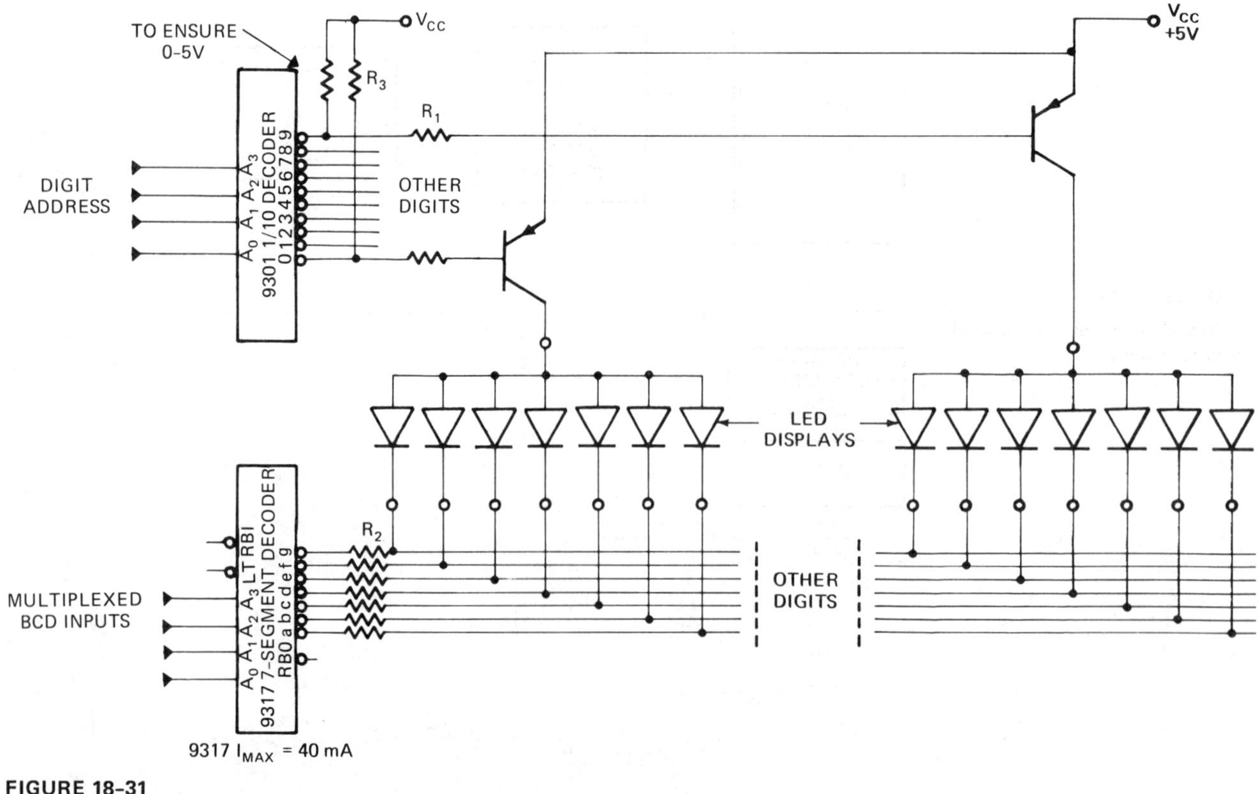

FIGURE 18-31

Fairchild multiplex system connection to common-anode LED seven-segment readouts.

$$P_{AV} = \frac{(\sqrt{N} \times 5V)^2}{N \times 125 \text{ ohms}} = \frac{N \times 25V^2}{N \times 125 \text{ ohms}}$$

$$= \frac{25V^2}{125 \text{ ohms}} = 200 \text{ mW}$$

The 5V readout in an eight-digit display must therefore receive:

$$\sqrt{N} \times 5V = \sqrt{8} \times 5V = 14.1V$$

The same power averaging is required by fluorescent and neon displays when they are multiplexed. The voltage applied during ON time must be $\sqrt{N}$ times the rated voltage of the readouts.

Power averaging is different for the LED display. At the point of light saturation, the forward-bias voltage across the diode goes up only slightly to produce a major increase in current. Therefore, only the current increase is significant and must be increased by the full value N.

The current is established by the shunt resistors, R_2, in Figure 18-31. The current is increased by dividing these values by N:

$$R_{2M} = \frac{R_{2S}}{N}$$

where:

R_{2M} = shunt resistor for a multiplexed display
R_{2S} = shunt resistor for a static display
N = number of readouts in the display being multiplexed

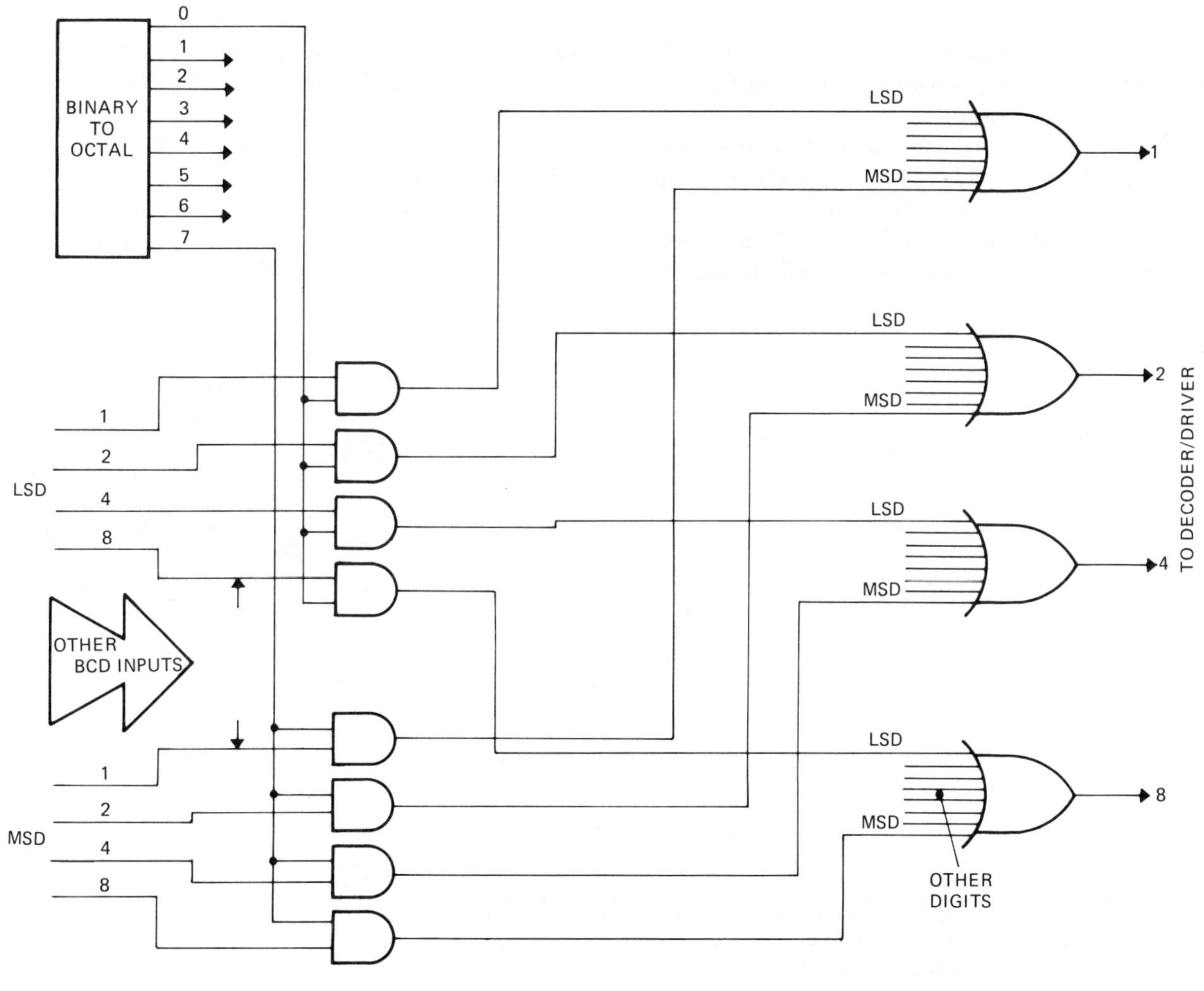

FIGURE 18-32

Typical eight-BCD-digit multiplexer logic.

18.10 Failsafe Circuits

If the clock circuit fails, the scan counter will stop in the state it was in at the time of the failure. The display addressed will be turned on and operating statically at the excess power levels needed for power averaging when multiplexed. This could result in burnout of the display. Figure 18-33 is a failsafe circuit used to off-address the scan counter in event of clock failure. If the clock pulses continue to arrive at the failsafe input, a sufficient charge will be maintained on C_2 to keep the transistor turned on. For failsafe operation, the scan decoder must have an extra input. A BCD-to-decimal decoder/driver could be used. When a 1 occurs on the A_3 input, the output is diverted to 8 or 9, keeping all eight displays off.

18.11 LED Dot Matrices

The use of lowercase numbers for B and D of a hexadecimal display creates some confusion. In particular, the lowercase b can easily be mistaken for a 6. A dot matrix, such as the Texas Instruments TIL 311 shown in Figure 18-34, is an improvement over the seven-segment bar-type display of Figure 18-13. The device comes with a latch decoder and driver built in. All but the corner and center vertical dots are connected as LED pairs in series. The constant current drivers supply the LED current so that the paired and single LEDs light at the same brilliance. The logic supply is +5V ±10%, while the LED supply can be from +4V to +7V. The LED supply lead connects to common anodes. The logic inputs are TTL compatible.

18.12 Summary

The end result of the function of many digital circuits is a visual display. We have seen this in the clock and counter circuits of Chapter 14. These readouts are usually numerical, but in many cases letters, decimal points, and other characters are included. Often lamps indicate the operating mode of the machine and on maintenance panels give indication of defects in the machine functions. All this presents a need for circuits to decode, control, drive, and in some cases multiplex the information to be read out.

In numerical controls and other automation machinery, the end result of complex digital functions is a mechanical action requiring application of high-current AC power to operate a motor or solenoid. This is generally done by energizing a relay. The relay coil is driven by a circuit very much like the lamp driver.

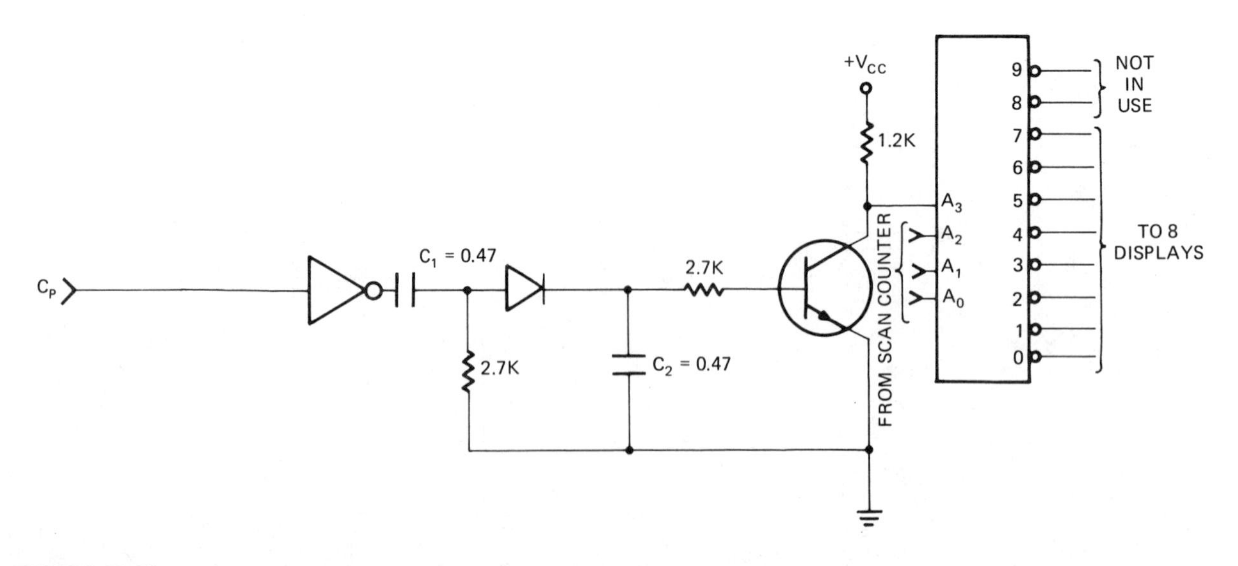

FIGURE 18-33

Fairchild failsafe circuit. Clock input keeps charge on C_2 high enough to keep transistor turned on. If clock fails, the logic 1 level on the collector is applied to A_3 of the scan decoder. The A_3 (8) input addresses off all 0 through 7 displays.

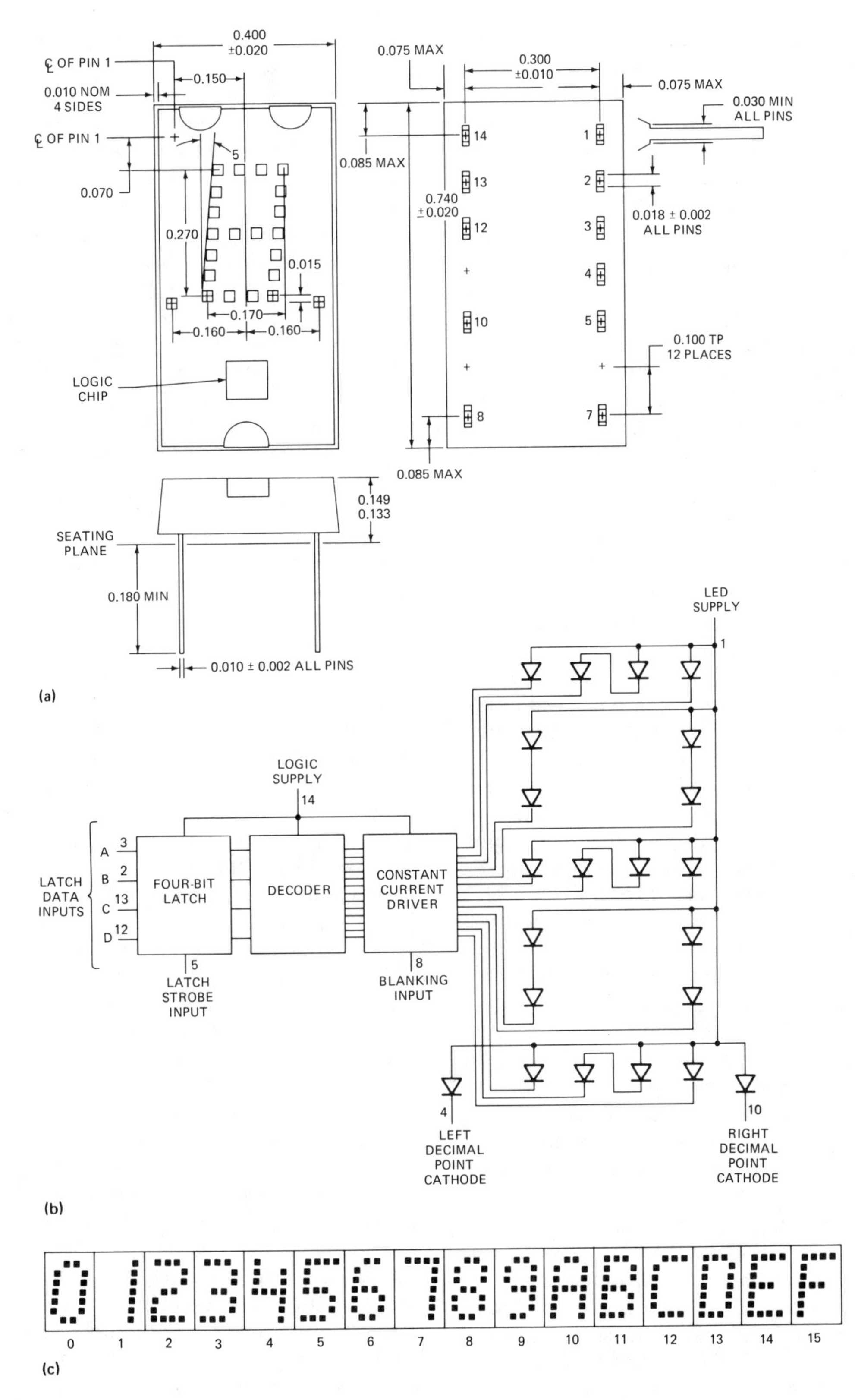

FIGURE 18–34

Texas Instruments TIL 311 hexadecimal display. The display contains both the LED dot matrix and the logic chip for the latch decode and driver functions. (a) Logic diagram of TIL 311. (b) Block diagram of TIL 311. (c) Dot displays for hexadecimal numbers.

An incandescent lamp usually requires more current, and in some cases more voltage than a logic circuit output can supply. In such cases, a lamp driver may be built using one transistor and a base resistor. The resistor size is computed as explained in Figures 18-1 and 18-2. If the base current required to operate a one-transistor lamp driver is more than the driving circuit can supply, the same design procedure can be used with two transistors in a Darlington pair, as explained in Figures 18-3 and 18-4. A driver circuit to energize a relay can be constructed by using a technique similar to that of a lamp driver. These are explained in Figures 18-6 through 18-8. If large numbers of lamp drivers are needed, monolithic transistor arrays are available to provide seven transistors to a single 16-pin DIP package. These are shown in Figure 18-9(a). They are also available in Darlington pairs, as shown in Figure 18-9(b). These are adequate for relatively low-power lamps. For somewhat higher power, a series of quad power drivers is available. Figure 18-10 shows a quad power driver with NAND gate inputs. A logic 0 would turn these on. The extra input is a convenience for lamp testing, as any number of drivers can have their extra input connected to the lamp test line. When this line is grounded through a pushbutton switch, all lamps should light, testing both lamps and drivers. When the switch is open, the NAND gate inputs of all drivers are enabled for normal use.

In some cases, readouts are direct and require no decoding. The output of a counter or register may be read out in binary form by using individual 8, 4, 2, and 1 lamps, as shown in Figure 18-11. Here experienced personnel can decode mentally, but if a large number of digits are involved, decoders may be used to decode from binary-coded decimal (BCD) to decimal for the purpose of driving in-line readouts. Integrated circuit decoders are available for this job, as Figure 18-12 shows.

In modern digital equipment, it is more common to use a single segmented readout for each decimal digit than ten individual lamps. At present, the seven-segment digit is the most popular. They are available in incandescent, fluorescent, neon, and LED form. The segments are lettered A through G, as shown in Figure 18-13. Any number 0 through 9 can be produced by lighting the correct selection of the seven elements. The numbers are shown in Figure 18-14. Decoding from BCD to the seven levels of A through G is complicated unless we first decode to straight decimal. From decimal to seven levels is easy if we consider that each segment except E has only two or three OFF states. Using NOR gates and the decimal lines to turn the segments off during the correct number is an easy solution. This is shown in Figure 18-17. For most applications, a BCD decoder/driver, like the TTL 9317 shown in Figure 18-18 can be used. Besides directing BCD to seven-segment decode, it provides a driver circuit for each element A through G. It also provides lamp test and ripple blanking leads. When the lamp test line is grounded, all seven elements will be turned on. The seven outputs show inversion symbols indicating that they provide grounds to turn the segments on in the same manner the transistor of a lamp driver grounds the end of a lamp. The ripple blanking leads are used in a multidigit fixed decimal point display to turn off unnecessary zeros. Figure 18-20 shows digits to the left of the decimal point wired to turn off all zeros to the left of the MSD. Figure 18-21 shows digits to the right of the decimal point connected to turn off all zeros to the right of the LSD.

Many single-board microprocessor systems need a hexadecimal display. The seven-segment LED readout is often used, but the lowercase letters B and D must be displayed to avoid confusion with 8 and 0. Decoder/drivers can now be purchased with hexadecimal capability. The confusion between B and 8 and D and 0 is eliminated by using the LED dot matrix display, which saves power and improves display readability. Figure 18-24 shows the usual filament wiring of an incandescent seven-segment readout. Figure 18-25 shows the fluorescent readout. It functions like a vacuum tube. The filament provides electrons that are attracted to whichever anodes A through G are positively charged. The anodes are coated with a material that fluoresces when bombarded by electrons. The grid can be negatively charged to control brilliance. The neon segmented readout—which has been in use for many years—has been available in ten segments, as Figure 18-26 shows. The neon readout is a gas tube that conducts by ions. When the anode is at ionization potential, conduction by ions occurs between a negative or grounded cathode and the positive anode. When this happens, the gas glows in the vicinity of the cathode. The ten-segment readout known as a NIXIE tube has ten cathodes bent in the shape of the numbers 0 through 9. They are selected by grounding the correct cathode. These are also available in seven segments.

The most modern of the seven-segment readouts is the LED (light-emitting diode). These have a definite advantage of low power requirements and long life but are currently available only in small size and limited brilliance. Figure 18–27(a) shows the characteristic curve of the light-emitting diode to be like that of any other diode in that a small forward-bias voltage drop must occur before there is a significant amount of current drawn. This means that, beyond the point V_D, very little voltage increase is needed for a large current increase. This fact becomes important to us in multiplexing displays. Figure 18–27(b) is a current-versus-candlepower curve. From this we see that brightness increases with diode current until the point of light saturation, and from that point on, further current increases have little effect. When arranged for seven-segment readout, LEDs may be purchased as either common anode or common cathode, as shown in Figure 18–28.

Figure 18–29 shows a typical circuit starting with a BCD output through decoder/driver circuits to the readout. A six-digit display would have six identical sets of these. For displays with more readouts, it is often economical to multiplex the readouts. In multiplexing, we switch the power supply to only one readout at a time. If there are twelve readouts, each readout is on one-twelfth of the time and off eleven-twelfths of the time. An immediate misconception is that this saves power, which is not true. To operate the multiplexed readouts at normal brilliance, the average power must be kept at the same level as an unmultiplexed display. To accomplish this, the voltage and/or current delivered to the readouts must be increased as described in Section 18.9. The real advantage in multiplexing is the saving in decoder/driver circuits. Only one is needed to serve all the readouts. To accomplish this saving, three other circuits are needed—a counter, a scan decoder, and a multiplexer. These are shown in Figure 18–30. Because of these added circuits, multiplexing is usually economical only for displays of eight digits or higher. In a multiplexed display, the segments A through G are all wired in parallel. This simplifies the wiring and may provide an advantage in addition to the circuits saved. In a multiplexed circuit, there is a danger that the loss of clock pulses would cause one readout to be lighted continuously at the peak power, which would cause a rapid burnout of the segments. To avoid this, some form of failsafe circuit must be used. Figure 18–33 shows one such circuit.

Glossary

Display: A visual readout of digital data.

Readout (as used in this chapter): A single digit of a display. In general, the term may be used to indicate any visual presentation of data, including the printout of a computer.

Incandescent Readout: An indicator in which an electric current passes through a wire or filament, heating it to a temperature at which it glows. See Figure 18–24.

Fluorescent Readout: An indicator that is constructed like a vacuum tube triode. A heated filament emits electrons that are attracted to a positive anode. The anode is coated with a fluorescent material that glows when bombarded by electrons. See Figure 18–25.

Neon Readout: A tube containing neon gas and two electrodes—the anode and the cathode. When a sufficiently high voltage is applied between the anode and the cathode, the gas ionizes and a conduction by ions occurs between the electrodes. Neon produces a bright red glow when it is ionized, making the tube a useful indicator.

NIXIE Tube: A registered trademark of Burroughs Corporation. The NIXIE tube is a neon tube with a common anode and ten separate cathodes, each bent to form one of the digits 0 through 9. See Figure 18–26.

LED (Light-Emitting Diode): A transparent diode that glows when conducting a forward-bias current. See Figure 18–27.

Lamp Test: A connection designed to test all or sets of indicators in a digital display by interrupting their normal function and causing all nondefective indicators to light. Lamp test connections must be made through diodes or logic gates in a manner that does not interfere with the normal function of the indicator.

Multiplexer: A circuit with a number of sets of input lines and a single set of output lines. It periodically, and one at a time, connects a set of input lines to the output so that each input shares the output for an equal time slot during every sampling period. See Figures 18–30 through 18–32.

Scan Decoder: A binary-to-decimal decoder used

to energize the common sides of a set of multiplexed readouts.

Hexadecimal Readout: A display that produces the hexadecimal digits.

Seven-Segment Display: A readout that contains seven distinct segments. See Figure 18-13. A number of letters is displayed by lighting one or more of the segments.

Decoder/Driver: An IC that can decode a binary number and produce an output capable of driving a display.

LED Dot Matrix: A digital readout whose display is produced by lighting several LED dots. See Figure 18-34.

Questions

1. What two elements in the lamp driver circuit tend to limit the collector current to a safe level?
2. What are two most important differences between the single-transistor driver and the Darlington pair driver?
3. When fully loaded, an RTL circuit has a minimum 1 level under 1V. For which lamp driver might this be a problem? Why?
4. Under what conditions is a resistor used in series with the lamp in a driver circuit?
5. Why is a diode used in the relay driver circuit? How is it biased?
6. If all the numbers 0 through 9 can be made with seven segments, why is an eighth segment sometimes used?
7. In encoding from decimal to seven segments, why is the NOR gate the ideal logic gate to use?
8. What are two advantages of ripple blanking?
9. Explain how ripple blanking connections differ between the left and right side of the decimal point.
10. What is a lamp test and how is it connected?
11. Describe the advantages and disadvantages between seven-segment readouts in incandescent, fluorescent, neon, and LED form.
12. There is a common connection for the incandescent and LED readouts. What is the common connection in the fluorescent and neon readouts?
13. What is light saturation of the LED?
14. What advantages are attainable by multiplexing a 12-digit display?
15. Would 60 Hz be satisfactory for a clock in a ten-digit multiplexed display? Explain.
16. When a 12-digit display is multiplexed, what circuits are saved? What new circuits must be supplied?
17. Approximately what voltage must be supplied to the filaments of a 5V incandescent display of eight multiplexed digits.
18. Approximately what peak current must be supplied to the segments of an LED readout in a nine-digit display if the LEDs reach light saturation at a current of 4 mA?
19. Explain the operation of a multiplexer.
20. In what way can multiplexing a display simplify the wiring?
21. If a seven-segment display is used for hexadecimal digits, what problems are created?
22. Under what condition may pull-up resistors be needed when the 14495 decoder/driver is used?
23. For hexadecimal digits, which type of LED display is an improvement over the seven-segment display?

Problems

18-1 Draw the schematic diagram of a single-transistor lamp driver.

18-2 Draw the schematic diagram of a Darlington pair lamp driver.

18-3 A lamp is to be used to indicate a parity error by lighting when a 2.4V logic 1 level appears on the output of a parity checker. A No. 47 lamp, 6.3V at 150 mA, is to be used. The transistor has a minimum β of 100; the lamp supply is 6.8V. Draw the lamp driver and supply the value for R_B.

18-4 The single-transistor lamp driver of Problem 18-3 was found to require an excessive base current. Redesign the driver using a Darlington pair.

18-5 Draw the logic diagram of a divide-by-N counter TTL 7493 (see Figure 14-7) connected to form a decade counter. Use one quad AND power gate and a hex inverter circuit to drive four No. 680 lamps (5V, 60 mA). Include a lamp test circuit.

18-6 The circuits of Figure 18-35 are those needed for the logic of Problem 18-5. Label the leads and draw the interconnections.

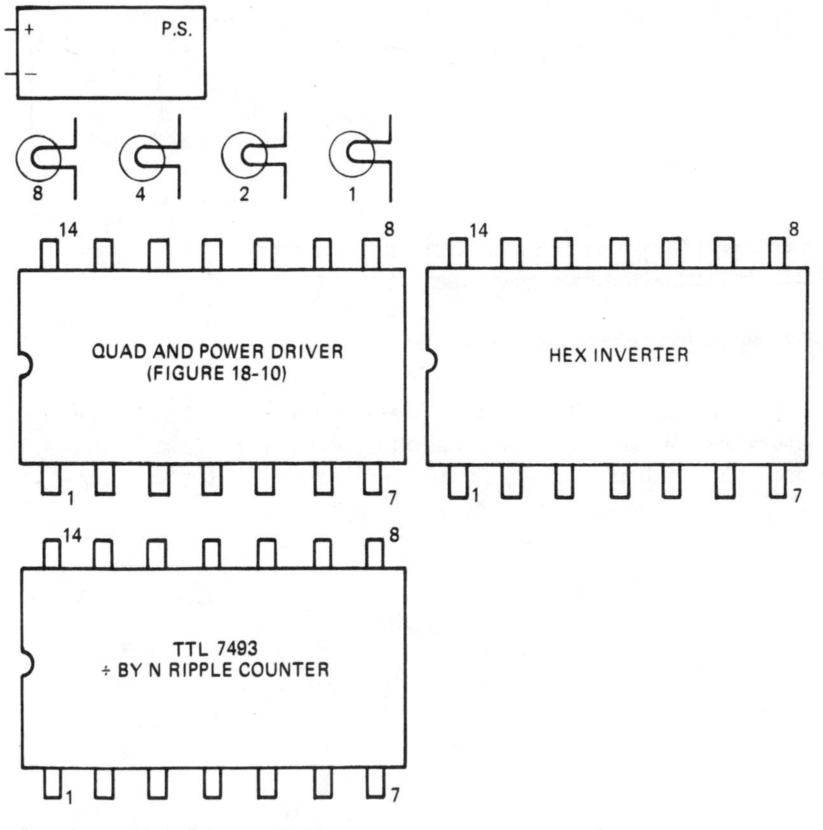

FIGURE 18-35. Problem 18-6

18-7 Draw a seven-segment readout and label the segments A through G.

18-8 Draw the numbers 0 through 9 in the form they would take as seven-segment readouts.

18-9 Draw the logic diagram of a divide-by-N ripple counter TTL 7493 connected to form a decade counter. Use a BCD-to-seven-segment decoder, TTL 9317 (Figure 18-18), to drive an LED seven-segment readout.

18–10 The circuits of Figure 18–36 are those needed for the logic of Problem 18–9. Label the leads and draw the interconnection.

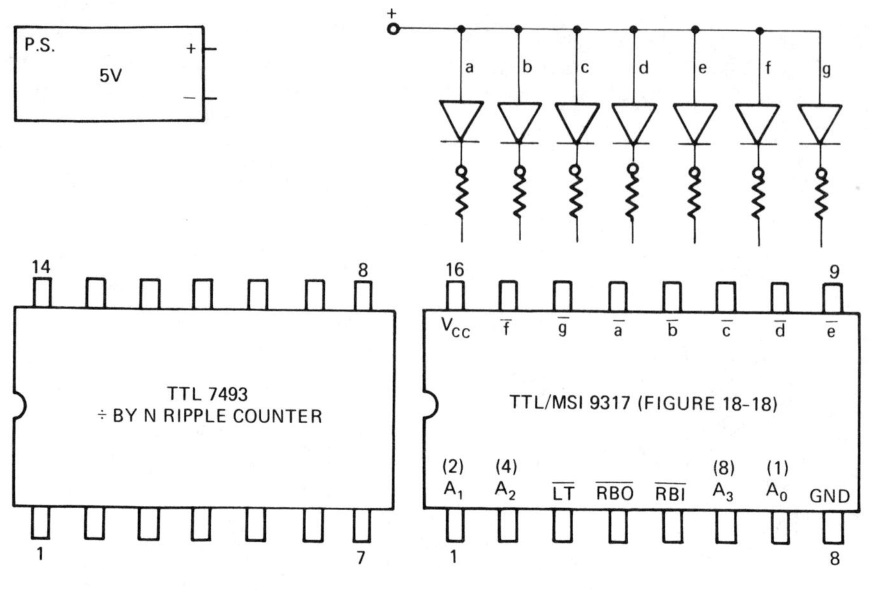

FIGURE 18–36. Problems 18–10 and 18–11

18–11 The LED in Figure 18–36 requires a static 20 mA at 1.65V to obtain light saturation. What size resistors should be used for a static display? For a multiplexed display of nine digits? (Allow 0.35V for the driver.)

CHAPTER

Control Waveform Generators

Objectives

Upon completion of this chapter, you will be able to:

- Generate clock-pulse-width control waveforms using a ring counter.
- Assemble a shift counter and use it to generate control waveforms.
- Use a shift counter as a decimal counting unit.
- Generate control waveforms using a binary counter.
- Produce random control and timing waveforms and pulses using shift counter outputs and logic gates.

19.1 Introduction

If one had to develop a simple block diagram that would apply to most digital electronic systems, Figure 19-1 would work reasonably well. The timing and control waveform generator of this diagram includes the clock and timing circuits discussed in Chapter 17. In addition to these, complicated systems require a large variety of control signals that begin and end on clock or delayed clock timing. These waveforms are generated in the control waveform generators and are used to enable or inhibit activities within the functional unit. Designers can proceed with the design of the functional unit, drawing in control signals wherever needed, having full confidence that they can later be developed in the control waveform section. The reason they can proceed confidently is that there are several logic circuits ideally suited to the production of control signals. The techniques described here apply without reservation to systems using small-scale integrated (SSI) circuits and medium-scale integrated (MSI) circuits.

For large-scale integrated (LSI) circuits, a prime objective is to minimize input and output leads. Because of this, control waveform generating circuits are included within the LSI itself to minimize input and output leads. This results in some duplication, such as is found in the BCD-to-octal decoders described in both the scan decoder and the multiplexer of Figures 18-31 and 18-32. There are just too few leads available to bring these signals in from an external circuit. Another deviation from the procedure described here is generation of timing sequencing and control signals with a minicomputer or microprocessor.

Computer routines can be used to provide control signals the same as or similar to those described in this chapter. To accomplish this requires a knowledge of programming and special knowledge of the computer and of the techniques needed for a digital interface with it.

The microprocessor, on the other hand, is a large-scale integrated circuit of about 40 pins. It was originally developed to provide timing and control routines for the digital calculator and has since proved useful in other digital systems. It is generally used in conjunction with one or more memory circuits, which are discussed in Chapter 22. Although the microprocessor itself has become relatively inexpensive, extensive labor is involved in programming and interfacing it with the functional unit. At present, it is thought that a system requiring more than 50 integrated circuits by conventional control methods would be economical to control by a microprocessor. This chapter explains the techniques used to generate control for the smaller systems and, at the same time, lays a background for the microprocessor by showing the circuit functions that the microprocessor must be programmed to duplicate.

19.2 Ring Counter

The ring counter of Figure 19-2 can be used to produce waveforms that are one clock period wide, as shown in the timing diagram. If the outputs are connected back to the inputs, they will

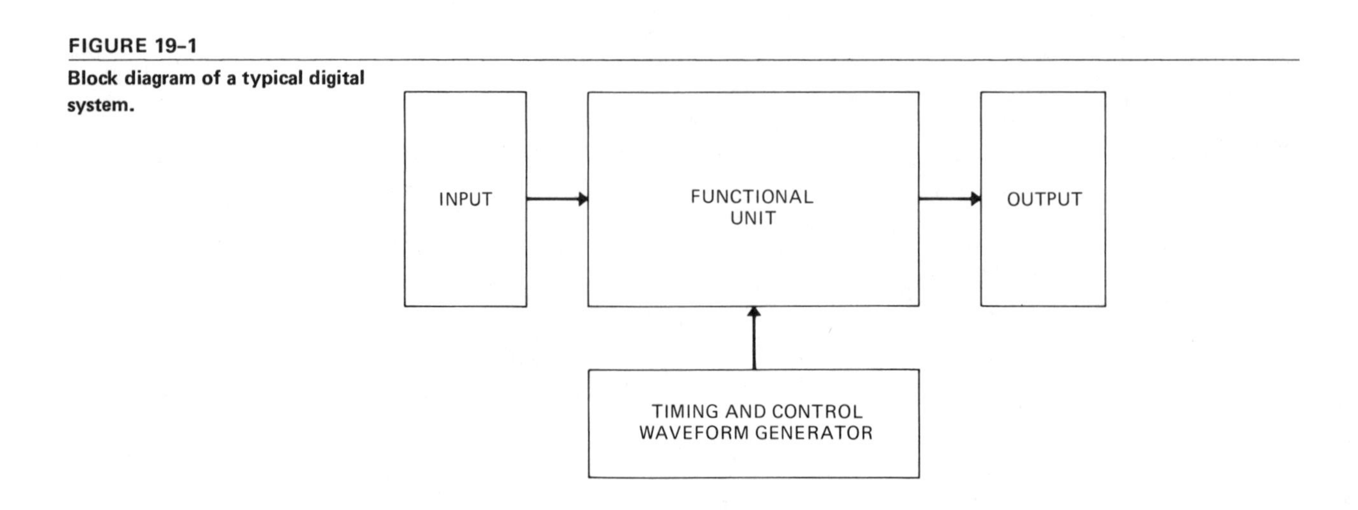

FIGURE 19-1
Block diagram of a typical digital system.

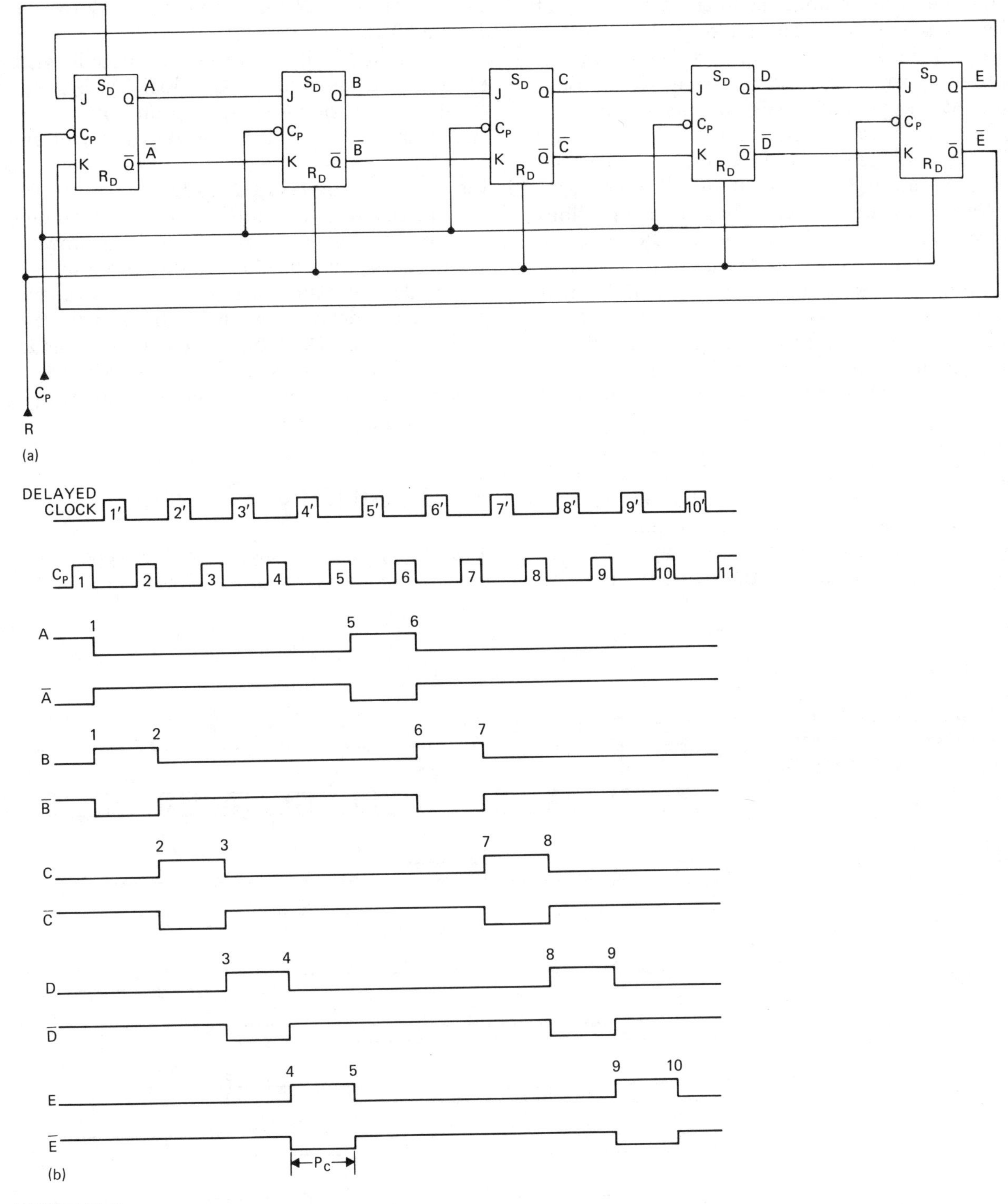

FIGURE 19–2

(a) Five-bit ring counter. (b) Five-bit ring counter waveforms.

continue to cycle until the next reset. Note that the reset line clears the last four flip-flops but sets a 1 into the first flip-flop. The 1 level is shifted down the register, moving to the right one bit on the trailing edge of each clock pulse. When the 1 level reaches the last flip-flop, that flip-flop will steer a 1 back into the first flip-flop, to repeat the cycle. If only one cycle of waveforms is needed, removing the connections between the output of the last flip-flop and the input to the first will result in all flip-flops remaining at 0 between clock pulse 5 and the next reset.

The advantage to the ring counter is that it can produce clock-period-wide control signals without use of decoding circuits. The number of waveforms can be increased by adding more flip-flops to the right of those shown in Figure 19-2. To obtain a 7-to-8 control waveform, an eight-bit ring counter would be needed. Various changes can be made to produce double-clock-period-width waveforms. The first two flip-flops may be set instead of one. In that case, the outputs of the five-bit counter will be as shown in Figure 19-3.

Another variation producing double-width pulse would be to divide the clock by toggling a flip-flop and using the flip-flop output to clock the ring counter. This produces waveforms identical to those of Figure 19-2(b) except that they are two clock periods wide.

Increasing the width of the ring counter outputs by connecting to OR gates, as shown in Figure 19-4, may produce an unwanted spike in the middle of the output caused by uneven propagation delay between flip-flops. The waveforms of Figure 19-3, however, can be expanded with an OR gate without the possibility of spikes because of the overlap in the time periods of the two input signals.

19.3 Binary Counter

The binary counter toggle or steer type can be used to develop control waveform voltages. The

FIGURE 19-3

Double-clock-period-width output waveforms produced when the first two bits of the ring counter of Figure 19-2 are preset at 0.

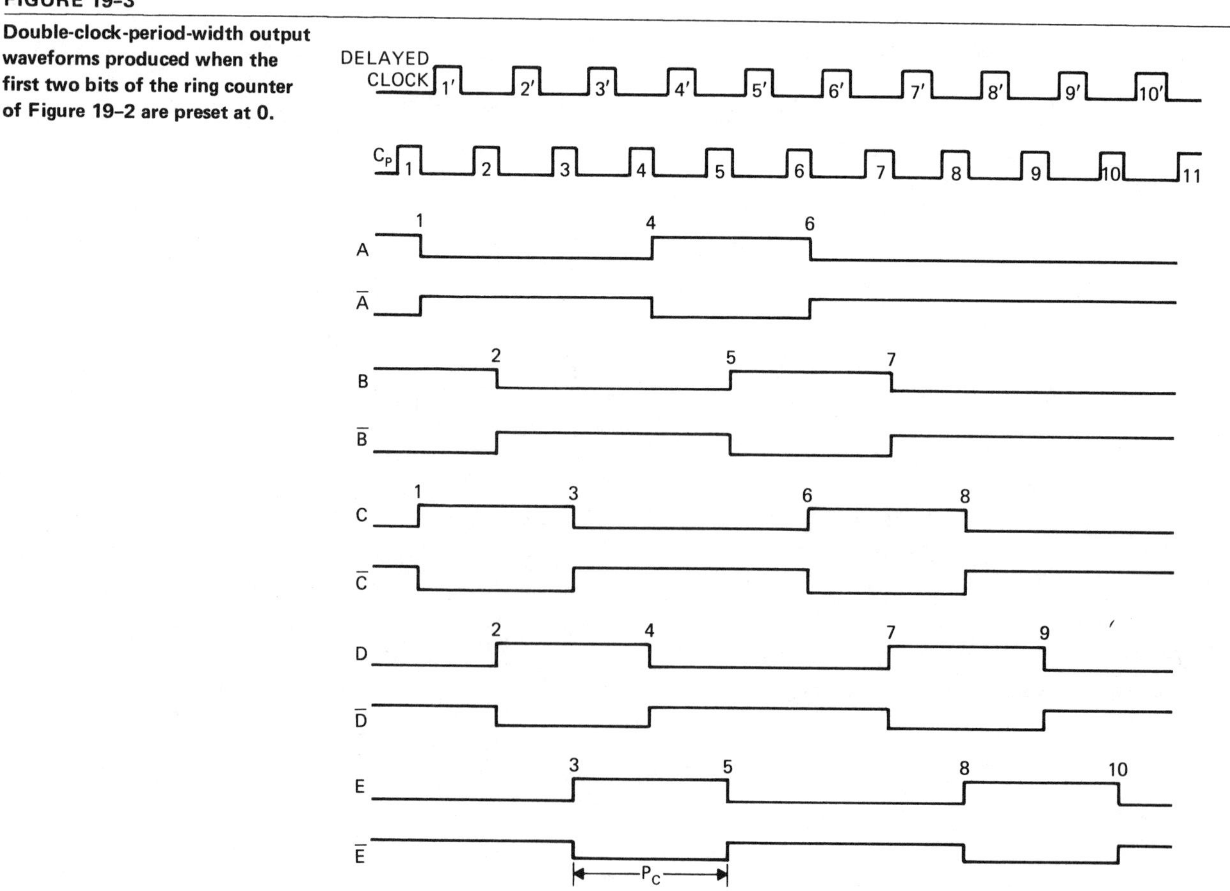

three-bit toggle counter of Figure 19-5 can produce waveforms that divide down the clock frequency. In addition, these can be decoded to produce clock-period-width gates like those of a seven-bit ring counter. Use of the three-bit counter produces the same clock-period-width gates as a seven-bit ring counter, but seven decoding gates are required for the positive, and seven more are required for the complements. If all seven waveforms and their seven complements were needed, we could simply compare the cost of a saving of four flip-flops with the cost of 14 gates and certainly the more complex wiring required by the decode circuit. On the surface, the use of the ring counter is favored, and if all or most of the available waveforms are needed, the ring counter is the better answer. However, when a few of the available waveforms are actually used, then the counter is better, because decode gates are required only for the

FIGURE 19-4

(a) Spike produced by extending the width of ring counter waveforms by connecting to OR gate. (b) Use of overlapping inputs to eliminate spikes in the output waveform.

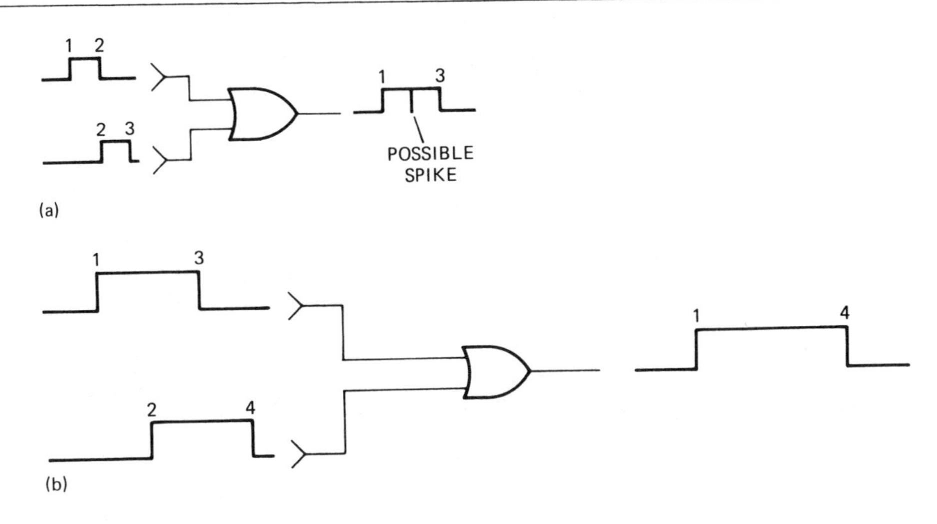

FIGURE 19-5

(a) Three-bit binary counter, which requires gates to produce clock-period-width waveforms. (b) Three-bit binary counter waveforms.

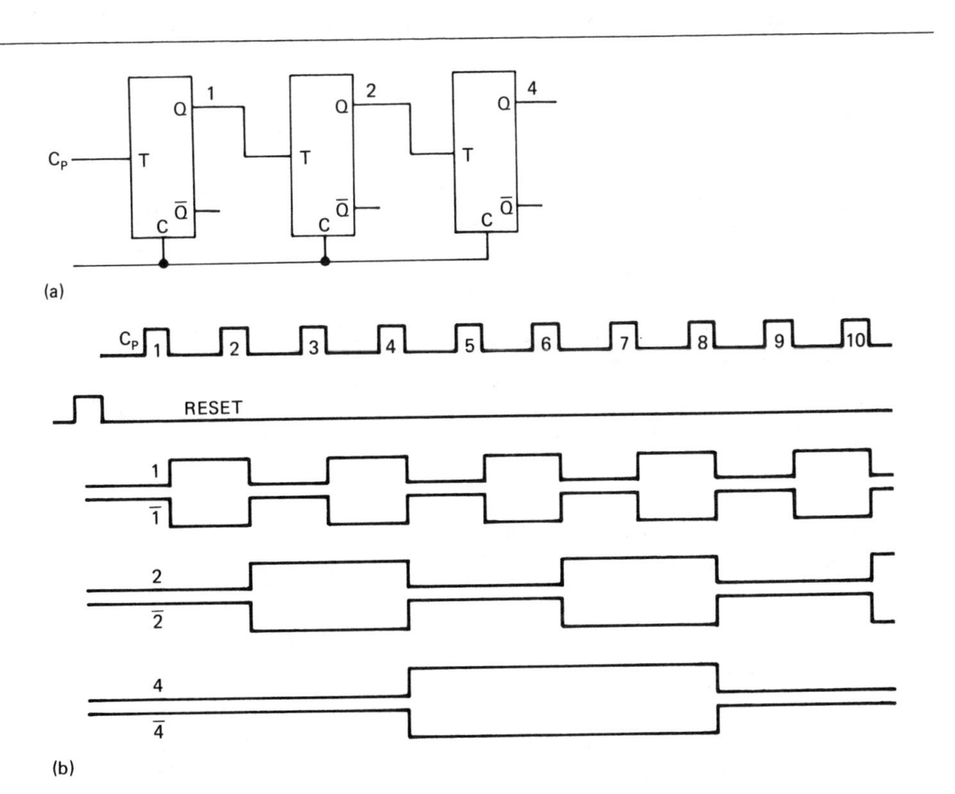

FIGURE 19-6

Five-bit toggle counter used to develop a control waveform from clock pulse 30 to 31.

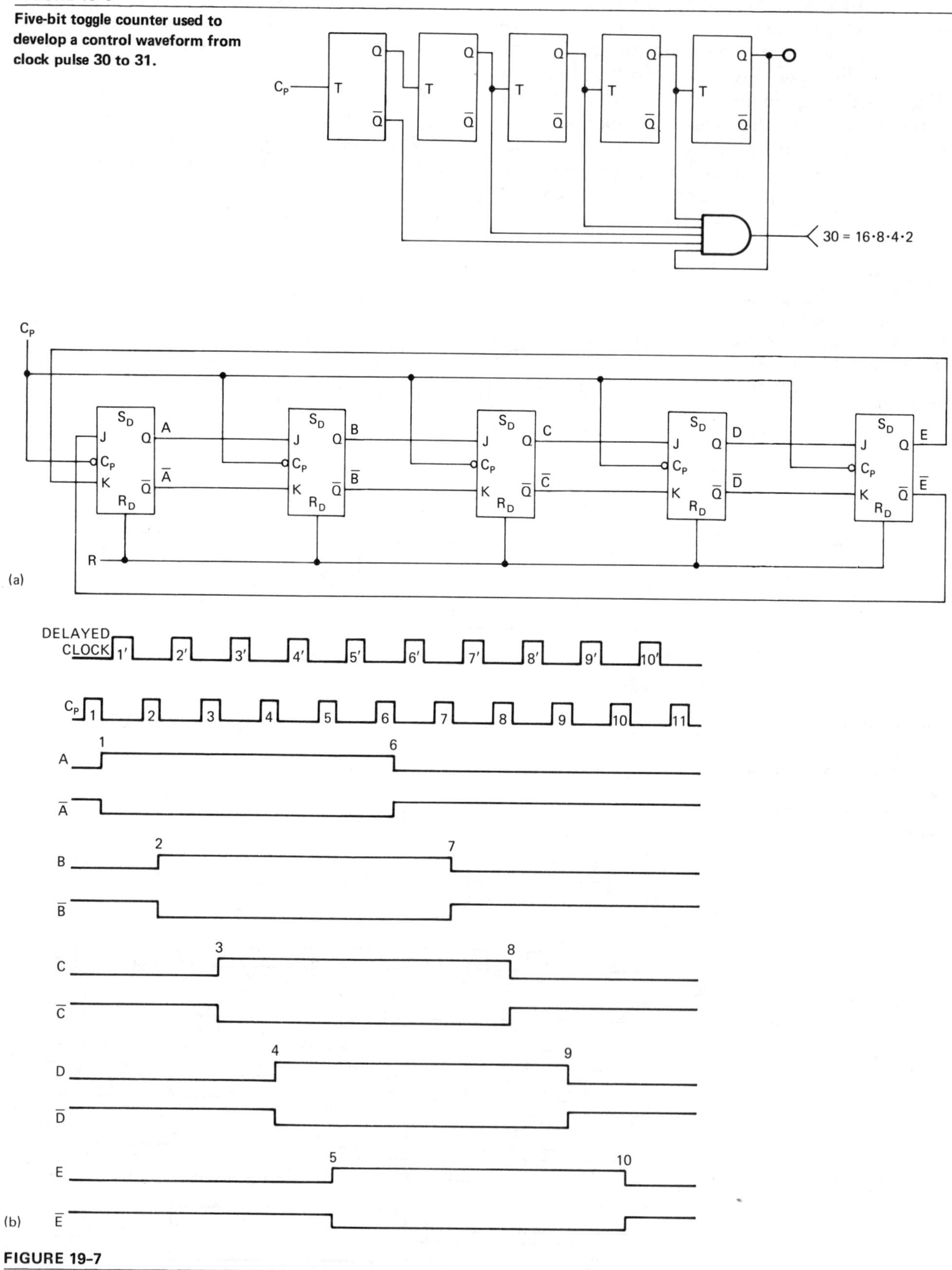

FIGURE 19-7

Examples 19-1 and 19-2. (a) Five-bit shift counter. (b) Shift counter output waveforms.

waveforms that are to be used. Figure 19-6 shows a counter used to produce a control waveform of 30 to 31. This would have required 31 flip-flops to produce with a ring counter. The binary counter needs only five flip-flops and the gate to decode.

19.4 Shift Counter

The most versatile of the control waveform generators, and the easiest to use, is the shift counter. As Figure 19-7 shows, the shift counter differs from the ring counter by the crossing over of the output leads and the fact that all flip-flops are cleared during reset. With all flip-flops in the clear state, all except the first are steered to 0. The crossover of the wiring between the first and last flop-flop results in the last flip-flop steering the first flip-flop to the 1 or set state. As the timing diagram indicates, the first clock pulse will set the first flip-flop, steering the second flip-flop to 1. On the second clock pulse, the second flip-flop clocks to 1 and begins to steer the third flip-flop to 1. The fifth flip-flop is clocked to 1 on the fifth clock pulse. This changes the steer on the first flip-flop. Because of the crossover, the first flip-flop is now being steered back to 0. Starting with the sixth clock pulse through the tenth, the flip-flops clock back to 0 one clock pulse at a time.

The shift counter can be used in conjunction with two-input logic gates to provide control waveforms of varying width by merely selecting the output that turns the gate on at the right time and a second output to turn it off at the right time.

■ **EXAMPLE 19-1**

Using the shift counter of Figure 19-7(a), provide waveforms that go high during times 2 through 4, 3 through 6, 2 through 7, 2 through 8, and 1 through 9.

Solution: See Figure 19-8. One input to the gate turns it on. The second input turns the gate off.

For waveforms less than five clock periods wide, AND and NOR gates are used. They can be used interchangeably by complementing the inputs. No gates are needed for waveforms five clock periods wide; these are available directly from a five-bit shift counter. For positive waveforms wider than five clock periods, OR gates or NAND gates are used.

FIGURE 19-8
Solution to Example 19-1.

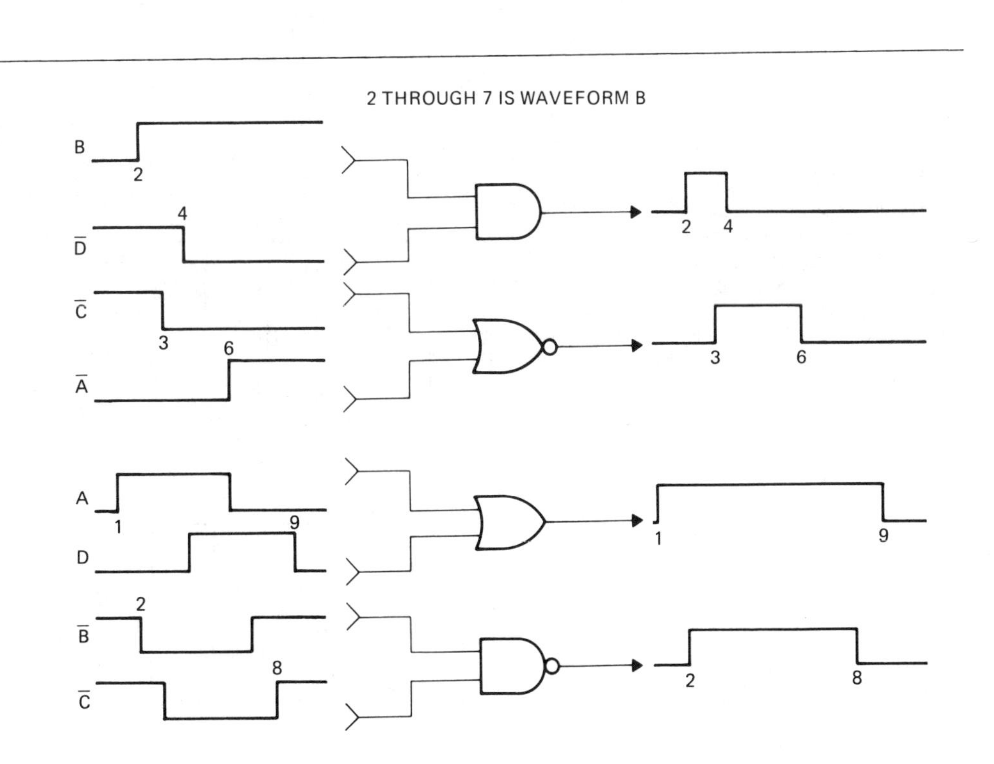

■ EXAMPLE 19-2

Using the shift counter of Figure 19-7(a), provide four waveforms that go low during times 1 through 3, 4 through 7, 1 through 6, 2 through 9, and 1 through 7.

Solution: See Figure 19-9.

The voltage waveforms developed in Examples 19-1 and 19-2 will repeat themselves every ten clock pulses until the next reset pulse. If a single voltage waveform is required, then another shift counter triggered by the last flip-flop of the first shift counter, as shown in Figure 19-10, will provide additional waveforms. Having two five-bit shift counters and working with four input logic gates, one can produce any width of waveforms between reset and clock pulse 100.

■ EXAMPLE 19-3

Provide voltage waveforms that go high between times 42 and 44, 33 and 36, and 22 and 27.

Solution: See Figure 19-11, which shows control waveforms generated from the outputs of two five-bit shift counters.

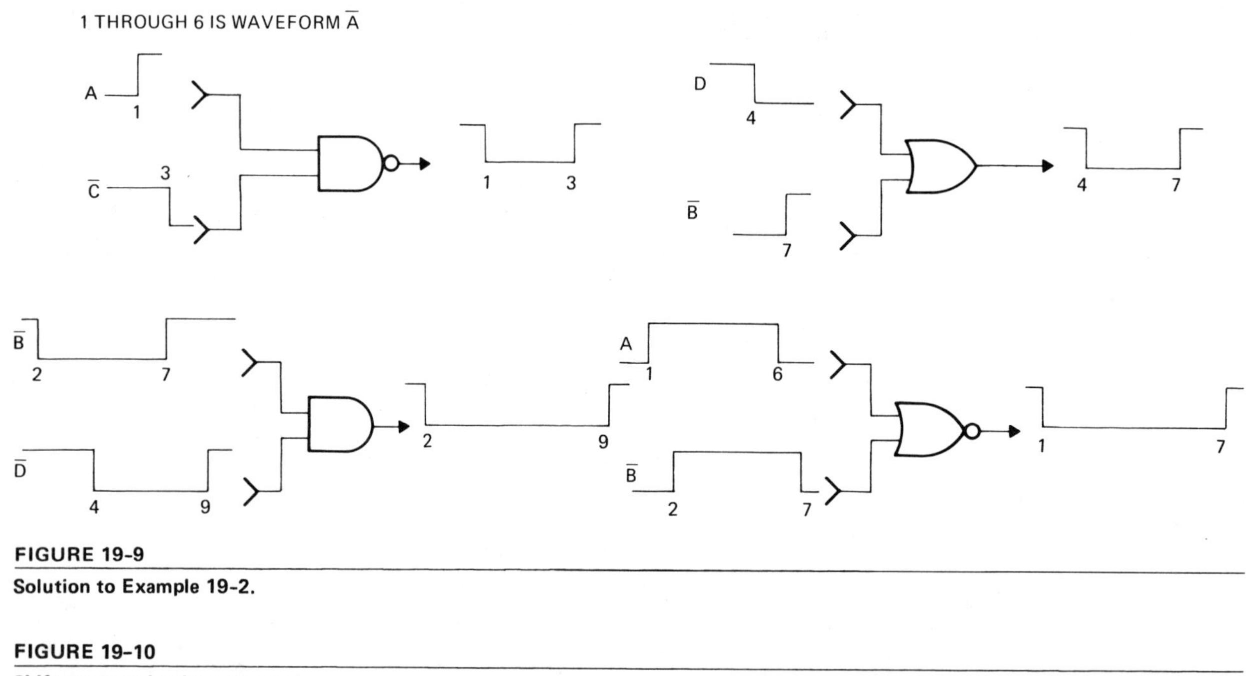

FIGURE 19-9
Solution to Example 19-2.

FIGURE 19-10
Shift counter circuits connected as two separate decades. The second decade is clocked by the E output of the first decade.

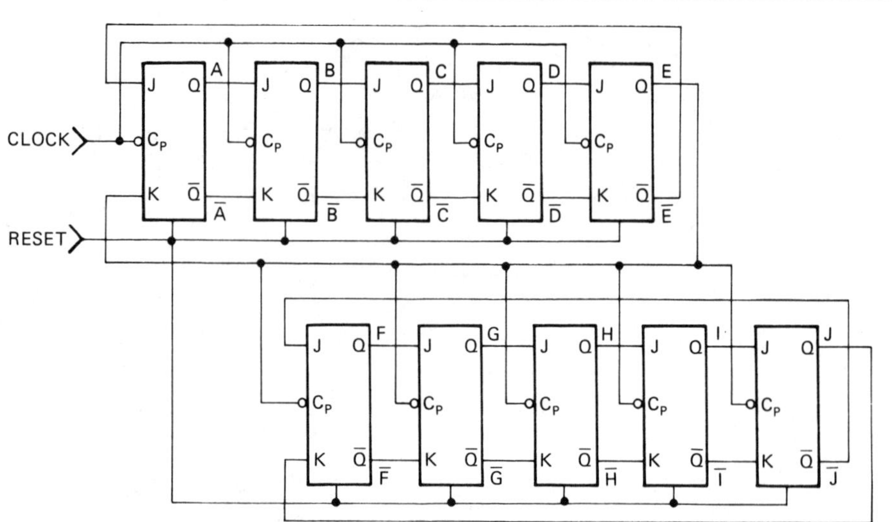

The shift counter can serve as a decimal counting unit by making connections as shown in Figure 19-12. The Q output of the fifth flip-flop is used to clock the next higher decade.

Although it is convenient in terms of our thinking to divide shift counters into sets of five flip-flops—in effect forming DCUs—it is not necessary to do this. If smaller- or larger-size shift counters prove economical, they should be used and the same techniques will apply.

Often a set of waveforms are needed during the first 10 or 20 clock periods, but the clock itself continues for hundreds of clock pulses after the last waveform before a reset occurs. In this case, a decode of the number after the last desired waveform can be used to inhibit the

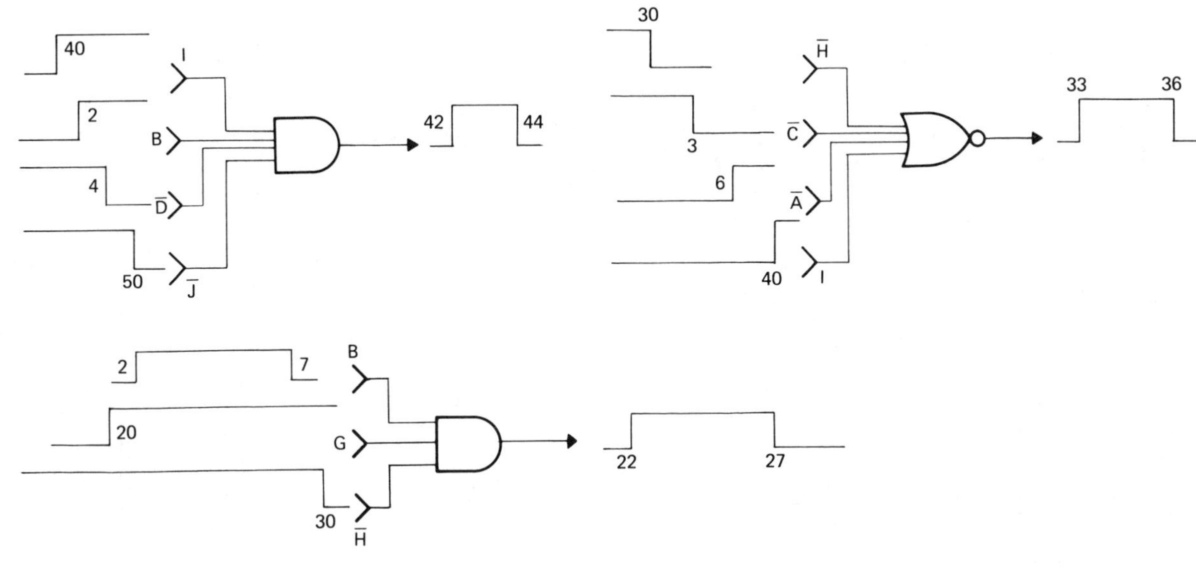

FIGURE 19-11

Solution to Example 19-3.

FIGURE 19-12

Shift counters forming three DCUs for a count to 1000.

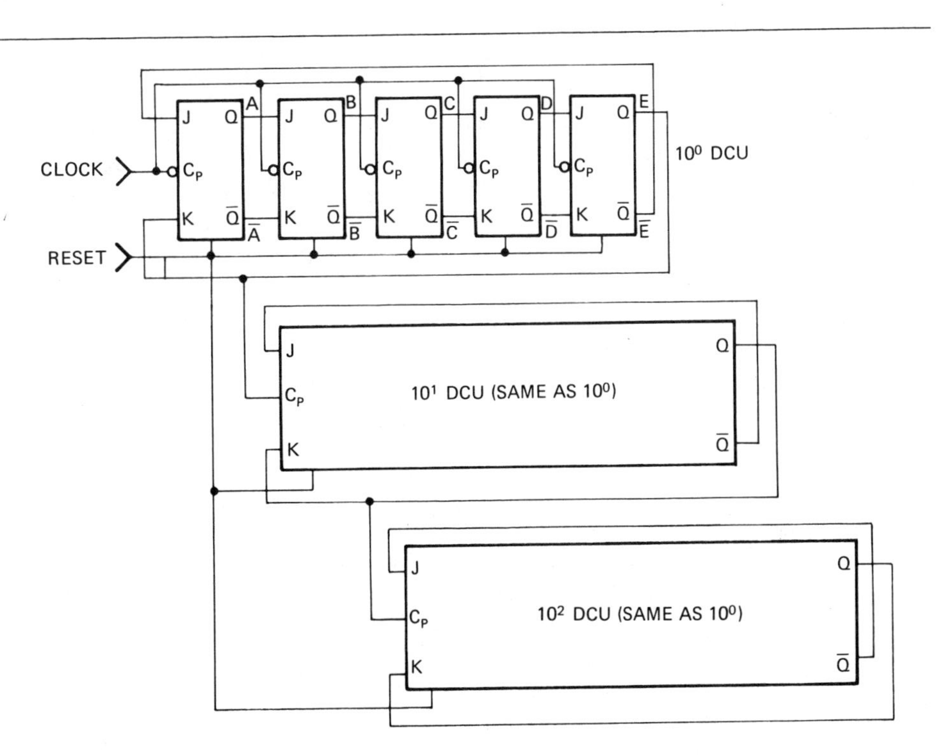

clock line. This will hold the shift counter in that state until reset, preventing any unwanted repetition of the lower-period waveforms. Figure 19–13 shows a shift counter that will stop at the count of 28. The primary advantage of the shift counter is its decoding. It requires only two input gates. One input turns the gate on at the correct time; the other turns it off. The 1.2.4.8 weighted DCU requires three- and four-input gates to decode.

19.5 Dual Clock Gating

In Section 17.7, we discussed the need for dual clock, or clock and delayed clock, systems. One method of timing or sequencing machine functions is to isolate clock pulses onto separate circuit lines and use them to start necessary functions at the correct clock time. In a dual clock system, the counters used to generate gating signals that isolate clock pulses should be triggered by delayed clock (C_P') pulses, as shown in Figure 19–14. This avoids gating with coincident leading or trailing edges. The pulse to be enabled through the gate is bracketed by the control signals, as Figure 19–15 shows.

19.6 Summary

Throughout our discussions of digital subassemblies, we have used waveforms to inhibit or enable some of the operations. We assumed

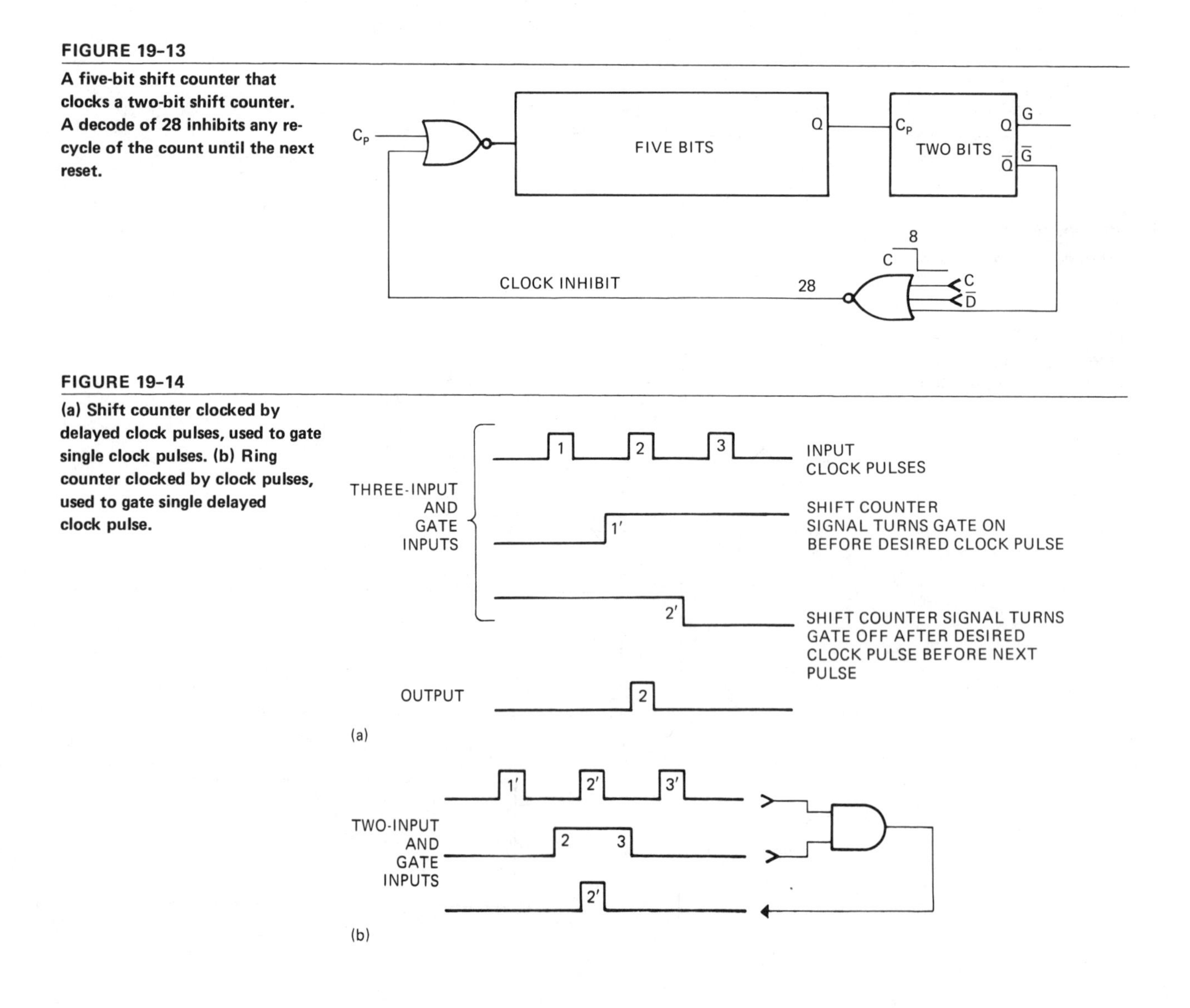

FIGURE 19–13
A five-bit shift counter that clocks a two-bit shift counter. A decode of 28 inhibits any recycle of the count until the next reset.

FIGURE 19–14
(a) Shift counter clocked by delayed clock pulses, used to gate single clock pulses. (b) Ring counter clocked by clock pulses, used to gate single delayed clock pulse.

these waveforms would be generated somewhere within the digital system. The subject of this chapter is the special circuits and methods used to generate the variety of control waveforms needed for timing, sequencing, enabling, and inhibiting the many functions that occur in a complex digital system.

If one were to use a general block diagram that would apply to any digital system, it would be like that of Figure 19-1. It would contain input, output, the functional unit, and a timing and control waveform generator. With this philosophy in mind, logic designers can proceed with the design of the functional unit, drawing in control waveforms where needed, in full confidence that they can be generated in the control waveform section because there are a number of logic circuits that are ideally suited to the production of control signals.

The *ring counter* is advantageous for generating clock-period-wide waveforms. Figure 19-2(a) is a logic diagram of a five-bit ring counter. It is wired like a normal shift register except that the first flip-flop is set while the others are reset by the reset line. The output of the last flip-flop may be wired back to the input flip-flop without crossover, Q to J, $\overline{Q}$ to K. When reset occurs, only the first flip-flop will be in the 1 state. As the counter is clocked, the single 1 level is passed one clock pulse at a time through the counter until it reaches the last flip-flop. If the output is wired back to the input, the last flip-flop will steer the 1 level back to the first flip-flop and the process will continue. The waveforms resulting from this are shown in Figure 19-2(b). The advantage of the ring counter is that it produces clock-period-wide control signals without the need for decoding gates. Some variations on the standard ring counter are often used. If a single set of clock-period-wide waveforms were needed between resets, the wiring back from the output flip-flop might be eliminated and the first flip-flop would be steered to 0.

If double-clock-period-width pulses are desired, the first two flip-flops of the counter will be set during reset. This results in the waveforms of Figure 19-3.

Extending the width of ring counter pulses by connecting them to OR gates, as shown in Figure 19-4(a), may produce spikes, which are often undesirable. These are caused by uneven propagation delay of flip-flops in the counter. The overlapping waveforms of Figure 19-3 can be combined by connection to OR gates without difficulty, as shown in Figure 19-4(b).

When waveforms far removed from reset are needed, the ring counter is no longer advan-

FIGURE 19-15

Ring counter clocked by delayed clock, used to enable single clock pulses through AND gates.

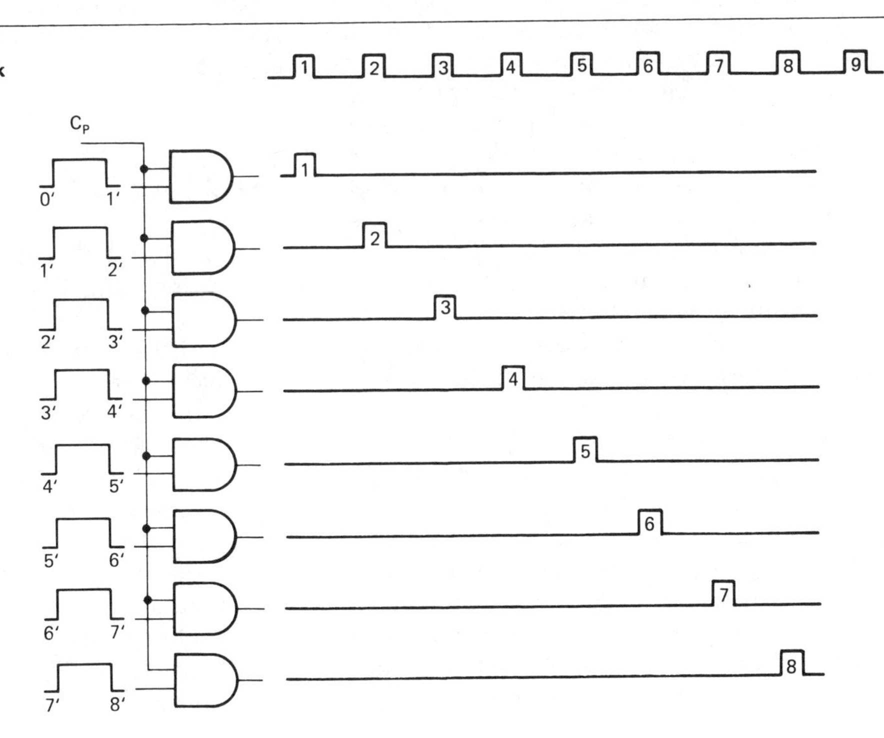

tageous, as it requires one flip-flop per clock period. To obtain a waveform of the single clock period 31 would require a ring counter of at least 31 bits. In this case, a *binary counter* and one or more decoding gates would be more economical. Figure 19-6 shows the five-bit binary counter and five-input AND gate used to develop a waveform at clock pulse 31.

The most versatile circuit for waveform production is the *shift counter*. It differs from the ring counter in that it has a normal (all 0) reset and the output is wired back with a crossover. The Q of the last flip-flop is wired to the K of the first flip-flop, the $\bar{Q}$ to the J. Figure 19-7 shows a five-bit shift counter and the waveforms it produces. At reset, all flip-flops are reset to the 0 state. The crossover leads steer the first flip-flop to 1; all others are steered to 0. On the trailing edge of the first clock pulse, the first flip-flop goes to 1 and remains in that state for five clock periods. Each clock pulse, until clock pulse 6, another flip-flop goes to the set state. When the rightmost flip-flop changes to the 1 state, it changes the steer of the first flip-flop. Each clock pulse thereafter, for five clock pulses, the flip-flops one at a time return to the 0 state. By clock pulse 10, the rightmost flip-flop changes to 0, its crossed-over leads steer the first flip-flop back to 1, and the cycle repeats. It repeats every ten clock pulses thereafter.

The shift counter, in conjunction with two-input decoding gates, can provide waveforms of varying clock-period widths. A shift counter of five bits can provide waveforms with trailing edges ten clock periods away from reset. This can be increased by two clock periods per flip-flop—twice the capacity of the ring counter. But the ring counter can provide clock-period-wide waveforms without any decoding gates. The shift counter's main advantage over the binary counter is that it requires only two input gates to produce its waveforms—one to turn the gate on, the second to turn it off. An example of control waveform generation using the five-bit shift counter is given in Examples 19-1 and 19-2 (Figures 19-8 and 19-9). Note that, in all cases, only two inputs are needed—one to turn the output on, the other to turn it off.

Besides serving as a control waveform generator, the shift counter is occasionally used as a decade counter. As a decade counter, it requires five flip-flops—one more than the binary counter—but decoding can be accomplished with two-input decoding gates. With the availability of integrated circuit decoders, this may no longer be a useful advantage where decoding of BCD digits is involved; but where individual waveforms far removed from reset are needed, the shift counter may be advantageous. Figure 19-12 shows three decades of shift counters. The Q output of the fifth flip-flop in each decade is used to trigger the next higher decade. Each decade can be decoded individually with two-input gates. If random waveforms far removed from reset are needed, three- or four-input gates may be used, as explained in Example 19-3 and Figure 19-11.

Thus far we have discussed the formation of control waveforms that are one or more clock periods in width. In many instances, we need single clock pulses isolated onto a separate line so that we can use them to time the starting of digital operations. When this is done, it is ideal to have a dual clock system. Otherwise, the pulses we are gating will have coincident trailing edges with the waveforms enabling them through the gates. This may result in an output with an erratic edge or pulses of reduced size. Ideally, the waveforms enabling the pulse through the gate bracket the pulse on both sides. This can be done easily if clock pulses are enabled onto separate lines by control waveforms produced by counters triggered with delayed clock pulses. Figure 19-14(a) shows the ideal shift counter waveforms for enabling clock pulse 2 onto a separate line. Figure 19-14(b) shows the ideal ring counter waveform for isolating 2′ onto a separate line.

Glossary

Ring Counter: A connection of flip-flops differing from that of a shift register in that only the first bit is set by the reset pulse and the output of the last flip-flop is wired back to the input of the first flip-flop without crossover (Q to J, $\bar{Q}$ to K). See Figure 19-2. The ring counter is ideal for generating clock-period-wide control waveforms without the need for decoding gates.

Shift Counter: A connection of flip-flops differing from that of a shift register only in that the output of the last flip-flop is wired back to the input of the first with crossover (Q to K, $\bar{Q}$ to J). See Figure 19-7. The shift counter outputs are ideal for producing control waveforms of various clock-period widths.

Microprocessor: An integrated circuit of 40 or more pins, originally developed to control the arithmetic processes of a digital calculator and later applied as a universal control circuit for medium-sized digital systems.

Questions

1. For what conditions is the ring counter the ideal control waveform generator?
2. What is unusual about the reset line connection for the ring counter?
3. How many flip-flops are required for a ring counter that must produce a clock-period-wide control waveform beginning with clock pulse 7?
4. If the connections between output and input flip-flops were removed, how would it change the operation of the ring counter of Figure 19-2?
5. For what condition is the binary counter advantageous as a control waveform generator?
6. How many flip-flops are required for a binary counter used to produce a clock-period-wide control waveform beginning with clock pulse 7? How many inputs are required on the decode gate?
7. Under what condition is the shift counter the ideal waveform generator?
8. How does the connection between input and output flip-flops on the shift counter differ from that on the ring counter?
9. How would removing the connections between the input and output flip-flops affect the operation of the shift counter?
10. How many flip-flops are required for a shift counter used to produce a clock-period-wide waveform beginning with clock pulse 7? How many inputs are required on the decode gate?
11. To isolate clock pulse 4 onto a separate line with the aid of a shift counter, how many inputs should the decode gate have? What would be the advantage of operating the shift counter with the delayed clock line?
12. If shift counters are to be connected as decimal counting units, how many flip-flops are needed for each DCU? To what should the clock line of the second DCU be connected?

Problems

19-1 Draw the logic diagram of a four-flip-flop ring counter.

19-2 Draw a timing diagram of the waveforms available from the ring counter of Problem 19-1.

19-3 Figure 19-16 shows two TTL dual J-K flip-flops. Draw in the interconnections needed to produce a four-bit ring counter. (*Hint:* The 7473 J-K flip-flop has no set input; but if we redefine the outputs, calling Q a $\overline{Q}$ and $\overline{Q}$ a Q, then the reset will function as a set. J and K will also reverse their functions.)

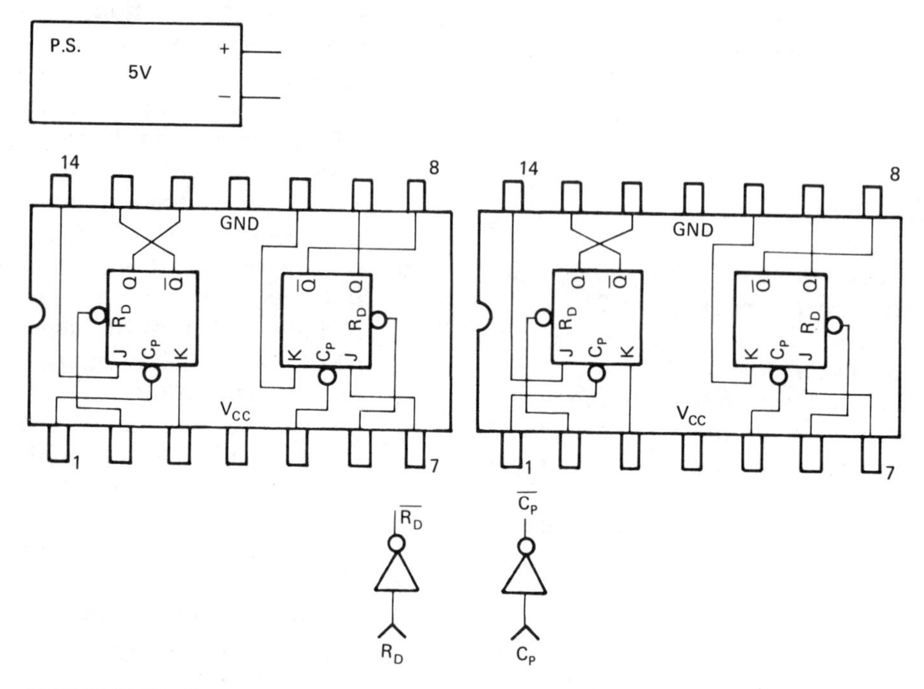

FIGURE 19-16. Problem 19-3

19-4 Draw a four-flip-flop shift counter clocked by a delayed clock line. Label the outputs A through D and $\overline{A}$ through $\overline{D}$.

19-5 Draw the waveforms that would occur at the outputs of the shift counter of Problem 19-4. (Assume a 0′ following reset.)

19-6 The eight AND gates of Figure 19-15 are enabled by ring counter waveforms to pass pulses 1-8. Use eight three-input AND gates to accomplish the same function, using the shift counter outputs of Problem 19-4. Label the inputs to the gates.

19-7 Repeat Problem 19-6 using one inverter and eight three-input NOR gates.

19-8 Show how the universal shift register of Figure 13-8, along with the necessary inverters, could be wired to function as a ring counter.

19-9 Repeat Problems 6-6 through 6-8.

19-10 Repeat Problems 6-10 through 6-14.

19-11 Repeat Problems 7-11 through 7-13.

19-12 Repeat Problems 7-20 through 7-23.

CHAPTER

Digital-to-Analog Converters

Objectives

Upon completion of this chapter, you will be able to:

- Construct a resistor or ladder network for a voltage source digital-to-analog converter.
- Select or construct a set of analog switches suitable for applying the voltage to the resistors in a ladder network.
- Test a digital-to-analog converter by using a counter and an oscilloscope.
- Construct a current source digital-to-analog converter.
- Use monolithic D/A converters.
- Use the digital-to-analog converter to construct a transistor curve tracer.
- Select and use a hybrid circuit digital-to-analog converter.

20.1 Introduction

In the control of manufacturing processes, the outputs of sensing devices that tell us temperature pressure, strain, weight, and so forth are analog voltages. The exact time that heat is to be turned off or on, or pressures are to be changed, is often a complicated formula of several such parameters. Because these voltages are already in analog form, analog computer methods are often used to make the calculations and apply the control signals. If, however, a high degree of precision or speed is required or if the computer is already available in digital form, an analog-to-digital converter may be used to prepare the signals for digital calculation. Conversely, the output of a digital computer may be useless as a control signal for the application of precise degrees of temperature or pressure and a digital-to-analog converter would be needed.

A digital-to-analog converter provides an analog voltage or current at its output. The output amplitude is proportional to the digital value applied to the input.

A D/A converter contains the three essential elements shown in Figure 20-1. These are a divider network, a set of bit switches, and a precision power supply. Ancillary items most often used with the converter are the register or latches used to store the digital number while it is being converted and an output amplifier that may be used to buffer or amplify the results.

The three most important characteristics of a D/A converter are resolution, accuracy, and speed. Resolution is primarily a matter of the size of the converter in bits, but the number of bits in the conversion does have an impact on speed and accuracy. The larger the number of bits, the longer the "settling time" needed for the conversion. Accuracy cannot exceed $\pm 1/2$ LSB, so the larger the number of bits the greater the accuracy. Imperfections in the switches, resistor networks, and precision power supply place a further limitation on speed and accuracy.

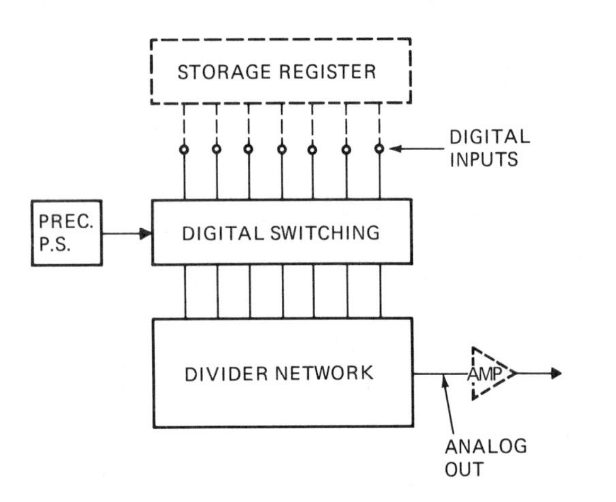

FIGURE 20-1

Block diagram of digital-to-analog converter. The reference voltage level is applied to the divider network through switches controlled by the digital inputs.

20.2 Resistor Network

Figure 20-2(a) shows the simplest form of digital-to-analog converter. In actual practice, the manual switches are replaced by electronic switches controlled by the outputs of a counter or storage register. The value of the resistors is equal to the resistance chosen for the 2^0 bit divided by the binary value of the bit for which it is used. If we were to add another bit (2^4), the resistor used would be $R_0/16$. As the switches are thrown to represent binary numbers, a different voltage divider relationship occurs for each binary number. The resistors of all switches in the 1 state form a parallel network between the output lead and V. The resistors of all switches in the 0 state form a parallel network between the output lead and ground. For a binary 5 (0101), the results are as shown in Figure 20-2(b). A similar voltage divider relationship could be worked out for any binary number set on the switches. In this case, the result would be 1V per bit. For any value of V:

$$V_A = \frac{V}{15} \times \text{(Number set on the switches)}$$

20.3 Binary Ladder Network

The ladder network of Figure 20-3 has several advantages over the resistor network, one being that only two values of resistance are needed, and if one considers the option of putting resistors in series or parallel, a single value of resistance can be used. With the resistor network, it is difficult to select a suitable standard value for resistor R_0 and still find that the necessary submultiples are available in standard values. We shall find also that with the ladder network the loading effect on the switches does not change radically as we increase the number of bits.

With the binary ladder, an equivalent divider network is easy to determine: Assume all switches at ground except 2^3. From point A to ground, a pair of 2R are in parallel, resulting in R, as shown above the short bracket in Figure 20-4. This R is in series with the R between A and B, as shown in the long bracket in Figure

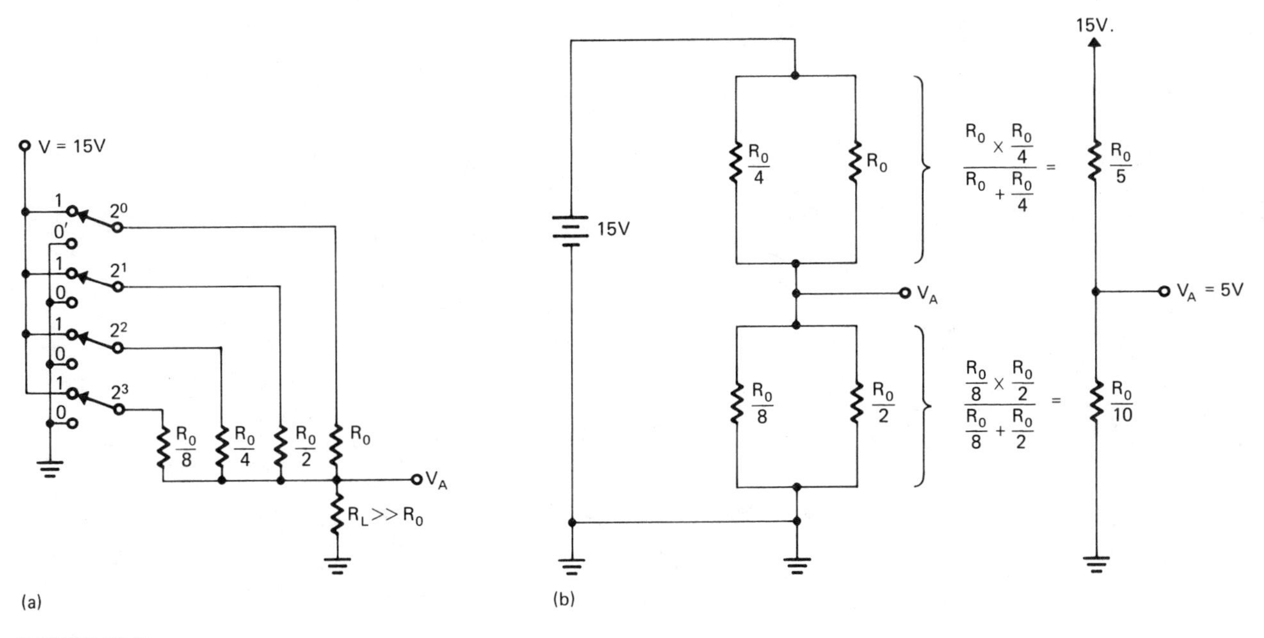

FIGURE 20-2

(a) Resistor network that will divide a voltage, V, in proportion to the value of the binary number set on the switches. (b) Equivalent divider formed when switches are set to 0101.

FIGURE 20-3

Ladder network with the LSB switch closed. This places 2R between point B and ground.

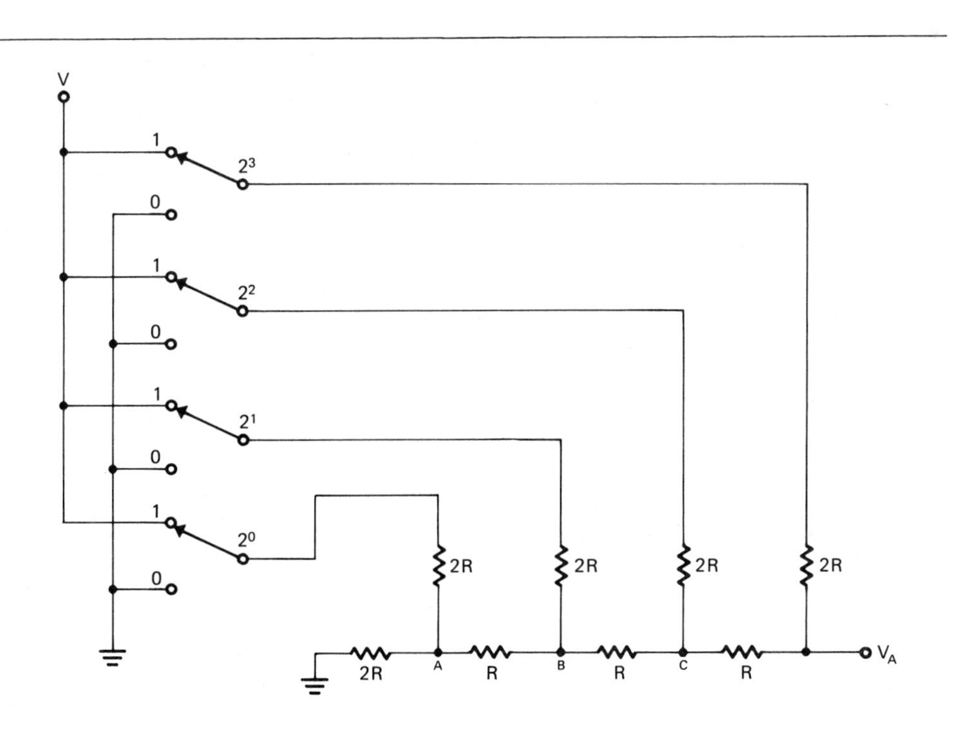

FIGURE 20-4

Ladder network with the LSB switch closed. This places 2R between point B and ground.

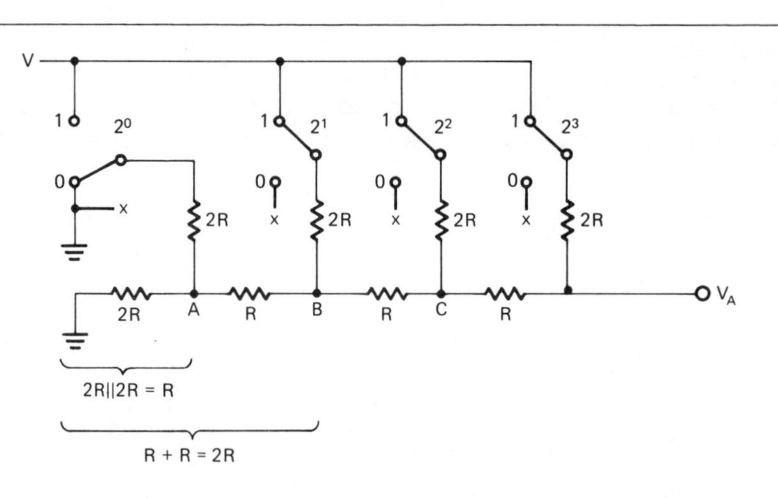

FIGURE 20-5

(a) Circuit with all switches closed except the MSB, resulting in an R to ground at each point A through C. (b) Circuit with only the MSB switch turned on, resulting in a 50 percent voltage divider.

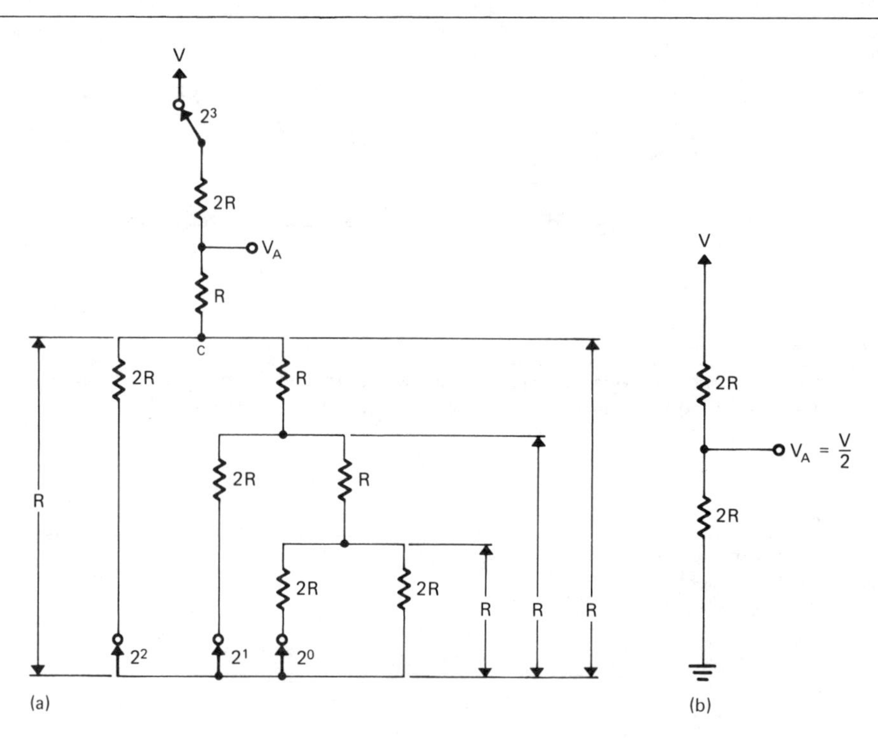

20-4. As we continue this analysis for the 2^1 and 2^2 switches being grounded, we find that for each switch the process repeats itself, forming the circuit of Figure 20-5(a), which, in the end, is equivalent to the 50 percent voltage divider of Figure 20-5(b).

With all switches except the MSB in the 0 or ground position, the results is V/2 at the output, V_A. If V were 16 volts, the output would be 8 volts, or 1 volt per count. With switches other than the MSB being 1, the analogy is more complicated, but the results are still 1 volt per count for a four-bit converter and a V of 16 volts. Figure 20-6 shows the divider relationship for a count of four, or 0100. The output at V_A is V/4, or 4V if V = 16. Again, it is 1 volt per count. This resolution can be changed by changing the value of V.

20.4 Voltage Source D/A Converter

We have shown that for single switches the voltage V_A will be V/16 per count. The superposi-

FIGURE 20-6

Ladder network divider relationship for a binary number (4) 0100.

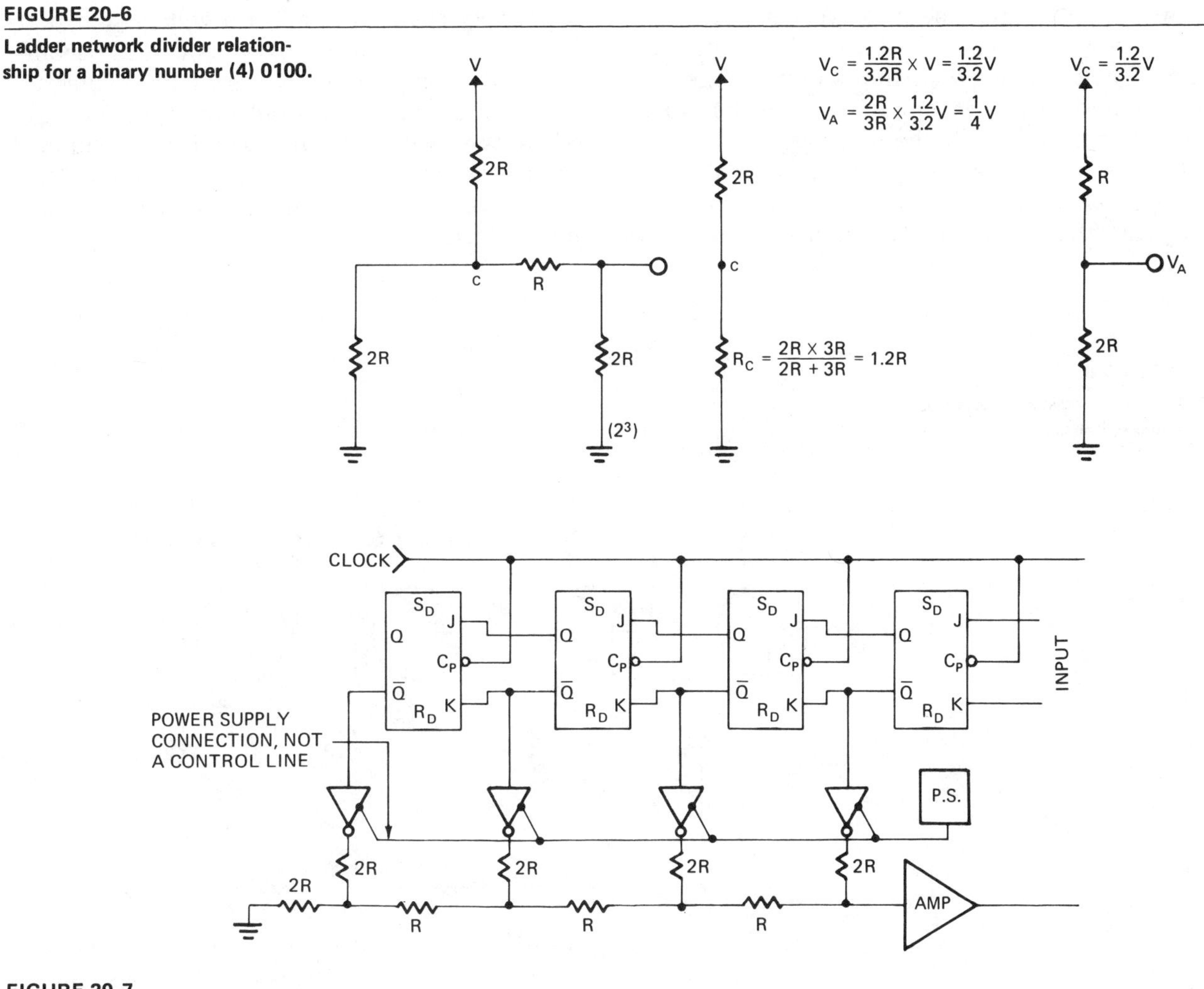

FIGURE 20-7

A complete digital-to-analog circuit including a register to store the number being converted, the ladder network and switches, a precision power supply, and an output amplifier.

tion theorem dictates that it will be V/16 volts per count for any number of switches that are on. Figure 20-7 shows the typical supporting circuitry for the digital-to-analog converter. The number would normally be stored in registers for the period of the conversion. If the switches are inverter types, they would be connected to the Q outputs of the register. If a high degree of accuracy is not needed, a simple transistor switch without the precision supply might be used, as in Figure 20-8. As these switches are inverting, the base leads will connect to the $\bar{Q}$ leads of the register. When the 0 output of a flip-flop is low, the switch will draw little or no current through R_C, placing the voltage $V \approx V_{CC}$ at the top of the 2R ladder resistor. If a register is in the reset state, the high level from the $\bar{Q}$ output will turn the transistor on, shorting the top of the 2R ladder resistor to ground through the saturated transistor. To operate properly, the load resistance R_L must be very large in comparison to R. This may require that an amplifier with high input impedance be used to buffer the output. The value of R_C must be very small in comparison to R so that the current drawn by the ladder network will not drop a significant amount of voltage across R_C.

It is seldom practical to seek accuracies of more than ±1/2 count or 1/2 LSB, but as the number of bits in the converter are increased the tolerance of the resistors, accuracy of the power supply, and errors created by the switches become a problem. Use of close-tolerance resistors and clamping the switch outputs to a precision power supply offer one improvement for accuracy, shown in Figure 20-9. The V_{CC} must be several volts higher than the ladder voltage, V. The precision power supply must be set to a value lower than V by the diode potential, V_D. When a switch transistor is not drawing collector current, a small current will be passing through R_C and the clamp diode—enough current to accurately clamp the collector voltage to exactly V. As the resolution or volts/count becomes lower, the single-transistor switch supplies too high an offset error resulting from $V_{CE\ SAT}$ when the switch is supposed to be grounded. More elaborate switching circuits with DC offsets of less than 5 mV are available in integrated circuit form.

FIGURE 20-8

D/A circuit using single-transistor inverting switches.

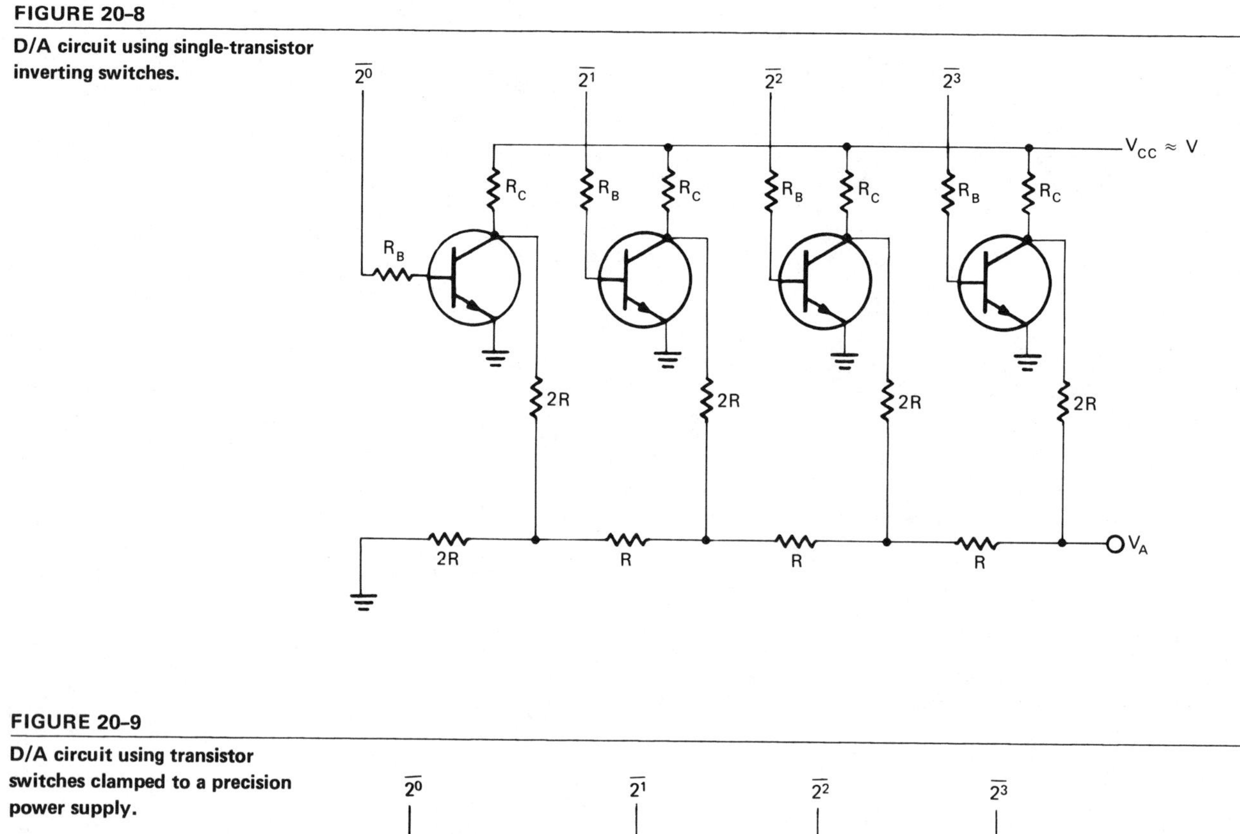

FIGURE 20-9

D/A circuit using transistor switches clamped to a precision power supply.

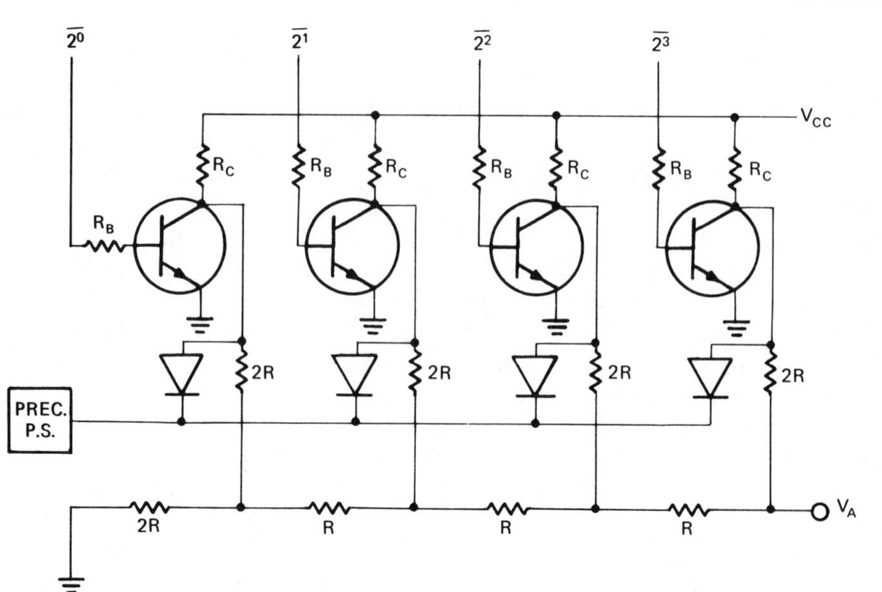

20.5 Test Waveforms of a Digital-to-Analog Converter

The digital-to-analog converter can be tested by statically setting numbers into the switches and measuring the output level. A dynamic and more reliable test can be made by connecting the switches to the output of a binary counter of equivalent number of bits.

As the counter counts, an oscilloscope connected to output V_A will, as Figure 20-10 shows, display a staircase waveform. The first step is the count of 0, and the last or highest is the count of 15. As Figure 20-11 shows, each step is exactly the same height, V/16. A change occurs after each clock pulse. After the count of 15, the counter returns to the count of 0 and

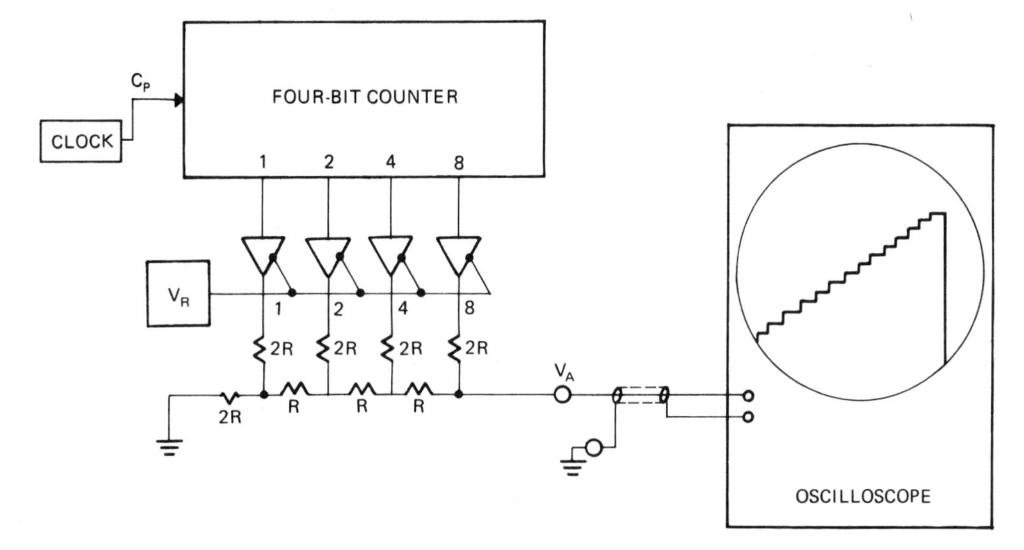

FIGURE 20-10
A binary counter attached to a digital-to-analog converter, producing a staircase waveform at the analog output.

FIGURE 20-11
Test waveform of a digital-to-analog converter. As the digital input from the binary counter changes from 0 through 15, the analog output changes in steps of V/16.

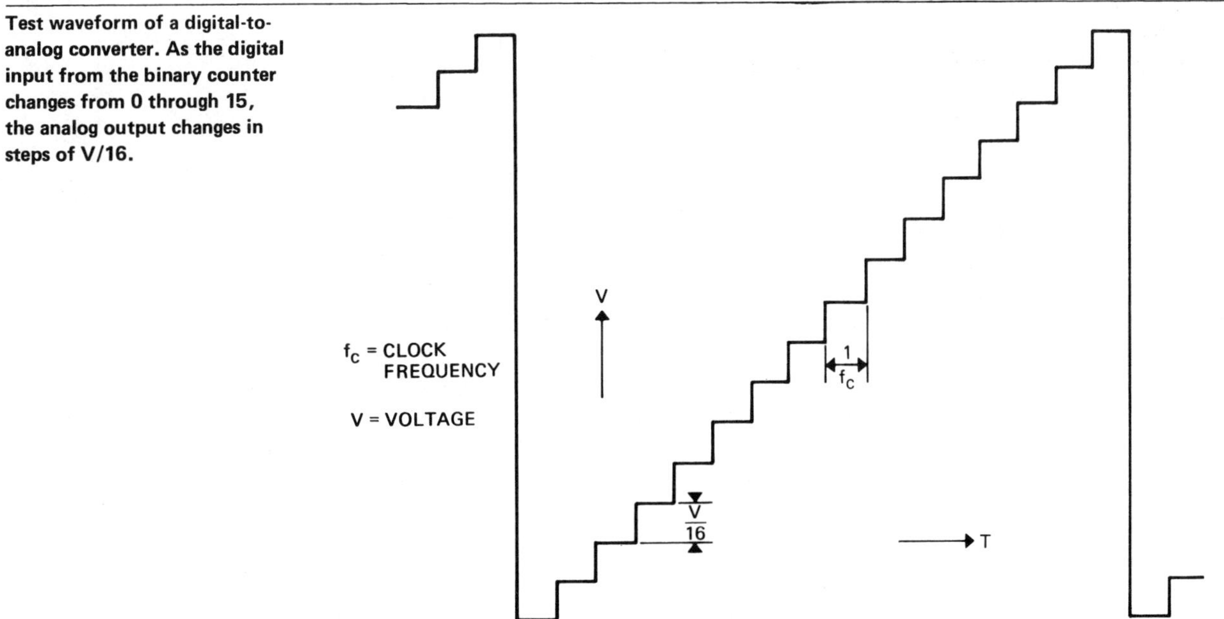

the analog voltage returns to 0 and begins the steps again.

20.6 Current Source D/A Converter

The methods discussed thus far use voltage divider techniques, and the load applied to the output, R_L, must be of very high resistance to avoid altering the divider relationship. The output, of course, can be buffered with an operational amplifier of high-input impedance, but if many bits are involved, the offset errors created by the switches become excessive. Use of a method that produces current sources proportioned to the binary value of the digital inputs is a favorable solution. The current source method also has the advantage of being faster than the voltage divider method. The major disadvantage is a full-scale conversion voltage much lower than the voltage applied to the network. This is, however, not objectionable for all cases.

Figure 20-12 shows an ideal equivalent

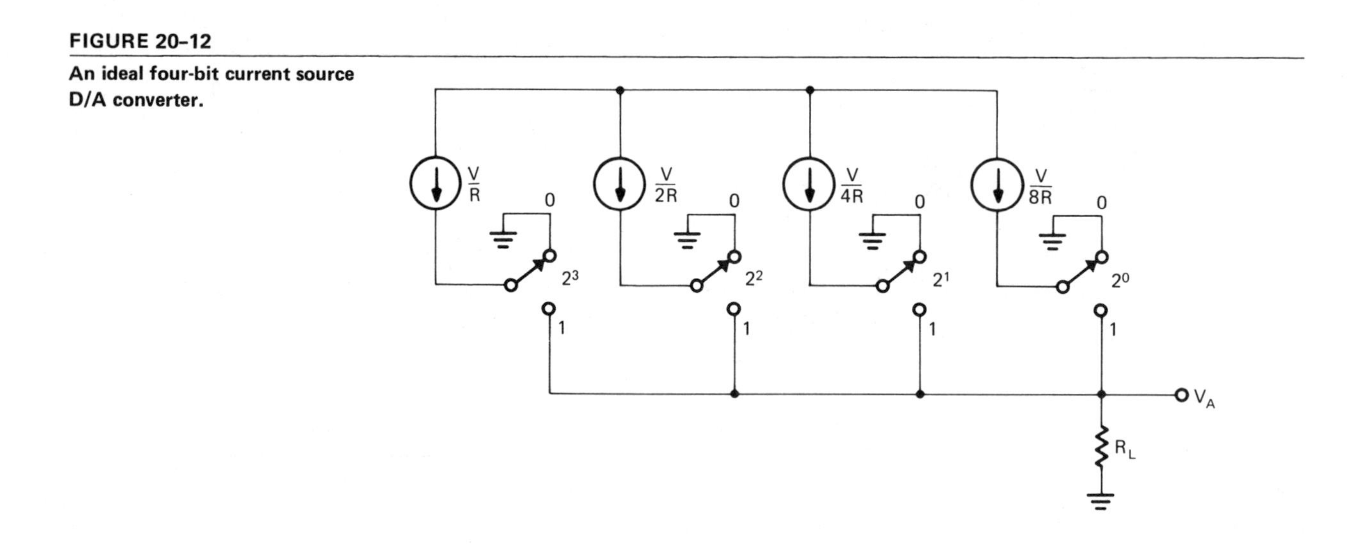

FIGURE 20-12
An ideal four-bit current source D/A converter.

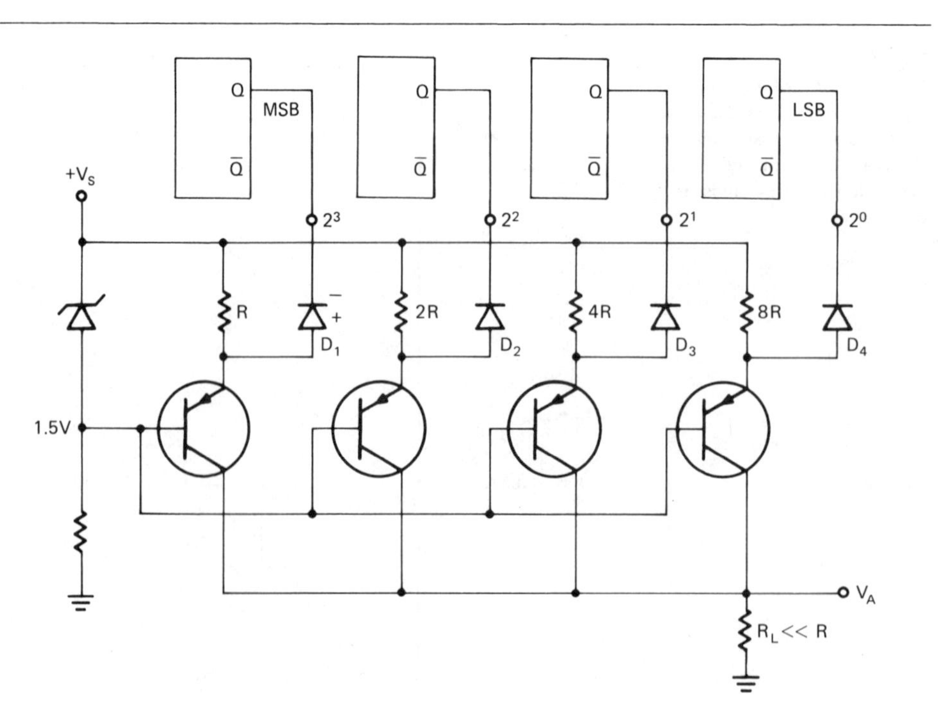

FIGURE 20-13
Current source D/A converter. When Q of a register is low, the current source is shorted to ground through the diode and Q, instead of passing through the transistor and R_L.

circuit for a current source D/A. In this circuit, the value of the current through R_L will be the minimum current V/(8R) times the binary number set on the switches. The current for any switch in the 0 state will be diverted to ground, but those in the 1 state will sum together and pass through R_L, resulting in a voltage drop proportional to both R_L and the count set on the switches. No difficulty is created by using a low value of R_L. Examination of the real circuit will show that a high value of R_L will tend to create inaccuracies. Figure 20-13 shows a four-bit current source D/A converter: $R_L << R$. If a binary 1-level voltage is applied to the cathodes of a diode, D_1 through D_4, that diode will appear like an open switch. The corresponding transistor will be turned on. If the binary count is at 15, a positive 1-level voltage will be applied to the cathode of the diodes D_1 through D_4. They will be reverse-biased and will have no effect on the circuit. Let us select an R_L such that with the maximum current flowing through it V_A is less than 1V. The voltage on the base leads should, therefore, be slightly above 1V. With the voltage $+V_S$ to the emitters through the resistors, a current will flow that will drop all but a few volts of $+V_S$ across the resistor. Instead of the ideal equivalent circuit of Figure 20-12, the more exact equivalent circuit of Figure 20-14 results.

Let us look at an individual bit, as shown in Figure 20-15(a), in the on state (binary 1). With the +2.6V binary 1 level on the diode cathode, the current will flow through the transistor, as the diode will be reverse-biased. Figure 20-15(b) shows the bias levels when a logic 0 is applied to the diode. The diode is forward-biased, allowing the current to flow through the diode and sink to ground through the logic circuit. Under this condition, V_E is less positive than V_B, a condition that turns off a PNP transistor. If a current source D/A converter were connected to a counter and its output checked on an oscilloscope, as described in Section 20.5, the same

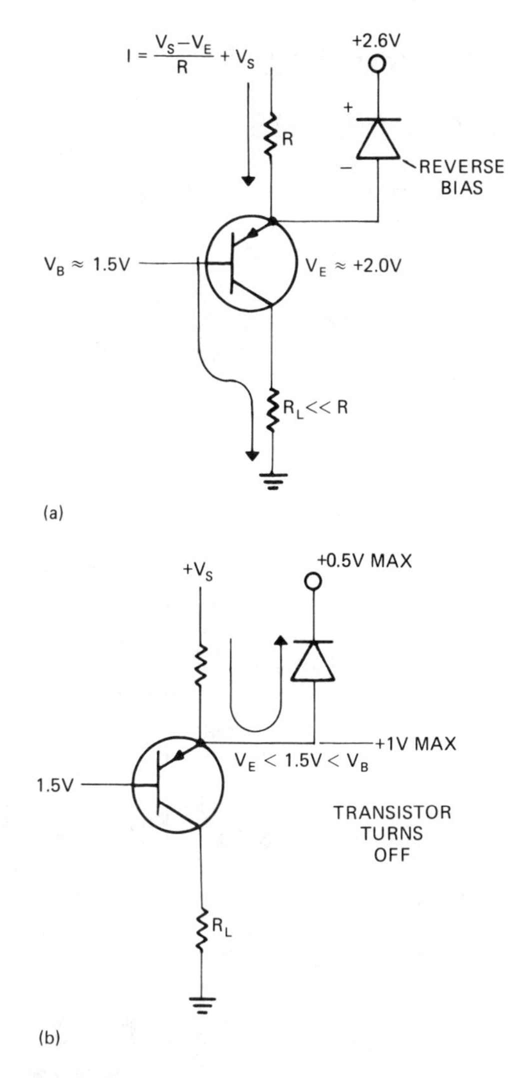

FIGURE 20-15

(a) Binary 1 level on the cathode of the diode. (b) Binary 0 level on the cathode of the diode.

FIGURE 20-14

Equivalent circuit of the current source D/A converter with full-scale input.

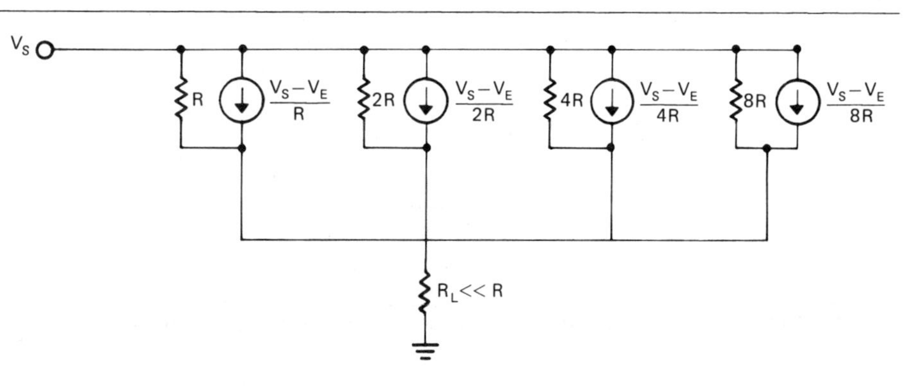

staircase voltage waveform would occur, but the voltage level at the top of the step would be very much lower than V_S.

20.7 Monolithic D/A Converters

Complete D/A converters are available in monolithic integrated circuit form. The most widely available of these is the DAC-08. The DAC-08 is an eight-bit high-speed converter supplied in a 16-pin DIP package. The pin-out and equivalent circuit are shown in Figure 20-16. The digital inputs B1 (MSB) to B8 (LSB) provide a digital range of 00000000 to 11111111. A current divider network is used, and the full-scale output current at I_{OUT} is about equal to the reference current supplied through V_{REF} (+) and V_{REF} (-).

Figure 20-17(a) shows the method used to establish an input reference current and a typical set of values used to provide a full-scale current I_{FS} of 2 mA. Figure 20-17(b) shows the relationship between the current through I_{OUT} as compared to $\overline{I_{OUT}}$. The output current flowing through the load resistance to ground establishes the value of full-scale voltage and voltage increment per bit. This is shown in Figure 20-18(a). Figure 20-18(b) shows the method used to provide a voltage output with positive and negative swing. Under this mode of operation, the MSB

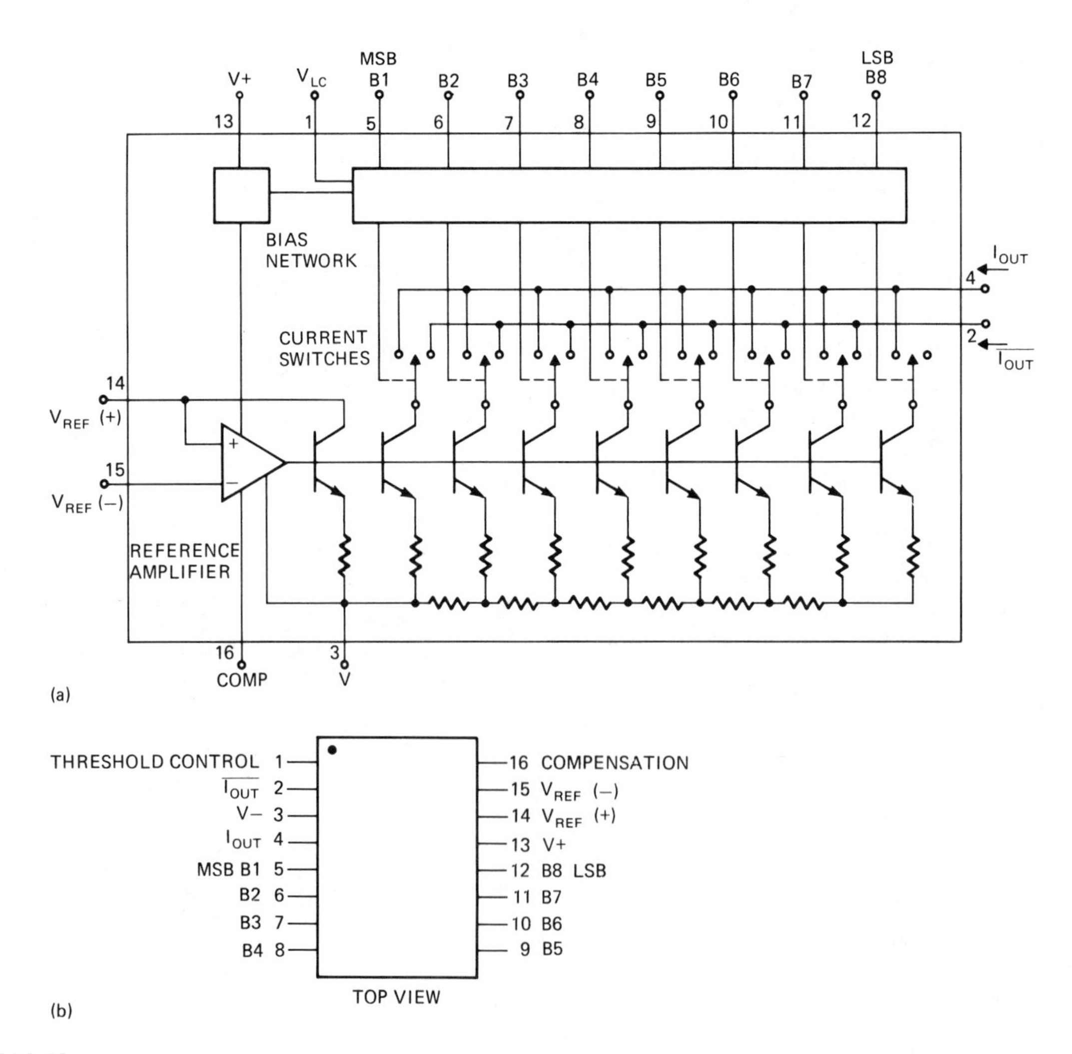

FIGURE 20-16

DAC-08, eight-bit monolithic digital-to-analog converter. (a) Equivalent circuit. (b) Pin connection.

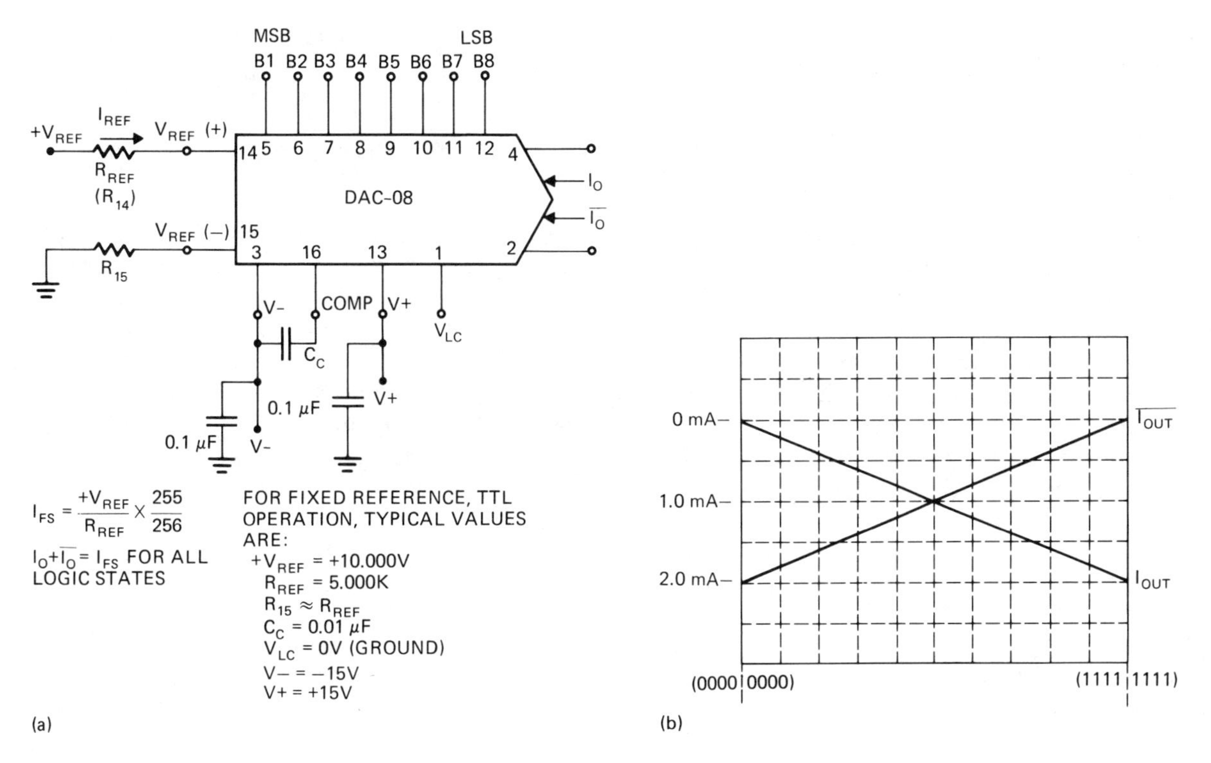

FIGURE 20-17

(a) Basic positive reference operation. Reference current is established by the value of the voltage V_{REF} and the resistor with the equation $I_{FS} = (+V_{REF}/R_{REF}) \cdot 255/256$. ($R_{15}$ is normally equal to R_{REF}.) (b) True and complementary output operation. Values of I_{OUT} and $\overline{I_{OUT}}$ for digital inputs between 00000000 and 11111111 and 2 mA I_{REF}.

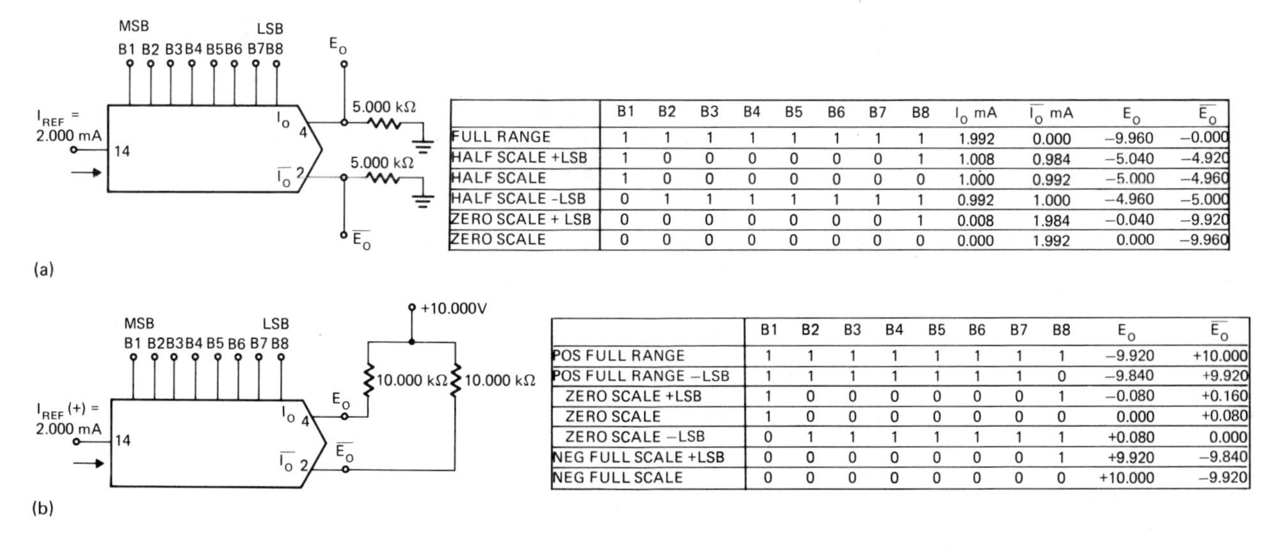

	B1	B2	B3	B4	B5	B6	B7	B8	I_O mA	$\overline{I_O}$ mA	E_O	$\overline{E_O}$
FULL RANGE	1	1	1	1	1	1	1	1	1.992	0.000	−9.960	−0.000
HALF SCALE +LSB	1	0	0	0	0	0	0	1	1.008	0.984	−5.040	−4.920
HALF SCALE	1	0	0	0	0	0	0	0	1.000	0.992	−5.000	−4.960
HALF SCALE -LSB	0	1	1	1	1	1	1	1	0.992	1.000	−4.960	−5.000
ZERO SCALE + LSB	0	0	0	0	0	0	0	1	0.008	1.984	−0.040	−9.920
ZERO SCALE	0	0	0	0	0	0	0	0	0.000	1.992	0.000	−9.960

	B1	B2	B3	B4	B5	B6	B7	B8	E_O	$\overline{E_O}$
POS FULL RANGE	1	1	1	1	1	1	1	1	−9.920	+10.000
POS FULL RANGE −LSB	1	1	1	1	1	1	1	0	−9.840	+9.920
ZERO SCALE +LSB	1	0	0	0	0	0	0	1	−0.080	+0.160
ZERO SCALE	1	0	0	0	0	0	0	0	0.000	+0.080
ZERO SCALE −LSB	0	1	1	1	1	1	1	1	+0.080	0.000
NEG FULL SCALE +LSB	0	0	0	0	0	0	0	1	+9.920	−9.840
NEG FULL SCALE	0	0	0	0	0	0	0	0	+10.000	−9.920

FIGURE 20-18

(a) Negative output voltage operation. (b) Bipolar output operation.

serves as a sign bit, and in the negative quadrant full-scale is with all input bits at zero.

When pin 1 (V_{LC}) or threshold control is grounded, the inputs B1 to B8 are set for a threshold voltage (V_{TH}) of 1.4V and are correctly set for TTL inputs. The threshold voltage equation is $V_{TH} = V_{LC} + 1.4V$. Figure 20-19 shows the means of establishing correct input thresholds for a variety of IC technologies. The use of active circuits, rather than simple voltage divider networks, ensures that the DAC-08 input thresholds will accurately track with any changes in supply voltage to the driving circuits.

The settling time of the DAC-08 is about 85 nanoseconds to obtain within 1/2 LSB of eight-bit accuracy. It settles to a six-bit accuracy in about 65 to 70 nanoseconds. The RC time constant of the output circuit has a significant effect on the settling time. The output capacitance of the DAC-08 is 15 pF. Increasing values of both load resistance and load capacitance will add to this effect.

20.8 Transistor Curve Tracer Using D/A Converters

Figure 20-20 shows a typical set of characteristic curves for a transistor. To display these on an oscilloscope, we must be able to sweep the horizontal of the oscilloscope with the voltage V_{CE} changing from 0 to a level above saturation but below the breakdown level. This can be accomplished by rectifying the output of a 12.6V filament transformer. The voltage for the vertical can be accomplished by applying the voltage drop across a small resistor in the collector circuit to the vertical deflection circuit. Figure 20-21 shows this. At each half cycle of the rectified sine wave, V_{CE} increases from 0 to 18V and back to 0V. The same voltage applied to the horizontal input will cause the trace to deflect from left to right and back. At the same time,

FIGURE 20-19

Using pin 1 of DAC-08 to establish the logic level for the digital inputs.

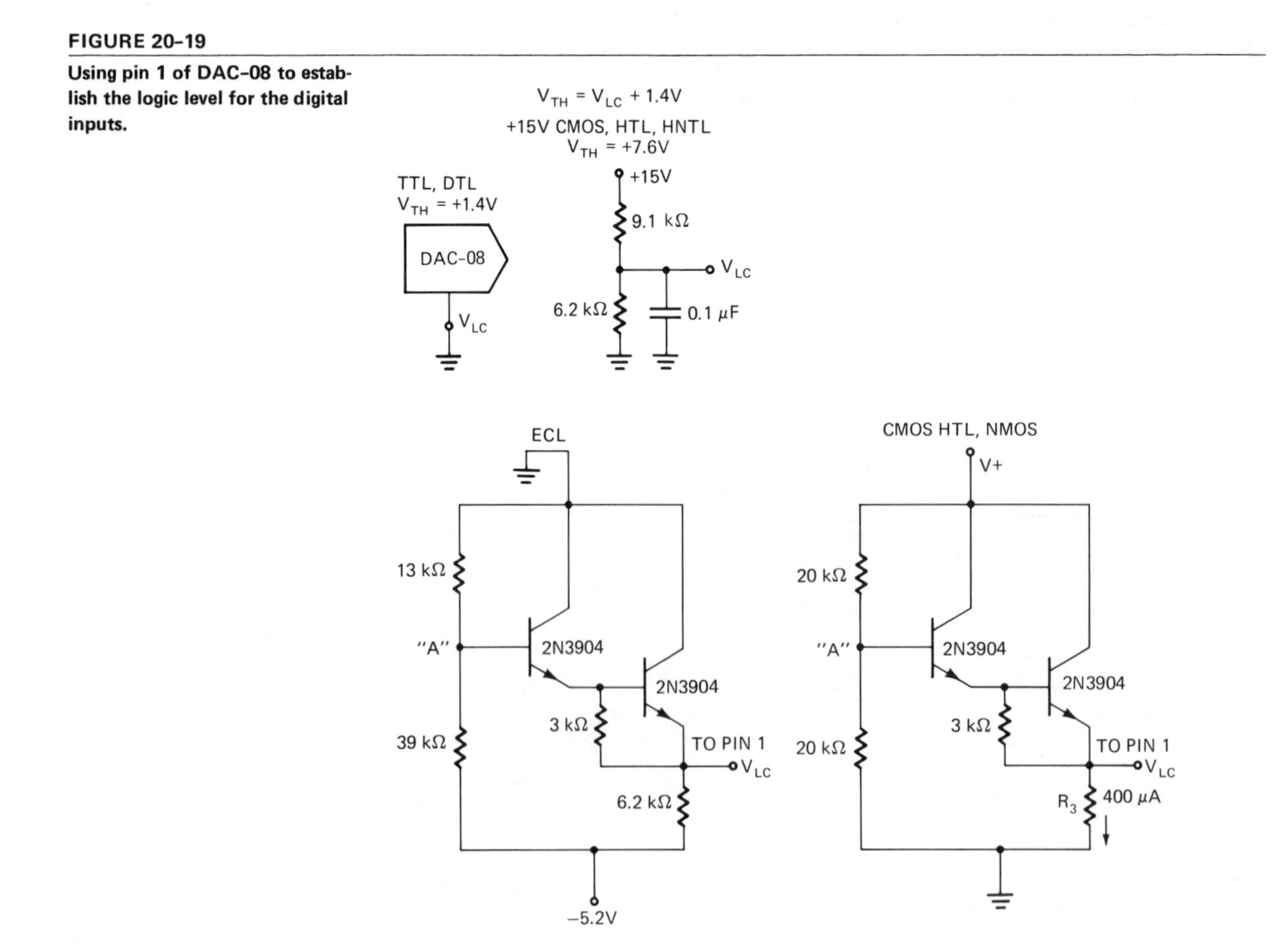

the current I_C (and the voltage $I_C R$) will increase with the V_{CE} increase. The scope will trace through a single characteristic curve for I_B = 10 μA. If, instead of a single curve, we would like to have 16 curves of increasing levels of I_B, the first addition is to square up the rectified sine wave so they can trigger a counter. This can be accomplished with a Schmitt trigger. Figure 20-22 shows the complete curve tracer. The output of the Schmitt trigger will cycle the counter continuously from 0 through 15.

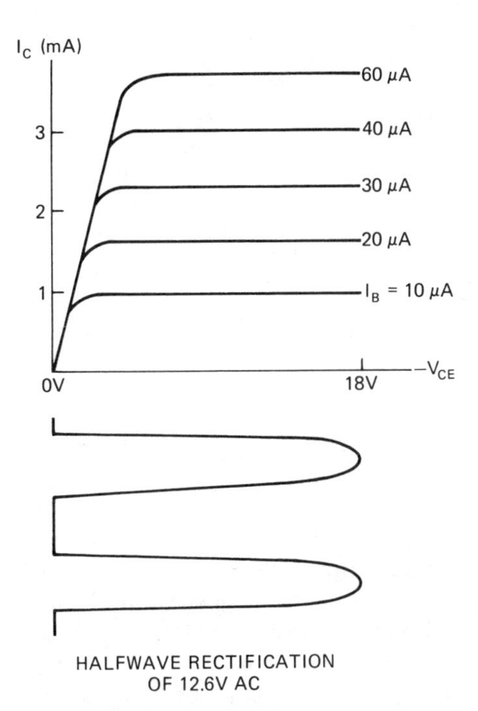

FIGURE 20-20

Halfwave rectified 12.6V AC used to sweep a transistor with V_{CE} of 0 to $18V_{PK}$.

A current source D/A can be readily adjusted for a particular level of μA per bit by varying V_S. If V_S were adjusted for 10 μA per bit, the tracer would produce characteristic curves for base currents (I_B), changing from I_B = 0 to I_B = 150 μA. The counter would increase one count and the base current would increase 10 μA after each half sine wave of the rectifier output, reaching the top curve of 150 μA at the count of 15 and returning to 0. The curves would retrace themselves again every 16/60 of a second.

20.9 Summary

There are many conditions in manufacturing and instrumentation that require a digital number to be converted to an analog equivalent. The device that performs this function is called a *digital-to-analog converter* (abbreviated D/A). The actual conversion circuit is a simple resistor network such as that shown in Figure 20-2. In this circuit, the LSB converts to a voltage of V/15—the resolution of the converter. If V = 15V, then the resolution is 1 volt per count. The range of the converter is the maximum number that can be set on the switches, or the maximum analog output voltage.

Another more popular form of resistor network is the *ladder network*. It has the advantage that only two values of precision resistors are needed. As Figure 20-3 shows, it requires only R and 2R. If a single value is available, the second

FIGURE 20-21

Transistor curve tracer for a single base current.

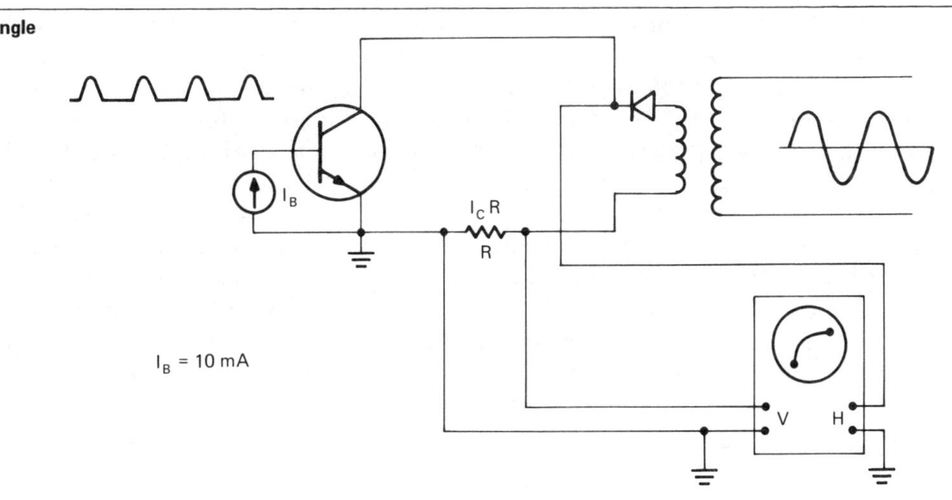

FIGURE 20-22

Curve tracer with a four-bit counter and D/A converter used to step the base current through 16 increasing levels.

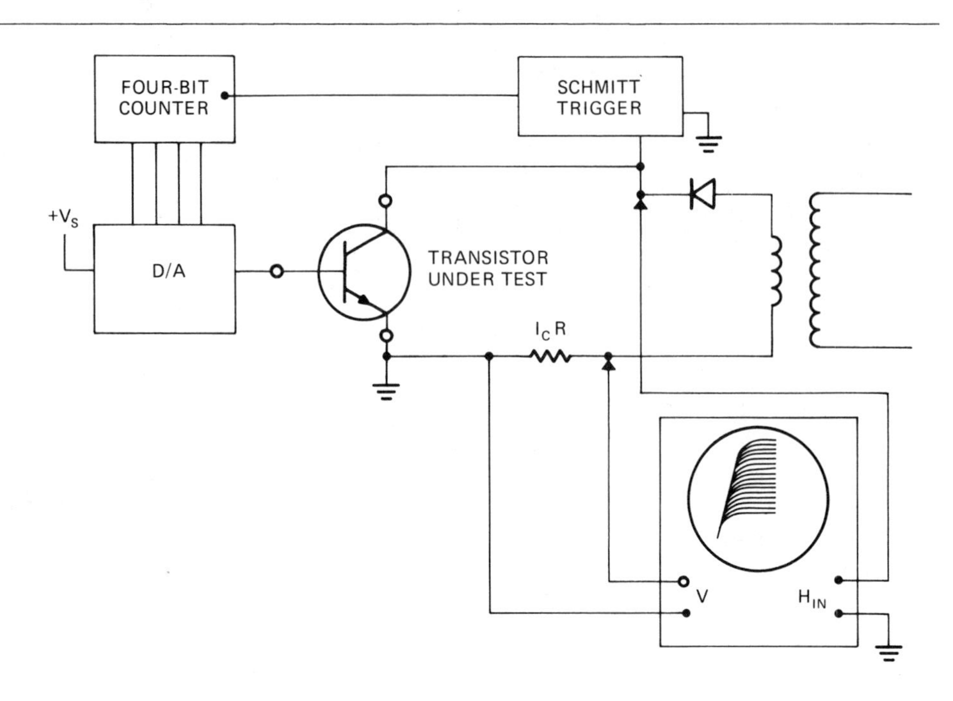

can be obtained by series or parallel connections. The voltage used with the ladder network must be one increment higher than with the resistor network.

At the end of each 2R resistor are switches that can be set in logic 1 or 0 position. Each value set on the resistors forms a different voltage divider between the applied voltage and ground. If the resistors are precise in value, the output between V_A and ground is always proportional to the binary number set on the switches.

Manual switches will rarely be used, as most conversions are made at a high rate of speed. Electronic switches of bipolar or FET construction must be connected in place of each toggle switch. The simplest form of electronic switching is a single transistor inverter switch, as shown in Figure 20-8. Because of the inversion, the complement of the number must be applied to the base resistor leads. This is seldom a problem, as the digital number is usually stored in a register from which complements are readily available. This switch introduces several sources of error. Errors due to variations in resistor R_C and variations in transistor β can be corrected by clamping the collectors to a precision power supply. This improves accuracy for the 1 levels applied to the top of the resistors but does not help the 0 levels, which are offset by $V_{CE\ SAT}$ of the transistor. Using both positive and negative supply voltages, precision switches are available with offsets as low as 5 mV.

A D/A circuit can be tested using a counter of the same bit size as the converter. Figure 20-10 shows a four-bit counter connected to a four-bit D/A converter. If an oscilloscope is connected to the output of the converter, then, as the counter cycles through the count 0 through 15, a staircase pattern should appear on the face of the oscilloscope, as shown in Figure 20-10. As Figure 20-11 shows, each step has a height equal to the resolution of the D/A converter. Each step has a width of a clock period. Any lack of uniformity in the height of the steps indicates an error or malfunction of the circuit.

The D/A circuits discussed thus far were *voltage source* D/A conversions. As the number of bits in the conversion increases, voltage source D/A either becomes inaccurate or requires voltages so high as to be impractical. *Current source* D/A conversion can provide better resolution and higher accuracy. As shown in Figure 20-12, this converter comprises current sources that are successively divided in half. The bit switches either apply the current to the output if it is 1 or ground it if it is 0. Whereas, in the voltage source D/A, the load resistor had to be very large in comparison with the network resistors ($R_L > R$), the current source D/A requires a low value of R_L to maintain accuracy. Figure 20-13 shows the transistor-diode switch and resistors that form the current source D/A.

Both current and voltage source D/A circuits are available in small modules. A popular eight-bit D/A converter is the DAC-08, shown in Figure 20-16. It uses a current divider network and can be used for either unipolar or bipolar operation.

An interesting application of the D/A converter can be found in the *transistor curve tracer*, shown in Figure 20-22. A rectified 12 volts is used to sweep the collector circuit of the transistor. This causes transistor current to increase from 0 through saturation; but at the same time, the rectified pulses are shaped by a Schmitt trigger and used to pulse a four-bit counter. The counter drives a D/A converter that steps the base current through 16 levels. Each level traces a characteristic curve on the scope.

Glossary

Digital-to-Analog Converter (abbreviated D/A): A circuit that converts a binary number to an analog voltage.

Questions

1. List the three essential elements of a digital-to-analog converter.
2. What are the three most important operating characteristics of a digital-to-analog converter?
3. What is an advantage of the R-2R ladder network of Figure 20-3 over the resistor network of Figure 20-2(a)?
4. Which specifications of a D/A converter can be tested statically and which must be tested dynamically?
5. What is the maximum number of steps obtainable with a four-bit D/A converter?
6. Which type of D/A converter achieves better resolution and higher accuracy—current source or voltage divider?
7. Does the DAC-08 D/A converter use a current or voltage divider network?
8. How is the reference current level into the DAC-08 established?
9. What relationship exists between input reference current, I_{REF}, and full-scale output current, I_{FS}?
10. As a digital input to a DAC-08 changes from 00000000 to 11111111, how does the value I_O change? How does the value $\overline{I_O}$ change?
11. Can the DAC-08 be used for ±1.0V operation as well as ±10V operation?
12. What function does the D/A converter perform in the curve tracer circuit of Figure 20-22?

Problems

20-1 A five-bit digital-to-analog resistor network has its largest resistor: R_0 = 10 KΩ. What would be the value of the smallest resistor in the network?

20-2 If the converter of Problem 20-1 were to have a resolution of 0.5 volts per count, what supply voltage would have to be used?

20-3 Draw a ladder network for a five-bit D/A converter using only 10 KΩ 1 percent resistors.

20-4 Describe the difference between voltage and current source D/A converters.

20-5 What errors result from using a single transistor inverter as a switch?

20-6 A three-bit D/A converter is being tested by a three-bit binary counter. Draw the oscilloscope pattern that should appear at the D/A output. Label the dimensions. The clock is 100 KPS; the resolution is 1 volt per count.

20-7 The current source D/A of Figure 20-13 has $R = 10\ K\Omega$, $+V_S = 10V$, $R_L = 600$ ohms. What is the resolution in mA per count? What is the resolution in mV per count?

20-8 If the load resistor of the converter in Problem 20-7, R_L, is reduced to 100 ohms, what would be the resolution in mV per count and mA per count?

20-9 Refer to Figure 20-17(a) and, for the typical values for TTL operation, calculate I_{FS}.

20-10 If the DAC-08 is wired for unipolar operation as shown in Figure 20-18(a), and the binary pattern is 10000011, determine E_O.

20-11 In Figure 20-21, why is the resistor not in the collector side of the circuit?

20-12 What are two functions of the rectified voltage in Figure 20-21?

20-13 What is the function of the counter and the D/A converter in Figure 20-22?

CHAPTER

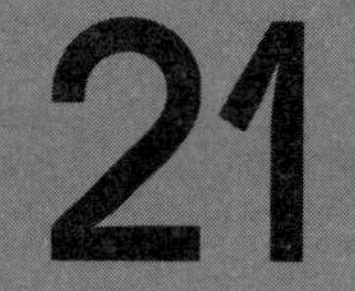

Analog-to-Digital Converters

Objectives

Upon completion of this chapter, you will be able to:

- Construct an analog comparator.
- Use integrated circuit comparators.
- Assemble analog comparators and logic circuits to form a simultaneous conversion A/D converter.
- Use a counter, D/A, and analog comparator to form an A/D converter.
- Use successive approximation registers.
- Assemble a register, D/A, and analog comparator to provide a successive approximation A/D converter.
- Assemble ring counters and other logic to generate the control signals for a successive approximation A/D converter.
- Assemble and use sample-and-hold circuits.

21.1 Introduction

The digital-to-analog converter discussed in Chapter 20 began with a digital number input and converted it to an analog voltage of so many volts or millivolts per count. For every increase of 1 in the digital number, there is an exact incremental increase in the analog output. For every decrease of 1 count in the digital number, there is an exact incremental decrease in the analog voltage. We can predict the analog output for a given input number by multiplying the volts per count times the binary input number. The analog-to-digital converter, on the other hand, receives an analog (DC) voltage on its input and provides a digital number on the output. For each increment of DC input voltage, there is 1 count added to the digital number on the output.

In Figure 21-1, a four-bit A/D converter with a resolution of 1 count per 200 mV receives an input of 2.1V. The voltage is high enough for a digital output of 1010, which requires 2V. We cannot be sure whether the leftover 100 mV will provide another count and raise the output to 1011. For normal accuracy, we expect that somewhere between 2.1V and 2.3V the digital number will change from 1010 (2.0V) to 1011 (2.2V). This sets a practical limit to the accuracy of conversion at ± 1/2 LSB. To this we must add other errors in the system design. This limitation, however, does not rule out the possibility of an operation converted to digital having an overall accuracy much higher than that of an equivalent analog system. The resolution of an A/D converter is the analog value of the LSB. It can be determined by dividing the full-scale voltage by $(2^n - 1)$, where n is the number of bits in the converter.

21.2 Analog Comparator

21.2.1 Function

The techniques used to convert analog voltages to digital numbers all use one or more analog comparators. Figure 21-2 shows the logic symbol for an analog comparator. The inputs are two analog voltages. The output is a binary 1 or 0 level. In this case, a binary 1 occurs on the output if $A > B$. This can, of course, be changed by merely interchanging the input leads to obtain a binary 1 for $B > A$.

Figure 21-3 is a typical schematic diagram of an analog comparator. Two analog voltages are applied to inputs A and B. If $B > A$, the current I_{C1} will be greater than I_{C2}. The emitter of Q_3 will be more positive than the base. There will be no current through either Q_3 or Q_4 and no voltage drop across the output load resistor. This produces a binary 0 output. If $A > B$, the current I_{C2} will be greater than I_{C1}. This will forward-bias the base-to-emitter junction of Q_3. The resulting voltage drop across R_{C3} will turn on Q_4. The collector current I_{C4} will cause a 1-level voltage drop across the resistor, R_4, producing a binary 1 output for $A > B$.

21.2.2 Integrated Circuit Comparators

Integrated circuit comparators come in both TO-5-type metal packages and standard 14-pin DIP. Typical comparators are shown in Figure 21-4. The LM319 is a dual comparator. The output is open-collector, rendering it compatible with either TTL or CMOS circuits. The inputs

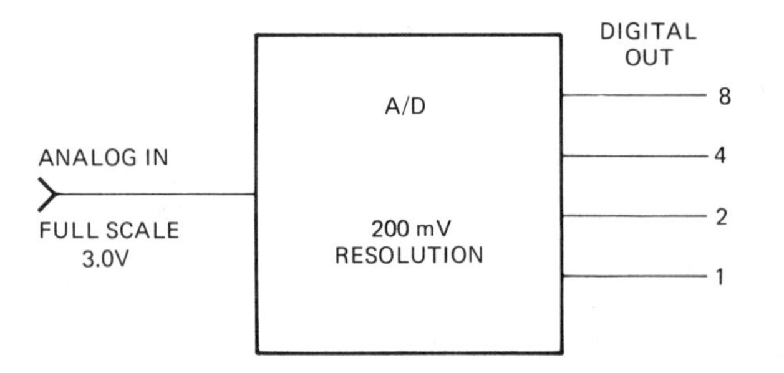

FIGURE 21-1

Analog-to-digital converter, which receives an analog voltage input and produces a proportional digital output.

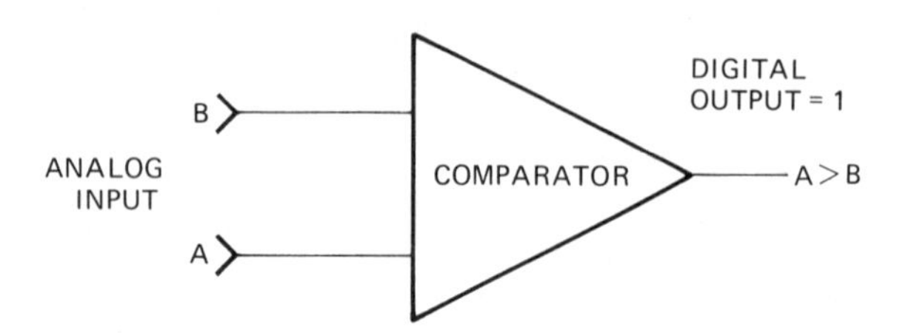

FIGURE 21-2

Analog comparator, which receives two analog inputs, A and B. The output will be a digital 1 level if $A > B$, a digital 0 if $A < B$.

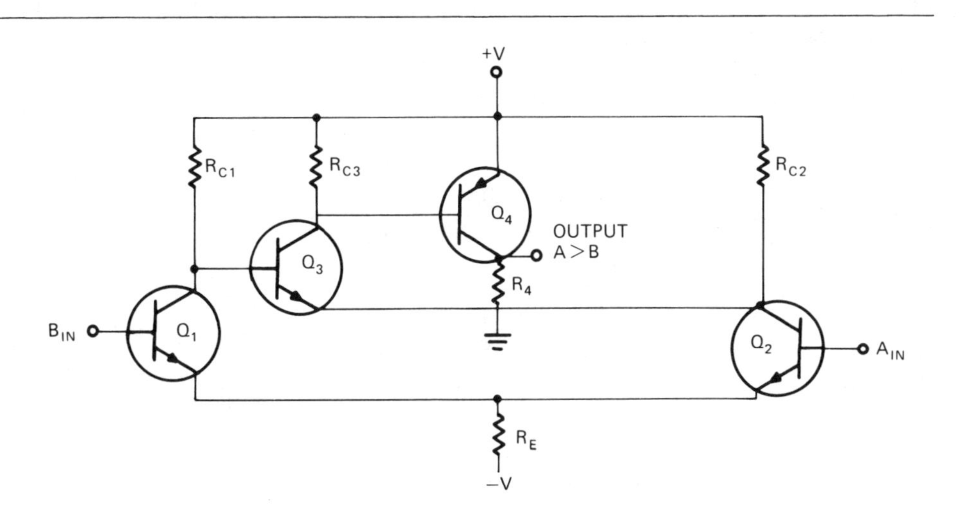

FIGURE 21-3

A typical bipolar comparator circuit.

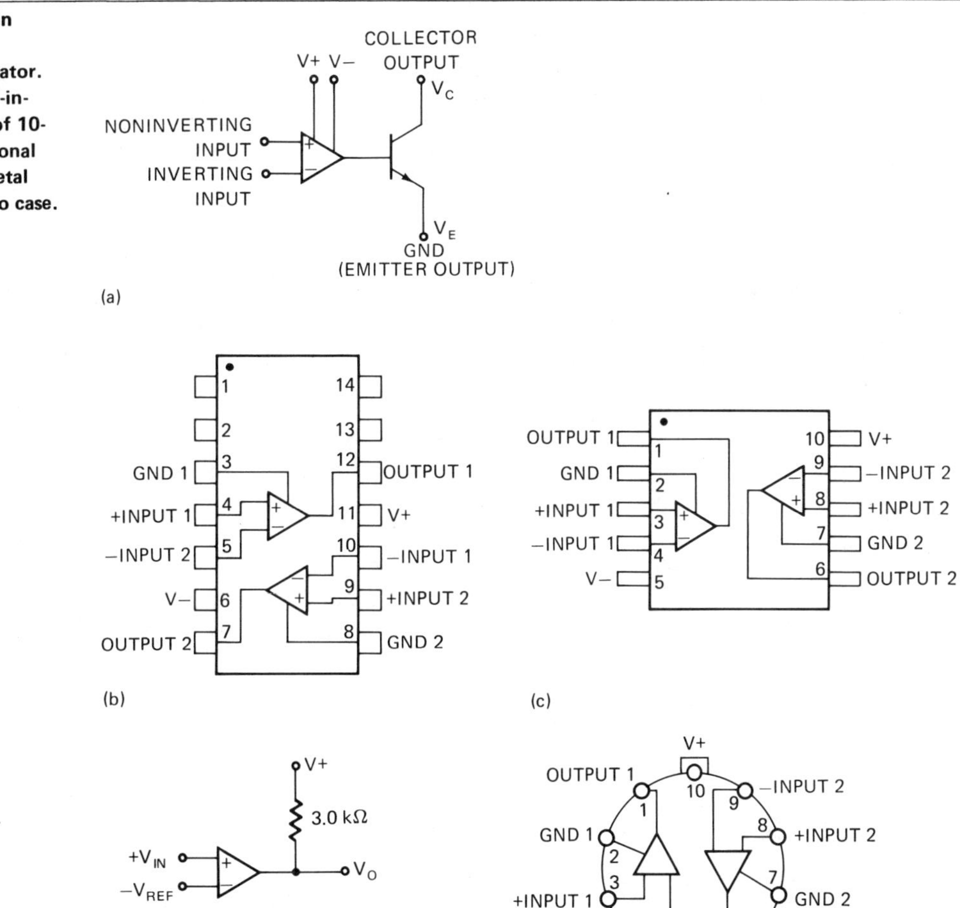

FIGURE 21-4

Basic comparators, housed in TO-5 14-pin DIP or 10-pin flat packs. (a) Basic comparator. (b) Top view of 14-pin dual-in-line package. (c) Top view of 10-pin flat package. (d) Functional diagram. (e) Top view of metal can, with pin 5 connected to case.

are marked + and –. The output remains high as long as the + input is slightly more positive than the – input. Some comparators use a pull-up resistor on the output as shown in Figure 21-4(d).

This is true regardless of which type of input, + or –, is used for reference. The LM319 has a minimum gain of 8V/mV so that the output will switch within a few millivolts of the reference voltage. The analog voltage range is dependent on the supply voltage used. With supply voltages V+ = 15V and V– = –15V, the input voltage range is ±13V. The output switching time is about 80 nanoseconds. Each comparator has a separate ground allowing for signal ground isolation from power ground.

The LM339 is a quad comparator. Its output is also open-collector. Although it was specifically designed for single-power-supply operation, it can be employed with dual supply such as ±15V. Under this condition, however, the output will switch between the voltage at the top of the pull-up resistor and the negative supply, as shown in Figure 21-5. This creates a problem of interfacing to logic circuit voltage levels normally 0V to 5V for TTL, or 0V to as high as 15V for CMOS. The solution may be as simple as an additional resistor, but the RC time constant between that resistor and the logic circuit input will have an effect on the speed of the circuit.

21.3 A/D Simultaneous Conversion

Simultaneous conversion is the most rapid method of A/D conversion. As Figure 21-6 shows, it requires an analog comparator for each count in the output. A three-bit A/D converter having an output of 0 through 7 requires seven analog comparators. A reference voltage, V_R, is divided repeatedly so that each successive reference input to the comparator is $V_R/8$ higher than the one below it. When the analog voltage is applied to V_A, every comparator with a reference voltage input below the analog input will produce a binary 1 output. Those comparators with V_R inputs higher than V_A will produce binary 0 outputs. The gates and inverters are arranged to steer the correct registers to set on the clock pulse.

For an A/D of low resolution, simultaneous conversion is the simplest and fastest method. If the A/D of Figure 21-6 were expanded by one

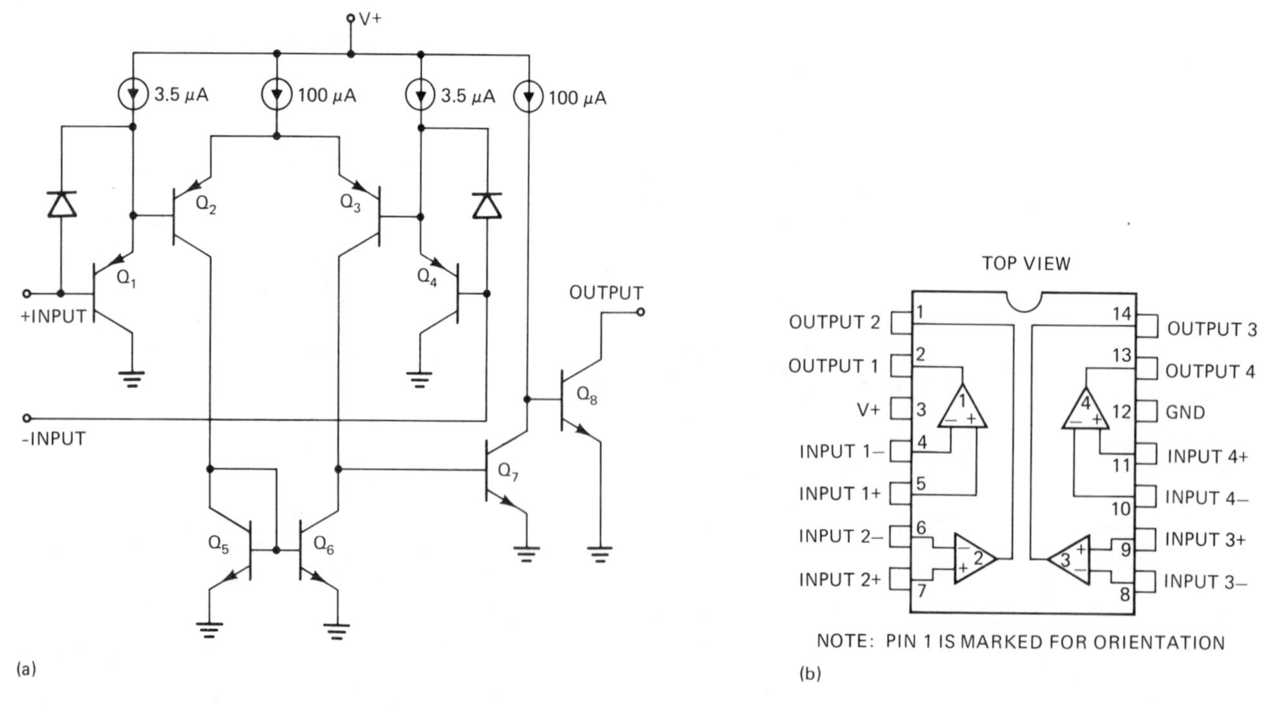

FIGURE 21-5

Integrated circuit quad comparator Am339 (LM339), designed for analog inputs between V+ and ground. (a) Schematic diagram. (b) Connection diagram.

digital bit, the division of the reference voltage would be $V_R/16$, and a total of 15 comparators would be needed along with the added digital circuits. Each added bit of resolution doubles the number of comparators required. The circuit of Figure 21-6 is also called a *flash converter.*

21.4 Counter-Controlled A/D Converter

The simplest method of high resolution A/D conversion uses three main elements, as the block diagram of Figure 21-7 shows—the counter, the digital-to-analog converter, and a single analog comparator.

The cycle starts with counter reset to 0. This produces a D/A output voltage of 0. If there is an analog voltage (V_A) applied to comparator input A, then B is not greater than A and the comparator output of binary 0 will enable the clock pulses through the NOR gate. The counter will count, and with each count the D/A output will increase one increment with each count. The count will continue until the increasing D/A output results in a binary 1 out of the comparator, indicating B > A. The 1 on the NOR gate will inhibit passage of clock pulses, stopping

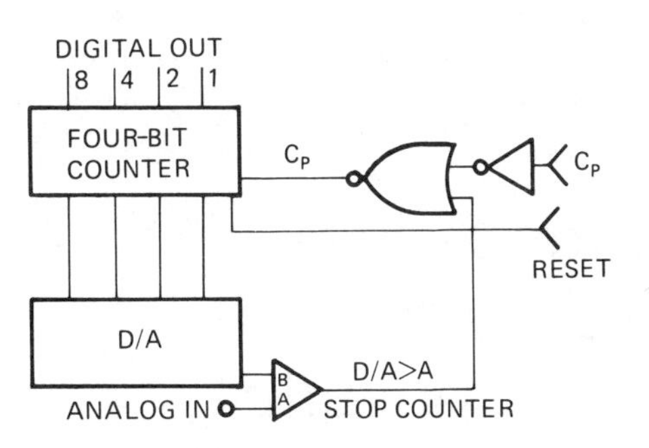

FIGURE 21-7

Analog-to-digital conversion, accomplished by having a digital-to-analog converter controlled by a counter.

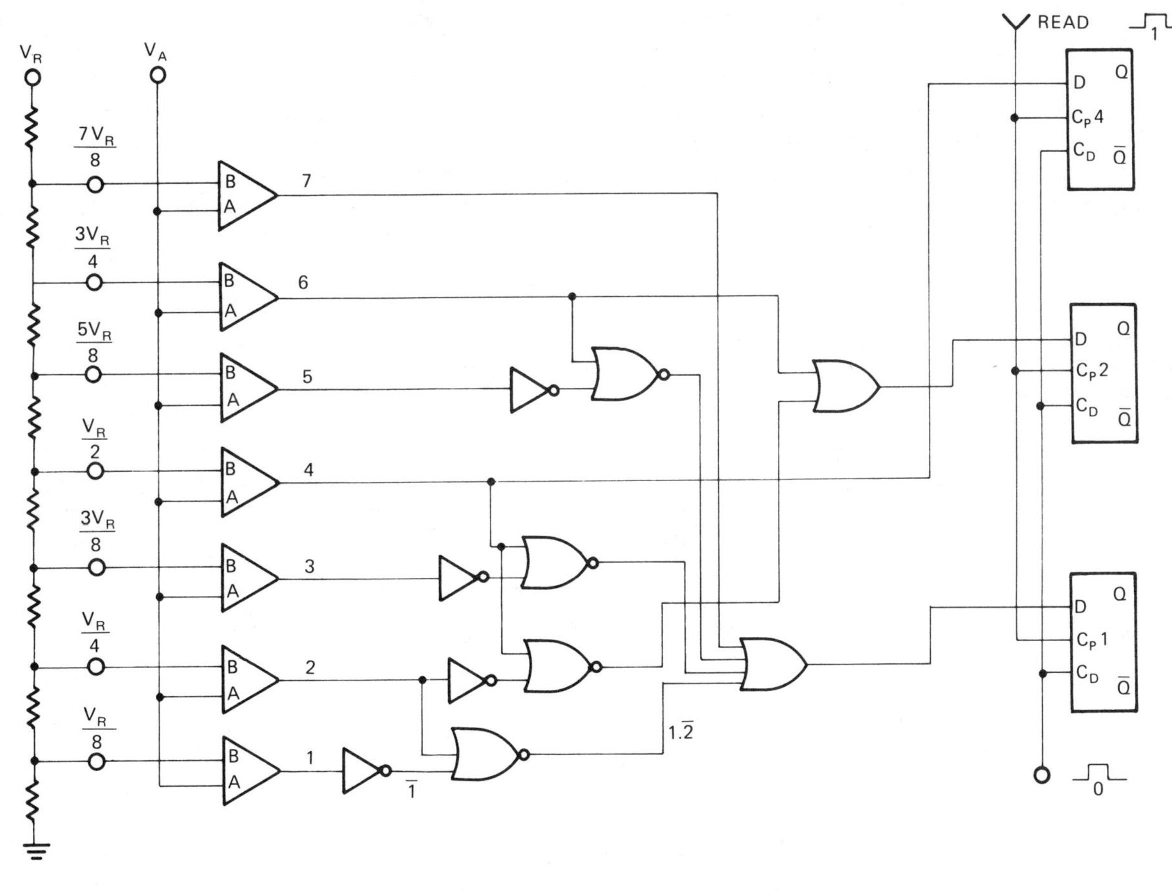

FIGURE 21-6

A three-bit simultaneous conversion analog-to-digital converter. Conversion requires only the time needed for a reset-and-read pulse.

the counter. The output is then available in parallel form from the Q outputs of the counter flip-flops. This method requires fewer circuits and control signals but requires as many as 2^n clock periods to complete a conversion using an n-bit counter. The conversion time can be reduced substantially if an up-down counter is used and the converter is allowed to track the analog voltage continuously by counting up when A > B and down when B > A. In this system, the comparator operates the up-down control of the counter.

21.5 Successive Approximation A/D Converter

21.5.1 Operation

Successive approximation is the method capable of the highest speed and accuracy. Figure 21-8 shows a block diagram of a four-bit successive approximation A/D converter. The three main elements of the converter are the four-bit register, which can be set and reset one bit at a time beginning with the MSB, a digital-to-analog converter, and the analog comparator. Besides these main elements, an elaborate clock and control waveform generator is needed. As we noted in Section 19.1, we will proceed with the design of the system, drawing in control signals where needed, and assuming the ability to generate any such waveforms in the control waveform section. The following steps show the order in which successive approximation occurs:

Step 1. (a) Turn on 2^3 (8) flip-flop. (This places 1/2 of the full-scale voltage on B.) (b) Compare output of D/A with the analog input. If D/A > A, then reset the 8 bit. If not, inhibit reset (leave 8 flip-flop set).

Step 2. (a) Turn on the 4 bit. (b) Compare output of D/A with the analog input. If D/A > A, then reset the 4 bit (reset of 8 is inhibited). If not, inhibit reset (leave 4 flip-flop set).

Step 3. Repeat Step 1, but use 2 flip-flop.

Step 4. Repeat Step 1, but use LSB flip-flop.

21.5.2 Clock and Delayed Clock Timing

For this conversion, use of clock and delayed clock is ideal. Flip-flops can be set with succeeding clock pulses and reset with corresponding delayed clock pulses. Figure 21-9 shows the 2^3 (8) flip-flop and the necessary inputs for its trial-and-error operation. Figure 21-10 shows the waveforms appearing at various points of Figure 21-9. The decision-making ele-

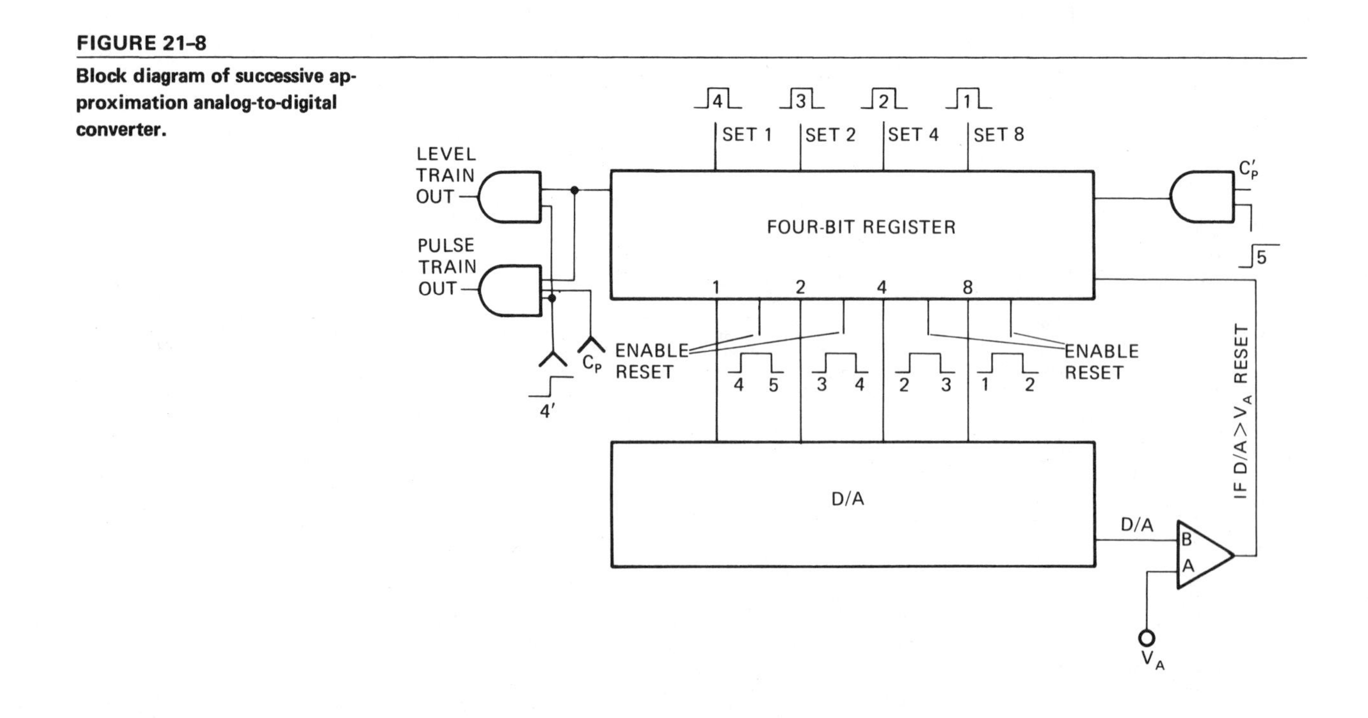

FIGURE 21-8

Block diagram of successive approximation analog-to-digital converter.

FIGURE 21-9

MSB flip-flop of a successive approximation A/D converter showing control waveforms needed. Flip-flop is set by clock pulse 1, producing V/2 at the D/A comparator input. If D/A < A, clock pulse 1 delayed is enabled to reset the MSB flip-flop.

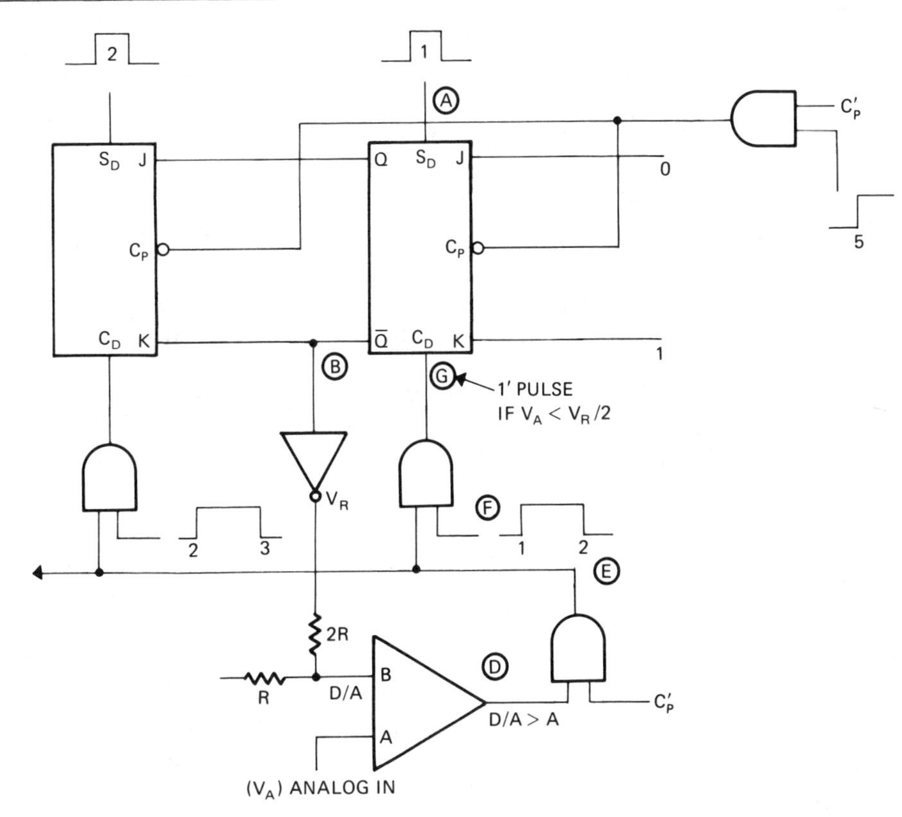

FIGURE 21-10

Waveforms of the MSB flip-flop shown in Figure 21-9. (a) For the condition $V_A > V_R/2$. (b) For the condition $V_A < V_R/2$.

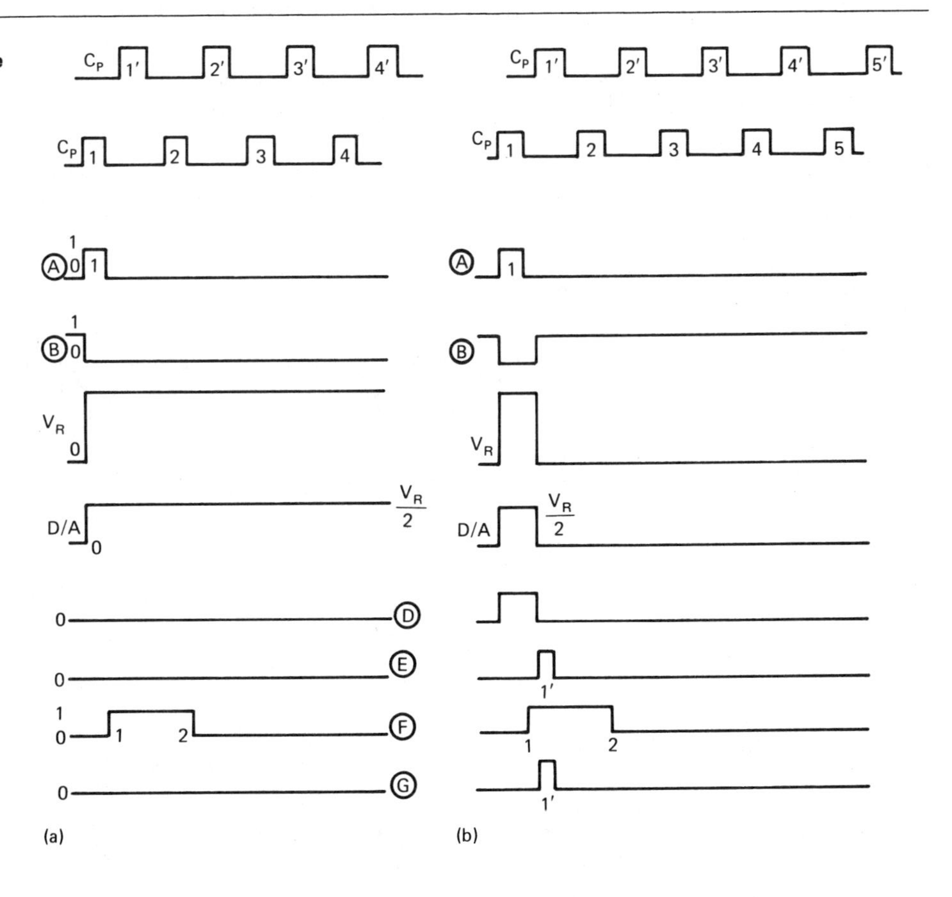

ment of this system is the analog comparator. It must determine whether the D/A voltage generated by each added register bit either is too much or is needed to represent the analog input. If, in comparing D/A with the analog input, it determines that D/A is greater than A, it produces a 1 level, which enables the clock prime pulse to reset the last flip-flop. If D/A remains less than or equal to the analog input, the comparator output remains at 0 level, inhibiting the clock prime pulse from resetting the flip-flop.

The output from successive approximation is available as a parallel output, or, unlike the other methods described here, it may be shifted out as a serial number. If it is shifted out in serial—an ideal method if it is to be transmitted to a remote point—there will be no need to reset at time 0, for after the last bit is shifted out, the register bits will all be at 0. If the output is taken in parallel, then a zero reset pulse must be ORed into the clear inputs of each flip-flop.

21.5.3 Timing and Control Circuits

The counter-controlled A/D of Figure 21-7 and the successive approximation A/D of Figure 21-8 compare closely in number of circuits, but there is a big difference in the number of timing signals and control waveforms. The counter-controlled A/D requires only a reset pulse and clock line. The successive approximation A/D must be supplied with not only reset and clock pulses but also delayed clock pulses, a set pulse for each bit, and an enabling waveform for each bit. Finally, if serial output is desired, a waveform is needed to enable the clock pulses to shift out the register after conversion is completed. We have, as described in Section 19.1, shown merely the active elements of the circuit, and we have drawn in the timing and control waveforms with full confidence that they can be generated elsewhere in the system. It would be instructive now for us to see exactly how these are to be generated.

The four successive clock pulses needed for successive setting of the A/D register bits are generated with the help of a register similar to a ring counter but not having the outputs crossed back to the inputs. This provides for only one set of clock-period-width gates between each 0 time reset. As Figure 21-11 shows, the register is clocked by delayed clock pulses, producing clock-period-width enable signals that bracket a single clock pulse. Normally, the first flip-flop of a ring counter is preset while the remaining flip-flops are reset, but the J-K ICs used in this circuit have no asynchronous preset input. That problem is solved by simply crossing over the output leads of the first flip-flop. A four-bit

FIGURE 21-11
Register and gates used to develop set pulses for successive setting of the four registers of the A/D converter.

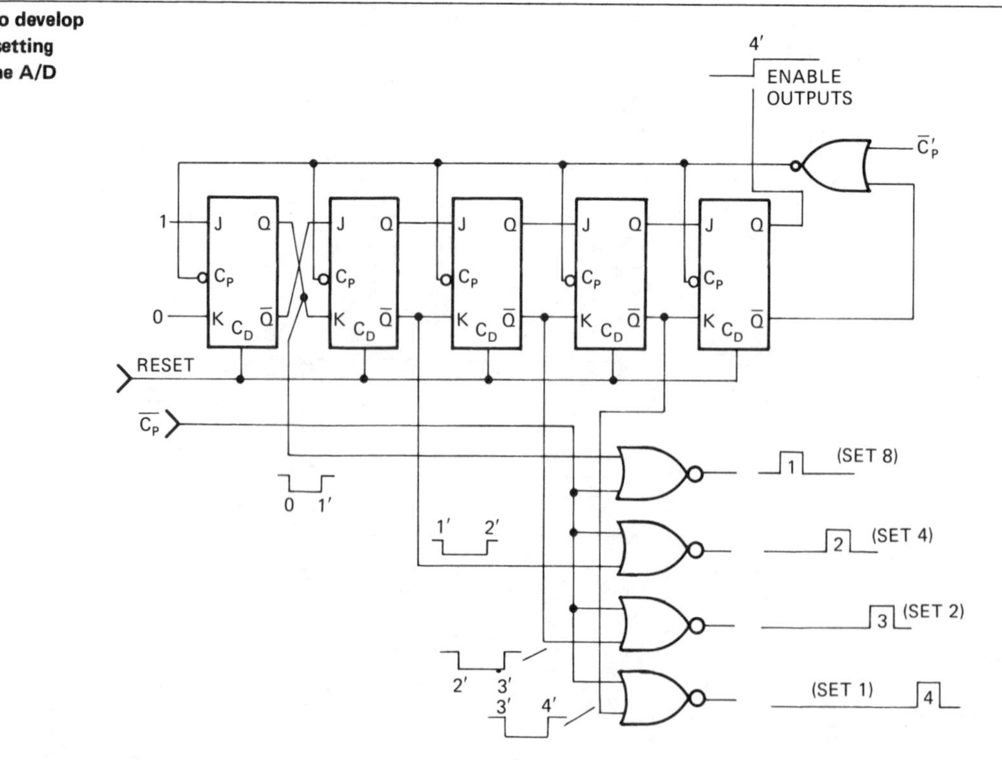

simultaneous conversion is complete on the leading edge of delayed clock 4′. The last flip-flop in the ring counter is set on the trailing edge of 4′. The outputs of this flip-flop are used to inhibit further clock pulses to the ring counter and also to enable the A/D converter outputs.

The clock-period-wide waveforms used to enable reset of the correct flip-flop when D/A > V_A are generated in a register like the one shown in Figure 21-12. This register also functions like a ring counter without the outputs crossed back to the inputs. It is triggered by the clock pulse line and, therefore, produces clock-period-width gates that are ideally timed for enabling delayed clock pulses to reset the correct register bits when the comparator determines that D/A > A. The last flip-flop, which sets on the trailing edge of clock pulse 5, produces a signal enabling the C_P' line to shift the A/D register through the serial outputs beginning $C_P'5$.

21.5.4 Conversion Waveforms

Figure 21-13 shows the timing diagram of the serial outputs and the Q outputs of each of the

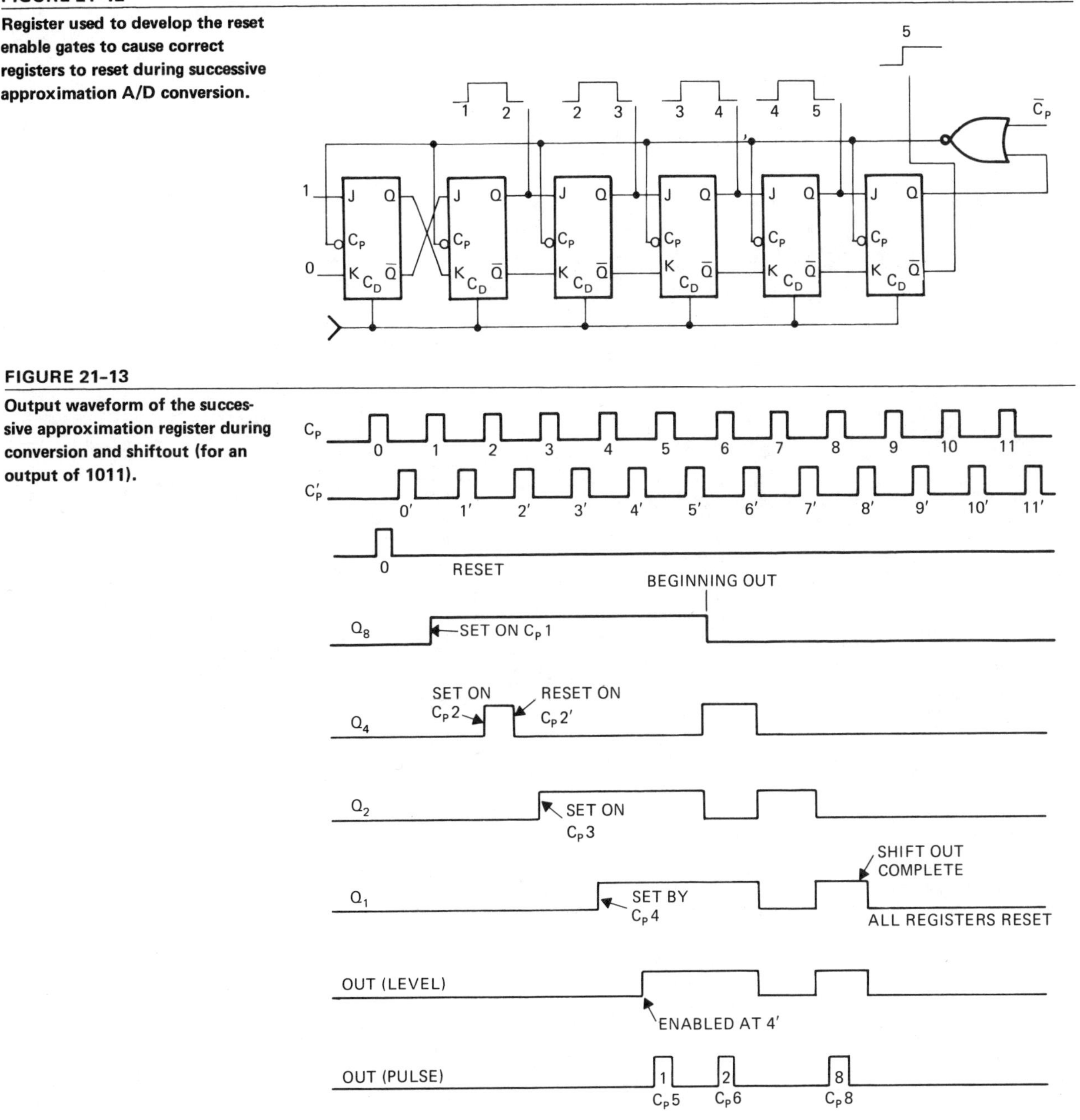

FIGURE 21-12
Register used to develop the reset enable gates to cause correct registers to reset during successive approximation A/D conversion.

FIGURE 21-13
Output waveform of the successive approximation register during conversion and shiftout (for an output of 1011).

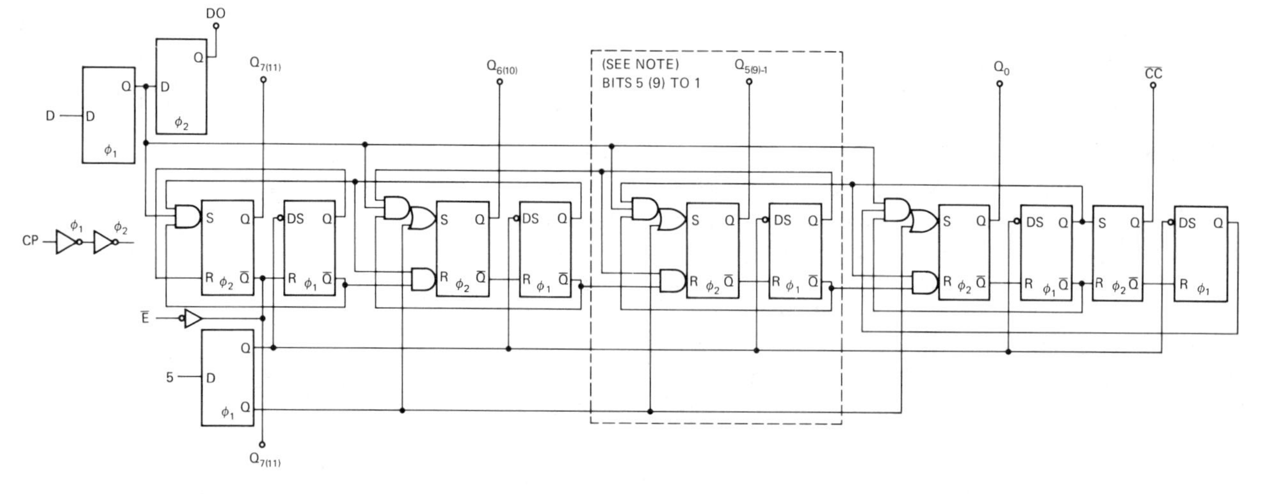

NOTE: CELL LOGIC IS REPEATED FOR REGISTER STAGES. Q_5 TO Q_1 Am2502, Q_9 TO Q_1 Am2504.

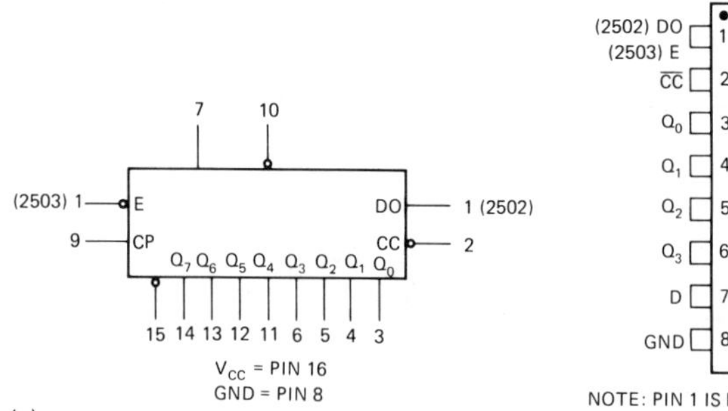

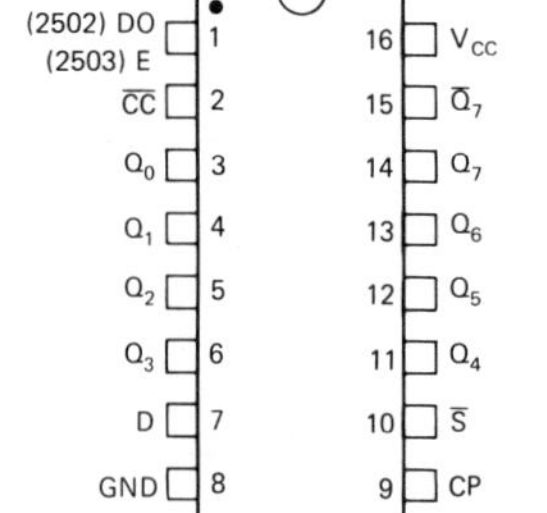

NOTE: PIN 1 IS MARKED FOR ORIENTATION.

(a)

TIME	INPUTS			OUTPUTS									
t_n	D	$\overline{S}$	$\overline{E}$	D_0	Q_7	Q_6	Q_5	Q_4	Q_3	Q_2	Q_1	Q_0	$\overline{CC}$
0	X	L	L	X	X	X	X	X	X	X	X	X	X
1	D_7	H	L	X	L	H	H	H	H	H	H	H	H
2	D_6	H	L	D_7	D_7	L	H	H	H	H	H	H	H
3	D_5	H	L	D_6	D_7	D_6	L	H	H	H	H	H	H
4	D_4	H	L	D_5	D_7	D_6	D_5	L	H	H	H	H	H
5	D_3	H	L	D_4	D_7	D_6	D_5	D_4	L	H	H	H	H
6	D_2	H	L	D_3	D_7	D_6	D_5	D_4	D_3	L	H	H	H
7	D_1	H	L	D_2	D_7	D_6	D_5	D_4	D_3	D_2	L	H	H
8	D_0	H	L	D_1	D_7	D_6	D_5	D_4	D_3	D_2	D_1	L	H
9	X	H	L	D_0	D_7	D_6	D_5	D_4	D_3	D_2	D_1	D_0	L
10	X	X	L	X	D_7	D_6	D_5	D_4	D_3	D_2	D_1	D_0	L
	X	X	H	X	H	NC	NC	NC	NC	NC	NC	NC	NC

H = HIGH VOLTAGE LEVEL
L = LOW VOLTAGE LEVEL
X = DON'T CARE
NC = NO CHANGE

NOTE: TRUTH TABLE FOR Am2504 IS EXTENDED TO INCLUDE 12 OUTPUTS.

INPUTS: CLOCK, START, DATA
OUTPUTS: Q_7, Q_6, Q_5, Q_4, Q_3, Q_2, Q_1, Q_0, CONVERSION COMPLETE, DO

(b)

FIGURE 21-14

Successive approximation register in integrated circuit form. (a) Logic diagram, logic symbols, and pin-out of the Am2502, an eight-bit SAR supplied in a 16-pin DIP circuit. (b) Truth table and timing diagram of Am2502 operation. Operation is with inverted outputs.

four A/D register flip-flops for a conversion of 2.2V DC to a binary number 1011. The bottom two waveforms would occur at the output of the two AND gates shown in Figure 21-8.

21.6 Successive Approximation Registers

Successive approximation registers for A/D converter application are available in integrated circuit form. Figure 21-14(a) shows the Am2502 SAR. Unlike the register explained in Section 21.5, the 2502 operates in complement form. Since most integrated circuit DACs work on active low functions or, like the DAC-08, have the option of inverted outputs (I_O and $\overline{I_O}$), the Am2502 can often be used without resorting to inverters on the outputs Q_0 through Q_7. A clock input controls the conversion rate and, although the 2502 may function with a clock frequency as high as 25 MHz, the maximum frequency used will normally be limited by the DAC with which the SAR is employed.

As Figure 21-14(b) shows, conversion begins with the start S input being held low for at least one clock period. On the rising edge of the first clock pulse following the active S input, the SAR resets, causing all outputs to go high except Q_7. After the settling time of the DAC, the comparator output of the A/D is fed back to the D input which, on the rising edge of the second clock pulse, is clocked into both D_0 (the serial output register) and D_7 (the MSB register). On the same clock edge, Q_6 goes low. Again the comparator output is fed back to D and is clocked into D_0 and D_6. This continues for eight clock pulses, with the comparator output being clocked into successively lower registers, resulting in the digital equivalent of the analog voltage being stored and available as a parallel output from Q_0 through Q_7. An equivalent serial output will have occurred on D_0 during the eight clock periods. Eight clock periods after the start pulse, conversion complete (CC) output goes low.

Figure 21-15 shows a typical connection of

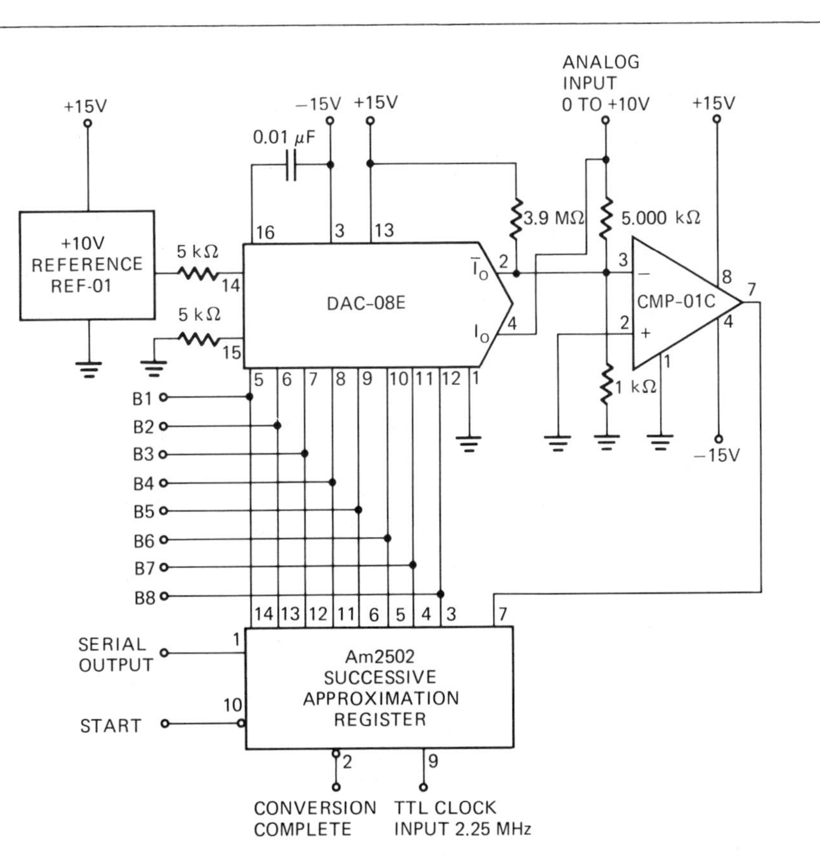

FIGURE 21-15

Complete eight-bit A/D converter using the Am2502 SAR and the DAC-08. The resistors are selected for an analog input of 0V to 10V. The digital output can be obtained in parallel from B_1 through B_8 or in serial from pin 1 of the DAC-08.

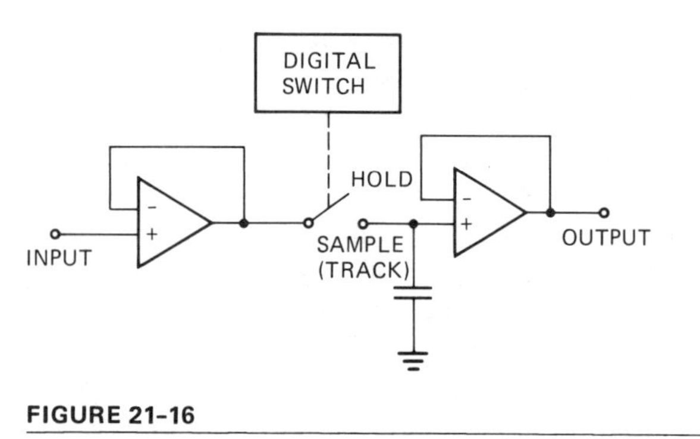

FIGURE 21-16

Simplified diagram of a sample-and-hold circuit.

IC components using the SAR, eight-bit DAC, and a comparator. The Am2504 is a 12-bit SAR that can be used with 12-bit DACs. For successive approximation in excess of 12 bits, these devices can be cascaded.

21.7 Sample-and-Hold Amplifier

Connecting the analog voltage to be converted directly to the A/D comparator may have several disadvantages, particularly if the analog voltage is changing rapidly. The stability of the conversion may be improved by using a sample-and-hold circuit. This type of circuit makes a rapid decision as to the analog level and stores that level on a capacitor throughout the time required for the A/D to make the conversion.

Figure 21-16 is a simplified diagram of a sample-and-hold amplifier. At the beginning of the conversion cycle, the switch is turned on and the capacitor is charged to the level of the input. The input buffer is a voltage follower that presents an extremely high impedance to the analog source yet has a very low output impedance and charges or discharges the capacitor with a very low time constant. When the circuit is switched to the hold mode, the capacitor is then loaded with the extremely high impedance of the output buffer. The result is that the capacitor approaches the ideal analog memory condition by charging or discharging instantly to the exact analog level during the sample period and receiving an imperceptible amount of discharge during the hold period. (The term *track* is commonly used when the sampling time is greater than the hold time.)

Figure 21-17 shows the SMP-10 sample-and-hold circuit. The sample-and-hold control operates with TTL threshold if pin 13 is grounded. Very high-speed switching is accomplished through the use of a "super charger," which supplies a high charging current until the voltage on the hold capacitor nears the final value, at which point the low current diode bridge completes the final charge. With supply voltages of ±15V, the input voltage range is about ±11V. The device is designed for use with a 5000 pF capacitor. The null inputs are used to cancel out both offset voltage errors and charge transfer errors. Figure 21-17(c) shows typical connections of the SMP-10.

21.8 Azimuth and Elevation Data Corrector

A typical application of the A/D converter is found in some radar sets. The information received from a large modern radar set is, in many cases, fed to high-speed digital computer circuits. An important part of this information is the radar antenna direction in the form of azimuth (compass direction) and elevation (angle above the horizon) pedestal readouts. If the target being tracked by the radar is stationary, the antenna direction and the target direction are the same. If the target is moving, the antenna will lag behind the target an amount dependent on the speed of the target. To track the target, error voltages must be created that are proportional to the amount that the antenna is off center from the target. This is accomplished by using a four-quadrant feed horn, as shown in Figure 12-18(a). If the energy reflected by the target comes back parallel to the focal axis, then each horn receives an equal amount of the reflected energy, and the error voltage formulas of Figure 21-18(b) equal 0. If the target is moving, the reflected energy will be offset from the focal axis and an error voltage for azimuth, elevation, or both will be created. These are used to control the pedestal motors that rotate and elevate the antenna. At the same time, these analog voltages can be converted to digital and added to the azimuth and elevation digital data from the pedestal to provide exact target direc-

FIGURE 21-17

Precision Monolithics SMP-10 sample-and-hold circuit. (a) Pin connections. (b) Functional diagram. (c) SMP-10 with null adjustment and hold capacitor connected.

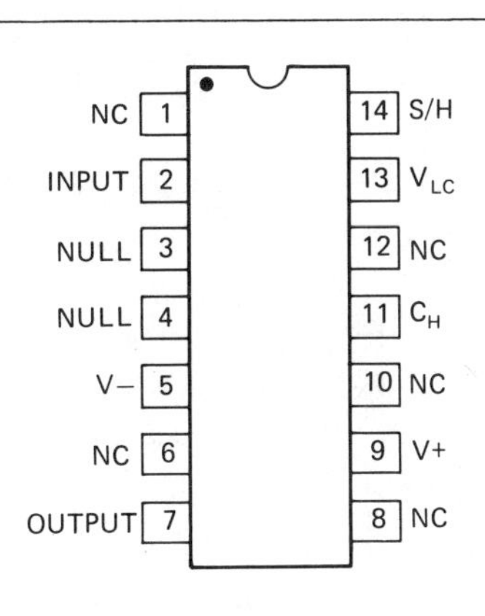

(a)

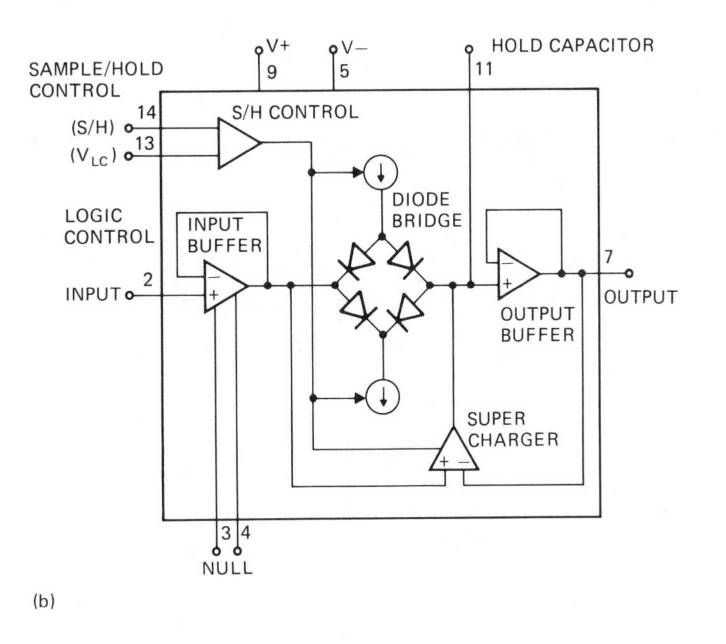

(b)

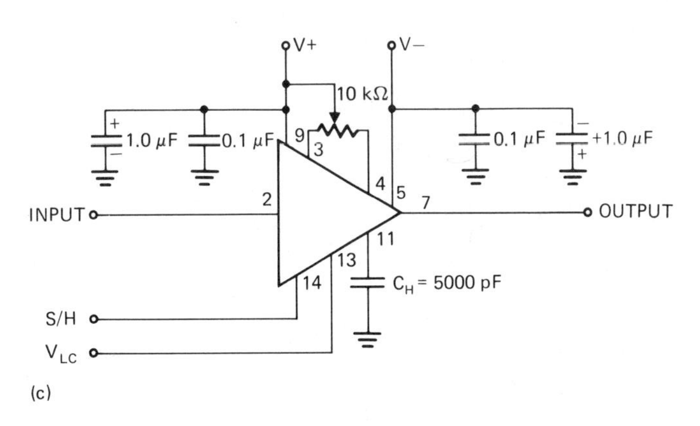

(c)

FIGURE 21–18

Typical application of A/D converter. (a) Radar antenna feed horn produces analog voltage proportional to antenna lag behind the target. (b) This voltage can be converted to digital for use in the radar computer using error voltage formulas.

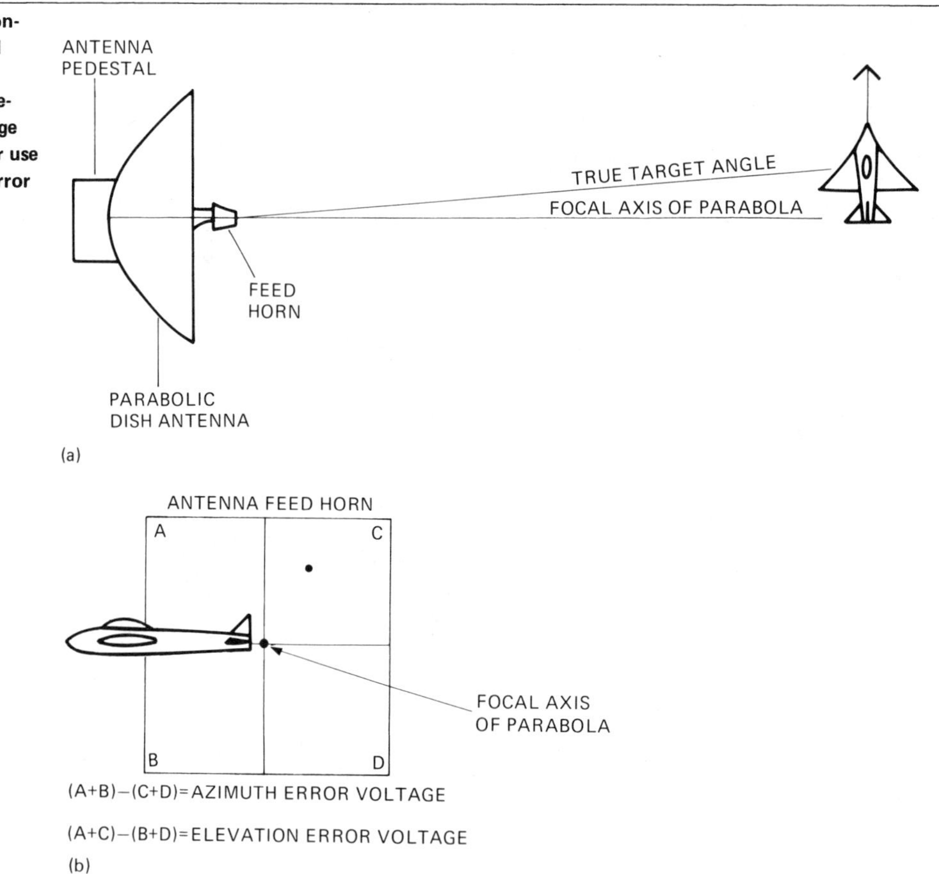

FIGURE 21–19

Block diagram of an azimuth digital data corrector.

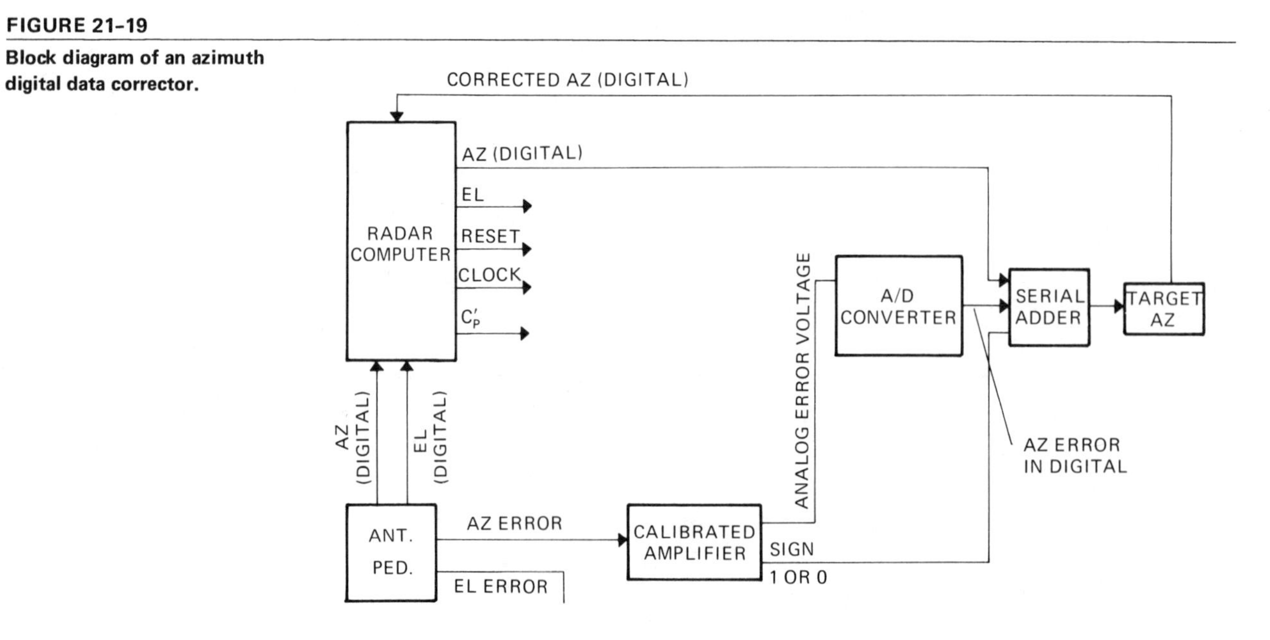

tion. They can also be used as target speed vectors. Figure 21-19 is a block diagram of an azimuth data corrector.

21.9 Summary

The *analog-to-digital converter* (A/D) performs a reverse function of the D/A we discussed in Chapter 20. The A/D receives an analog or DC voltage at its input and converts it to a digital number. The digital number output is proportional to the analog voltage input.

An A/D converter is designed or adjusted for a resolution of so many volts or millivolts per count. The converter in Figure 21-1 has a resolution of 200 mV per count—meaning that for every increase of 200 mV at the analog input, the digital number at the output will increase by 1. Likewise, for every decrease of 200 mV at the analog input, the digital output will decrease by 1. The converter in Figure 21-1 is a four-bit output. Therefore, the highest output is 15, which means that at an input of 3V, we have reached full scale of the converter. If we apply to the input a voltage of 2.1V, we cannot be sure whether the converter will read 1010 (2.0V) or 1011 (2.2V). This sets a practical limit to the accuracy of ± 1/2 LSB. To this error—known as the *quantizing error*—must be added the errors caused by the electronic components in the system.

The *analog comparator* shown in Figure 21-2 is a key element in all A/D converters. It receives two analog voltages on its inputs and provides a logic 1 or 0 at its output, depending on whether $A > B$ or $B > A$. The comparator in Figure 21-3 produces a logic 1 for $A > B$, but the input leads can be rotated to obtain a 1 level for $B > A$.

Integrated circuit comparators can be purchased in a variety of cases, as shown in Figure 21-4. The output of a comparator will remain high as long as the + input is slightly more positive than the - input.

Simultaneous conversion is the most rapid form of A/D conversion. As Figure 21-6 shows, it requires one clock pulse to clear the output register and a second to clock the levels that are on the D inputs to the output. The analog input is applied to the A inputs of all seven comparators. The reference voltage V_R is divided into eight precisely equal parts, applied to $V_R/8$ through $7V_R/8$ comparator inputs. When the analog input is applied, all comparators having reference inputs of lower voltage than the analog input will produce a logic 1 output. The logic between the comparators and the register D inputs is designed to give the highest comparator control over the output. The highest comparator output not only steers the correct D input but must also inhibit the effect of the lower comparators.

The *counter-controlled* A/D converter is the simplest and cheapest means of conversion. It uses three main elements, as Figure 21-7 shows—a counter, a comparator, and a D/A converter. As the counter increases its count from reset, the output of the D/A (digital-to-analog) converter increases one increment per count. The comparator compares the D/A output with the analog input voltage. As long as $A > B$, the count continues. The moment the counter arrives at a count for which $B > A$, the 1 level from the comparator inhibits the clock line. The digital output can then be read at the parallel Q outputs of the counter.

This is a slow conversion, as enough time must be allowed between resets for the counter to reach its full count. The conversion time may be improved by using an up-down counter and allowing the converter to track the analog voltage continuously by counting up when $A > B$ and down when $B > A$.

Another widely used instrument is the *successive approximation* A/D converter. Like the counter method, it uses a D/A converter and a comparator, but in place of the counter is a storage register. It is faster than the straight counter, as it requires only one clock period for each flip-flop in the register. Unfortunately, as Figure 21-8 shows, it requires an elaborate set of control waveforms.

The conversion begins with the number 1 clock pulse setting the MSB flip-flop. This causes an output from the D/A equal to 1/2 the full-scale voltage of the converter. If the analog input is below that level, the comparator output enables the MSB flip-flop to be rest by $C_P'1$. If the analog input is above that level, it inhibits resetting of the flip-flop. Next, clock pulse 2 sets the next-highest-bit flip-flop. This adds another voltage (1/4 full scale) to the output. If $A > A/D$, the comparator inhibits reset. If $A < A/D$, the comparator enables reset of the second flip-flop. This process continues until the LSB register has been tried. At that time, the output is available as a parallel number or it may be shifted out in serial.

Figure 21-9 details the first two flip-flops of the register, showing how clock and delayed clock pulses are used to accomplish the trial turn-on in a single clock period per flip-flop. From Figure 21-8 we can see that four individual clock pulses are needed to operate a four-bit converter. These are obtained using the ring counter of Figure 21-11. The counter is clocked by C_P' pulses producing waveforms that bracket the clock pulses and enable them through the NOR gates. Figure 21-12 shows a second ring counter, which produces gates that enable the C_P' pulse to reset the proper flip-flop when the comparator determines the trial voltage is too high.

Successive approximation registers (SARs) are available in integrated circuits such as the Am2502 shown in Figure 21-14. This device is a high-speed device, and the A/D conversion rate will normally be limited by the DAC and not by this SAR. Figure 21-15 shows how an SAR, eight-bit DAC, and a comparator are interconnected to form an analog-to-digital converter.

When the analog input voltage is changing rapidly, a *sample-and-hold circuit* is necessary to keep the voltage stable for conversion. Figure 21-16 shows a simplified diagram, while Figure 21-17 illustrates the SMP-10.

Glossary

Analog-to-Digital Converter: A circuit that receives an input analog voltage and produces a binary number at its output.

Resolution: The analog value of the least significant bit (LSB). It can be determined by dividing the full-scale voltage by $(2^n - 1)$, where n is the number of bits in the converter.

Analog Comparator: A circuit that compares two analog voltages at its + and - inputs. If the voltage at the + input is slightly greater than the voltage at the - input, the output is high.

Sample-and-Hold Amplifier: A circuit that "reads" the input analog voltage and stores that level on a capacitor. See Figure 21-16.

Questions

1. What is the function of an analog comparator in an A/D circuit?
2. If the analog comparator of Figure 21-3 produces a logic 1 output when $A > B$, what modification is needed to obtain a logic 1 output for $A < B$?
3. Can the LM319 be used with either TTL or CMOS circuits?
4. What is the most rapid method of A/D conversion?
5. What would happen to the performance of the A/D converter of Figure 21-7 if the input leads to the comparator were reversed?
6. What can be done to reduce the conversion time of a counter-controlled D/A converter?
7. Which A/D circuit can provide the output in serial?
8. An A/D converter is controlled by an up-down counter. What operates the up-down control?
9. In successive approximation A/D conversion, which register bit is tried first—the MSB or the LSB?
10. In successive approximation A/D, what happens when the output of the D/A converter exceeds the analog voltage?
11. If the output of a successive approximation A/D converter is taken in serial, is an initial reset needed for the A/D register? If the output is taken in parallel only, is an initial reset needed?
12. What is the ideal logic circuit for generating the control waveforms for the successive approximation A/D converter?
13. In most successive approximation A/D converters, what components limit the maximum frequency of operation?
14. What type of circuit can be used to prevent the analog input voltage from varying during conversion?
15. What feature of a sample-and-hold circuit prevents the loading down of the analog input during the sample time?
16. What feature of a sample-and-hold circuit prevents excessive discharge of the hold capacitor during A/D conversion?

Problems

21-1 If an A/D converter has a resolution of 100 mV per count, what analog voltage is represented by a logic 1 in the third significant digit of the output?

21-2 How many additional analog comparators are needed to expand the A/D circuit of Figure 21-7 to a four-bit output?

21-3 A counter-controlled A/D converter uses a five-bit counter. The reference voltage of its D/A circuit is 16V. What is the resolution of the A/D? What analog voltage does the MSB represent?

21-4 Complete the diagram of Figure 21-9 for a four-bit A/D converter.

21-5 Change the timing diagram of Figure 21-13 to an output of 1101.

21-6 If the resolution of the converter in Problem 21-7 is 50 mV per count, what analog voltage does the output represent?

21-7 The A/D of Figure 21-15 has an analog input of 3.85V. Draw a set of waveforms like those of Figure 21-14(b), indicating the A/D conversion of this analog input.

CHAPTER

22

Memory Technology

Objectives

Upon completion of this chapter, you will be able to:

- Use the accepted language and terminology in describing computer memories.
- Select the proper form of memory for a given application.
- Test the operation of ferrite core memory.
- Use semiconductor memories.

22.1 Introduction

Upon completion of a high school physics course, a student can solve problems that are very complex in comparison with the problems we solve in our daily lives. Seldom in our daily routine must we resort to writing down facts and figures, since our short-term memory can handle most of them. In physics problems, so many facts and figures are generated so rapidly that we must write them on a scratch pad so that we can later find them and put them to use. For the problem solver, the scratch pad is an auxiliary memory. Likewise, the computer can handle many routine functions for which the registers within its arithmetic and control units provide sufficient storage of data; but, for more complex routines, memory circuits must be employed.

In a physics class, students sit a long time observing and remembering a routine for solving a problem. Similarly, a program for solving a problem must be stored in the computer's memory—hence, the name *general-purpose stored-program digital computer.*

A computer can operate in cycles of microseconds; yet it must often receive data from terminals that are mechanically operated and hundreds of times slower. If those terminals work first into a memory circuit, many terminals can use one computer in a procedure known as *time share.*

All this points to a need for memory circuits. Not long ago a computer contained one working memory of limited capacity. Today's computer may have numerous types of memories, which we can categorize as *working memories* or *auxiliary memories.* Working memories are contained within the computer's central processing unit, and they are usually random-access. A large central processor may contain one large memory, along with several smaller memories, referred to as scratch pad memories, buffer memories, and read-only memories.

The buffer memory stores input data that are arriving at too slow a rate for the high-speed processing they are to receive or data that are not complete enough for the computer to process. If the working memory were used for this storage, it would keep the central processor from doing other work. A buffer memory can operate on a first-in, first-out basis and can perform this function much faster than the main memory.

The scratch pad memory is a small-capacity memory built into the arithmetic unit. For certain arithmetic operations, it provides a faster memory cycle time than is available from the main memory.

The read-only memory (ROM) is programmed during manufacture or it can be programmed—only once—by the user. The data in the memory can be read out whenever needed, but the data are not erasable. There is no way to remove the data and write new data in their place. This type of memory is ideal for fixed routines or for special display readouts that can be called from ROM, instead of coding or decoding circuits.

Auxiliary memories—in most cases, peripheral devices such as magnetic tapes and magnetic disks—are ideal for storing inventory records and other bookkeeping data. They may also be used to hold compiler programs and computer subroutines. The auxiliary memory has a much longer cycle time than the working memory, but its capacity is almost unlimited.

22.2 The Memory Bit

The capacity of computer memories, like the capacity of a storage register, is measured in number of bits. A single bit of memory can be supplied by a minute ferrite core, which can be magnetized in the clockwise direction to store a 1 or in the counterclockwise direction to store a 0. It can be supplied by an integrated circuit flip-flop. The charged or discharged state of a minute capacitor can be used as a memory bit. The foremost requirement of a memory bit is that it have two distinct electrical states that can be easily translated into the 1 and 0 voltage levels that the computer uses.

A single bit of memory is a small amount indeed. It is estimated that the human brain contains billions of bits of memory. Computer designers discuss the need for obtaining a billion-bit memory capacity. A compiler or translator program for Fortran or Cobol uses over 20,000 bits of memory.

22.3 Words and Bytes

Mere assembly of tens of thousands of memory cells does not make a useful memory. We need some system for locating and retrieving all the information—which is not economical on a single-bit basis. The memory must, therefore, be organized into words that have an addressable location in the memory.

In many cases, a word is the same size as the total number of bits in the arithmetic logic unit of the computer. However, one or more overhead bits are often included for error detection or correction. In recent years, memory words have been organized in bytes. A byte is eight bits. The byte, therefore, has the capacity for two BCD or hexadecimal digits. It can also store letters, numbers, and special characters in ASCII code. The ASCII code explained in Chapter 24 requires seven bits. The eighth bit may be used as parity.

22.4 Memory Terms

Before discussing some of the many types of memories currently in use, let us first define a few of the terms that describe memories and their performance:

1. To *write* means to enter a data word into memory.
2. To *read* means to retrieve a data word from memory.
3. *Access time*—the time required to read one word out of memory—is usually identical to the time required to write one word into memory. Modern memory devices have access times lower than 50 nanoseconds.
4. *Memory cycle time*—a more exact measure of computer operating speed in that it includes the time required to transmit and store the memory data—can be as low as 100 nanoseconds.
5. The *address* is built into each word location so that data words entered into a large memory can be retrieved later from thousands of others that are simultaneously in the same memory. The address is usually transmitted to the memory as a binary number in parallel form.

Listed next are some comparative characteristics of memories. The italicized terms represent the more desirable or important of these characteristics:

Random access (sequential access or serial access),
Destructive readout or *nondestructive readout*,
Static or dynamic,
Volatile or *nonvolatile*,
Erasable or nonerasable.

In selecting a memory where cost, power consumption, or size considerations did not dictate otherwise, the ideal memory would be one having all the italicized characteristics. Ferrite core memories, used almost exclusively as the working memory of computers built during the 1960s, and still in wide use, have all but one of the italicized desirable characteristics. Their only shortcoming is destructive readout, a problem easily overcome by reading out into a register and immediately writing it back in while that same location is still being addressed.

Random access (serial access or sequential access): The most useful form of memory is the random access memory (RAM). As Figure 22–1 shows, the address is sent to the memory as a parallel binary number (octal or hexadecimal numbers can often be used for ease of description). Any location in the memory can be addressed without waiting or sorting through items in numerical sequence. The memory cycle time is the same for any location addressed.

The serial access memory operates as shown in Figure 22–2. A shift register of N bits can store either words of N bits or n words N/n bits long. A data word is written into the left side, but it must arrive at the correct time so that the LSB will be in the correct position. The word is shifted through the register, requiring N clock pulses for the LSB to reach the output. After reaching the output, it may be read out. At the same time, it is recirculated back to the input. Recirculation continues until it is to be replaced by a new word; then recycle is inhibited and a new word is entered. A particular word is not addressable at random but can be read out only every Nth clock pulse.

As shown in Figure 22–3(a), sequential access applies to auxiliary memories such as the magnetic tape and, to a lesser extent, the magnetic disk. Data are inserted on the tape in conjunction with an address number. As Figure 22–3(b) shows, the tape is divided into records. The magnetic tape memory uses thin plastic tape that is coated with ferrite. The ferrite can be magnetized in small domains, which means that minute segments of 1- and 0-level magnetization can be placed side by side during the write period and later be distinguished by the read heads. As data are written on the tape, they are automatically assigned addresses or record numbers. The records are numbered sequentially on the tape. When a particular record is addressed, the address number is applied to a comparator, where it is compared to the record numbers in the read registers and a decision is made, as Figure 22-4 shows. The reader can distinguish between address numbers and data by assigning one an odd parity and the other even parity. The time required to retrieve data depends on how far through the sequence of records the tape

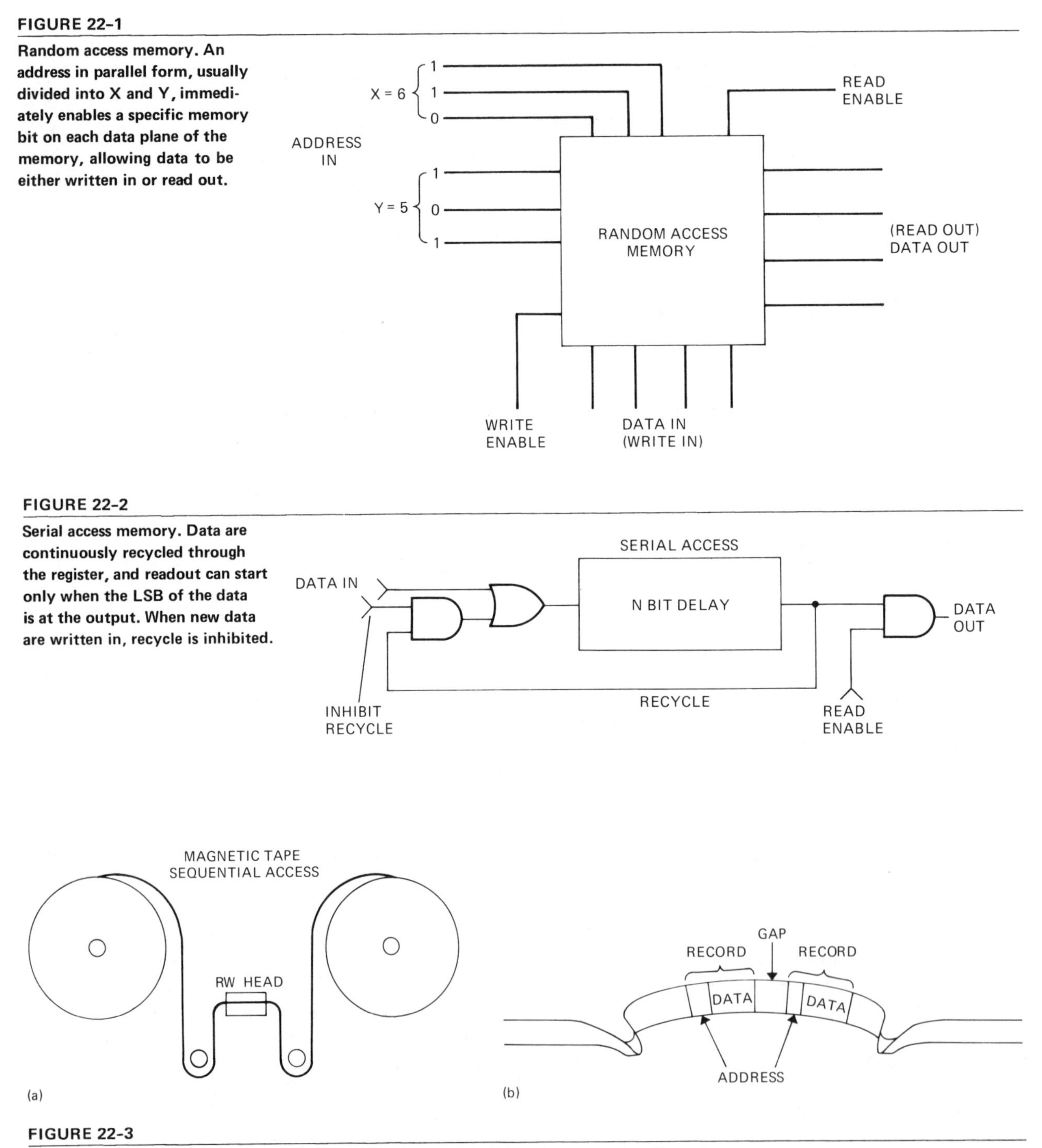

FIGURE 22-1

Random access memory. An address in parallel form, usually divided into X and Y, immediately enables a specific memory bit on each data plane of the memory, allowing data to be either written in or read out.

FIGURE 22-2

Serial access memory. Data are continuously recycled through the register, and readout can start only when the LSB of the data is at the output. When new data are written in, recycle is inhibited.

FIGURE 22-3

(a) Magnetic tape sequential access memory. (b) Magnetic tape divided into records, with each record numbered sequentially.

reader must search before it arrives at the record addressed.

Destructive readout or nondestructive readout: Some memory bits can be read out only in a fashion that returns their state to 0. This shortcoming can usually be overcome by writing the data back in while the memory location is still addressed.

Static or dynamic: Data stored in dynamic memory are in continual motion—for example, being shifted through a register or transmitted through a delay line or acoustic tube. The data levels are continuously refreshed during transmission and will vanish shortly after transmission shifting or cycling has been discontinued. In a static memory, the data are entered as a charge, logic, or magnetic state that will remain in one memory position until a new number is written in its place.

Volatile or nonvolatile: If the power is removed from a computer or digital device, a nonvolatile memory will not lose the data stored in it. Magnetic memories are nonvolatile. Semiconductor read-write memories are volatile.

Erasable or nonerasable: Many read-only memories are nonerasable. The program data is built into them during manufacture. The programs or data in them cannot be changed. If a change is needed, the ROM is discarded and replaced by a new one. Some read-only memories, however, can be erased. The most widely used erasable ROM is the semiconductor EPROM or UVPROM, which can be erased with ultraviolet light. Read-write memories are erasable. They do not, however, require an erase function in that new data can be written directly into an address without regard to the existence of previous data.

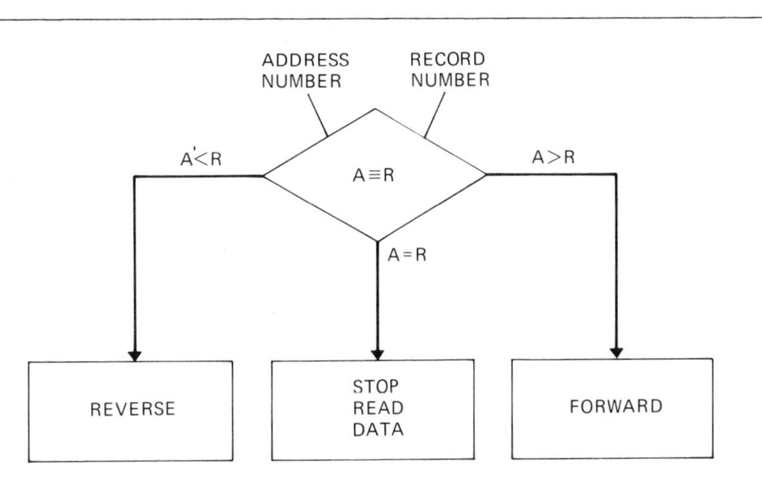

FIGURE 22-4

Flowchart showing possible decisions when digital comparator compares record numbers read off the tape with the number in the address register.

22.5 Ferrite Core Memory

The ferrite core memory bit is a small washer about equal in diameter to the lead in a lead pencil. The ferrite material can be magnetized in either a clockwise (CW) or a counterclockwise (CCW) direction. It also has the capacity to retain magnetism after the magnetizing force—which comes from a wire threaded through the hole of the ferrite washer—has been removed. As Figure 22-5 shows, the current

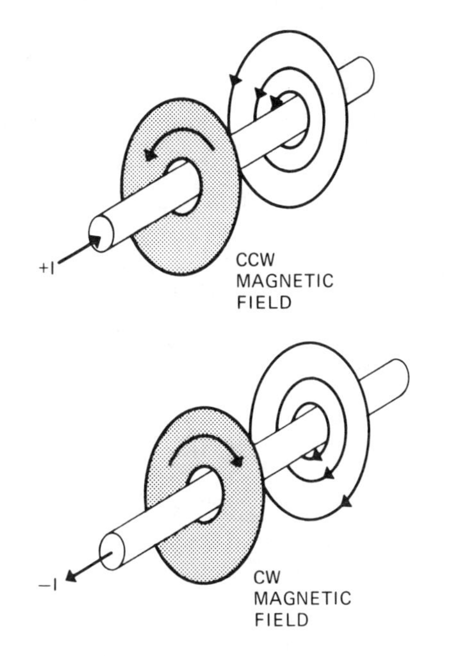

FIGURE 22-5

Ferrite cores. Direction of magnetization depends on direction of current through the wire.

through the wire creates a magnetic field of concentric lines of force around the conductor. (The direction of the magnetic field, CW or CCW, can be predicted by the left-hand rule.) The ferrite core will be magnetized in the same direction as the field around the conductor.

Figure 22-6 shows the hysteresis curve of the ferrite core. If a current running through the wire is increased to +I, the magnetic flux, H, in the core will reach saturation at the magnetic 1 level (point 1). As shown by the graph at point 2, reducing the current to 0 does not remove the magnetism. Once magnetized, a core will retain its 1 or 0 level indefinitely. At point 3 on the curve, we find that a negative current of -I is required to switch the core from the 1 to the 0 state. At point 4 on the curve, the current returns to 0, but the core retains its magnetization in the 0 state. An additional advantage is that the current switching the core can come from several wires, each supplying a current of I/2.

Figure 22-7 shows 14-bit word lines con-

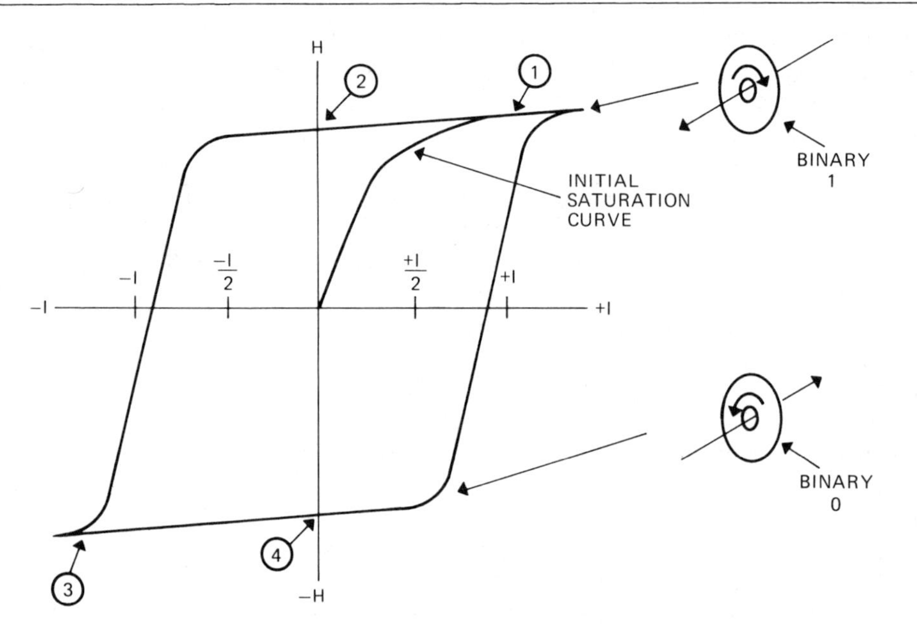

FIGURE 22-6

Hysteresis curve of the ferrite core showing retention of magnetic states in spite of a reverse magnetizing current, ± I/2. A full magnetizing current, ±I, however, does switch the magnetic state of the core.

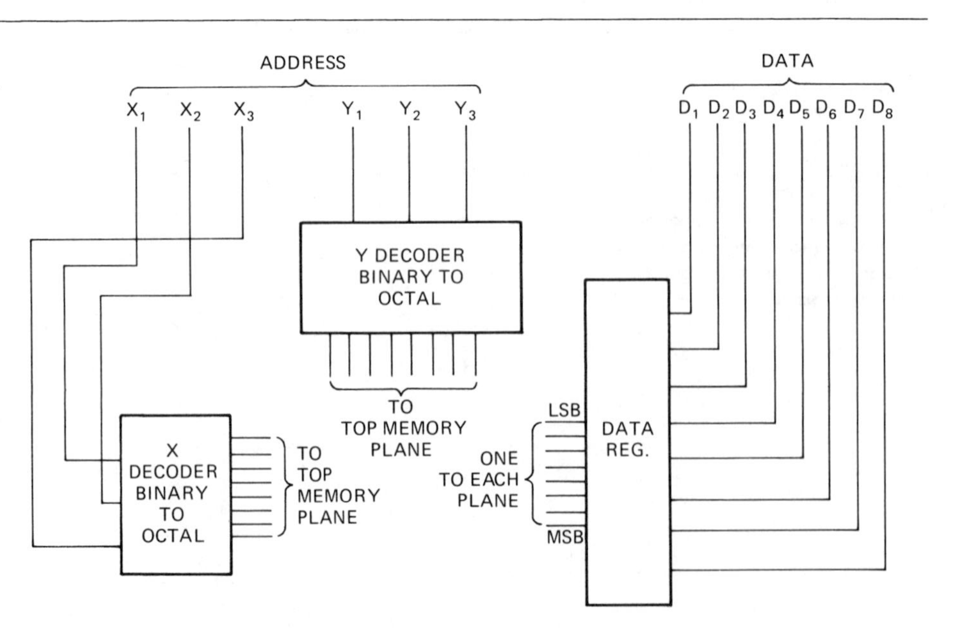

FIGURE 22-7

14-bit word lines connected to a 64-word-by-8-bit ferrite core memory. The 3 X and 3 Y binary address bits each decode into 8 X lines and 8 Y lines. The data are stored in an 8-bit memory register before being written into the memory.

nected to a 64-word-by-8-bit ferrite core memory. Six bits are address; the remaining eight are the data. The address is decoded into 8 X lines and 8 Y lines, each of which is connected to a current driver that drives a current line. When activated, it drives a current of +I/2, only half the amount needed to switch a core to the 1 state. As Figure 22-8 shows, each X wire threads through a row of eight cores on the top plane before extending downward to the next plane below. Each Y wire does the same. As each current driver provides a current of only +I/2, only that core for which both the X and Y lines are energized will be switched; hence, the term *coincident current memory.*

The usual ferrite core memory is addressed as a coincident current memory. To understand its functioning, let us use a small eight-cube memory as an example. This memory would have 512 bits, arranged in eight-bit words. The cores are arranged in eight planes of 64 cores each. When an eight-bit word is stored in this memory, one bit is stored on each plane and it will have the same XY location on each plane.

FIGURE 22-8

Coincident current memory. A decoded address X_3, Y_5 activates current lines that pass through each plane of the memory. The X_3, Y_5 core on each plane can be switched when a current of + I/2 is received from each current driver.

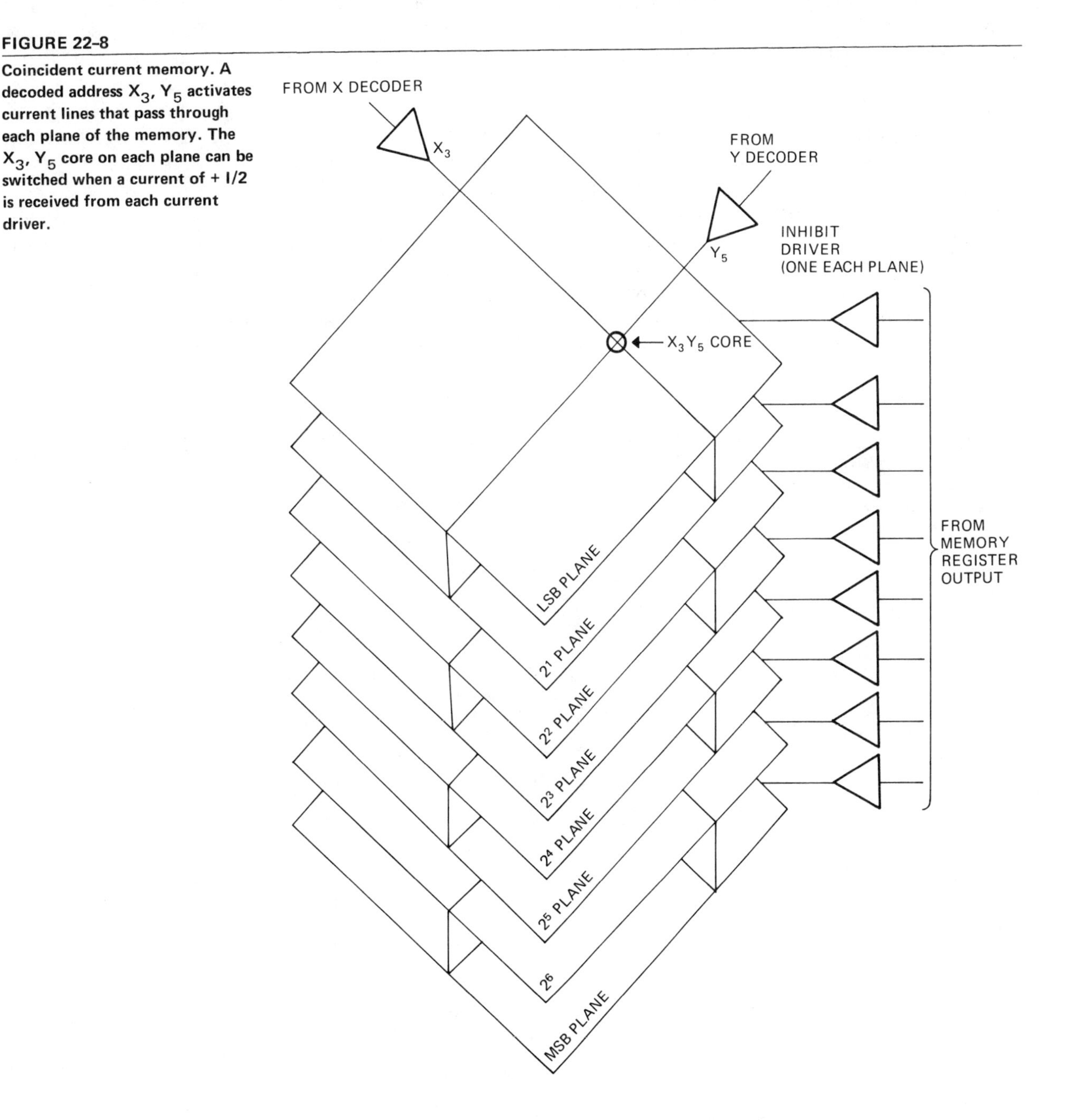

Figure 22-8 shows an address of 011101 (X_3, Y_5). The X_3 line threads through eight cores on each plane and is grounded at the far end. The Y_5 line threads through eight cores on each plane and is grounded at its far end. This means that the X_3, Y_5 core on each plane will be switched. It does not allow for the fact that some of the data bits may be 0. This is provided for by another wire, called the *inhibit line.* There are separate inhibit wires and an inhibit current driver for each plane of the memory—hence, for each bit of the data. The inhibit driver is energized only for the data bits that are 0; it provides a current of -I/2. This again is not enough to switch a core, but it cancels out one of the +I/2 currents of the addressed core and prevents it from switching. Whereas each X and Y line threads through a row of cores on every plane in the memory, the inhibit line threads through every core on one plane only.

Thus far we have what is needed for writing data into the core memory. To provide for a readout of the memory, another wire must be threaded through the cores, one wire on each plane. This wire is called the *sense wire.* When a core is switched from a 1 state back to 0, it induces a current in the sense wire, which causes the output of the sense amplifier to set a 1 into the memory buffer register for that bit. If the core is already at 0, then no current is induced into the sense line and a 0 is stored in that memory register bit. Figure 22-9 shows a single plane of the memory with X, Y, and sense wires (inhibit wires are left out for clarity). To read a word out of memory, it is addressed in the same coincident current method, but the addressed cores will this time receive a -I/2 current from an X line and a -I/2 current from a Y line. If an address core on any of the planes is in the 1 state, it will switch to 0 when subjected to the coincident -I/2 currents. A core that switches from 1 to 0 induces a current on the sense line and through the sense amplifier sets a 1 in its memory register bit. On those planes for which the core is already at 0, no current is induced in the sense wire, leaving the corresponding memory register bit reset. In reading the data, all eight cores become 0. This amounts to a destructive readout except that immediately after the read pulse the data word is in the memory buffer register. The read pulse is automatically followed by a write pulse, which enters the data back into the memory location.

With the exception of the XY current

FIGURE 22-9

8-by-8-bit memory plane, showing X, Y, and sense wires. Inhibit wires are left out for clarity.

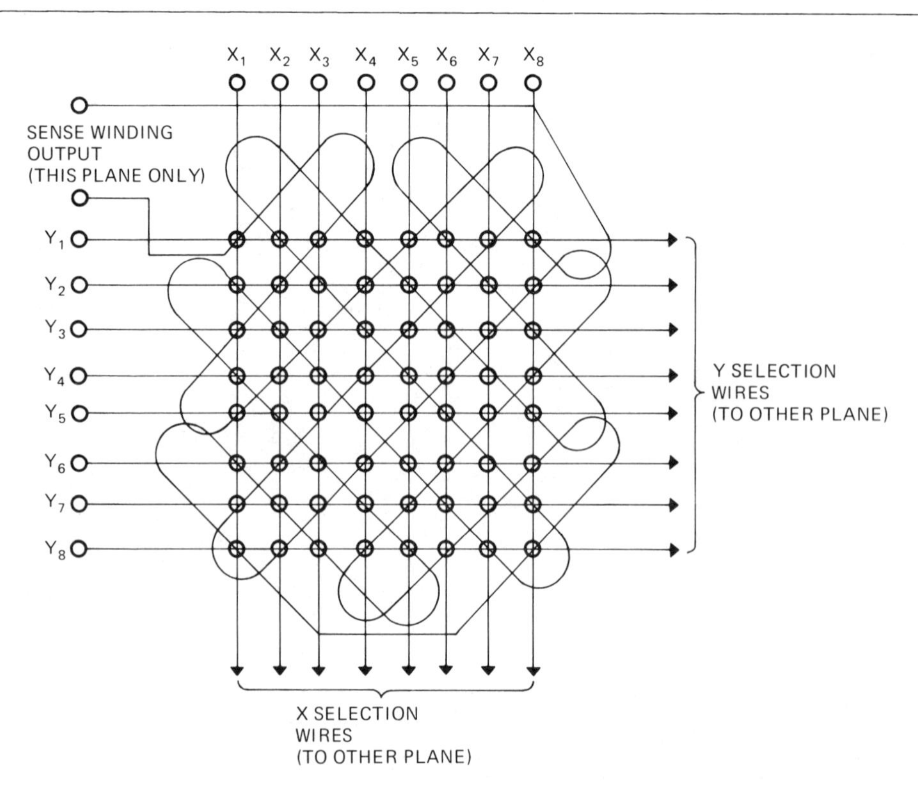

drivers, Figure 22-10 shows the logic needed for each plane of the ferrite core memory. The memory register handles the data in both reading and writing data into the memory. Figure 22-11 is the timing diagram of two memory cycles. As far as the current drivers are concerned, each memory cycle includes both a read and a write operation. The difference lies in whether the write enable allows a new data word to be entered into the memory register, which results in new data in the memory also, or whether the read enable allows the output of the sense amplifiers to be entered into the register, resulting in a read operation with the old data word being set back into the same location. Figure 22-11(a) shows waveforms for a read operation. Figure 22-11(b) shows those for a write operation.

Although the ferrite core memory offers all the ideal operating characteristics for a memory, its cost and size are serious drawbacks. A great deal of expensive manual work goes into producing it. It does not lend itself well to batch fabrication or miniaturization. The plated wire memory is an attempt to overcome these defects. It functions much like the ferrite core memory; but, instead of a core, the magnetic material is plated on the X wires. The Y, inhibit, and sense wires are etched on printed circuit cards that sandwich the plated wires. Although the full length of the wire is plated, the magnetic material is such that it can be magnetized in narrow domains confined to the points of X and Y crossing.

22.6 Semiconductor Memory Circuits

The semiconductor memory is composed of interconnected DIP-type circuits, instead of the interconnected planes that are typical of the ferrite core memory. Figure 22-12 shows five typical memory circuits in different size DIP packages. The EPROM 2732, a 32K NMOS memory, shown in Figure 22-12(a), is organized as 4096 words, with each word eight bits wide. It comes in a 24-pin package with a window that permits erasure of the contents by ultraviolet light. The bipolar RAM 82S09, shown in Figure 22-12(b), is a high-speed, 64-word-by-9-bit memory. It is commonly used for high-speed memories requiring eight bits plus parity. It uses the not-so-common 28-pin package. The data is in separate I/O. The CMOS RAM 5101, shown in Figure 22-12(c), is a very low-power RAM commonly used for battery or battery backup operation. The package is 22-pin. The NMOS static RAM 2114, shown in Figure 22-12(d) is organized as 1K words by four bits. The data is common I/O. An 18-pin package is used. The NMOS dynamic RAM 4116, shown in Figure 22-12(e), is organized as 16K by one bit. It

FIGURE 22-10

Logic circuits required for each plane of a ferrite core memory.

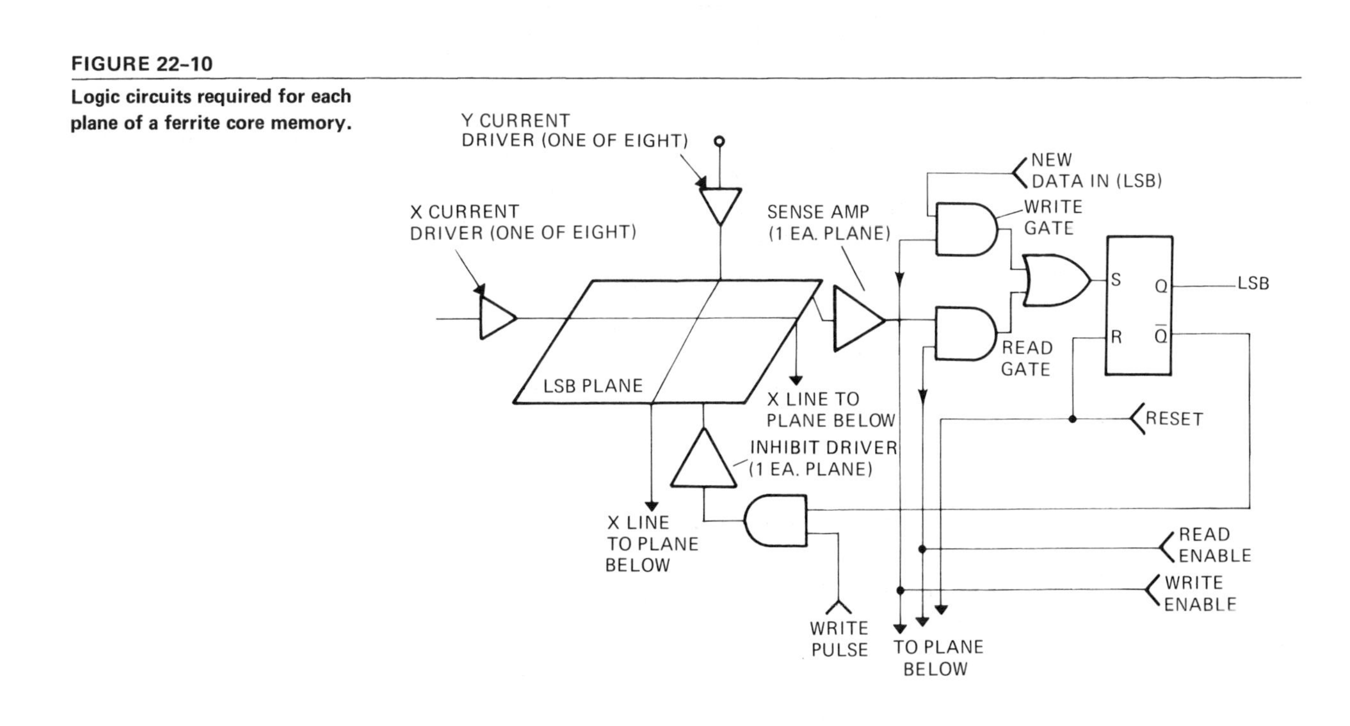

FIGURE 22–11

(a) Waveforms of a read operation. (b) Waveforms of a write operation.

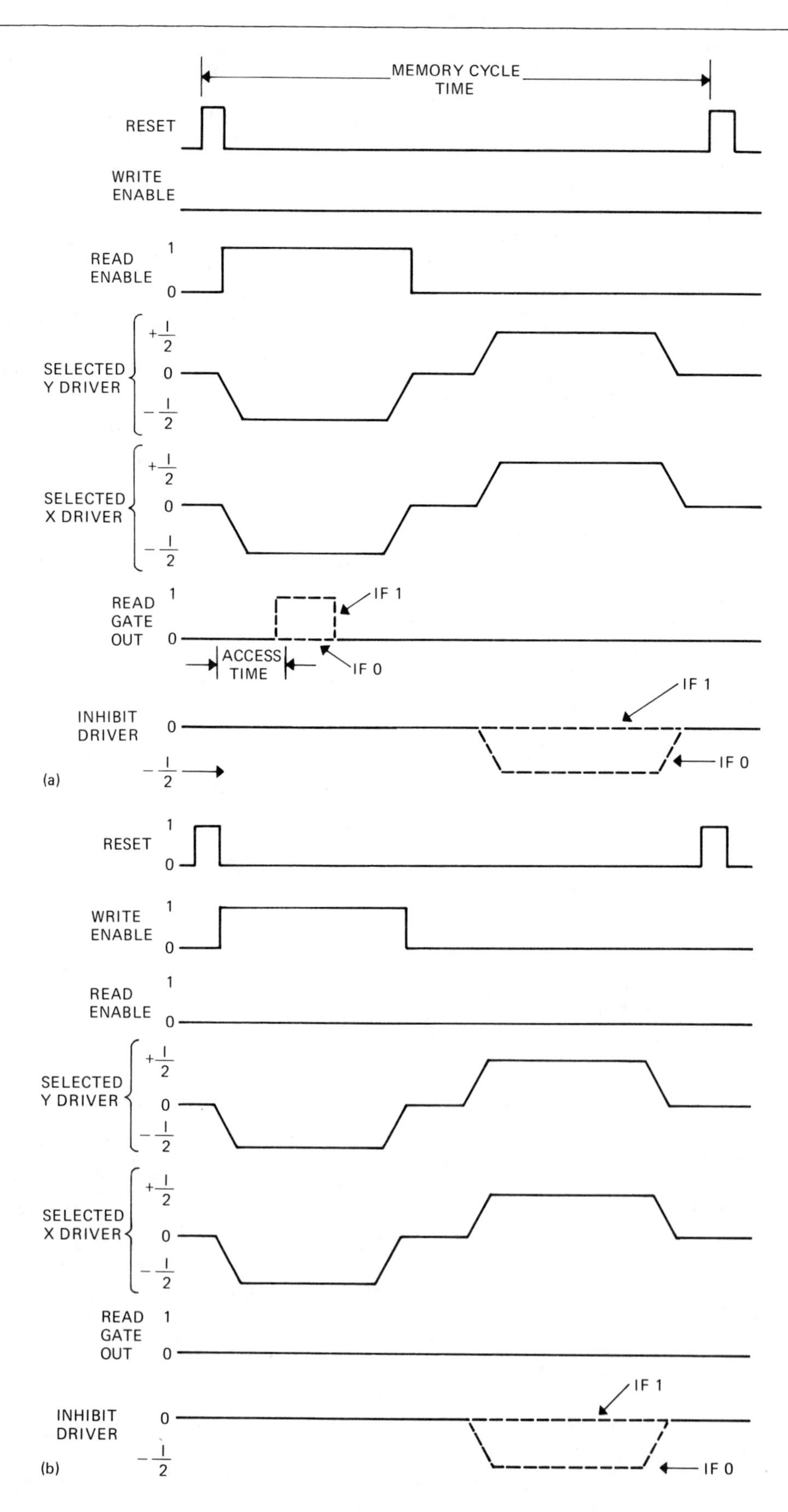

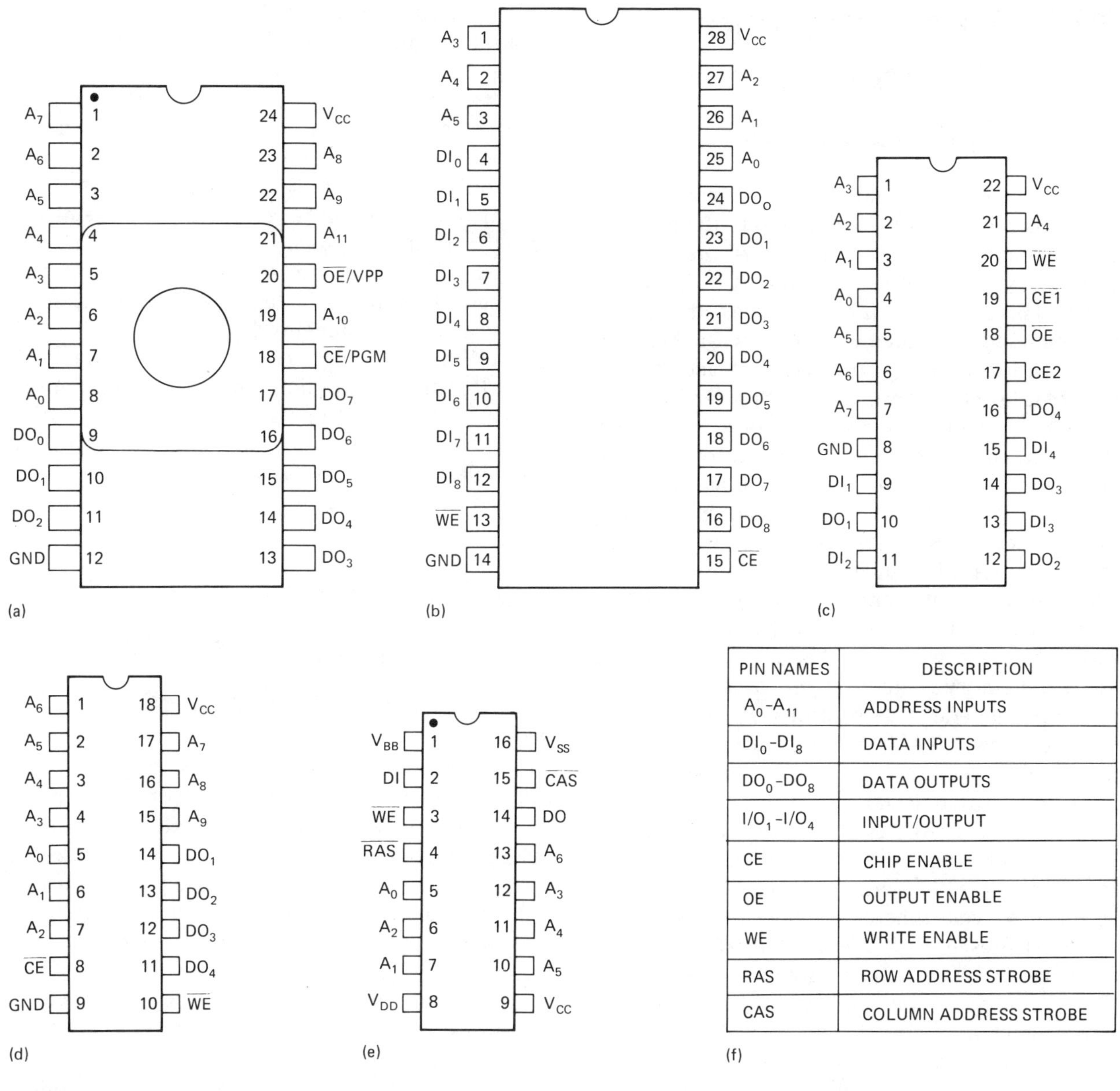

PIN NAMES	DESCRIPTION
A_0-A_{11}	ADDRESS INPUTS
DI_0-DI_8	DATA INPUTS
DO_0-DO_8	DATA OUTPUTS
I/O_1-I/O_4	INPUT/OUTPUT
CE	CHIP ENABLE
OE	OUTPUT ENABLE
WE	WRITE ENABLE
RAS	ROW ADDRESS STROBE
CAS	COLUMN ADDRESS STROBE

FIGURE 22-12

Semiconductor memories. (a) EPROM 2732, a 32K NMOS memory. (b) Bipolar RAM 82S09. (c) CMOS RAM 5101. (d) NMOS static RAM 2114. (e) NMOS dynamic RAM 4116. (f) Pin names and descriptions.

comes in a 16-pin package. The 4116 uses a multiplexed addressing scheme.

In addition to the different package sizes, semiconductor memories vary in terms of technology (NMOS, CMOS, bipolar, ECL). They can be bit organized or word organized. For example, the 4116 is bit organized in that it provides only one data bit per circuit. For a memory system of eight-bit words, therefore, eight of these circuits would be wired with addresses in parallel. The 2732, on the other hand, is already eight data bits wide.

The data can be separate or common I/O (input/output). While the 82S09 uses separate pins for data-in and data-out, the 2114 handles input and output data through the same four pins.

22.7 Semiconductor Memory Applications

The cost per bit of semiconductor memory is so much lower than the cost per bit of magnetic core memory that the use of core memory in new designs of digital machines is becoming a rarity. This is in spite of the fact that, at present, no semiconductor memory exists with the ideal combination of read-write, random access, and nonvolatile characteristics of the ferrite core memory. Semiconductor memories that are read-write and random access are, unfortunately, volatile. Those that are nonvolatile and random access are either "read-only" or require an erase cycle before being rewritten. Because of this, several types of semiconductor memories are generally used for working memory functions in a digital system. Where the function must necessarily be read-write, RAMs are used. Where volatile memory is inconvenient, ROMs are used.

The general classifications of semiconductor memories—RAM (random access memory) and ROM (read-only memory)—are somewhat imperfect in that ROMs are also by circuit function random access. A better distinction would be to classify those into which data can be written in circuit with a write cycle time about equal to its read cycle time as "read-write" memories, but it is present-day practice to call read-write memories RAMs and read-only memories ROMs.

A computer with nothing in its memory is like a newborn baby. It can do nothing. If the computer's memory consists of only volatile RAM, then each time the power is turned off, it returns to its infantile state. The most essential program needed by any computer is a program for loading other programs. A program that can take another program from an input device and store it correctly in the computer's memory is called a *bootstrap loader.* For computers using core memory, the bootstrap loader is usually entered into RAM using manual switches on the control panel of the computer. It is a laborious task, but once the program is entered into the nonvolatile ferrite core memory, the loading of the program need not be repeated unless a malfunction is suspected. The program will remain valid even after power has been off for a long period of time. It will be there when the computer is powered up again.

Where semiconductor memory is used, loader programs are supplied by read-only memories. The program is built into the memory during manufacture and remains valid no matter how often power is turned off and on. A change to the loader program, no matter how minor, can only be accomplished by discarding the old ROM and installing a revised version. Semiconductor read-only memories are useful only for program and data listings that are not subject to change. Chapter 24 deals with the different types of read-only memories.

With the exception of a few simple dedicated computer applications, a program cannot be run without the use of read-write (RAM) memory. Read-write memory is needed to store variables that are used in the program or generated during the running of the program.

22.8 Integrated Circuit Memory Technologies

Another consideration in the selection of the ideal integrated circuit for a memory application is the technology. Memory circuits are made in the four technologies listed in Figure 22–13. ECL provides the highest speed in terms of either access or cycle time. In all other respects except noise immunity, ECL is the least advantageous technology. Bipolar memories are not quite as fast as ECL. They have the advantage of single-power-supply operation and are lower in cost per bit than ECL.

MOS memories are by far the most widely used because the cost per bit is lowest. Older versions of MOS RAMs are much slower than equivalent bipolar RAMs, but as MOS circuits of smaller geometrics have been developed, the gap between MOS and bipolar memories has been closing. Access times of MOS RAMs have improved from a speed of about 350 nanoseconds in the mid 1970s to below 100 nanoseconds in 1981. This compares with about 50 nanoseconds for bipolar RAMs of comparable size.

A similar story of improved speed has occurred for CMOS RAMs. The major advantage of CMOS RAMs is their low power consumption at medium clock speeds. They have the slowest speed of the four technologies. That speed, however, has changed from about 500 nanoseconds access time for a 1024-bit RAM in the mid 1970s to less than 200 nanoseconds access time for a 4096-bit RAM in 1981.

22.9 Summary

The register circuits we discussed in Chapter 12 function to supply a computer or other digital machine with rapid, short-term, low-capacity memory. For many applications, this is not enough, and data must often be transferred from registers into memory circuits that supply large-capacity storage for indefinite periods. To fill this need, various types of memory circuits have been devised. There are two general classifications of memories—the *working memory* and the *auxiliary memory*. The working memory is found inside the computer's central processing unit. Auxiliary memories are high-capacity peripheral devices used to store data such as inventory records and computer subroutines. In the more recent computer models, the working memory capacity has been reinforced by special memories, such as scratch pad memories and buffer memories.

The basic element of the memory is the *memory bit*, which must have two states—designated 1 and 0—and it must be possible to switch rapidly between the 1 and 0 states. Memory bits must be very small and consume little power in order to be practical in quantities of 60,000 or more to a single memory. The ferrite core, magnetic film, and integrated circuit flip-flop have these characteristics and form the bulk of the memory bits in use today.

Mere assembly of tens of thousands of memory bits does not make a useful memory. We need some system of locating and retrieving the information. Therefore, the bits are organized into *words*, each having an *addressable location*. Words of 8, 16, or 32 bits are common. In many cases, the words are further divided into *bytes*. A byte equals eight bits.

The most useful and versatile memory is the *random access memory* (RAM). The access time, or time required to retrieve a word from the memory, is the same for any location in the random access memory. When the address is sent to the memory, the data word becomes available within a few clock periods. The ferrite core memory has been used as a random access memory for several decades.

The ferrite core is a minute washer with an outside diameter slightly larger than the lead in a pencil. As Figure 22-5 shows, an electric current in a wire running through the core can magnetize it in a clockwise or counterclockwise direction. As shown by the hysteresis curve of Figure 22-6, when the current in the wire increases to a level of +I (point 1), the core is magnetically saturated in the clockwise direction; but, as indicated at point 2, when the current is returned to 0, the clockwise magnetic state is retained. To remove the magnetic state, the current must be reversed to a -I. This reverse current switches the core to the opposite, or CCW, magnetic state. These opposite magnetic states are designated logic 1 and 0. As the curve indicates, currents of +I and -I in magnitude can switch the core between the 1 and 0 states, but the half currents +I/2 or -I/2 do not cause it to switch. This is an important point, for the system of addressing the memory takes advantage of the fact that the half currents from two different wires running through the core can add or subtract to control the switching of the correct memory bits.

The cores are usually arranged in XY

FIGURE 22-13

Comparison of semiconductor memory technologies according to speed, power consumption, capacity, and cost per bit.

TECHNOLOGY	SPEED (ACCESS TIME)	POWER CONSUMPTION	CAPACITY	COST PER BIT
ECL	HIGHEST (30 ns)	HIGHEST	LOW (1K BY 1)	HIGHEST
BIPOLAR	HIGH (45-50 ns)	HIGH	LOW (1K BY 8)	HIGH
NMOS	MEDIUM (55 ns) (120 ns)	LOW	HIGHEST (64K BY 1) (2K BY 8)	LOWEST
CMOS	LOW (200 ns)	LOWEST	HIGH (2K BY 8)	LOW

planes, with an X and Y wire running through each core. There is a separate plane for each data bit. Figure 22–7 shows the arrangement for a 64-word-by-8-bit memory. Of 14 bits transmitted to the memory, six are address bits, three X and three Y. These are decoded into X and Y coordinates, which send +I/2 current down one of the X lines and one of the Y lines. Only the core at the intersection of the lines will receive enough current to switch. The data bits are stored in the memory register. Each output of this register operates an *inhibit driver.* If a logic 0 is stored as a data bit, the inhibit driver sends a -I/2 current through a third wire, to cancel out one of the half currents, preventing the core from switching. Figure 22–8 shows the overall eight-cube memory receiving an X_3, Y_5 address. There is one plane for each data bit, and each plane has a separate inhibit driver. The X and Y lines, however, wind through each of the eight planes, addressing the same X_3, Y_5 core on each plane. It becomes a 1 or 0 data bit depending on the inhibit driver, which is operated by the memory register output.

To read data out of the memory, a fourth wire is threaded through each core—the *sense wire.* There is a separate sense wire on each plane. In reading data out of the memory, the addressed X and Y drivers send a current of -I/2, which will switch all cores in that address to 0. If a core is in the 1 state and switches to 0, it induces a pulse onto the sense wire. If the core is already at 0, no pulse is induced onto the sense line. The sense line operates a driver that sets the data into the memory register. After the reading of data from a particular memory location, the data at that address have been switched to 0, which constitutes a destructive readout, but immediately following the read operation, while the same location is addressed, a +I/2 current is transmitted through the addressed XY lines and the data that are in the memory register are written back into the same memory location. Figure 22–11 is the timing diagram for this operation.

Today's computers use *semiconductor memories.* These memory circuits are housed in IC packages and manufactured with the same technologies as other digital devices. The cost per bit for semiconductor memories is much lower than for magnetic core memories. Unfortunately, semiconductor memories that are *read-write* are volatile, and those semiconductor memories that are nonvolatile are *read-only.*

The most essential program needed by a general-purpose computer is a program for loading other programs; it is called a *bootstrap loader.* Most computers with semiconductor memories have their bootstrap program stored in read-only memory because ROM is nonvolatile. Except for some dedicated control application, computers need read-write (RAM) memory to store data that is generated during the running of programs. MOS memories are most often used because they have the lowest cost per bit and consume less power than bipolar memories.

Glossary

Random Access Memory (RAM): A memory for which every address has the same cycle time. The data become available within the same access time for all addresses.

Serial Access Memory: A memory in which the data circulate in serial form. Readout of the data must start when the LSB arrives at the output terminal. See Figure 22–2.

Sequential Access Memory: A memory in which data stored on a tape receive record numbers in numerical order. To obtain a particular number, the reader must search through the tape in numerical sequence until the addressed record is found. See Figure 22–3.

Semiconductor Memory: A solid state memory circuit that is manufactured using either NMOS, CMOS, bipolar, or ECL technology.

Bootstrap Loader: A computer program that loads other programs. It is most often used just after the computer is turned on. A computer using primarily volatile semiconductor read-write memory will normally have its bootstrap loader program in nonvolatile ROM.

Read-Only Memory: Memory that is programmed during manufacture. The data can be read from it but cannot be changed during the computer operation.

Buffer Memory: Memory used to store data arriving too slowly to take advantage of the high-speed capabilities within the computer.

Scratch Pad Memory: Small-capacity memory, used within the arithmetic unit to provide higher-speed operation than is available from the main memory.

Memory Byte: An arrangement of memory bits into 8-bit segments to provide BCD, hexadecimal, or alphanumeric storage. A memory word may contain eight or more bytes.

Access Time: Time interval between the address-to-memory location and the time the data become available at the output of the memory registers.

Cycle Time: Total time required for the computer's complete read-write cycle. See Figure 22–11.

Destructive Readout: A memory that loses the data from storage when they are read out. This compares with nondestructive readout, in which the data can be read out any number of times without being altered. Some memory systems for which the readout is destructive are set up with a write cycle automatically following the read cycle, to write the data back into the memory location, thereby giving the effect of nondestructive readout.

Volatile: A memory that will lose the data stored in it if the power is removed. A nonvolatile memory will retain the data stored in it even if the power is removed.

Hysteresis Curve: A curve showing the extent to which the magnetic flux in a core lags behind the magnetizing force as the magnetizing force is changed from 0 to saturation in both the positive and negative directions.

Ferrite Core: A small, washer-shaped memory element, made from a compressed, powdered, ferrimagnetic material, having a square hysteresis curve, making it ideal for a memory element.

Questions

1. How many bits are contained in a byte?
2. Can the ferrite core be magnetized in both the clockwise and counterclockwise directions?
3. If power is removed, will a ferrite core hold its data?
4. List four semiconductor technologies used in the manufacture of memories.
5. What type of semiconductor memory is normally used for bootstrap loader program?
6. List four semiconductor technologies that can be used to make memory circuits in order of speed.
7. Explain the differences between random, sequential, and serial access.
8. What is a destructive readout? How is the ferrite core memory rendered nondestructive readout?
9. Describe a volatile and a nonvolatile memory.
10. What form of memory is nonerasable?
11. Explain the procedure used to address a record on magnetic tape.
12. In the eight-cube memory of Figure 22–8, through how many cores does the current from the Y_5 current driver pass? Why is only the X_3, Y_5 core on each plane likely to switch?
13. If the X_3, Y_5 core on each plane receives two $+I/2$ currents, how do those destined to be 0 avoid switching?
14. What is the purpose of the sense wire on each memory plane?
15. How does the plated wire memory differ from the ferrite core memory?
16. Explain the difference between access time and cycle time.
17. How many ASCII code characters can be stored in each memory byte?
18. Which semiconductor memory technology supplies memory circuits requiring the lowest power per bit?
19. Which semiconductor memory technology supplies memory circuits with the highest number of bits per circuit?
20. What two types of input/output arrangements are available with semiconductor memories?
21. How does the semiconductor memory designer overcome the problem that semiconductor read-write memories are volatile?

Problems

Use the following list of memory characteristics to solve Problems 22-1 and 22-2:

Erasable or nonerasable,
Destructive readout or nondestructive readout,
Volatile or nonvolatile,
Dynamic or static,
Read-write or read-only,
Serial or sequential or random access.

22-1 Choosing one characteristic from each line in the preceding list, classify each of the following memories: (a) Semiconductor ROM, (b) ferrite core, (c) magnetic tape, and (d) semiconductor RAM.

22-2 Choosing one characteristic from each line in the preceding list, classify each of the following memories: (a) 2732 EPROM, (b) dynamic shift register, (c) 4116 RAM, and (d) 2114 RAM.

CHAPTER

Semiconductor Read/Write Memories

Objectives

Upon completion of this chapter, you will be able to:

- Connect semiconductor read/write memory chips to a microprocessor.
- Decode a 64K memory map address system.
- Use timing diagrams for address access time, chip select access time, and memory read cycle time.
- Use data-in and data-out lines.
- Avoid bus contention problems.
- Understand write cycle and read cycle timing.
- Compare different random access memory cell designs.
- Compare the advantages and disadvantages of different types of RAMs.
- Compare static versus dynamic MOS RAMs.
- Organize RAM chips to assemble memory systems.
- Study the sense amplifiers used in dynamic RAMs.
- Use an interface chip to multiplex the microprocessor's address bus lines for dynamic RAMs.
- Study refresh, read cycle, write cycle, and read modify write timing for dynamic RAMs.
- Calculate the power drain of dynamic RAMs.
- Study causes of dynamic RAM errors.
- Use error detection and correction circuitry.

23.1 Introduction

An integral part of any computer system is semiconductor memory. As we discussed in Chapter 22, there are two types—volatile and nonvolatile. Semiconductor read/write memory is volatile. Semiconductor read-only memory is nonvolatile. There are, however, several types of read-only memory: mask programmed ROMs, PROMs, EPROMs, and EEROMs. This chapter deals only with semiconductor RAMs, while Chapter 24 covers the different types of read-only memories.

Within a computer, the central processing unit (CPU) and memory communicate over a set of conductors called *buses*. There are address, data, and control buses. Each location in memory has a unique address. The CPU sends out a binary pattern on the address bus, and when it is received by memory, it is decoded. The address is then ready to send or receive data over the data bus. The control bus is used to transfer timing signals to synchronize all communication between the CPU and memory.

This chapter shows how address, data, and control buses are connected to RAMs and how the CPU selects particular memory chips. As with other digital devices, memory chips are primarily manufactured using either TTL, ECL, NMOS, or CMOS technology. The advantages and disadvantages of each type of technology are covered in this chapter. Semiconductor RAMs fall into two categories—static and dynamic. Which type of RAM to use may depend upon the application. In this chapter, we will analyze both types. Dynamic RAMs require additional circuitry, as will be discussed in Section 23.6.

23.2 Memory Buses

23.2.1 Address Bus

The pin-out of RAM memories can be divided into four elements: power, address bus, data bus, and control bus. The address inputs of a RAM are normally designated with the letter A, with the least significant bit being A_0, as shown in Figure 23-1. In microcomputer applications, memory addresses are generated by the microprocessor. The low-order bits of the address bus normally connect in parallel to every memory

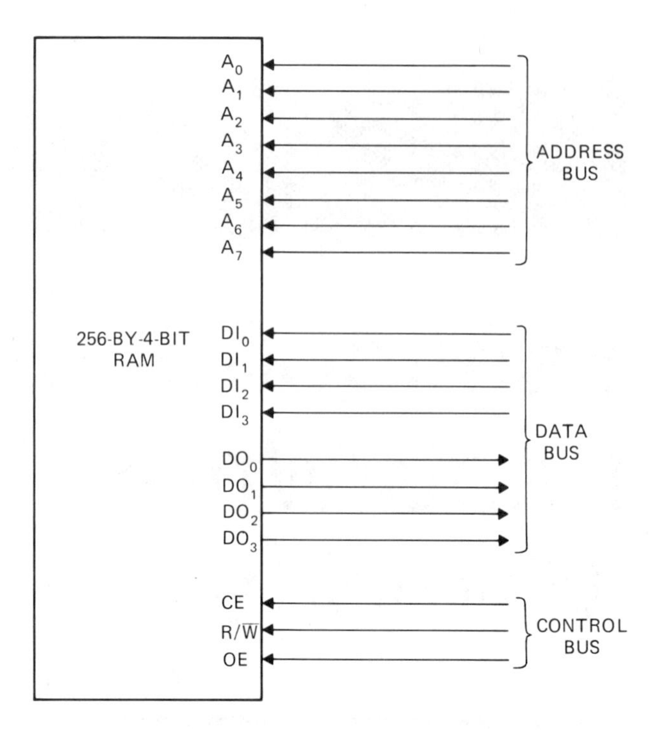

FIGURE 23-1

Memory buses. The low-order bits of the address bus connect to each memory chip in the system. In the modern memory system, data-in and data-out may be combined on a bidirectional data bus. Control bus signals may include the chip enable, the read/write control, and the output enable.

chip in the system. The chip select inputs of the memory may serve several functions but are normally a part of the address system. The chip select or chip enable functions differ from the address in that the chip select either selects or deselects the entire memory chip, while the address inputs determine which word or bit is addressed within the chip. The read/write control permits the RAM to be switched between read and write modes. The output enable is often needed to turn the data-out buffers off when other devices are using the data bus.

Chip selects may connect directly to a high-order memory address lead or to an address decoder output that decodes the upper address bits of the memory system. This is shown in Figure 23-2, where the twelve address leads of each device are connected in parallel to the address bus. The single one-of-eight decoder divides the lower half of the memory map into eight 4K blocks.

Address maps are often used to keep track of the location of numerous circuits in a memory address system. Figure 23-3 shows a 64K memory system for which an address map has

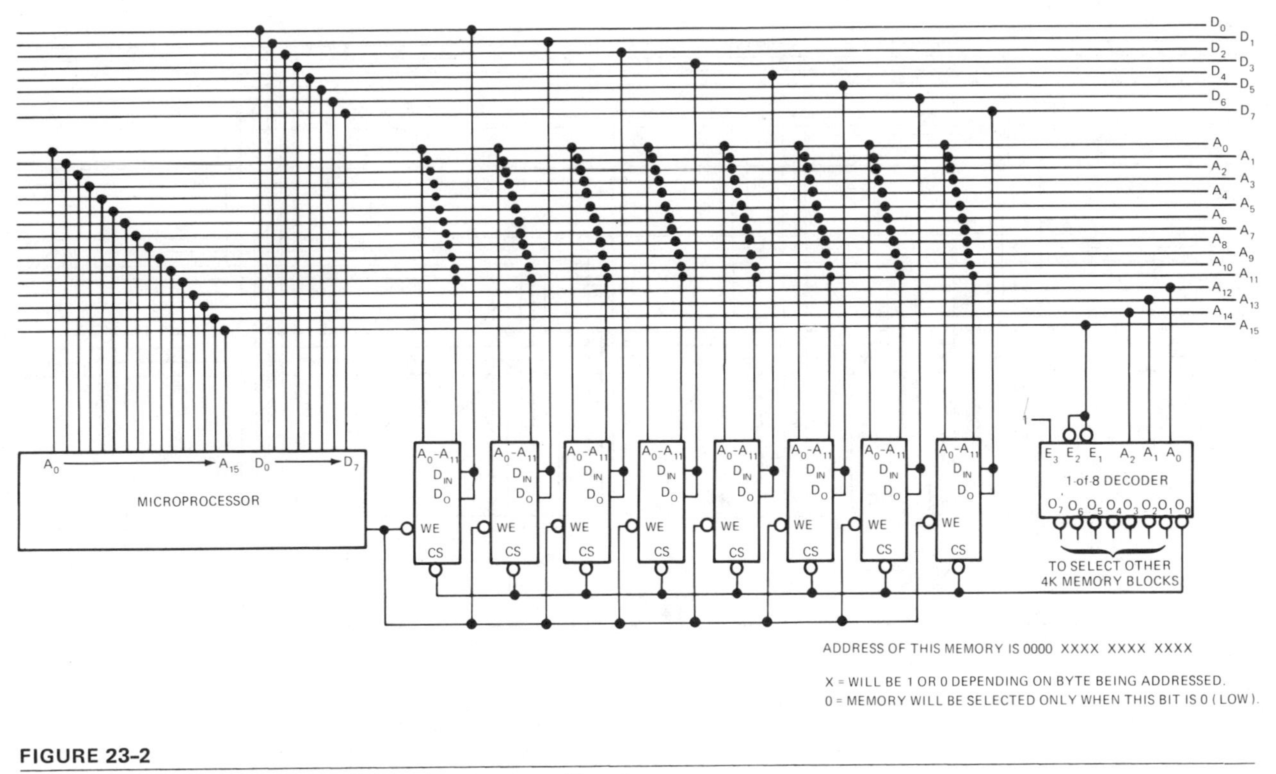

FIGURE 23-2

4K-by-1 static RAM connected to a 16-bit microprocessor address bus.

been drawn. In the example shown, address bit A_{11} is connected directly to the $\overline{CS}$ lead of the 2K-by-8 memory. As the memory map of Figure 23-3(b) shows, the addresses of this RAM will repeat themselves whenever A_{11} is not high. That is, the lower address bits, A_0 through A_{10}, will select any one of the 2048 bytes within the memory during both read and write operations, but only when A_{11} presents a low level to the $\overline{CS}$ input. If other memory devices were to occupy any portion of the extra memory blocks, it would be necessary to deselect this memory chip whenever other memory devices were being addressed. If there were no intention of using one or more of the extra address blocks, the deselection for these blocks would not be needed.

Figure 23-4(a) shows a simple decode gate used to select this 2K block of memory for the lower address block. A_{11} through A_{14} connected through an OR gate will cause the RAM to be addressed only when all four of these upper address lines are low. Failure to include A_{15} in the deselect decoding causes a repetition on the upper half of the memory, as shown in Figure 23-4(b). This image can be easily removed by increasing to a five-input OR gate with A_{15} connected to added input. If desired, the 2K-by-8 RAM can be moved to any of the other 2K blocks in the memory by using inverters at one or more of the OR gate inputs. The object of chip select decoding is to avoid bus contention. Bus contention occurs when intended or image addresses of two or more memories occupy overlapping address blocks, causing contention on the data bus when one device tries to drive data bits high as the other tries to drive them low. However, there is no need to deselect for memory images at addresses that are not going to occur.

The 1K convention: Note that $3FF_H = 1023_{10}$. The reference to this RAM as a "1K memory," although it has a total of 1024 bytes (counting 000), has become a convention. A memory described as "2K bits" is really 2048 (2 × 1024) bits. The same is true of a memory of 2K bytes or 2K words. By this convention, there would be 2048 bytes or 2048 words. At 32K, the actual size is 32,768. At 64K, the actual size is 65,536. At this size, the convention is commonly violated with occasional reference being made to "65K memory." The term *page* is commonly used to indicate 256 memory bytes or words. Each 1K byte or word is four pages.

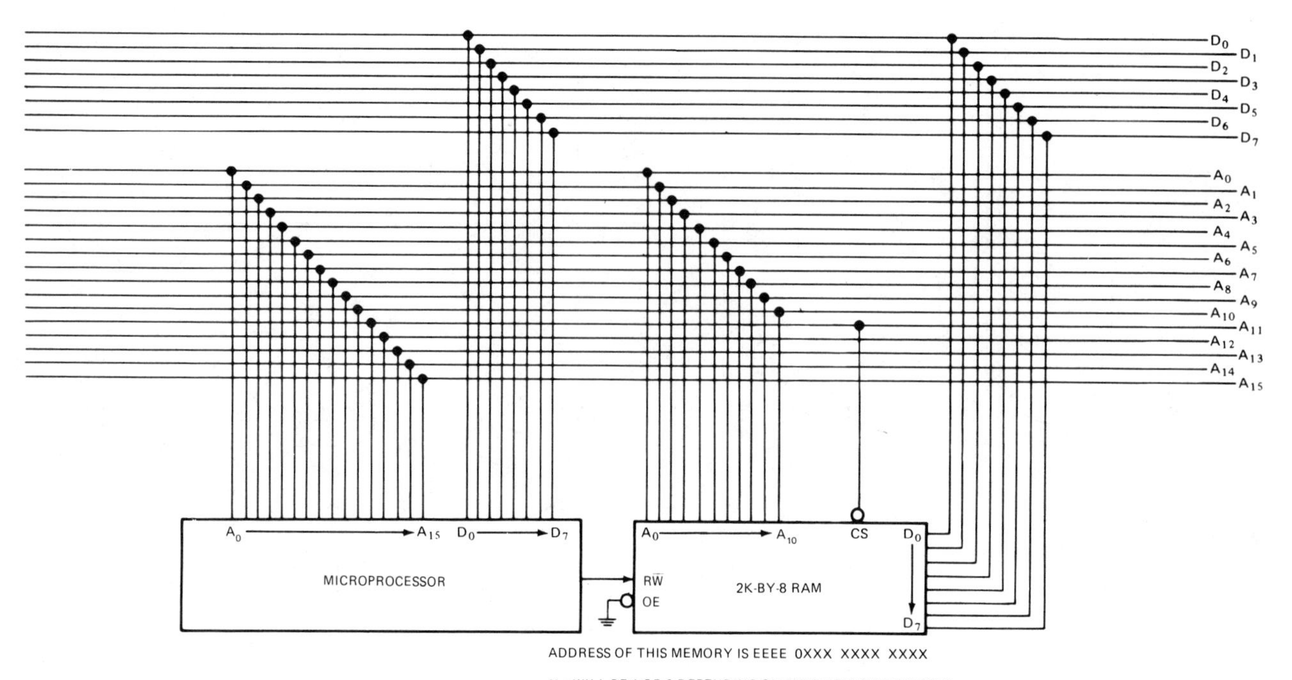

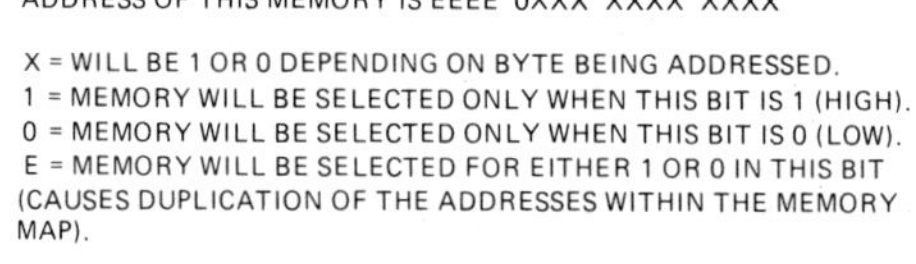

(a)

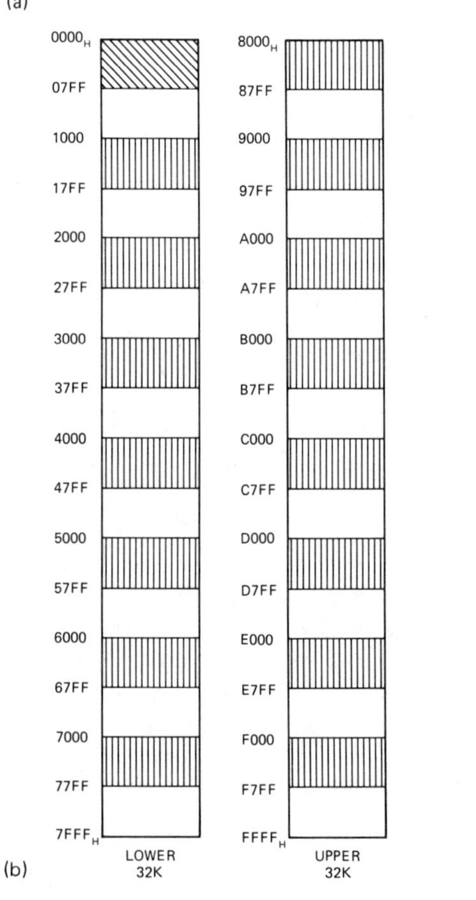

(b)

FIGURE 23-3

(a) 2K-by-8 static RAM connected to a 16-bit microprocessor address bus. (b) Address map of a 64K memory system showing that the addresses of this RAM repeat themselves whenever A_{11} is not high.

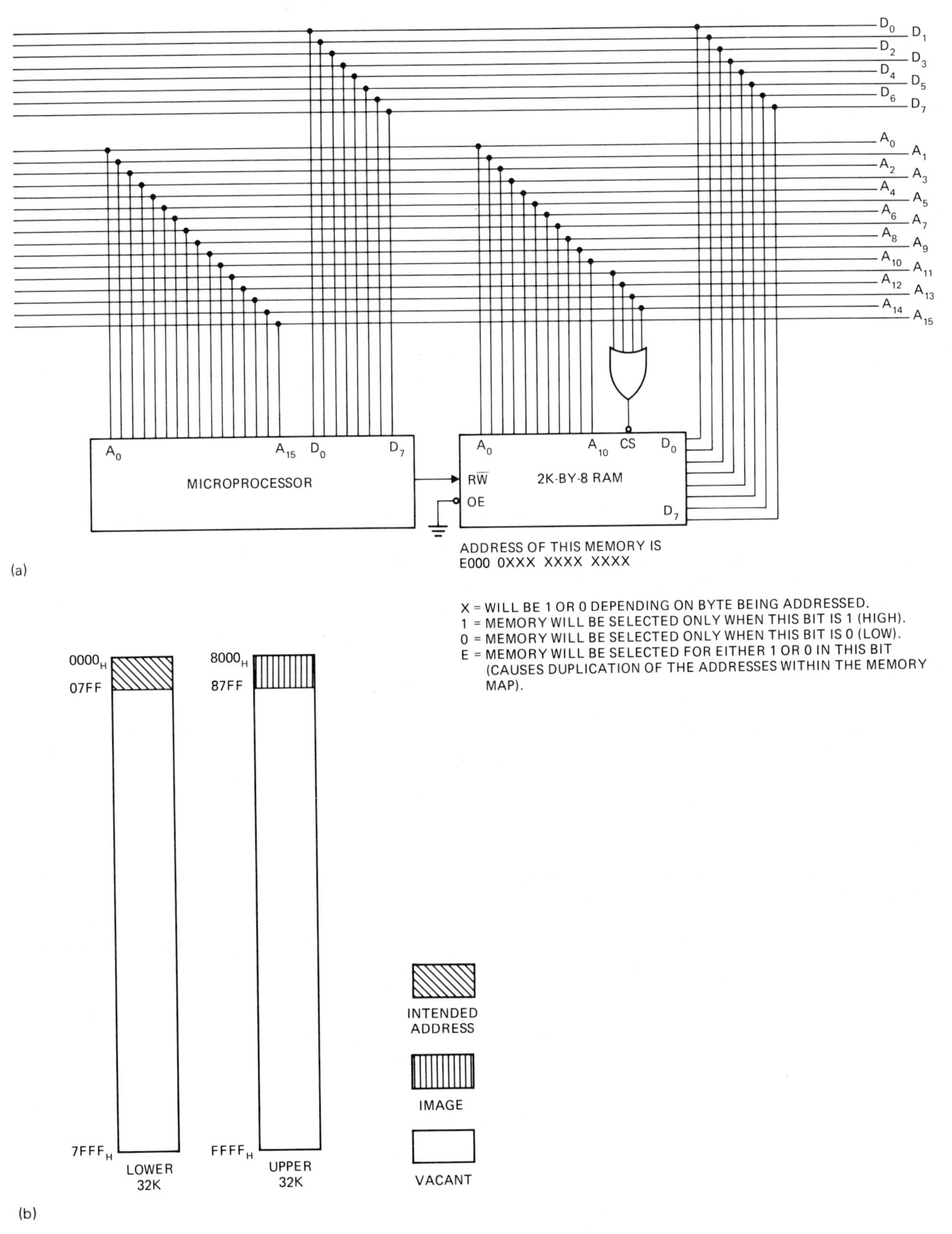

FIGURE 23–4

(a) 2K-by-8 static RAM connected with chip select decoding. (b) Address map of a 64K memory system showing the lower half of the memory addresses vacant for all but the intended address and a repetition on the upper half caused by failure to include A_{15} in the deselect decoding.

FIGURE 23-5
Definition of terms related to the size of memory address and data and their relation to a hypothetical 64K-by-16-bit memory system. *Note:* **The hexadecimal number 1000_H is equal to 4K (4096).**

TERM	DEFINITION	64K-BY-16-BIT MEMORY
K	1024 (2^{10}) (BITS, BYTES, OR WORDS)	64K WORDS 128K BYTES
PAGE	256 WORDS	256 PAGES
WORDS	THE DATA CONTENT OF EACH MEMORY ADDRESS	65,536 WORDS (64K WORDS)
BYTES	8 BITS	131,072 BYTES (2 BYTES PER WORD)

A 64K memory system contains 256 pages. The definitions of terms related to the size of memory address and data and their relation to a hypothetical 64K-by-16-bit memory system are shown in Figure 23-5.

Figure 23-6(a) shows a memory system with less than 32K of an available 64K being used. The lower 16K is occupied by a ROM. A 2K RAM occupies the address above the ROM and is deselected for all addresses used by the ROM. When A_{14} is low, the ROM is selected. When A_{14} is high, the RAM is selected. The ROM, being 16K, has sufficient address to occupy the lower 16K of the address system with 16K separate bytes of memory. The RAM, on the other hand, repeats itself, having image addresses at 2K intervals above the ROM as shown in Figure 23-6(b). The image of the ROM addresses can be vacated by using a two-input OR gate. Other decoding gates can be used to vacate the images of the 2K RAM.

Any one of the intended and image addresses could legitimately be used to address bytes of data in RAM. Normally the programmer is concerned with one block of these addresses. Concern for the images occurs if the memory is to be expanded. Then it is necessary only to add sufficient decoding to deselect the memories in any address block to be occupied by the new devices. If two memory devices occupy the same address block, and if the memory devices supply the same data bits, a contention on the data bus will occur. If two memory devices occupy the same address block, and one supplies lower data bits while the other supplies upper data bits, then there is no contention. (Such a scheme is shown later in Figure 23-10.)

23.2.2 Address Read Timing

Important timing specifications of a memory read cycle are the address access time, chip select access time, and memory read cycle time. These are shown in Figure 23-7. The zero point in timing is the time during which the last address bit becomes valid. The address access time is the period of time, after the system supplies the valid address, that it will take the memory to retrieve the data and supply it at the output buffer pins, assuming the buffers are turned on by the chip select.

Chip select access times are normally shorter than address access times. Although in past examples we have shown CS inputs to be connected directly to address bus leads, they are often ORed with other timing signals to delay the data output. This may be done to avoid gliches that occur on the data bus while the address is changing or to prevent bus contention with other devices that are still outputting to the bus after the previous cycle. This delay of chip select may not be necessary where the memory has an output enable to supply this function.

A few memories have been designed with address systems that are activated by the leading edge of a chip enable. Read cycle timing for

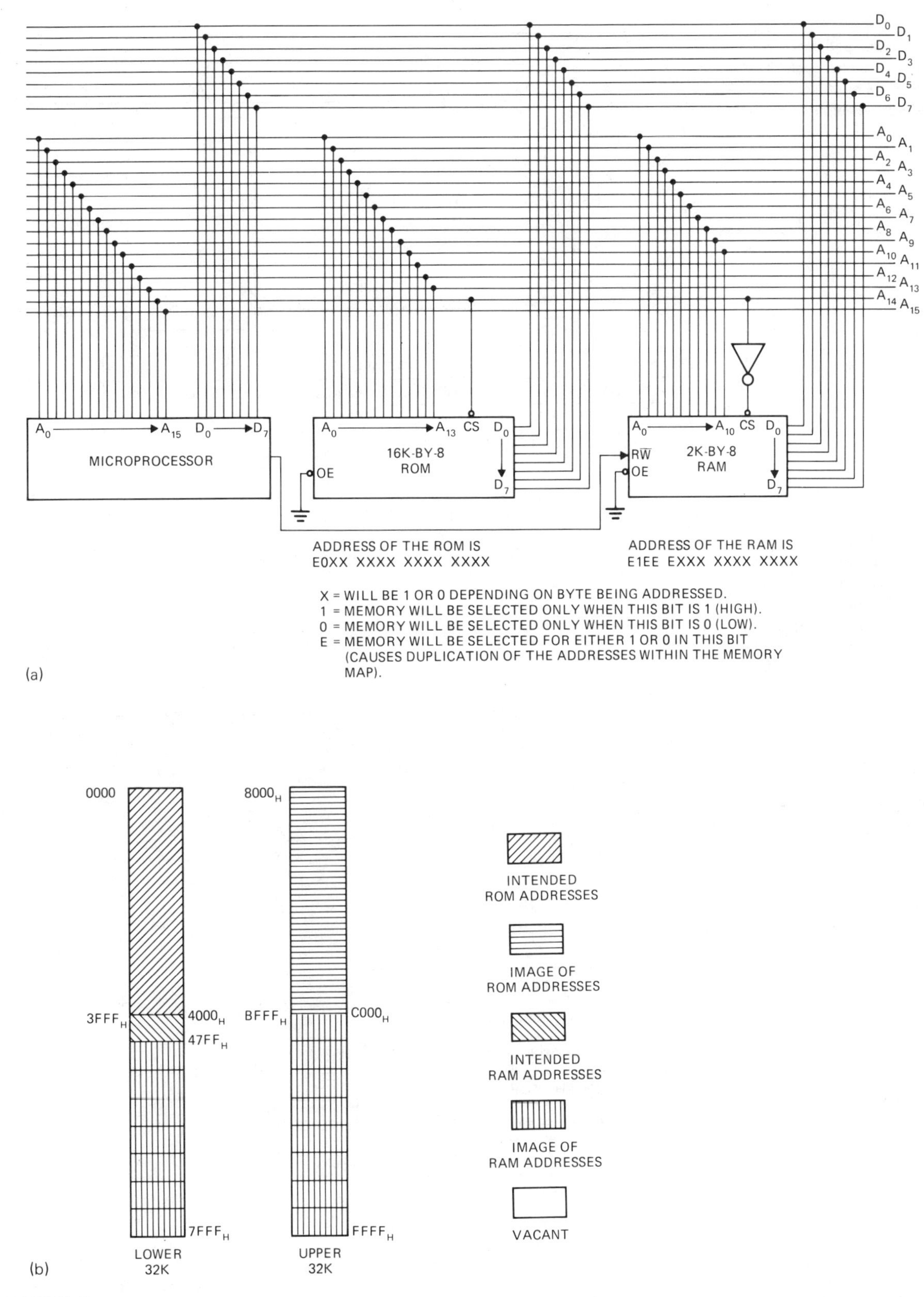

FIGURE 23–6

(a) Memory system containing a 16K-by-8 ROM and a 2K-by-8 RAM. (b) Address map showing intended address and images. The ROM is deselected when A_{14} goes high, causing it to be addressed in the lowest 16K of memory address. The RAM is selected when A_{14} goes high, causing it to occupy the address blocks beginning at 4000_H just above the ROM.

FIGURE 23-7

Timing diagram and specifications of the memory read cycle. The memory cycle normally begins when all address inputs become valid.

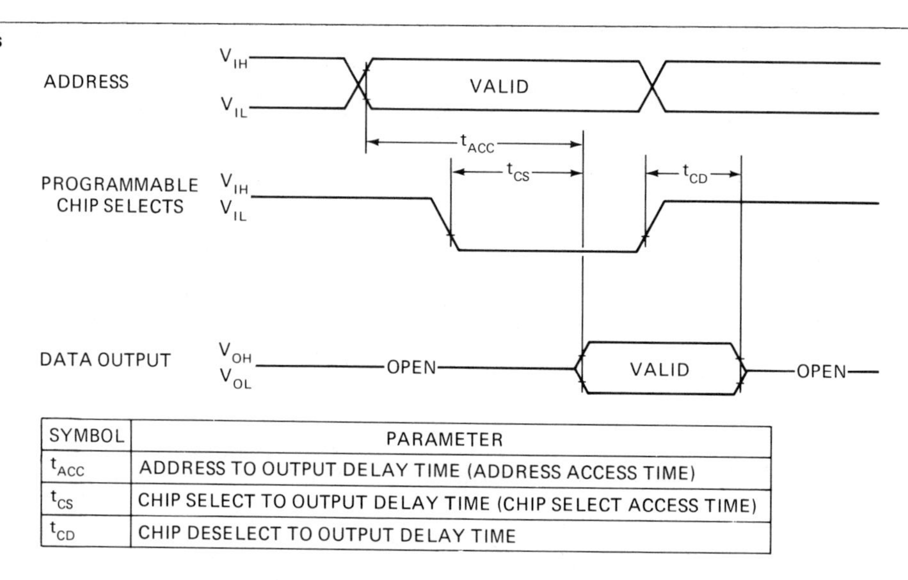

SYMBOL	PARAMETER
t_{ACC}	ADDRESS TO OUTPUT DELAY TIME (ADDRESS ACCESS TIME)
t_{CS}	CHIP SELECT TO OUTPUT DELAY TIME (CHIP SELECT ACCESS TIME)
t_{CD}	CHIP DESELECT TO OUTPUT DELAY TIME

FIGURE 23-8

Timing diagram and specifications of a read cycle for memories with dynamic addressing. Timing is referenced to the leading edge of the chip enable rather than valid address time.

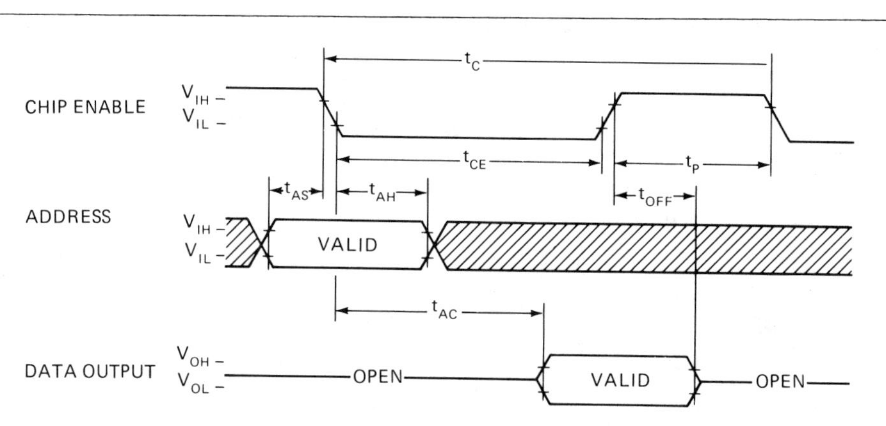

SYMBOL	PARAMETER
t_C	CYCLE TIME
t_{CE}	$\overline{CE}$ PULSE WIDTH
t_{AC}	$\overline{CE}$ ACCESS TIME
t_{OFF}	OUTPUT TURN OFF DELAY
t_{AH}	ADDRESS HOLD TIME REFERENCED TO $\overline{CE}$
t_{AS}	ADDRESS SETUP TIME REFERENCED TO $\overline{CE}$

such a memory is shown in Figure 23-8. For such memories, the address must be valid (t_{AS} = address setup time) before the edge of chip enable and must remain valid (t_{AH} = address hold time) after the leading edge of chip enable. For this type of memory, timing is referenced to the leading edge of chip enable rather than valid address time.

FIGURE 23-9
Write cycle with bus contention occurring.

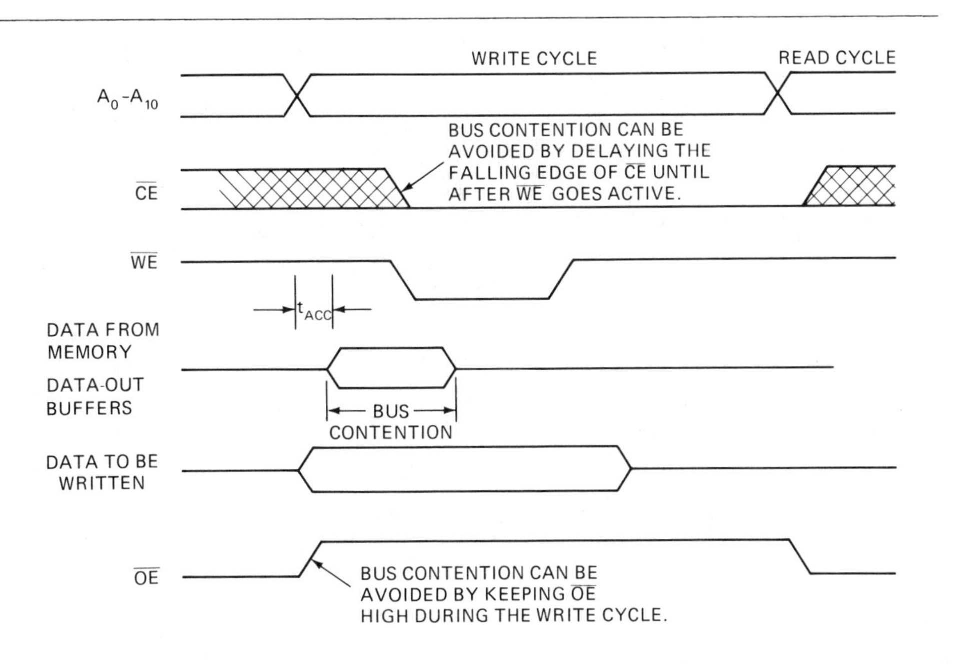

23.2.3 Data Bus and Data Bus Expansion

Several variations exist in memory data bus organization. They can be bit- or word-organized. They may have separate or common I/O. These differences were shown in Figure 22-12. Because of pin-out limitations, common I/O is usual for word-organized memories. Bit-organized memories normally have separate I/O, but the data-in and data-out leads can often be shorted externally for use on common I/O buses. The use of a common I/O bus leads to a problem of bus contention between the output buffers of the memory device (data-out) and the data-in being supplied by the system for storage in the addressed memory.

For a memory write to occur, the address must be valid, the chip must be selected, and the write enable must go active. This timing is shown in Figure 23-9. When $\overline{CE}$ goes low, the memory device, not realizing it is in a write cycle, will output to the data bus. If the intended input data are already on the bus, a bus contention will occur. When write enable goes active, this signal turns off the output buffers, but a bus contention has occurred between the output buffers of the memory being addressed and the data to be written. If the duration of the bus contention is short enough, there may be no detrimental effect. If the duration is lengthy, the bus may not recover in time to write valid data, overheating of circuits may occur, or excess power dissipation may result. Bus contention of this type may be prevented by keeping the chip select inactive until after write enable becomes active. If the memory circuit is equipped with an output enable, this control can be used to keep the output buffers off through the write cycle.

Computer data buses may vary in width from as few as four bits to as many as 16 or more bits. The majority of present-day microprocessor-based systems are eight data bits wide while minicomputer buses may be 12 or 16 data bits wide. Where the memory chip is bit-organized or its word size is shorter than the word size of the system data bus, they are connected with addresses in parallel, but each chip will output to different data bits. This is shown for 1K-by-4 memory in Figure 23-10 and for 4K-by-1 memory in Figure 23-2. In Figure 23-10, the four data bits of one device connect to data bus D_0 through D_3. The data leads of the other device connect to D_4 through D_7. A triple three-input OR gate provides expansion of the chip select.

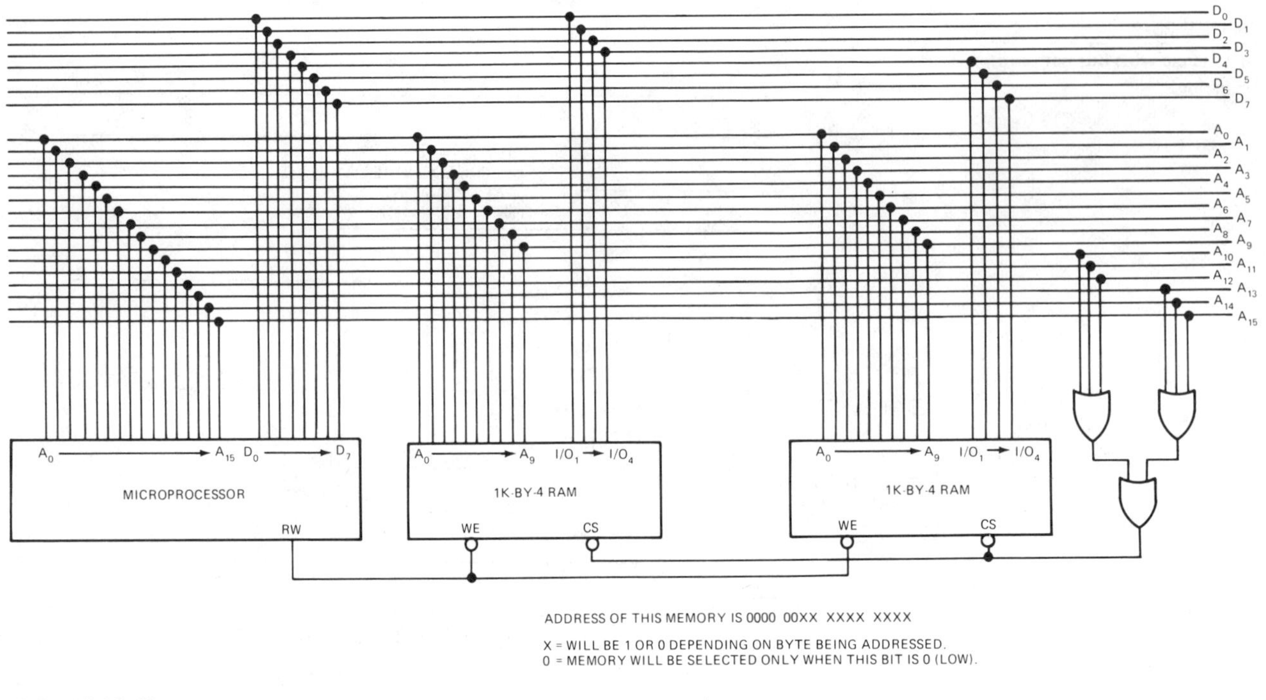

FIGURE 23–10

Two 1K-by-4 static RAMS (2114) with the addresses connected in parallel to provide four pages of microprocessor memory.

23.2.4 Write Cycle Timing

Figure 23–11 shows a typical waveform diagram for write cycle timing. The write cycle time, t_{WC}, is for most memories identical to the read cycle time. For most memories, a write occurs when both chip select and write enable are active. To avoid writing into an erroneous address, the address must be stable at least t_{AS} (address setup time) before the write time. The end of write time occurs when either $\overline{WE}$ or $\overline{CS}$ goes high. For valid data to be written, the data must be stable at least t_{DW} (data setup time) before the end of write time and remain stable for at least t_{DH} (data hold time) after the write time.

For memories with separate I/O, t_{CW}, a chip select to end of write time, may be specified, but for those with common I/O, chip select may come after write enable to avoid bus contention between data-out and data-in. The two most common timing parameters of the data-in are setup time, t_{DW}, and hold time, t_{DH}. The data-in must be valid t_{DS} before the end of $\overline{WE}$ and must remain stable t_{DH} after $\overline{WE}$. Write enable must usually remain active a minimum period of time to accomplish an effective write. Therefore, a minimum write pulse width t_{WP} is also common.

23.3 Bipolar Random Access Memories

23.3.1 TTL Random Access Memories

The static cell of a TTL memory is a flip-flop of the type shown in Figure 23-12. The memory can be bit-organized or word-organized. If it is bit-organized, each integrated circuit is like a single plane of a ferrite core memory, providing only one data bit for any address in the memory. To obtain words of N bits in length, N circuits must be wired with the address going in parallel to each circuit, but with the data outputs on separate leads. When the cell is not addressed, either X or Y or both will supply a ground to the emitter. When the cell is addressed, the X and Y emitters are high, causing the emitter current to flow through the read/write lines. To write into the cell, one of the read/write lines is

FIGURE 23–11

Timing diagram and specifications of the memory write cycle. A memory write occurs when both $\overline{CS}$ and $\overline{WE}$ become active. The end of write time occurs when either $\overline{CS}$ or $\overline{WE}$ goes high.

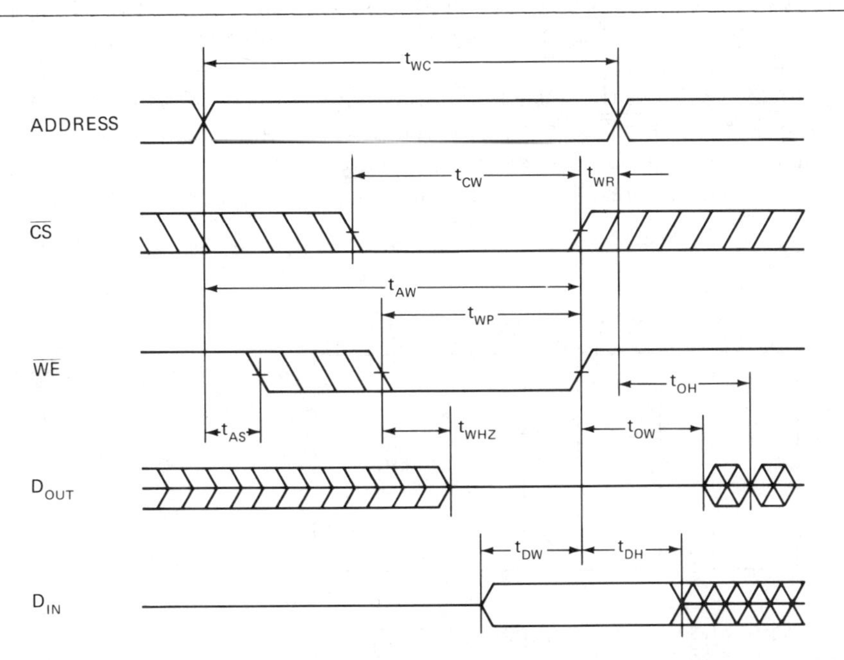

SIGNAL MAY BE CHANGING FROM LOW TO HIGH

SIGNAL MAY BE CHANGING FROM HIGH TO LOW

SIGNAL MAY BE CHANGING FROM HIGH IMPEDANCE TO LOW OR HIGH

SYMBOL	PARAMETER
t_{WC}	WRITE CYCLE TIME
t_{CW}	CHIP SELECTION TO END OF WRITE
t_{AW}	ADDRESS VALID TO END OF WRITE
t_{AS}	ADDRESS SETUP TIME
t_{WP}	WRITE PULSE WIDTH
t_{WR}	WRITE RECOVERY TIME
t_{OHZ}	OUTPUT DISABLE TO OUTPUT IN HIGH Z
t_{WHZ}	WRITE TO OUTPUT IN HIGH Z
t_{DW}	DATA TO WRITE TIME OVERLAP
t_{DH}	DATA HOLD FROM WRITE TIME
t_{OW}	OUTPUT ACTIVE FROM END OF WRITE

FIGURE 23–12

Bipolar memory cell.

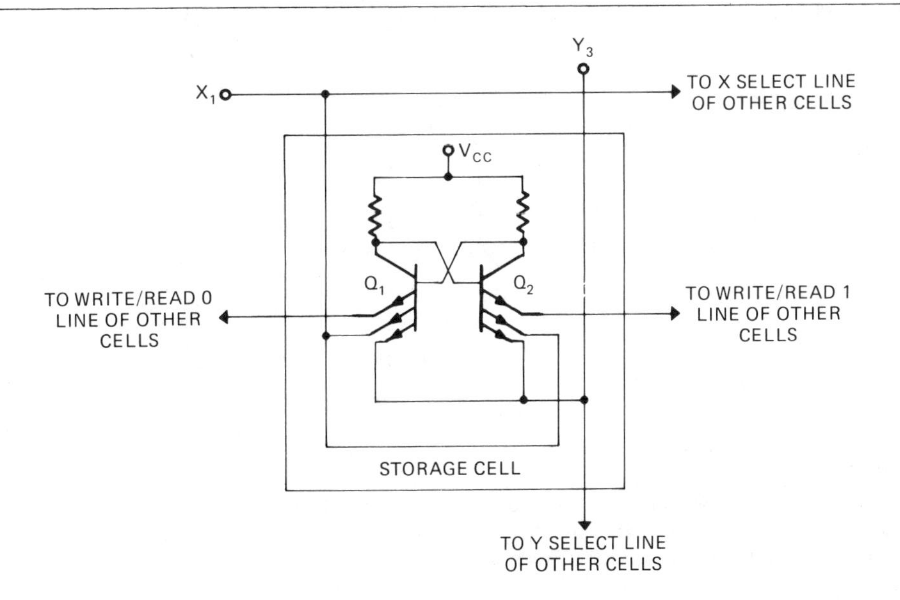

held high during the address while the other is grounded, causing one transistor to conduct while the other turns off.

Figure 23–13 shows the basic block diagram of a 256-by-1 memory. The address is handled in the coincident X Y fashion as in the ferrite core memory. The address bits A_0 through A_3 are decoded internally into 16 X leads, and A_4 through A_7 are decoded into 16 Y leads. If the address were 10010111, we would be enabling each of the 16 cells in the X_9 line to transfer the 1 or 0 state to its corresponding Y line, but only the Y_7 line would be enabled to transfer the data bit to the output. The addressed data

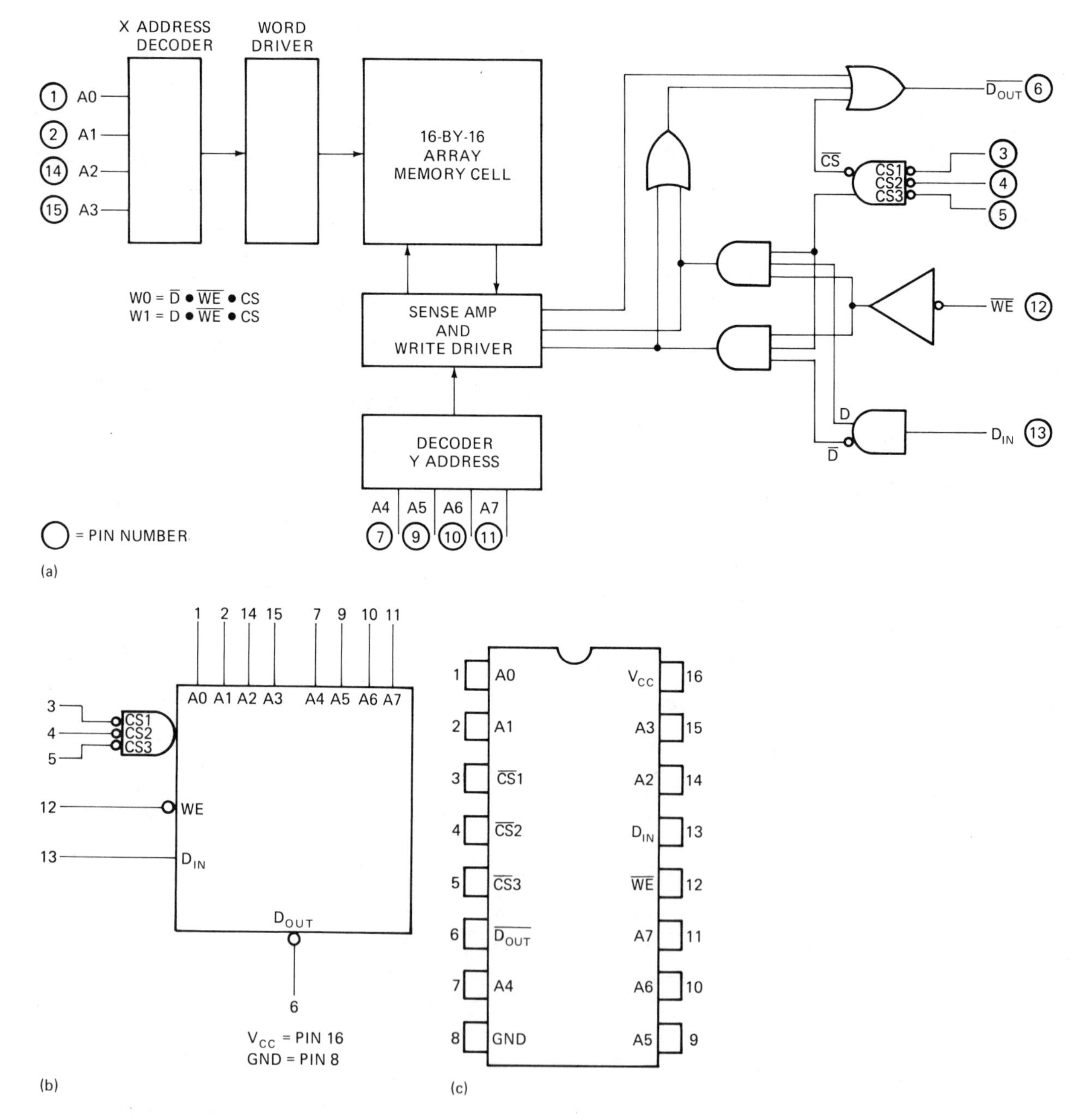

FIGURE 23–13

The Fairchild 93411, a 256-word-by-1-bit memory. The eight-bit address inputs decode to 16 X and 16 Y lines. There is a single data input (D_{IN}) and a single data output lead ($\overline{D_{OUT}}$). (a) Logic diagram. (b) Logic symbol. (c) Connection diagram.

would appear in inverted form at the $\overline{D_{OUT}}$ provided all $\overline{CS}$ inputs were low and the $\overline{WE}$ was high.

To write data into an addressed location, the $\overline{WE}$ must go low. When this happens, the level appearing at the D_{IN} lead will be written into the addressed location. The operation occurs in accordance with the truth table of Figure 23-14. If any $\overline{CS}$ inputs are high, neither read nor write operations will occur. Figure 23-15 shows the timing for a read cycle. The data-out will occur a maximum of (t_{ACS}) 30 nanoseconds after chip select goes low, and a maximum of (t_{AA}) 55 nanoseconds after the address. The data-in (D_{IN}) must be low and the write enable ($\overline{WE}$) must be high during read cycle. The data-out will turn off a maximum of 25 nanoseconds after removal of the chip select.

Figure 23-16 shows a more advanced version of TTL RAM, the 93L422. It is organized as 256 by 4 and has separate I/O. Figure 23-17 shows two of these devices wired in parallel for a 256-by-8-bit page of memory occupying addresses 400 through 4FF. Note that address and chip selects are wired in parallel so that both devices read and write at the same addresses. One device supplies memory for the lower four bits of data, and the other supplies memory for the upper four bits. Timing for this RAM is specified in a fashion similar to that of the 93411. The 93L422 has a maximum address access time of 65 nanoseconds with a typical power drain of 275 mW. A slightly faster version, the 93422, has a maximum address access time of 45 nanoseconds with a typical power drain of 475 mW. It has the same pin-out but consumes more power.

INPUTS			OUTPUTS	MODE
$\overline{CS}$	$\overline{WE}$	D_{IN}		
1	X	X	0	NOT SELECTED
0	0	0	0	WRITE 0
0	0	1	0	WRITE 1
0	1	0	$\overline{D_{OUT}}$	READ

FIGURE 23-14

Truth table of the 93411 256-by-1 memory. The truth table applies for any given address.

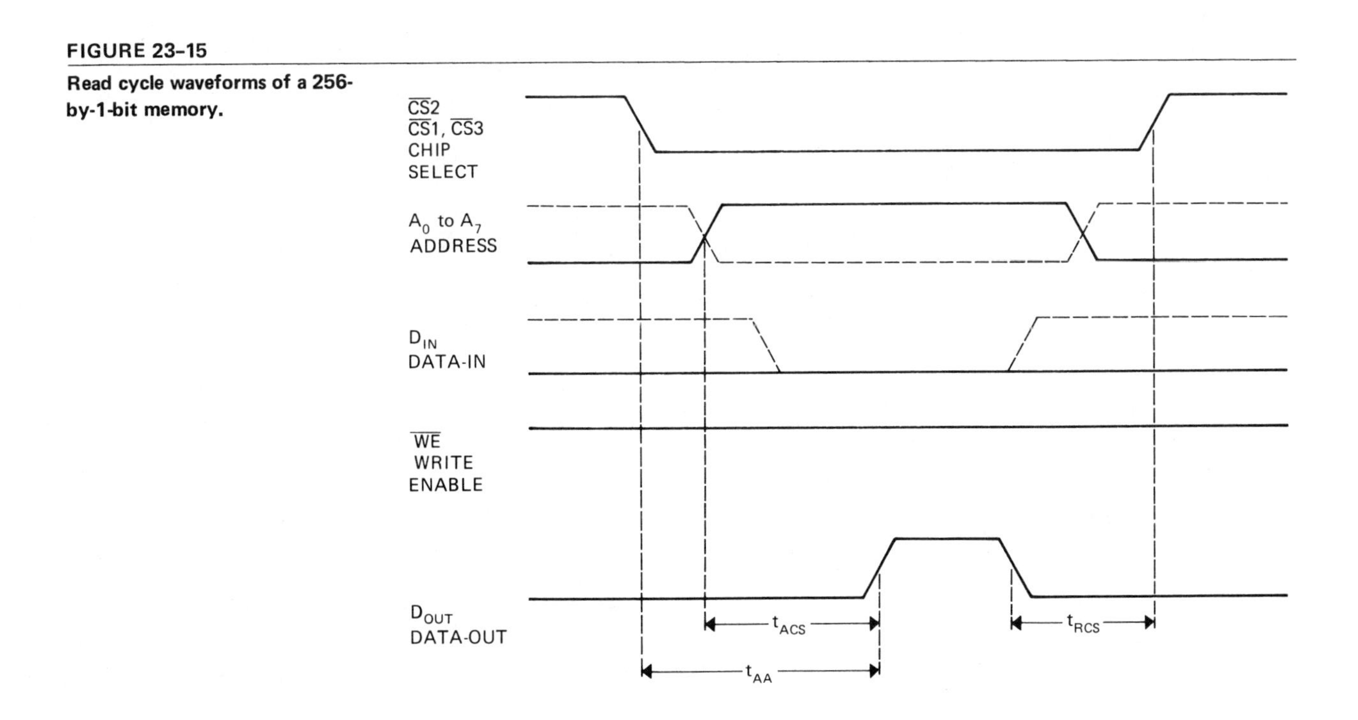

FIGURE 23-15

Read cycle waveforms of a 256-by-1-bit memory.

FIGURE 23-16
256-by-4 high-speed bipolar RAM, the 93L422. (a) Logic diagram. (b) Connection diagram of 22-pin DIP package.

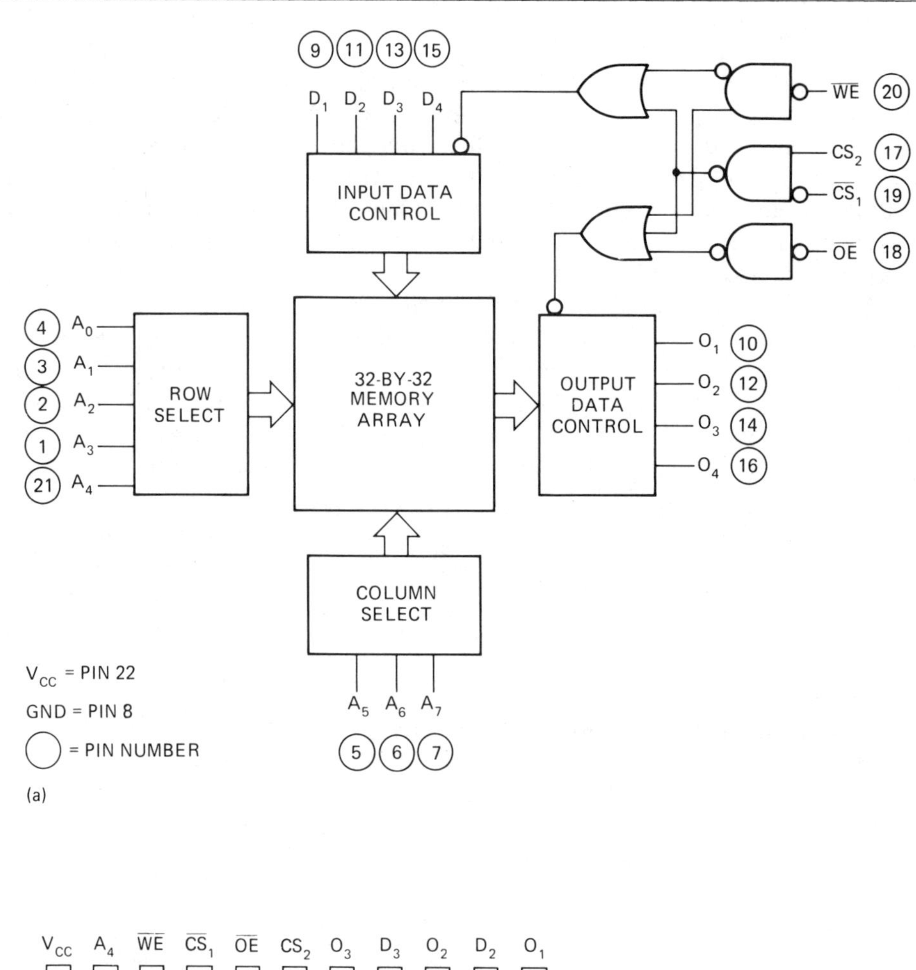

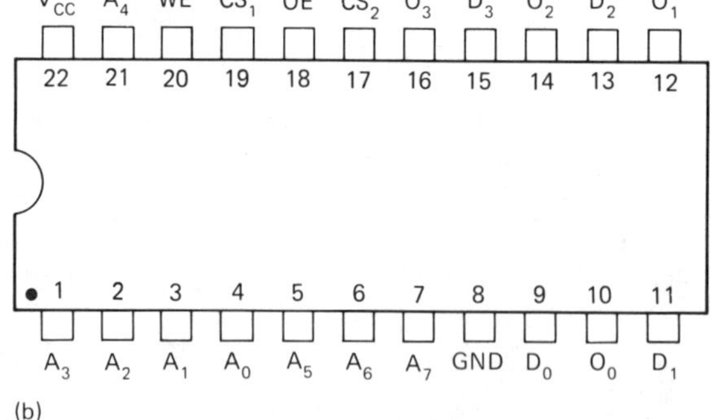

(b)

23.3.2 Emitter-Coupled Logic Memory

In selecting a memory, speed is an important factor. The 256-by-1-bit TTL memory of Figure 23-13 has a typical access time of 45 nanoseconds.

Figure 23-18 is a Fairchild F10415 ECL 1024-by-1 memory. The D_{OUT} lead is open-emitter, which differs from open-collector in that a resistor must be tied between the D_{OUT} and $-V_{EE}$ instead of $+V_{CC}$. This does, however, provide wired OR connection of the output. It has an access time of 35 nanoseconds. This

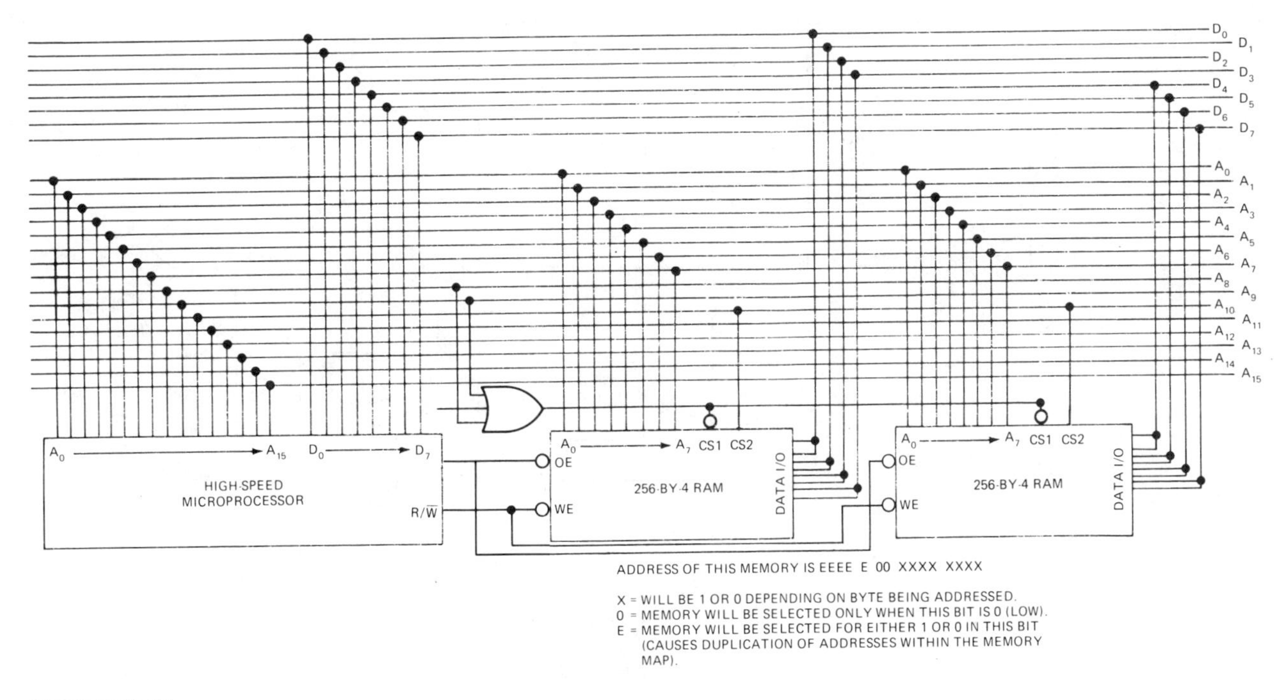

FIGURE 23-17

Two 93L422 RAMs wired in parallel for a 256-by-8-bit page of memory occupying addresses 400 through 4FF.

memory is, like the 256-by-1-bit memory, a 16-pin circuit, but it differs in that a single chip select input is provided. The two pins vacated are then used for address leads; this provides a 1024-bit address. It can be expanded to N bits by paralleling the address lines using the method shown in Figure 23-13.

The CS inputs are additional control inputs and, in large memory arrays, can be used to expand the address. Expansion of the number of bits per word, however, is accomplished by using a separate chip for each data bit and paralleling the address lines. Figure 23-19 shows the timing diagram for a write cycle of this memory. $\overline{WE}$ is not effective until 5 nanoseconds (t_{WSCS}) after $\overline{CS}$. $\overline{WE}$ pulse should not occur until 5 nanoseconds (t_{WSA}) after the address is complete. D_{IN} must be at correct level at least 5 nanoseconds (t_{WSD}) before $\overline{WE}$. $\overline{WE}$ must have a minimum pulse width of 30 nanoseconds (t_W). D_{IN} must be held at least 5 nanoseconds (t_{WHD}) after $\overline{WE}$ goes low. The address and $\overline{CS}$ must be held a minimum of 5 nanoseconds (t_{WHA} or t_{WHSC}) after $\overline{WE}$ goes low.

23.4 CMOS Static RAMs

A typical CMOS static RAM cell is shown in Figure 23-20. The P-N pairs of transistors are held in state by opposite voltages on the gate inputs. In one pair, the upper transistor is turned off; in the other, the lower transistor is off. This blocks any significant current flow between V_{DD} and V_{SS}. As is typical of CMOS, current flows only when the cell is being switched. The cell is switched to the state forced on the bit line when the word select goes active during a write cycle. The bit lines are floating during a read cycle, and when the word select goes active, the state of the cell is transmitted through the bit lines to the output.

The prime advantage of the CMOS RAM is the low power dissipation that occurs when the device is in standby. For a large memory composed of numerous CMOS devices, power is consumed primarily in the devices that are selected. The amount of power consumed in a deselected CMOS RAM is very little. Many CMOS RAMs

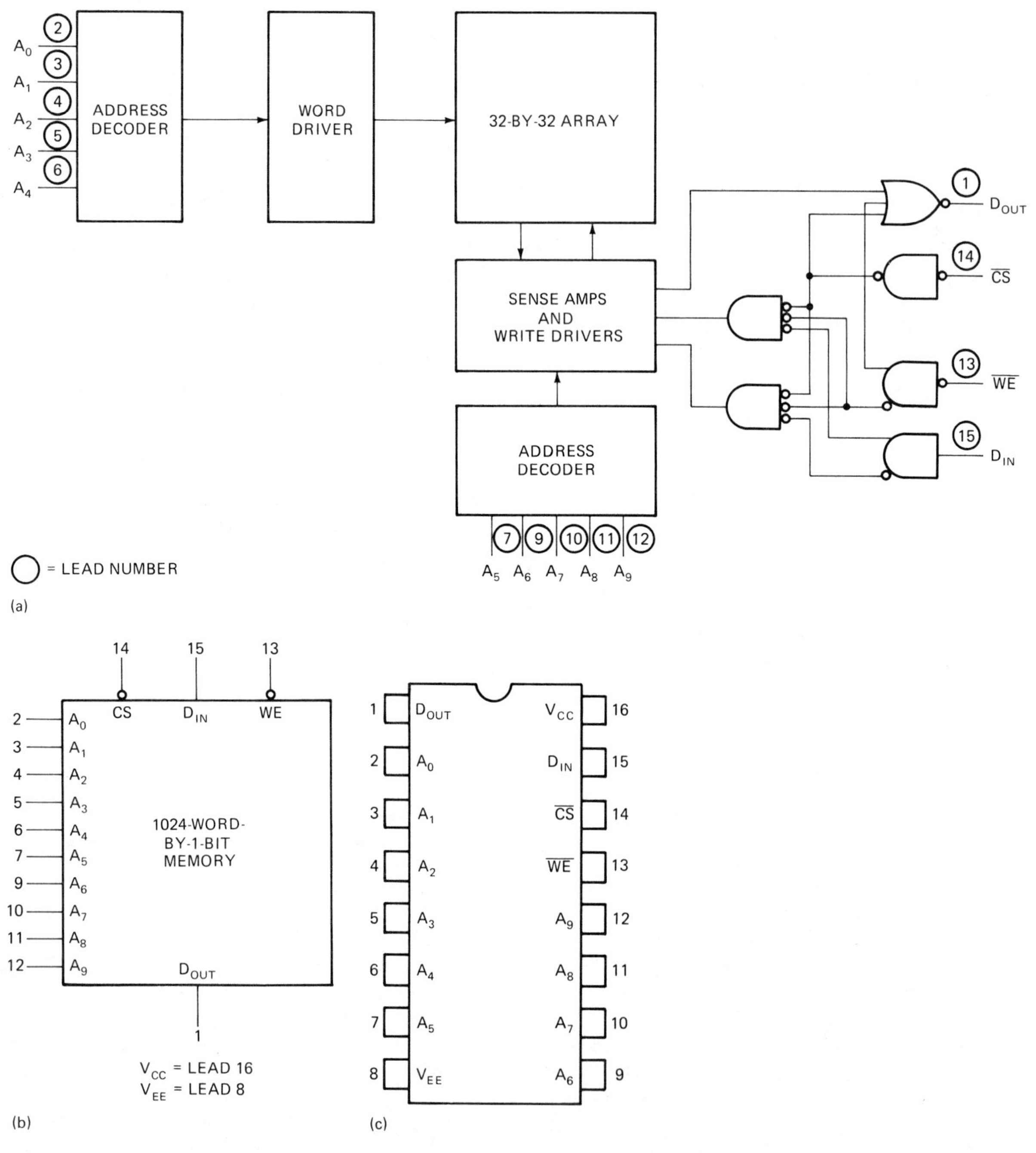

FIGURE 23-18

A Fairchild F10415 1024-by-1-bit RAM ECL memory. (a) Logic diagram. (b) Logic symbol. (c) Connection diagram.

can retain data with supply voltage levels reduced to two volts. This results in even lower power drain and makes it possible to use small batteries to retain the data when main power is removed.

Figure 23-21 compares the power drain of 1K-by-1 RAMs in four technologies. At maximum cycle time, comparisons yield similar results for newer and larger RAMs. However, as the table indicates, for newer RAMs, the speed gap between CMOS and other technologies is narrower.

Figure 23-22 shows the 6116 CMOS RAM, which is configured as a 2K-by-8. It provides high operating speeds, low operating power,

FIGURE 23-19

Write cycle waveforms of 1024-by-1-bit memory.

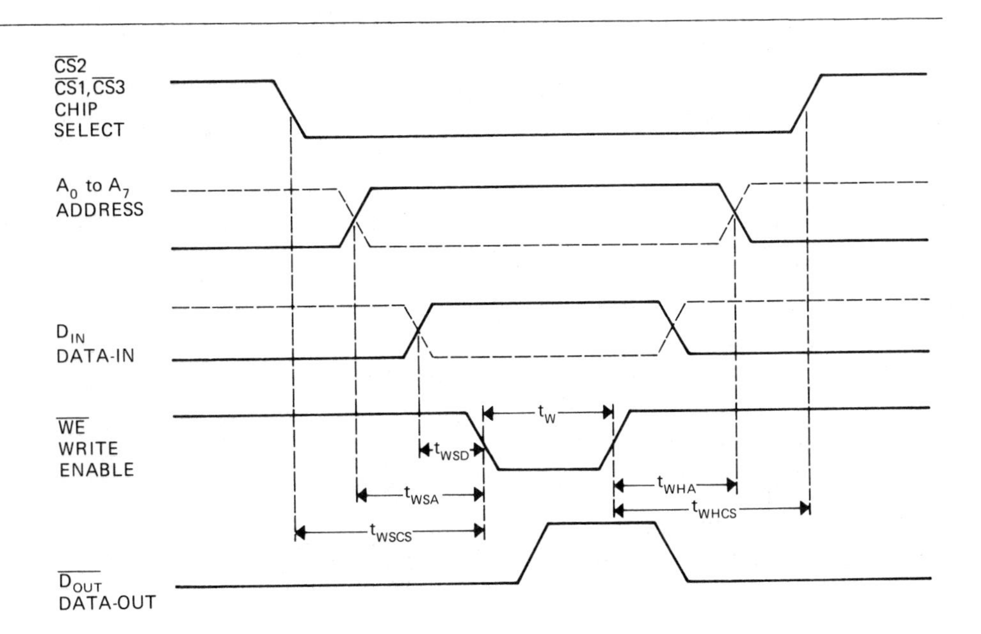

FIGURE 23-20

Typical CMOS static RAM cell. Like the bipolar memory cell, the CMOS cell is a flip-flop, but unlike the bipolar cell, it draws little current.

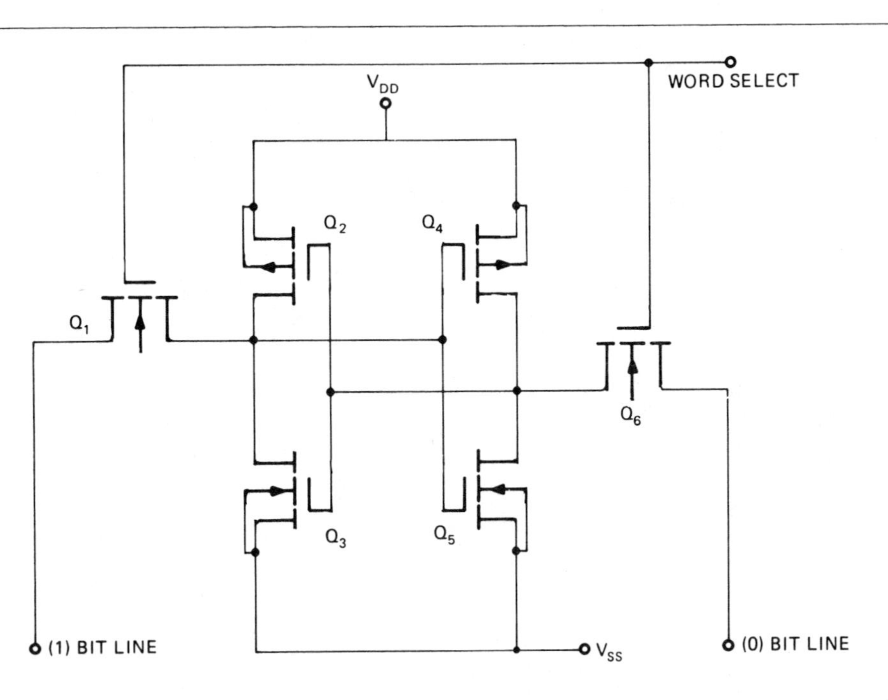

FIGURE 23-21

Speed/power relationships of RAMs in four integrated circuit technologies.

TECHNOLOGY	DEVICE NUMBER	DEVICE SIZE	SPEED (ACCESS TIME)	POWER DISSIPATION (PER BIT)	POWER DOWN FEATURES AUTOMATIC WITH DESELECT
CMOS	6518 6116	1K × 1 2K × 8	90 ns 120 ns	12.5 μW 11 μW	0.5 μW PER BIT
NMOS				300 μW	(ON SOME DEVICES) 20 μW PER BIT
(STATIC)	2115 4802	1K × 1 2K × 8	30 ns 200 ns	4.7 μW	(DYNAMIC RAM WITH $\overline{RAS}$ ONLY REFRESH) 0.36 μW PER BIT
(DYNAMIC)	4164	64K × 1	120 ns		
BIPOLAR (TTL)	93425A	1K × 1	30 ns	475 μW	NONE
ECL	10425	1K × 1	12 ns	525 μW	NONE

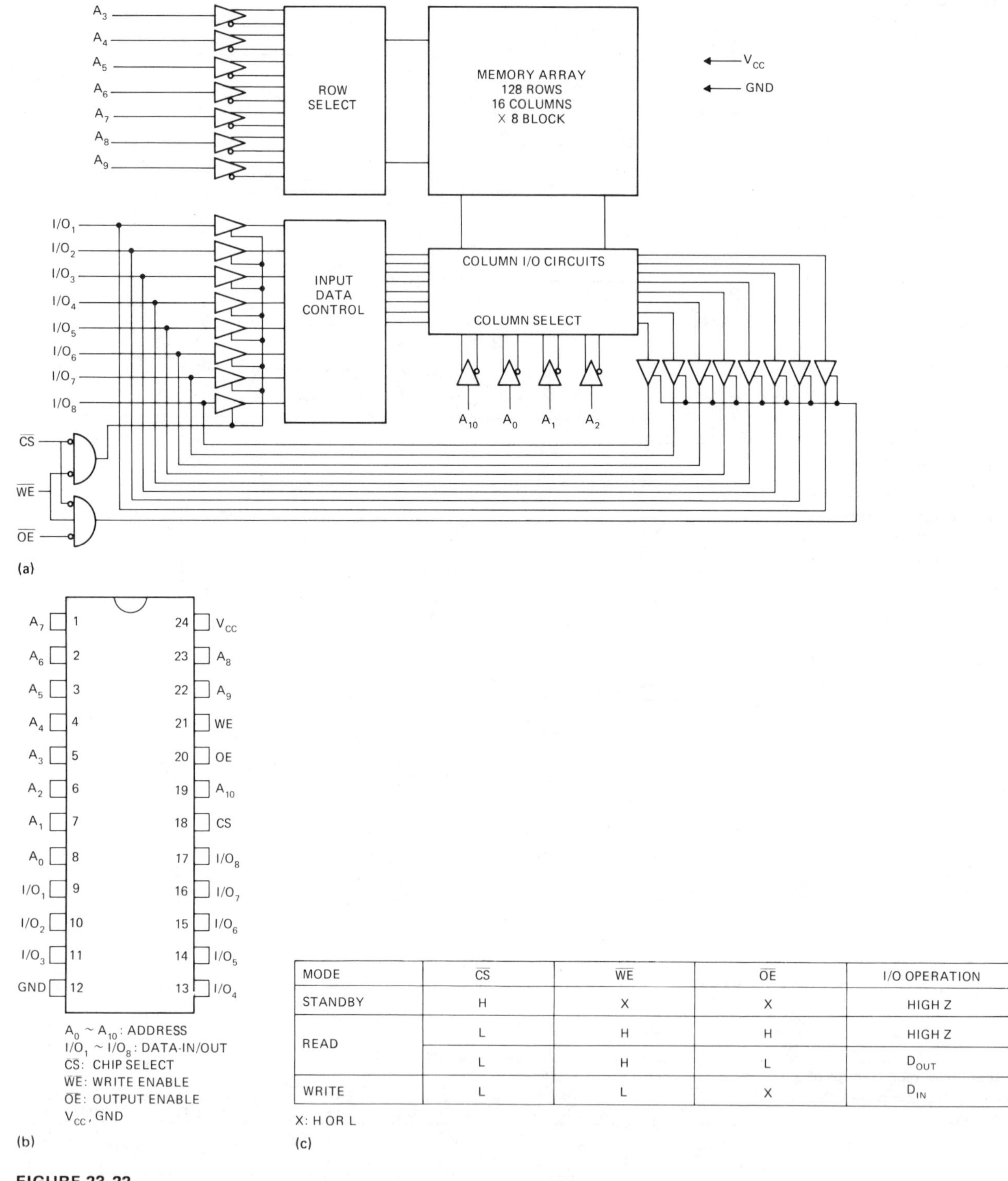

MODE	$\overline{CS}$	$\overline{WE}$	$\overline{OE}$	I/O OPERATION
STANDBY	H	X	X	HIGH Z
READ	L	H	H	HIGH Z
	L	H	L	D_{OUT}
WRITE	L	L	X	D_{IN}

X: H OR L

(c)

FIGURE 23-22

The 6116 CMOS RAM, configured as a 2K-by-8 RAM. (a) Functional block diagram. (b) Pin arrangement. (c) Truth table.

automatic power down with deselect, and battery backup capability. It is supplied in a 24-pin package. Operating current at minimum cycle time is typically 35 mA. When deselected ($\overline{CS}$ lead high), current drain is typically 4 mA. In the data retention mode with V_{CC} at only 3V, current drain would be no greater than 50 μA. Maximum address access time is 120 nanoseconds.

Figure 23-23(a) shows a typical circuit for automatic application of battery voltage in event of power loss. The V_{CC} to the 6116 is connected

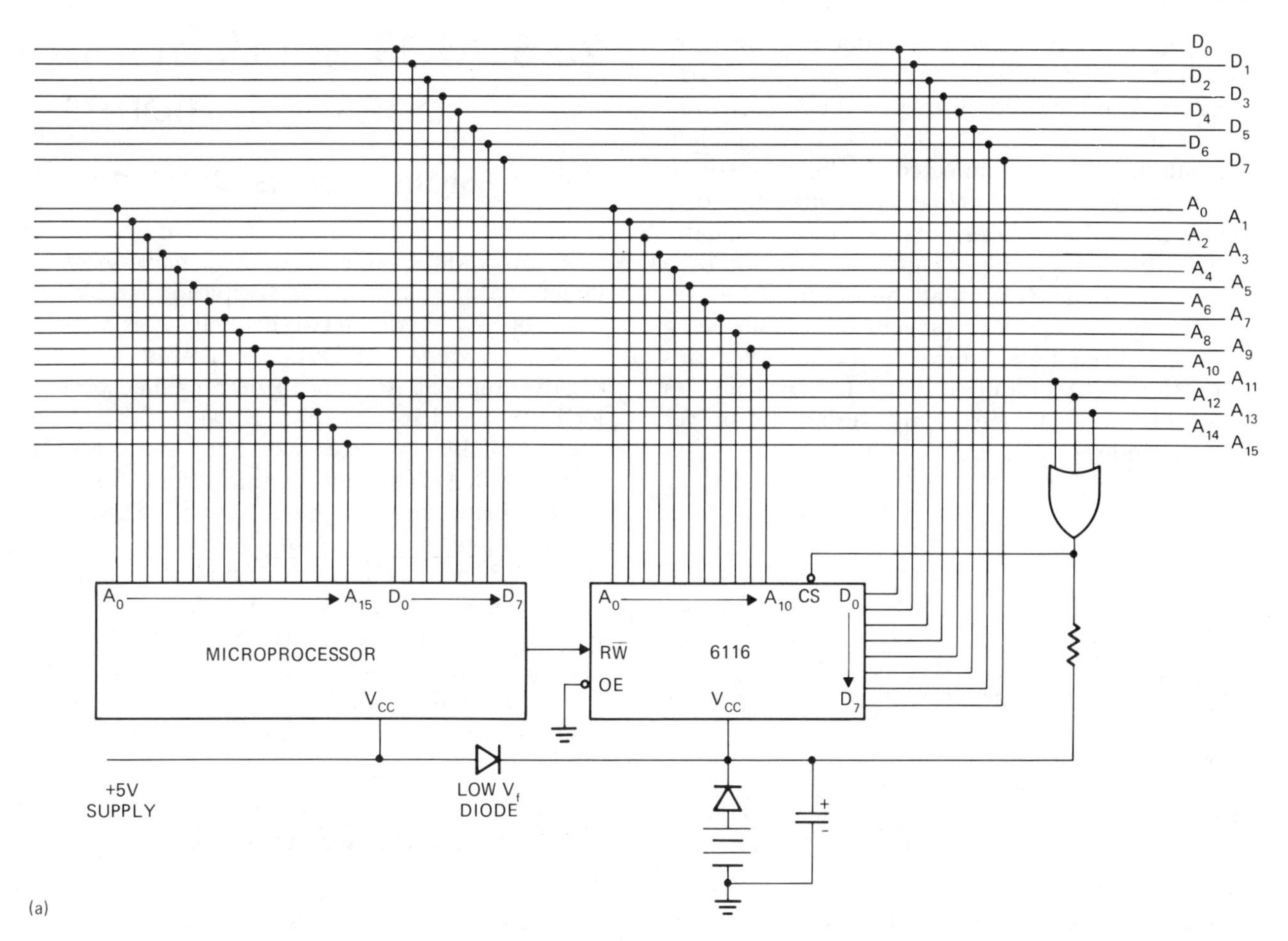

LOW V_{CC} DATA RETENTION CHARACTERISTICS (T_A = 0 to +70°C)

ITEM	SYMBOL	TEST CONDITIONS	MIN	TYP	MAX	UNIT
V_{CC} FOR DATA RETENTION	V_{DR}	$\overline{CS} \geqslant V_{CC} - 0.2V$ $V_{IN} \geqslant V_{CC} - 0.2V$ or $V_{IN} \leqslant 0.2V$	2.0	–	–	V
DATA RETENTION CURRENT	I_{CCDR}	$V_{CC} = 3.0V$, $\overline{CS} \geqslant 2.8V$ $V_{IN} \geqslant 2.8V$ or $V_{IN} \leqslant 0.2V$	–	–	50	μA
CHIP DESELECT TO DATA RETENTION TIME	t_{CDR}	SEE RETENTION WAVEFORM.	0	–	–	ns
OPERATION RECOVERY TIME	t_R		t_{RC} *	–	–	ns

*t_{RC} = READ CYCLE TIME

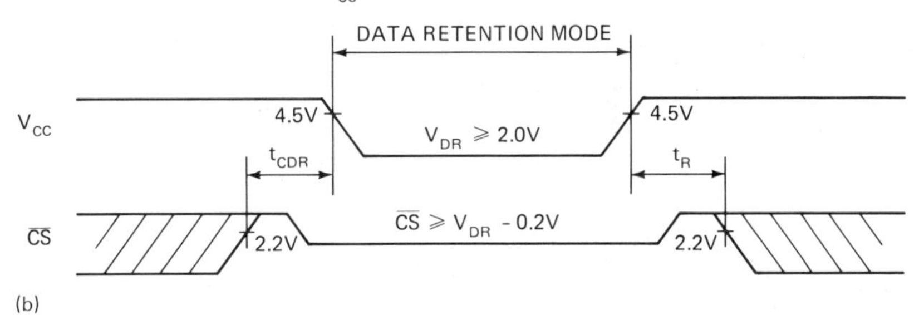

FIGURE 23-23

(a) 6116 CMOS battery backup circuit. (b) Low V_{CC} data retention characteristics and waveforms.

through a low V_f diode (Schottky diode or transistor with base and collector shorted). The 3-volt battery is connected through a diode such that it is turned off when the normal V_{CC} is applied. A pull-up resistor connected to $\overline{CS}$ causes it to go to a level slightly below V_{DR} when the logic gate goes dead from loss of power. Figure 23-23(b) shows the waveforms for the data retention mode. Note that the $\overline{CS}$ must be held at the deselect level to prevent data from being written during power down. Some older versions of CMOS RAMs required that $\overline{WE}$ (write enable) be switched high before and during the power down mode.

DATE AVAILABLE	BIT-ORIENTED	WORD-ORIENTED	SPEED (CYCLE TIME)
1975	1K × 1 (2102)	256 × 4 (2101)	500 ns
1978	4K × 1 (2141)	1K × 4 (2114)	250 ns
1981	16K × 1 (2167)	2K × 8 (4802)	70 ns

FIGURE 23-24

Advancement in speed and capacity for NMOS static RAMs.

23.5 MOS Random Access Memories

23.5.1 Static versus Dynamic

In general, the MOS RAM is used more widely than the CMOS, bipolar, or ECL because of its larger capacity, low power dissipation, and lower cost per bit. There are two general types in use—static and dynamic. Dynamic RAMs have the disadvantage of needing to be refreshed periodically to avoid loss of data. However, they have been available in larger sizes than static RAMs. Dynamic RAMs are bit-organized and, for most small capacity memory needs, are more awkward to use than static RAMs. Therefore, we will discuss static RAMs first. Figure 23-24 shows the advance in NMOS static RAMs from the 1K-by-1 level available in 1975 to the 2K-by-8 level available in 1981. In this same time period, speed of NMOS static RAMs advanced from 500 nanoseconds access time to under 100 nanoseconds.

23.5.2 Static RAM Cell

Figure 23-25 shows an NMOS static RAM cell. It is similar to the CMOS RAM cell in that it is

FIGURE 23-25

NMOS static RAM cell.

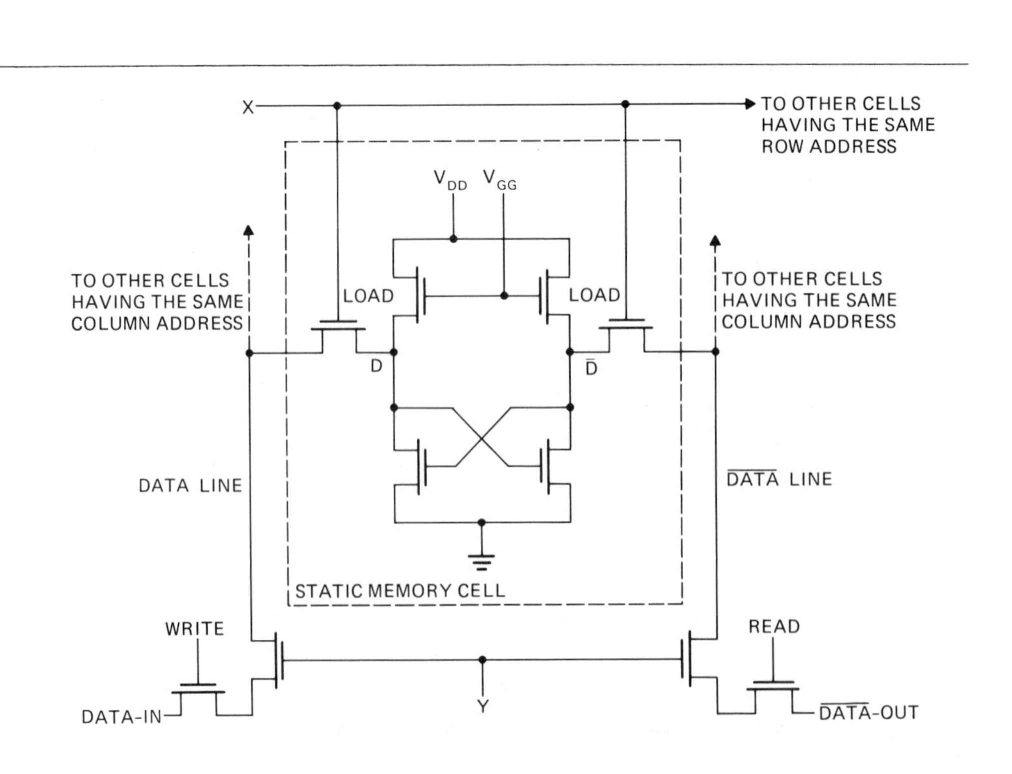

composed of a pair of cross-coupled inverters forming a flip-flop. The upper transistors, however, act as load transistors and a small amount of current flows through the side that is turned on in order to provide the voltage difference between D and $\bar{D}$. The cell size of the NMOS is smaller than that of the CMOS, allowing for higher density.

23.5.3 Word-Organized Static RAMs

The 2114 1K-by-4 static RAM was the first common I/O static RAM to be offered in the 300-mil-wide DIP package. At the time, it was ideal for eight-bit microprocessor application in that it took a width of only two devices to supply the eight data bits. It was widely sourced and was produced in quantities and at prices that made its use widespread. Figure 23-26 shows the block diagram and pin-out of the 2114.

Because the majority of static RAMs were used with eight-bit microprocessors, a byte-wide RAM was an ideal configuration. The 4801 1K-by-8 and the 4802 2K-by-8 static RAMs have become widely sourced. Figure 23-27 shows the pin-out of these RAMs. The 4802 differs from the 4801 in that pin 19 is an NC (no connect) on the 4801, and a higher-speed version is available for the 4801.

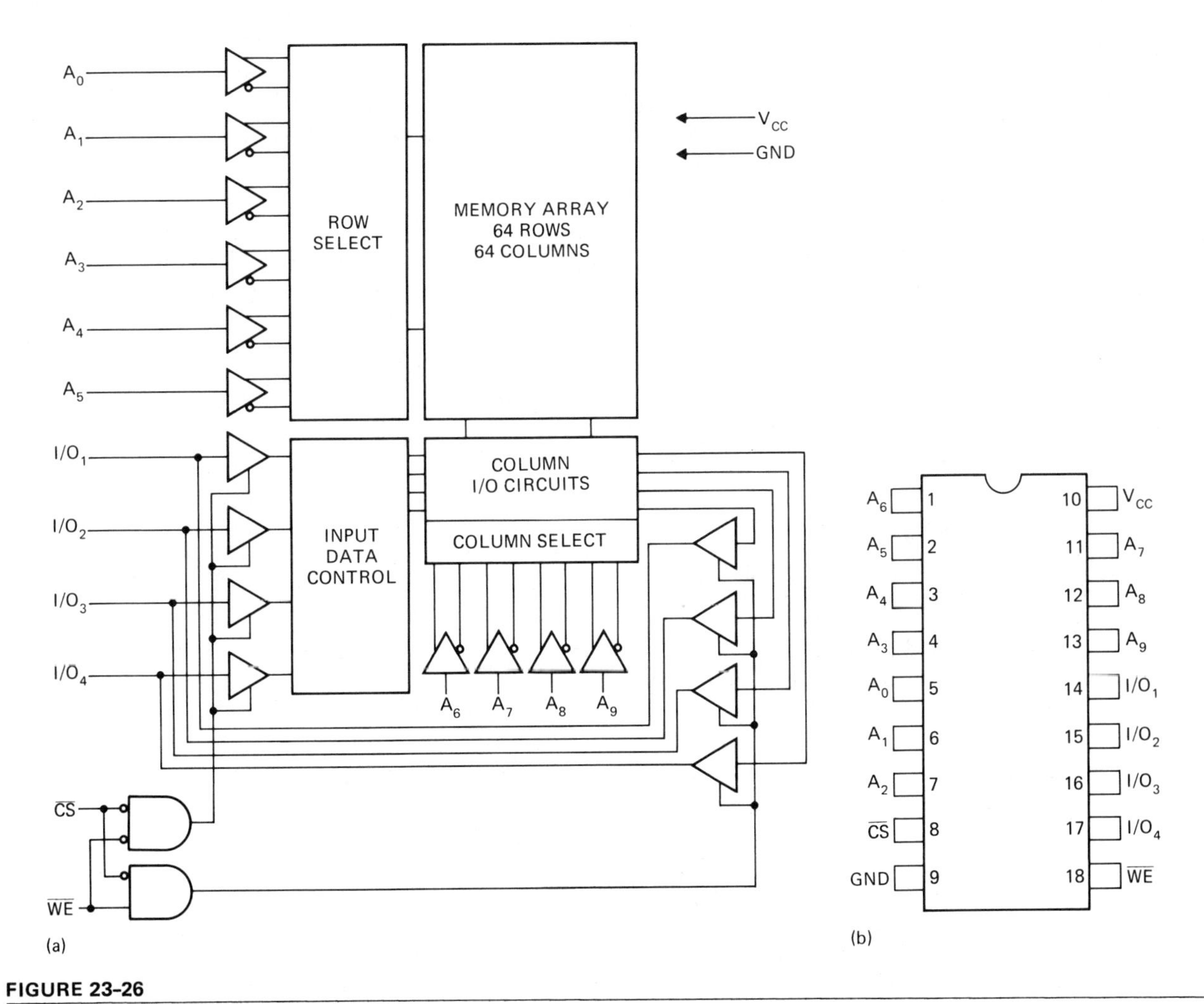

FIGURE 23-26
2114 1K-by-4 static RAM. (a) Block diagram. (b) Pin configuration of 18-pin DIP package.

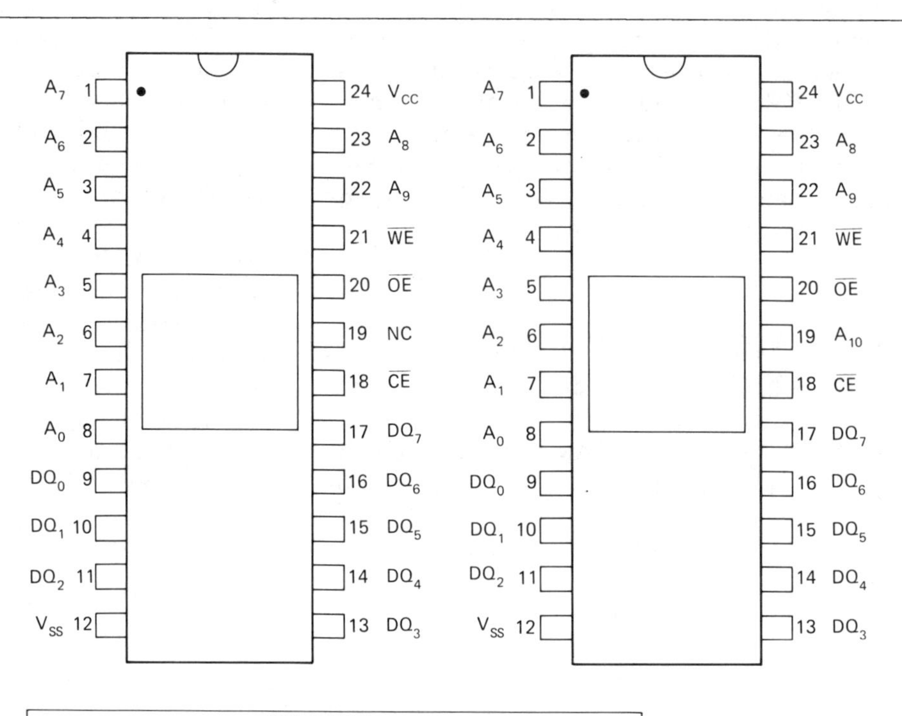

PIN NAMES			
A_0-A_9	ADDRESS INPUTS	$\overline{WE}$	WRITE ENABLE
$\overline{CE}$	CHIP ENABLE	$\overline{OE}$	OUTPUT ENABLE
V_{SS}	GROUND	NC	NO CONNECTION
V_{CC}	POWER (+5V)	DQ_0-DQ_7	DATA-IN/ DATA-OUT

FIGURE 23-27
Pin-out of the 4801 1K-by-8 and the 4802 2K-by-8 static RAMS.

23.6 Dynamic RAM

23.6.1 Dynamic RAM Cells

The static memory cell shown in Figure 23-25 requires six transistors per cell. The stored value (1 or 0) in the state of the cell flip-flop is retained indefinitely, as long as power is not removed or its state is not switched by a write function. On the other hand, the dynamic cell stores 1 or 0 states as the presence or absence of a charge on a minute capacitor in the memory cell. This charge will leak off within a few milliseconds. The circuit is designed to refresh the cell when it is read. As long as every cell in the memory is read at least once every 2 milliseconds, no data will be lost.

When originally developed, dynamic RAM cells used four transistors and used the normal gate to source capacitance of these FET transistors for storage. Efforts to reduce the size of the memory cell led to development of a three- and finally a single-transistor cell. These cells are shown in Figure 23-28. It is somewhat misleading to call the cell in part (c) of the figure a single-transistor cell because the capacitor connected to the source terminal of the single transistor is more than the equivalent of the transistor in the space it occupies.

Although the size of the storage capacitor relative to the cell transistor is large, the actual capacitance of the storage cell is minute (0.04 pF to 0.07 pF). Difficulty in reading the trivial amount of charge stored in the cell is a limita-

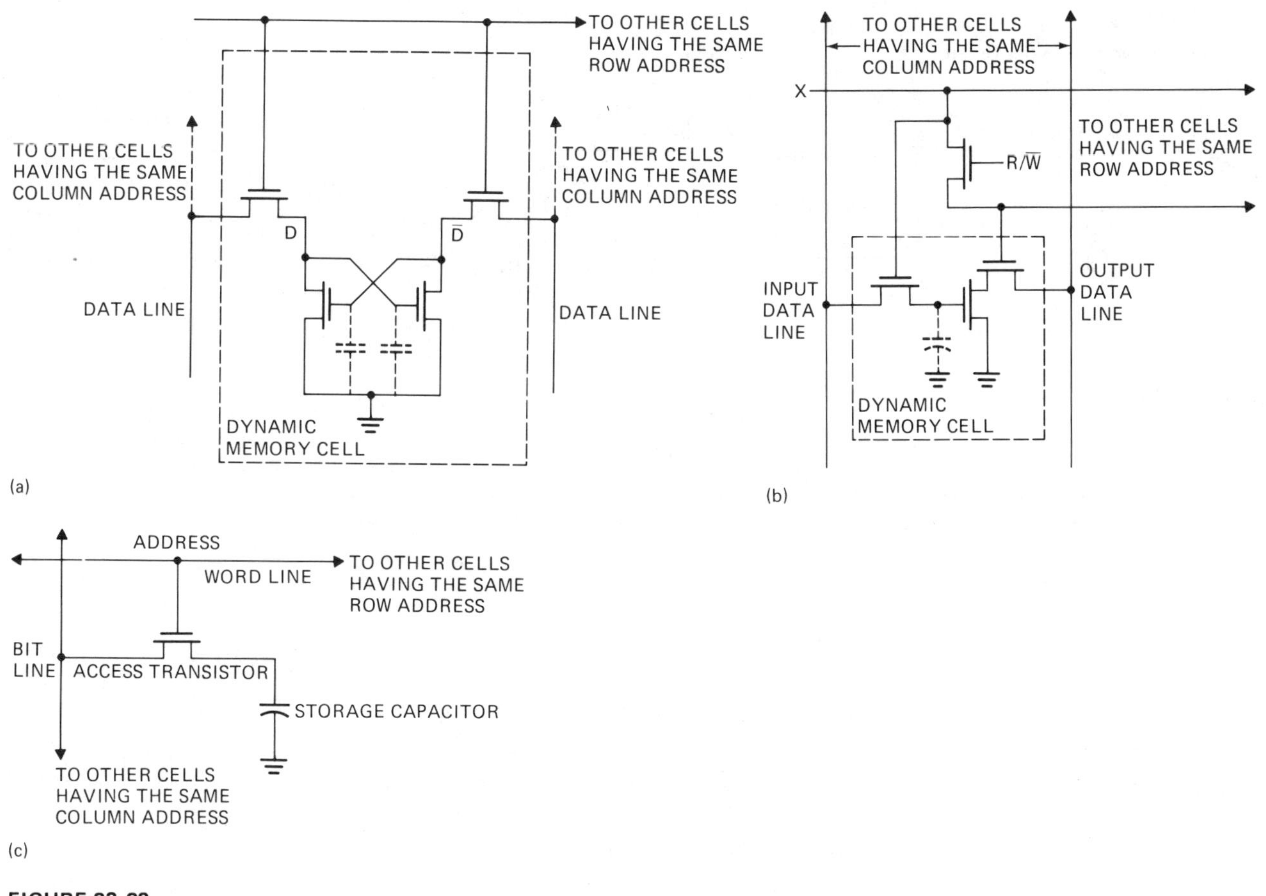

FIGURE 23-28

Dynamic RAM storage cells. (a) Four-transistor cell. Each cell is its own sense amp. (b) Three-transistor cell. A cell transistor buffers the output during a read or refresh cycle. (c) Single-transistor cell. The cell transistor does not amplify the output during a read. The cell is read or refreshed using a differential sense amp.

tion to increasing the density of the dynamic RAM.

23.6.2 Sense Amplifier

Three- and four-transistor cells contain transistors that amplify the charge on the capacitor with each read or refresh cycle. This does not happen in the single-transistor cell, and so a sense amplifier must be used to amplify the results of reading the cell. The sense amplifier occupies "real estate" on the chip that could otherwise be allocated to storage cells. In addition, the sense amplifier is a high-power consumer in comparison with other components on the chip. For this reason, the number of sense amps is minimized by connecting a large number of cells to each sense amp. Figure 23-29 shows a typical layout for a 16K dynamic RAM in which 128 cells are connected, 64 on each side of a sense amp. The unfortunate result of connecting large numbers of cells to the same sense amp is a lengthy bit line, resulting in a high bit line capacitance that attenuates the signal from the cell.

Bit line capacitances of 16K RAMs may vary from 0.7 pF to 1.1 pF. When a storage cell is connected to the bit line during a read or refresh cycle, and the ratio of bit line to cell capacitance is as high as 20 : 1, the result is a signal attenuation of 20 : 1. With the signal attenuated, the voltage imbalance between opposite nodes of the sense amps is barely enough to guarantee switching. Extending the bit line to

FIGURE 23-29

Simplified sense amplifier circuit.

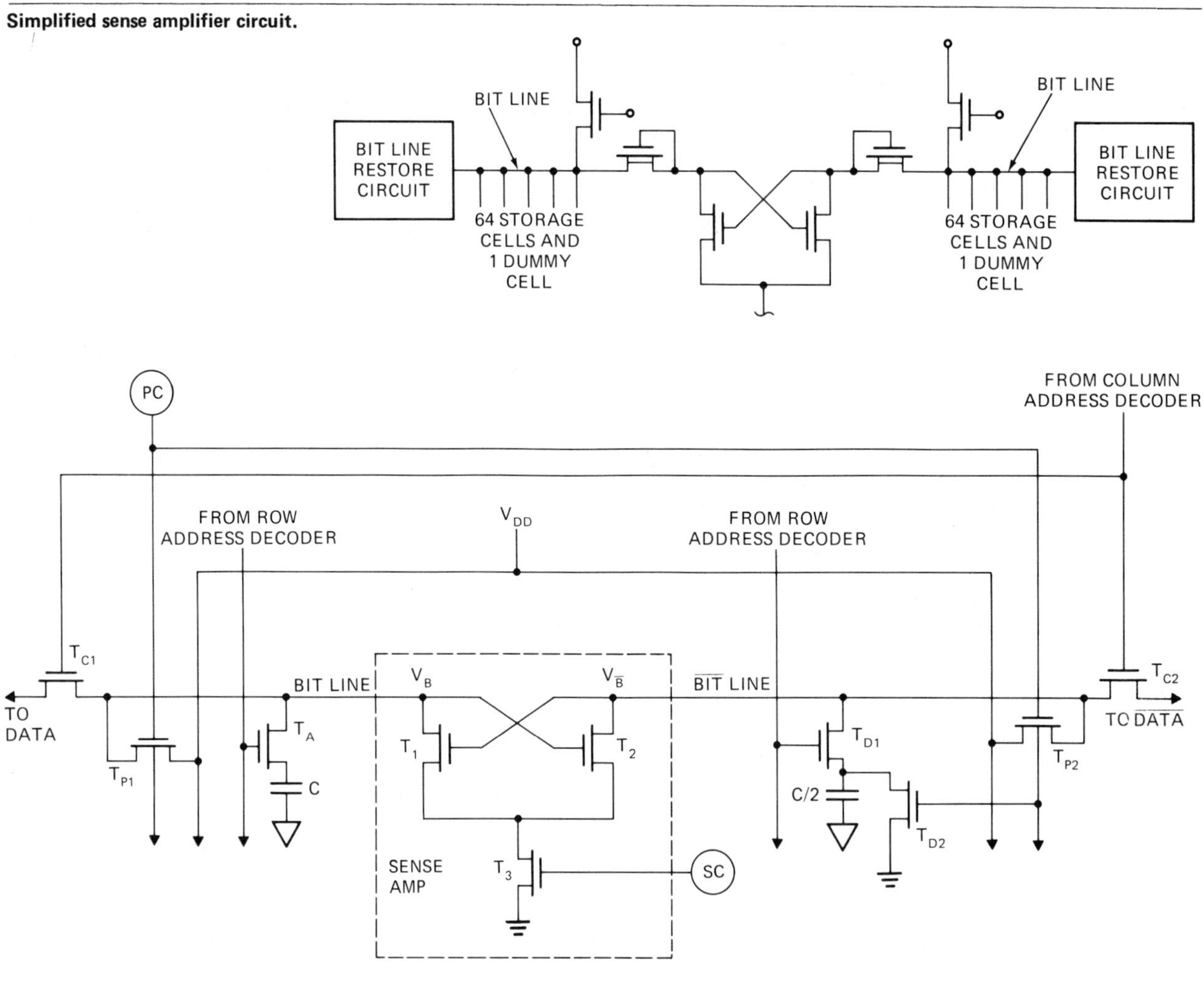

FIGURE 23-30

Simplified diagram of a sense amp showing provisions for precharge, cell selection, sensing, and column select functions. (The 127 unselected storage cells and BIT line dummy cell are not shown for simplicity.)

add more cells per sense amp increases the attenuation ratio and jeopardizes the reliability of sense amp switching.

The three-power-supply, 16K dynamic RAM is a mature product, widely sourced by both United States and foreign manufacturers, with only minor differences in the designs used by the various manufacturers. A square layout with 128 cells connected to each of 128 sense amps is used. The attenuation ratio between bit line and storage cell is typically 15 : 1. This ratio requires sense amps that switch reliably with a voltage imbalance of 200 mV. Judging from the success of the 16K RAM, the state-of-the-art of NMOS technology was sufficient to meet the above delicate conditions.

To see how the sense amp functions in the reading of the cell, let us look at a typical cycle in which a cell connected to the left-hand bit line is addressed. The dummy cell on the opposite side is automatically selected. Figure 23-30 shows these components of the RAM. There are four specific phases in the read-write or refresh cycles of the dynamic RAM. These are described in Figure 23-31. The first is the precharge. This occurs when both $\overline{RAS}$ and $\overline{CAS}$ are high. The precharge clock turns on T_{P1} and T_{P2} charging BIT and $\overline{BIT}$ lines to V_{DD}. During this same phase, the dummy cells are charged to ground through T_{D2}. When $\overline{RAS}$ goes low, the address decoder selects one of 128 cells connected to each bit line and turns on T_A, the access transis-

PHASE OF CYCLE	STATE OF TRANSISTORS							FUNCTION
	T_P	T_{D2}	T_A	T_{D1}	T_3	T_1/T_2	T_C	
PRECHARGE	ON	ON	OFF	OFF	OFF	FLOAT	OFF	PRECHARGE OCCURS WHEN BOTH $\overline{RAS}$ AND $\overline{CAS}$ ARE HIGH. BOTH BIT AND $\overline{BIT}$ LINES ARE CHARGED TO V_{DD} (+12V).
ROW ADDRESS SELECT	OFF	OFF	ON	ON	OFF	FLOAT	OFF	WHEN $\overline{RAS}$ GOES LOW, ROW ADDRESS DECODER SELECTS ONE OF 128 CELL ACCESS TRANSISTORS AND CONNECTS CELL TO BIT LINE AND DUMMY CELL TO $\overline{BIT}$ LINE.
SENSING	OFF	OFF	ON	ON	ON	ON/OFF OFF/ON	OFF	T_3 TURNS ON THE SENSE AMPLIFIER IN A STATE THAT REFRESHES THE CHARGE ON THE CELL.
COLUMN ADDRESS SELECT	OFF	OFF	ON	ON	ON	ON/OFF OR OFF/ON	ON	WHEN $\overline{CAS}$ GOES LOW, COLUMN ADDRESS DECODER SELECTS ONE OF 128 BIT-LINE TRANSISTOR PAIRS AND TURNS IT ON. THIS CONNECTS BIT LINES TO COMPLEMENTARY DATA LINES.

FIGURE 23–31

Operation phases occurring during memory read or write cycles of a dynamic RAM. (See Figure 23-30.) A refresh cycle will include only the first three phases. Column address select need not occur during a refresh cycle.

tor. At the same time, T_{D1}, the access transistor of the dummy cell, is turned on.

Let us assume a BIT line capacitance to storage cell capacitance ratio of 15:1. The $\overline{BIT}$ line capacitance to dummy cell capacitance ratio is, therefore, 30 : 1. If a high level is stored, then when the access transistors turn on, the BIT line and storage cell have the same charge, causing the BIT line to remain near V_{DD}. The dummy cell, however, is at ground, and therefore, when the zero charge in the dummy cell equilibrates with the charge on the $\overline{BIT}$ line, the results are as follows:

$$Q_{\overline{B}} = C_{\overline{B}}V_{DD} = (C_{\overline{B}} + C_D)V_{\overline{B}}$$

$$= \left(C_{\overline{B}} + \frac{C_{\overline{B}}}{30}\right)V_{\overline{B}}$$

$$C_{\overline{B}}V_{DD} = \left(C_{\overline{B}} + \frac{C_{\overline{B}}}{30}\right)V_{\overline{B}}$$

$$V_{DD} = \left(1 + \frac{1}{30}\right)V_{\overline{B}}$$

Therefore,

$$V_{\overline{B}} = \frac{30}{31}V_{DD} = 11.6V$$

and $V_B > V_{\overline{B}}$

where:
Q_B = combined charge of $\overline{BIT}$ line and cell
$C_{\overline{B}}$ = capacitance of $\overline{BIT}$ line
C_D = capacitance of dummy cell
$V_{\overline{B}}$ = voltage on $\overline{BIT}$ line after equilibration
V_{DD} = voltage on $\overline{BIT}$ line before equilibration (12V)

When the sense clock turns on T_3, the sense amp will come alive with T_2 turned on, causing the $\overline{BIT}$ line to ground through T_2, T_3, while the BIT line remains near V_{DD} refreshing the charge on the cell.

If a zero level is stored, then when the access transistors turn on, the dummy cell will again reduce the BIT line to 11.6V, but with the storage cell at ground, as shown here:

$$Q_B = C_B V_{DD} = (C_B + C)V_B$$

$$= \left(C_B + \frac{C_B}{15}\right)V_B$$

$$C_B V_{DD} = \left(C_B + \frac{C_B}{15}\right)V_B$$

$$V_{DD} = \frac{16}{15}V_B$$

Therefore,

and $V_B = \frac{15}{16}V_{DD} = 11.25\text{V}$

$$V_B < V_{\overline{B}}$$

Q_B = combined charge of BIT line and cell
C_B = capacitance of BIT line
C = capacitance of storage cell
V_B = voltage on BIT line after equilibration
V_{DD} = voltage on BIT line before equilibration (12V)

When the sense clock turns on T_3, the sense amp will come alive with T_1 turned on, causing the BIT line to ground through T_1, T_3, thus refreshing the charge.

If the cycle is an $\overline{\text{RAS}}$ only refresh, then the last phase of the cycle does not occur. If it is a read cycle, then the sense phase is followed by a decode of column address. The column decode selects one of the 128 BIT/$\overline{\text{BIT}}$ line pairs by turning on the T_{C1}/T_{C2} transistors, transmitting the state of the BIT lines to the output through the data lines. If the cycle is a memory write, then the low impedance data input buffer overrides the sense amplifier and stores the state of the data-in during the time T_{C1} and T_{C2} are turned on.

23.6.3 Dynamic RAM Organization and Timing

Dynamic RAM Versions and Pin-Outs. At the date of this writing, most manufacturers of 16K dynamic RAMs have succeeded in developing a 64K dynamic RAM using cell and sense amp techniques similar to those used on 16K dynamic RAMs. Along with the increase in capacity has come a reduction to single 5V power supply operation. There are, however, significant differences in layouts between vendors. Some use 512 sense amps with 128 cells per sense amp, while others use 256 sense amps with 256 cells to each sense amp. The 256-sense-amp versions have lower power drain. Other quality differences between these layouts have yet to be determined. The 256K dynamic RAM is already in existence and within a few years may dominate the dynamic RAM market. Figure 23-32 compares the pin-out of these three dynamic RAMS—the 16K, the 64K, and the 256K—as they exist today in the 16-pin DIP package. All three of these devices use an address multiplexing scheme to economize on pins. Note that although a 16K address requires 14 address inputs, the 4116 has only pins A_0 through A_6. These seven pins receive the row address that is latched in by row address strobe $\overline{\text{RAS}}$. This is followed by a change to column address on pins A_0 through A_6, which is latched in by column address strobe $\overline{\text{CAS}}$.

Like the static RAM, the dynamic RAM

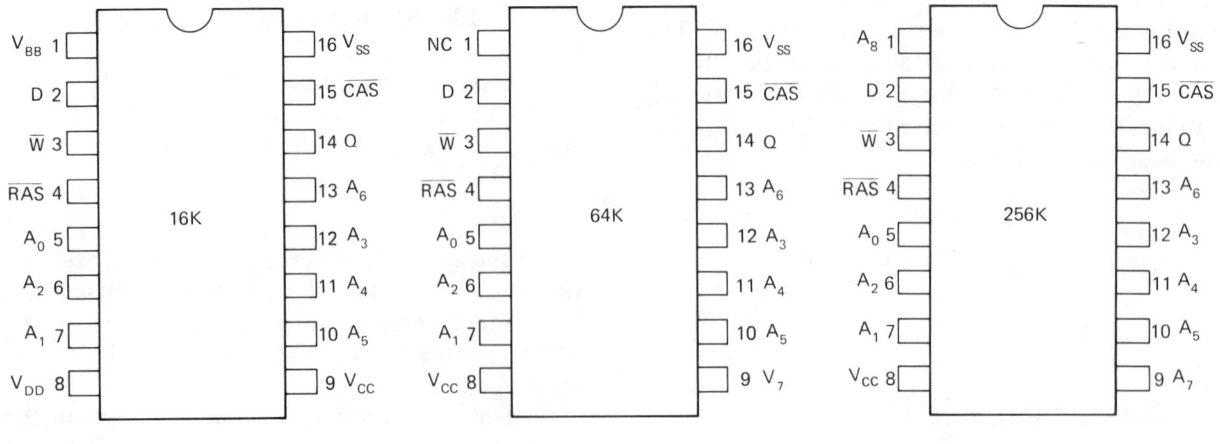

Pin	16K	64K	256K
1	V_{BB}	NC	A_8
2	D	D	D
3	$\overline{W}$	$\overline{W}$	$\overline{W}$
4	$\overline{RAS}$	$\overline{RAS}$	$\overline{RAS}$
5	A_0	A_0	A_0
6	A_2	A_2	A_2
7	A_1	A_1	A_1
8	V_{DD}	V_{CC}	V_{CC}
9	V_{CC}	V_7	A_7
10	A_5	A_5	A_5
11	A_4	A_4	A_4
12	A_3	A_3	A_3
13	A_6	A_6	A_6
14	Q	Q	Q
15	$\overline{CAS}$	$\overline{CAS}$	$\overline{CAS}$
16	V_{SS}	V_{SS}	V_{SS}

FIGURE 23-32
Modern versions of the dynamic RAM. The 16K dynamic RAM uses three power supplies. The 64K and 256K versions use only +5V for power. All three, however, use a similar multiplexed address and cell sensing scheme.

has memory read and memory write cycles. But along with these, there is a refresh cycle that is used to prevent loss of data due to leakage of charge from the storage cells. There are also several modes of operation available from a dynamic RAM that do not exist for the static RAM. These are the read-modify write and page mode operation.

Refresh Timing. The timing of the 4116 dynamic RAM is primarily referenced to the row address strobe ($\overline{\text{RAS}}$). Figure 23-33 shows a typical $\overline{\text{RAS}}$ waveform. The cycle time (t_{RC}) of the memory can be measured from one falling edge of $\overline{\text{RAS}}$ to the next. There is a minimum negative pulse width ($t_{\overline{RAS}}$) needed to allow for charge transfers to and from the storage cells. A minimum positive pulse width or precharge time (t_{RP}) is needed to allow time for precharge of the BIT lines before beginning a new cycle. The multiplexed address input pins must contain valid memory row address information at least t_{ASR} before the falling edge of $\overline{\text{RAS}}$. The row address must remain valid for at least t_{RAH} after the falling edge of $\overline{\text{RAS}}$.

Read Cycle Timing. The row address timing specifications are the same for normal read/write functions as they are for refresh functions. As Figure 23-34 shows, the column address strobe

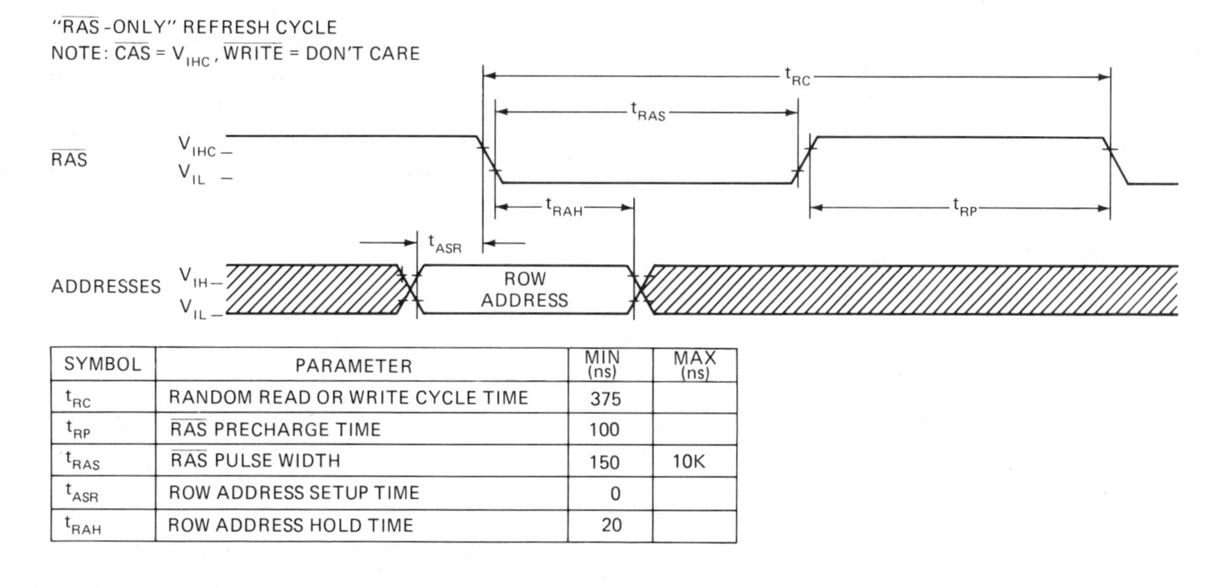

SYMBOL	PARAMETER	MIN (ns)	MAX (ns)
t_{RC}	RANDOM READ OR WRITE CYCLE TIME	375	
t_{RP}	$\overline{\text{RAS}}$ PRECHARGE TIME	100	
t_{RAS}	$\overline{\text{RAS}}$ PULSE WIDTH	150	10K
t_{ASR}	ROW ADDRESS SETUP TIME	0	
t_{RAH}	ROW ADDRESS HOLD TIME	20	

FIGURE 23-33

Row address strobe ($\overline{\text{RAS}}$) timing parameters. Although shown here for the refresh cycle, the same specifications hold for both read and write cycles.

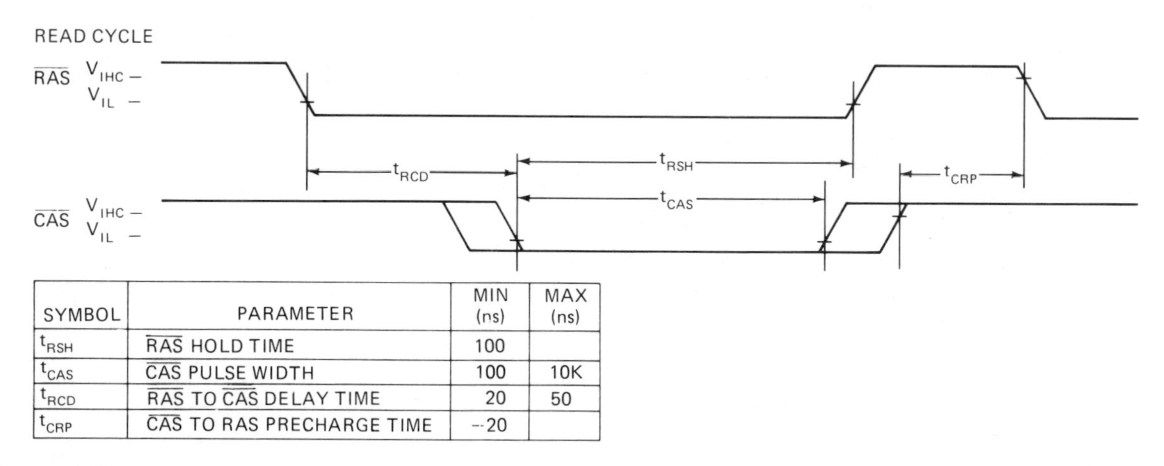

SYMBOL	PARAMETER	MIN (ns)	MAX (ns)
t_{RSH}	$\overline{\text{RAS}}$ HOLD TIME	100	
t_{CAS}	$\overline{\text{CAS}}$ PULSE WIDTH	100	10K
t_{RCD}	$\overline{\text{RAS}}$ TO $\overline{\text{CAS}}$ DELAY TIME	20	50
t_{CRP}	$\overline{\text{CAS}}$ TO RAS PRECHARGE TIME	−20	

FIGURE 23-34

Column address strobe ($\overline{\text{CAS}}$) timing parameters with respect to $\overline{\text{RAS}}$. Although shown here for the read cycle, the same specifications hold for the write cycle.

($\overline{CAS}$) must occur at least t_{RCD} after the falling edge of $\overline{RAS}$ and at least t_{RSH} before the rising edge of $\overline{RAS}$. The rising edge of $\overline{CAS}$ must occur at least t_{CRP} before the end of the cycle, and $\overline{CAS}$ must have a negative pulse width of not less than t_{CAS}.

The column address information must be valid at least t_{ASC} before the falling edge of $\overline{CAS}$. The column address information must remain valid at least t_{CAH} after the falling edge of $\overline{CAS}$.

The write must be high during the read cycle or, as Figure 23-35 shows, must go high at least t_{RCS} before the falling edge of $\overline{CAS}$ and remain high for at least t_{RCH} after the rising edge of $\overline{CAS}$.

With the inputs timed correctly, valid data-out will occur t_{RAC} after the falling edge of $\overline{RAS}$ or t_{CAC} after the falling edge of $\overline{CAS}$ (whichever comes later). Valid data will remain on D_{OUT} for t_{OFF} after the rising edge of $\overline{CAS}$.

Write Cycle Timing. Write cycle timing can take either of two forms. In one, write goes active before the falling edge of $\overline{CAS}$ and D_{OUT} remains in high impedance throughout the cycle. This makes possible the use of common I/O with D_{IN} and D_{OUT} shorted together. Figure 23-36 shows the timing specifications of write and D_{IN} waveforms for an early write cycle. The write must go low at least t_{WCS} before the falling edge of $\overline{CAS}$ and at least t_{CWL} before the rising edge of $\overline{CAS}$. It must remain low at least t_{WCH} after the falling edge of $\overline{CAS}$.

Read Modify Write. In a system with separate I/O, the existing data can be read from a memory address, and immediately following, in the same cycle, new data can be written in its place. The minimum cycle time would be the same as for an ordinary read cycle. Because of the fact that read data and write data are active during

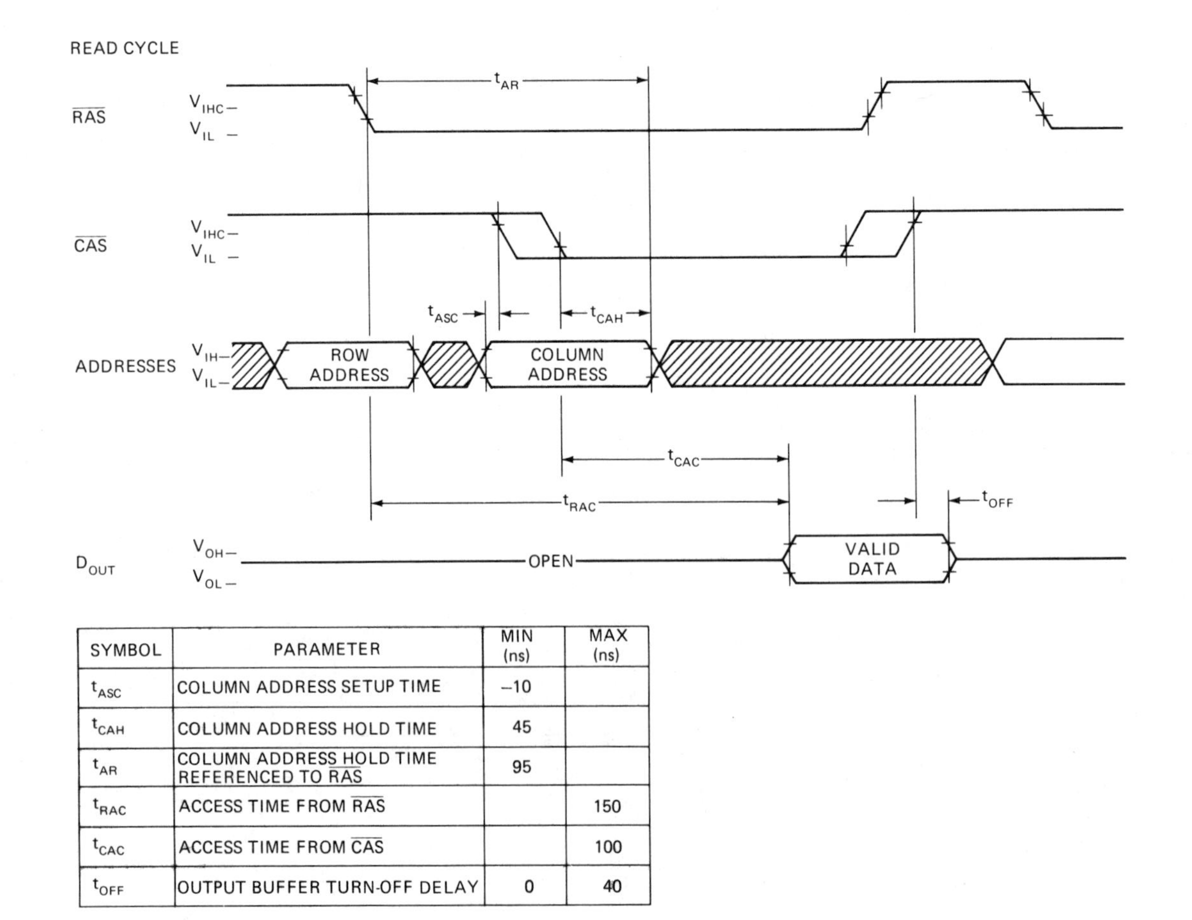

SYMBOL	PARAMETER	MIN (ns)	MAX (ns)
t_{ASC}	COLUMN ADDRESS SETUP TIME	−10	
t_{CAH}	COLUMN ADDRESS HOLD TIME	45	
t_{AR}	COLUMN ADDRESS HOLD TIME REFERENCED TO $\overline{RAS}$	95	
t_{RAC}	ACCESS TIME FROM $\overline{RAS}$		150
t_{CAC}	ACCESS TIME FROM $\overline{CAS}$		100
t_{OFF}	OUTPUT BUFFER TURN-OFF DELAY	0	40

FIGURE 23-35

Column address inputs and data-out timing parameters with respect to $\overline{RAS}$ and $\overline{CAS}$.

the same period, this option should not be used with common I/O.

The timing of a read-modify write cycle is shown in Figure 23-37. Note that the specification t_{CWD} calls for the write to become active some time after the falling edge of $\overline{CAS}$. In addition, write must not go active sooner than t_{RWD} after the falling edge of $\overline{RAS}$.

Page Mode Operations. In reading a block of data for which the ROW address remains the same and only the column addresses are different, $\overline{RAS}$ can be held low while read cycles that use only column addresses are initiated. This provides read cycle times of less than half the normal read cycle. The same is true of page mode write cycles. Page mode read and write cycles are shown in Figure 23-38. The flexibility needed in memory systems timing in order to take advantage of page mode operation, without interfering with normal read/write and refresh timing, complicates memory system design. In some applications, however, the benefits may warrant the complication.

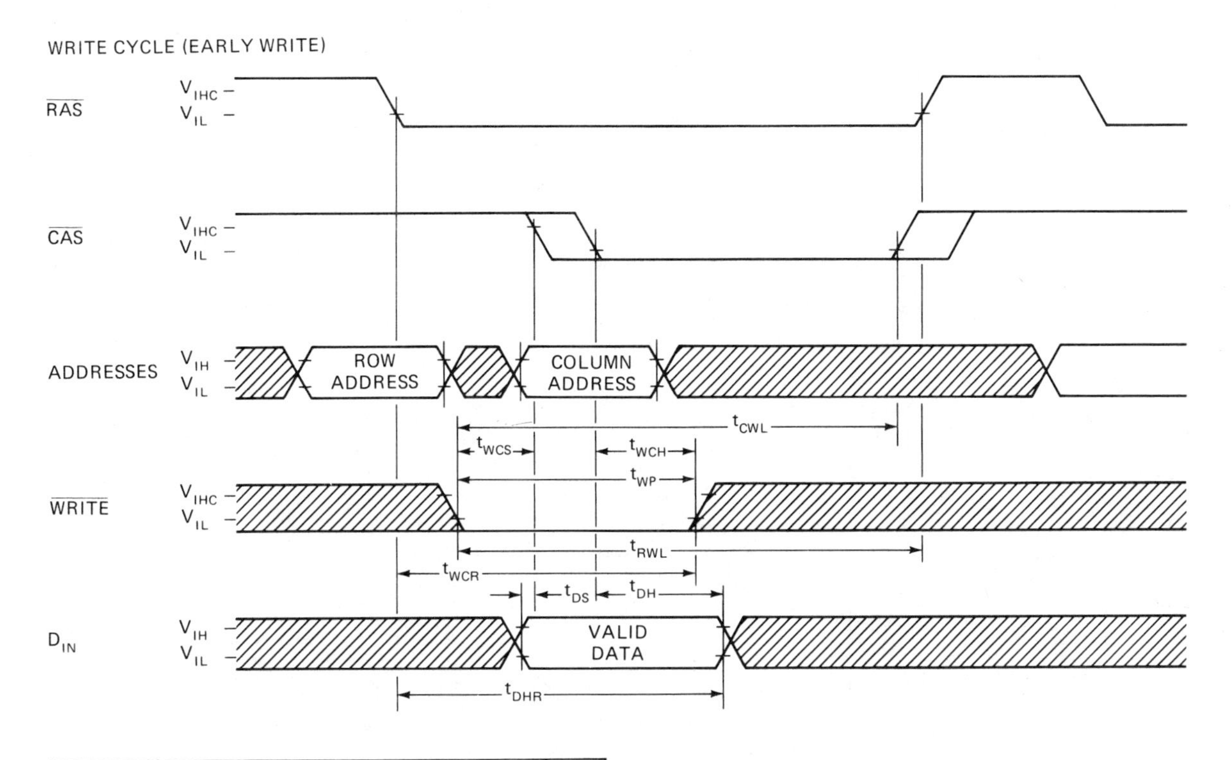

SYMBOL	PARAMETER	MIN (ns)
t_{WCH}	WRITE COMMAND HOLD TIME	45
t_{WCR}	WRITE COMMAND HOLD TIME REFERENCED TO $\overline{RAS}$	95
t_{WCS}	WRITE COMMAND SETUP TIME	−20
t_{WP}	WRITE COMMAND PULSE WIDTH	45
t_{RWL}	WRITE COMMAND TO $\overline{RAS}$ LEAD TIME	60
t_{CWL}	WRITE COMMAND TO $\overline{CAS}$ LEAD TIME	60
t_{DS}	DATA-IN SETUP TIME	0
t_{DH}	DATA-IN HOLD TIME	45
t_{DHR}	DATA-IN HOLD TIME REFERENCED TO $\overline{RAS}$	95

FIGURE 23-36

Timing parameters of the write and data. For a memory write cycle, the timing parameters for $\overline{RAS}$, $\overline{CAS}$, and address bus are the same as for a read cycle.

READ-WRITE/READ-MODIFY WRITE CYCLE

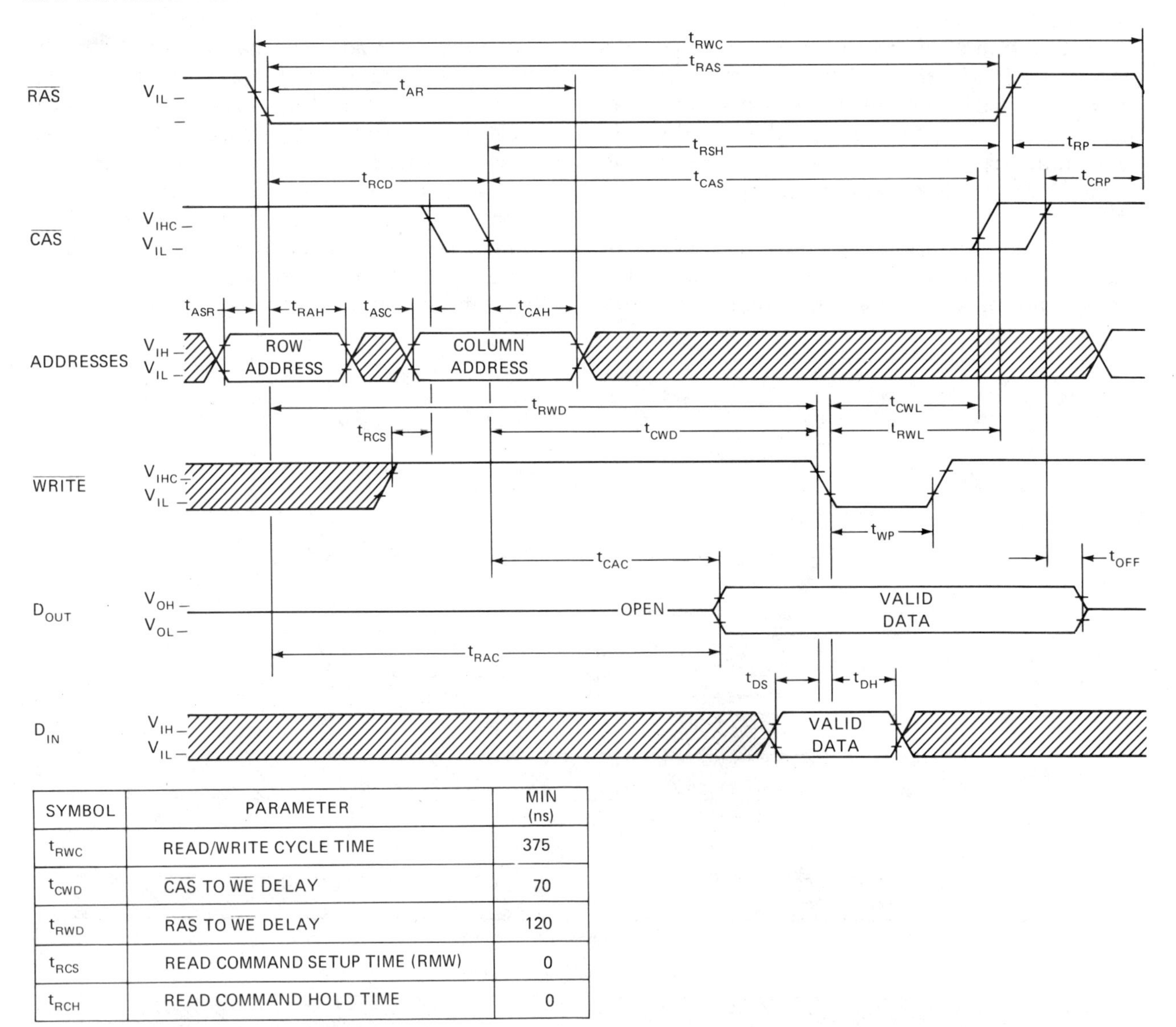

SYMBOL	PARAMETER	MIN (ns)
t_{RWC}	READ/WRITE CYCLE TIME	375
t_{CWD}	$\overline{\text{CAS}}$ TO $\overline{\text{WE}}$ DELAY	70
t_{RWD}	$\overline{\text{RAS}}$ TO $\overline{\text{WE}}$ DELAY	120
t_{RCS}	READ COMMAND SETUP TIME (RMW)	0
t_{RCH}	READ COMMAND HOLD TIME	0

FIGURE 23-37

Timing parameters of the read-modify write cycle. Note that leading edge of write pulse occurs after the leading edge of $\overline{\text{CAS}}$, instead of before (as in an early write cycle).

PAGE MODE WRITE CYCLE

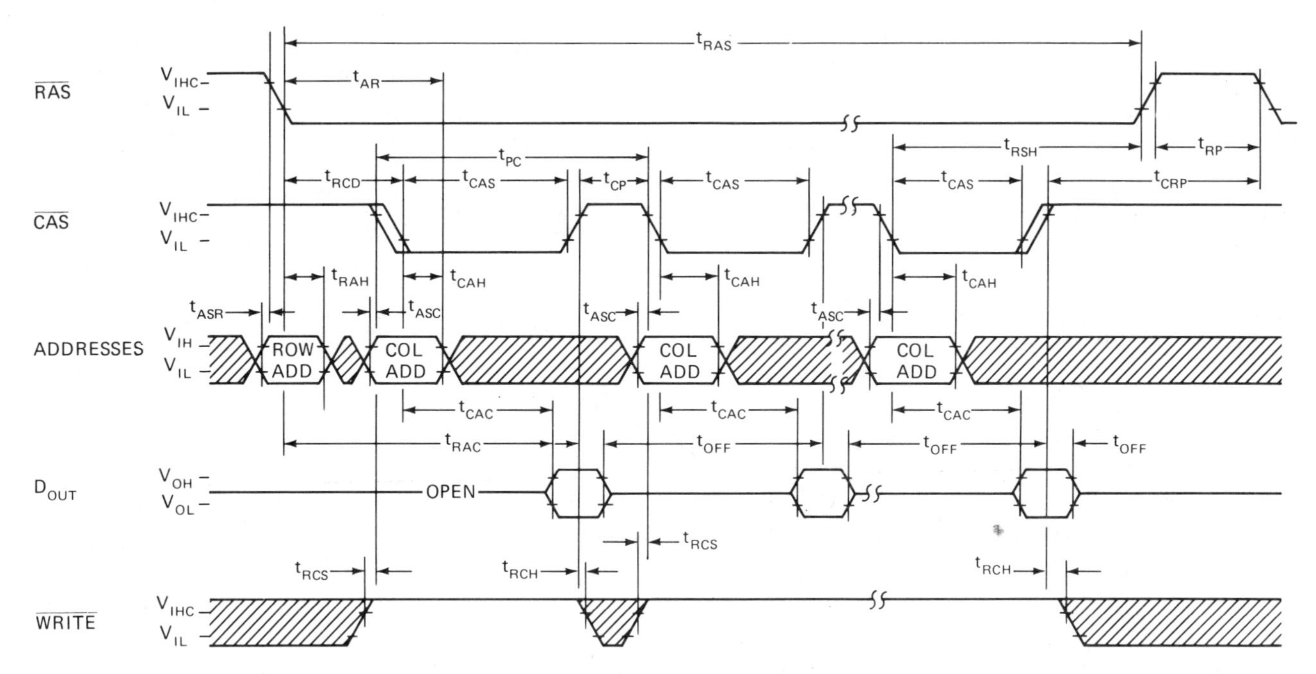

PAGE MODE READ CYCLE

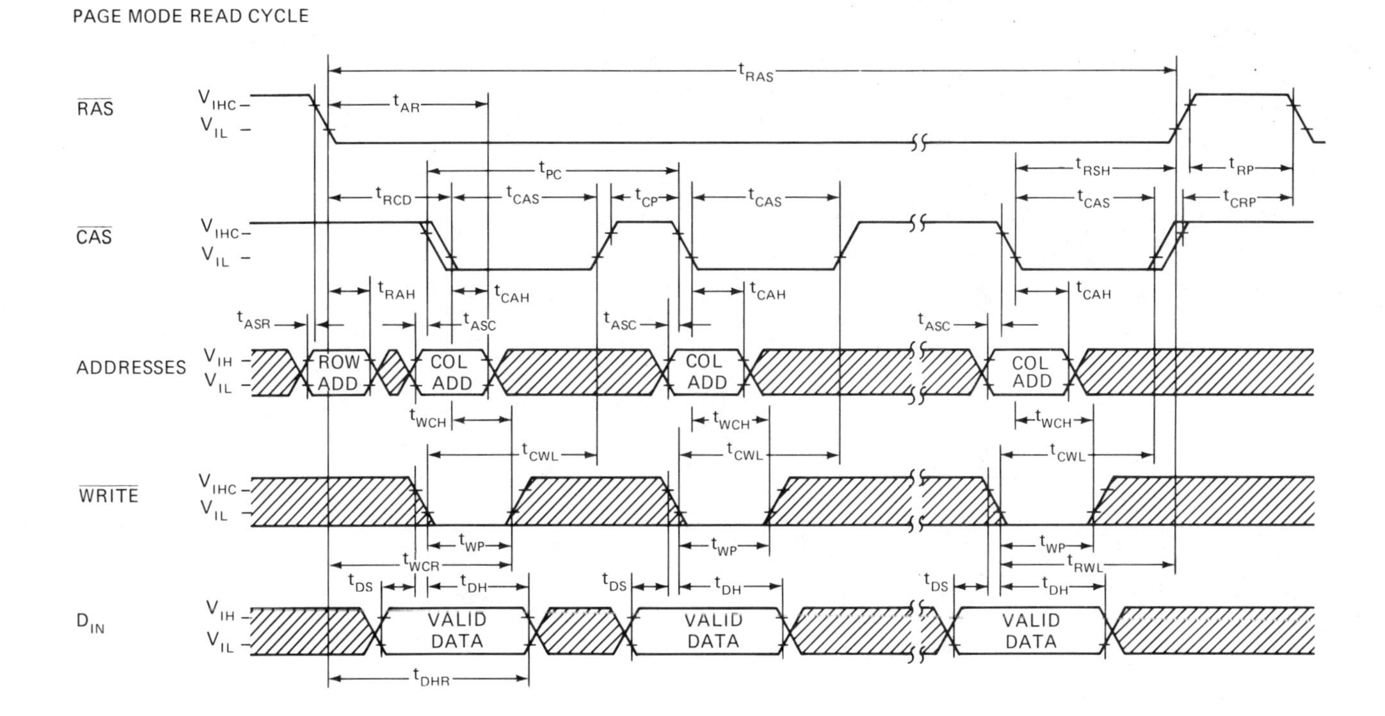

SYMBOL	PARAMETER	MIN (ns)
t_{CP}	CAS PRECHARGE TIME (PAGE MODE)	60
t_{PC}	PAGE MODE CYCLE TIME	170

FIGURE 23-38

Page mode operation of the dynamic RAM. While holding $\overline{RAS}$ low, a series of column addresses can occur causing a read or write of a series of data from memory locations having the same row address but different column addresses.

23.6.4 Hardware for Refresh and Multiplexing

Figure 23-39 shows the Intel 3242, a 28-pin IC designed to provide address multiplexing and refresh hardware for a 16K dynamic RAM. Fourteen address inputs, A_0 through A_{13}, are multiplexed out to seven outputs, $\overline{O}_0$ to $\overline{O}_6$. As the inversion occurs for both read and write cycles, there is normally no reason to reinvert the outputs. In addition to multiplexing the address inputs, a seven-bit refresh counter is included and its outputs are also multiplexed to the memory address pins. The user must, however, supply correctly timed refresh and ROW enable signals. Note that a refresh enable overrides a ROW enable. A clock may be supplied for the refresh counter, or the refresh and clock inputs may be tied together. A zero detect signal is supplied too, in case there is a need to synchronize the refresh operations with normal memory read and write functions.

When a microprocessor initiates a memory read or memory write cycle, arbitrarily over-

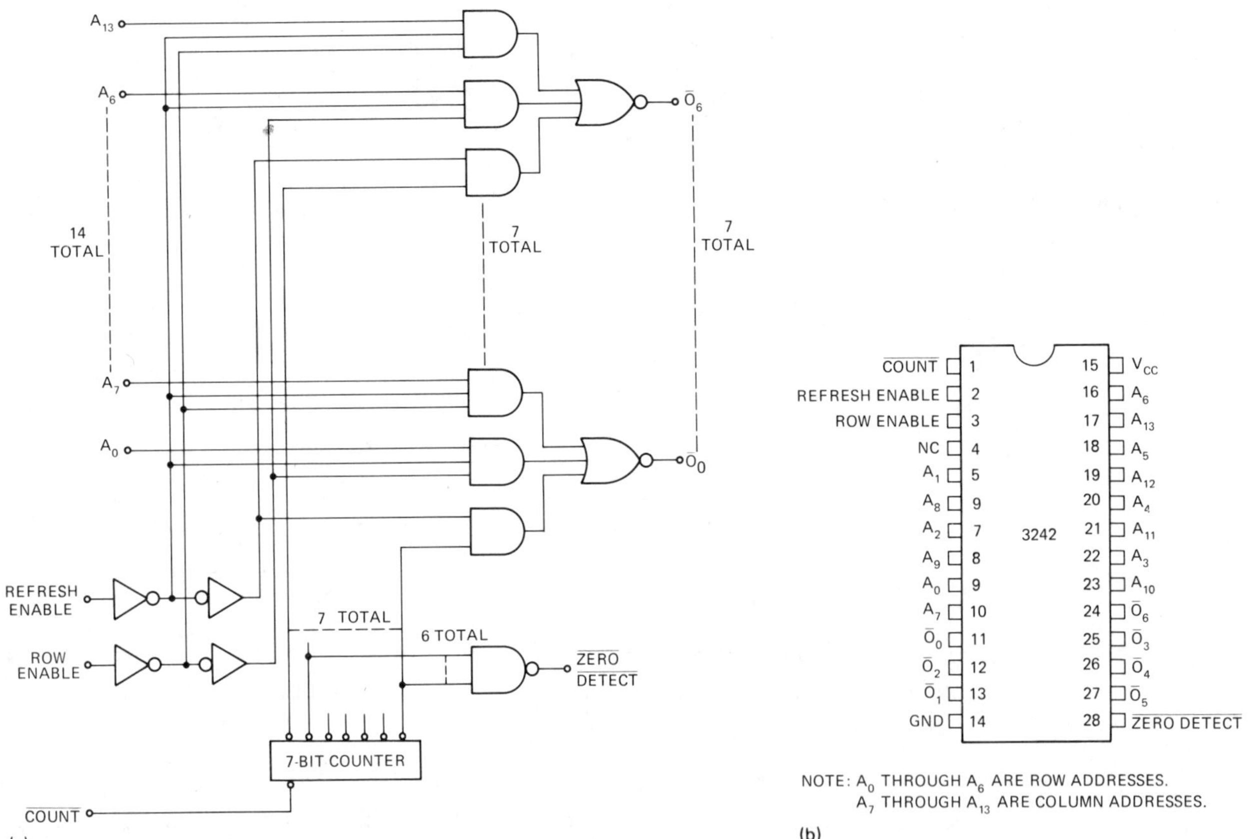

REFRESH ENABLE	ROW ENABLE	OUTPUT
H	X	REFRESH ADDRESS (FROM INTERNAL COUNTER)
L	H	ROW ADDRESS (A_0 THROUGH A_6)
L	L	COLUMN ADDRESS (A_7 THROUGH A_{13})

$\overline{\text{COUNT}}$ ADVANCES INTERNAL REFRESH COUNTER.
$\overline{\text{ZERO DETECT}}$ INDICATES ZERO IN THE FIRST SIX SIGNIFICANT REFRESH COUNTER BITS (USED IN BURST REFRESH MODE).

(c)

FIGURE 23-39

The Intel 3242 refresh controller circuit. The circuit supplies multiplexing functions for the address inputs and, in addition, contains a seven-bit counter for generating refresh addresses. (a) Logic diagram. (b) Pin configuration. (c) Truth table and definitions.

riding that cycle for the sake of refresh would be catastrophic to the running of the program. Obviously, refresh cycles must be carefully worked into system timing. There are many timing options for refresh. The one chosen may be a sell-off between speed, hardware, and software economics. The minimum refresh cycle is only two-thirds of the minimum time required for a normal read or write cycle, but it is seldom convenient to design different periods into a memory timing system. If the time saving is desirable, however, it may be done. Since refresh occurs one row at a time and there are 128 rows, one refresh cycle must occur at least every 15.6 μs. In a system where memory cycle time is 500 μs, one refresh must occur every 30 memory cycles. In a microprocessor system of 1 μs instruction time, which handles a row at a time refresh by software, a jump to refresh subroutine would have to be inserted every fifteenth instruction. To say the least, this would be inconvenient. For such cases, refresh, in which all 128 ROW addresses are completed at one time, would be more software convenience. It would require a jump to refresh subroutine once every 1800 instructions. Some systems, however, cannot allow a complete shutdown of interrupt functions for the 128 or more instruction cycles required to complete a total refresh. Other microprocessor techniques that may be used to interrupt a program for the sake of refresh involve the use of interrupt, ready, or hold signals. All allow completion of the instruction in progress prior to entering a refresh routine or releasing the memory buses for the sake of hardware-controlled refresh. The time lost in these delays tends to favor the use of refresh burst rather than row-at-a-time. Where, due to other system memory limitations, cycles are much longer in duration than the minimum cycle time of the dynamic RAM, it may be possible to insert refresh cycles at the beginning or end of normal read-write cycles.

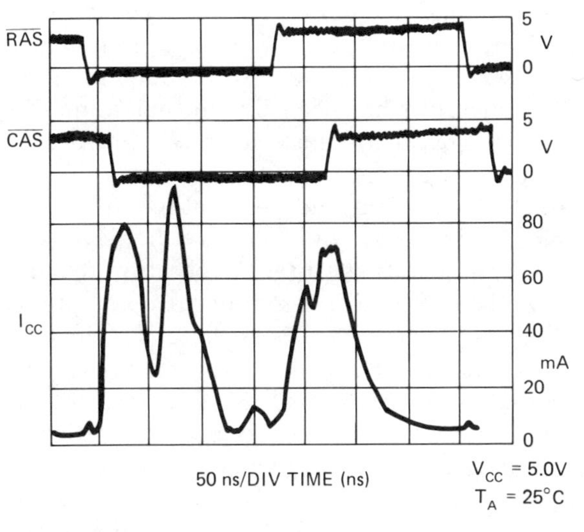

FIGURE 23-40

Motorola oscillograph of I_{CC} current waveform of the 64K dynamic RAM showing effect of the edges of $\overline{RAS}$ and $\overline{CAS}$ on the supply current.

23.6.5 Power Drain of Dynamic RAMs

Because memory systems using dynamic RAMs are usually massive in size, the power drain resulting from these large numbers of RAM devices is an important concern. Figure 23-40 shows the relationship of currents at the V_{CC} lead with the edges of the address strobes $\overline{RAS}$ and $\overline{CAS}$. Much of the activity that causes power drain within the device occurs just after the edges of the strobes. It logically follows that the more frequent the occurrence of the strobes, the higher the power drain. For that reason, current drain is specified for three conditions, as shown in Figure 23-41. I_{CC1}, operating current, represents the highest condition of

FIGURE 23-41

Supply current specifications for the 64K dynamic RAM.

DC ELECTRICAL CHARACTERISTICS (0°C ≤ T_A ≤ 70°C)(V_{CC} = 5.0V ±10%)

SYMBOL	PARAMETER	MIN	MAX	UNITS
I_{CC1}	OPERATING CURRENT AVERAGE POWER SUPPLY OPERATING CURRENT (RAS, CAS CYCLING; t_{RC} = t_{RC} MIN)	–	54.0	mA
I_{CC2}	STANDBY CURRENT POWER SUPPLY STANDBY CURRENT ($\overline{RAS}$ = V_{IH}, D_{OUT} = HIGH IMPEDANCE)	–	4	mA
I_{CC3}	$\overline{RAS}$ ONLY REFRESH CURRENT AVERAGE POWER SUPPLY CURRENT, REFRESH MODE ($\overline{RAS}$ CYCLING, $\overline{CAS}$ = V_{IH}; t_{RC} = t_{RC} MIN)	–	45	mA

current flow that could occur. It occurs only for a memory operating continuously with read-write cycle of minimum cycle time. If the cycle time is longer than the minimum, then the current is computed as though part of the cycle is at current I_{CC1} and the remainder at I_{CC2}. For example, if the 64K RAM specified were operating at a 650 nanosecond cycle time, while t_{RC} minimum of the RAM were 350 nanoseconds, then,

$$I_{CC} = I_{CC1} \times \frac{350\ \text{ns}}{650\ \text{ns}} + I_{CC2} \times \frac{(650 - 350)\ \text{ns}}{650\ \text{ns}}$$

$$= 54\ \text{mA} \times 0.538 + 4\ \text{mA} \times 0.462$$

$$= 29.05 + 1.85\ \text{mA} = 30.9\ \text{mA}$$

One of every 24 cycles must be a refresh cycle in order to ensure a refresh of all cells. This is determined by the following:

$$\frac{15.6\ \mu\text{s between refresh cycles}}{0.650\ \mu\text{s per cycle}}$$

The effect of this on operating current is usually not considered significant but,

$$I_{CRF} = I_{CC3} \times \frac{350\ \text{ns}}{650\ \text{ns}} + I_{CC2} \times \frac{(650 - 350)\ \text{ns}}{650\ \text{ns}}$$

$$= 45\ \text{mA} \times 0.538 + 1.85\ \text{mA}$$

$$= 24.2\ \text{mA} + 1.85\ \text{mA} = 26\ \text{mA}$$

$$I_{CC} = \frac{1}{24} \times 26\ \text{mA} + \frac{23}{24} \times 30.85$$

$$= 1.08\ \text{mA} + 29.6\ \text{mA} = 30.68\ \text{mA}$$

Though the calculation shows a lack of significance in the effect of the refresh on operating current, refresh current calculation does become significant for very large memories when it is known that some portion of the memory is being subjected to refresh only while the remainder is being read or written. The current in the refresh-only RAMs is computed as follows:

$$I_{CRF} = I_{CC3} \times \frac{350\ \text{ns}}{15600\ \text{ns}} + I_{CC2} \times \frac{(15600 - 350)\ \text{ns}}{15600\ \text{ns}}$$

$$= 45\ \text{mA} \times 0.022 + 4\ \text{mA} \times 0.978$$

$$= 0.99\ \text{mA} + 3.91\ \text{mA} = 4.9\ \text{mA}$$

The operating and refresh currents of I_{DD} for the 16K RAM are determined by a method similar to the above computation. The I_{CC} and I_{BB} are taken as specified.

23.6.6 Errors in Dynamic RAMs

Errors in semiconductor memories are generally classified as hard errors, which occur repeatedly at the same address, and soft errors, which occur on rare occasions at random addresses. A hard error indicates a defective device, and the defect can usually be detected by testing the device to the limit of its specifications. Some hard errors, on the other hand, may be easily missed if they are the result of a pattern sensitivity. Such errors occur only when the device is subjected to a particular pattern of reading and writing data.

Soft errors can be caused by low noise immunity, which is a factor in selecting the type of memory circuit design. Use of a good memory board layout, including adequate decoupling capacitors near the power pins of each memory device, can help keep noise errors to a minimum.

Recently it has been shown that some soft errors are bound to persist in spite of efforts to prevent noise errors. In fact, a study of the effects of alpha particle penetration of the dynamic RAM array by Timothy C. May and Murray H. Woods of Intel Corporation demonstrated that some soft errors occur due to alpha particle penetration. Soft error rates for the 64K dynamic RAM are a vast improvement over the initial designs of the 16K dynamic RAMs simply because the cause has been identified as alpha particles emanating from trace amounts of uranium and thorium found in the packaging material close to the die. Soft errors cannot be accelerated by high temperature, but an accelerated test can be conducted by exposing the die to highly radioactive sources of known α flux density, measuring the error rate per hour, and interpolating the results that could be expected from the low level of radiation of the packaging or die-coating material. Reduction in the uranium/thorium content of die-coat material has made it possible for vendors to guarantee a mean time between errors of 10^6 hours. To date, most vendors of 64K dynamic RAMs are using a polyimide die coat. The die coat is thicker than the penetrating range of α particles. The uranium/thorium content of polyimide is extremely low in comparison to other packaging materials (7 ppb as compared to 500 to 5000 ppb for aluminous ceramics).

23.6.7 Error Detection

Errors caused by α radiation set a finite limit on the failure rate attainable with semiconductor memories. For a memory in which errors rarely occur, it is often sufficient to merely know when they occur so that the operation can be aborted and repeated. It is hoped that in the second attempt the error will not occur.

The most common method of error detection is parity, in which a parity bit must accompany each word when it is written. Prior to writing, the data word is examined by a parity generator, which determines whether the number of 1 bits in the word is odd or even. In an odd parity system, a 1 is placed in the parity bit if the number of 1s in the data word is even. A 0 will be placed in the parity bit if the number of 1s in the data word is odd. The result must be a data plus parity, which is odd. For even parity, data plus parity must be even. When a word is read from memory, a parity checker determines if parity is odd or even. If incorrect parity is detected, an alarm is energized. Parity was discussed in detail in Chapter 8. The fact that a parity bit must be stored in memory with each word adds additional overhead to the memory. For an eight-bit data word, nine bits would be needed, or 12.5% overhead. For a 16-bit data word, only $6\frac{1}{4}$% overhead would be needed.

As the size of the memory increases, error detection becomes more desirable. As the size of the data word increases, the problem of overhead becomes less critical.

23.6.8 Error Correction

In a binary system, correcting an error is a trivial matter of changing a 1 to a 0 or a 0 to a 1. This can be done quite simply with an exclusive OR gate. We have just shown that, with the sacrifice of some overhead, a parity system can detect an error. The problem remains to determine which bit in the word is in error. To do this there is an even more severe sacrifice in overhead and, as a result, even further advantage to a larger data word.

For example, let us use a 16-bit memory word. A parity generator receives all 16 bits of the data word, from which it generates the parity bit. With the parity bit included, a 17-bit word is stored in each address of the memory. When the 17-bit word is read, a parity checker determines whether there is an error. If the parity is in error, it could be any one of the 16 data bits that we wish to correct by inverting it. If we add an 8-bit parity generator at the input to record only the parity of the odd-numbered bits, and a corresponding parity checker at the data-out to determine the parity of the odd-numbered data bits when they are read, we then have to store 18 bits for a 16-bit word, but when an error occurred, we would know whether the error was in the odd or even bits. Next we divide the data bits into pairs and generate a parity bit from only the odd-numbered pairs. With this parity bit added to the stored memory word and an equivalent eight-bit parity check added at the output, we come closer to locating the error. Next we divide the data word into quarters and, with another eight-bit parity system, determine whether the error is in the odd or even quarters. Next we add an eight-bit parity check of the upper half of the data word.

Figure 23–42(a) is a block diagram of the error-correcting system. For a 16-bit data word, the memory word stored is 21 bits. As the diagram shows, during a memory write the data-in is examined and parity bits (check bits) are generated. The check bits are stored with the data word. During a memory read of that same address, an identical set of parity generators examines the data portion of the word and generates another set of check bits. If 16-bit parity is in error, a syndrome generator compares the newly generated check bits with those taken from memory. Where the bits differ, the syndrome generator outputs a 1; where they are the same, it outputs a 0. The syndrome word is then decoded as shown in the truth table in Figure 23–42(b), and a high level is transmitted to exclusive OR with the data bit that is in error. The parity and check bits that are stored in memory with the data word are as susceptible to error as the data bits themselves. Soft errors in check bits can be controlled by enabling error correction only when overall parity is in error. This leaves the system vulnerable only to soft errors in memory used to store the parity bit. Another check bit used to monitor for error of the parity bit would then provide a system with capacity to detect and correct single bit errors per word. Double bit errors, which are many times less frequent than single bit errors, could be detected but not corrected. These would be indicated by a nonzero output from the syndrome generator in conjunction with no parity error.

The method just explained is error detection and correction in a straightforward and

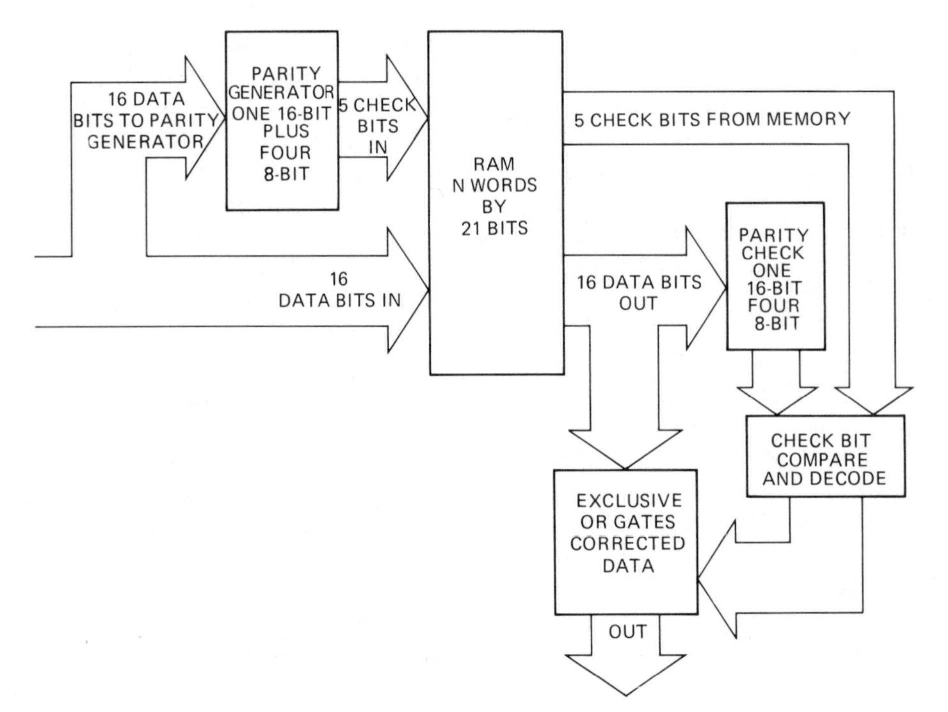

(a)

<table>
<tr><th>DATA BITS</th><th>FULL</th><th>HALVES</th><th>QUARTERS</th><th>PAIRS</th><th>ODD/EVEN</th></tr>
<tr><td>D_0</td><td rowspan="16">1</td><td rowspan="8">0</td><td rowspan="4">0</td><td rowspan="2">0</td><td>0</td></tr>
<tr><td>D_1</td><td>1</td></tr>
<tr><td>D_2</td><td rowspan="2">1</td><td>0</td></tr>
<tr><td>D_3</td><td>1</td></tr>
<tr><td>D_4</td><td rowspan="4">1</td><td rowspan="2">0</td><td>0</td></tr>
<tr><td>D_5</td><td>1</td></tr>
<tr><td>D_6</td><td rowspan="2">1</td><td>0</td></tr>
<tr><td>D_7</td><td>1</td></tr>
<tr><td>D_8</td><td rowspan="8">1</td><td rowspan="4">0</td><td rowspan="2">0</td><td>0</td></tr>
<tr><td>D_9</td><td>1</td></tr>
<tr><td>D_{10}</td><td rowspan="2">1</td><td>0</td></tr>
<tr><td>D_{11}</td><td>1</td></tr>
<tr><td>D_{12}</td><td rowspan="4">1</td><td rowspan="2">0</td><td>0</td></tr>
<tr><td>D_{13}</td><td>1</td></tr>
<tr><td>D_{14}</td><td rowspan="2">1</td><td>0</td></tr>
<tr><td>D_{15}</td><td>1</td></tr>
<tr><td>CHECK BIT NUMBER</td><td>MSB C_4</td><td>C_3</td><td>C_2</td><td>C_1</td><td>LSB C_0</td></tr>
</table>

1 = INDICATES DATA BIT USED IN GENERATING THIS CHECK BIT.
0 = INDICATES DATA BIT NOT USED IN GENERATING THIS CHECK BIT.

RESULT OF CHECK BIT COMPARE IF BIT IN ERROR				
1	0	0	0	0
1	0	0	0	1
1	0	0	1	0
1	0	0	1	1
1	0	1	0	0
1	0	1	0	1
1	0	1	1	0
1	0	1	1	1
1	1	0	0	0
1	1	0	0	1
1	1	0	1	0
1	1	0	1	1
1	1	1	0	0
1	1	1	0	1
1	1	1	1	0
1	1	1	1	1
MSB				LSB

(1 BIT IS A NO COMPARE.)

(b)

FIGURE 23–42

(a) Error correction for an N-word-by-16-bit memory system. (b) Truth table showing results of check bit comparison.

most understandable form. It does not, however, take full advantage of the error information that could possibly be obtained from six check bits. Several integrated circuit manufacturers are supplying integrated circuits for error detection and correction. Figure 23–43 is the Advanced Micro Devices 2960 error detection and correction circuit. The device supplies circuits for both check bit generation during a memory write and error detection and correction during a memory read.

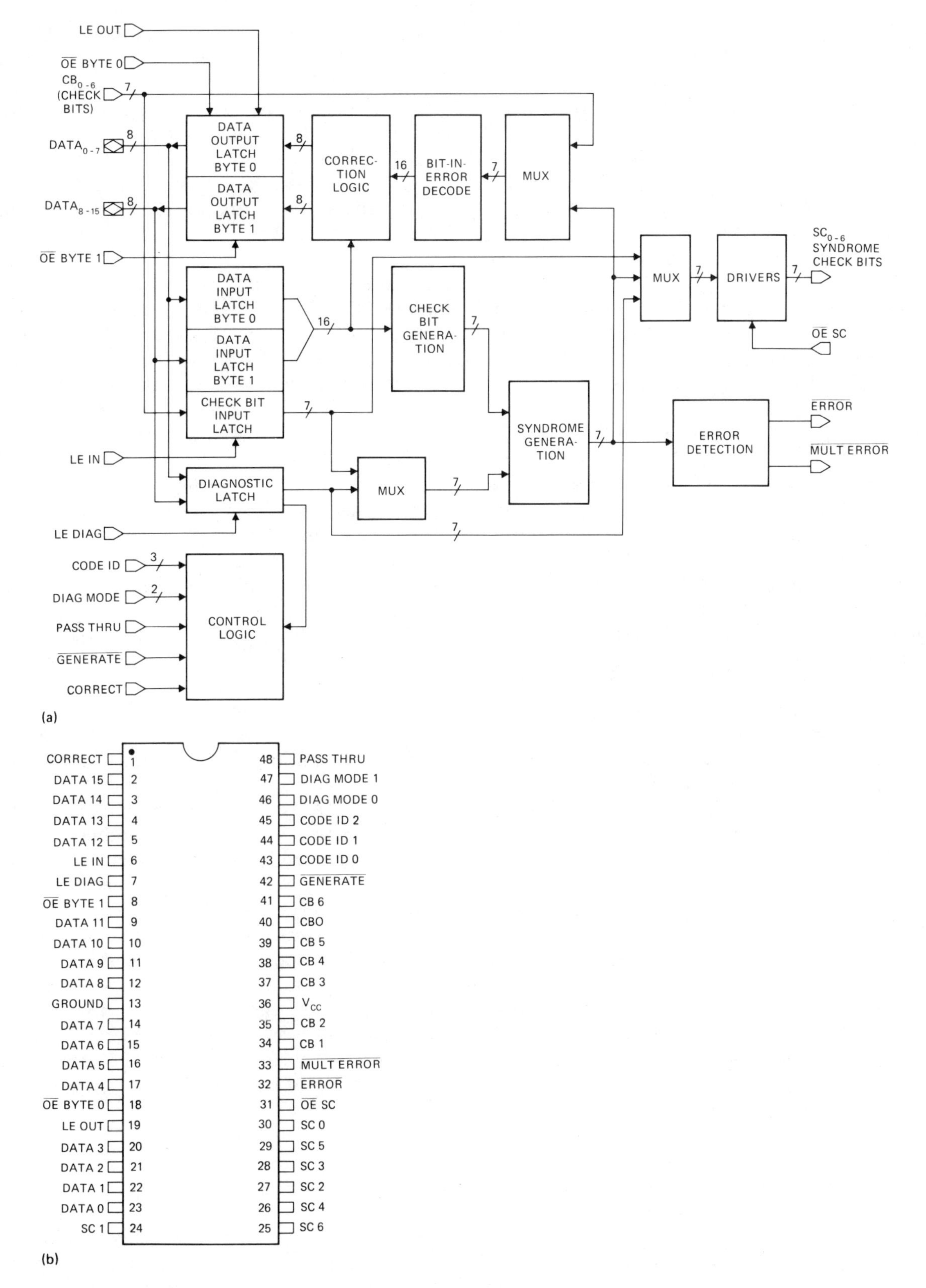

FIGURE 23-43

Advanced Micro Devices Am2960 16-bit error detection and correction unit. (a) Block diagram. (b) Connection diagram.

23.7 Summary

RAM semiconductor memories are used in all general-purpose computers. They are connected to the microprocessor by address, data, and control buses. Every memory chip receives the low-order address lines, while the high-order address lines are used for chip selection. Sometimes, in large systems, the address lines used for chip select are wired to decoders. The output of the decoder is connected to the memory chip select pins, as shown in Figure 23-2. A memory map shows which addresses RAM, ROM, and I/O occupy. The object of chip select decoding is to avoid bus contention.

A memory's address access time is the amount of time it takes for the memory to receive a valid address and output the data to its output buffer.

Memory chips that are word-organized normally have common data I/O lines. Bit-organized memories usually have separate I/O lines; these lines can often be externally tied together. The steps involved for a microprocessor to write data to RAM are: (1) send a valid address to memory; (2) select the particular chip or chips; and (3) activate the write enable line. This timing is shown in Figure 23-9. To prevent bus contention, it may be necessary to keep the chip select inactive until after the write enable or to use the output enable if available. To avoid writing data into an erroneous address, the address must be stable before the write time. The microprocessor must send valid data to the memory chips before the end of write time and remain stable after the write time.

There are two types of random access memories—static and dynamic. The static cell is a flip-flop, as shown in Figure 23-12. Figures 23-11 and 23-15 show the timing diagrams for the write and read cycles, respectively. Figure 23-17 shows how two memory chips organized as 256-by-4 are wired in parallel for a 256-by-8 memory system.

When fast access times are needed, a user may choose emitter-coupled logic. Typical access time is 45 nanoseconds. Figure 23-2 shows how 1K of ECL RAM are interconnected.

Chapter 9 dealt with the advantages and disadvantages of CMOS devices. CMOS RAMs have lower power drain and are obviously a practical solution for portable equipment.

MOS RAMs are the most widely used memory chips. They are manufactured with high density, low power dissipation, and low cost per bit. There are two types—static and dynamic. Dynamic RAMs have higher density than static RAMs but have the distinct disadvantage of having to be refreshed periodically. Figure 23-25 shows an NMOS static RAM cell. The 2114, shown in Figure 23-26, is a popular static memory chip. The single transistor cell of a dynamic cell is shown in Figure 23-28. A 16K dynamic RAM layout is shown in Figure 23-29. Figure 23-31 gives the operation phases occurring during memory read or write cycles of a dynamic RAM. Dynamic RAMs use a multiplexing scheme for the address lines. The multiplexing scheme allows the manufacturer to produce high-density RAMs in 16-pin dual-in-line packages. Timing waveforms are critical for dynamic RAMs. The memory chips must receive the row and column addresses at the right times. The refresh timing must meet the manufacturer's specifications. These waveforms are illustrated in Figures 23-33 through 23-37.

An IC called a refresh controller such as the Intel 3234 can be used to provide address multiplexing and refresh control. This IC also includes a refresh address counter.

Current drain of dynamic RAMs is specified for three conditions, illustrated in Figure 23-41.

There are two types of errors that occur in dynamic RAMs—hard errors and soft errors. Hard errors occur at the same address and indicate a defective device. Soft errors occur at random addresses and can be caused by low noise immunity. Soft errors also occur because of a phenomenon known as alpha particle penetration. The most common method of error detection is by parity. This requires additional memory space. Advanced Micro Devices 2960 error detection and correction circuit, shown in Figure 23-43, can be used. It requires an additional six bits of memory for each data word.

Glossary

Address Bus: A group of conductors over which a memory chip receives a binary pattern identifying a memory location.

Data Bus: A group of conductors over which a memory chip can send and receive binary data.

Control Bus: A conductor or group of conductors used to synchronize the transfer of data.

Chip Select: A line or lines that are used to enable or disable the memory chip.

Memory Map: A method of keeping track of the location of numerous circuits in a microcomputer address system.

Bus Contention: The situation that arises when two devices try to use the data bus at the same time. One device tries to drive the data bus high as the other tries to drive it low.

Memory Page: 256 words of memory occupying consecutive addresses.

Address Access Time: The time it takes a memory device to retrieve the data and supply it to the output bus buffer after a valid address is received.

Chip Select Access Time: The time from chip select going active until valid data occurs at the data-out pins of a memory circuit.

Common I/O: The use of the same pins or bus lines for both data-in and data-out functions.

Write Cycle Time: The time required for a memory to complete a valid write cycle and be ready for the next read or write.

Read Cycle Time: The time required for a memory to complete a valid read cycle and become available for the next read or write cycle.

Bit-Organized: Memory data bus organization in which each address in a memory chip stores a single bit.

Word-Organized: Memory data bus organization in which each address in a memory chip stores several bits—usually four or eight bits.

Static Ram: A memory device that retains its data indefinitely as long as power is supplied to the chip. It does not require a periodic refresh cycle.

Dynamic RAM: A memory device whose data has to be periodically refreshed by performing a redundant read or write operation.

Sense Amplifier: A circuit that senses the state (1 or 0) of a dynamic RAM cell and amplifies it for the purpose of having it read or refreshed.

Bit Line: A conductor within a dynamic RAM to which a number of single transistor cells are connected.

Refresh Cycle: A memory cycle executed for the purpose of refreshing the data stored in a dynamic RAM cell.

Page Mode: Reading a block of data from a dynamic RAM by which the row address remains the same and the column addresses change.

Hard Error: An error that occurs on every occasion that the address of the defective bit is read.

Soft Error: A memory bit written or read at an incorrect state due to noise or α particle penetration. Soft errors occur at random address or data bits and are not repeating.

Alpha Particle Penetration: Alpha particles from uranium and thorium used in the packaging material penetrating the chip. The effect is to alter the charge being read or stored causing a soft error.

Error Detection: A method, such as using parity, to determine whether or not an error has occurred.

Questions

1. Into what four groups can the pin-out of a semiconductor RAM be divided?
2. What is the functional difference between a chip select input and an address input?
3. What is the purpose of chip select decoding?
4. What is the actual size of 4K bytes of memory?
5. Which is normally the shorter time interval—address access time or chip select access time?
6. If the most significant address bit of a memory address system is not used in chip select decoding, what form will the upper half of the address map take?
7. If a memory chip has a WE pin, must this line go high or low for the microprocessor to write data to memory?
8. Could bus contention occur if the chip select line is activated before the write enable line is activated?
9. If a memory map shows that a 1K memory has three image locations, under what condition must address decoding be expanded?
10. What type of digital circuit is a static TTL RAM cell?
11. How many address pins are there on a 256-by-4 memory chip?
12. What is a prime advantage of a CMOS memory chip?
13. What type of memory chips should be used for battery-operated equipment?

14. Typically does CMOS RAM have faster or slower access time from that of NMOS devices?
15. What are the two types of MOS random access memories?
16. Are dynamic RAMs bit- or word-organized?
17. List an advantage and a disadvantage of dynamic RAMs over static RAMs.
18. Is a static NMOS RAM cell similar to a CMOS RAM cell?
19. How many address pins are there on a 2114 RAM chip?
20. How many 2114 packages are required to assemble 4K bytes of RAM?
21. Draw a single transistor dynamic RAM cell.
22. Are sense amplifiers needed in dynamic RAMs that use a three-transistor cell arrangement?
23. Is a row of dynamic memory cells refreshed during a microprocessor read operation?
24. How do manufacturers economize on address pins on dynamic memories?
25. Can a 256K dynamic RAM be housed in the same size package as a 16K dynamic RAM?
26. Assume that a dynamic RAM is selected. What function is occurring when (a) $\overline{RAS} = 1$ and $\overline{CAS} = 1$, (b) $\overline{RAS} = 0$ and $\overline{CAS} = 1$, and (c) $\overline{RAS} = 0$ and $\overline{CAS} = 0$?
27. During a refresh cycle, is each bit in a row refreshed at the same time?
28. If a dynamic RAM is wired for common I/O, can a read modify write operation be performed?
29. What is the type of operation called when a block of data is read from memory and the row address remains the same and only column addresses are different?
30. What is the purpose of Intel's controller chip 3242?
31. A 16K dynamic RAM has how many actual memory bits? What is the address of the highest bit if the lowest is address 0000_H?
32. What are the two classifications of errors in dynamic RAMs?
33. Is alpha particle penetration a hard or soft error?
34. Can decoupling capacitors eliminate some types of soft errors?
35. Is a noise error a hard or soft error?
36. What method other than decoupling capacitors is used to reduce or eliminate soft errors?
37. What is the purpose of American Micro Device's 2960 IC?
38. Under what condition can two memory devices be selected for the same address blocks without causing a bus contention?
39. What is the most common point of zero reference for memory timing?
40. What is the usual point of zero reference for a memory with a dynamic address system?
41. For a memory with both chip enable and output enable functions, what two methods can be used to avoid bus contention during a write cycle?
42. What features of CMOS RAMs make them ideal for battery backup operation?
43. What feature of dynamic RAM makes them more awkward to use than static RAMs?
44. What advantages do dynamic RAMs have in comparison to static RAMs?
45. What limitation exists for connecting large numbers of dynamic RAM cells to the same sense amp?
46. What happens to the bit line of a dynamic RAM during precharge?
47. When the cell addressed during a dynamic RAM memory cycle contains a zero charge, what charge will exist on the dummy cell before equilibration?
48. During a dynamic RAM memory cycle, when the cell addressed contains a high level charge, what charge will exist on the dummy cell during equilibration?
49. For common I/O operation of the dynamic RAM, what must be the relative timing between $\overline{CAS}$ and the write pulse?
50. What is a read-modify-write cycle? What must be the relative timing between the $\overline{CAS}$ and the write pulse in order to have an RMW cycle?
51. How is page mode operation accomplished with the dynamic RAM?
52. Why is the current drain of dynamic RAMs relative to the memory cycle times?
53. Why do refresh cycles consume less power than read or write cycles?
54. What are three operating current specifications of dynamic RAMs, and when will these currents occur?
55. What is the relationship between error correction circuits and parity generators?
56. What is the purpose of the polyimide die coat on the 64K RAM chip?

Problems

23-1 The 1-of-8 decoder of Figure 23-2 is rewired as follows: E_3 to A_{12}, E_2E_1 to GND, A_0 to A_{13}, A_1 to A_{14}, A_2 to A_{15}, and memory chip selects to 05. What is the address of the memory?

23-2 Draw an address decoder that will select the 4802 memory of Figure 23-3 for address $E1E1_2$. Draw a memory map showing the memory address and all images.

23-3 If a NAND gate were substituted for the OR gate in Figure 23-4, what would be the new memory address? Draw the memory map.

23-4 For the memory system of 23-6, draw the decoding logic needed to give the ROM an intended address of $COOO_H$ to $FFFF_H$ while keeping the RAM at the same intended address of 4000_H to 47FF. Use added decoding needed to avoid bus contention. Draw the new memory map.

23-5 Draw a timing diagram like that of Figure 23-7. The address becomes valid at time 0. Chip select goes low at 120 nanoseconds. Data becomes valid at 250 nanoseconds. Chip select goes inactive at 450 nanoseconds, and data leaves the bus at 500 nanoseconds. Label the actual values of t_{ACC}, t_{CS}, and t_{CD}.

23-6 Draw a timing diagram like that of Figure 23-8. The address becomes valid at time 0. Chip select is active from 100 to 290 nanoseconds. WE is active from 120 to 300 nanoseconds. D_{OUT} goes to high impedance from 130 to 300 nanoseconds. D_{IN} contains valid data from 140 to 320 nanoseconds. Determine the parameters listed in Figure 23-11.

23-7 A dynamic RAM using only 5V power supply is configured like the RAM in Figure 23-30 (V_{DD} = 5V). The bit line to cell capacitance ratio is 10. Determine the voltage differences for operation of the sense amp in reading both 1 and 0 levels.

23-8 Draw the timing diagram of an $\overline{RAS}$-only refresh cycle like that of Figure 23-34. Row address is valid from time 0 to 100 nanoseconds. $\overline{RAS}$ goes active from 20 to 200 nanoseconds. Determine the timing parameters given in Figure 23-10.

23-9 A 64K RAM is in standby with a refresh cycle occurring every 15.6 μs. From the specifications given in Figure 23-41, compute I_{CC} for the standby period.

23-10 Draw the timing diagram of a dynamic RAM read cycle like that of Figure 23-35. Row address is valid from 0 to 100 nanoseconds. $\overline{RAS}$ is active from 20 to 200 nanoseconds. Column address is valid from 110 to 180 nanoseconds. $\overline{CAS}$ is active from 120 to 240 nanoseconds. Valid data occurs from 160 to 260 nanoseconds. Compute the parameters given in Figure 23-35.

23-11 Draw the timing diagram of a dynamic RAM write cycle like that of Figure 23-35. Row address is valid from time 0 to 100 nanoseconds. $\overline{RAS}$ is active from 20 to 200 nanoseconds. Column address is valid from 110 to 180 nanoseconds. $\overline{CAS}$ is active from 120 to 240 nanoseconds. Write is active from 80 to 160 nanoseconds. D_{IN} is valid from 100 to 170 nanoseconds. Compute the timing parameters given in Figure 23-35.

23-12 The dynamic RAM of Problems 23-10 and 23-11 has average cycle times of 750 nanoseconds. From the specifications in Figure 23-41, compute the supply current, I_{CC}. Consider the effect of refresh as insignificant.

23-13 The 3242 refresh controller is adequate for 16K dynamic RAM. Draw the external logic necessary to expand this circuit for use as a 64K refresh controller with 256 row refresh capability. Use a block symbol for the 3242 and show needed logic gates and flip-flops to make the expansion.

23-14 For the error correction described in Figure 23-42(b), draw the logic diagram of the circuit needed to decode the check bits and correct the data.

CHAPTER

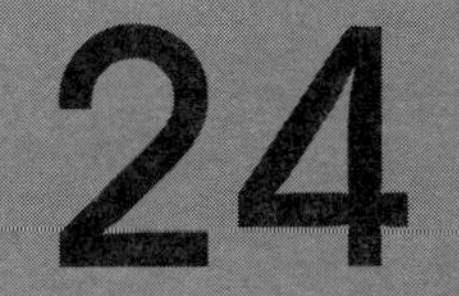

Serial Memory and Read-Only Memory

Objectives

Upon completion of this chapter, you will be able to:

- Connect and use a dynamic shift register as a serial memory.
- Connect and use a static shift register as a serial memory.
- Write a program for a read-only memory.
- Use a read-only memory character generator in a CRT display circuit.
- Compare fusible link PROMs to mask programmed ROMs.
- Use fusible link PROMs.
- Assemble logic gates using a programmable array logic.
- Use field programmable logic arrays.
- Use ultraviolet erasable PROMs.

24.1 Introduction

Chapter 23 dealt with one type of semiconductor memory—RAM. This chapter covers the following other types of memories and logic arrays: serial memories, mask programmed memories, fusible link PROMs, ultraviolet erasable read-only memories, programmable array logic, and field programmable logic arrays. Each device has specific advantages and provides a designer with powerful tools for different applications. This chapter shows how and where these devices can be used.

24.2 Serial Memory

24.2.1 Identity to Shift Registers

This serial memory is essentially a shift register of many bits. In fact, a serial memory is more often referred to as either a static or dynamic shift register. As memory circuits, these devices can be connected to recirculate to provide for a nondestructive readout. They are volatile. They are ideal for handling serial data, which might otherwise have to be twice converted if stored by RAM. Those used for memory purposes may differ from those used for operations registers, in that reduced power drain is a more important aspect in their design or selection. They commonly use two or more clock phases to accomplish the reduced power drain.

24.2.2 Dynamic Shift Registers

One way to store a 1 voltage level is to charge a capacitor to the 1 level and then isolate it so that the charge cannot drain off. By using such a method, a shift register could be constructed, as Figure 24-1 shows. If C_1 were charged to the 1 level when the ϕ_1 switches were closed, then most of that voltage would be passed on to C_2 when the ϕ_2 switches were closed. If the switches were alternately opened and closed a second time, a depleted 1 level would reach C_4 after the second closing of the ϕ_2 switches. This appears somewhat like the operation of a shift register, except that the signal level decays in its progress through the register. If we were to try to follow down this 1 level with a 0 by discharging C_1 when the ϕ_1 switches closed the second time, then on the second closing of the ϕ_2 switch, the C_2 capacitor would lose most of its charge to the larger C_1, storing a binary 0 in C_2 along with the binary 1 in C_4. This register requires two capacitors per bit of storage. Unless the input capacitors are extremely large and the output capacitors extremely small, there is a serious decay of the binary levels as data pass through the register. The table in Figure 24-1 shows a decay to 73 percent of the input 1 level at the output of the second bit, even if succeeding capacitors decline in value by a factor of 10. Such a system seems possible but impractical until we apply MOS semiconductor techniques to overcome its weak points.

Referring to Chapter 5, we know that the common emitter or common source switch can restore signal levels. The common source switch is ideal for this function. Because of the extremely high input impedance, the MOSFET can be turned on with the charge on a small capacitor without rapidly discharging it. It is no drawback that an inversion occurs, in that each register stage will require two phases, as was the

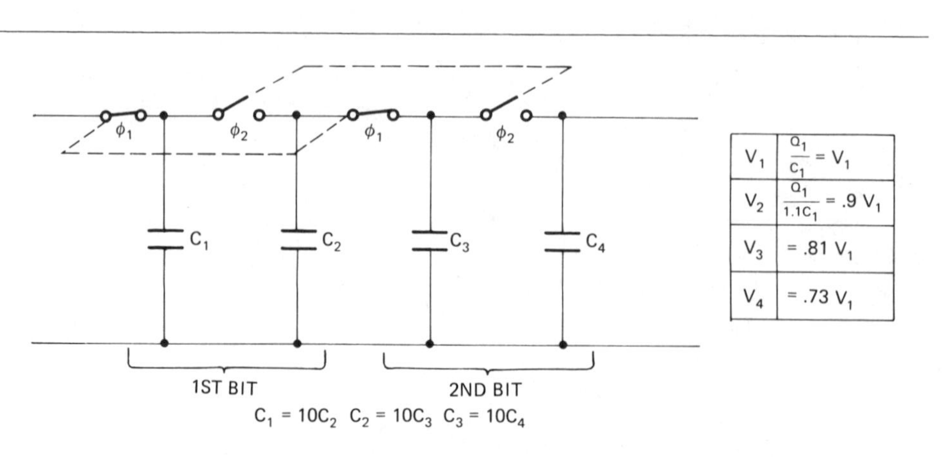

FIGURE 24-1
Manual equivalent of an AC MOSFET shift register. If C_1 is charged to a logic level, that level can be passed to C_4 by alternately opening and closing the switches ϕ_1 and ϕ_2.

case with the master-slave flip-flop. Figure 24–2 shows one stage for a shift register of this type. This is a simplified version with manual switches. Let us assume the transistors to be N channel, so that we may proceed with the explanation in positive logic. If a positive 1 level is placed on the input, the first MOSFET turns on and all the V_{DD} drops across R_1. If a 1-level charge exists on C_1, a momentary closing of the ϕ_1 switch discharges C_1 to ground through Q_1. The 0 level now on the gate of MOSFET 2 turns it off, making a 1-level voltage available at the terminal of the ϕ_2 switch. A momentary closing of the ϕ_2 switch will charge capacitor C_2 to the 1 level.

Because the manual switch is not practical for an automatic system, a MOSFET is used in place of the switch. The MOSFET is ideal for this function because the drain and source terminals can be interchanged without materially altering the conductivity. Figure 24–3 shows half a register stage with the switch replaced by a MOSFET. When a positive 1 level is applied to the input, the input MOSFET is turned on and the drain current drops the V_{DD} across R_1. With a positive 1-level charge on the capacitor, C_1, the switch MOSFET is biased to discharge the capacitor when the positive pulse, ϕ_1, turns it on, as Figure 24–3(a) shows. When a 0 is applied to the input, the input MOSFET is turned off. R_1 may now operate as the drain resistor of the switch MOSFET. When the pulse ϕ_1 is applied to the gate, capacitor C_1 will charge to a positive 1 level through R_1 and the switch MOSFET, as Figure 24–3(b) shows.

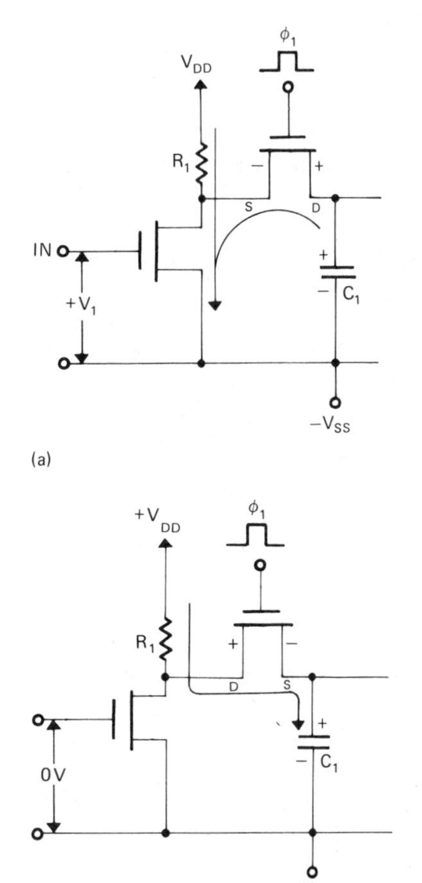

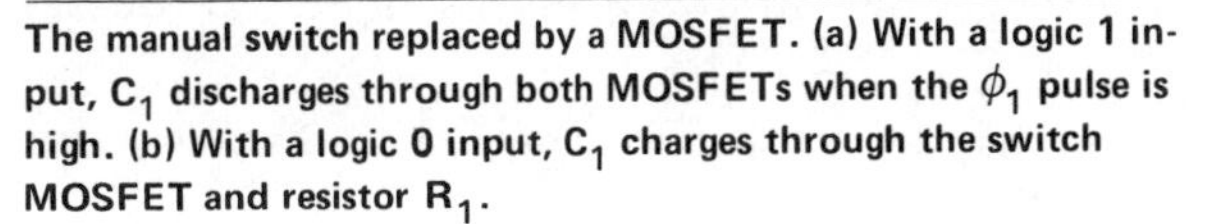

FIGURE 24–3

The manual switch replaced by a MOSFET. (a) With a logic 1 input, C_1 discharges through both MOSFETs when the ϕ_1 pulse is high. (b) With a logic 0 input, C_1 charges through the switch MOSFET and resistor R_1.

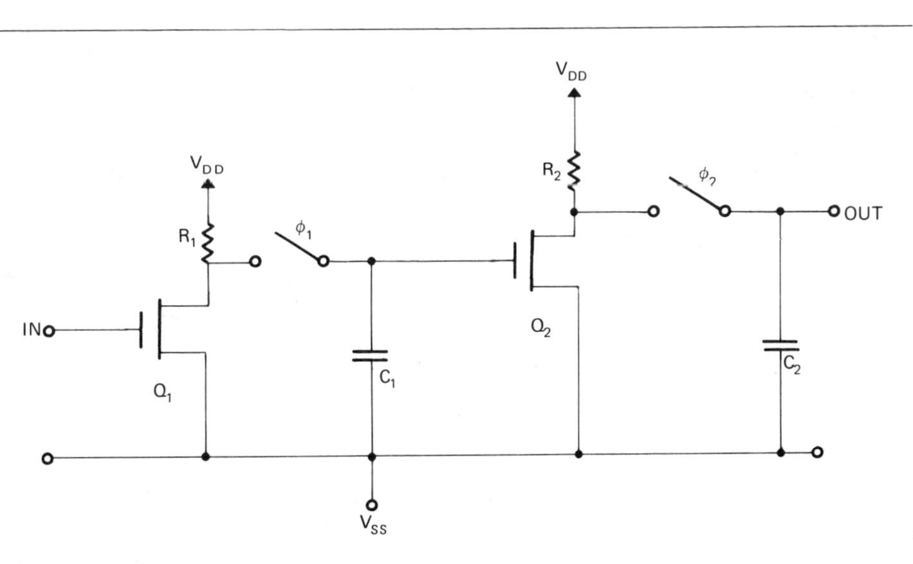

FIGURE 24–2

A MOSFET inverter used to charge the capacitors and prevent degrading of the logic levels as the number is shifted through the register. The inversion is canceled because there are two inverters per bit of storage.

In discrete components, a resistor or capacitor may be lower in price than a transistor, but in the manufacture of integrated circuits, this is not the case, and MOSFETs acting as resistors replace the drain resistors. Figure 24-4 shows the usual form of this AC register. The MOSFET acting as the drain resistor is turned on along with the switch MOSFET. The high turn-off resistance helps further to prevent a change in charge during turn-off. The turn-off and input resistances of the MOSFETs are so extremely high that the mere capacity of the gate circuit can in itself hold sufficient charge to activate the input gates. This capacitance, unfortunately, can store a useful charge for only a few milliseconds. Because of this, the data cannot be shifted in and held in static storage by inhibiting the clock line after shift-in and enabling it later for shift-out.

The data will disappear if not continuously moved through the register. This limits its use to serial access buffer memories. The timing must be such that the first bit of the data can be used exactly when it reaches the output half of the last stage in the register; but data can be recirculated from the output back to the input.

The dynamic shift register just described lends itself readily to use in digital calculators, display circuits, and other serial devices. Figure 24-5 shows the Signetics* MOS dual 100-bit register. The input bit has an added buffer circuit. The output has a driver circuit. Bits 2 through 99 resemble Figure 24-4 except that P-channel MOS circuits are used, and therefore a negative V_{DD} power supply is used. For use with TTL circuits, $+V_{CC}$ is applied to the positive power terminal, and, as indicated in the timing diagram, the inputs and outputs are compatible with TTL. Dual-phase negative-going clock pulses must be supplied. Two 100-bit registers are contained in one 8-pin package with a power drain of only 400 μW per bit at 1 MHz clock rate.

Figure 24-6 shows an even larger dynamic shift register with recirculation circuits already included. They are available in 512- or 1024-bit sizes. The power drain is 150 μW per bit at 1 MHz clock. The maximum clock rate is 5 MHz.

*Signetics® is a registered trademark of Signetics Corporation.

24.2.3 Static Shift Registers

The static shift register used for serial memory purposes uses a cell that requires several clock phases to attain low power operation. Figure 24-7 is a 1024-bit static shift register. The circuit is a P-channel silicon gate MOS, but the three negative-voltage clock phases are generated internally. All inputs and outputs, including the clock, are TTL logic levels. As shown in the block diagram, logic for recirculation is included on the chip, requiring only an external jumper. Unlike the dynamic shift register, this register has no minimum clock speed and can operate to 1 MHz. Power consumption is typically 160 μW per bit.

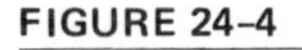

FIGURE 24-4

A MOSFET used to supply the drain resistance. The gate-to-source capacitance is high enough to eliminate the need for a discrete capacitor.

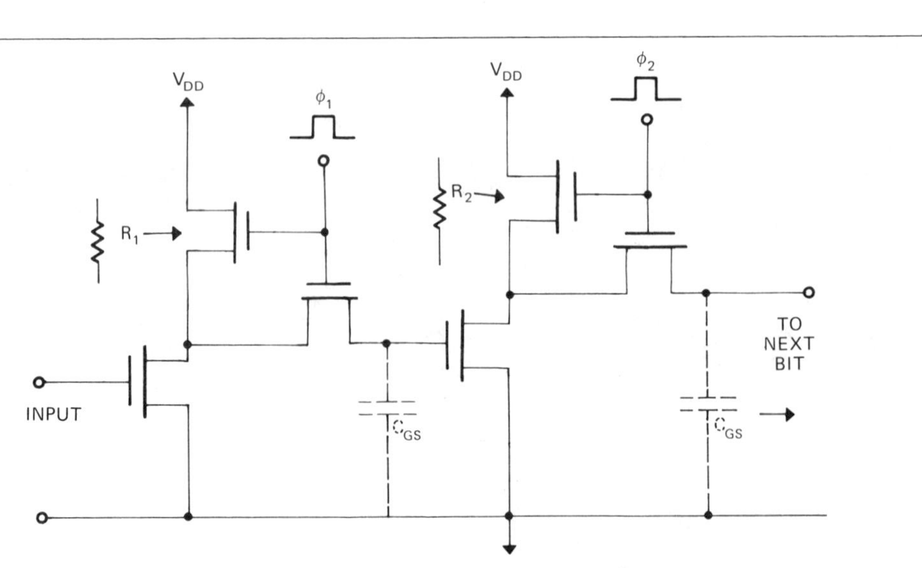

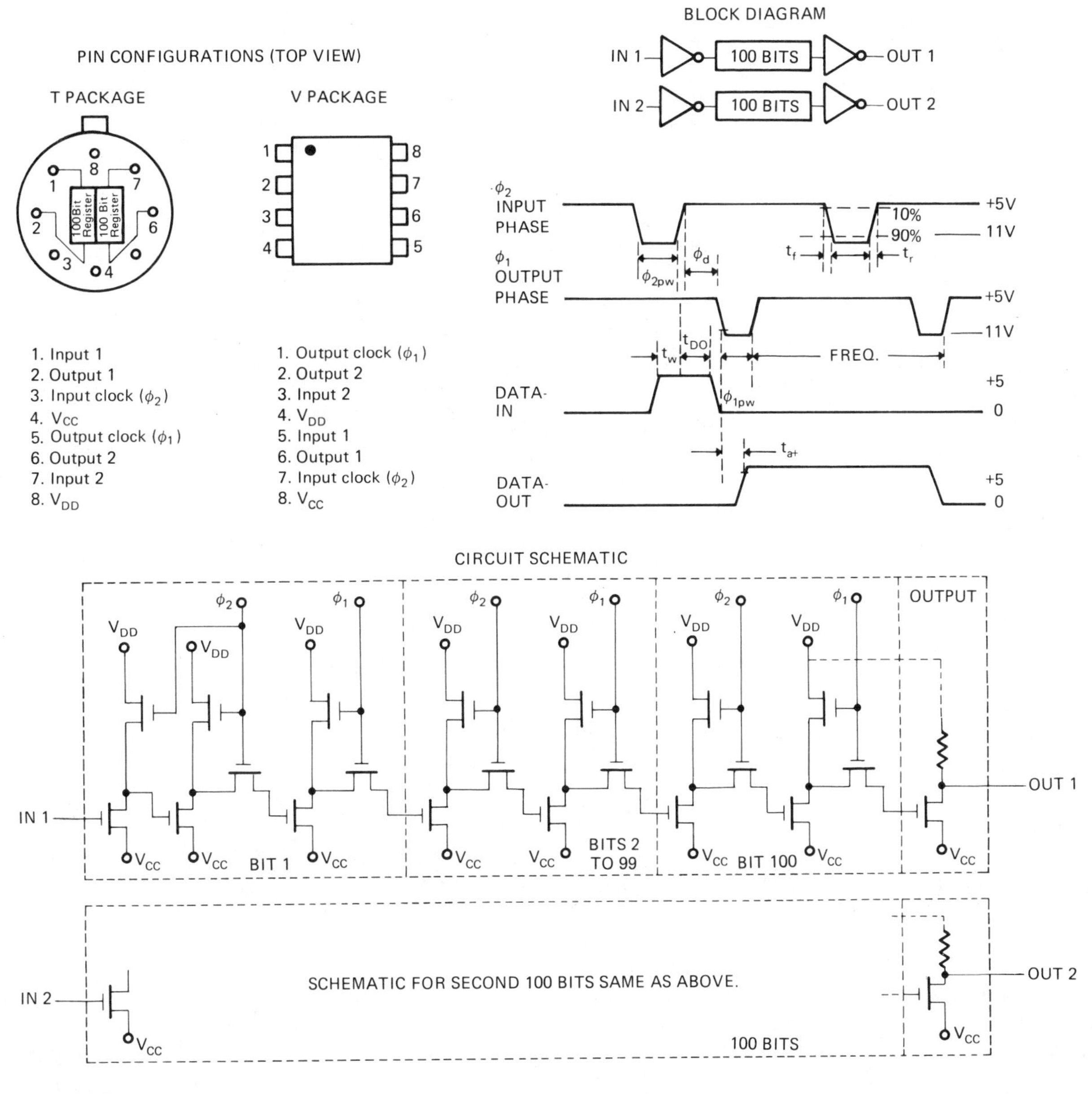

FIGURE 24-5

Signetics MOS dual 100-bit dynamic shift registers.

24.3 Read-Only Memory

24.3.1 Types of ROMs

The read-only memory is as useful in digital application as the read-write RAM. The ROM has the advantage of being nonvolatile and does not have to be loaded each time the system is started up. There are several types available that are erasable and several that are nonerasable. The mask programmable, or MROM, is programmed by the manufacturer to a program supplied by the user. The initial charge for the mask may be as high as $1000, and unless a large number of devices are purchased to a given program, it is an expensive memory. If even a minor change becomes necessary, the mask charge is repeated. The MROM has the highest long-term reliability of all the ROMs. They may be bipolar or MOS, depending on size, and they are fast enough for use with microprocessors.

The fusible link PROM, sometimes referred to as the program once memory, is programmed by blowing minute metallic fuses to produce either 1 or 0 levels. Some manufacturers supply PROMs that are originally 0, and fuses must be

FIGURE 24-6

Signetics 512- or 1024-bit dynamic shift registers with logic for recirculation included.

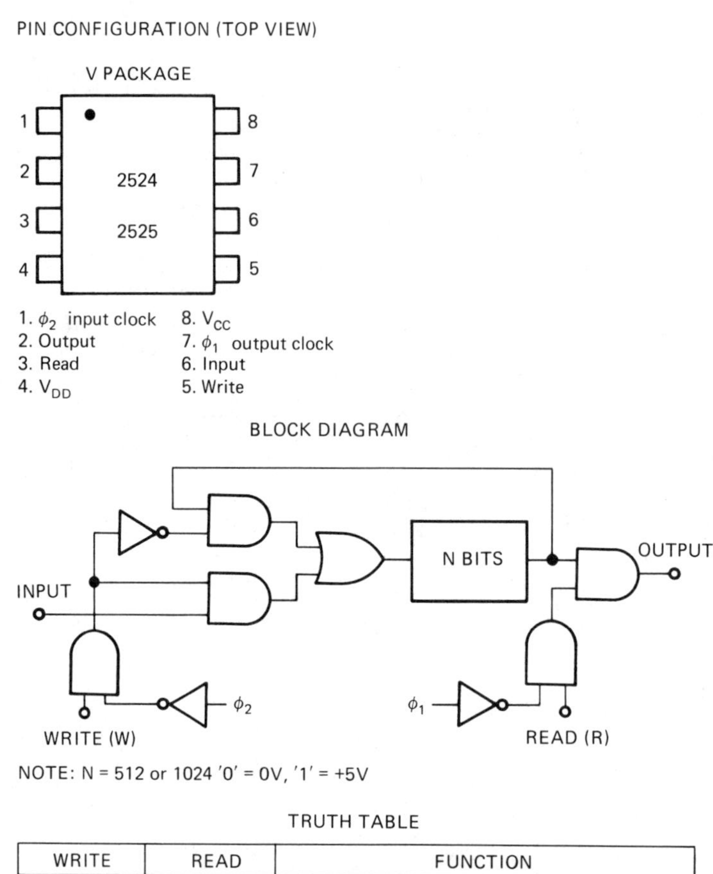

TRUTH TABLE

WRITE	READ	FUNCTION
0	0	RECIRCULATE, OUTPUT IS '0'
0	1	RECIRCULATE, OUTPUT IS DATA
1	0	WRITE MODE, OUTPUT IS '0'
1	1	READ MODE, OUTPUT IS DATA

blown to produce 1s. Others supply PROMs that are originally 1, and fuses must be blown to produce 0s. A number of standard configurations are available, such as 256-by-4, 512-by-4, and 2K-by-8, for which the parts supplied by different manufacturers are interchangeable once they are programmed, but the method of programming is different for each manufacturer. If programmed according to specification, the fusible link PROMs are highly reliable. They are bipolar devices having access times fast enough for present microprocessor applications.

The ultraviolet erasable EPROM can be programmed by the user. It has two operating modes—the read mode and the program mode. The voltages applied differ for the two modes. In the program mode, each word is addressed and voltages are applied at the data-out pins to program 1 or 0 levels. In the read mode, normal operating voltages are applied, and the memory outputs the data that were programmed. The inputs and outputs are at normal TTL levels. The EPROM has a transparent lid over the silicon chip and can be erased by exposing it to ultraviolet light of the correct wavelength and intensity.

The electrically erasable EEPROMs come close to functioning like RAMs, with certain important exceptions. They are like RAMs in that they can be read and written in the same circuit, but unlike RAMs, an erase mode must precede the write cycle and, equally important, the write cycle takes many times longer than the read cycle. However, they are the closest to a nonvolatile read-write semiconductor memory that the industry has thus far produced. They have many interesting applications as peripheral devices into which the contents of RAM can be transferred prior to a loss of power. In this respect, they are an alternative to a battery backup system. Figure 24-8 compares the four commercially available types of ROMs with re-

FIGURE 24–7

Signetics 1024-bit static shift register with internal clock generator. All inputs and outputs are TTL-compatible.

PIN CONFIGURATION (TOP VIEW)

V PACKAGE

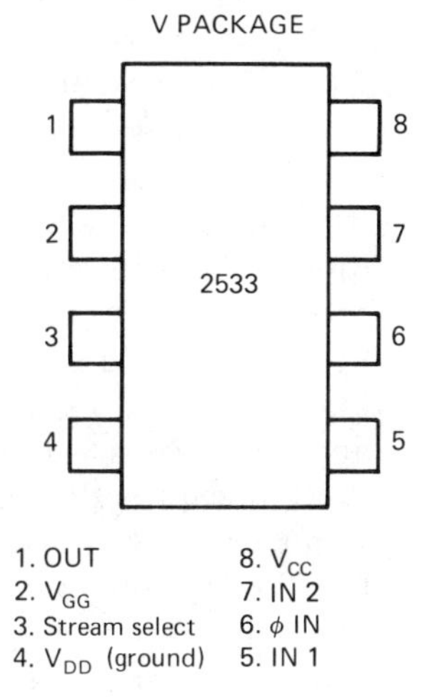

BLOCK DIAGRAM

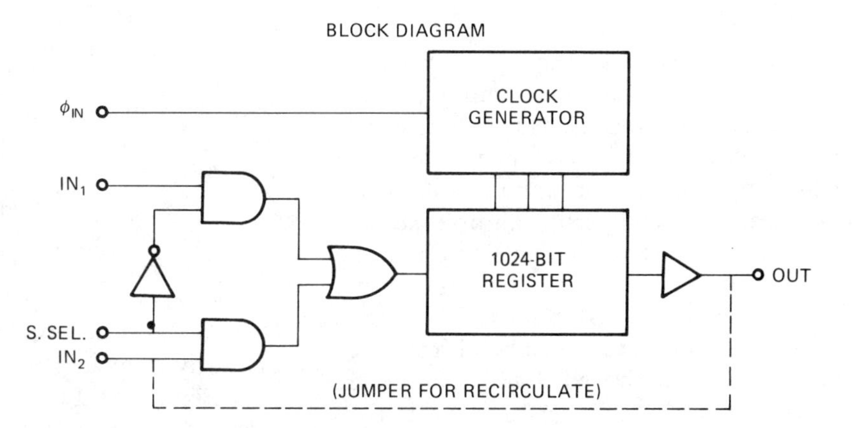

TRUTH TABLE

STREAM SELECT	FUNCTION
0	IN 1
1	IN 2

NOTE: '0' = 0V, '1' = +5V

TYPE	PROGRAMMED	ERASABLE OR NONERASABLE	PRICE	TECHNOLOGY	TYPICAL READ ACCESS TIME (ns)	SIZES AVAILABLE (1982)
MASK PROGRAMMED (MROM)	DURING MANUFACTURE	NONERASABLE	HIGHEST FOR SMALL QUANTITIES	TTL	30	32 × 8 TO 2K × 8
			LOWEST FOR VERY LARGE QUANTITIES	NMOS	350	2K × 8 TO 16K × 8
FUSIBLE LINK (PROM)	AFTER MANUFACTURE BY MANUFACTURER, DISTRIBUTOR, OR USER	NONERASABLE	LOW	TTL	60	32 × 8 TO 2K × 8
ULTRAVIOLET ERASABLE (EPROM)	BY USER OUT OF CIRCUIT	ERASABLE	HIGH	NMOS	200	1K × 8 TO 8K × 8
ELECTRICALLY ERASABLE (EEPROM)	BY USER IN CIRCUIT	ERASABLE	HIGHEST	NMOS	250	1K × 8 TO 2K × 8

FIGURE 24–8

Types of ROMs and characteristics that affect their applications.

spect to programming, speed, erasability, and price.

24.3.2 Mask Programmed ROMs

Read-only memories (ROMs) are memories for which the data are programmed into the memory during manufacture or are programmed manually by the user before assembly into the digital system. The circuitry of the read-only memory is much simpler than that of a RAM. The usual structure is a matrix of lines like that shown in Figure 24-9. The lines are connected by conductive elements at some of their intersections. The conducting element may be left out in manufacture to form a 0 or left in to produce a 1.

As in a decoder circuit, a given combination of 1s and 0s applied to the inputs (address lines) produces the programmed combination of 1s and 0s on the output leads. The address is decoded to energize a single line in the matrix. This single matrix line produces the 1s and 0s on the output leads, depending on the presence or absence of the conductive elements at its intersections with the output lines. The conductive element must be a diode or transistor, so that the inactive input lines can be isolated from the single line that is activated by a particular address. In Section 6.8, we discussed the decimal-to-BCD encoder. In Figure 6-35, OR gates were used to produce the encoding, but this could be accomplished in diode matrix style, as shown in Figure 24-10(a). From the ROM or matrix point of view, the encoder seems to be arranged with the diodes connected with a single input and multiple outputs, as shown for input 6 in Figure 24-10(b). But if we draw those diodes that are connected to a single output, as shown for output 2 in Figure 24-10(c), we have, in fact, a diode OR gate, and the identity of the logic gate encoder and the ROM matrix becomes apparent. Thus we may conclude that the ROM does not provide a new capability but can instead be considered a more standard means for handling coding and decoding—one that promises large economies in manufacturing digital machines, for much of what goes on in digital logic is a matter of coding and decoding.

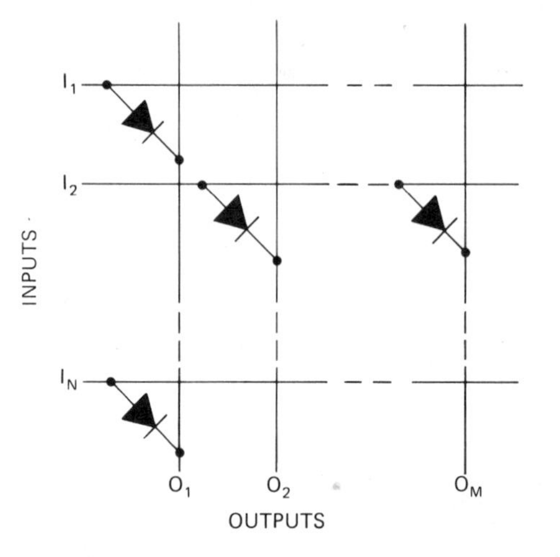

FIGURE 24-9

The read-only memory—a matrix of lines with logic 1 or 0, depending upon the presence or absence of a conducting element at the intersection of input address lines with the output lines.

The conducting elements that join the lines at intersections may be diodes, bipolar or MOS transistors in integrated circuit form. The manufacturer may preprogram these by masking to leave out the connections during processing. For those that are programmable by the user, an intersection destined to be a 0 is addressed and a specified voltage level high enough to burn out the connection is applied through that address. Once a junction is programmed, there is no practical way to change it.

Figure 24-11 shows three types of arrays that have been used to produce read-only memories. In Figure 24-11(a), an addressed line produces a forward-bias current through each diode along the line. The output lines with diodes at the intersections will receive the current. Output lines for which the diodes were left out (or burned out) will receive no current.

For the transistor and MOS arrays shown in Figure 24-11(b) and 24-11(c), a 1 level on the address line turns on those devices having a gate or base lead. This shorts the output line to ground, producing a 0 out. Figure 24-12 is a memory of this type. This chip is organized into an 8K-word-by-8-bit memory. The thirteen inputs, A_0 through A_{12}, are decoded internally to provide the word lines of the memory matrix.

These contents of the SY2364 memory are permanently programmed to customer order. The number of bits per word can be expanded by paralleling the address leads, as Figure 24-13 shows. Each of the programmed chips in Figure

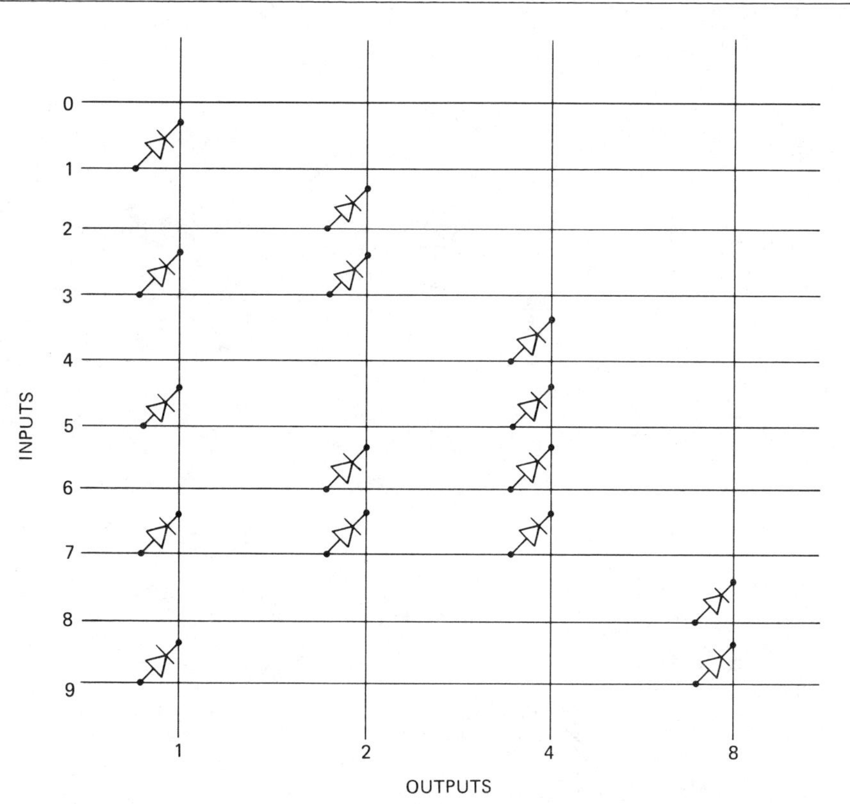

(a)

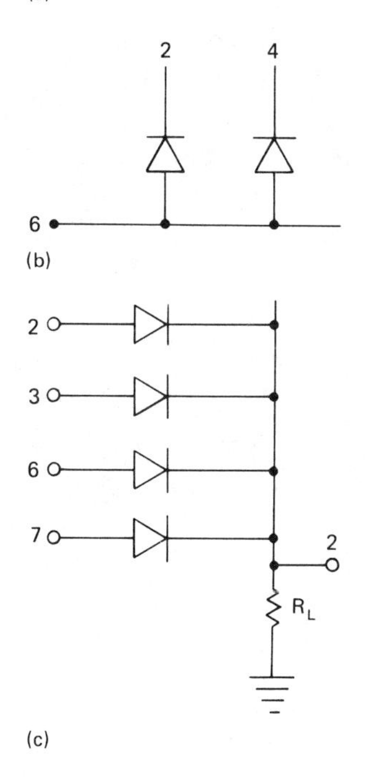

FIGURE 24–10

Encoding accomplished in diode matrix style. (a) ROM forming a decimal-to-BCD encoder. (b) Active elements along input line 6. (c) Active element along output line 2, showing identity to four-input diode OR gate.

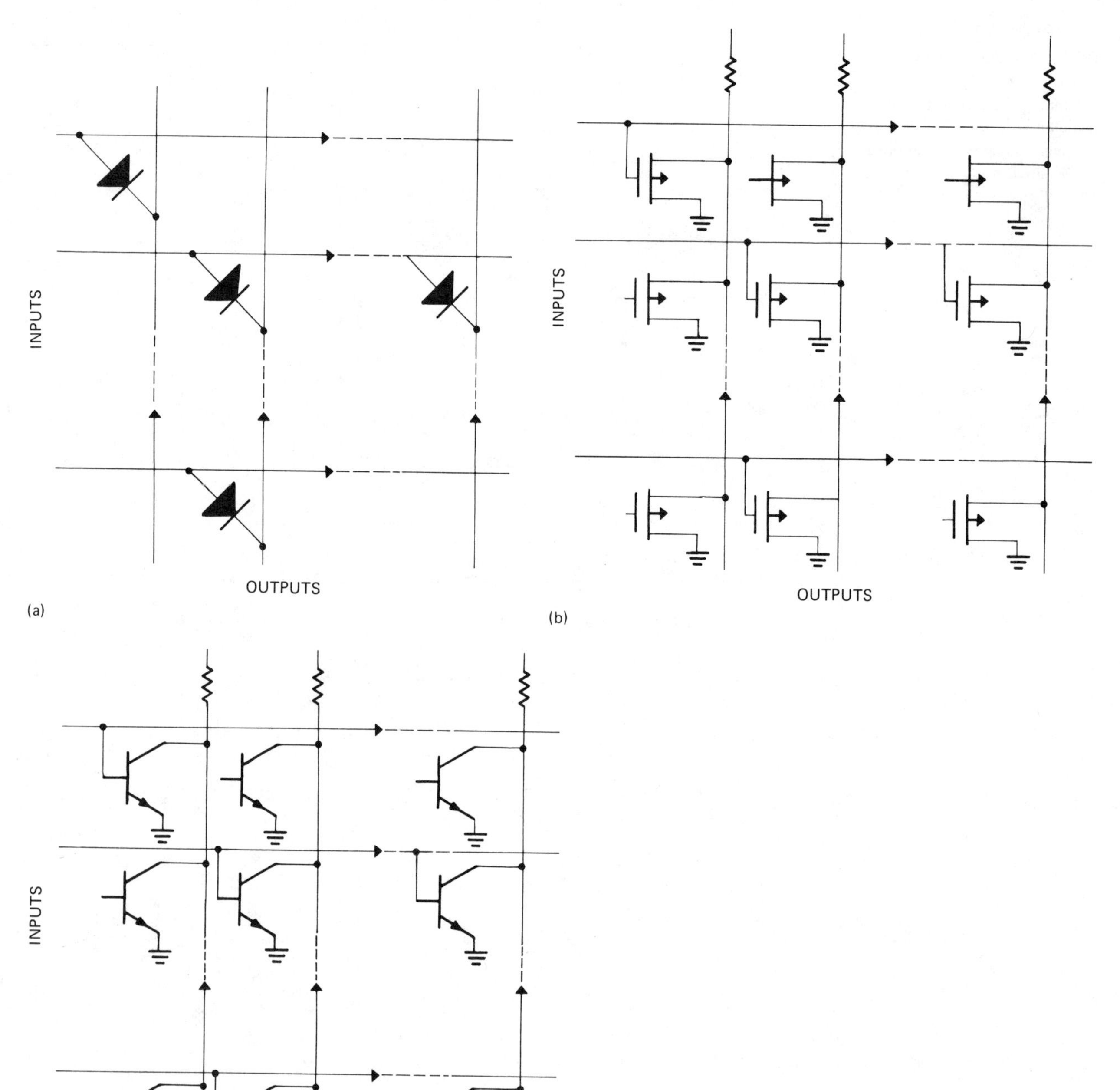

FIGURE 24–11

Three types of arrays used to produce read-only memories. (a) Diode ROM matrix. (b) Transistor ROM matrix. (c) MOS ROM matrix. Memory may also include decode of input address lines, output driver circuits, and chip select circuit.

(a)

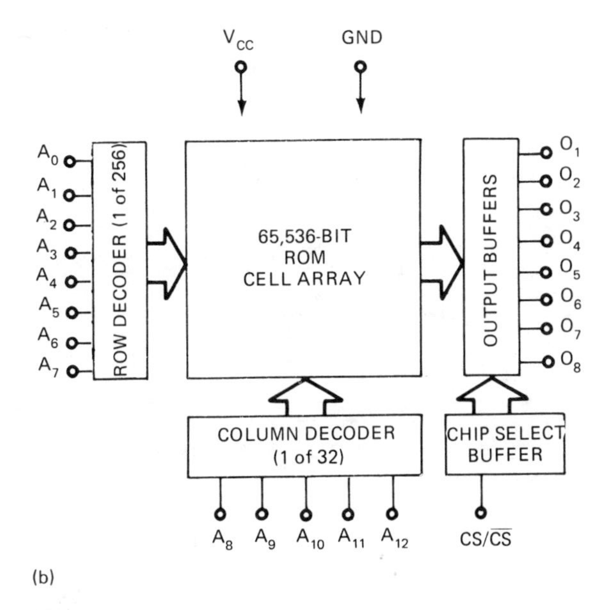

(b)

FIGURE 24-12

Synertek TTL 8K-word-by-8-bit read-only memory. (a) Pin configuration. (b) Block diagram.

24-13 may be programmed differently. Expansion of the address is accomplished by using the chip select and the three-state output capabilities, as Figure 24-14 shows.

24.3.3 ROM as a Mathematical Table

One of the numerous applications of read-only memories is using them as mathematical tables in calculator and computer arithmetic units. Figure 24-15 shows the ROM used as a sine table. The angle in degrees or radians is used as the address. The output provides the binary or BCD equivalent of the sine of the angle. Let us take a simplified example by using only two significant digits 0° to 90°, which calls for a 90-word address. A TTL 93406 ROM, which provides 256 words by four bits, can be used. No address expansion will be needed. Even though the device is designed for straight binary, with eight address lines we can use a BCD input without external decoding. A four-place table would require a BCD output of 16 bits. This table could be constructed by connecting four 93406 circuits, as in Figure 24-16.

It might not seem logical to accept a table of two significant digits at the input and twice that many digits at the output, but to provide a third significant digit would require a minimum of 900 words by 16 bits. With extensive BCD-to-binary encoding, we would need sixteen 93406 circuits. An alternative here would be to provide two-digit constants for use with a mathematical routine for interpolation. This would require only two additional 93406 circuits. In fact, given the simple ROM 0° to 90° sine table, we have other trigonometric functions of all angles

FIGURE 24-13

An 8K-word-by-32-bit ROM using four SY2364 memory chips. Expansion in bits per word is accomplished by paralleling address lines. Each SY2364 chip may have a different program.

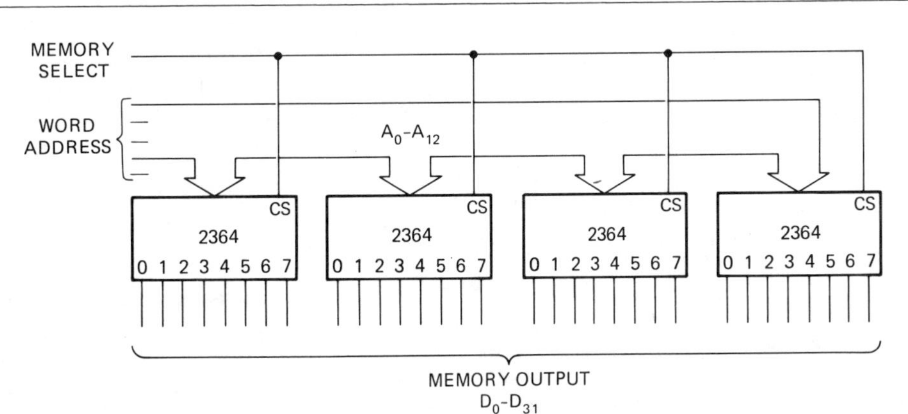

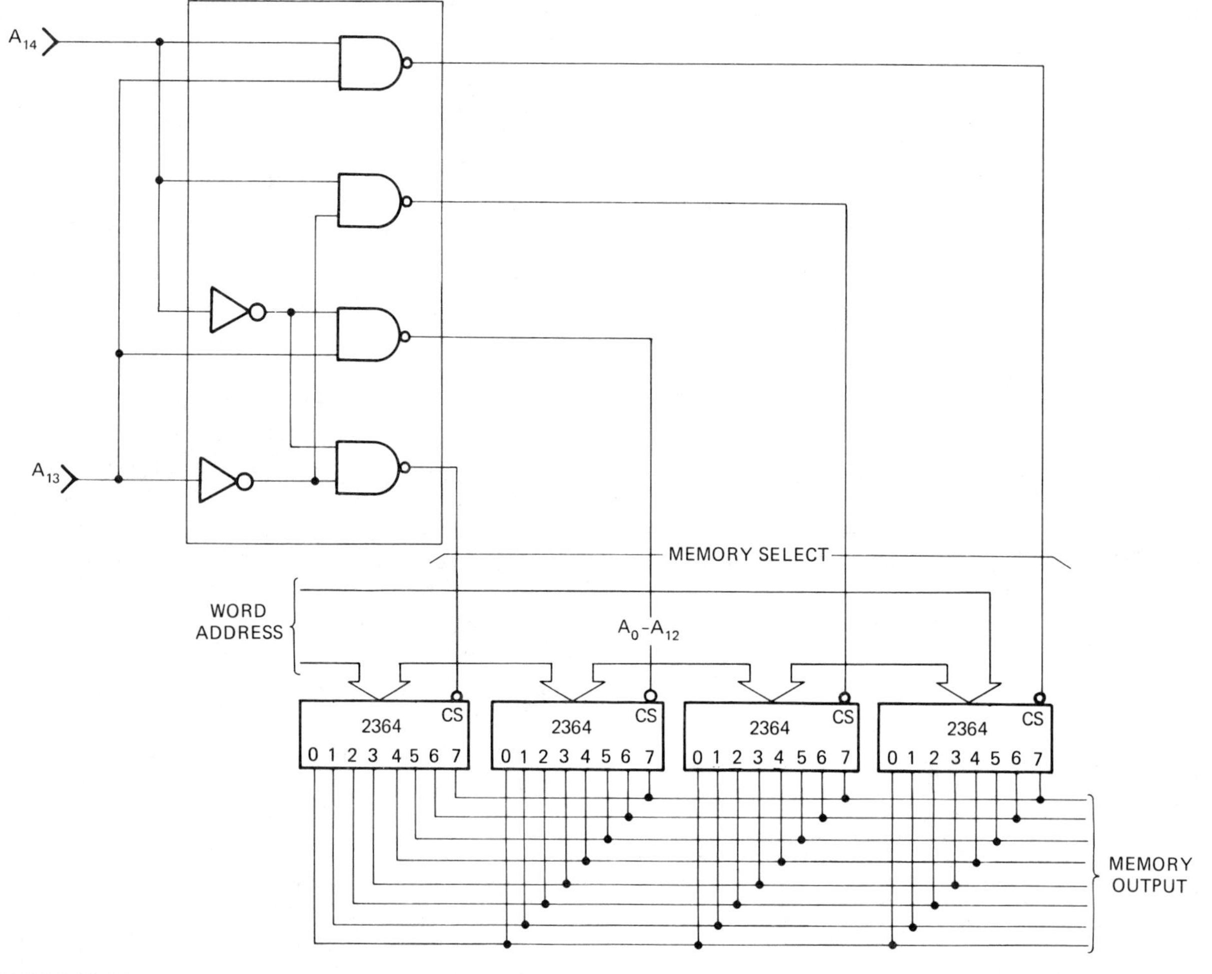

FIGURE 24-14

A 32K-word-by-8-bit ROM using four SY2364 memory chips. Address expansion is accomplished by using the chip select and three-state output capabilities.

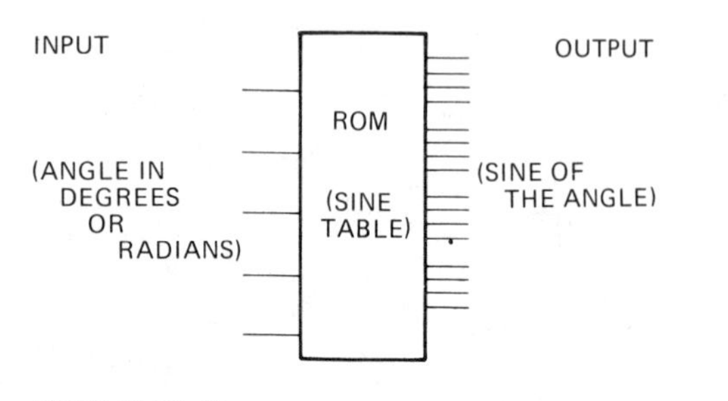

FIGURE 24-15

Block diagram of a read-only memory used to provide a sine table. The input is the angle in degrees or radians. The output is the sine of the angle.

from 0° to 360° available to us by mathematical routines. For example,

$$\cos^2 \theta = 1 - \sin^2 \theta$$

$$\tan \theta = \frac{\sin \theta}{\cos \theta}$$

Preceding routines like these with a simple interpolation of the sine table would not be adding very much to the operating cycle.

Obtaining the necessary programming for the sine table would be a matter of converting

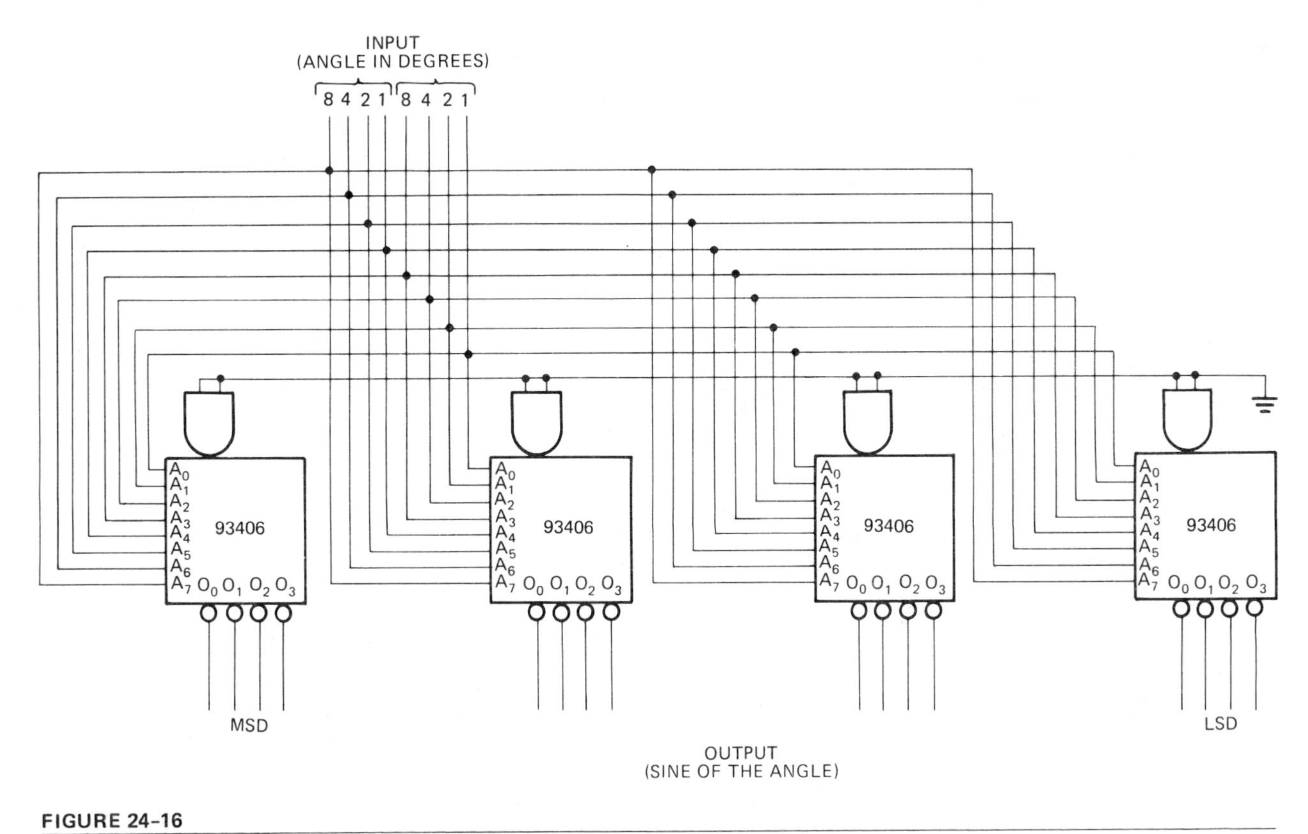

FIGURE 24-16

Four 93406 256-word-by-4-bit read-only memory circuits connected to provide a sine table with two digits (0° to 90°) and four output digits.

the four most significant digits of a 0°-to-90° sine table from decimal to BCD and filling in the data on a customer coding form. Figure 24-17 shows the format of this form covering only the first 15 words of the 256-word memory. A different coding form would be needed for each of the four digits of the output.

24.3.4 CRT Display Circuit

A more impressive use of the ROM is that of character generator for cathode ray tube (CRT) readouts. As an output device for computers and other digital terminals, the CRT display has several distinct advantages. It consumes nothing in the way of paper, cards, or other stationery. It has no mechanical parts, so it is easy to maintain and operate. It is easy to read and can be read at much greater distance than conventional output devices like page printers or teletype outputs. The reader who is not familiar with the functioning of CRT circuits should review this subject. (See also the Glossary at the end of this chapter.)

24.3.5 Character Generator

The American Standard Code for Information Interchange has been widely accepted for use in teletype, computer, and other functions that require digital representation of numbers, letters, and special characters, such as punctuation, mathematical, and other symbols found on a keyboard. The ASCII code is a seven-bit or seven-level code that allows for 128 characters, but, as some of these are carriage commands and some are reserved for future use, they are not all needed for CRT display purposes.

Figure 24-18 is a table of the ASCII code showing only those characters normally used for display purposes. The top row represents the three most significant digits of the seven-bit code. The left column contains the four least significant digits. The numbers below 0100000 and those above 1011111 are not needed for display. This leaves 64 characters that must be generated in the display. Only six of the seven ASCII bits need be used. There is an exclusive code for each of the 64 characters without including the ASCII

FIGURE 24-17

Fairchild customer coding form—format used to provide the manufacturer with programming information for a TTL 93406 1024-bit read-only memory.

CUSTOMER CODING FORM

CUSTOMER ______________________ Location ______________________

Cust. P/N ______________________ Cust. Dwg. # ______________________

Function ______________________ SL # ______________________

Chip Select Code — CS_1 (13) = ___, CS_2 (14) = ___.*

*If not specified, chip select code will be '00.' Package pin numbers are shown in parentheses.

CUSTOM ROM TRUTH TABLE

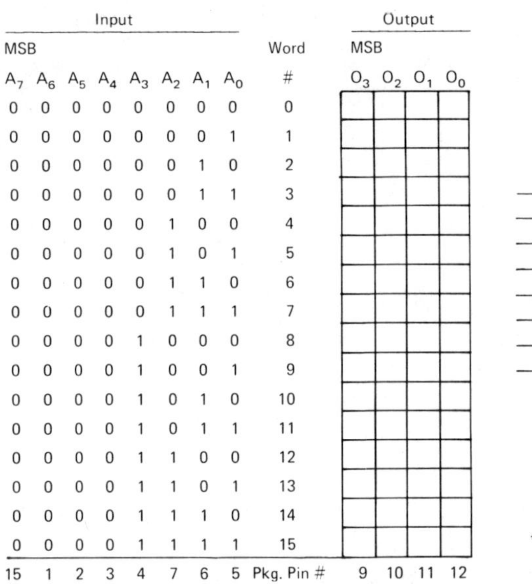

Input MSB A_7	A_6	A_5	A_4	A_3	A_2	A_1	A_0	Word #	Output MSB O_3	O_2	O_1	O_0
0	0	0	0	0	0	0	0	0				
0	0	0	0	0	0	0	1	1				
0	0	0	0	0	0	1	0	2				
0	0	0	0	0	0	1	1	3				
0	0	0	0	0	1	0	0	4				
0	0	0	0	0	1	0	1	5				
0	0	0	0	0	1	1	0	6				
0	0	0	0	0	1	1	1	7				
0	0	0	0	1	0	0	0	8				
0	0	0	0	1	0	0	1	9				
0	0	0	0	1	0	1	0	10				
0	0	0	0	1	0	1	1	11				
0	0	0	0	1	1	0	0	12				
0	0	0	0	1	1	0	1	13				
0	0	0	0	1	1	1	0	14				
0	0	0	0	1	1	1	1	15				
15	1	2	3	4	7	6	5	Pkg. Pin #	9	10	11	12

LOGIC SYMBOL

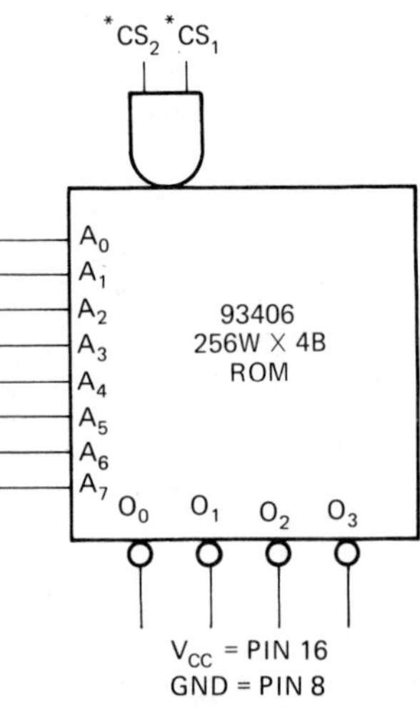

V_{CC} = PIN 16
GND = PIN 8

*Chip selects active level may be programmed per customer requirements. If not specified, both CS will be active low.

CONNECTION DIAGRAM

Signal	Pin	Pin	Signal
A_6	1	16	V_{CC}
A_5	2	15	A_7
A_4	3	14	$\overline{CS_2}$
A_3	4	13	$\overline{CS_1}$
A_0	5	12	$\overline{O_0}$
A_1	6	11	$\overline{O_1}$
A_2	7	10	$\overline{O_2}$
GND	8	9	$\overline{O_3}$

FIGURE 24-18

The American Standard Code for Information Interchange (ASCII). Only the center four columns are shown. The left two columns contain carriage controls. The right columns contain lowercase letters and sometimes are not used.

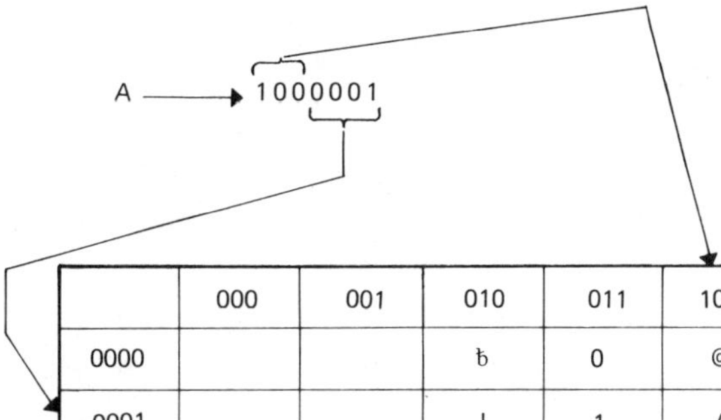

A → 1000001

	000	001	010	011	100	101	110	111
0000			ƀ	0	@	P		
0001			!	1	A	Q		
0010			"	2	B	R		
0011			#	3	C	S		
0100			$	4	D	T		
0101			%	5	E	U		
0110			&	6	F	V		
0111			'	7	G	W		
1000			(	8	H	X		
1001			)	9	I	Y		
1010			*	:	J	Z		
1011			+	;	K	[		
1100			,	<	L	/		
1101			–	=	M	]		
1110			.	>	N	↑		
1111			/	?	O	←		

MSB. These six bits are used as a portion of the address to a read-only memory that contains data bits needed to form the characters. Several standard character formats are available.

Figure 24-19 shows a five-by-eight-character format for the letter S. A single-dimension address, of course, cannot provide us with a two-dimensional figure. The six bits of the ASCII code for S will take us only to one of the eight rows set aside for the letter S. The second dimension is provided by a three-bit row counter that counts through the eight rows. The five output lines change for each of the eight states of the counter. For CRT application, the data must be in serial form. A parallel-to-serial conversion in a converter operating at six times the clock rate of the row counter is needed. The extra clock bit provides the space between letters.

Figure 24-20(a) shows the block diagram of the circuits needed to extract the data bits from the character generator and convert them to serial form. Figure 24-20(b) shows the waveforms for this operation performed on the letter S. The ASCII input addresses us to the rows in the memory for the S character. The row counter changes the parallel outputs from the 000 row to 111. The parallel-to-serial converter gives the rows of data bits in serial form. A space is provided between characters by using blank bits in the parallel-to-serial converter or the first of six clock pulses to clock parallel data in and the remaining five to shift out in serial.

FIGURE 24–19

Signetics 2513 high-speed 64-by-7-by-5 static character generator. It provides CRT character display data if addressed by ASCII code and a mod 8-row counter.

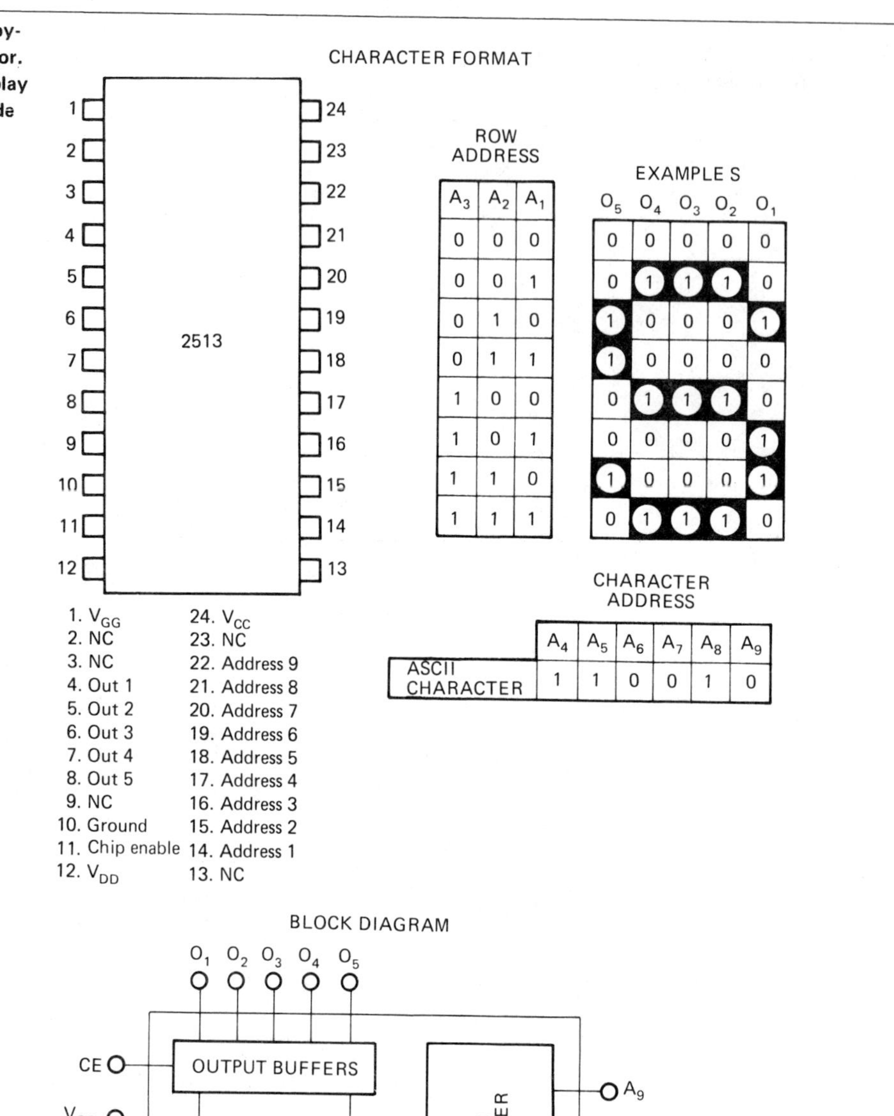

CE	OUTPUT
0	DATA
1	OPEN

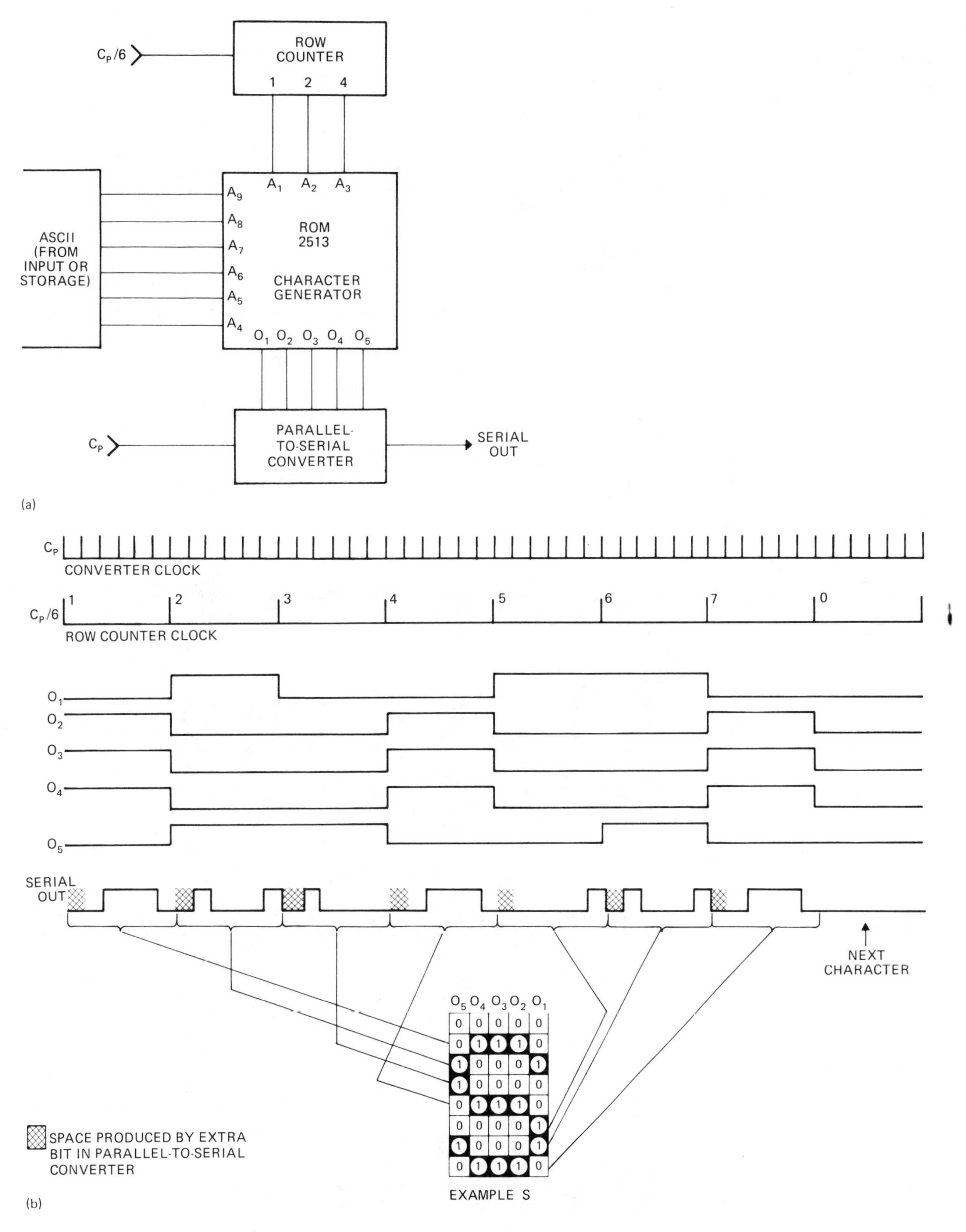

FIGURE 24-20

(a) Block diagram of the circuits needed to extract the data bits from the ROM character generator and convert them to serial form. (b) Waveforms produced from ASCII input 110010 (S).

COLUMN ADDRESS

A_3	0	0	0	0	1	1
A_2	0	0	1	1	0	0
A_1	0	1	0	1	0	1

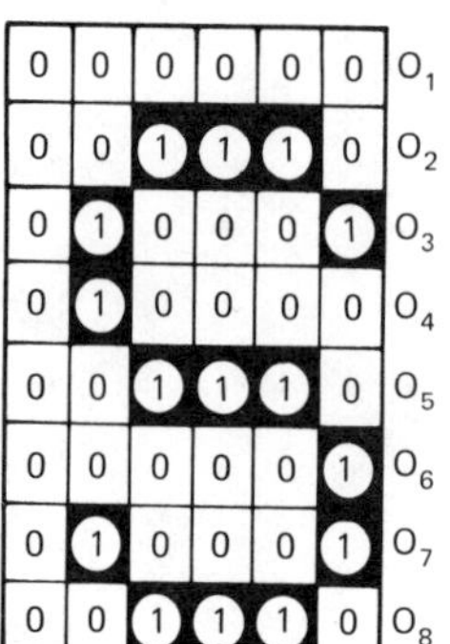

ADDRESS CHARACTER

	A_4	A_5	A_6	A_7	A_8	A_9
ASCII CHARACTER	1	1	0	0	1	0

FIGURE 24-21

Signetics 2516 character generator format for the letter S. The eight-bit output is in column form instead of row form. The counter used is a modulo 6-column counter.

Another popular character format is shown in Figure 24-21. The outputs in this format provide one column for each count of the column counter instead of counter addressing by rows. The circuitry needed to extract and convert these data is much like that of Figure 24-20(a) except that a modulo 6 counter would be used instead of an 8. The parallel-to-serial converter would need eight bits.

Figure 24-22 shows the complete ASCII character font provided by the 2516 character generator. That obtained from the 2513 is the same except for the blank column on the left. Note the identity to the four middle columns of the ASCII code chart of Figure 24-18.

24.3.6 Single-Character Deflection

In Figure 24-20, we saw the circuitry needed to produce the serial intensity bits for the display

FIGURE 24-22

Complete ASCII character font provided by a single 24-pin ROM circuit, Signetics MOS 2516.

of the ASCII characters on the CRT. The CRT obviously must be intensified with the trace or electron beam in the correct position. This means that deflection voltages accurately synchronized with the intensity bits must also be generated. Figure 24–23 shows the horizontal and vertical staircase voltages that must be generated. The trace must move horizontally the width of the character once for each of the eight lines or rows. Each time the trace is reflected back to the left, the vertical voltage brings the trace down one step. The intensity voltage intensifies the electron beam at the correct time and in the correct location to form the letter T. The intensity waveform shown produces white letters on a dark background. A simple inversion of this voltage produces dark letters on a white background.

To accomplish the perfectly synchronized staircase voltages, we must add to the circuit of Figure 24–20 a divide-by-6 counter and two 3-bit D/A converters. Figure 24–24 shows these additions. The divide-by-6 counter operates a digital-to-analog converter to produce the staircase voltage for the horizontal circuit. This counter may also be used in generating the $C_P/6$ clock pulse used to trigger the row counter. The row counter output, besides addressing the character generator, operates a D/A converter that generates the vertical deflection staircase. The eight steps of the vertical staircase are six times wider than those on the horizontal staircase. The

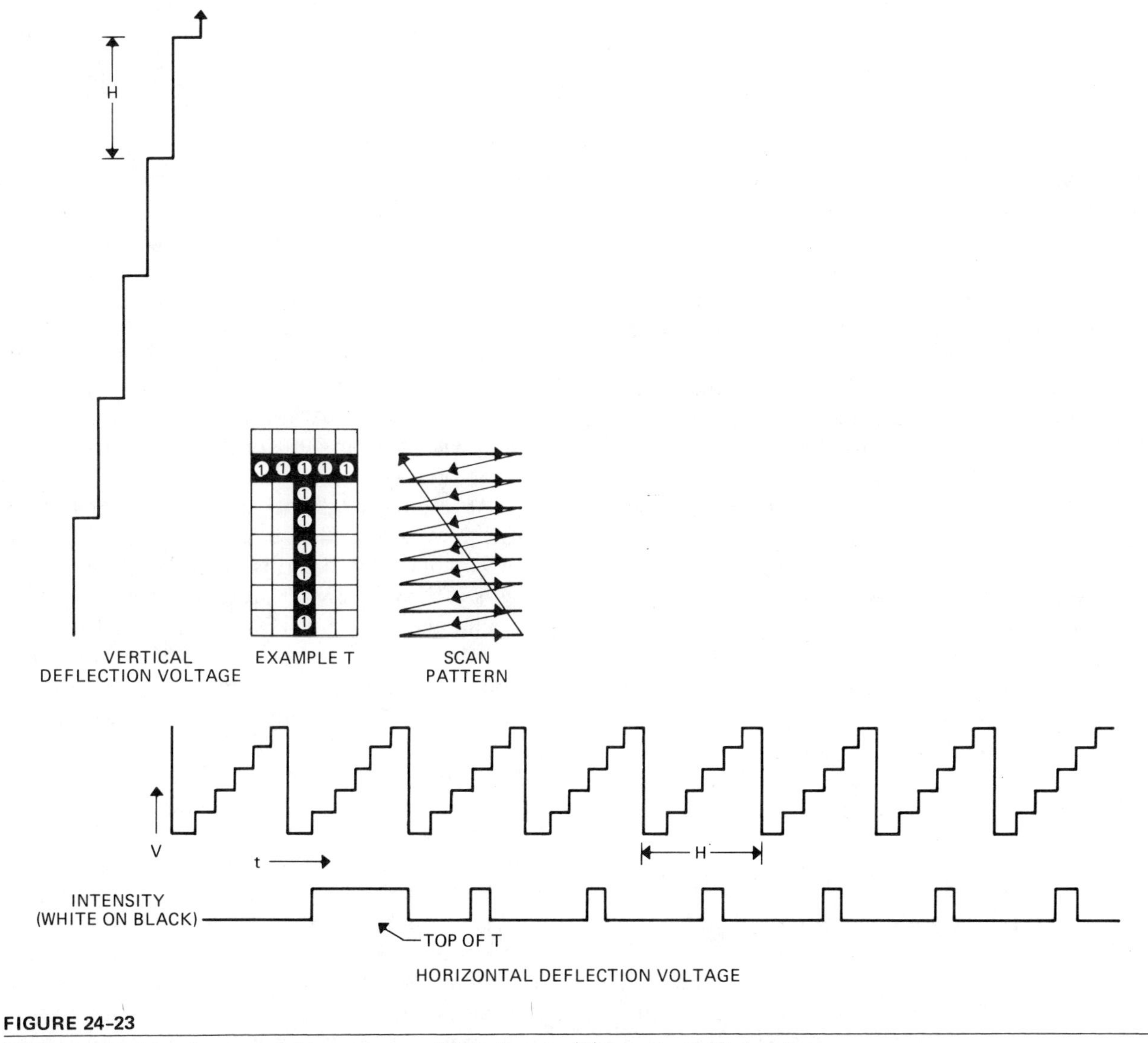

FIGURE 24–23

Staircase deflection voltages needed to produce an ASCII character (T) from the 2513 character generator.

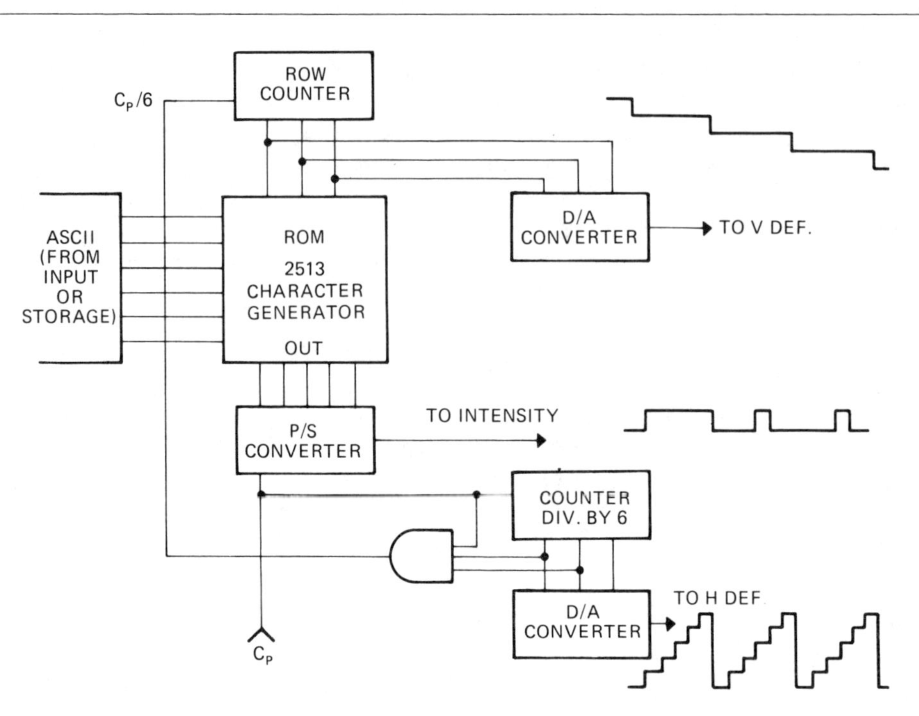

FIGURE 24-24

A mod 6 counter connected to a D/A converter to produce the horizontal staircase deflection voltage. The output of the row counter connected to another D/A converter produces the vertical deflection staircase.

circuit of Figure 24-24 is correct for a 2513 or equivalent character generator putting out one 5-bit row at a time.

Using a 2516 or equivalent column-oriented character generator with the output coming one 8-bit column at a time would require waveforms as shown in Figure 24-25. The trace starts on the left and sweeps vertically upward once with each cycle of the parallel-to-serial converter. Note how the intensity bits differ from those in Figure 24-23. The circuit used to generate these waveforms would differ slightly from those of Figure 24-24. The parallel-to-serial converter would be eight bits instead of six. The divide-by-6 counter and its D/A circuit would change to divide by 8 and would operate the vertical instead of the horizontal. The divide-by-8 row counter would change to a divide-by-6 column counter and its D/A circuit would operate the horizontal instead of the vertical. The $C_P/6$ clock would change to $C_P/8$. The actual number of circuits remains the same.

24.3.7 Display and Refresh of a Line of Characters

When the last row of a character is completed, at the instant the row counter clocks to 0, a new ASCII character should appear at the ROM input. At the same instant, a horizontal voltage increment must be added to place the second character directly to the right of the first. This can be accomplished by storing the ASCII data in serial memories or shift registers, each bit of the character in a separate register. This register must be clocked at $C_P/48$. The $C_P/48$ clock will also operate a counter and D/A converter. The output of this D/A converter is summed together with the previously generated horizontal staircase voltage.

As Figure 24-26(a) shows, each of the six ASCII bits is stored in a separate register. Each time a character scan is completed, a $C_P/48$ pulse shifts all six N-bit shift registers, starting a new character at the ROM inputs. At the same time, the $C_P/48$ pulse clocks a counter that operates a third D/A converter. The voltage staircase from this converter is summed in an analog adder with the bit staircase. The new D/A output moves the sweep to the right the distance of one character with each count of the divide-by-N counter. After the count of N, this staircase returns to the bottom step and a refresh of the illumination of the line is started.

A single sweep through a line of characters is not enough to make them visible for a reasonable time. Refresh of the illumination is easily accomplished by recirculating the data in the serial memory. Figure 24-26(b) shows the wave-

FIGURE 24-25

Staircase deflection voltages needed to produce an ASCII character (T) from the 2516 character generator.

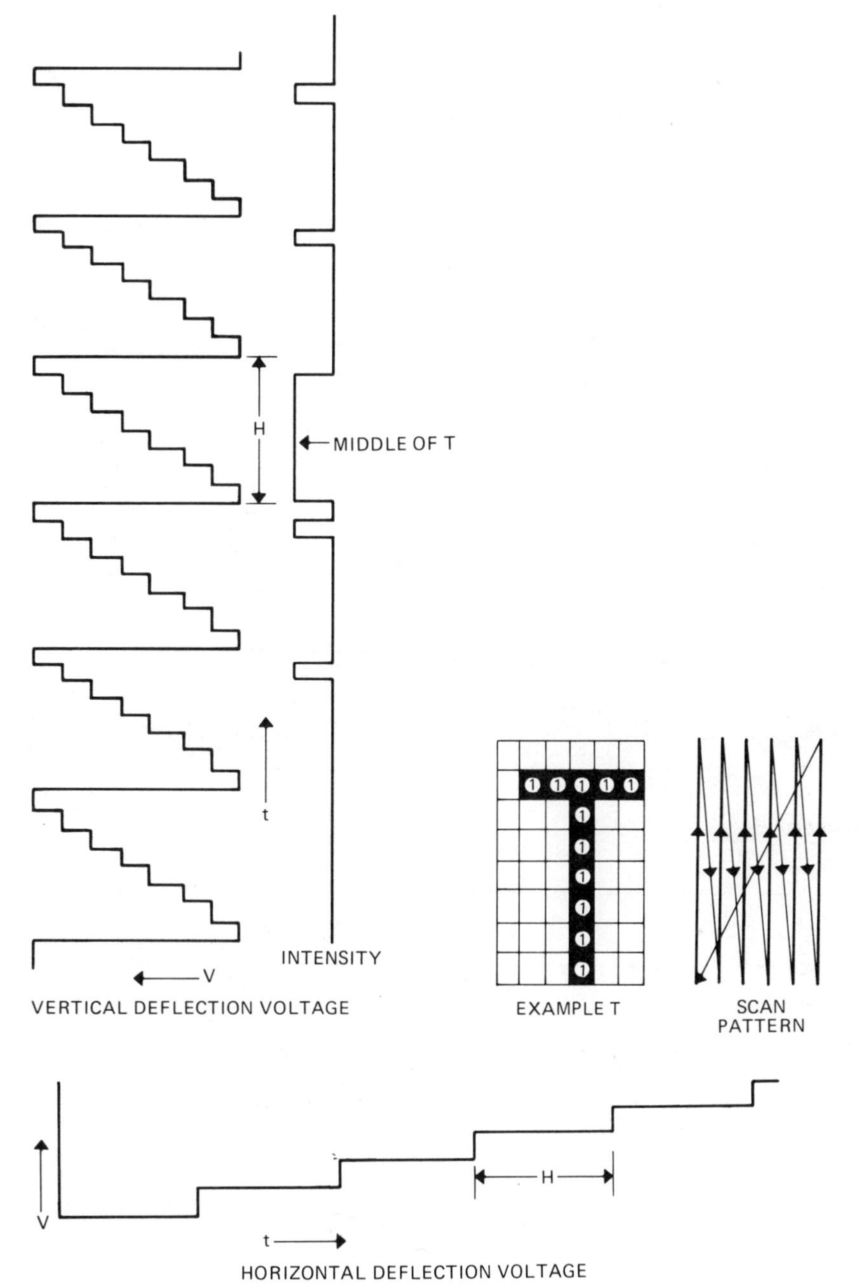

forms resulting from the letters STO being displayed on a CRT. The horizontal deflection input is obtained by combining the bit steps and the character steps in an analog adder. The bit steps come from the divide-by-6 counter and its D/A converter. The character steps come from the divide-by-N counter and its D/A converter.

24.3.8 Display of a Page

To expand the circuit of Figure 24-26 for a display of M lines would require serial memory expansion to M × N bits. A clock pulse of $C_P/48N$ would be generated to clock a divide-by-M counter. A D/A converter connected to

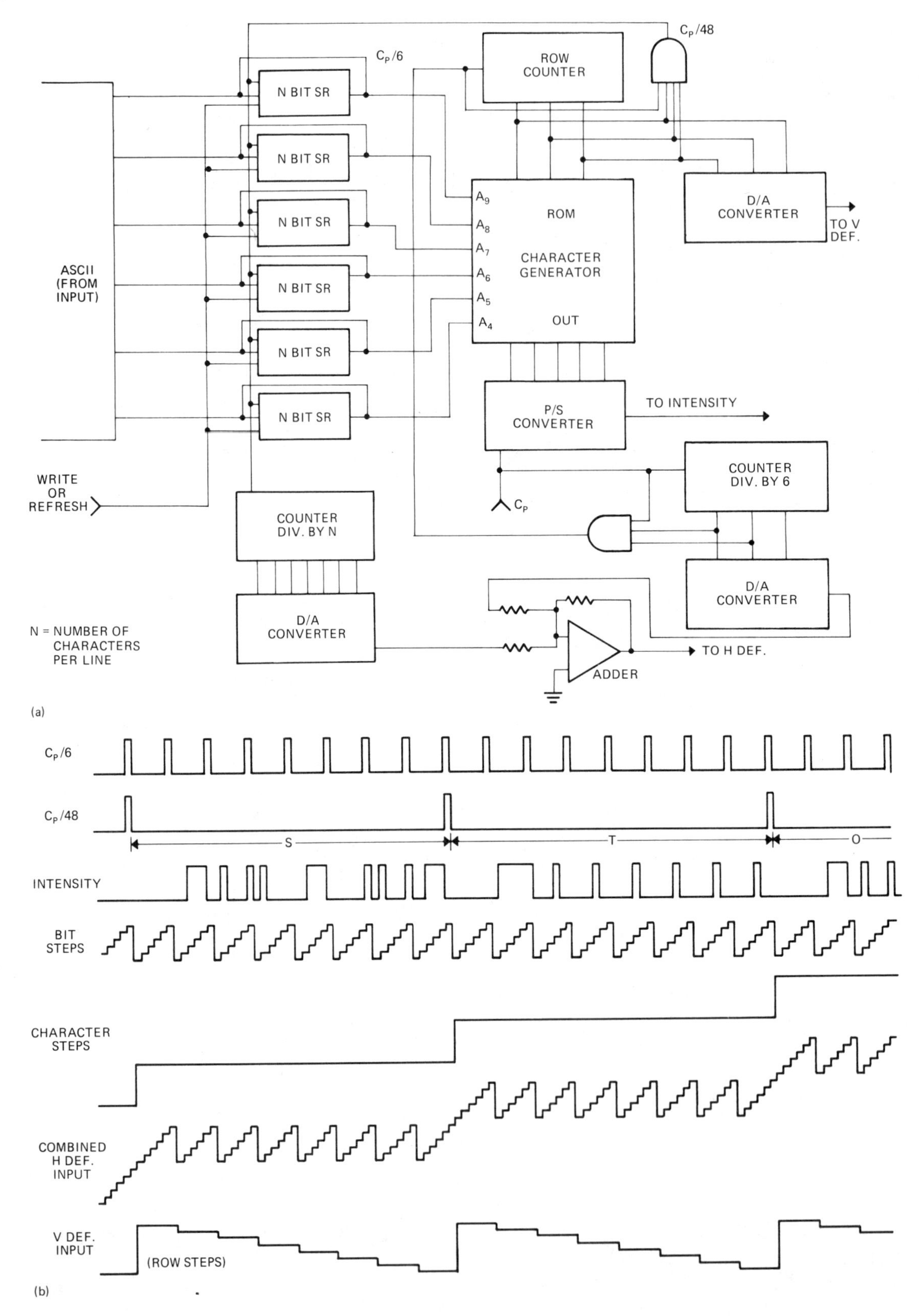

FIGURE 24-26

Display and refresh of a line of characters. (a) N-bit shift registers or serial memories used to store the data (each bit stored in a separate register). (b) Waveforms resulting from the letters STO being displayed on a CRT.

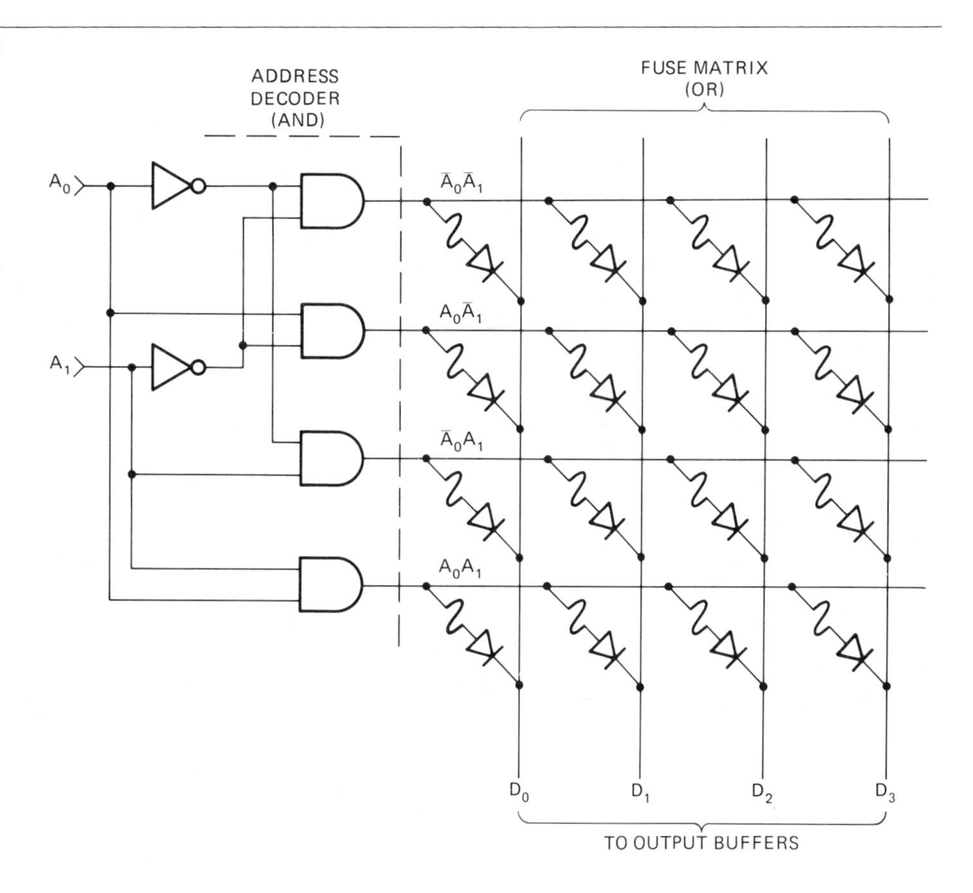

FIGURE 24-27

Simplified version of fusible link PROM.

the divide-by-M counter would produce a line step staircase. The line step staircase would be combined in a second analog adder with the row step staircase to form the vertical deflection input.

24.4 Fusible Link PROM

As shown in Figure 24-10, the diodes in the matrix of the ROM are included or left out during manufacturing to provide the desired outputs from each memory address. This operation can be performed by the manufacturer only at a substantial cost. For very large quantities of identical circuits, the cost per circuit may be very low, but where small quantities are needed the cost per circuit is high. For ROMs for which the content may be subject to change, the mask programmed ROM may not be practical because each time a change is required, a new mask charge is involved. The fusible link PROM is more practical where smaller quantities of devices are involved or where the program content may be subject to change.

The fusible link PROM differs from the mask programmed ROM in that the connecting elements of the program array contain fuses. This is shown in Figure 24-27. The fuses can be blown or can be left intact to provide the desired 1 and 0 conditions at each address in the memory. The blowing of the fuses is accomplished by a special programming circuit that applies carefully controlled voltages and currents while the device is being addressed. Once fuses are blown, they cannot be replaced. Therefore, a change in program requires that the old PROM be replaced with another that has been programmed to a new listing. Blank PROMs are readily available, relatively easy to program, and much less expensive than having a mask made for a new ROM.

Fusible link PROMs come in a variety of sizes and configurations. Those from different manufacturers are interchangeable after programming but usually program differently. Three-state versions and open-collector revisions are available.

Figure 24-28 shows the 7611 PROM. It is configured as 256-by-4, and two of these PROMs must be connected to fill an eight-bit microprocessor data bus. One circuit supplies the lower

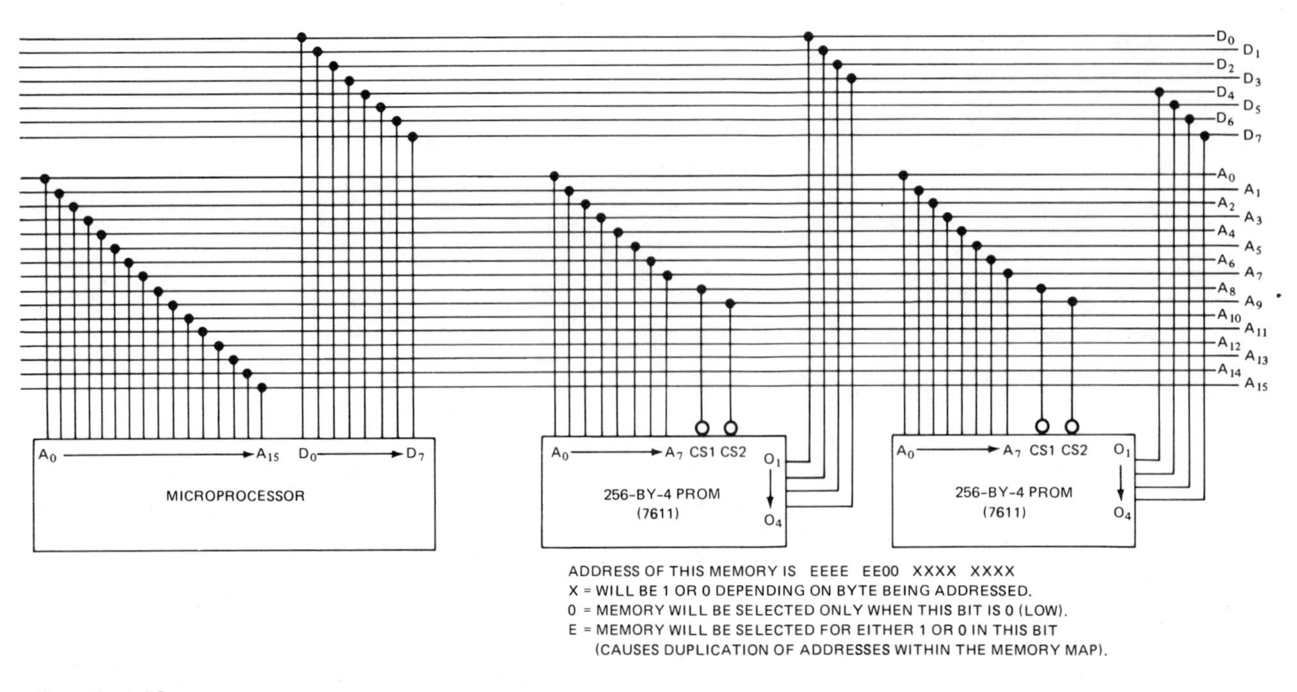

FIGURE 24-28
Two 256-by-4 PROMs connected to provide one page of ROM.

four data bits; the second supplies the upper four data bits. The 7641 is a more advanced model of this memory with a 512-word-by-8-bit configuration. It supplies two full pages of memory in a single 24-pin circuit. This family of fusible link memories, supplied by Harris Semiconductor, Inc., is among the easiest to program because the programming voltage and average currents used are within the specifications of the 7400 series TTL high-voltage open-collector buffers. This means they can be programmed using standard digital logic circuits without the aid of linear or discrete components.

Figure 24-29 shows a typical circuit used at the memory data pins. The open-collector circuits are turned on, effectively grounding the end of the pull-up resistors and dropping 12V across the resistor. Those data bits to be programmed with a 0 level will receive a program pulse that turns off the open-collector output, allowing the voltage to rise to a level that will blow the fuse. Only one data fuse can be blown at a time.

There is no quick, simple way to transfer data from a program sheet to a PROM. There are at least four principal operations involved, as follows:

1. Render data into machine-readable form.
2. Verify the machine-readable data.
3. Transfer the data to the PROM.
4. Verify that the PROM is correctly programmed.

The need for verification must not be overlooked, because the probability of human errors in punching a thousand or more keys to a program is very high, and usually a single error is enough to make a PROM useless. Verification after programming is necessary, since not all fuses blow on the first attempt.

There are a number of commercial PROM blowers that facilitate these four essential operations. Typical of these is the Data I/O Model 5 programming system. This system provides a random access memory of 4K bytes. The program can be loaded into RAM using a keyboard. Then it can be verified by stepping through each address and checking the display. Once the program is loaded and verified, it can be transferred automatically from RAM to PROM. The blowing of the PROM is verified as each address is blown. An optional paper tape punch can provide a tape for future use of the same program. The tape not only can be used for programming by the originator but also can be sent to a PROM manufacturer or distributor for the purchase of preprogrammed PROMs. The Model 5 programs PROMs from the data stored in its RAM. The RAM can be loaded from one of three sources: the keyboard, the tape reader, or a master PROM.

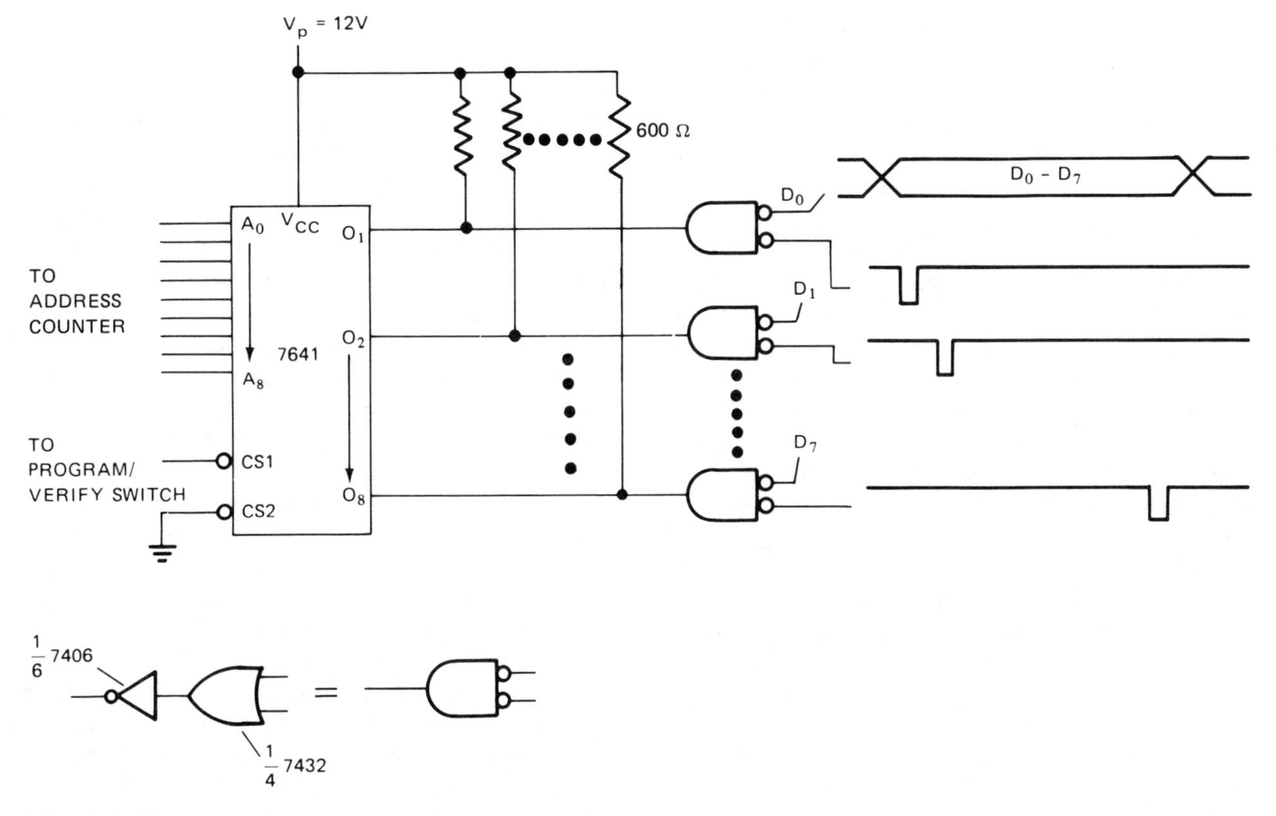

FIGURE 24–29

Fusing pulses of the correct width and timing applied to a Harris 7641 PROM through NOR gates. The standard 7433 NOR buffer has too low a voltage rating to accommodate V_p. The NOR function can be made using a 7432 OR gate with a high-voltage inverting buffer 7406.

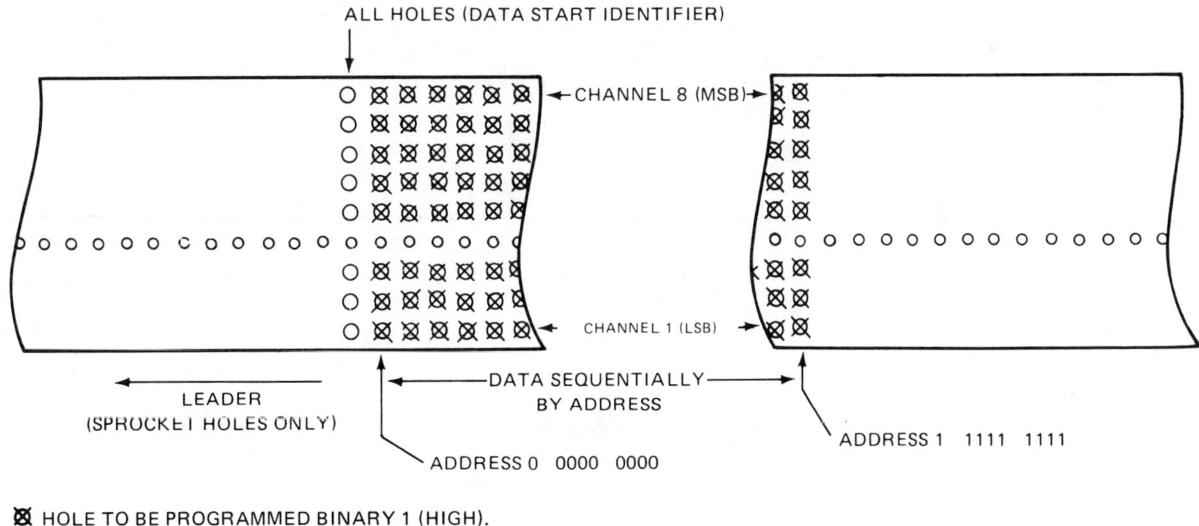

FIGURE 24–30

Perforated paper tape in binary format for a 512-by-8 PROM. The programmer recognizes address zero as the address following the first rub-out (all holes).

For most commercial PROM blowers, the program in machine-readable form is a perforated paper tape. There are various formats used for the tape. The simplest of these is the binary format, shown in Figure 24–30. The tape shown in the figure is for a 512-by-8 PROM. A tape for a 512-by-4 PROM would have no holes in channels 5 through 8 following the rub-out.

Unfortunately, paper tape equipment that provides a binary format is somewhat specialized and not readily available to all laboratories. Most PROM suppliers can program from a tape punched in ASCII hexadecimal code. A tape containing this code is shown in Figure 24-31. As the name implies, the data to be programmed are expressed as hexadecimal digits that are punched on the tape as ASCII characters. *Note:* in this discussion, PROMs that store eight data bits per address are referred to as eight-bit PROMs; those that store four data bits per address are referred to as four-bit PROMs.

Using a teletype with a paper tape punch, the data bytes are typed MSD followed by LSD and then a space, such that each address of an eight-bit PROM will result in two hexadecimal characters, 0 through F, punched in ASCII code on the tape followed by a space character. If the PROM has only four data bits, then only a single character is punched before each space. A single "SOH" character (control A) is punched prior to address zero, and an ETX (control C) is typed at the end of the tape. The PROM programmer accepts the characters following the "SOH" as being address zero.

If a mistake is made in typing the hexadecimal characters, it is very easy to correct it as long as the space bar has not been depressed. The space bar is an execute command for the PROM programmer. For an eight-bit PROM, it programs the address based on two characters preceding the space bar. Additional characters preceding those are ignored. For a four-bit PROM, additional characters preceding the one character before the space are ignored. However, if the space bar has been depressed, and a mistake exists in the one character preceding for a four-bit PROM, or either of two characters preceding for an eight-bit PROM, then the tape must be discarded and a new one started.

The PROM programmer is designed to ignore the ASCII characters for carriage return and line feed. The reader function of the teletype is useful when listing the contents of the tape in a convenient format for verification. If all addresses in a PROM are not to be programmed, the ETX character after the last character to be programmed will prevent any fuses from being blown in the remaining addresses of the PROM. All addresses preceding the ETX must contain the correct hexadecimal

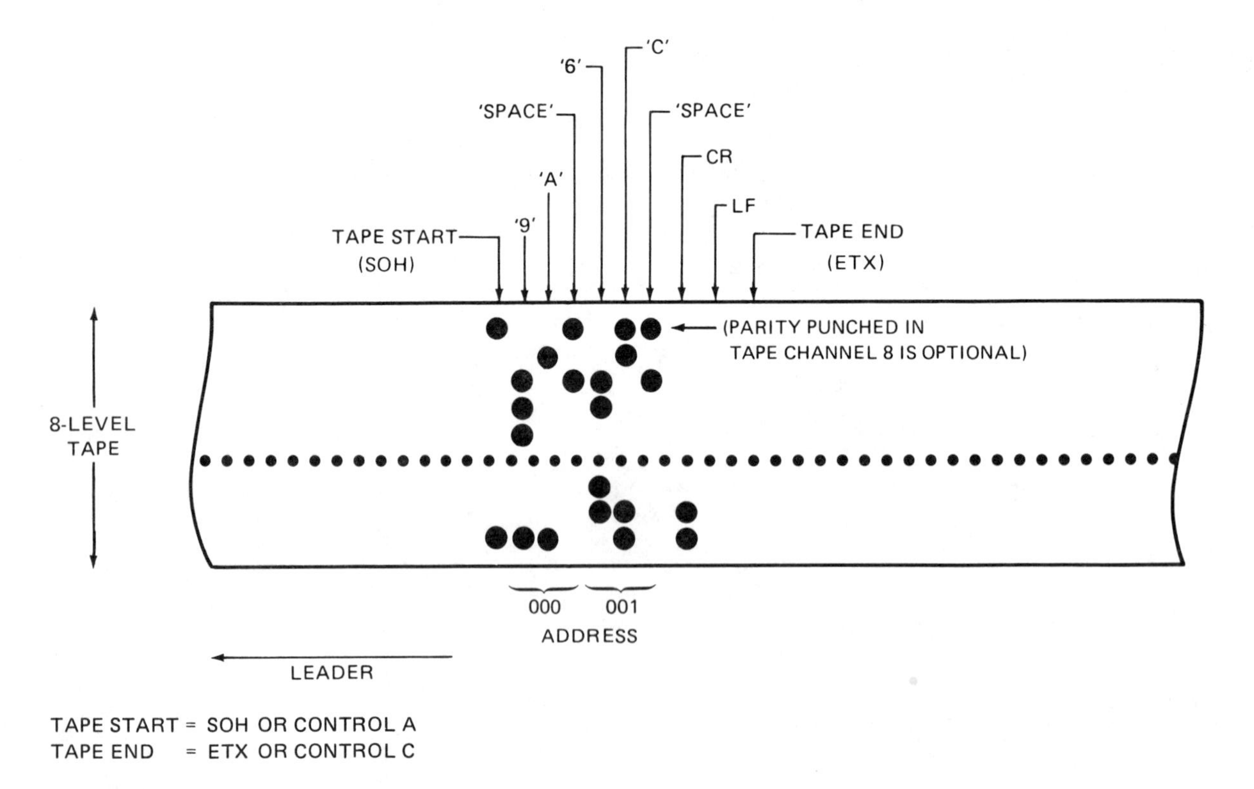

FIGURE 24-31

ASCII hexadecimal paper tape. The programmer accepts the characters following the "SOH" as being address zero.

characters plus space bar. No addresses can be skipped. If an unprogrammed section is to remain between two programmed segments of the PROM, then the addresses of the unprogrammed section must be programmed all F or all 0, depending on whether the unprogrammed device outputs all high levels or all low levels.

The price charged by distributors for programming PROMs to a user's tape is relatively low in comparison to the price of the unprogrammed PROM. This is particularly true for the large, higher priced PROMs. Figure 24-32 is a distributor's step-by-step instruction for generating an ASCII hexadecimal programming tape.

24.5 Programmable Array Logic Devices

Programmable array logic devices are similar to fusible link PROMs but are specifically designed to replace random logic. The PROM is essentially a binary-to-decimal decoder in which each line of the decoded output is connected or not connected to a set of output OR functions. The connect or disconnect with the output OR is accomplished by the fuse being blown or not being blown. (See Figure 24-27.) The input AND function is fixed as an address decoder, while the output OR function is selected by the blowing of fuses to supply the data for each decoded memory address.

The programmable array logic, PAL*, on the other hand, permits the input AND function to be varied by the blowing of fuses while the output OR function is fixed. This can be seen in Figure 24-33. Note that the connection from the input buffer goes to the cathode of the array diodes rather than to the anode, as is the case for the PROM in Figure 24-27. The PAL is designed to implement AND/OR logic. Fuses are

*PAL® is a registered trademark of Monolithic Memories, Inc.

FIGURE 24-32
PROM programming tape generation procedure.

Step 1	Go to TWX machine, place machine on local loop, turn on tape punch, and turn on main power (TWX machine—eight-level tape output).
Step 2	Press and release "HERE IS" key about 20 times to run off a 1-foot leader.
Step 3	While holding "CTRL" key down, type the letter A once. This places the programmer into the "read data" configuration.
Step 4	Starting with PROM address 0, enter the data word desired in hexadecimal code one digit at a time, MSD first.
Step 5	After typing the data digits, if no mistake has been made, hit the space bar once and only once.
	(If a typographical error occurs, you can correct it by typing in the correct digits before you hit the space bar. The programmer is set up to ignore all but the last two keys entered before the space bar is hit. If you have already hit the space bar, it is too late and you must begin all over again from Step 2.)
Step 6	Repeat Steps 4 and 5 for each word in the PROM. *Do not* skip any addresses. If you desire all output bits to be 0, type 00 at those locations.
Step 7	When the line you are on has run out of space, at the end of a word and after the space, you may hit "LINE FEED" and "CARRIAGE RETURN" to start a new line of paper, after which you enter the data for the next word. The programmer will ignore the two paper commands.
Step 8	When you have finished entering the data for all the addresses in the PROM, hold the "CTRL" key depressed and type the letter C once. The CTRL C on the tape will take the programmer out of the "read data" configuration.
Step 9	To complete the tape, press and release the "HERE IS" key 10 times to provide a trailing leader.
Step 10	If you desire, you can check your tape by playing it back and reading the printout to check it for errors. Tape is now ready to be sent in for programming the PROM.

FIGURE 24-33

Hypothetical type 2H4 PAL.

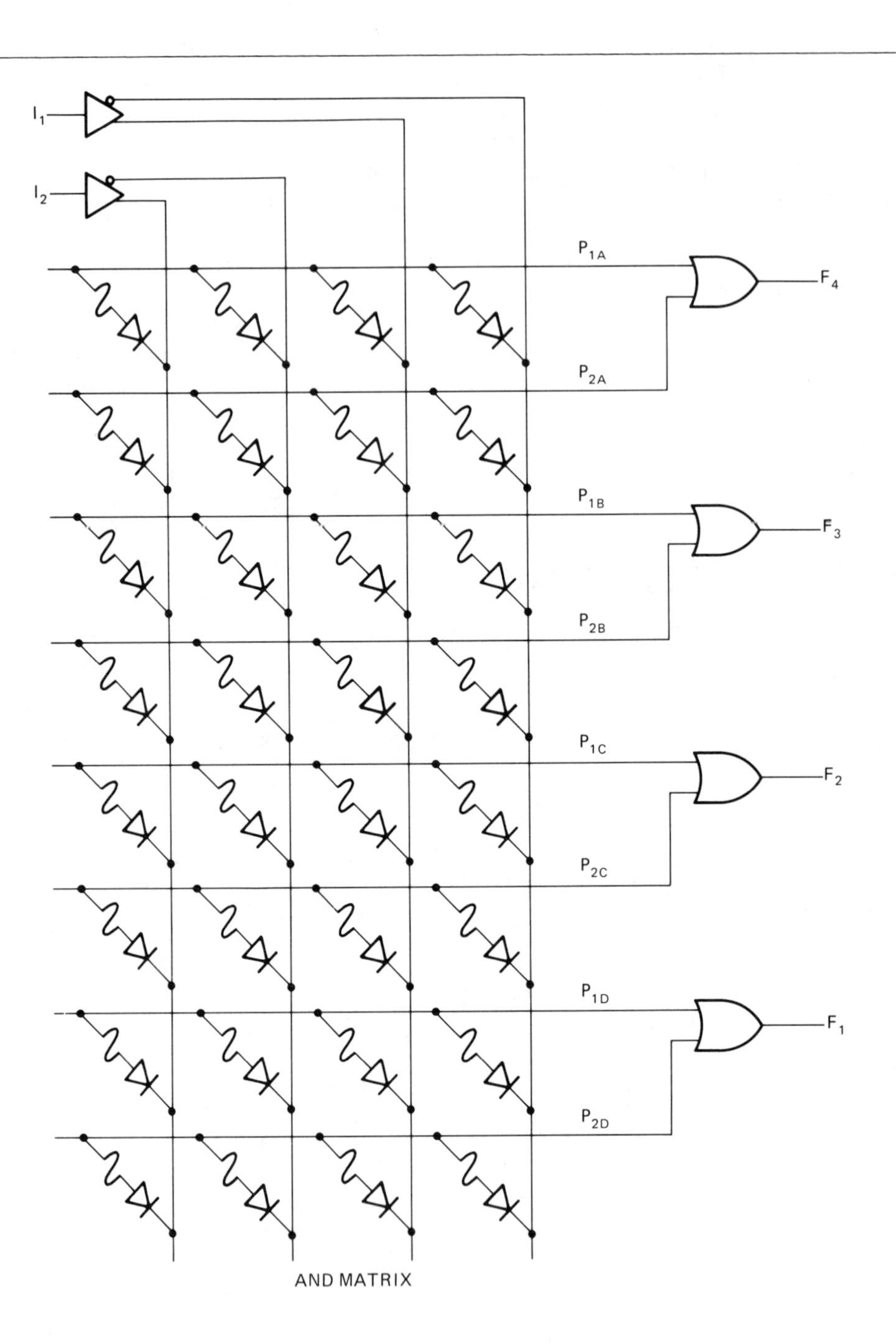

retained to connect any desired input in either true or inverted form to any of the array of AND gates. The AND gate outputs are connected to OR gate inputs. Devices can be selected to provide from dual input OR up to an OR gate with all 16 AND gates connected to it.

Figure 24-34 shows five devices of the series 20 PALs. Series 20 indicates 20-pin DIP devices. The first number of the part number is a count of the inputs. The final number is the number of OR gates. The AND gate array is the common element among the devices, and this allows for a uniformity in programming.

Figure 24-35 shows the AND gate array of a PAL12H6. The vertical lines numbered at the top represent input and inverted input buses. The horizontal lines numbered at the left are product terms and represent AND gates shown in a simplified form. Each intersection of a horizontal line with a vertical line represents a separate input to the AND gate through a fuse that can either be blown to remove the input or

FIGURE 24-34

(a) Explanation of PAL designations. (b) Five series 20 PALs providing AND/OR logic. The number of the PAL describes the number of inputs and outputs available and whether the outputs are high (AND/OR logic) or low (AND/NOR logic) or with output registers.

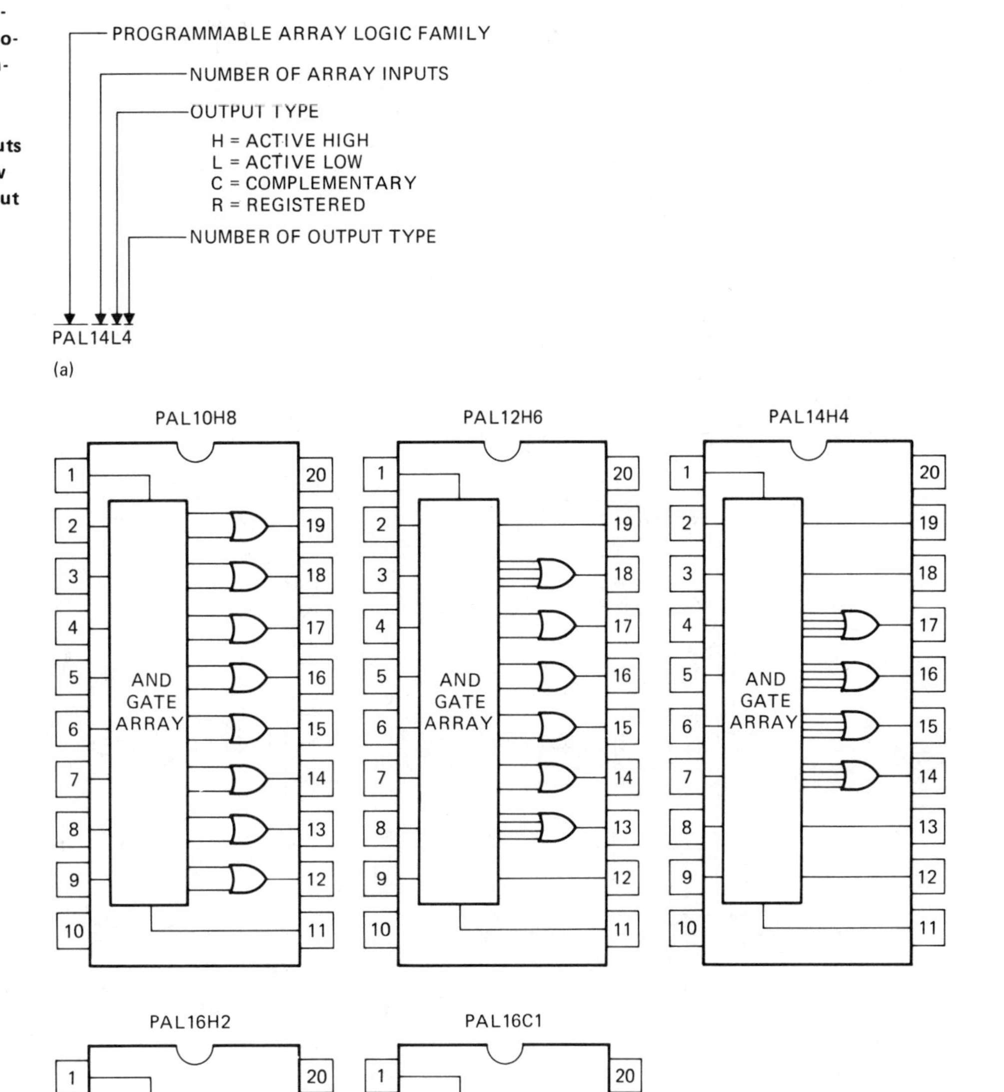

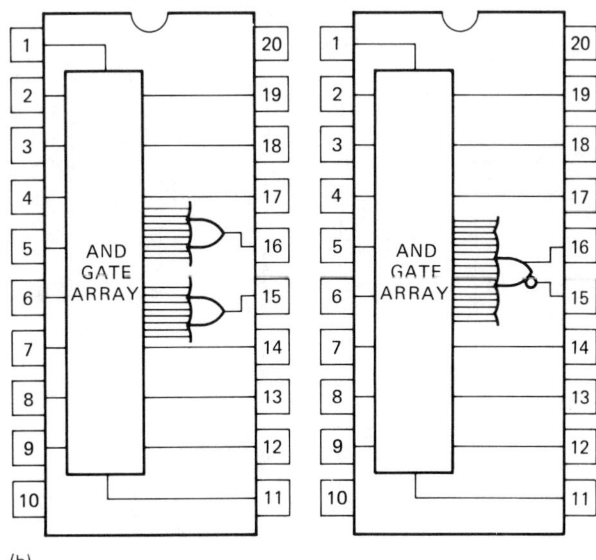

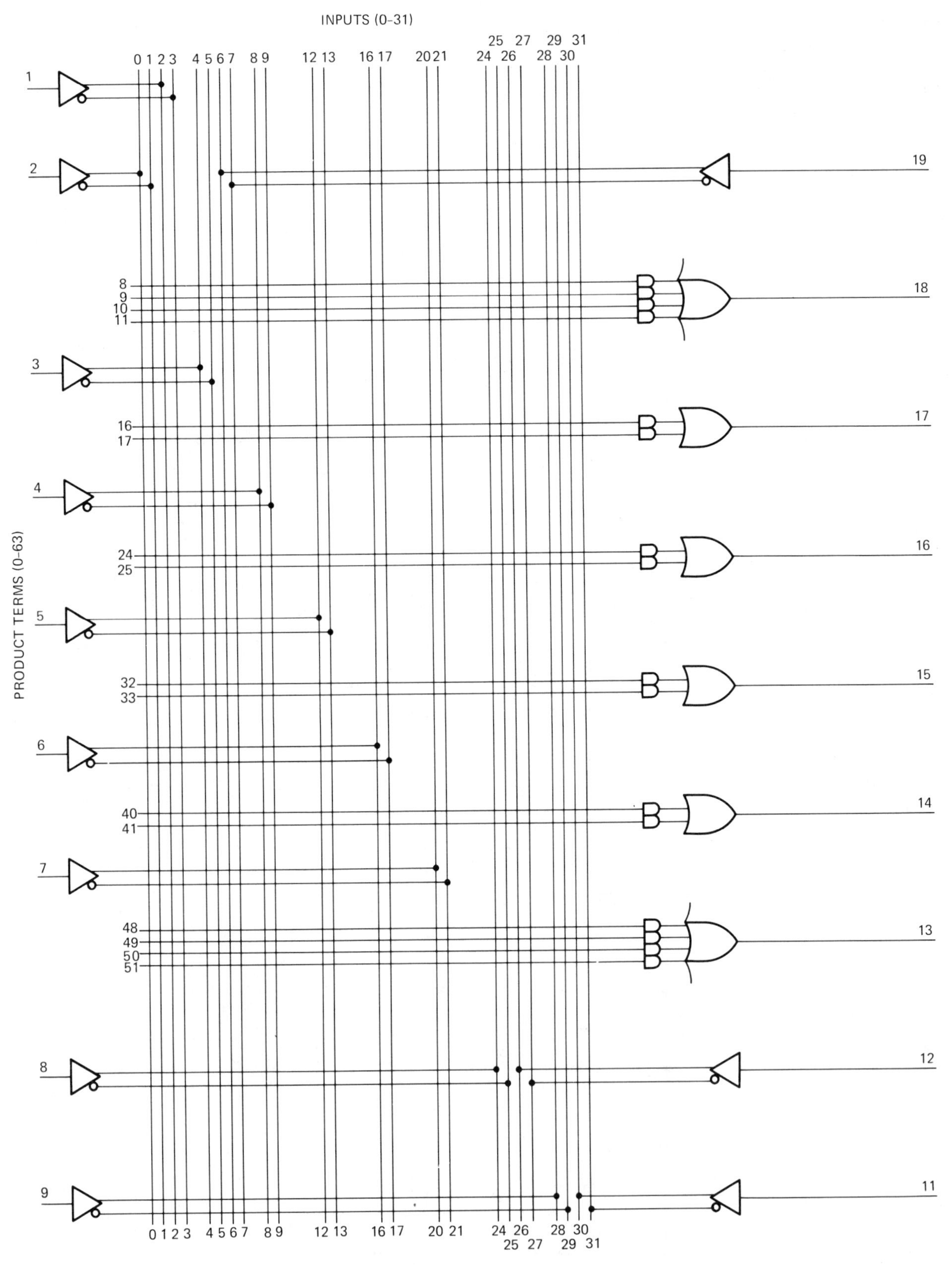

FIGURE 24-35

Logic diagram of a PAL12H6 with all fuses intact (blank). Each product line is connected to every input line through diodes and fuses. By leaving the correct fuses intact, the user can form any desired AND function of true or inverted inputs. The AND gates are connected to OR gate inputs as shown.

not blown to maintain it. The user places dots or crosses at the intersection where fuses are to be maintained.

Figure 24-36(a) compares a single programmed line to the actual AND function it creates. The user places an X at the intersection of product term line and input bus to indicate a fuse to remain intact, thus providing an input to the AND gate. The product line as marked in this figure is equivalent to the AND function shown. Leaving all fuses to a product term intact produces a low level or logic 0 to the OR gate enabling it. As shown in Figure 24-36(b), the shorthand method of indicating no fuses to be blown for a product term is an X in the AND symbol.

It may appear that the logic available from the PAL is limited because the output function is, in all cases, an OR, while the input functions must, in all cases, be ANDs. Inversions can occur only at the input terms. This fact is not as limiting as it first appears. The degree to which logic functions can be converted to AND/OR form is not always obvious, but with the aid of Boolean algebra it is seldom difficult.

Consider the simplest of applications. Figure 24-37 shows a set of simple logic gates we would like to provide from a single PAL. Use of standard SSI logic would require a minimum of three circuits, and to economize on space, we would like to implement these same functions with one PAL. As six outputs and twelve inputs are needed, we will try a 12H6.

Producing the inverter is simply a matter of leaving a single fuse intact on the first product

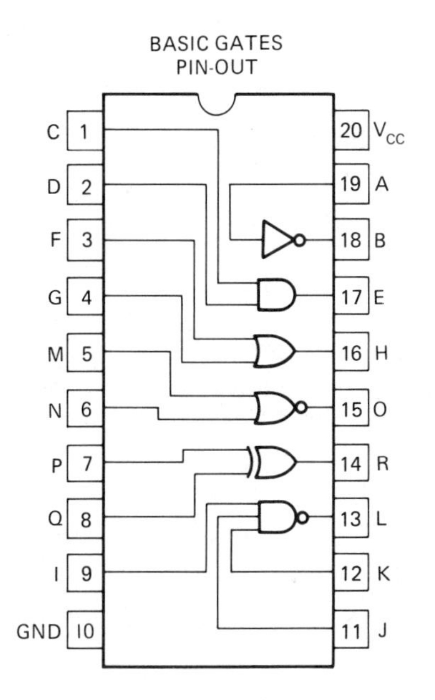

FIGURE 24-37

Simple logic gates to be implemented with a PAL. Six outputs and twelve inputs are needed. A 12H6 has the correct capacity.

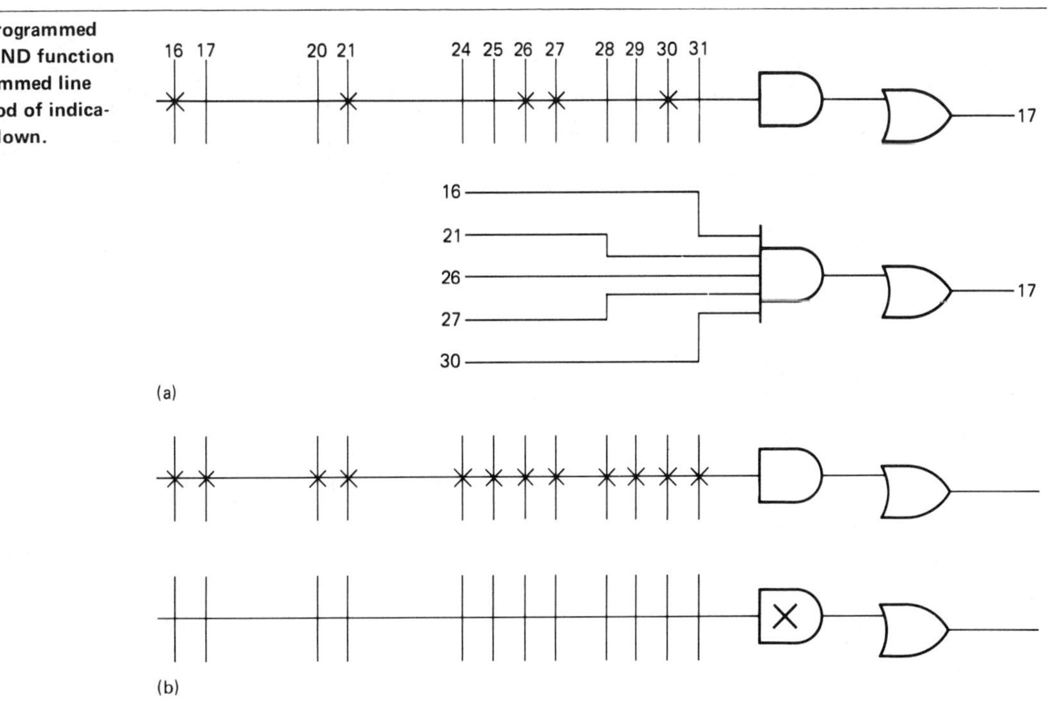

FIGURE 24-36

Product terms. (a) Programmed line and the actual AND function it creates. (b) Programmed line and shorthand method of indicating no fuses to be blown.

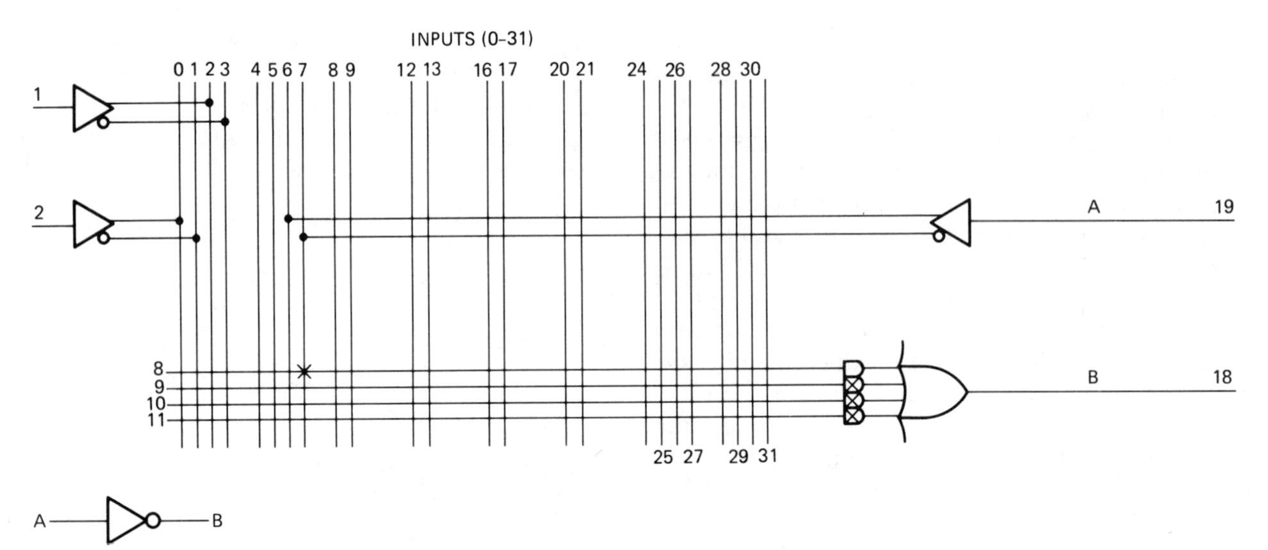

FIGURE 24-38

An inverter, B = $\overline{A}$. The fuse retained is between inverted input for pin 19 and product line 8. All fuses along product lines 9, 10, and 11 are left intact to enable the OR gate.

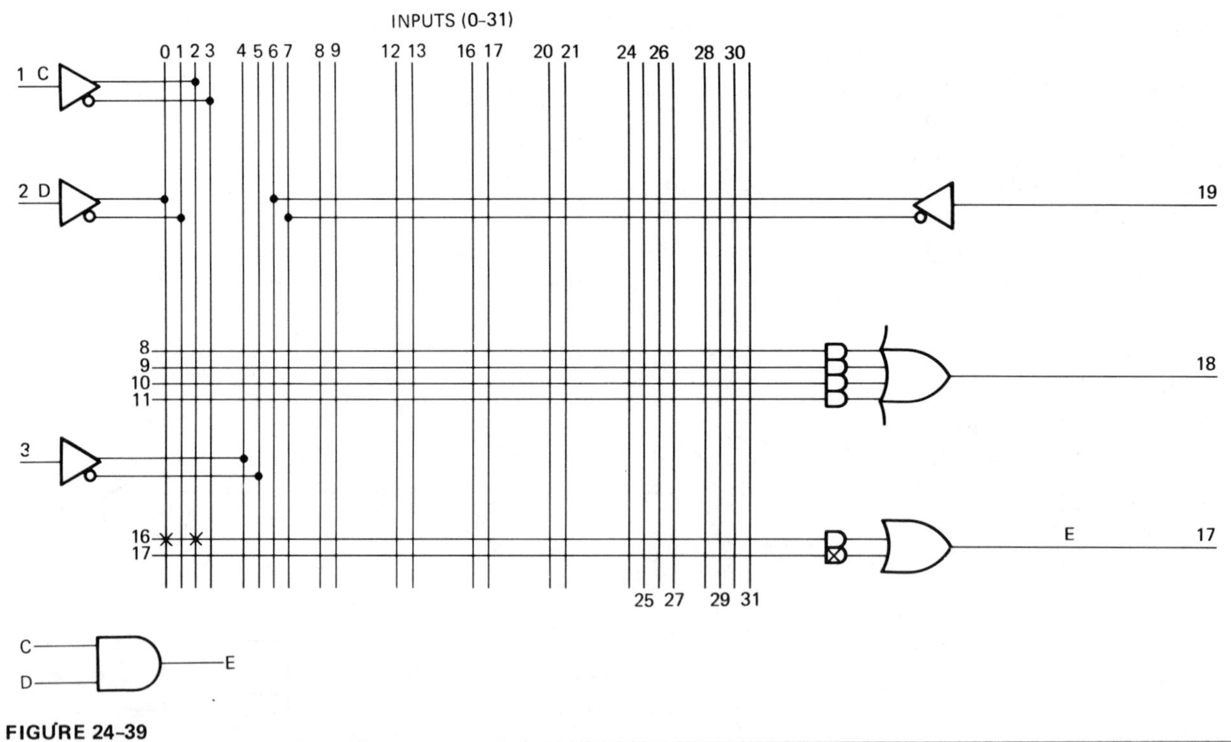

FIGURE 24-39

Two-input AND gate, E = C • D. Fuses are retained for input pins 1 and 2 at product line 16, and the second OR gate input is enabled by retaining all fuses along product line 17.

term line. As Figure 24-38 shows, the fuse retained is the inverted bus of input 19. The remaining fuses on product term 8 are to be blown. The other inputs to the OR gate must be enabled with 0 levels. This is accomplished by leaving all fuses of product terms 9, 10, and 11 intact.

Next, the two-input AND gate will be accomplished using product term line 16. Fuses will be retained on the noninverted bus for input pins 1 and 2. All fuses will be retained on product term line 17, which enables the second input of the OR gate. This is shown in Figure 24-39.

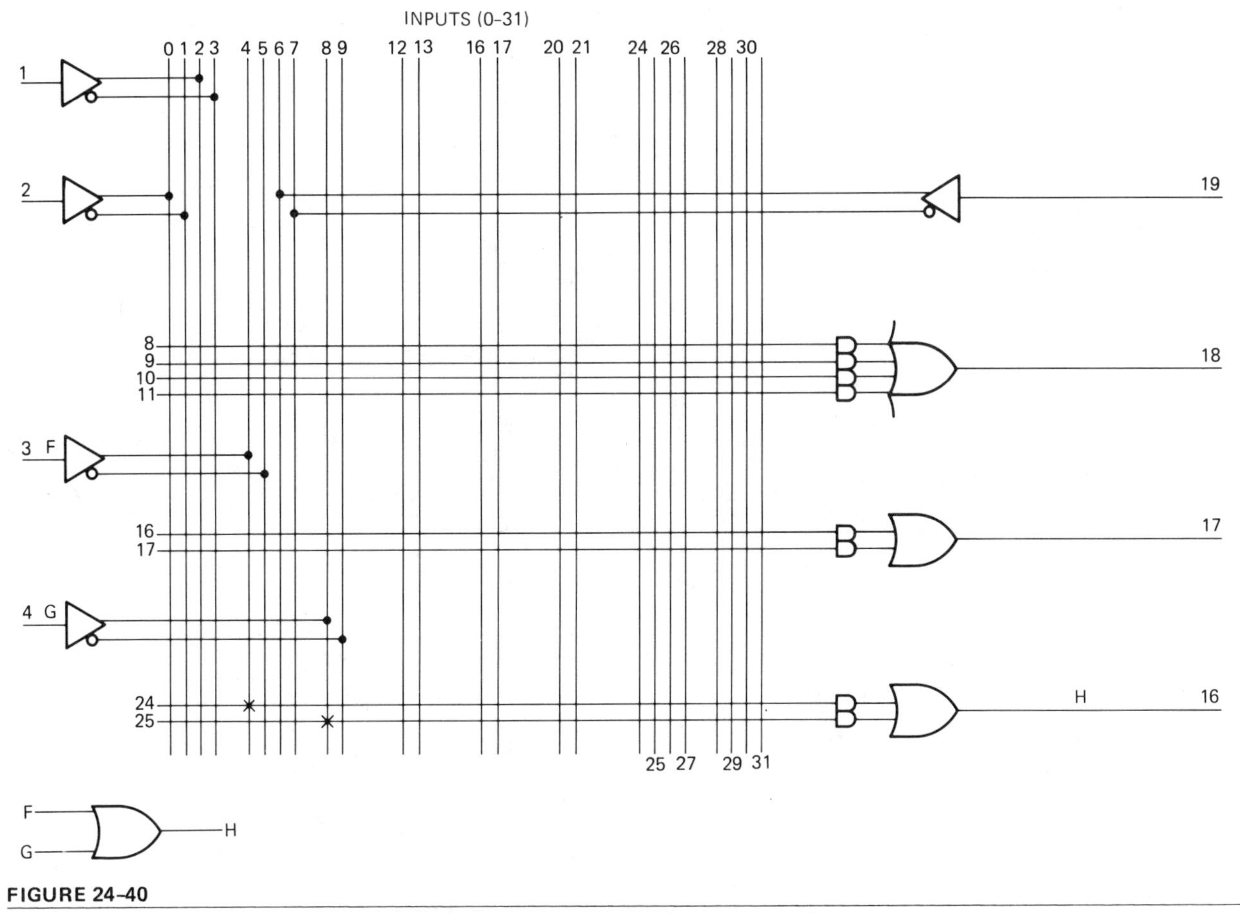

FIGURE 24-40

Two-input OR gate, H = G + F. A single fuse is retained on product term line 24, where the input for pin 3 connects, and on product term 25, where the input for pin 4 crosses.

To provide a two-input OR, we retain the fuse for the noninverted input pin 3 at product term line 24 and for the noninverted input pin 4 at product term line 25. The remaining fuses along product term lines 24 and 25 are blown. Thus, these product term lines become straight-through buffers to the two-input OR gate. This is shown in Figure 24-40.

Providing a two-input NOR gate calls for a conversion of the symbol to an AND gate with inputs inverted as shown in Figure 24-41(a). Fuses are retained where the inverted bus for input pin 5 crosses product term line 32 and, on product term line 32, where the inverted bus for input pin 6 crosses. Fuses along product term line 33 are retained in order to enable the extra OR gate input. This is shown in Figure 24-41(b).

Next, the exclusive OR is considered as a sum of product terms (the output is high when the inputs are different): $R = (\overline{P} \cdot Q) + (P \cdot \overline{Q})$. Figure 24-42(a) shows the logic symbol. Product term line 40 provides one AND gate. The fuses to the inverted pin 7 input and the noninverted pin 8 input are retained on product line 40. Product line 41 provides the second AND gate. The fuses for the noninverted input pin 7 and the inverted input pin 8 are retained on product term line 41.

It might seem that the three-input NAND gate could be provided using a single product term line, but there would be no way to provide an output inversion, so again we resort to the DeMorgan equivalent symbol as shown in Figure 24-43(a). Product lines 48, 49, and 50 are used to provide inverted inputs from I, J, and K, respectively. The remaining input to the OR gate is enabled by leaving all fuses along product term 51 intact.

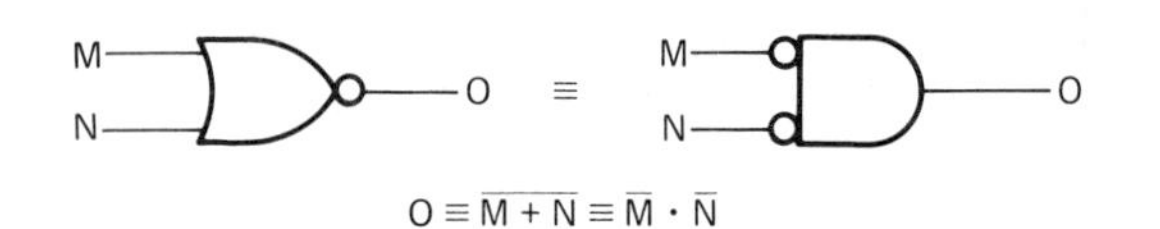

(a)

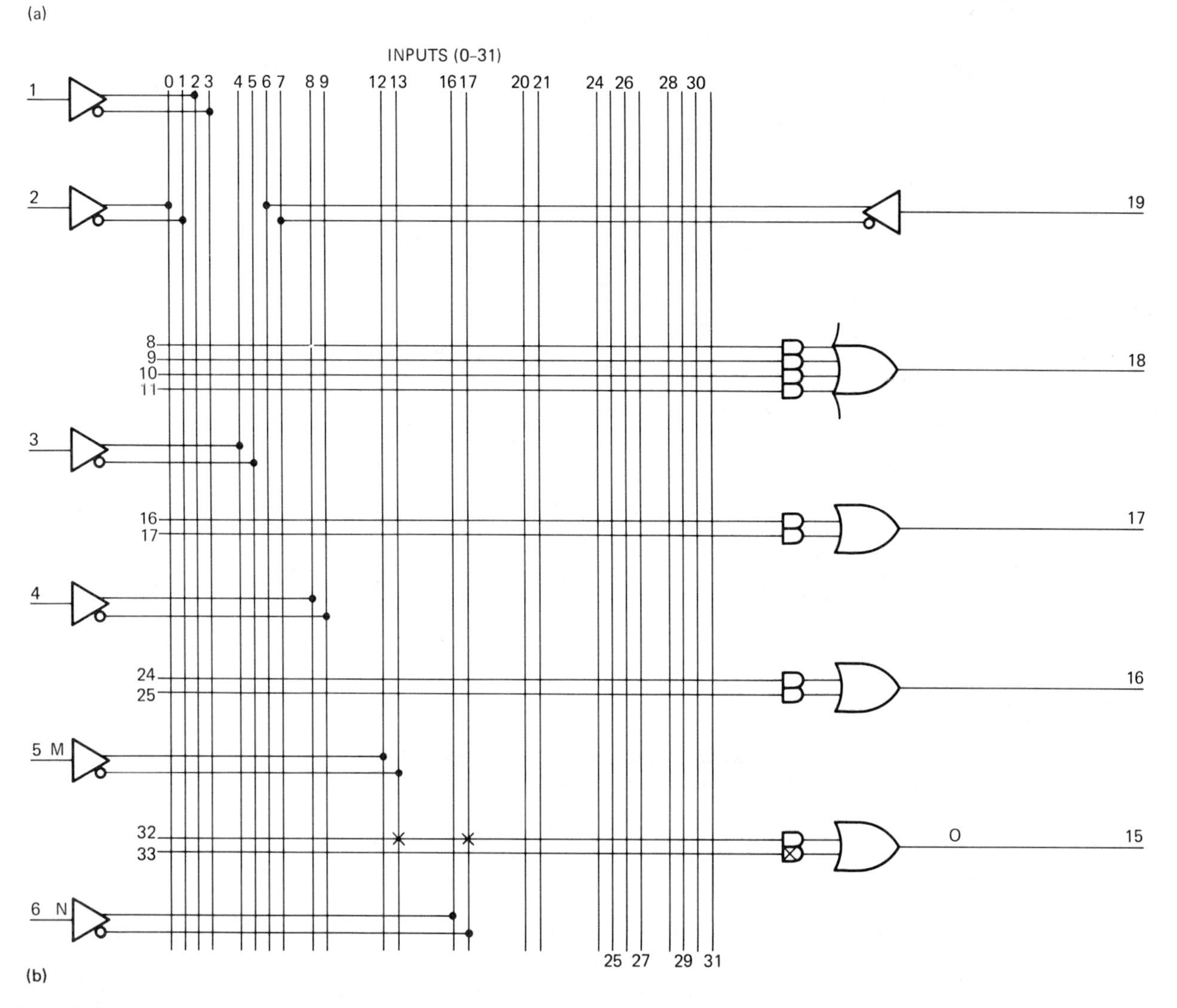

(b)

FIGURE 24–41

Two-input NOR gate, O = $\overline{M + N}$. (a) A two-input NOR gate is converted to an AND gate with inverted inputs. (b) Fuses are retained on product line 32 where inverted inputs for pins 5 and 6 cross. All fuses on product term 33 are left intact to enable the OR gate.

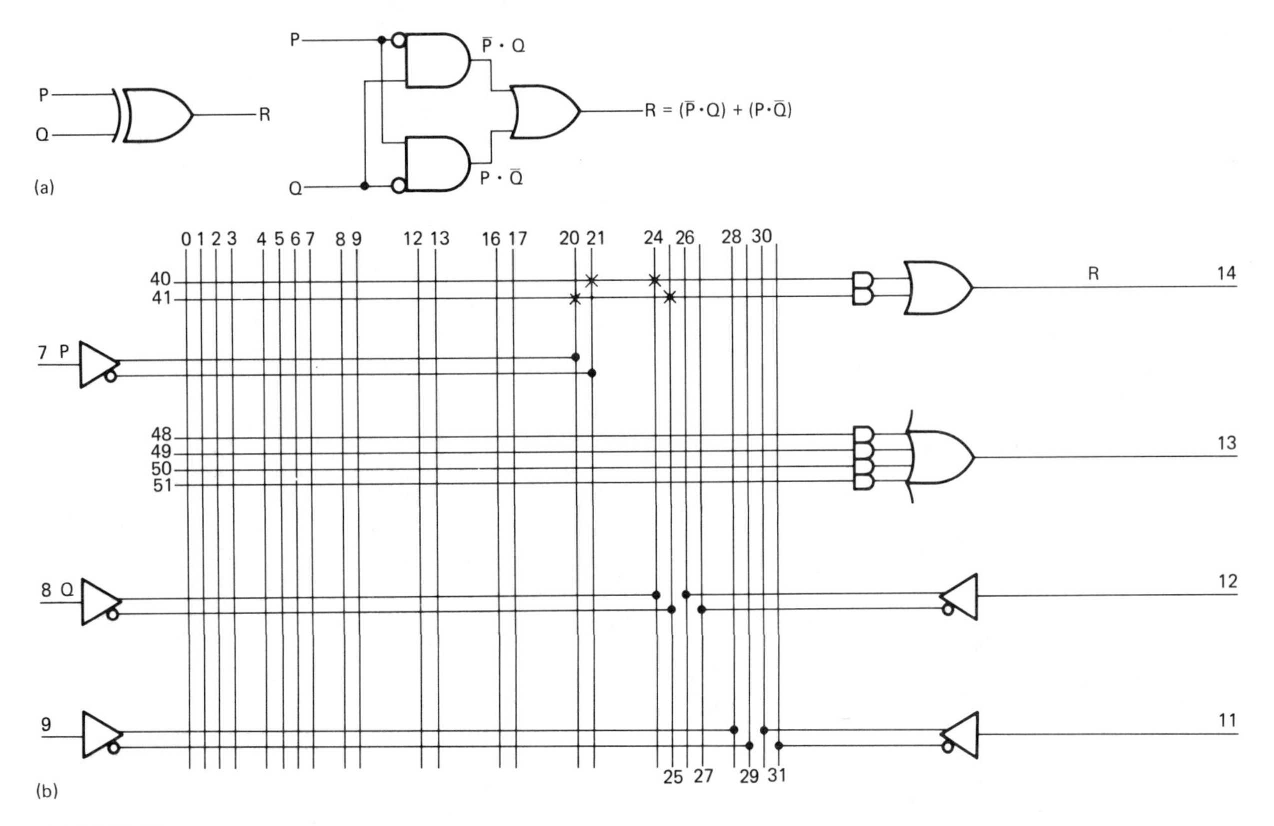

FIGURE 24-42

Exclusive OR gate, $R = (\bar{P} \cdot Q) + (P \cdot \bar{Q})$. (a) An exclusive OR gate is converted to two AND gates and an OR gate. (b) Fuses are left intact on product term line 40 where the inverted input for pin 7 and the true input for pin 8 cross. Fuses are left intact on product term line 41 where the true input for pin 7 and the inverted input for pin 8 cross.

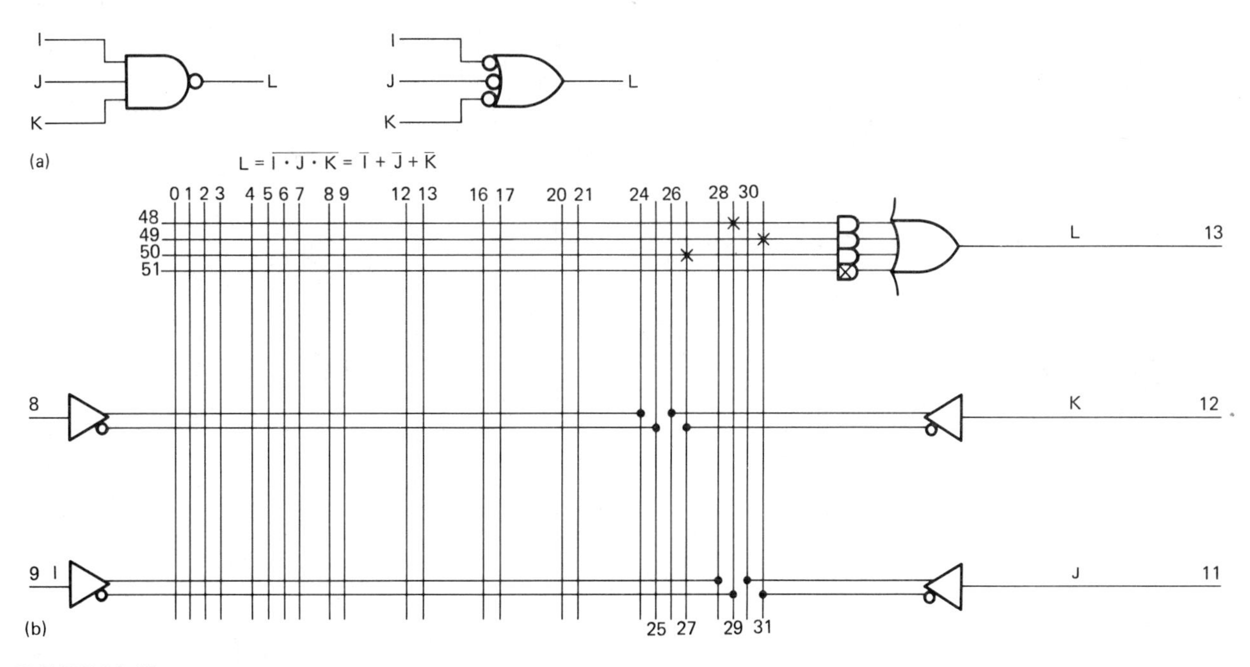

FIGURE 24-43

Three-input NAND gate, $L = \overline{I \cdot J \cdot K}$. (a) A three-input NAND gate is converted to an OR gate with inverted inputs. (b) A fuse is left intact where the inverted input for pin 9 crosses product term line 48, where the inverted input for pin 11 crosses product term line 49, and where the inverted input for pin 12 crosses product term line 50. All fuses along product term line 51 are left intact to enable the OR gate.

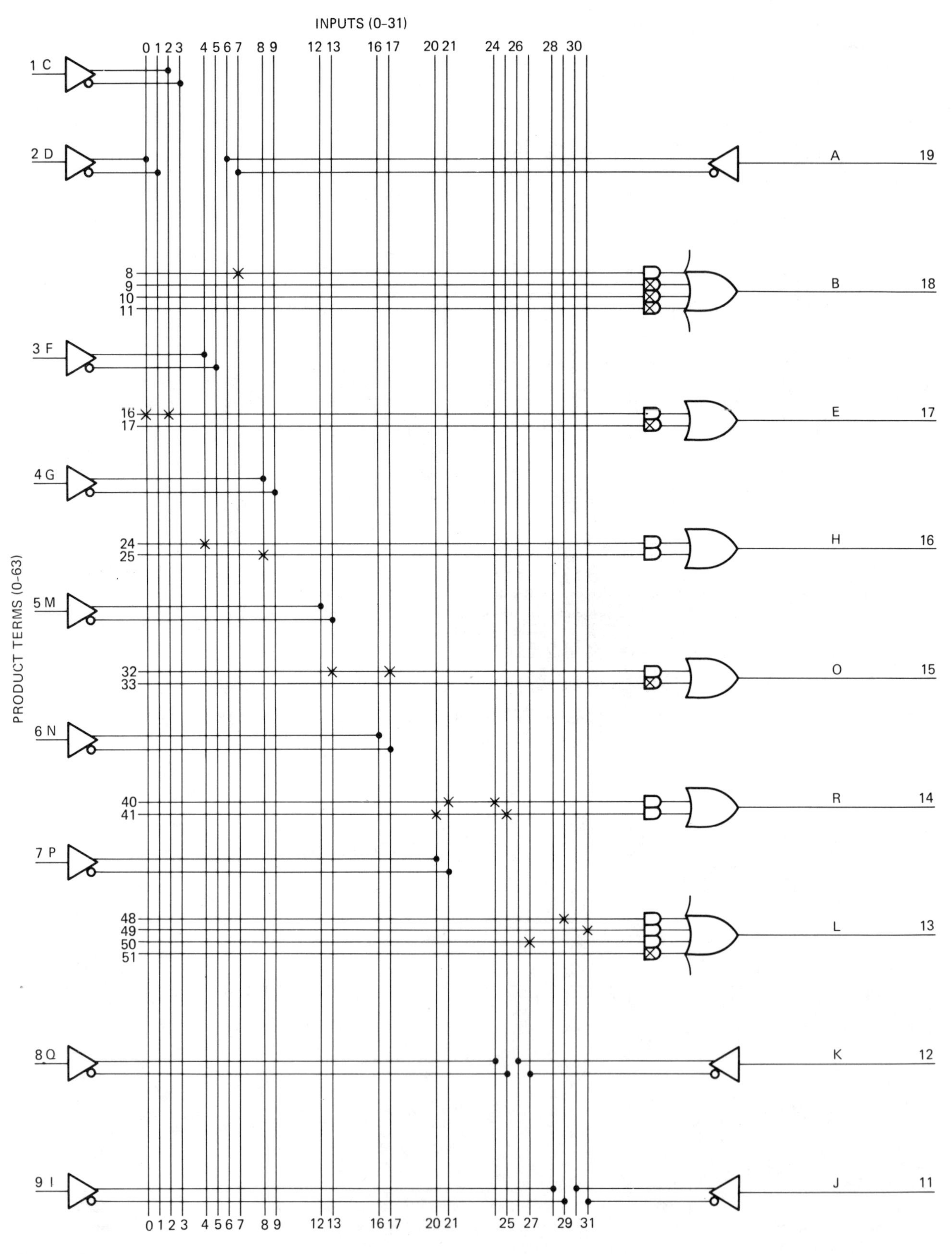

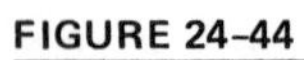

FIGURE 24-44

PAL12H6 logic diagram marked for fuses to keep intact to provide the simple logic of Figure 24-37.

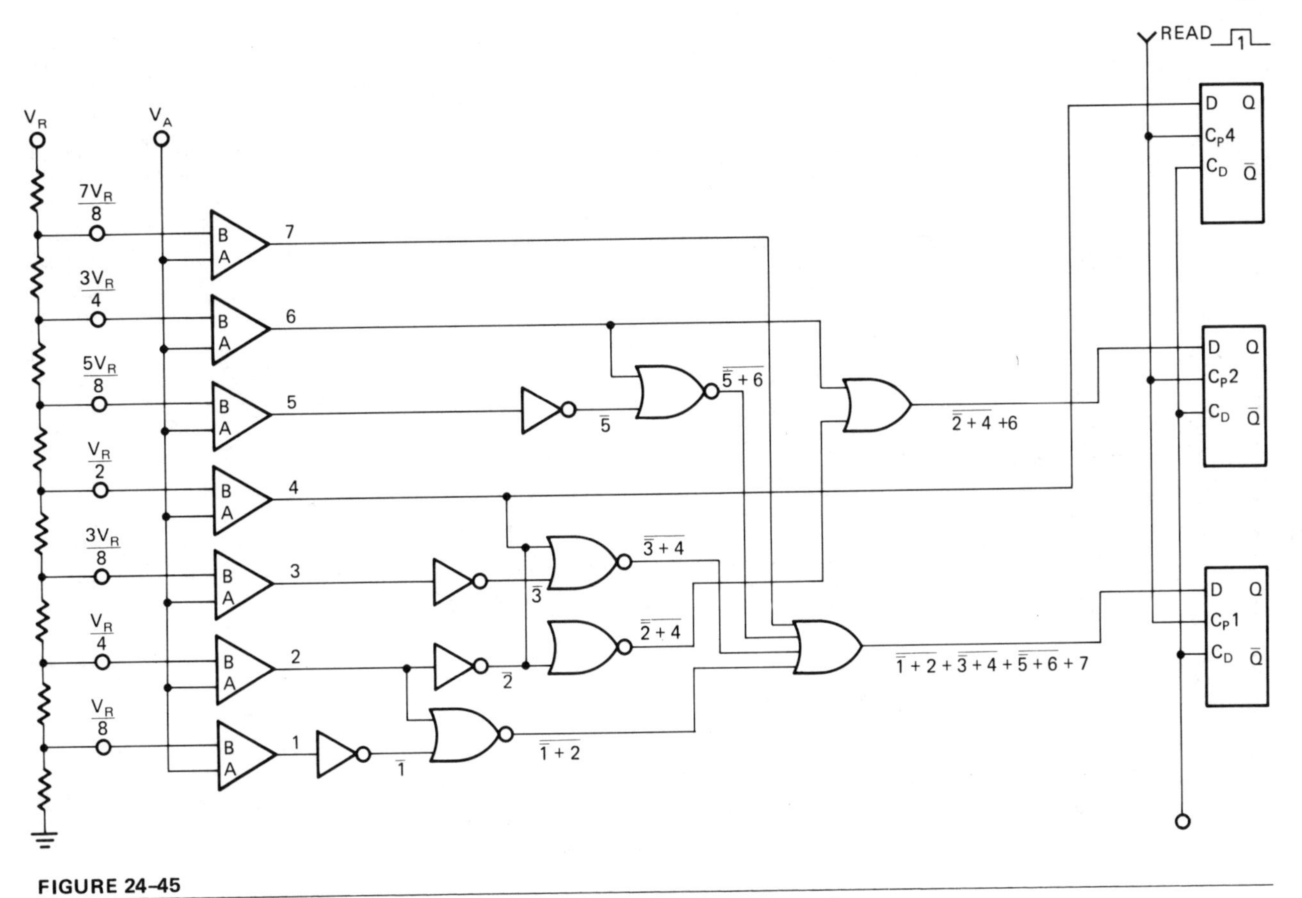

FIGURE 24-45

Simultaneous A/D logic gates to be implemented by a PAL. The Boolean equations are first converted to AND/OR logic form.

Figure 24-44 shows the logic diagram of the PAL12H6 marked for fuses to be left intact in implementing the logic of Figure 24-37.

Let us progress from the simple gates to a more complex application. One drawback of the A/D simultaneous conversion circuit pictured in Figure 21-6 is the logic needed between the comparator and the registers. Logic for this three-bit conversion and even a larger converter can be provided by a single PAL. First, we will write a Boolean equation for the output of the two OR gates by describing the signals as they pass through each gate. This is shown in Figure 24-45. The signal to the 1 register is:

$$R_1 = \overline{\overline{1} + 2} + \overline{\overline{3} + 4} + \overline{\overline{5} + 6} + 7$$

We manipulate the equation until inversion exists over only the individual input terms. By DeMorgan's theorem,

$$R_1 = \overline{\overline{1}} \cdot \overline{2} + \overline{\overline{3}} \cdot \overline{4} + \overline{\overline{5}} \cdot \overline{6} + 7$$

Double inversions cancel:

$$R_1 = 1 \cdot \overline{2} + 3 \cdot \overline{4} + 5 \cdot \overline{6} + 7$$

For the second OR gate,

$$R_2 = \overline{\overline{2} + 4} + 6 \equiv \overline{\overline{2}} \cdot \overline{4} + 6 \equiv 2 \cdot \overline{4} + 6$$

The logic reduced to AND/OR is shown in Figure 24-46.

To supply this function, we need a PAL with seven inputs and two outputs. If this were to be the only function, then a 12H6, 14H4, or 16H2 would do equally well. For this example, let us use the 16H2. As Figure 24-47 shows, pin 15 has been assigned as the output to the LSB register and pin 16 as the output to the 2 register. The 4 register will have a direct connection. Inputs 1 through 7 are assigned to comparator outputs 1 through 7, respectively. The fuses to be retained are also shown in Figure 24-47.

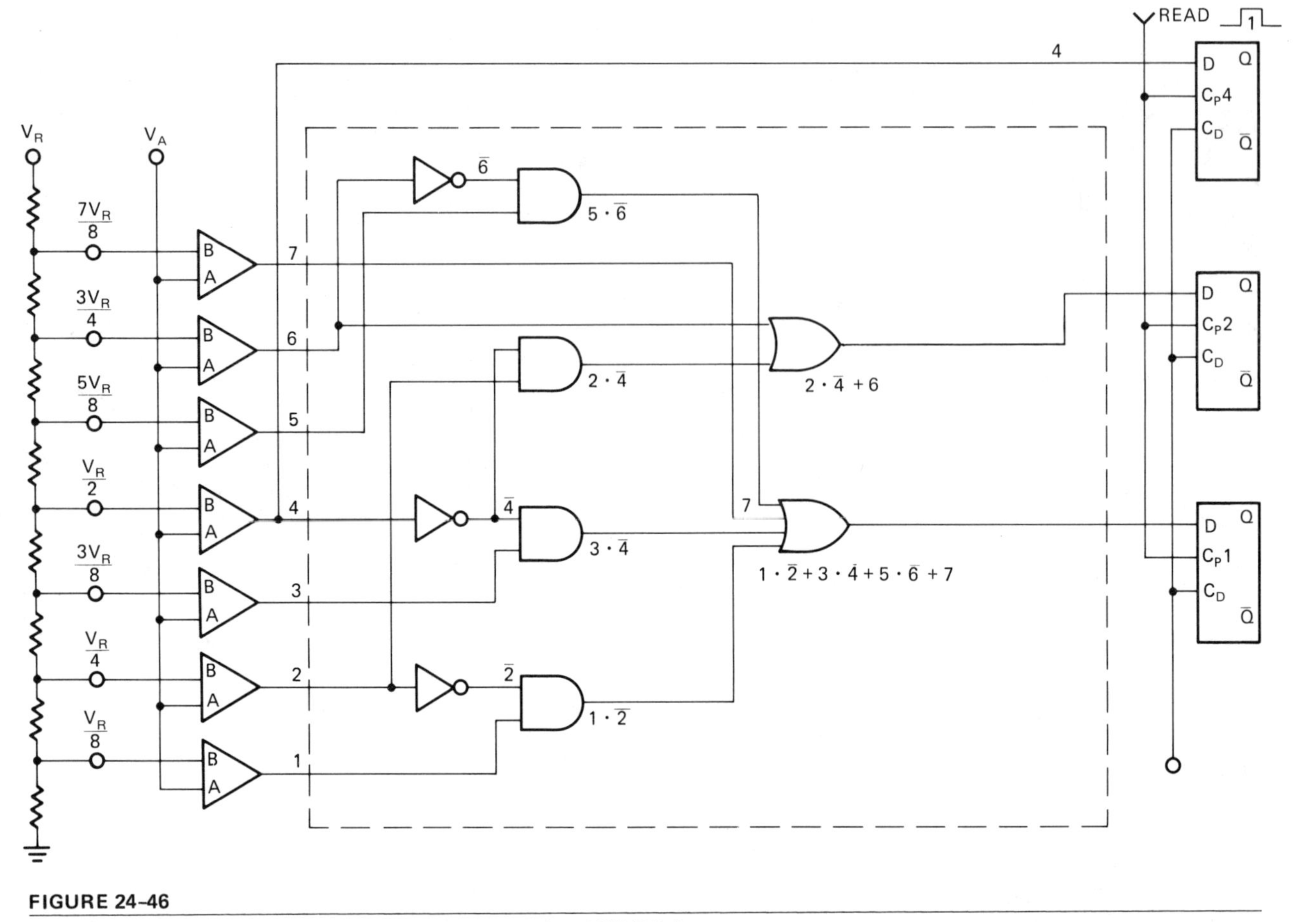

FIGURE 24-46

Simultaneous conversion circuit with logic gates converted to AND/OR form.

Another form of PAL in this series is the L type. Examples of these PALs are shown in Figure 24-48. The L devices differ from the H devices in that the output gates are NOR instead of OR. This means that the output provides an inversion whether it is advantageous or not, and instead of using DeMorgan equivalent circuits to provide for an output inversion, we may resort to a DeMorgan equivalent circuit to compensate for an output inversion. Looking again at the simple gates in Figure 24-37, the inverter changes in that the noninverted input is used and the output NOR provides the inversion instead.

The AND gate is more difficult. This conversion is shown in Figure 24-49(a). The output function desired has no inversions, but we want to write it so that an inversion exists over the entire right-hand member of the equation without changing its identity.

The final form of the equation must have an inversion over the entire right-hand member, but below that inversion, the rules are the same as for the H-type PAL. The equation is manipulated until inversions exist over individual input variables only. For the AND gate, the end result must be $E = C \cdot D$. Knowing that it must be produced with an output inversion and still be identical, we write $E = \overline{\overline{C \cdot D}}$ under the output inversion. We manipulate the equation in the same fashion we used for the H-type PAL (inversions may exist on input variables only):

$$E = C \cdot D \equiv \overline{\overline{C \cdot D}} \equiv \overline{\bar{C} + \bar{D}}$$

← output inversion
← input inversion

Note that in Figure 24-39, we used a single product term line, while in Figure 24-50, we use two product term lines.

The OR gate must also be considered from the point of view that the inversion is there

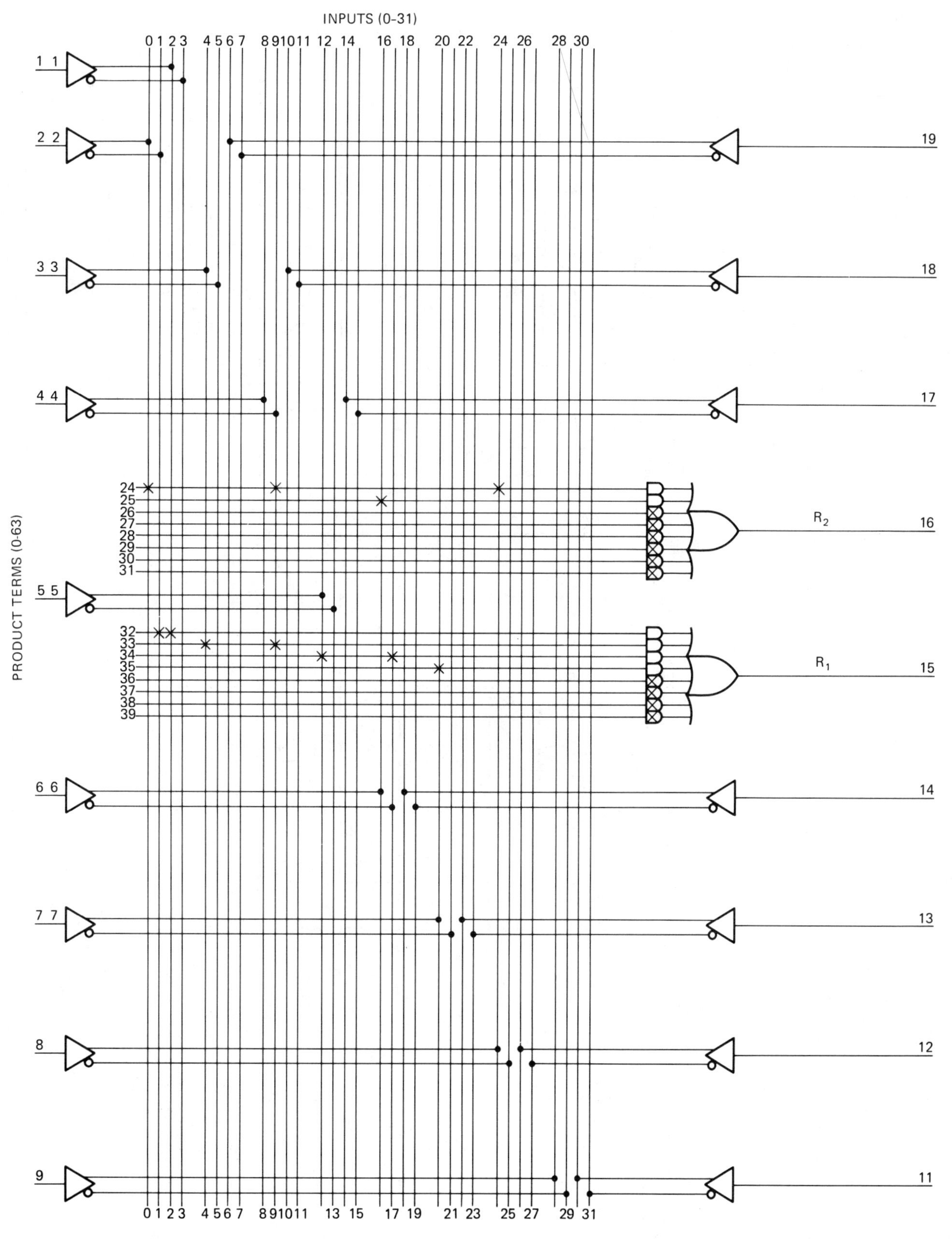

FIGURE 24-47

PAL16H2 logic diagram marked for fuse retentions needed to implement the logic gates for the circuits of Figure 24-45.

FIGURE 24-48

PALs with inverted outputs providing AND/NOR logic.

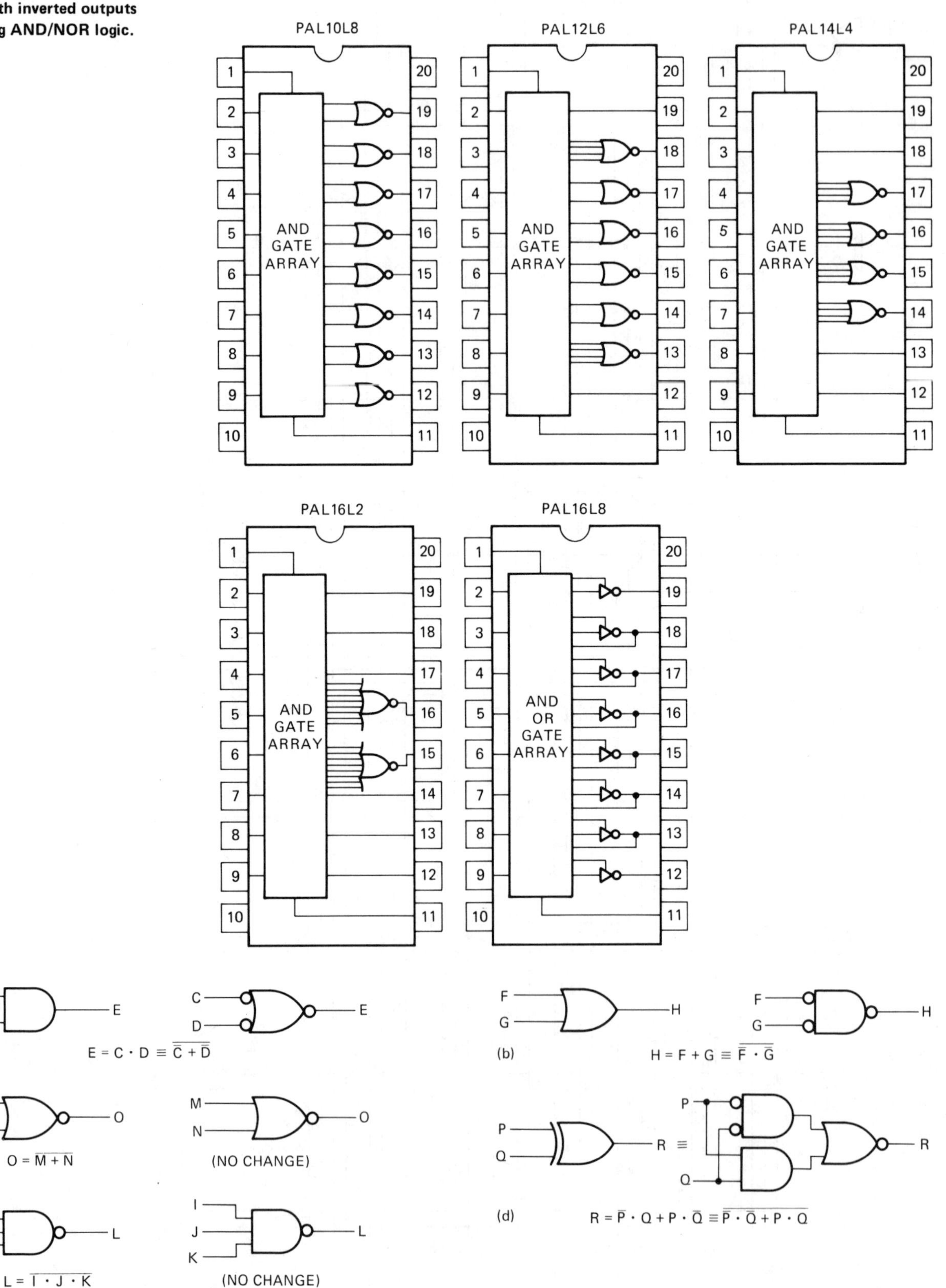

FIGURE 24-49

Simple logic gates and conversions to ideal form for implementation with AND/NOR logic.

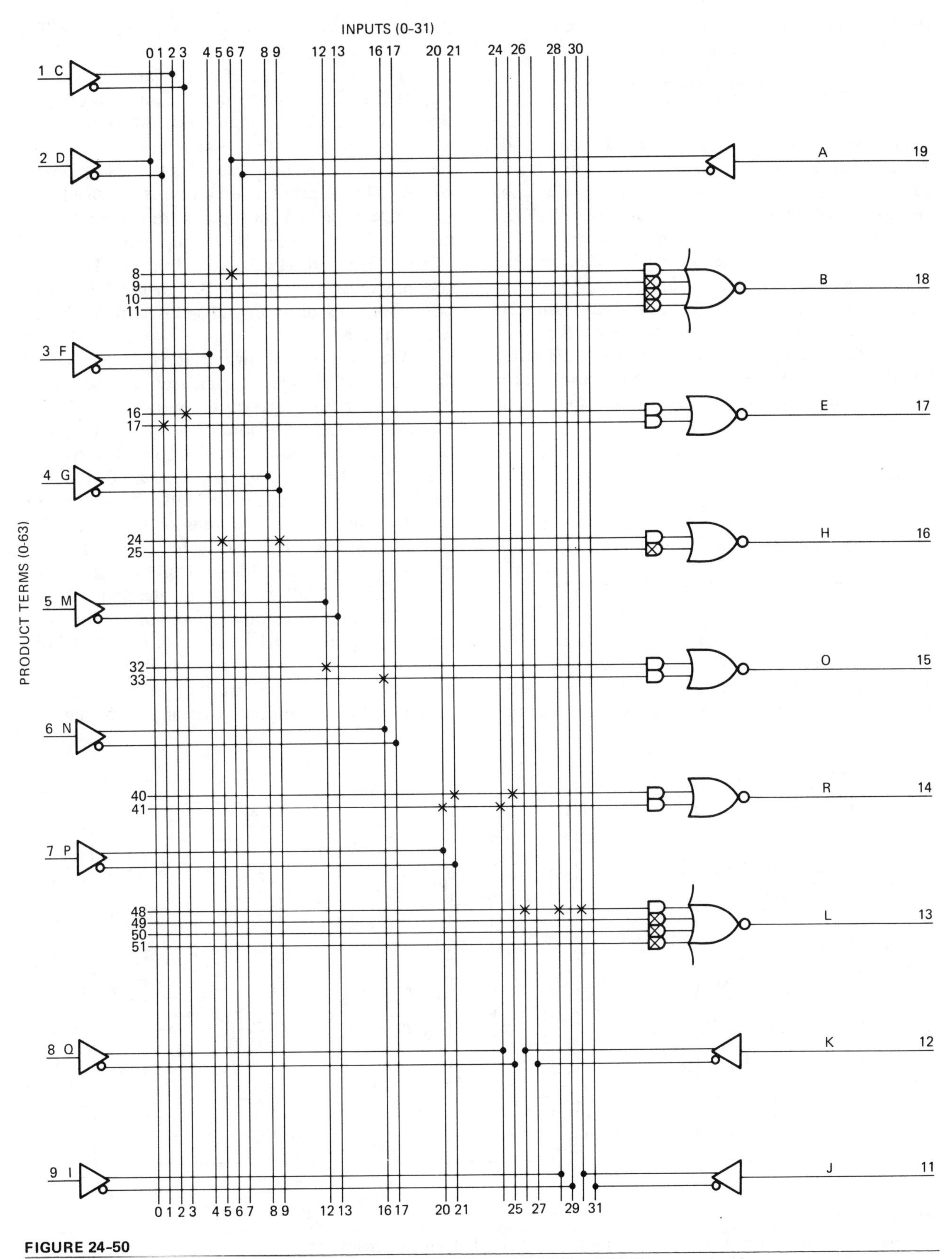

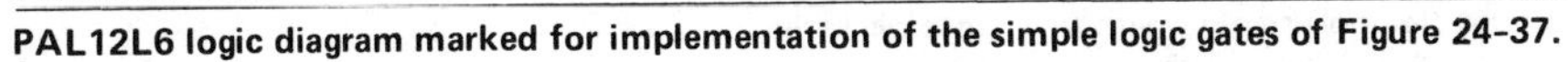
FIGURE 24-50

PAL12L6 logic diagram marked for implementation of the simple logic gates of Figure 24-37.

whether we need it or not and that the results we must have are the same as with the 12H6:

$$H = F + G \equiv \overline{\overline{F + G}} \equiv \overline{\bar{F} \cdot \bar{G}}$$

Again, we put the inversion over the right-hand member of the equation and invert the terms and change the operation to maintain the identity. This time we change from OR to AND. The results are shown in Figure 24-49(b).

The NOR gate, as shown in Figure 24-49(c), is already in correct form, $O = \overline{M + N}$. As Figure 24-50 shows, the noninverted M and N inputs provide separate inputs to the NOR gate.

The exclusive OR gate, as shown in Figure 24-49(d), follows the same pattern as before (we wish the output to be high when the inputs are different):

$$\begin{aligned} R = \bar{P} \cdot Q + P \cdot \bar{Q} &\equiv \overline{\overline{\bar{P} \cdot Q + P \cdot \bar{Q}}} \\ &\equiv \overline{\overline{\bar{P} \cdot Q} \cdot \overline{P \cdot \bar{Q}}} \equiv \overline{(\bar{\bar{P}} + \bar{Q}) \cdot (\bar{P} + \bar{\bar{Q}})} \\ &\equiv \overline{(P + \bar{Q}) \cdot (\bar{P} + Q)} \equiv \overline{\bar{P}P + PQ + \bar{Q}\bar{P} + Q\bar{Q}} \\ &\equiv \overline{0 + P \cdot Q + \bar{P} \cdot \bar{Q} + 0} \equiv \overline{P \cdot Q + \bar{P} \cdot \bar{Q}} \end{aligned}$$

For the exclusive OR gate, the results are the same as in Figure 24-42. While Figure 24-42 describes an output that is high when the inputs are different, Figure 24-49 describes an output that is low when the inputs are the same. The results are identical.

The three-input NAND gate, shown in Figure 24-49(e), is already in correct form for the L-type PAL. A three-input AND function, along one product line that is inverted through an enabled NOR gate, provides this. No manipulation of the symbol or equation is required.

Figure 24-51 compares the simple logic gates and the methods of implementing them using 12H6 and 12L6. Note that for 12H6, inversions exist only at inputs. For the 12L6, output inversions exist on all gates and on inputs where needed.

We have just shown that H devices and L devices can produce identical results, even though their fuse patterns are radically different. This raises the question of why both sets of devices are needed. The selection of a specific member of the PAL series is governed not only by the number of inputs and outputs needed but also by the required number of OR and NOR gate inputs. If an H device selected on the basis of inputs and outputs does not provide the necessary number of OR gate inputs, changing to an L device will alter the equation and allow the more numerous inputs to be supplied by AND functions or product terms, for which there are as many as 16 inputs per product term. A 16-input NAND gate is available with any L device, but

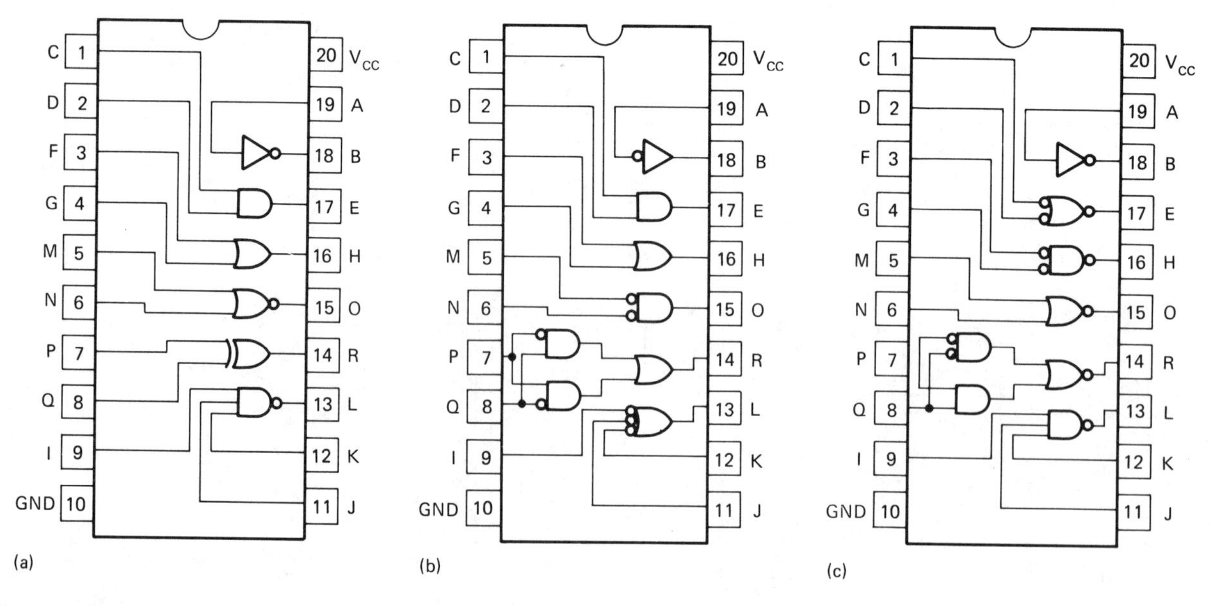

FIGURE 24-51

Implementation of simple logic gates. (a) The basic logic gates can be built using programmable array logic devices. (b) Using 12H6, the logic must be implemented with inversions occurring only on the inputs. (c) Using 12L6, the logic must be implemented with output inversions.

no higher than eight-input NAND gates are available from the 16H2. The same is true of an OR gate. On the other hand, the H devices can supply a 16-input AND or a 16-input NOR that is not available from the L devices.

Let us try a more complex application of the L-type PALs by implementing the logic of Figure 24-52. The circuit requires four outputs and nine inputs. The 14L4 will fill these requirements. The logic equations, as described at each gate output, are manipulated into suitable form. The correct equations are given in the block below the logic diagram. Figure 24-53 shows a logic diagram of the 14L4 marked for fuse retention needed to provide the logic of Figure 24-52.

Another version of the PAL series includes the registered devices shown in Figure 24-54. These devices have feedback buffers and are capable of providing sequential logic.

Let us return to the simultaneous A/D converter for an example of an application of the registered devices. Figure 24-55 is the logic diagram of a converter with a resolution of $V_R/10$. The PAL16R4 provides both logic gates and registers. The logic is already shown in correct AND/OR form. The outputs are active low used to drive a display circuit.

Figure 24-56 shows the logic diagram of the 16R4 marked for fuse retention needed to implement the logic of Figure 24-55. Note that pin 19 is used as input 1. The output buffer is disabled at that pin by retaining fuses along product line 1. For inputs 2 through 9, pin numbers and variable numbers are in agreement.

24.5.1 Programming of PALs

PALs can be programmed manually on most commercial PROM programmers. All PALs in the 20 series program as if they were 512-by-4 fusible link PROMs. The fuse data marked on the logic diagram of the PAL must first be transferred onto a programming format sheet. The format sheets supplied by the manufacturer have the state of phantom fuses already marked. Figure 24-57 shows a typical programming format for the PAL12H6. Note that phantom fuses marked as H and L correspond to missing input and product term lines on the 12H6 logic diagram. The four-bit data words are equated to product terms marked in the left column. The addresses are numbered in consecutive rows from 0 to 1FF. The user fills in the blanks by inserting an L where the X occurs on the logic

FIGURE 24-52

Logic circuit to be implemented with a 14L4. Equations in the block are in correct form for L-type PALs.

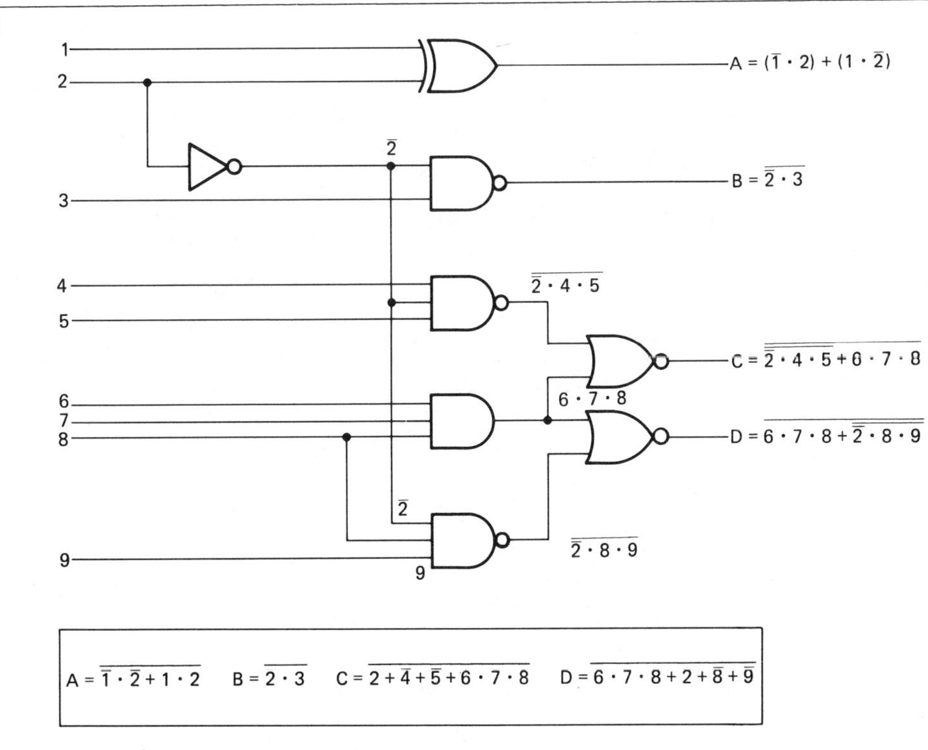

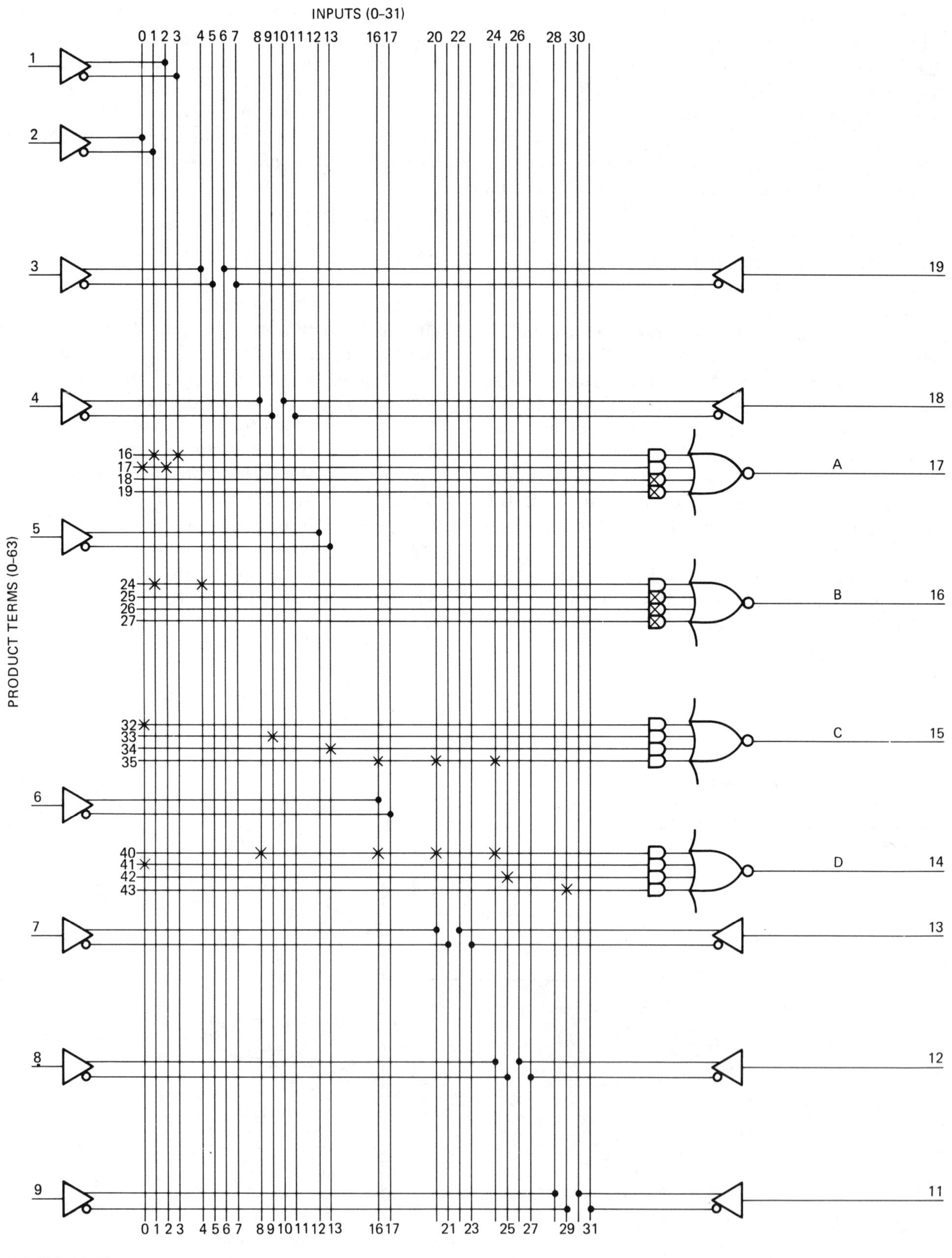

FIGURE 24-53

Logic diagram of 14L4 marked for fuse retention needed to implement the logic of Figure 24-52.

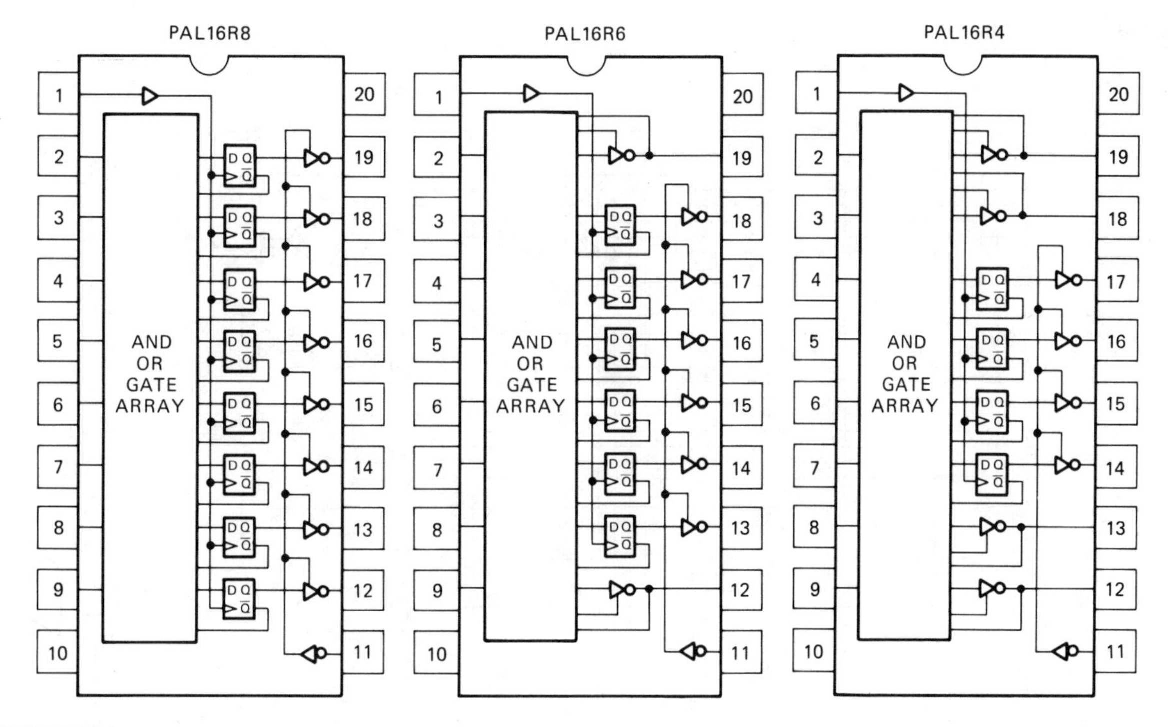

FIGURE 24-54

Series 20 PALs with registers and three-state outputs.

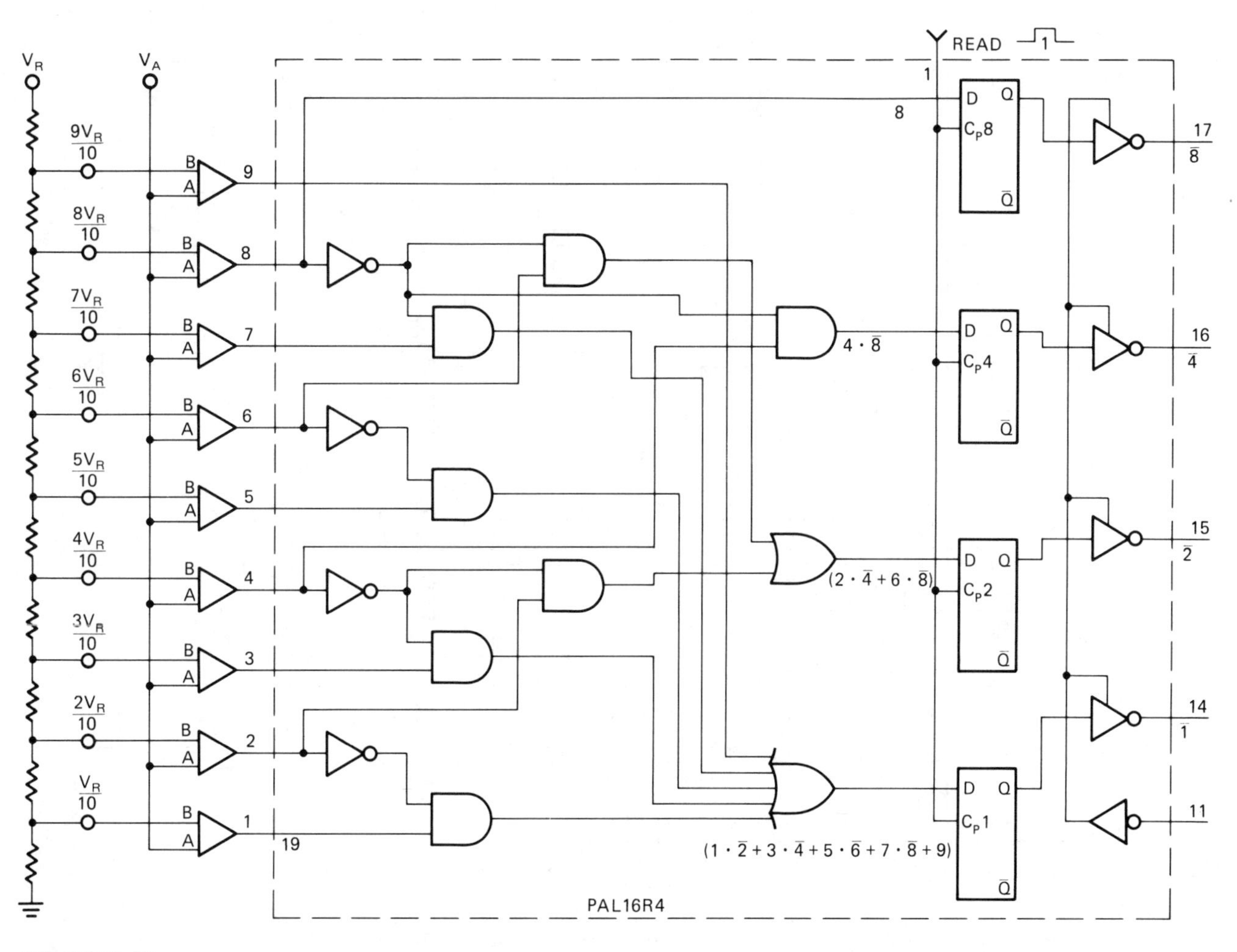

FIGURE 24-55

Simultaneous conversion A/D using registered PAL for digital logic.

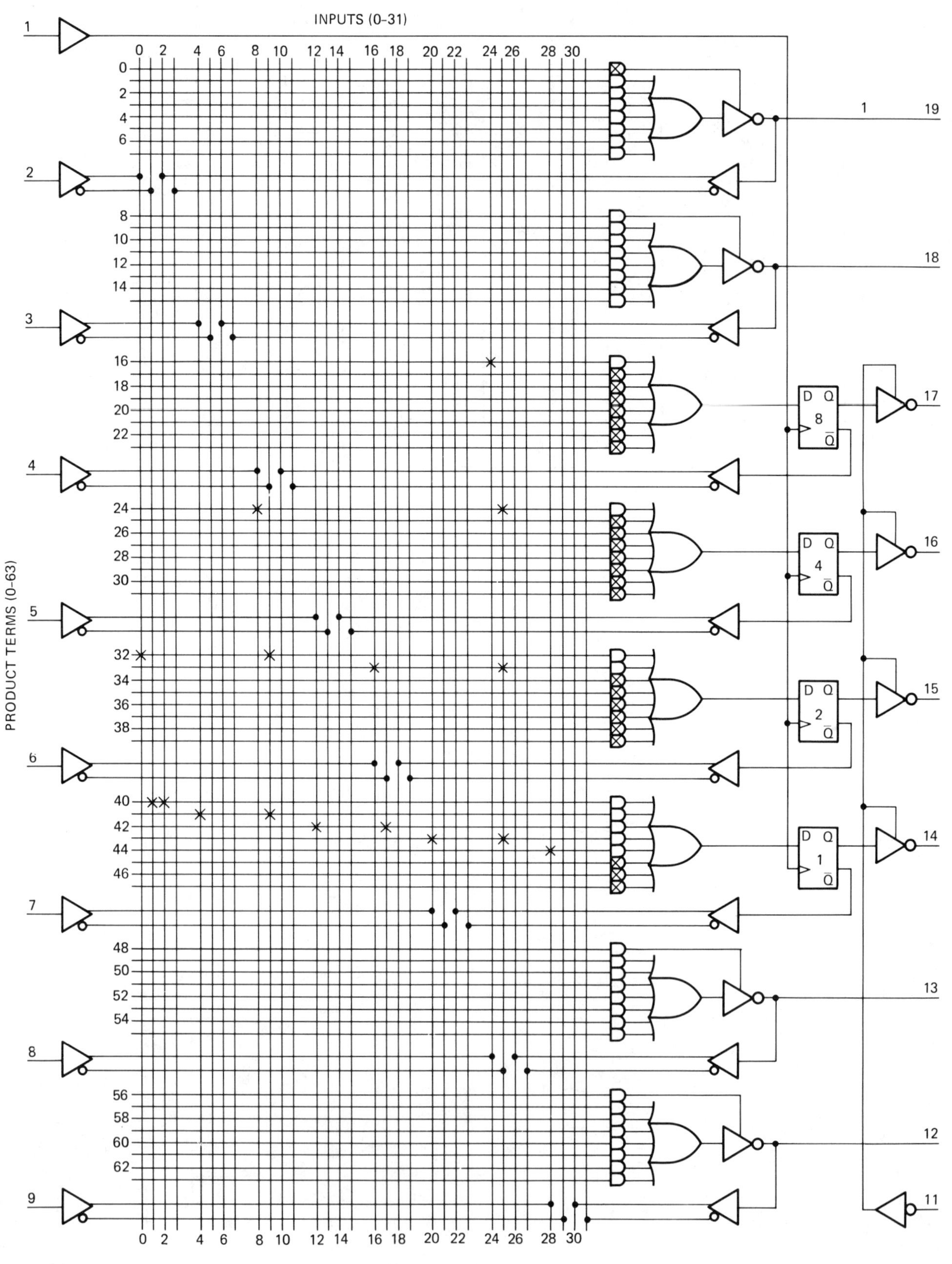

FIGURE 24-56

Logic diagram of PAL16R4 marked for fuse retention needed to implement the logic of Figure 24-55.

PROGRAMMING FORMAT CONTAINING
PHANTOM FUSES
INPUTS (0-31)

PROGRAMMING FORMAT

PAL12H6
PATTERN: ____
NAME: ____

PRODUCT TERMS (0-63)

DATE: ____

	0	1	2	3	4	5	6	7	8	9	10	11	12	13	14	15	16	17	18	19	20	21	22	23	24	25	26	27	28	29	30	31
WORD	0	1	2	3	4	5	6	7	8	9	A	B	C	D	E	F	10	11	12	13	14	15	16	17	18	19	1A	1B	1C	1D	1E	1F
O_4 24											H	H			H	H			H	H			H	H								
O_3 16											H	H			H	H			H	H			H	H								
O_2 8											H	H			H	H			H	H			H	H								
O_1 0	H	H	H	H	H	H	H	H	H	H	H	H	H	H	H	H	H	H	H	H	H	H	H	H	H	H	H	H	H	H	H	H
WORD	20	21	22	23	24	25	26	27	28	29	2A	2B	2C	2D	2E	2F	30	31	32	33	34	35	36	37	38	39	3A	3B	3C	3D	3E	3F
O_4 25											H	H			H	H			H	H			H	H								
O_3 17											H	H			H	H			H	H			H	H								
O_2 9											H	H			H	H			H	H			H	H								
O_1 1	H	H	H	H	H	H	H	H	H	H	H	H	H	H	H	H	H	H	H	H	H	H	H	H	H	H	H	H	H	H	H	H
WORD	40	41	42	43	44	45	46	47	48	49	4A	4B	4C	4D	4E	4F	50	51	52	53	54	55	56	57	58	59	5A	5B	5C	5D	5E	5F
O_4 26	L	L	L	L	L	L	L	L	L	L	L	L	L	L	L	L	L	L	L	L	L	L	L	L	L	L	L	L	L	L	L	L
O_3 18	L	L	L	L	L	L	L	L	L	L	L	L	L	L	L	L	L	L	L	L	L	L	L	L	L	L	L	L	L	L	L	L
O_2 10											H	H			H	H			H	H			H	H								
O_1 2	H	H	H	H	H	H	H	H	H	H	H	H	H	H	H	H	H	H	H	H	H	H	H	H	H	H	H	H	H	H	H	H
WORD	60	61	62	63	64	65	66	67	68	69	6A	6B	6C	6D	6E	6F	70	71	72	73	74	75	76	77	78	79	7A	7B	7C	7D	7E	7F
O_4 27	L	L	L	L	L	L	L	L	L	L	L	L	L	L	L	L	L	L	L	L	L	L	L	L	L	L	L	L	L	L	L	L
O_3 19	L	L	L	L	L	L	L	L	L	L	L	L	L	L	L	L	L	L	L	L	L	L	L	L	L	L	L	L	L	L	L	L
O_2 11											H	H			H	H			H	H			H	H								
O_1 3	H	H	H	H	H	H	H	H	H	H	H	H	H	H	H	H	H	H	H	H	H	H	H	H	H	H	H	H	H	H	H	H
WORD	80	81	82	83	84	85	86	87	88	89	8A	8B	8C	8D	8E	8F	90	91	92	93	94	95	96	97	98	99	9A	9B	9C	9D	9E	9F
O_4 28	L	L	L	L	L	L	L	L	L	L	L	L	L	L	L	L	L	L	L	L	L	L	L	L	L	L	L	L	L	L	L	L
O_3 20	L	L	L	L	L	L	L	L	L	L	L	L	L	L	L	L	L	L	L	L	L	L	L	L	L	L	L	L	L	L	L	L
O_2 12	L	L	L	L	L	L	L	L	L	L	L	L	L	L	L	L	L	L	L	L	L	L	L	L	L	L	L	L	L	L	L	L
O_1 4	H	H	H	H	H	H	H	H	H	H	H	H	H	H	H	H	H	H	H	H	H	H	H	H	H	H	H	H	H	H	H	H
WORD	A0	A1	A2	A3	A4	A5	A6	A7	A8	A9	AA	AB	AC	AD	AE	AF	B0	B1	B2	B3	B4	B5	B6	B7	B8	B9	BA	BB	BC	BD	BE	BF
O_4 29	L	L	L	L	L	L	L	L	L	L	L	L	L	L	L	L	L	L	L	L	L	L	L	L	L	L	L	L	L	L	L	L
O_3 21	L	L	L	L	L	L	L	L	L	L	L	L	L	L	L	L	L	L	L	L	L	L	L	L	L	L	L	L	L	L	L	L
O_2 13	L	L	L	L	L	L	L	L	L	L	L	L	L	L	L	L	L	L	L	L	L	L	L	L	L	L	L	L	L	L	L	L
O_1 5	H	H	H	H	H	H	H	H	H	H	H	H	H	H	H	H	H	H	H	H	H	H	H	H	H	H	H	H	H	H	H	H
WORD	C0	C1	C2	C3	C4	C5	C6	C7	C8	C9	CA	CB	CC	CD	CE	CF	D0	D1	D2	D3	D4	D5	D6	D7	D8	D9	DA	DB	DC	DD	DE	DF
O_4 30	L	L	L	L	L	L	L	L	L	L	L	L	L	L	L	L	L	L	L	L	L	L	L	L	L	L	L	L	L	L	L	L
O_3 22	L	L	L	L	L	L	L	L	L	L	L	L	L	L	L	L	L	L	L	L	L	L	L	L	L	L	L	L	L	L	L	L
O_2 14	L	L	L	L	L	L	L	L	L	L	L	L	L	L	L	L	L	L	L	L	L	L	L	L	L	L	L	L	L	L	L	L
O_1 6	H	H	H	H	H	H	H	H	H	H	H	H	H	H	H	H	H	H	H	H	H	H	H	H	H	H	H	H	H	H	H	H
WORD	E0	E1	E2	E3	E4	E5	E6	E7	E8	E9	EA	EB	EC	ED	EE	EF	F0	F1	F2	F3	F4	F5	F6	F7	F8	F9	FA	FB	FC	FD	FE	FF
O_4 31	L	L	L	L	L	L	L	L	L	L	L	L	L	L	L	L	L	L	L	L	L	L	L	L	L	L	L	L	L	L	L	L
O_3 23	L	L	L	L	L	L	L	L	L	L	L	L	L	L	L	L	L	L	L	L	L	L	L	L	L	L	L	L	L	L	L	L
O_2 15	L	L	L	L	L	L	L	L	L	L	L	L	L	L	L	L	L	L	L	L	L	L	L	L	L	L	L	L	L	L	L	L
O_1 7	H	H	H	H	H	H	H	H	H	H	H	H	H	H	H	H	H	H	H	H	H	H	H	H	H	H	H	H	H	H	H	H
WORD	100	101	102	103	104	105	106	107	108	109	10A	10B	10C	10D	10E	10F	110	111	112	113	114	115	116	117	118	119	11A	11B	11C	11D	11E	11F
O_4 56	H	H	H	H	H	H	H	H	H	H	H	H	H	H	H	H	H	H	H	H	H	H	H	H	H	H	H	H	H	H	H	H
O_3 48											H	H			H	H			H	H			H	H								
O_2 40											H	H			H	H			H	H			H	H								
O_1 32											H	H			H	H			H	H			H	H								
WORD	120	121	122	123	124	125	126	127	128	129	12A	12B	12C	12D	12E	12F	130	131	132	133	134	135	136	137	138	139	13A	13B	13C	13D	13E	13F
O_4 57	H	H	H	H	H	H	H	H	H	H	H	H	H	H	H	H	H	H	H	H	H	H	H	H	H	H	H	H	H	H	H	H
O_3 49											H	H			H	H			H	H			H	H								
O_2 41											H	H			H	H			H	H			H	H								
O_1 33											H	H			H	H			H	H			H	H								
WORD	140	141	142	143	144	145	146	147	148	149	14A	14B	14C	14D	14E	14F	150	151	152	153	154	155	156	157	158	159	15A	15B	15C	15D	15E	15F
O_4 58	H	H	H	H	H	H	H	H	H	H	H	H	H	H	H	H	H	H	H	H	H	H	H	H	H	H	H	H	H	H	H	H
O_3 50											H	H			H	H			H	H			H	H								
O_2 42	L	L	L	L	L	L	L	L	L	L	L	L	L	L	L	L	L	L	L	L	L	L	L	L	L	L	L	L	L	L	L	L
O_1 34	L	L	L	L	L	L	L	L	L	L	L	L	L	L	L	L	L	L	L	L	L	L	L	L	L	L	L	L	L	L	L	L
WORD	160	161	162	163	164	165	166	167	168	169	16A	16B	16C	16D	16E	16F	170	171	172	173	174	175	176	177	178	179	17A	17B	17C	17D	17E	17F
O_4 59	H	H	H	H	H	H	H	H	H	H	H	H	H	H	H	H	H	H	H	H	H	H	H	H	H	H	H	H	H	H	H	H
O_3 51											H	H			H	H			H	H			H	H								
O_2 43	L	L	L	L	L	L	L	L	L	L	L	L	L	L	L	L	L	L	L	L	L	L	L	L	L	L	L	L	L	L	L	L
O_1 35	L	L	L	L	L	L	L	L	L	L	L	L	L	L	L	L	L	L	L	L	L	L	L	L	L	L	L	L	L	L	L	L
WORD	180	181	182	183	184	185	186	187	188	189	18A	18B	18C	18D	18E	18F	190	191	192	193	194	195	196	197	198	199	19A	19B	19C	19D	19E	19F
O_4 60	H	H	H	H	H	H	H	H	H	H	H	H	H	H	H	H	H	H	H	H	H	H	H	H	H	H	H	H	H	H	H	H
O_3 52	L	L	L	L	L	L	L	L	L	L	L	L	L	L	L	L	L	L	L	L	L	L	L	L	L	L	L	L	L	L	L	L
O_2 44	L	L	L	L	L	L	L	L	L	L	L	L	L	L	L	L	L	L	L	L	L	L	L	L	L	L	L	L	L	L	L	L
O_1 36	L	L	L	L	L	L	L	L	L	L	L	L	L	L	L	L	L	L	L	L	L	L	L	L	L	L	L	L	L	L	L	L
WORD	1A0	1A1	1A2	1A3	1A4	1A5	1A6	1A7	1A8	1A9	1AA	1AB	1AC	1AD	1AE	1AF	1B0	1B1	1B2	1B3	1B4	1B5	1B6	1B7	1B8	1B9	1BA	1BB	1BC	1BD	1BE	1BF
O_4 61	H	H	H	H	H	H	H	H	H	H	H	H	H	H	H	H	H	H	H	H	H	H	H	H	H	H	H	H	H	H	H	H
O_3 53	L	L	L	L	L	L	L	L	L	L	L	L	L	L	L	L	L	L	L	L	L	L	L	L	L	L	L	L	L	L	L	L
O_2 45	L	L	L	L	L	L	L	L	L	L	L	L	L	L	L	L	L	L	L	L	L	L	L	L	L	L	L	L	L	L	L	L
O_1 37	L	L	L	L	L	L	L	L	L	L	L	L	L	L	L	L	L	L	L	L	L	L	L	L	L	L	L	L	L	L	L	L
WORD	1C0	1C1	1C2	1C3	1C4	1C5	1C6	1C7	1C8	1C9	1CA	1CB	1CC	1CD	1CE	1CF	1D0	1D1	1D2	1D3	1D4	1D5	1D6	1D7	1D8	1D9	1DA	1DB	1DC	1DD	1DE	1DF
O_4 62	H	H	H	H	H	H	H	H	H	H	H	H	H	H	H	H	H	H	H	H	H	H	H	H	H	H	H	H	H	H	H	H
O_3 54	L	L	L	L	L	L	L	L	L	L	L	L	L	L	L	L	L	L	L	L	L	L	L	L	L	L	L	L	L	L	L	L
O_2 46	L	L	L	L	L	L	L	L	L	L	L	L	L	L	L	L	L	L	L	L	L	L	L	L	L	L	L	L	L	L	L	L
O_1 38	L	L	L	L	L	L	L	L	L	L	L	L	L	L	L	L	L	L	L	L	L	L	L	L	L	L	L	L	L	L	L	L
WORD	1E0	1E1	1E2	1E3	1E4	1E5	1E6	1E7	1E8	1E9	1EA	1EB	1EC	1ED	1EE	1EF	1F0	1F1	1F2	1F3	1F4	1F5	1F6	1F7	1F8	1F9	1FA	1FB	1FC	1FD	1FE	1FF
O_4 63	H	H	H	H	H	H	H	H	H	H	H	H	H	H	H	H	H	H	H	H	H	H	H	H	H	H	H	H	H	H	H	H
O_3 55	L	L	L	L	L	L	L	L	L	L	L	L	L	L	L	L	L	L	L	L	L	L	L	L	L	L	L	L	L	L	L	L
O_2 47	L	L	L	L	L	L	L	L	L	L	L	L	L	L	L	L	L	L	L	L	L	L	L	L	L	L	L	L	L	L	L	L
O_1 39	L	L	L	L	L	L	L	L	L	L	L	L	L	L	L	L	L	L	L	L	L	L	L	L	L	L	L	L	L	L	L	L
	0	1	2	3	4	5	6	7	8	9	10	11	12	13	14	15	16	17	18	19	20	21	22	23	24	25	26	27	28	29	30	31

FIGURE 24-57

Monolithic Memories typical programming format for the PAL-12H6.

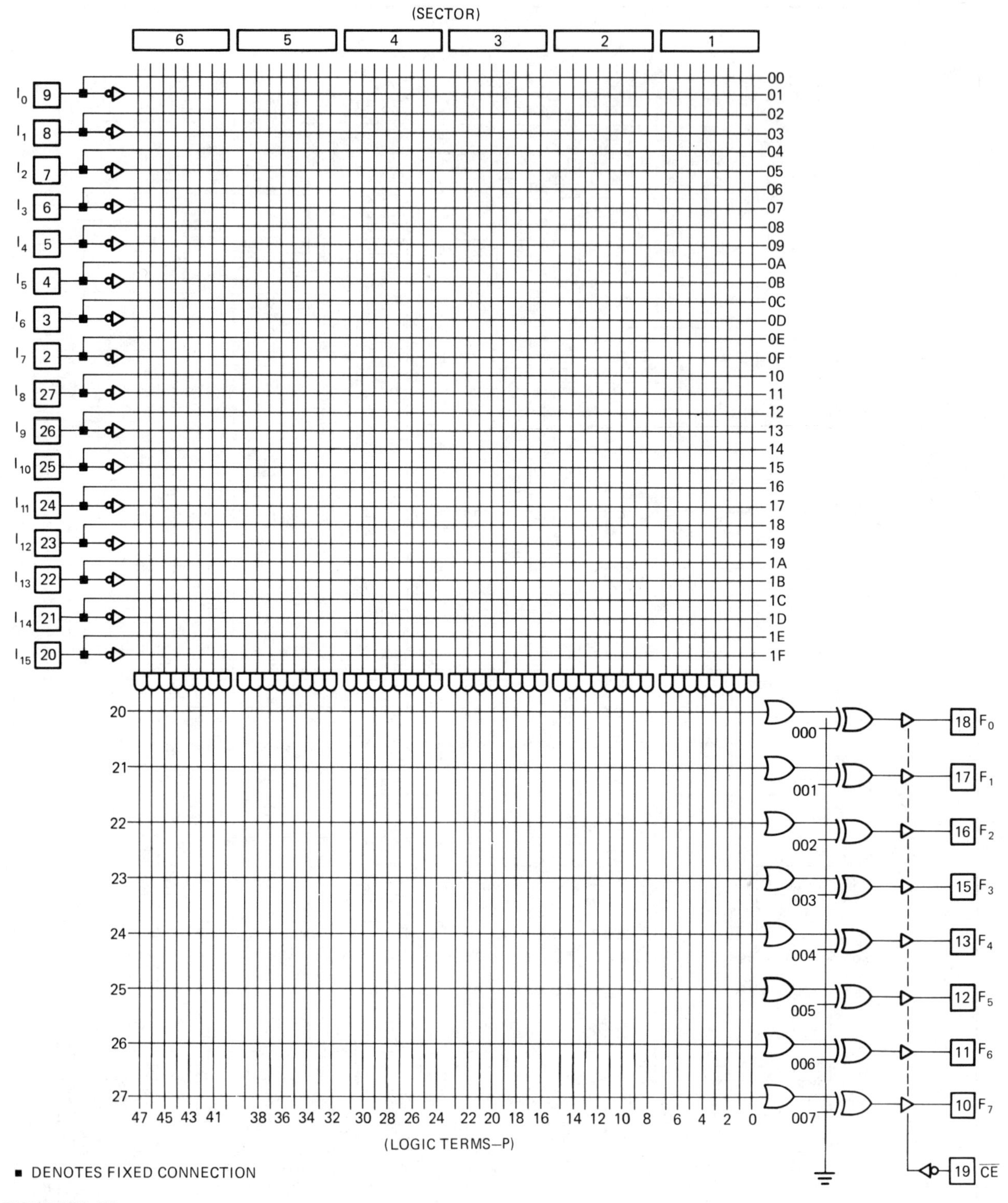

FIGURE 24-58

Complete logic diagram of the 82S100 FPLA (16×48×8).

diagram and L along entire product term lines where the AND gate is X. The remaining blanks are filled with H indicating fuses to be blown. As most PROM programmers receive data-in from a hexadecimal keyboard, the data are usually further translated to a hexadecimal listing.

Manual programming is practical for only small-scale and prototype operations. Special development systems for PALs allow the operator to program using Boolean algebra expressions.

24.5.2 Integrated Fuse Logic (Field Programmable Logic Arrays)

Thus far we have seen the fusible link technique used for PROMs and PALs—the PROMs using a fixed AND array at the inputs and a programmable OR array at the outputs, and the PALs using a programmable AND array at the inputs and a fixed OR array at the outputs. A further development in this area is the field programmable logic array. This device allows programming of both the input AND array and the output OR array.

Figure 24-58 shows the complete logic diagram of the 82S100, which is a typical member of this family of parts. The outputs are tri-state (82S100) or open-collector (82S101). It is an FPLA (16 × 48 × 8). The numbers indicate 16 input variables and eight output functions. The 48 indicates 48 possible product terms (AND gates), and each product term can exist as an input to any of the eight OR gates. Fuses are retained along the product lines to form the AND functions. Fuses are retained along the sum lines to form the output OR functions. In addition, the OR outputs can be switched to NOR by blowing fuses at the exclusive OR leads.

24.6 Ultraviolet Erasable PROMs (EPROMs)

The EPROM is an attractive device for use as a program storage memory. In addition to being random access and nonvolatile, it is erasable.

The EPROM storage cell is shown in Figure 24-59. Each cell is a field effect transistor with a floating gate. When a cell is addressed during a normal read cycle, it conducts (turns on), provided there is no charge on the floating gate. Programming is accomplished by charging the floating gate to store a binary 0 and leaving it uncharged to store a binary 1. A high-voltage pulse on a program pin, in combination with normal logic inputs applied to the address and data, accomplishes the programming. Once a cell is programmed to produce a logic 0 (V_{OL} output), it cannot be changed without the entire EPROM being erased. After programming voltages are removed, the silicon dioxide surrounding the floating gate is a sufficient insulator to hold the charge for an extended period. Some loss of charge (less than 10%) occurs during the first few days after programming. From then on, a normal cell loses no more than 10% of the remaining charge in ten years (at 125° C).

Figure 24-60 shows several present-day versions of EPROMs. The 2708 1K-by-8 has the disadvantage of requiring three power supplies. New versions such as the 2K-by-8 (2716) and 4K-by-8 (2732) require only a +5V power supply for operation after programming. Beginning with the 8K-by-8 (2764) and the 16K-by-8 (27128), a 28-pin circuit is required to provide the necessary pin-out. The 27128 is the largest version available to date.

FIGURE 24-59
EPROM storage cell.

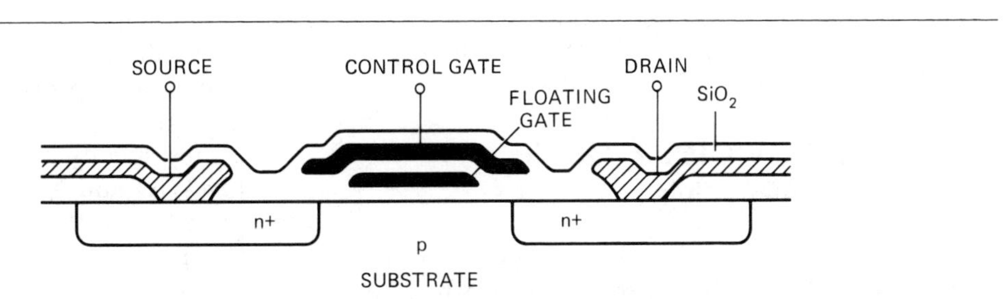

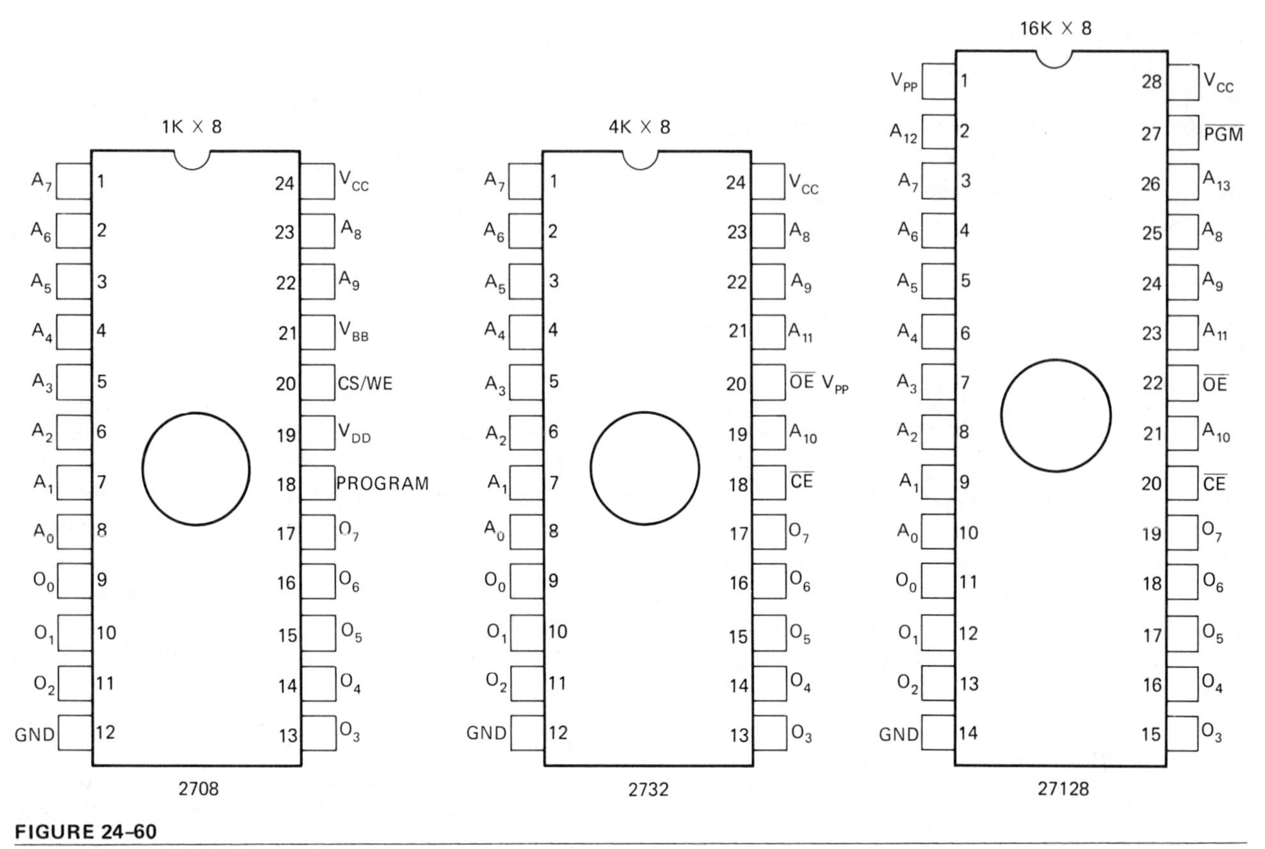

FIGURE 24-60

Three versions of EPROMs.

FIGURE 24-61

Mode selection truth table of the EPROM 2732. Program mode requires V_{PP} = 25 and normally occurs in a special programming circuit rather than in the operating circuit.

PINS / MODE	$\overline{CE}$/PGM (18)	$\overline{OE}/V_{PP}$ (20)	V_{CC} (24)	OUTPUTS (9-11, 13-17)
READ	V_{IL}	V_{IL}	+5	D_{OUT}
STANDBY	V_{IH}	DON'T CARE	+5	HIGH Z
PROGRAM	V_{IL}	V_{PP}	+5	D_{IN}
PROGRAM VERIFY	V_{IL}	V_{IL}	+5	D_{OUT}
PROGRAM INHIBIT	V_{IH}	V_{PP}	+5	HIGH Z

An additional advantage to the newer, single-power-supply EPROMs is a standby mode power level that is considerably lower than operating power. The power reduction occurs automatically when the device is deselected. Figure 24-61 shows the mode selection truth table of the 2732. Programming is accomplished with a high-level pulse (V_{PP} = 25V) on pin 20.

24.6.1 Ultraviolet Erasure of EPROMs

The EPROM can be erased by exposing it to ultraviolet light of the correct wavelength. The ultraviolet light adds photon energy to the electrons that form the charge on the floating gates of the storage cell. In the high energy state, the electrons are capable of passing through the SiO_2 insulation. Given sufficient light intensity and exposure time, the programmed cells of the EPROM will discharge.

The erase procedures recommended by various manufacturers of EPROMs are all similar in that they specify the wavelength of ultraviolet light (2537A) and energy required in W-sec/cm^2 at the window of the device being erased.

Lamps for erasing EPROMs are available from several sources. Figure 24-62 gives a list of

FIGURE 24-62
Standard models of ultraviolet lamps, their intensities, and their erasure times.

MODEL	POWER RATING	MINIMUM ERASE TIME FOR INDICATED DOSAGE WITHOUT A FILTER OVER THE BULB
R-52	13,000 μW/cm²	19.2 min
S-52	12,000 μW/cm²	20.7 min
S-68	12,000 μW/cm²	20.7 min
UVS-54	5,700 μW/cm²	43.8 min
UVS-11	5,500 μW/cm²	45.6 min

ultraviolet lamp models, their intensities, and their minimum erasure times. For a given device type, we would expect erase exposure times to vary only by the intensity of the UV source. Experimental results on 2708 EPROMs from five different vendors show a lack of uniformity in the exposure time required to erase these devices. This lack of uniformity could result in the use of an unreliable erase procedure. The only practical test for an erased condition is to determine that the outputs are all high (FF_H) at all addresses. This condition first occurs when the last memory cell of the EPROM discharges to the switching threshold. In this initial erase state, some memory cells are so close to the switching threshold as to exhibit little or no noise immunity. The use of these marginally erased, low noise immunity devices may result in intermittent failures.

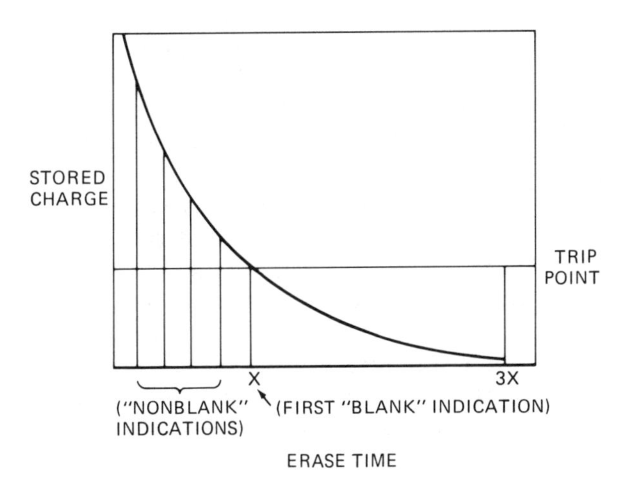

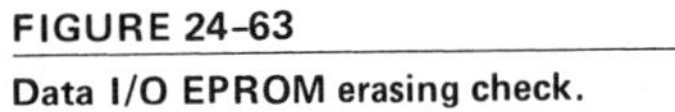

FIGURE 24-63
Data I/O EPROM erasing check.

Data I/O instructions for determining the exposure time needed to erase an EPROM with a particular UV source are as follows: EPROMs are erased by placing them approximately one inch from a source of high-intensity 254 nm (UV) radiation. The amount of time required for erasure is specified in terms of minimum integrated dosage, expressed in W-sec/cm², and is a function of the power rating of the UV source.

As the UV source ages, its intensity diminishes, which means that an EPROM left to erase for "the usual" time may not be completely erased if the tube is old. Incomplete erasure may cause increased access time or unstable bits, which can switch a long time after programming. The UV source should be periodically checked by the following procedure, which is illustrated in Figure 24-63.

1. Place a programmed EPROM under the UV source for a measured, short period of time.
2. Put the EPROM in the programmer and check to see if it is blank.
3. If it is not blank, put it under the source again for another measured period. Repeat steps 2 and 3 until the EPROM is blank.
4. Multiply "time-to-blank" by 3 and use this figure as a minimum for an erase time.

As Figure 24-63 shows, the EPROM being erased is periodically tested for first blank indication. Since the discharge rate of the cells is exponential, twice as much exposure time is required to discharge from the trip point to a reliable 0 level than was required to discharge to the trip point. This procedure would produce a reliable exposure time regardless of the age or strength of the UV source.

Use caution with UV sources and avoid unnecessary exposure. Radiation of the recommended intensity can cause burns and impair vision. Finally, guard against unwanted erasure of an EPROM after programming by covering its quartz lid with a UV filtering label material.

24.7 Summary

The serial memory and the read-only memories have a number of special applications in digital systems. The serial memory is a shift register of many bits. Both dynamic and static shift registers are used as serial memories. As memories, they are serial access and erasable and can be wired to recirculate the data for a nondestructive readout. They are volatile. In design, they differ from normal operations registers in that they must be very economical in power consumption in order to be used in such large numbers of bits. The techniques used to save power require use of two or more clock phases.

The dynamic shift register, like the dynamic memory cell, uses the gate-to-source capacitance of a MOSFET to store the logic levels. Figure 24-4 shows a typical dynamic register stage. It can be divided into identical half stages, which operate in two phases. As Figure 24-3(a) shows, a 1 level at the input will discharge the capacitor C_1 when clock ϕ_1 turns on the switch transistor. A 0 level on the input causes the capacitor to charge when clock ϕ_1 turns on the switch transistor. For each half stage of the register, there is an inversion, so that a 1 level at the input stores a 1 level on the capacitor at the output of the second stage. The switch transistors isolate each gate-to-source capacitance, causing the data to move through the register a half bit on clock pulse ϕ_1 and a half on ϕ_2. The data stored in the small gate-to-source capacitance will leak off if a clock rate of at least 500 cycles per second is not maintained.

Figure 24-5 shows a Signetics Corporation 2500 series dual 100-bit dynamic shift register. It uses P-channel devices, but the input and output are at TTL logic levels. Two negative voltage clock phases must be supplied. Figure 24-6 is a 512- or 1024-bit dynamic shift register with logic for recirculation included on the chip. It has a power drain of only 150 μW per bit at 1 MHz clock. It has a maximum clock rate of 5 MHz.

High-capacity registers suitable for serial memory application are also available as static shift registers. These use MOS flip-flop circuits, but they are operated by several clock phases to reduce power consumption. Figure 24-7 is a Signetics Corporation 2500 series 1024-bit static shift register. The circuit has TTL-compatible inputs and outputs, including the single clock input, which synchronizes a three-phase clock generator contained on the chip. Recirculation can be accomplished by an external jumper. Power dissipation is only 160 μW per bit.

Another important semiconductor memory is the read-only memory (ROM). The ROM is programmed into the digital system before assembly and cannot be written into during operation. The memory is formed by a matrix of lines, the address lines running in the X direction, the output lines in the Y direction. One address line is energized at a time. As Figure 24-11 shows, the 1 or 0 level at the output lines depends on the presence or absence of a conducting element at the intersection of the address line with the output lines. The conducting elements may be diodes, bipolar, or MOS transistors, as Figure 24-11 shows. Figure 24-12 shows an 8-K-word-by-8-bit read-only memory. The number of bits per word can be expanded by paralleling the address lines, as Figure 24-13 shows. Expansion of the address can be accomplished, as Figure 24-14 shows, by using the chip select inputs and combining the outputs in a three-state connection.

One application of read-only memories is their use as mathematical tables, such as the sine function table shown in Figure 24-16. The input is the angle in degrees or radians. The output is the sine of the angle. A more impressive use of ROMs is found in their application as character generators, particularly in CRT displays.

The American Standard Code for Information Interchange (ASCII) is a digital code widely used in addressing character generators. As shown in Figure 24-18, 64 characters of this code provide most of the characters appearing on a telegraphic keyboard. Only the first six bits of the code are needed. The MSB may be excluded. Figure 24-19 shows a MOS 2513 character generator. The addresses for the ASCII characters are the six ASCII bits at inputs A_4 through A_9. The remaining address is provided by the three bits of a row counter. As the row counter counts 0 through 7, a different row of the eight-row character appears at the output leads as five parallel data bits, 0_1 through 0_5. The parallel data bits are converted to serial in a parallel-to-serial converter, as shown in Figure 24-20(a) and (b).

Other ROM character formats are available. Figure 24-21 shows the letter S of a MOS 2516 character generator. It differs from the MOS 2513 in that the outputs are a column of eight bits. Instead of a row counter, a mod 6 column counter is used to address one column of the output at a time. Figure 24-22 shows the complete set of ASCII characters available from the MOS 2516. Note the resemblance of this set of characters to the ASCII code shown in Figure 24-18. To cast the characters on the CRT screen, a set of perfectly synchronized deflection voltages must be generated. Figure 24-23 shows these for the MOS 2513. The horizontal deflection starts at the right and moves the beam to the right one step for each bit of the data. As the scan pattern indicates, at the end of each row the vertical deflection voltage steps the beam down to the next row.

Figure 24-24 shows the circuits that must be added to produce the staircase deflection voltages. A digital-to-analog (D/A) converter controlled by a divide-by-6 counter produces the six-step staircase for the horizontal deflection. The counter is clocked by the same clock as the parallel-to-serial converter producing the intensity input. The row counter operates a D/A converter that generates the eight-step vertical deflection staircase. To display a line of characters and periodically refresh the intensity, additional circuits are supplied to those shown in Figure 24-24.

As Figure 24-26(a) shows, to display a line of N characters, an N-bit shift register is used to store each bit of the ASCII input. As the scan of each character is completed, the shift registers are clocked, placing a new character at the inputs A_4 through A_9. It requires 48 clock pulses for each character. Therefore, the registers are clocked, with a $C_P/48$ pulse. At the completion of each character, the sweep should be moved to the right the width of one character. This is accomplished by clocking a divide-by-N counter that operates a D/A converter. The output of this converter produces a character-step staircase that is added in an analog adder with the bit-step staircase. Figure 24-26(b) shows the waveforms of this addition. When the divide-by-N counter returns to 0, the trace returns to the right. At the same time, the first character of the line has been recirculated, and the scan of each character in the line is repeated periodically to maintain a desirable intensity of the display.

For low-volume applications, the fusible link PROM, as shown in Figure 24-27, is a more practical device than the mask programmed ROM. The fusible link devices are purchased blank and are programmed by the user. Once the fuses are blown, they cannot be replaced. There are a number of commercially available PROM programmers, which can be used to program these devices. Data can be entered into a PROM programmer by using hex keyboards or paper tape or by down-loading from a computer.

A device similar to the fusible link is the programmable array logic (PAL). PALs are designed to replace random logic. These devices are designed so that the input is a fusible AND function and the output is a fixed OR function, as shown in Figure 24-33. With the aid of Boolean algebra and DeMorgan's theorem, different logic functions can be obtained, as illustrated in Figures 24-37 through 24-44. More complex functions, such as $A/\overline{D}$ conversion, can be done as shown in Figure 24-45. PALs are available with registers and three-state outputs. See Figures 24-54 through 24-56.

Programmable array logic can be programmed manually on most PROM programmers. Manufacturers provide programming format sheets such as the one shown in Figure 24-57. Some special development systems allow the user to program PALs using Boolean algebra.

Field programmable logic arrays, FPLAs, allow a user to program both the input AND function and the output OR function. See Figure 24-58.

The ultraviolet erasable PROM (EPROM) is another type of nonvolatile memory. It stores charge in a low-leakage cell, which is a field effect transistor with a floating gate. Programming is accomplished by charging the floating gate to store a binary 0 and leaving it uncharged to store a binary 1. After it is programmed, it cannot be changed without the entire EPROM being erased. Erasure is accomplished by placing the device under an ultraviolet lamp.

Glossary

Matrix: A circuit composed of parallel sets of lines running in perpendicular directions. The lines may be joined by conductive elements at their intersections.

ASCII: The American Standard Code for Information Interchange—a seven-bit digital code used to represent the letters, numbers, and special characters found on the telegraphic keyboard. It is widely used for other digital communication and display.

CRT (Cathode Ray Tube): Electron tube used

for television, oscilloscope, and digital displays.

Phosphor: A material used to coat the inner surface of a CRT screen. A phosphor substance will fluoresce or glow when bombarded by a high-intensity electron beam. It will retain its glow for some period after the electron beam has passed.

Electron Beam (Trace): High-velocity electrons, which are focused into a narrow beam by a form of electron optics commonly called an electron gun. The electron beam is used to trace images on the phosphor-coated screen of the CRT.

Horizontal Deflection Plates: A set of parallel plates in the CRT through which the electron beam passes. Voltage waveforms applied to these plates cause the trace or electron beam to be deflected horizontally.

Horizontal Deflection Voltage: Voltage applied to the horizontal deflection plates of the CRT.

Horizontal Deflection Staircase: Voltage with a staircase waveform, used to deflect the trace horizontally in steps.

Vertical Deflection Plates: A set of parallel plates in the CRT through which the electron beam passes. Voltage waveforms applied to these plates cause the trace or electron beam to be deflected vertically.

Vertical Deflection Voltage: Voltage applied to the horizontal deflection plates of the CRT.

Vertical Deflection Staircase: Voltage with a staircase waveform, used to deflect the trace vertically in steps.

Intensity Input: Input to the CRT that causes an increase or decrease in the electron beam intensity. In the digital display system, 0 produces black (no luminance); 1 produces full brilliance. A digital inversion of the input can change a black-on-white display to white on black.

Fusible Link PROM: A programmable semiconductor read-only memory device in which the connecting elements in the program array are fuses.

PROM Programmer: A piece of equipment used to "blow" or program a semiconductor read-only memory.

Integrated Fuse Logic: See *Field Programmable Logic Array.*

Field Programmable Logic Array: A fusible link device that allows programming of both the input AND array and the output OR array.

Ultraviolet Erasable PROM: A semiconductor read-only memory device that can be programmed by the user and erased by placing the device under ultraviolet light.

Programmable Array Logic (PAL): Semiconductor devices designed to replace random logic with equivalent AND/OR and AND/NOR functions. The AND portion of the function is accomplished by blowing the correct fuses in a programmable AND array. The OR/NOR function is determined by selection of the correct member of the PAL series of device.

Questions

1. What is done to convert a shift register to a nondestructive readout memory?
2. For what applications are serial memories more advantageous than RAMs?
3. Why are two clock phases needed for the dynamic shift register of Figure 24-5?
4. In Figure 24-6, what form will the output data be in?
5. In the circuit of Figure 24-4, explain the effect of the clock pulse width and frequency on the power drain.
6. What may be contained in an integrated circuit ROM besides the memory matrix?
7. Describe the procedure for expanding the number of bits per word on a ROM like the SY2364 of Figure 24-12.
8. Describe the procedure for expanding the number of words in a read-only memory using the SY2364.
9. A read-only memory circuit is programmed to perform like a BCD adder-subtractor. How many input and output leads are required?
10. How many 93406 ROMs would be needed for a BCD adder-subtractor?
11. What advantages does the CRT display have over other readout terminals?
12. What is the meaning of ASCII?
13. How many bits are in the ASCII code? How many are used in the character generator?
14. Write the numbers 0 through 9 in ASCII code.
15. Write the letters ABC and XYZ in ASCII code.
16. Write the number 9 in BCD, in ASCII,

and as it would appear in data bits from the MOS 2513 character generator and from the MOS 2516 character generator.

17. ASCII display data are stored in a read-only memory in two dimensions. The six bits of the ASCII code form one dimension. What is the second?
18. The 5×7-character format includes no space between letters. How can this space be provided?
19. In Figure 24-20(b), label the individual serial out bits that originate from the second row, O_1 through O_5.
20. Describe the difference between the MOS 2513 and the MOS 2516 formats.
21. Describe the difference between the scan patterns of Figures 24-23 and 24-25.
22. Why are the intensity voltages for producing the letter T not the same for both the MOS 2513 and the MOS 2516?
23. The row counter addresses the top row at the count of 0, requiring a descending staircase: How can a positive-voltage D/A converter be used for this function?
24. How would the N-bit shift register and divide-by-N counter circuit of Figure 24-26 differ between use for a MOS 2513 and a MOS 2516?
25. How are the bit-step and character-step voltages combined in Figure 24-26?
26. In Figure 24-26(b), describe how each of the following would differ if a MOS 2516 were being used:

Clock pluses	Row steps
Bit steps	Combined horizontal deflection
Character steps	

27. Explain how the circuit of Figure 24-26(a) would be expanded for a 16-line display with N bits to the line.
28. How is a refresh of the display intensity accomplished?
29. How does a fusible link PROM differ from a mask programmed ROM?
30. Once fuses are blown in a fusible link PROM, can they be erased?
31. What four steps are necessary for transferring data from a program sheet to a PROM?
32. What are the ways by which data can be entered into a PROM programmer?
33. How does a programmable array logic differ from a fusible link PROM?
34. What is the common element in PALs?
35. Is DeMorgan's theorem useful in the writing of PAL programs?
36. What do the letters H and L represent in PAL numbering?
37. Describe the difference between the fusible link PROM and the PAL.
38. How often will the addresses of the PROMs in Figure 24-28 repeat themselves in a 64K address system?
39. Are the fuses of the 7641 PROM blown one address at a time or one bit at a time?
40. How does the programmer recognize address zero when a binary tape is used?
41. For an ASCII hexadecimal program tape, how will the tape for a 512-by-4 PROM differ from that for a 512-by-8 PROM?
42. Explain the numbering system of the PAL 20 series.
43. What is the largest size AND gate available from H series PALs?
44. What is the largest size AND gate available from L series PALs?
45. What are the horizontal lines appearing on the PAL logic diagram?
46. What does an X placed in the AND gate of a PAL schematic diagram indicate?
47. What are the numbers of three PALs with registers? How does integrated fuse logic differ from PALs?
48. What is the nature of the gate circuit of an EPROM storage cell?
49. How do EPROMs store their data?
50. When a 2708 EPROM is erased, are the cells storing a binary 0 or a binary 1?
51. Why does it take years for EPROMs to be erased with ordinary sunlight?
52. Can the amount of ultraviolet light exposure needed to totally erase an EPROM be exceeded by more than 20% without damaging an EPROM?
53. How does the EEPROM differ from a nonvolatile RAM?

Problems

24-1 Draw a diode matrix that produces octal-to-binary conversion.

24-2 Draw a microprocessor bus diagram like that of Figure 23-6 showing the SY2364 connected with a 4802 RAM. The RAM is to occupy the lower 2K addresses. Show the select decoding needed to prevent bus contention.

24-3 Draw a set of 93406 ROMs wired to make a BCD adder-subtractor.

24-4 Draw a ROM sine table to convert 0 to 1.57 radians (three significant digits input) with four-digit (BCD) output. Use SY2364.

24-5 In the format of Figure 24-16, write the program for the BCD adder-subtractor of Problem 24-3.

24-6 The sine table of Figure 24-16 is in two significant digits, 0° to 90°. Draw a block diagram (single block) of that memory, showing only inputs and outputs. Combine all CS1 leads of the four chips and show as one output. Leave all CS2 inputs on ground.

24-7 Using two of the block diagram symbols created in Problem 24-6, a single inverter, and the necessary pull-up resistors, draw a ROM sine table for 0 to 1.57 radians.

24-8 Expand the drawing of Figure 24-20 by drawing the logic for the parallel-to-serial converter.

24-9 Draw the logic diagrams of the divide-by-6 counter, its D/A converter, and the $C_P/6$ generator.

24-10 Redraw the circuit of Figure 24-24 with modifications needed for use with the MOS 2516 character generator.

24-11 Draw a set of waveforms like those of Figure 24-26(b) that would apply to use of a MOS 2516.

24-12 Draw a block diagram using all of Figure 24-26(a) as one block. Add the logic needed for a display of 16 lines.

24-13 Draw a microprocessor address bus like that of Figure 24-28 but show two 7641 PROMs connected to supply two pages of ROM at both the upper and lower extremes of the address bus. Show chip select decoding needed to prevent the occurrence of images from either device.

24-14 Which high output (AND/OR type) PAL could be used to provide the logic for the comparator shown in Figure 8-23. What is the equation for the output in AND/OR terms?

24-15 (a) Using DeMorgan's theorems, convert the equations for the A/D logic gates of Figure 24-45 to AND/NOR form.
(b) Mark the 14L4 logic diagram of Figure 24-55 for the necessary fuse retention needed to produce this logic.

24-16 (a) Using DeMorgan's theorems, convert the equations for the logic of Figure 24-52 to AND/OR form suitable for implementation with H-type PALs.
(b) Of the PALs shown in Figure 24-34, which device could not supply all of these functions without the use of external feedback?

24-17 Using pin 18 as the Σ out and the upper D flip-flop as carry register, mark the 16R4 logic diagram of Figure 24-56 for the necessary fuse retention needed to provide a serial adder. Use pins 2 and 3 as the A and B inputs.

24-18 Using pin 16 as $A > B$, pin 17 as $A < B$, and pin 18 as $A = B$ outputs, mark the 16R4 logic diagram of Figure 24-56 for correct fuse retention needed to provide the serial comparator shown in Figure 15-8.

Use the following information to solve Problems 24-19 through 24-23. The BCD adder of Figures 10-20 and 10-21 is to be implemented by using the 74S283 for the binary adder portion. The decimal carry generator and add 6 circuit is to be implemented using the 14L4. The decimal carry, C_O, is the only output to be used as an external feedback. The circuit will have five inputs: 2_{IN}, 4_{IN}, 8_{IN}, 16_{IN}, C_O. It will have four outputs: 2_O, 4_O, 8_O, C_O.

24-19 Determine the correct equation for the AND/NOR logic of C_O.

24-20 Determine the correct equation for the AND/NOR logic of 2_O.

24-21 Determine the correct equation for the AND/NOR logic of 4_O.

24-22 Determine the correct equation for the AND/NOR logic of 8_O.

24-23 Mark the 14L4 logic diagram of Figure 24-55 for the correct fuse retention to provide the functions determined in Problems 24-19 through 24-22.

Selected Answers

Chapter 2 (Exercises)

1. 1, 512, 8, 128, 32, 64, 16, 256, 2
5. .101001, 000100, .001011, .010100, .101011, .110100
7. .01111, .010101, .011, .1101, .11
9. 45, 119, 22, 69, 21, 15, 17
15. 165, 155, 121, 3315, 753
17. 79, 168, 23, 65, 155

Chapter 3 (Exercises)

1. 10111, 100000, 101101, 110010
3. 100001.0101, 101000, 10001.0111
5. 01001010, 0001100, 010.010
7. 00110, 00101, 01010, 1001000, -101101, 1010010
11. 2745, .124, 83.274, 861.37
13. 100010001, 10111011, 11110011, 101101000, 100.111, 10110.1, 10010.01001, 111.10111

Chapter 4 (Problems)

4-1. 2 μs, 12 μs

4-4. 10 μs

4-6. 2 μs, 8 μs, parallel

4-7. $X = A \cdot \overline{B}$

4-8. $X = A \cdot \overline{B}$

4-18. $X = AB + C + B = B + C$
$X = AB + BC + D$
$X = A \cdot \overline{B} + \overline{A} \cdot B$

4-21. $AB\overline{C}$
$(\overline{A} + B) \cdot \overline{C}$
$\overline{A}B + CD$

4-23. 01001100, 00010110

4-24. 11111101, 10111111

Chapter 5 (Problems)

5-1. (a) +5V; (b) 0V

5-2. bottom diode

5-5. 4.3V to 4.4V in both circuits

5-6. 1.06 mA

5-7. 4 inputs

5-9. 2.27 mA

5-10. 60 μA

Chapter 7 (Problems)

7-2. $X = \overline{A+B} + \overline{C+D} = (\overline{A} \cdot \overline{B})+(\overline{C} \cdot \overline{D})$

7-3. $X = \overline{A \cdot B + \overline{A+B}} = \overline{A \cdot B + \overline{A} \cdot \overline{B}}$

7-8. $X = \overline{A \cdot B} \cdot (A+B)$

Chapter 8 (Problems)

8-1.

1110101	1111011
0011010	0111001
1110011	0100111
0101010	1100101

8-6.

36	58	73
49	92	25

8-8. C = 0: Output is the same as the input data.
C = 1: 00110100 11110000
10001010 01011000

Chapter 9 (Problems)

9-1. Sink loads—Figures 6-15 and 7-3
Source loads—Figures 6-8 and 7-14

9-3. 1-level noise immunity 1.3V typical
0.4V minimum
0-level noise immunity 0.58V typical
0.4V minimum

9-5. 7.5 ns

9-6. $V_{NL} = 2V$, $V_{NH} = 2V$

9-8. 500 mW

Chapter 10 (Problems)

10-1.

111111	1010111
10100110	10100000

10-14. 4 four-bit adders, 4 two-bit adders, 1 quad exclusive OR, 1 hex inverter, 2 quad two-input NAND, 2 triple three-input NAND

Chapter 11 (Problems)

11-3. 3 quad exclusive OR gate, 2 dual AND OR invert, 1 hex inverter

Chapter 13 (Problems)

13-12. 001

13-13. 111

13-14. 0011

Chapter 15 (Problems)

15-3.
```
  11000      011000
-111111 → +000001
           ------
           011001 → -100111 = -39
```

```
 111111      111111
- 11000 → +101000
          ↙100111 = +39
         1

 101011      101011
- 10101 → +101011
          ↙010110 = +22
         1

  10101      010101
-101011 → +010101
            101010 → -10110 = -22
```

Chapter 16 (Problems)

16-4. Inverter

16-6. $t = 200\ \mu s$
Sampling frequency = 1250 periods per second

16-7. V_{SS} = GND
V_{EE} = -3V to -5V
V_{DD} = +5V

Chapter 17 (Problems)

17-1. 642 kohms

17-6. MV 1 approx. 220 pF
MV 2 approx. 27 pF

17-8. Use Figure 17-6(b).
$RC = 2.27 \times 10^{-5}$
@ $t = 2200$ pF, $R \approx 10K$
$R_S = 10R \approx 100K$

17-12. $R_T = 1800\ \Omega$
$C_T = 120$ pF
$C_S = (C_Y - C_{IN}) = 33$ pF

17-15. (5 oscillator frequencies could be used.)
For F_0 at Q_4, $\frac{f_0}{2}$ at Q_5, $\frac{f_0}{8}$ at Q_7
$$f_{OSC} = \frac{f_0}{16} = 4\text{ kHz} = \frac{1}{2.2\ R_T C_T}$$
$R_T C_T = 1.14 \times 10^{-4}$
@ $C_T = 1000$ pF, $R_T \approx 120\ k\Omega$

Chapter 18 (Problems)

18-4. R_b = 26.6 kohms

18-11. R_s = 150 ohms (static)
R_s = 16.7 ohms (multiplexed)

Chapter 20 (Problems)

20-1. 625 ohms

20-8. 100 μA, 10 mV

20-9. 1.992 mA

20-10. -5.08V

Chapter 21 (Problems)

21-1. 400 mV

21-3. 500 mV per count, MSB = 8V

Chapter 22 (Problems)

22-1. (a) Semiconductor ROM: nonerasable, nondestructive readout, nonvolatile, static, read-only, random access; (c) magnetic tape: erasable, nondestructive readout, nonvolatile, static, read-write, sequential

22-2. (b) Dynamic shift register: erasable, destructive readout, volatile, dynamic, read-write, serial; (d) 2114 RAM: erasable, nondestructive readout, nonvolatile, static, read-write, random access

Chapter 23 (Problems)

23-1. Memory address is $B000_H$ to $BFFF_H$.

23-3. Memory will occupy one intended and one image block in the address map, at addresses 7800_H through $7FFF_H$ and $F800_H$ through $FFFF_H$.

23-7. 239 mV when sensing a 1 level
216 mV when sensing a 0 level

23-9. @ t_{RC} min = 375 ns, $I_{C(REF)}$ = 5.54 mA

Chapter 24 (Problems)

24-15 (a) $\overline{R}_1 = \overline{1} \cdot \overline{3} \cdot \overline{5} \cdot \overline{7} + \overline{1} \cdot 4 \cdot \overline{5} \cdot \overline{7} + 2 \cdot \overline{3} \cdot \overline{5} \cdot \overline{7} + 2 \cdot 4 \cdot \overline{5} \cdot \overline{7} + \overline{1} \cdot \overline{3} \cdot 6 \cdot \overline{7} + \overline{1} \cdot 4 \cdot 6 \cdot \overline{7} + 2 \cdot \overline{3} \cdot 6 \cdot \overline{7} + 2 \cdot 4 \cdot 6 \cdot \overline{7}$
$\overline{R}_2 = \overline{2} \cdot \overline{6} + 4 \cdot \overline{6}$

24-16 (a) $A = \overline{1} \cdot 2 + 1 \cdot \overline{2}$
$B = 2 + 3$
$C = \overline{2} \cdot 4 \cdot 5 \cdot \overline{6} + \overline{2} \cdot 4 \cdot 5 \cdot \overline{7} + \overline{2} \cdot 4 \cdot 5 \cdot \overline{8}$
$D = \overline{2} \cdot 6 \cdot 8 \cdot 9 + \overline{2} \cdot \overline{7} \cdot 8 \cdot 9$

24-20. $2_O = \overline{\overline{2_{IN}} \cdot \overline{C_O} + C_O \cdot 2_{IN}}$

24-21. $4_O = \overline{\overline{4_{IN}} \cdot \overline{C_O} + \overline{2_{IN}} \cdot C_O \cdot 4_{IN}}$

24-22. $8_O = \overline{8_{IN} \cdot C_O \cdot 4_{IN} + 8_{IN} \cdot C_O \cdot 2_{IN}} + \overline{\overline{8_{IN}} \cdot \overline{C_O} + \overline{8_{IN}} \cdot \overline{2_{IN}} \cdot \overline{4_{IN}}}$

Index of Digital Circuits

```
   XX      NNNNN    X      X
  /         /       |       \
Manufacturer  Type of  Package  Temperature
              device            range
```

X = letters
N = numbers

Digital integrated circuits are commonly coded as shown above. The prefix letters identify the manufacturer. The four or five numbers identify the type of device. For small- and medium-scale TTL circuits, the 5400 and 7400 series differ only by supply voltage and temperature range. A 5402 and a 7402 will have the same logic drawing and pin connections. Occasionally the letters S, L, or H will appear in the middle of the four type numerals, signifying Schottky, low-power, or high-speed devices; but drawings and pin connections are the same as those without the letters.

TTL Circuits

CMOS Circuits

ECL Circuits

Static RAMs

Index

Note: Page number in **boldface** type indicates end-of-chapter glossary definition for the term.